D1207560

PSSC PHYSICS Fourth Edition

PSSC PHYSICS FOURTH EDITION

Uri Haber-Schaim Judson B. Cross
John H. Dodge James A. Walter

D. C. HEATH AND COMPANY

Lexington, Massachusetts Toronto

Published simultaneously in Canada.

Printed in the United States of America.

Library of Congress Catalogue Number 75-18974

International Standard Book Number: 0-669-97451-X

Preface

This textbook and its correlated teaching materials represent the latest products of a pioneering effort that was launched in 1956. Perhaps the best way to describe the motivation, philosophy, and development of PSSC Physics is to quote from the prefaces to the first three editions.

The Physical Science Study Committee is a group of university and secondary school physics teachers working to develop an improved beginning physics course. The project started in 1956 with a grant from the National Science Foundation, which has given the main financial support. The Ford Foundation and the Alfred P. Sloan Foundation have also contributed to the support of the program.

This textbook is the heart of the PSSC course, in which physics is presented not as a mere body of facts but basically as a continuing process by which men seek to understand the nature of the physical world. Besides the textbook there are the following closely correlated parts: a laboratory guide and a set of new and inexpensive apparatus; a large number of films, standardized tests; a growing series of paperback books by leaders in related fields; and a comprehensive teacher's resource book directly related to the course.

The PSSC physics course is the work of several hundred people, mainly school and college physics teachers, over a period of four years. A brief account of this collaboration is given at the end of the book. Here it is appropriate, however, to recognize two of these collaborators. Professor Jerrold R. Zacharias, of the Department of Physics of the Massachusetts Institute of Technology, called together a committee of leaders in physics and in education from which this project sprang. He has been active in all phases of the project. Professor Francis L. Friedman, also of the Department of Physics at M.I.T. and a member of the Committee from the beginning, has played the major role in developing the textbook and has contributed significantly to all parts of the program. . . .

James R. Killian, Jr.
Chairman, Board of Trustees
Educational Services Incorporated
September 1960

From the Preface to the First Edition

Revisions for the second edition of PSSC Physics have been carried out in the light of the experience accumulated over the five years since the publication of the first edition. During this time, the course has been taught by some 6000 teachers to 640,000 students. We are deeply indebted to many of these teachers, and to some of their students, who have kept us informed of their successes and their tribulations through the "feedback" channels maintained for that purpose. We are also grateful to numerous colleagues from colleges and universities all over the world who have given us the benefit of their detailed criticisms of various chapters and sections. Throughout this period, the PSSC Planning Committee has been active in critically re-examining the existing materials and in both suggesting and producing new materials for trial in schools. This tested material has been particularly valuable in producing the second edition. . . .

It has been our earnest effort to retain both the spirit and the content of the original course, making changes primarily for the purpose of strengthening or clarifying the arguments. It is only through actual use in the schools that we shall be able to estimate the results of this effort.

Byron L. Youtz
August 1965

From the Preface to the Second Edition

Five years have passed since the appearance of the second edition. During this time science education has made further advances. In particular, new courses on the junior high school level provide the entering physics student with a much better background and approach than was the case five years ago. This development was a major factor affecting this revision, along with the persistent evidence that many teachers have to rush to complete the PSSC course in one year.

The course now starts directly with the study of light, the former Part II. We have deleted most of Part I of the previous edition. The chapters on kinematics and vectors from Part I were completely rewritten and moved to the beginning of the study of dynamics.

Although the former Part III, Dynamics, was left basically intact, a number of sections were rewritten, in particular in the chapter on Work and Kinetic Energy, and the one on Heat, Molecular Motion, and Conservation of Energy.

Significant rearrangements were made in the former Part IV, Electricity and Atomic Structure, to provide a clearer view of the problems which led to the development of the basic ideas of quantum physics.

Throughout the book we made editorial changes where we found ways to present the material more concisely without damaging the general spirit and style of the earlier editions.

We have added a new feature to the third edition: a new Appendix containing a detailed discussion of about 30 problems from the Home, Desk, and Lab pages throughout the book. The problems were selected for their value in presenting approaches to problem solving in general, and are indicated in the text by a dagger (†). The analysis of the problems was written specifically for student use and not merely as Teacher's Guide solutions. We have also retained, in another Appendix, the short answers to starred (*) problems which appeared in the second edition.

Uri Haber-Schaim
Judson B. Cross
John H. Dodge
James A. Walter
August 1970

From the Preface to the Third Edition

FOURTH EDITION

Just as in previous editions, the priorities for revisions were strongly influenced by teachers using the course. This time we concentrated on Chapters 17–21. Except for a few corrections, the rest of the text was left intact.

Chapter 17 has been rewritten to focus more on the accomplishments and limitations of applying classical mechanics to atoms in gases and solids, i.e., the description of internal energy in classical terms.

Chapters 18–21 now form two groups of chapters. Chapters 18 and 20 provide an extension of particle dynamics to include electric forces. Chapter 21, on the other hand, provides a brief introduction into circuitry based on voltage-current relations without trying to tie these relations to fundamental principles. We hope that the distinction between these two parts of electricity has been made clearer.

The fact that there is a demand for a Fourth Edition of PSSC constitutes, we feel, a vote of confidence for the original team that put the course together.

Uri Haber-Schaim
Judson B. Cross
John H. Dodge
James A. Walter
August 1975

Table of Contents

PHYSICS

Introduction

The Growth of Physics
The Tools of Physics
The People of Physics

The Growth of Physics

When a flash of lightning shatters the dark, the radio crackles and your eyes are dazzled for a few seconds. A moment later you hear the roll of thunder, and a loose windowpane rattles. Three states away in a storm-location center, a radio-locater can pinpoint the lightning stroke; and downtown the weather forecaster, hearing a faraway rumble, nods as if he had been expecting the storm.

Here is a chain of events, different events, taking place at different places and different times. They are all linked. How are they tied together, and just what is happening to eye, to ear, to the radio, and in the air itself?

Take another chain of events. In the test shops of a steel mill you can watch a thin bar of a new alloy pulled at its ends by the powerful jaws of a testing machine as big as a house, which is driven by a fast-whirring electric motor no bigger than a football. The solid bar slowly yields to the pull. It stretches out like a piece of taffy, and breaks a little later with a jangle of sound. How does it all work? What holds the bar together? Why did it finally give way? And, by the way, why is steel stronger than glass? Why is it more dense than aluminum? Why does it rust?

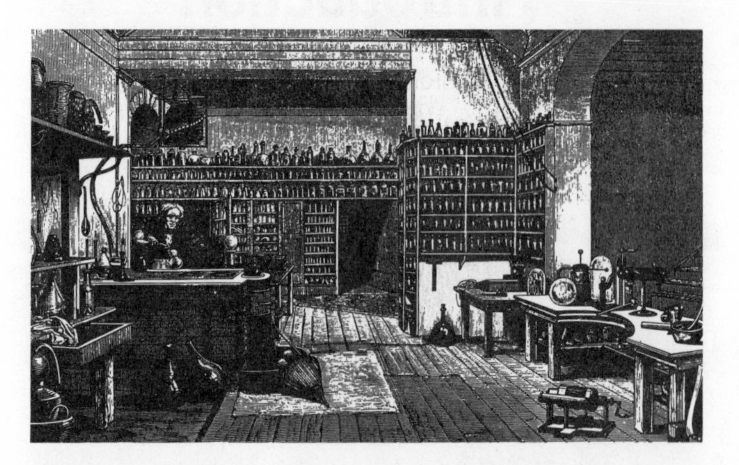

An engraving of the laboratory of Michael Faraday at the Royal Institution in London about 100 years ago. In the lower right-hand corner is some of the apparatus he used in his work on electricity. (From "Life and Letters of Faraday," by Dr. B. Jones, Longmans, Green.)

A typical modern American physics laboratory. Compare it with the picture of Faraday's lab.

Men once feared the "illness of the sun," when the sun disappeared and the earth darkened. Then we learned about the complex motion of the moon. Eclipses became far easier to predict than tomorrow's weather. The moon has been circling our planet since long before the first dinosaur walked the earth. A man-made satellite can circle the globe for a long period without propeller, jet, or wings—a tiny synthetic moon. How do satellites move? How can we design our own? How are we able to send men to the moon?

Physics enables us to answer such questions. It gives us the power to predict and to design, to understand and to adventure into the unknown. From what we learn in physics, new things are made. With new answers in physics, new questions are always arising. Many of these questions would never have been asked if physics itself had not been put to use.

Before the time of Galileo there were no astronomical telescopes. Once Galileo had put together two lenses to make an astronomical telescope and had discovered four moons revolving around Jupiter, more and better telescopes were designed and made. With their aid, new heavenly bodies were found, such as the many small planets called asteroids that move between the orbits of Jupiter and Mars.

New questions now arose. How could the complex motions of these moons and asteroids be explained? To answer questions like this, much of the special mathematical branch of physics called *mechanics* was developed. Beginning in the eighteenth century, rapid advances were made in this study of how objects move when subjected to complex forces. The new knowledge of mechanics led to the better design of machines. So we see that without the telescope, mechanics would have taken a slower course.

More recently, around fifty years ago, the beginnings of the understanding of atoms made it clear how to build much better air pumps than ever before. With these new pumps, a good vacuum was easy to obtain, and with a good vacuum experiments that formerly were impossible started researchers delving into the nature of electrons and atoms. The ability to produce a high vacuum, and the knowledge it brought, led to such varied practical achievements as radio and television tubes, concentrated orange juice, and atomic energy. This atomic research also clarified the very foundation of chemistry by making it possible to determine what holds two atoms together or what keeps them apart.

So the subject grows. It is like a great building under construction, not a finished structure around which you have only to take a guided tour. Though some parts are pretty well complete, and both useful and beautiful, others are only half done. Still others are barely planned. New parts will be started and completed by the men and women of your generation, possibly by you or your classmates. Once in a while, a finished room in this structure known as physics is found unsafe, or no longer large enough for new discoveries, and the room is abandoned or rebuilt. But the great foundations are well laid and stand on pretty solid ground. These remain unchanged though changes go on above them. It is the intent of this book to let you see the plan of the building, to show you what the builders have done, to look at some of the parts they are working on now, and to notice occasionally where the design is still incomplete.

The Tools of Physics

Physics requires tools—tools of every kind. As with nearly all activities of human beings, the key tool of the physicist is his mind. Next, he needs language to make clear to himself and to others what he thinks and has done, and what he wants to do. Mathematics, which can be thought of as a special, supremely clear and flexible international language of relation and quantity, is also an important tool in his kit—and his own eyes, ears, and hands are very important indeed. He regards these as the first instruments for collecting information about the events of the world, a world which he tries to understand and to control. Then, to aid his senses, and to produce the special circumstances he sometimes wants to study, he must make use of a rich variety of other tools, instruments, machines, or contrivances.

Sometimes the tools of physics may be simple. In 1896 Henri Becquerel discovered the strange radioactive properties of uranium, and began the branch called *nuclear physics,* with no other equipment than a photographic

In 1896 the French physical chemist Henri Becquerel discovered radioactivity by exposing a photographic plate to a uranium salt. The photographic plate was wrapped in black paper, and crystals of uranium acetate were laid on top.

plate, wrapped in black paper, and a few crystals of a special chemical salt. In 1934 Fermi and his associates in Rome discovered the slow-neutron basis of atomic energy. They used simple apparatus—a hospital radium supply, a marble fountain basin of water, some pieces of silver and cadmium, and an instrument made of little pieces of thin metal leaf mounted

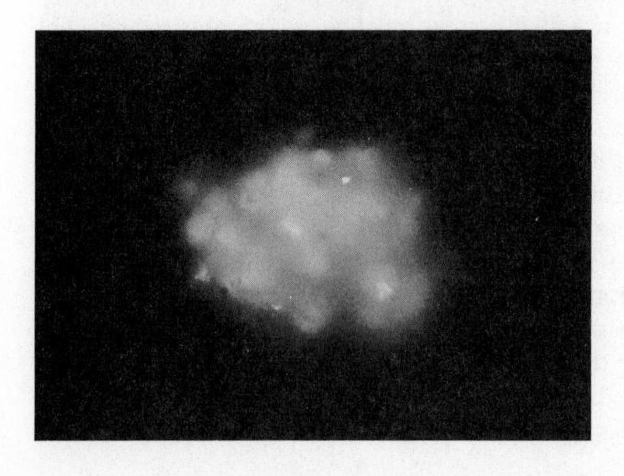

The resulting exposure, like the one reproduced here, could only result from some invisible yet penetrating emanations from the salt, since light was never permitted to strike the photographic plate. Here is an example of a simple means used to obtain an important fundamental result.

The bevatron at the University of California Lawrence Radiation Laboratory is an example of a large and complex tool. The man in the lower right gives an idea of the dimensions of this giant accelerator. The huge ring-shaped magnet weighs as much as a ship and requires as much power as a small town. This large tool is used in obtaining evidence of the behavior of particles too small to be seen with any device so far developed by man. (Courtesy: University of California Lawrence Radiation Laboratory.)

on a small microscope. Following up the work of Fermi, five years later Hahn and Strassmann discovered the fission of uranium. They worked with simple chemical equipment and a commonplace Geiger counter. Who will next found a branch of science with simple tools and a really good idea? We do not know, but it is bound to happen.

Sometimes the tools of physics may become wonderfully complex. The instrument-filled satellite and the space probe are tools of the physicist who seeks to understand the rain of particles falling upon the earth from outer space, or the properties of that space itself. The giant bevatron at Berkeley, California, and its accompanying liquid-hydrogen bubble chamber are the tools of the physicist interested in studying the properties of the tiny particles which somehow constitute the atom.

Here again is a very complex instrument to extend our senses. This liquid-hydrogen bubble chamber is used to make visible trails (seen in the next picture) of the tiny subatomic particles coming from the bevatron.

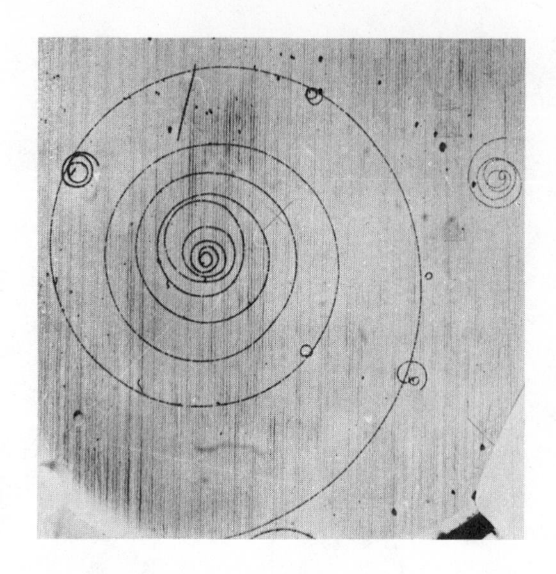

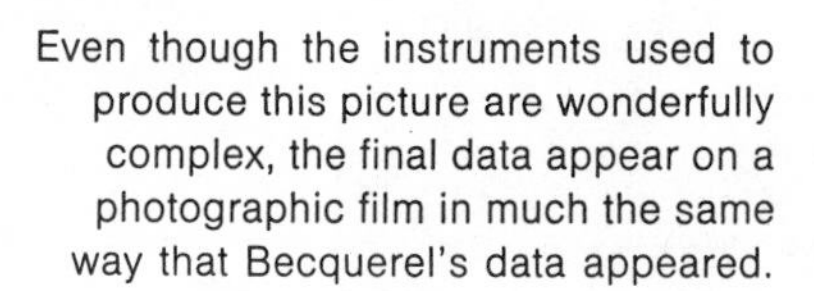

Even though the instruments used to produce this picture are wonderfully complex, the final data appear on a photographic film in much the same way that Becquerel's data appeared.

The People of Physics

The people who design and use the kind of equipment you have been reading about are physicists. When their skills are primarily those of designing and performing experiments, they are called experimental physicists. When, on the other hand, they are skilled primarily in the use of mathematics in the problems of physics, they are called theoretical physicists. Benjamin Franklin and Mme. Marie Curie were experimental physicists. Isaac Newton and Albert Einstein were theoretical physicists—perhaps the greatest.

In the earlier days, the tools, both experimental and mathematical, were so simple that a single man or woman could become skilled in the use of both kinds. Isaac Newton not only made the thrilling experiment of breaking sunlight into colors with a prism, but actually invented for his own purposes one of the most useful forms of mathematics, the calculus. Franklin contributed to electrical theory, besides inventing and performing many key experiments. Nowadays some of the tools are so complex that few physicists are versatile enough to become masters of all of them. But whether theorists or experimenters, the people who build physics are all physicists.

However, most of the people who study the fundamentals of physics do not go on to become physicists. Many go into related fields, such as engineering or other sciences, and many will leave science altogether. But whether you go on or not, you can find in the physicists' story of nature a great deal that will help you in understanding the changing and exciting world in which we live. For physics lies behind the headlines, behind the gadgets that create the new jobs, and behind the new problems every citizen has to face. In studying this growing subject, one of the most significant in the history of man, you will have a chance to nourish that curiosity about the world which marks us humans off so sharply from the other animals, that wonderful feeling of *wanting to know* which can be a deep satisfaction throughout a whole lifetime.

How Light Behaves

1-1 Sources of Light
1-2 Transparent, Colored, and Opaque Materials
1-3 Reflection
1-4 Light-Sensitive Devices
1-5 How Light Travels
1-6 The Speed of Light
1-7 Shadows
1-8 Light Beams, Pencils, and Rays
1-9 How We Locate Objects

CHAPTER

1

From the very beginning of history, men have puzzled about the nature of the light that affects our eyes. The questions they asked were probably the same that have occurred to you. What is light? How does it travel, and how fast? Is seeing always believing? Why are some objects colored, some white, and some black?

1-1 Sources of Light

Anyone who has spent a moonless night in the country, in a forest, or at sea knows how very dark it can be when the sun is on the other side of the earth. At early dawn, objects that we could not see a few minutes earlier begin to take shape. Then details sharpen, colors appear and brighten, and daylight begins. It is the light from the sun, rising above the horizon in the east, that gives shape, detail, and color to our world.

The sun, the stars, lamps, even lightning bugs, give off light. They are called luminous bodies (from the Latin word *lumen,* meaning light). Other objects—trees, grass, the pages of this book, for example—are nonluminous. They are visible only when they receive light from some luminous source and reflect it to our eyes.

Whether a body is luminous or nonluminous depends as much on its condition as on the material of which it is made. By changing the conditions we can make many familiar substances luminous or nonluminous at will. The filament, or fine wire, inside an electric light bulb is nonluminous unless it is heated by an electric current passing through it. We can take a cold piece of iron and make it glow red, yellow, or white by heating it in a bed of burning coals or over a gas flame. When solids and such liquids as melted metals are heated to temperatures above 800°C (about 1500 degrees Fahrenheit), they become sources of light. Such heated materials are known as incandescent bodies.

Careful observation shows that the light from candle flames comes from many small, hot particles of carbon heated by the burning gases from the candle wax. Thus the flame is another incandescent source of light. Many of the carbon particles are not completely burned in an ordinary flame. They cool off as they are carried above the flame by the air, become nonluminous, and make up the main part of the rising smoke.

Not all light sources are incandescent. Neon tubes and fluorescent lamps, like electric light bulbs, give off light as long as an electric current passes through them. Touching each of them, however, convinces us at once that there must be a difference in the way the light is produced. The neon and fluorescent tubes remain quite cool, whereas the incandescent bulb soon becomes too hot to touch. Pursuing this difference further, we find that by gradually increasing the current in the filament of the incandescent bulb we can increase its brightness, but that there is also an accompanying change in color. At first we see a dull red glow which changes to a bright yellow and, with sufficient current, can become "white hot" like the heated iron. On the other hand, if we increase the current through a neon tube to increase the brightness, we observe no change in color. Thus there is a basic difference between incandescent sources and other sources. In the former, changes in brightness, temperature, and color seem to be closely linked, while in the latter the color of the source depends mainly on the nature of the material and does not vary with brightness.

A great deal of light reaches our eyes from nonluminous surfaces. To convince ourselves, we need only imagine how the average room would appear if the walls and other surfaces were covered with a paint so black that it reflected none of the light reaching it. The lights would appear as bright glares against dark backgrounds. White ceilings or bright walls reflect and diffuse much of the light they receive and so increase the brightness within the room. In fact, when we use indirect lighting we hide the lamps from sight, and all of the light reaches us after being reflected from the walls and ceiling. On a larger scale the moon, which we often think of as a source of light at night, is really an indirect-lighting device that reflects sunlight.

1-2 Transparent, Colored, and Opaque Materials

When you look through a clean window at a brightly illuminated scene outside, you are hardly conscious of the fact that the glass is there. A substance that transmits light in this way is said to be transparent. Later in the day, when dusk has come, look out through the same window from inside a lighted room. In addition to the world outside, you now see in the glass a reflection of yourself and of the room. The light by which you see yourself must have started from within the room. Instead of going through the glass to the outside, this light was returned to you. It was reflected.

Does the thickness of a transparent body have any effect on the amount of light it transmits? A single piece of the glass seems to transmit light almost perfectly. But, if you pile up a thick stack of ten or twenty pieces of clear glass, some light is *absorbed* and the light passing through it dims and appears somewhat colored. Evidently we become aware of clear materials like plastics, glass, and water partly because they reflect as well as transmit light, and partly because some of the light is absorbed.

Such materials have another important effect on light. When light enters or leaves them, its direction changes in an interesting way. In Fig. 1–1 the apparent bending of the stick at the point at which it enters the water indicates that something happens to the light which passes from the stick to the camera. Another illustration of the same effect is shown in Fig. 1–2.

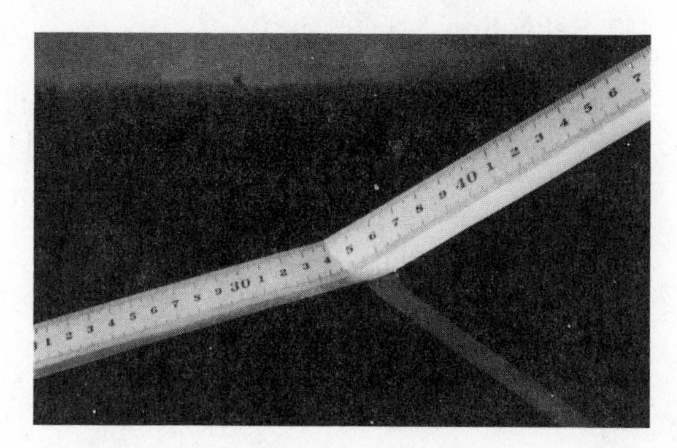

Figure 1–1
This photograph shows the apparent bending of a meter stick at the point where it enters water. The gray shape sloping down from the stick is a reflection.

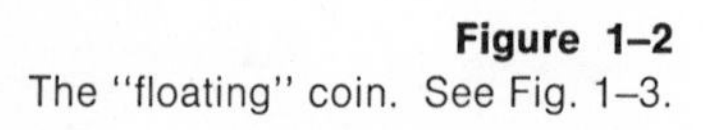

Figure 1–2
The "floating" coin. See Fig. 1–3.

The coin on the right is in an empty container, and an identical coin is in a water-filled container on the left. The coins were photographed simultaneously with the apparatus shown in Fig. 1–3. The coin in the water looks nearer to the camera and bigger than the other one. This "floating" coin effect and the illusion of the "bent" stick occur because light changes direction when it goes from one material, such as glass or water, into another material, such as air. This bending of a light path is called *refraction.* A detailed study of refraction contributes greatly to our understanding of what light is. We shall return to it several times.

Figure 1–3
The position of camera and containers used in photographing the "floating" coin in Fig. 1–2.

1-3 Reflection

All objects, whether transparent or opaque, reflect some of the light that falls upon them. Most surfaces send light off in many directions. This is called *diffuse reflection.* It is by the help of this diffused light that we see illuminated bodies, observe their texture and color, and distinguish them from their surroundings.

A few materials, such as highly polished sheets of silver, aluminum, or steel, absorb little from white light and also reflect in a much more regular manner than do rougher surfaces. An ordinary mirror consists of a thin film of silver placed on the back of a plate of glass. In Fig. 1–4, light from

Figure 1–4
Regular and diffuse reflection. The white paper (far left) sends more of the light from the lamp toward the camera than does the mirror.

a lamp shines on both a mirror and a piece of white paper. The paper appears well illuminated and is white against the dark background. On the other hand, the mirror appears quite dark. The same amount of light is reaching the two, and this mirror reflects light very well, as is indicated by the reflection of the white candle. Why, then, is the paper brighter than the mirror in the photograph? The mirror reflects light from the lamp away from the camera, while the paper reflects light in all directions so that some of it reaches the camera.

It is the regularity of reflection from smooth surfaces that allows the formation of images. Figure 1–5 shows a waterfront scene with boats and piers reflected in the water. Is the picture printed right-side up? Turn the

Figure 1–5
Nearly perfect reflection. Is the picture right-side up? Which are the actual boats? Which are their reflections?

page upside down and examine it again. There are several clues in the picture that tell you if it is right-side up or not.

In the next chapter we shall learn something about the laws of reflection that will allow us to discuss "why," as well as "what" happens when images are formed by mirrors or other smooth reflecting surfaces.

1-4 Light-Sensitive Devices

We have not yet mentioned how we know that light is present or what its color may be. We have taken for granted that marvelous instrument, the human eye. A thorough study of the eye would include discussions of how light is refracted to fall on the nerve cells of the retina at the back of the eyeball, how the eye adjusts itself to see clearly objects at different distances or of different brightnesses, how the effects of color are produced, and also how things can go wrong with our vision. Such a study could easily occupy us for a full year or more. Here we shall have to be content with the observation that in passing through structures in the front of the eye light is refracted in such a way that an image is formed on the retina. Chemical changes take place at the retina, and as a result electrical impulses are sent along the nerves to the brain.

There are many instruments, other than the eye, that respond chemically or electrically to the action of light. These are often used to study light by methods that are more convenient or more satisfactory than is visual observation. The best-known substances that are *photosensitive* are some of the chemical compounds containing silver. Such compounds are used in photographic films. The full process involved in the exposure and development of the film is complex. But the essential thing is that those portions of the film on which light has fallen are left with a deposit of finely divided silver after chemical treatment. In this state the silver does not appear as bright, shiny metal, but as dull, opaque particles, the familiar black portions of a photographic negative. The portions not exposed to light become clear and transparent.

There are also a number of devices that give an electric current when light falls on them, without any chemical changes taking place. Chief among these are photoelectric cells, or photocells, which we shall discuss later. The electric current through a photocell is proportional to the intensity of the light falling on it.

1-5 How Light Travels

The sun and the stars are so familiar that we seldom think of the vast stretches of almost empty space that separate them from us. Yet we know that the sun is about 1.5×10^{11} meters away from the earth and that the nearest star is about three hundred thousand times as far. Countless stars have been seen at distances so remote that comprehension almost fails us. All the information from which we have learned about this vast universe has come to us "riding swiftly astride beams of light." It must therefore be true that light can travel over very great distances and that it can travel freely in empty space.

As familiar as the sun itself are the shadows it casts. What can we learn from them about light? As we walk or run along on a sunny day, our shadows keep pace with us. This simple experience shows that light must travel much faster than we can run, for the shadows of our heads must lag behind us by the distance we move while light goes from our heads to the ground. Shadow shapes, too, reveal something about the nature of light. In Fig. 1–6 a set of lines connects various points on a shadow to the corresponding points on the object which casts the shadow. The lines are almost parallel and they all point to the source of the light. Evidently light travels in straight lines.

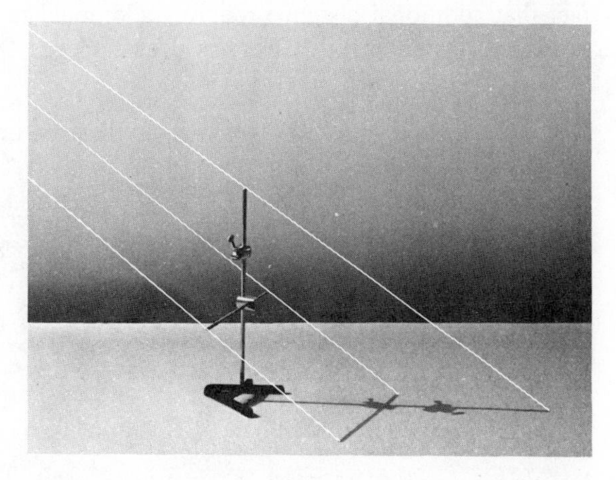

Figure 1–6
The formation of shadows. Lines connecting points on the shadow to points on the object casting the shadow are nearly parallel; all point to the light source.

1-6 The Speed of Light

All of us have, at one time or another, heard the roar of a jet plane high in the sky and instinctively looked for it in the direction of the sound. We finally saw it far ahead of the sound. Then in deciding where the airplane actually was, we believed our eyes rather than our ears. Why did we trust our eyes? We know that sound takes some time to travel so great a distance and that light travels far faster than sound. We did not assume that the plane was beyond the point where we saw it, because we believe that light travels fast indeed!

Galileo suggested a method for finding the speed of light, similar to the method he used to measure the speed of sound. Two men with lanterns were placed at a measured distance apart. The first uncovered his lantern and started a clock, and the second uncovered his lantern when he saw the light from the first. When the first man saw the second's light, he stopped the clock, thus measuring the time for light to travel from the first man to the second and back—or so Galileo hoped. This experiment failed to give the speed of light because light travels so fast. But it was not a complete failure. It showed that the speed of light was too great to allow its passage to be measured over short distances with the crude timing mechanisms then available.

The first evidence for the finite speed of light was recognized by Olaf Roemer in 1676 from astronomical observations on the motion of the moons of Jupiter. Once every revolution these moons disappear in the shadow of Jupiter and then emerge again (Fig. 1–7). The time between consecutive eclipses of any one moon equals its period of revolution. It was observed that these periods were not constant: they appeared slightly longer when the earth in its orbit around the sun was moving away from Jupiter (for example, from A to B in Fig. 1–7), and shorter when the earth was coming closer to Jupiter. Roemer reasoned that the earth does not affect the motion of Jupiter's moons, but that when it recedes from Jupiter the light from consecutive emergences of a moon from behind the shadow has to travel an additional distance to reach the earth (l in Fig. 1–7). The fact that additional distance requires additional time shows that light does not spread instantaneously but has a finite speed.

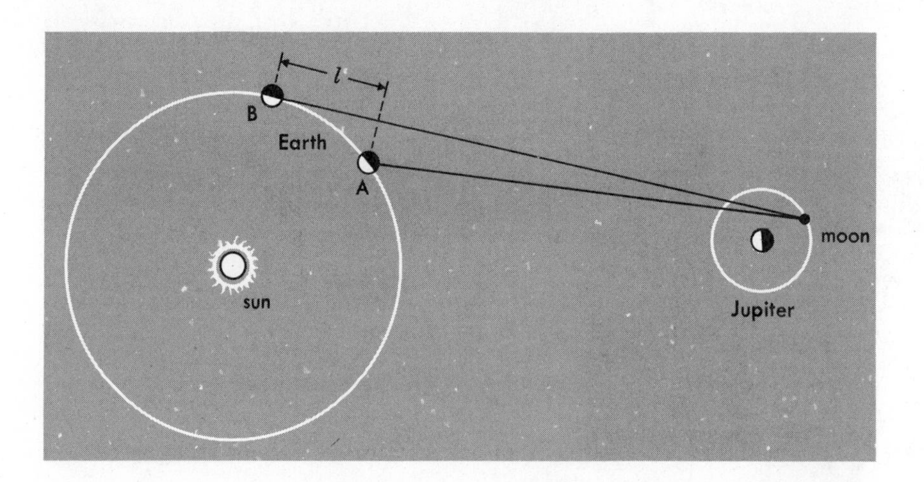

Figure 1–7
A schematic diagram illustrating Roemer's reasoning that light travels with finite speed. Sizes and distances are not drawn to scale.

This was the great contribution of Roemer. The numerical value of the speed of light could not be calculated accurately at that time for two reasons: his measurements of the time it would take light to cross the diameter of the earth's orbit were in error, and the diameter of the earth's orbit itself was not accurately known.

Later measurements of the delays in eclipse times showed that the time required to cross the earth's orbit is 16 min 20 sec. The average distance of the earth from the sun is now known to be 1.47×10^{11} meters; hence the speed of light is

$$c = \frac{2 \times 1.47 \times 10^{11}\ \text{m}}{980\ \text{sec}} = 3.00 \times 10^{8}\ \text{m/sec}.$$

The first determination of the speed of light over a distance short enough to be practical on the surface of the earth was made by Armand Fizeau in 1848. It required the invention of a timing device that could measure very short intervals with accuracy. We shall discuss Fizeau's method in Chapter 4.

Look at your own shadow, cast by the sun on a smooth floor or pavement. Notice the difference in sharpness between the shadow of your feet and that of your head. This difference is even more noticeable in the shadow of a thin vertical stick or rod as shown in the top half of Fig. 1–8. Apparently the shadow becomes wider and less sharp as the distance from the object to the edge of its shadow increases. The fuzziness of shadows also depends on the source of light. The bottom half of Fig. 1–8 shows the shadow cast when the source is very small. This shadow is quite sharply defined over its entire length. The shadows cast by tiny light sources are generally quite sharp, consistent with the straight-line propagation of light.

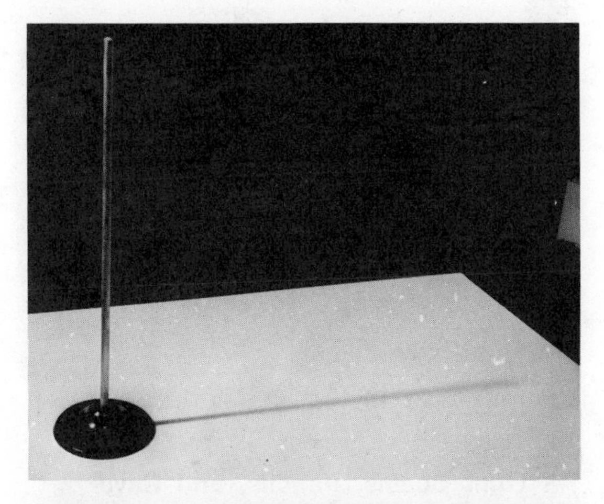

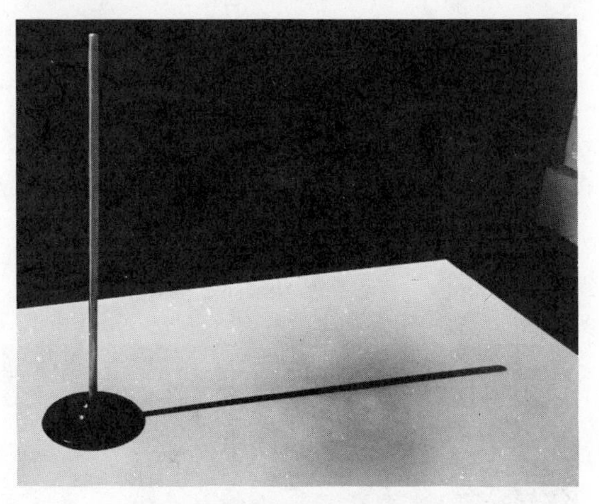

Figure 1–8
Shadows cast by an extended source of light (top) and by a point source of light (bottom). Notice the difference in their sharpness.

The sharp shadows cast by a source of light so small that it may be considered a point give us a hint about the reason for the less sharp appearance of shadows in sunlight. Every point on the surface of the sun sends out light. The shadow, therefore, is not really a single shadow but is the combination of a very large number of individual shadows cast by light from each point on the sun's surface. Figure 1–9 indicates how the shadow is formed when the source of light is large, like the sun. Traveling in straight lines, no light can reach the circular region between *c* and *d*; hence this region of the shadow is black. In the dotted region between the circle *ab* and the circle *cd*, light from some parts of the source gets past the object and onto the screen. Hence the shadow is less dark, and it finally fades off to an indistinct edge at the circle *ab*. Outside the shaded region between *a* and *b*, light arrives from all places on the source.

The dark part of a shadow, which no light reaches, is called the umbra. The less black parts make up the penumbra. When we are in the umbra of the shadow of the moon, the sun is in total eclipse. When we are in the penumbra, we can see part of the sun, and we say the eclipse is partial.

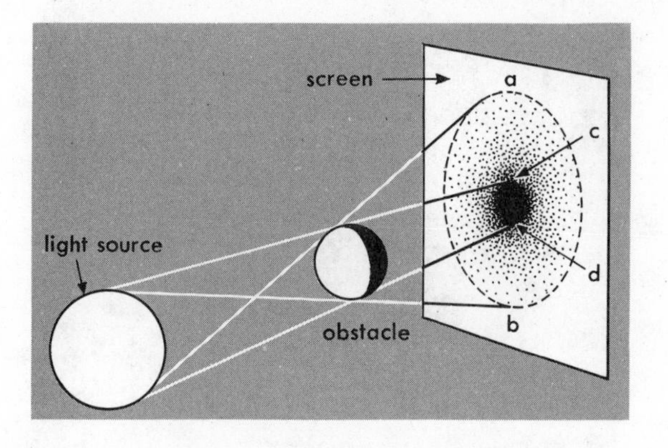

Figure 1–9
Formation of a shadow from an extended light source. Sketch is a cross section through the light source and obstacle in a plane perpendicular to the screen.

1-8 Light Beams, Pencils, and Rays

Figure 1–10 shows a light source which is pointed at a piece of paper. We know that the light comes from the source and we see it reflected from the paper, but we do not see it between the two objects. Common sense suggests, however, that the light must travel between the two. In Fig. 1–11 fine smoke particles have been introduced into the air and now the beam shows clearly all the way. The straight edges of the beam connect the source with the edges of the illuminated spot, confirming our belief that light travels in straight lines. The fact that we saw nothing of the beam until the smoke

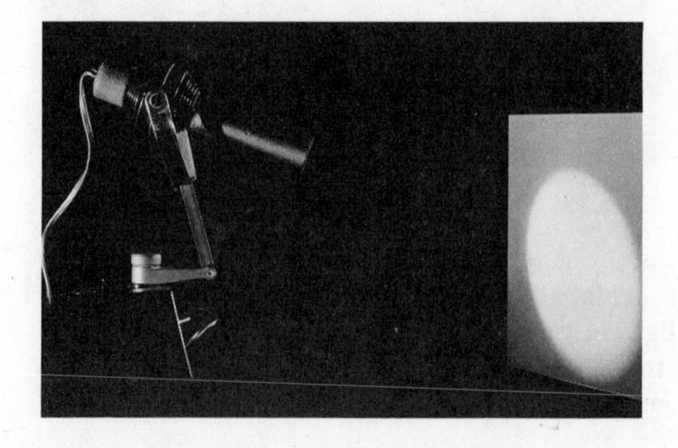

Figure 1–10
Light beams are invisible. Here we see the light source and the reflection from the white paper, but nothing in between source and target.

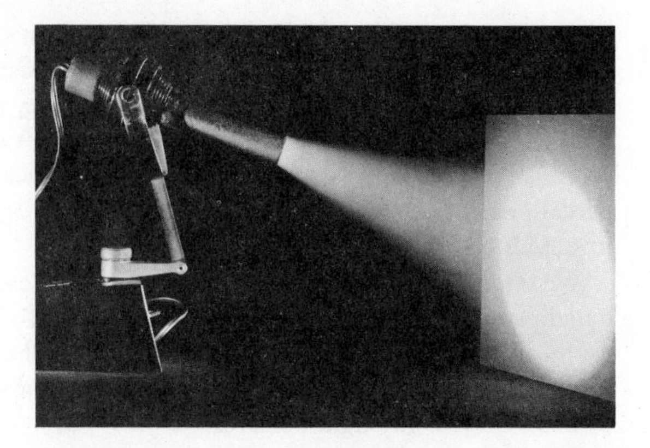

Figure 1–11
If fine particles are introduced into the air, the light beam shows clearly all the way from the source to the cardboard target. Compare with Fig. 1–10.

was introduced tells us that light enters our eyes only when we are looking directly toward a source or when there are illuminated bodies that can reflect light directly to our eyes. In this case the light was reflected from the paper and from smoke particles, making them act like new sources to send the light on new straight-line paths toward our eyes.

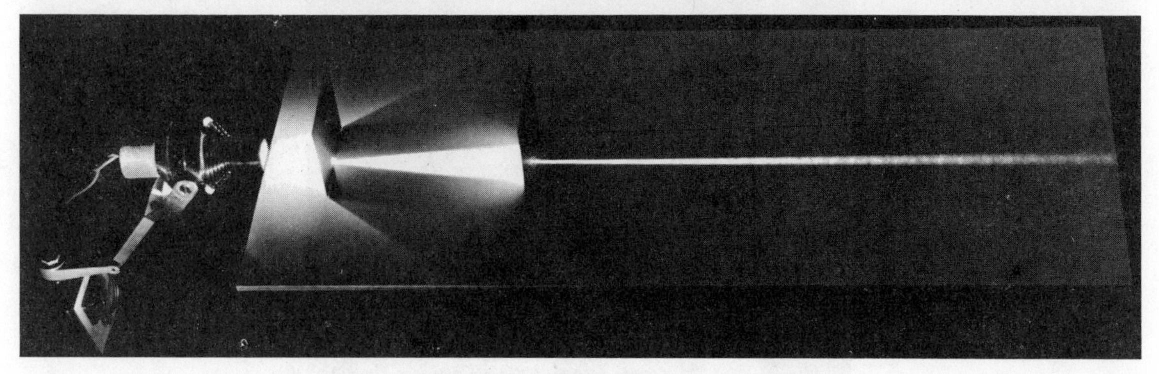

Figure 1–12
Beams and pencils of light. In the top picture, light coming from the source through a pinhole forms a cone-shaped beam. Below, a second pinhole in line with the first produces a narrow pencil of light.

In the upper picture of Fig. 1–12 we see a cone marked out by the beam. The axis of the cone (the line joining the center of the pinhole to the center of the base of the cone) passes through the center of the source and the pinhole.

By putting two pinholes in line, as shown in Fig. 1–12 (bottom), we get a very narrow *pencil* of light. It is often convenient to imagine such a pencil of light that is smaller and finer than any pencil that we can produce in practice. The finest pencil that one can imagine is just a straight line. We call this extreme pencil a *ray* of light. Of course, we can never produce single light rays, but the idea of a ray is a very useful one. It allows us to draw on paper lines that represent the directions in which light is traveling. A pencil of light can be found in nature; a ray of light is something we have invented to represent a very small pencil.

We often draw a few rays, for example, to indicate the limits of the illuminated region, the umbra and penumbra in Fig. 1–9. The rays we draw are useful like the lines on an architect's plan, but they are no more light than his lines are walls or windows. Light does not make rays; we do. They help us describe the way light behaves.

We know that light beams or pencils striking reflecting objects illuminate them so that they act as new sources. Light thus interacts with the material on which it falls: the light beam influences the material, and the material influences the behavior of the light beam. Can light beams also interact with each other? We can answer this question by letting the light beams from two different sources pass through a single pinhole, as is shown in Fig. 1–13. Each beam passes through the hole as if the other beam were not there.

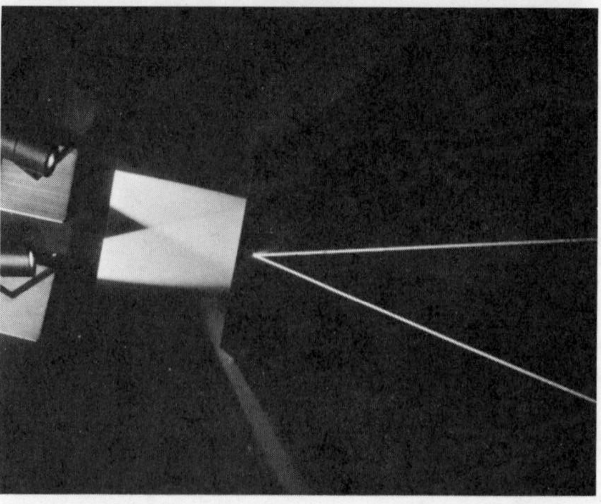

Figure 1–13
An experiment showing that two beams of light can pass through each other.

In general, each light beam from two or more sources behaves as if the other beams were not present, and this independence of action is very important. The procedure of studying the pencils or rays from different sources or different parts of a source would otherwise be of little value. Tracing the rays that show the directions of two jets of water, for example, does not tell us much about where they go when they hit each other. It is because the light from each of two sources acts as if it were alone that we can trace rays to find the regions of light and shadow when both sources are present.

1-9 How We Locate Objects

We can always locate the position of a point source of light if we know the directions of several of the rays that come from this source. We simply draw two or more of these rays backward until they meet. The point of intersection is the location of the source. When a cone of rays comes to your eye from such a source, you automatically change the shape of the eye to focus the diverging rays from the source, so that you see an image of the source. This process of focusing your eyes gives you information equivalent to that of tracing back the rays, and in certain circumstances, without being aware of it, you use this information to estimate the distance to the source.

We can estimate distances more easily with two eyes than we can with one. To convince yourself of this, have a friend hold a thin piece of wire away from other objects (Fig. 1–14). With another similar wire in your hand, try to touch the end of his wire when it is about as far away as you can reach. You will find that you can do this with considerable accuracy. Now try the same experiment with one eye closed. You will find it much more difficult to bring the two wires together.

One of the reasons you judge distances better with both eyes open is indicated by this experiment. Light coming from the end of an object must

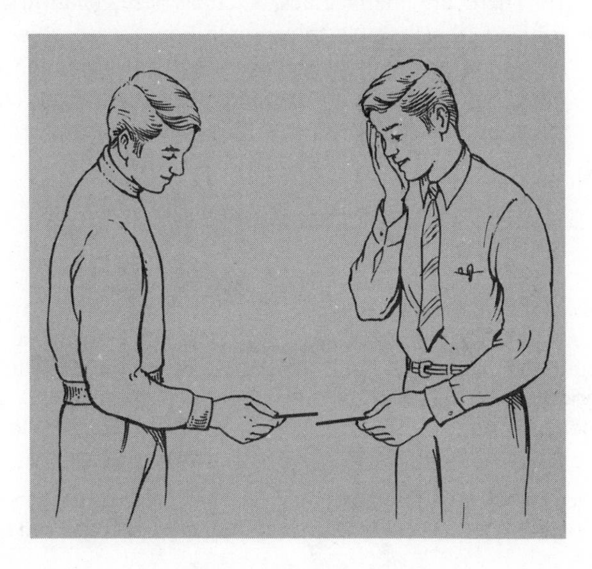

Figure 1–14
An experiment for testing your ability to judge distance with one eye.

travel in different directions to reach the two eyes (Fig. 1–15). The act of converging your eyes on the right sight lines and focusing them to the right cones of rays has been calibrated by your past experience. Your acquired knowledge of the distances associated with the particular act of converging and focusing tells you directly where the object is—and it tells you with more accuracy than just the act of focusing with one eye.

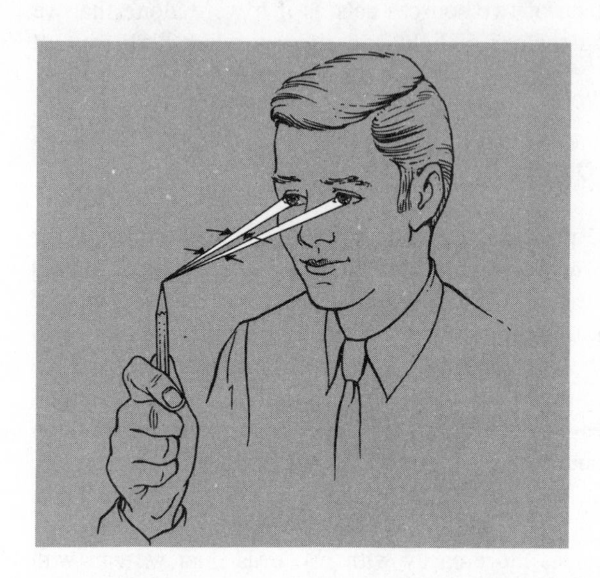

Figure 1–15
Binocular vision. We judge distance by the angle at which the eyes are converged on the subject, as well as by focusing them on the cones of rays.

If you leave other objects around when you try to measure distance, or if you use blocks instead of thin wires, you will be able to touch one object to another still more easily. In such circumstances your brain takes advantage of other clues, for example, the stereo effect that arises from the slightly different view of the faces of the blocks seen by each eye. These two views differ more as the object is moved closer. The size and position of nearby objects also help the brain to judge the distance of an object.

There are many clues, such as size, shadows, and motion of familiar objects, that help us to estimate how far an object is from us. When these clues are present, the direct interpretation of the rays in the cones of light entering the eyes and of the angle between the cones becomes less important, but at short distances these obvious physical clues are definitely used.

For Home, Desk, and Lab

1.* Which of these objects are luminous (when in normal operation)? (Section 1.)

Camera	Chrome trim on car
Firefly	Electric stove heating element
Flash bulb	Diamond
Mirror	

2.* How can you tell whether an object is luminous or nonluminous? (Section 1.)

3. We have seen that glass, although transparent, does not transmit all of the light that enters it, some of the light being absorbed. Is this also true of clear water? Be prepared to discuss what evidence you would look for to support your answer.

4. The color of the sun appears to change. For example, it looks redder at sunset than at noon.

What can you conclude from this regarding the passage of light of different colors through the air? Can you relate this phenomenon to the fact that the sky is blue?

5. Figure 1–11 shows clearly that smoke scatters light toward our eyes, even though the light was originally traveling in such a direction that it would not reach us. In view of this, why does a dense cloud of smoke overhead appear dark, rather than light? Write briefly. Be prepared to discuss.

6. We have seen that when light from a white object falls on a photographic film it produces a deposit of black silver after development. This results in the familiar negative image on the film. If you shine light through the negative onto another photographic film, after development what kind of an image will you have?

7. Move this book toward you, with one eye closed or covered. When the book has just reached the point at which the print blurs, have someone measure the distance of the book from your eye. Repeat the experiment with the other eye. Are the two distances approximately equal? Try this with a few people of different ages, and record the results along with their ages.
   (a) Is there a limit to the ability of the eye to adjust itself for clear vision?
   (b) Is this limit the same for all persons, or for the two eyes of any one person?
   (c) Does it vary, in general, with the age of the person? Compare your answers with those found by other people in the class.

8. Consider the changing appearance of the moon in the course of a month. Can you explain why a full moon rises close to sunset and a new moon rises close to sunrise?

9.† Artificial satellites can often be seen as bright objects high in the sky long after sunset. What must be the minimum altitude in meters of a satellite moving above the earth's equator for it to be still visible directly overhead two hours after sunset? (The radius of the earth is 6.38×10^6 m.)

10. (a) How long does it take light to reach the earth from the sun?
    (b) If the light from the nearest star takes 4.3 years to reach us, how far away is the star?

11. Radio waves travel at the same speed as light in empty space or in air.
    (a) How long does it take a radio signal to travel from New York to San Francisco, a distance of about 4.8×10^3 km?
    (b) A radar transmitter, which sends out radio signals of a particular type, when pointed at the moon receives a reflection 2.7 sec after the signal is sent. What does this experiment give as the distance of the moon from the earth?

12. (a) How does the 2.7 sec required for a radio wave to travel to the moon and back affect the conversation between an astronaut on the moon and ground station personnel?
    (b) Extend (a) to astronauts on Mars when it is at its nearest or farthest position from the earth. (The distance from the earth to the sun is 1.5×10^{11} m and from Mars to the sun is 2.3×10^{11} m.)
    (c) Extend (a) to an astronaut traveling close to the next nearest star, from which the light requires 4.3 years to reach us.

13.* In Fig. 1–9, is it possible to place the screen in such a way as to obtain the penumbra without the umbra? (Section 7.)

14. Figure 1–9 shows how a fuzzy shadow is produced. How could you produce a sharper shadow of the same obstacle?

15. We sometimes see total eclipses of the sun by the moon, and sometimes annular eclipses. In the latter, a ring of light from the sun is seen around the edge of the moon.
    (a) By drawing a diagram of the earth, moon, and sun, explain why two different kinds of solar eclipses occur. Do the distances of the moon from the earth and of the earth from the sun remain the same?
    (b) The moon is about a quarter of a million miles from the earth, and the sun is about 93 million miles away. If the moon has a diameter of two thousand miles, what is the approximate diameter of the sun?

16.* In the upper photo of Fig. 1–12 what would happen to the cone of light if the source were moved farther away? (Section 8.)

Further Reading

These suggestions are not exhaustive but are limited to works that have been found especially useful and at the same time generally available.

BRAGG, SIR WILLIAM, *The Universe of Light.* Macmillan, 1933; Dover, 1959. The entire book is interesting as well as easy reading.

BROWN, SANBORN C., *Count Rumford.* Doubleday Anchor, 1962: Science Study Series. (Chapters 10 and 18)

GRIFFIN, DONALD R., *Echoes of Bats and Men.* Doubleday Anchor, 1959: Science Study Series.

MINNAERT, M., *Light and Color in the Open Air.* Dover, 1954. A classic, describing many light effects such as rainbows, mirages, sunset colors, and blue sky.

WEISSKOPF, VICTOR F., *Knowledge and Wonder.* Doubleday Anchor, 1963: Science Study Series. (Chapter 3)

Reflections and Images

CHAPTER

2

2-1 The Laws of Reflection
2-2 Images in Plane Mirrors
2-3 Parabolic Mirrors
2-4 Astronomical Telescopes
2-5 Images and Illusions
2-6 Real and Virtual Images

2-1 The Laws of Reflection

All that we have said in Section 1–9 applies to finding the apparent position of a source of light. This, however, need not be the actual position of the source. If rays of light have had their direction changed by reflection or refraction, tracing such rays straight back to their point of intersection will not lead to the actual point from which the rays started. Nevertheless you will see the source apparently located at the point of intersection. You see an image of the source, not the source itself.

In Chapter 1 we briefly discussed the images seen in mirrors. Can we find laws of reflection that explain the position of such images? Figure 2–1 shows a pencil of light falling on a polished metal plate; the pencil is made visible by smoke. Notice that the *reflected* pencil *PR* is as sharply defined as is the *incident* pencil *IP*. Polished metals, liquid surfaces, and mirrors that reflect sharply defined pencils are said to be *specular reflectors.* Materials such as white paper give *diffuse reflection.* [See Fig. 2–2.]

To study specular reflection we place a sheet of white cardboard so that the two pencils just skim along its surface as in Fig. 2–3. We find that the cardboard must be held perpendicular to the reflecting surface. If we do the experiment over and over again, changing the direction in which the incident pencil strikes the surface, we always get the same result—the cardboard (and therefore the plane of the two pencils) is always perpendicular to the reflecting surface.

As a further step we draw a line through the middle of the cardboard at right angles to one edge. Then we place the cardboard in a position perpendicular to the reflecting surface as before, with the end of the line just touching the point at which the incident pencil strikes. Since the line is on the cardboard, it is in the same plane as the two pencils of light and also perpendicular to the reflecting surface. This observation gives us the first law of specular reflection:

> When light is reflected from a plane specular surface, the incident ray, and the normal (the perpendicular) to the surface at the point of contact all lie in the same plane.

Now suppose that we measure the angles between the normal and the two rays, as indicated in Fig. 2–3. We find that the angle between the reflected ray and the normal, called the *angle of reflection,* is equal to the angle between the incident ray and the normal, called the *angle of incidence.* This result is confirmed whenever we compare these two angles; hence we can state the second law of specular reflection:

> The angle of reflection is equal to the angle of incidence.

2-2 Images in Plane Mirrors

We can use these two laws of reflection to help us locate and explain the nature of images in an ordinary mirror. We start with the diagram shown

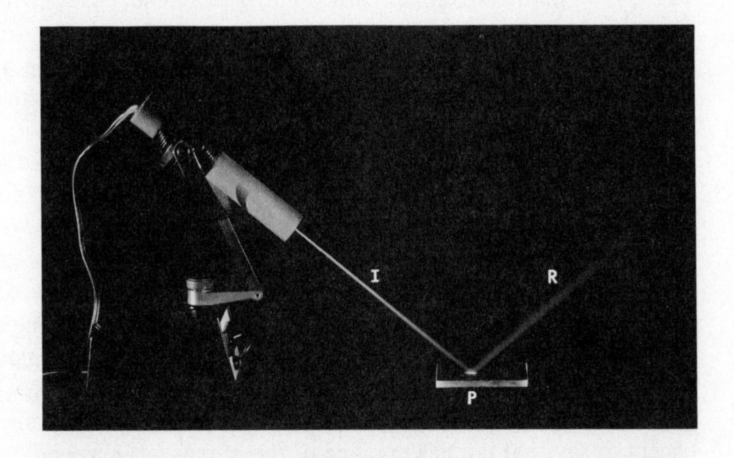

Figure 2–1
Specular reflection. A beam of light striking a polished metal surface produces a sharply defined reflected beam.

Figure 2–2
Diffuse reflection. When a beam of light strikes a piece of white paper, light is reflected in all directions.

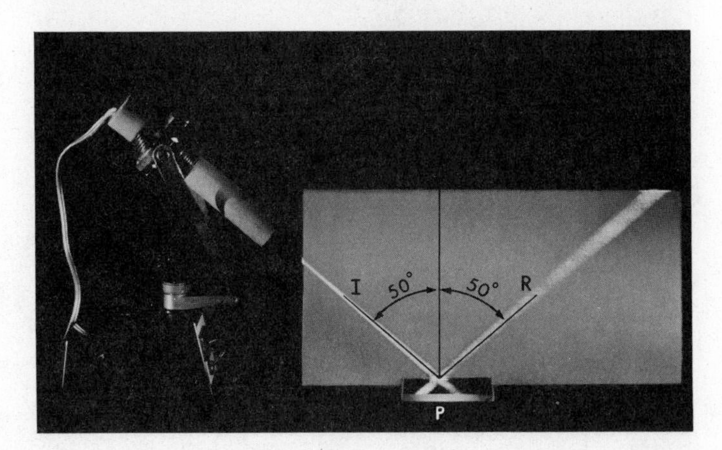

Figure 2–3
An experiment that demonstrates the first and second laws of specular reflection. The beams seem to cross because you see in the mirror the reflections of the lighted regions on the cardboard.

in Fig. 2–4, in which the horizonal line LM represents a line on the reflecting surface of a plane mirror which is perpendicular to the plane of the page. The point H represents a point on an object viewed in the mirror. Two rays are drawn in the plane of the paper diverging from H. One, HM, is perpendicular to LM and the other, HL, makes the angle i with a line, at L, that is perpendicular to LM. MH' and LH' are extensions behind the mirror of the two reflecting rays.

In accordance with the second law of reflection, we have made angle r equal to angle i. Angle i' and angle i are equal, and angle r' and angle r are equal. Therefore, angle i' equals angle r', and so the large triangle HLH' is isosceles. Because it is isosceles, $H'M = HM$.

In showing that $H'M = HM$ we did not specify the size of the angle i. The large triangle is isosceles no matter what the value of angle i; and $H'M$ is always equal to HM. Only the apex L of the large triangle changes. Its base HH' remains the same, as shown in Fig. 2–5, and $H'M = HM$ for any angle of incidence. Furthermore, we did not specify the plane of the paper except to say that it and the plane of the mirror are perpendicular. It could be any plane that includes the line HM, as Fig. 2–6 shows.

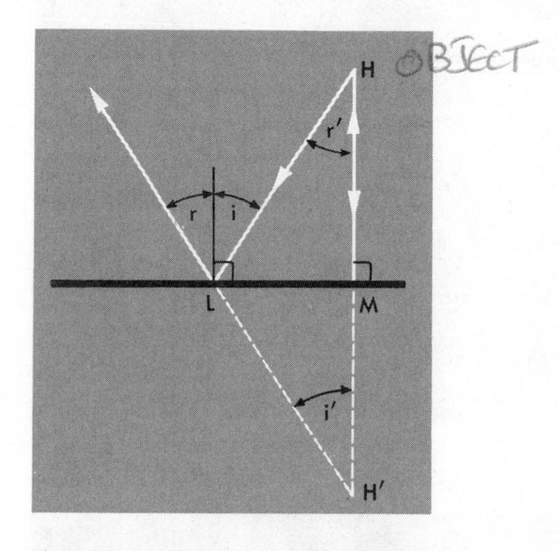

Figure 2–4
The geometry of two rays reflected from a plane mirror. Both rays originate at *H*. One is reflected back on itself from *M*, and the other is reflected from *L*. Angle *i* equals angle *r*. *MH'* and *LH'* are extensions of the two reflected rays.

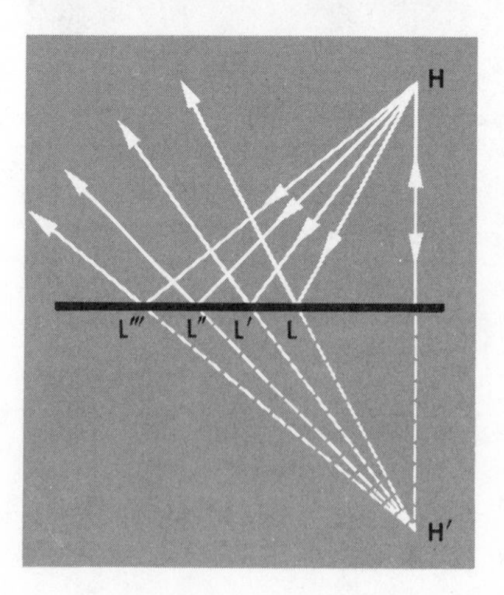

Figure 2–5
The geometry of more than two rays reflected from a plane mirror.

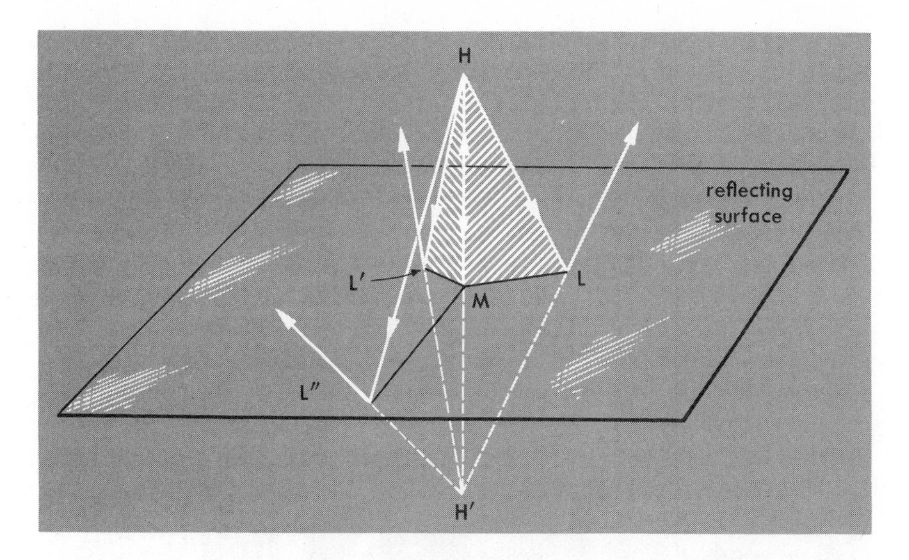

Figure 2–6
Four rays from H, three chosen at random and one perpendicular to the reflecting surface, strike the reflecting surface at the points M, L, L', and L''. The three triangles HML and HML' (both shaded) and HML'' (unshaded) lie in different planes that are perpendicular to the plane of the reflecting surface. All three reflected rays act as though they were diverging from the point H'.

Summing up our findings, we have shown that all the rays which start at a point H and are reflected by a plane mirror appear to diverge from a point located on the extension of the perpendicular from H to the mirror; this point is as far behind the mirror as H is in front of it. As far as the eye is concerned, then, light appears to be diverging from this point. We speak of this point as the image of point H.

By the same reasoning we can locate the image of any point on the arrow shown in Fig. 2–7, and thus find the complete image of the arrow.

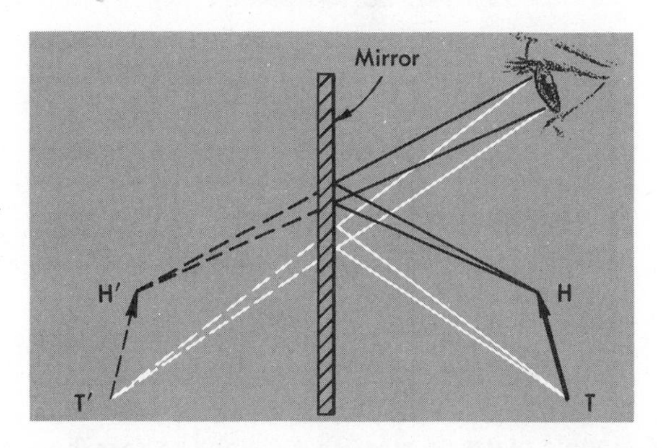

Figure 2–7
The two rays from H reach the eye as if they came from the same point H', so the head of the arrow appears to be at H'. Similarly, the white rays in the figure indicate that the tail of the arrow T appears as if it were at T'.

Every point on the image is exactly opposite the corresponding point on the object and equally far from the mirror surface. The image and object are exactly similar and of the same size.

In our construction we have extended the reflected rays behind the mirror, but we could not find the light from the arrow by going behind the mirror. Because the mirror causes the rays to appear as if they come from the image, although they actually do not, we say that a *virtual image* is formed. We use this term to distinguish it from a *real image,* which we shall meet later.

We have obtained the location of the virtual image by using the two laws of reflection. It is easy to verify the results by a simple experiment. A small mirror is mounted perpendicular to a tabletop, and an object having a height somewhat greater than the mirror is placed in front of it, as shown in Fig. 2–8. An identical object is then placed behind the mirror and moved about until its top portion, seen over the mirror, appears as a continuation of the image when it and the image are viewed from any possible position. The real object behind the mirror then is in the same position as the virtual image. Measurements show that this is the position given by our construction in Fig. 2–4.

Figure 2–8
Location of the virtual image. This experiment verifies our calculations based on the laws of reflection.

2-3 Parabolic Mirrors

On the top of the Pic du Midi in the French Pyrenees research workers curious about the effects of ultrahigh temperatures have been able to use mirrors to concentrate enough sunlight to melt steel. In India, using inexpensive "mirror ovens," housewives cook dinner with concentrated sunlight. Astronomers use the large mirrors in their telescopes to concentrate the faint light from distant stars; they can thus produce images on photographic plates of stars that cannot be seen by the eye alone. What shape mirror has this ability to concentrate light?

We cannot concentrate light with a plane mirror, for the light always diverges, appearing to come from the virtual image behind the mirror. But by using several plane mirrors, we can cause several pencils of light from the same source to cross each other in a small region of space. To be specific let us assume that the source is far away—a star, perhaps—and therefore the pencils of light that come from it to our mirrors are almost parallel. The farther the source, the smaller the angles between the pencils of light, until for a star all the pencils reaching the earth are practically parallel. In Fig. 2–9 twenty-five mirrors cocked at slightly different angles are arranged to reflect parallel light so that it passes through a single small

Figure 2–9
Construction of an approximately parabolic mirror from plane mirrors.

region. We see the region because the smoke there scatters some of the light in the many crossing pencils. Other regions are equally smoke-filled but appear darker because they contain fewer pencils of light. The pencils of light pass through the small region of overlap and then diverge from one another.

We can concentrate the light further if we use more and smaller mirrors. Cut each one of our twenty-five mirrors into four pieces, for example. Then each pencil has only one-fourth of the previous cross-sectional area, and we can make them converge into a smaller region. Now cut the mirrors again, making them still smaller. The region of overlap is cut down still more. Do it again, time after time. Eventually the mirrors are infinitely small, or at least as small as we can make them; but they intercept the same amount of light that our original twenty-five mirrors did. They reflect the same amount and send it all through a single tiny spot.

To construct a mirror which focuses light from a broad beam of parallel rays to a single spot by successively dividing and adjusting the orientation of plane mirrors is easy in theory but difficult in practice. We can illustrate the procedure quite easily, however, by concentrating parallel light on a line instead of a spot. Two stages in this process which you can carry out without too much difficulty are shown in Fig. 2–10. It is easy to imagine the similar processes carried out in space to produce the single spot.

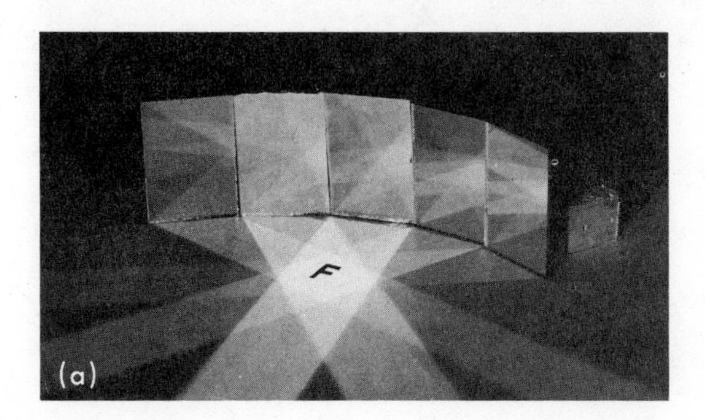

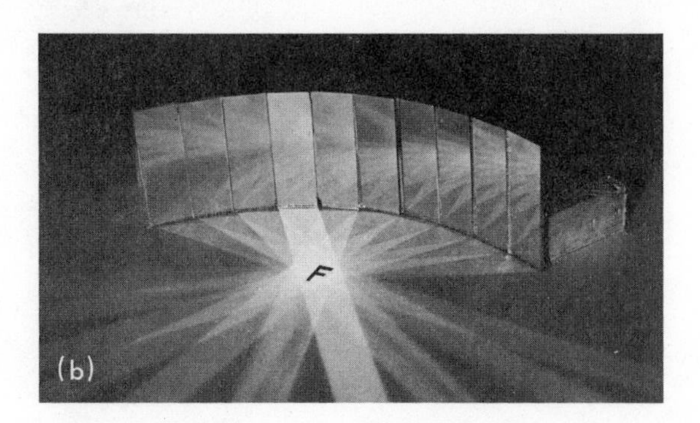

Figure 2–10
(a) The converging of light by five plane mirrors. (b) As the number of mirrors is increased to ten, the light is converged into a smaller region.

As we subdivide and reorient plane mirrors in space, we construct closer and closer approximations to a mirror with a continuously curved surface. We can imagine that the indefinite continuation of the process would result in a smooth mirror which focuses the parallel light. The shape of the smooth surface (called a paraboloid of revolution or parabolic mirror) is determined exactly by our imaginary procedure. The spot F to which all the reflected light converges is called the *principal focus* of the parabolic mirror, and its distance from the center of the mirror is called the *focal length* or *focal distance*. It is important to know the focal length, and we shall represent it by the letter f. In Fig. 2–11 a cross section is shown through the principal focus parallel to the direction of the incident light. Each of the rays we draw indicates how the light is reflected. It stays in the plane of the figure according to the laws of reflection; and the angle of reflection of each ray equals the angle of incidence. The curve shown, which is the intersection of the parabolic mirror with the plane of the figure, is called a parabola. The surface of the mirror is called a paraboloid of revolution because it can be generated by rotating the parabola of Fig. 2–11 about the incident ray through the principal focus F. This incident ray that passes through, and is reflected back through, the principal focus lies on the *axis of revolution* of the paraboloid.

When we make parabolic mirrors, we usually start by making a smooth surface that is approximately parabolic; then, if necessary, we improve it by distorting, grinding, and polishing until light sent in parallel to the axis of revolution is brought to an accurate focus.

We have seen that all of the light that falls on a parabolic mirror in a direction parallel to the axis is reflected in such a way that it passes through the principal focus. Since light can travel in either direction over a given path, we can interchange the incident and reflected rays by putting a tiny, intense source of light at the principal focus of a parabolic mirror. It follows, then, that any light starting from the principal focus will, after reflection, be traveling parallel to the axis of the mirror. Searchlights and some flashlights are constructed on this principle, as shown in Fig. 2–12. All light from the source that strikes the parabola travels outward in parallel paths to form a narrow, intense beam that penetrates to great distances through space.

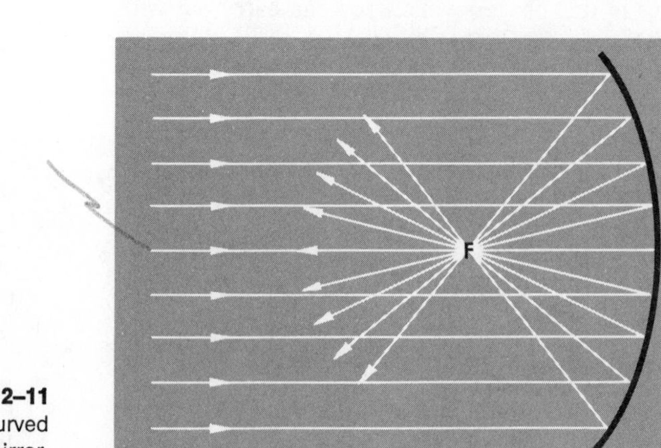

Figure 2–11
The converging of light by a curved mirror.

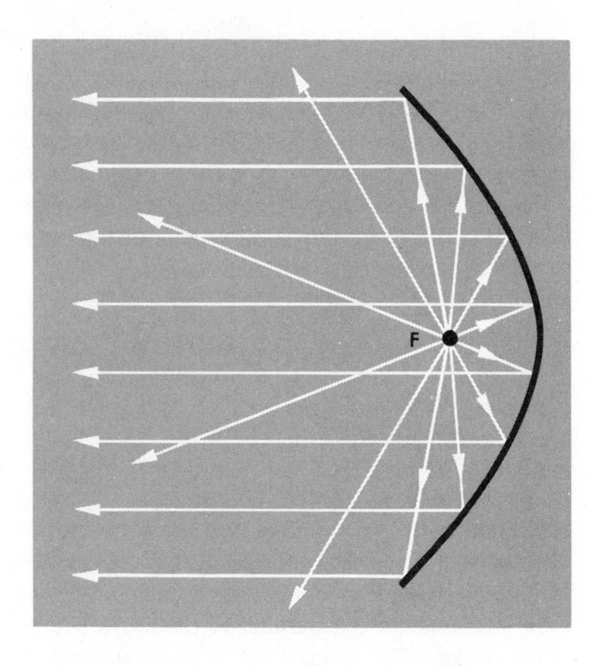

Figure 2–12
A flashlight. More than half of the light from a flashlight bulb located at the focal point F of the paraboloid is reflected as a parallel beam.

2-4 Astronomical Telescopes

We have seen that a parabolic mirror brings to a focus all the light that arrives parallel to the axis of revolution; also, when we place a tiny source at the focus of such a mirror it sends out a parallel beam. What happens to light that does not come in parallel to the axis or to light that diverges from a source which is not at the principal focus? The answers to these questions will lead us to understand such varied devices as astronomical telescopes and shaving mirrors.

Two ways are open to answer our questions. We can do experiments with parabolic mirrors, or we can decide logically what should happen according to the two laws of reflection. Indeed, we can use both, or we can mix the two methods; but we certainly must check our logical conclusions by appropriate measurements.

Let us start again with the light from a star hitting a parabolic mirror. Suppose this time that the parallel pencils of light do not come in along the axis of the mirror; instead, the direction of the rays is at an angle α to the axis. In Fig. 2–13 a plane is shown which includes the axis of the mirror

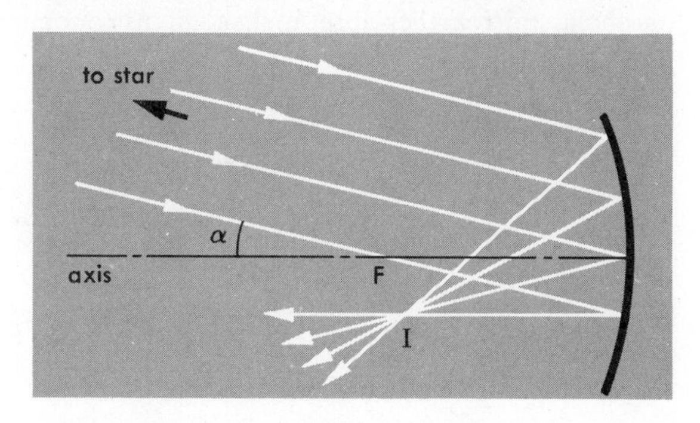

Figure 2–13
Formation of the image of a star that is not on the axis of the telescope.

and some of the rays representing the light from a star. By constructing the reflected rays so that the angle of reflection of each ray equals the angle of incidence, we can find out how we expect the mirror to work. As you see in the figure, the various reflected rays pass through a small region just below the principal focus on the side of the axis opposite the star. They do not pass through a single point; but very careful construction (or calculation based on geometry) shows that they do come close to one point even when we include those rays that are not in the plane of the figure. The mirror should therefore form a blurred spot of light at the position I where the light almost focuses.

We can find the approximate position of I by tracing just two of the many rays that converge on I. These we choose so as to make the job of ray tracing as easy as possible. In particular we know that a ray through the principal focus will go out parallel to the axis. Also a ray that comes in parallel to the axis will go out through the focus. A ray that hits the center of the symmetric mirror will leave it on the other side of the axis, going out on a symmetric path. Two of these rays will always allow us to find the image I. In Fig. 2–14 we locate I by tracing the ray that comes in through F (and therefore must go out parallel to the axis) and the ray that hits the mirror at the center. The image I lies at approximately the same distance from the center of the mirror as F, but it is at the same angle on one side of the axis that the star is on the other. Consequently, the images of several stars as seen from the center of the mirror form the same pattern as the stars themselves do in the sky; they are just turned top for bottom and right for left.

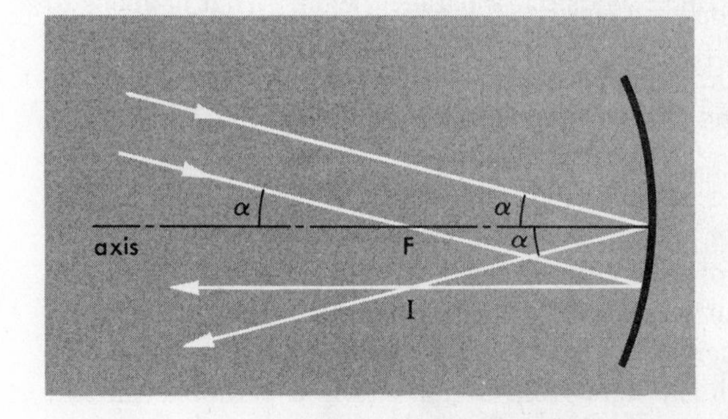

Figure 2–14
Simplified version of Fig. 2–13, showing how to find the position of the image.

Also, a photograph taken on a photographic plate placed perpendicular to the axis at F shows the spacing of the stars correctly (Fig. 2–15). The parabolic mirror therefore makes an astronomical telescope. It gathers

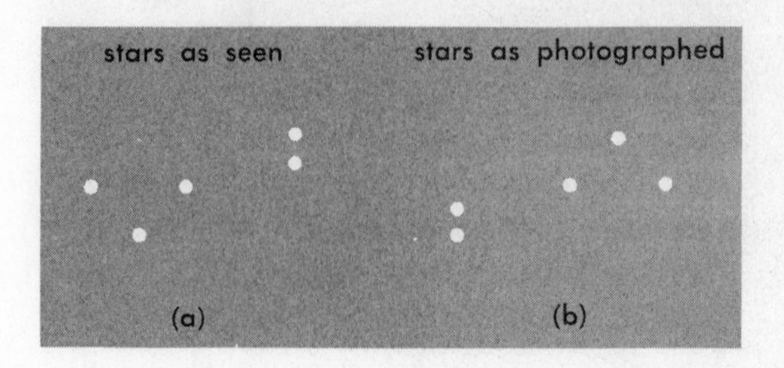

Figure 2–15
A sketch of stars as seen in the sky and as photographed with a telescope. Try looking at the page upside down.

light from the stars and concentrates it on nearby spots in positions corresponding to the positions of the stars on the celestial sphere. Nearly all modern large telescopes use parabolic mirrors.

The heart of the Hale telescope at Mt. Palomar is a parabolic mirror 5 meters in diameter. A metal framework which supports the parabolic mirror is mounted on axles and bearings in such a way that it can be pointed at various spots in the heavens. Figure 2–16 shows the actual reflecting telescope and gives an idea of the mounting required. This mounting can be controlled so that it follows the apparent motion of a star from east to west while a picture is being taken. With such telescopes photographs of the sky are taken, and every detail of our predictions and of the more careful calculations that predict the extent of blurring is confirmed.

Figure 2–16
The Hale telescope at Mt. Palomar Observatory. The arrow (bottom, center) points to the mirror.

A photographic film, exposed for a long time, will gather more light and record a fainter image than can our eyes. It also gives a permanent record of the relative intensity and positions of star images. Large astronomical telescopes are seldom used for direct viewing, because the eye is too insensitive and unreliable as a recording instrument.

Even with long exposure times, we need as much light as we can get to record images of faint stars and galaxies on a photographic plate. Therefore, telescopes with large-diameter mirrors are used to increase the amount of light collected. The largest telescopes gather enough light so that, with very sensitive photographic plates and long exposure times, faint galaxies as far away as 10^{22} km have been recorded.

2-5 Images and Illusions

You can shoot bullets through the light bulb of Fig. 2–17 without marring its surface or causing the light to flicker. You can pass your hand right through the bulb without feeling a thing. It's the little lamp that isn't there. The light is there, all right; and it was perfectly visible to someone near the camera when the picture was taken. The ruler in the picture was there too, made of genuine wood. If your bullets hit it, it would splinter; your hand could not pass through it.

Figure 2–17
Real image formed by a parabolic mirror. Notice the size measured on the centimeter scale.

The little bulb is an image, the image of an honest light bulb located well out of sight about 17 meters in front of the image. The image was formed by a parabolic mirror almost 2 meters behind the illusory bulb.

Thus we see that parabolic mirrors do form images of extended nearby objects. Indeed, this experiment can tell us more. We know that the genuine light bulb stood upright on its base while its image was suspended in space upside down. If we change the distance to the light bulb, the image moves, and it changes size. When we move the genuine bulb toward the mirror, the image moves away from the mirror toward the incoming source. It also grows bigger.

From a series of such experiments, using parabolic mirrors of different focal lengths and moving the source through a whole set of distances in front of each mirror, we can discover simple relations between the distances and between the size of the image and of the object, as the source is usually called in this kind of work. To discover these relations from the results of such experiments without using the laws of reflection is possible, but it requires many experiments and a good guess about how to represent the results simply and accurately.

On the other hand, once we know from experiment that images are actually formed, tracing a few rays in accord with the laws of reflection will lead us to the relations between object and image with very little work. A few checks with experimental results will then show that the laws of reflection are not leading us astray. We can then predict confidently what will happen with any parabolic mirror and an object at any distance in front of it.

Suppose that an object is located in front of a parabolic mirror a distance S_o beyond the principal focus (Fig. 2–18). The height of the object above the axis is H_o. Let us consider the light coming from the top of the object that hits the mirror. In particular follow the two rays drawn in the figure. One represents light passing through the focus, F, to the mirror and back out parallel to the axis. The other starts parallel to the axis and therefore must be reflected through the focus. Where they intersect after reflection must be the place where the light comes together again to form the image of the top of the object. These two rays are all we need, because any other reflected rays from the top of the object will pass through a very small region of space surrounding the intersection of the two rays we chose. We found the same thing was true for light coming from a star when the incident light was not quite parallel to the mirror axis.

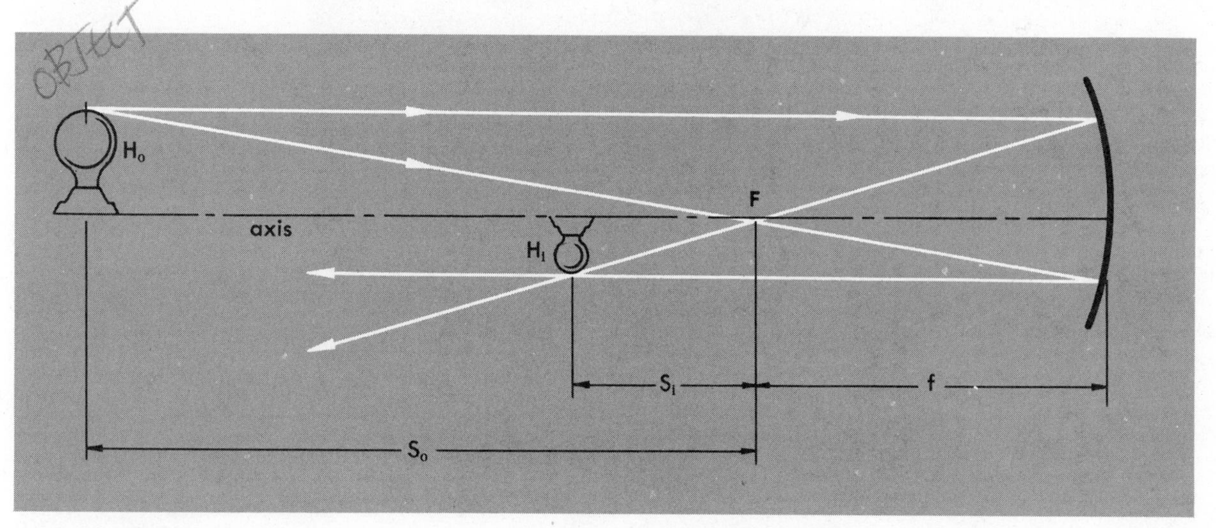

Figure 2–18
Finding the size and location of an image of an object at finite distance. The image is located by tracing two rays.

We see an image of the top of the object because the light moving along various paths is sent through an image point and diverges from there to our eye or camera just as if a small version of the object were located there. The intersection of the two rays tells not only how far away from F the image of the top is, but also determines the height H_i of the whole image as it extends down from the axis. Anytime we want to draw such a two-ray picture to scale, we can find out what the laws of reflection predict about the image.

For example, consider the first experiment we discussed in this section. We used the following setup to obtain the photograph in Fig. 2–17. The genuine light bulb was placed 17.50 m in front of the principal focus of the mirror. The focal length of the mirror, determined by using parallel light from the sun, was 1.75 m. So the distance between source and mirror was 19.25 m. The image is seen 0.175 m in front of the focal point or 1.92 m from the mirror. Furthermore, an ordinary large electric light bulb 19 cm high was the source; and the image, as measured on the centimeter scale beside it, is only 1.9 cm high as well as upside down. Tracing our two rays on a scale drawing of this setup shows that the image should lie 17.5 cm, or 0.175 m, in front of F, and that it should be only $\frac{1}{10}$ as tall as the object, in agreement with the experimental facts. With similar drawings you can predict the size and position of the image for other arrangements of object and mirror, or for mirrors of different focal length.

The geometry of these two-ray drawings is so simple that it hardly seems necessary to go further in specifying the relations. But it is a nuisance to draw diagrams all the time, and the simplicity of the geometry suggests that we can express our results in extremely simple forms for numerical computation. Let's start with the relation of image to object size.

In Fig. 2–19 we have left out everything but the ray from the top of the object going through the principal focus. Because this ray is reflected parallel to the axis and passes through the image, the "top" of the image must lie the same distance away from the axis as the reflected portion of this ray.

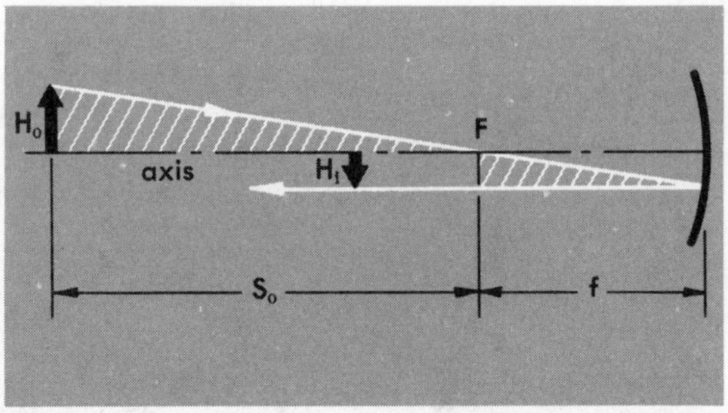

Figure 2–19
Finding the ratio of image size to object size by similar triangles. The ray from the top of the object through the principal focus is used to relate the ratio of the sizes to the ratio of the focal length and object distance.

Therefore, H_i has the same height as the perpendicular from F to the reflected ray. Now we can get the ratio of H_i and H_o immediately from the shaded similar triangles. Because f and S_o are the bases and H_i and H_o the corresponding altitudes, we get

$$\frac{H_i}{H_o} = \frac{f}{S_o}.$$

In fact, the base of the little triangle is a bit shorter than f because of the curvature of the mirror; but, as long as the object isn't too big, the ray will cut the axis at a small angle. The point where it hits the mirror will be almost exactly the distance f away from the principal focus.

In this excellent approximation, then, to determine the ratio of image size to object size all we need to know is the ratio of focal length to the distance of the object from the principal focus. Using the example of Fig. 2–17 with $f = 1.75$ m and $S_o = 17.5$ m, we get

$$\frac{H_i}{H_o} = \frac{1.75}{17.5} = \frac{1}{10}.$$

In the experiment it was also 1/10.

Now let's see what additional information we can obtain from the other ray, which comes in parallel to the axis and is reflected through the principal focus. We can use it to construct the shaded similar triangles in Fig. 2–20.

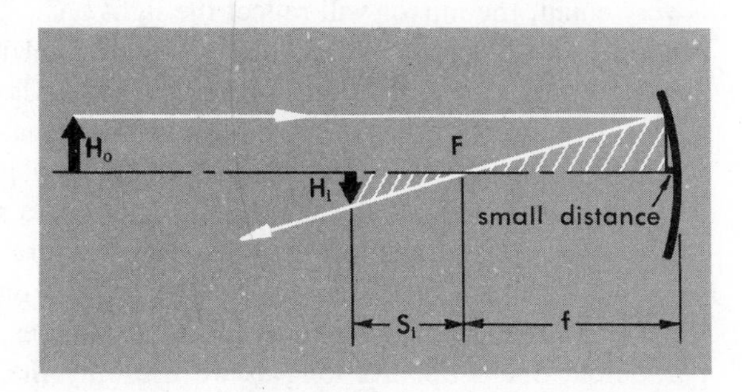

Figure 2–20
Finding another relation between image and object size. The ray from the top of the object parallel to the axis is used to relate the ratio of the sizes to the ratio of focal length and image distance.

From these triangles we find

$$\frac{H_i}{H_o} = \frac{S_i}{f}$$

again with a small error because in place of f we should use f minus the small distance pointed out in the figure. When this relation is combined with

$$\frac{H_i}{H_o} = \frac{f}{S_o}, \text{ we get } \frac{S_i}{f} = \frac{f}{S_o}$$

$$\text{or} \quad S_iS_o = f^2.$$

The distances of image and object from the focus are in inverse proportion. As the object comes toward the mirror, the image must move out so that the product S_oS_i stays equal to f^2. This, of course, is just what we found when we moved the light bulb at the beginning of this section. Originally S_o was 17.5 m and S_i was 0.175 m, giving the product $S_oS_i = (17.5)(0.175) = (1.75)^2$, which is the square of the focal length. You can work out what S_o and S_i were after we moved the light 8.75 m closer to the mirror. Use the new S_o and the equation $S_i = f^2/S_o$ to find S_i. Experimentally the illusory little bulb moves 0.175 m and doubles in height. Does your calculation agree?

Using $S_i = \dfrac{f^2}{S_o}$ and $H_i = \dfrac{f}{S_o} H_o$, we can quickly predict the positions and sizes of images in any parabolic mirror.

One example of the application of $S_oS_i = f^2$ is too attractive to resist. Let us ask: at what distance from the principal focus must the object be placed so that the image will be at the same place? That is, when are S_o and S_i equal? Clearly the answer is: when both S_o and S_i are equal to f; that is, when object and image are both located at a distance $2f$ from the central point of the mirror surface. Furthermore, image and object are the same size, for

$$H_i = \frac{f}{S_o} H_o \text{ gives } H_i = H_o.$$

When the object is on the axis a distance f in front of the focal point, and is very small, the mirror will reflect the light back into the small region surrounding the object itself. Similarly, a spherical mirror will bounce light coming from the center of the sphere of which it is a part, right back to the center. Apparently, the effective portion of a parabolic mirror must closely resemble a portion of a sphere. In particular, a parabolic mirror of focal length f must be almost the same as a section of a sphere of radius $2f$.

The relation is shown in Fig. 2–21, where a circle with center at C and a parabola with focus at F are superimposed. It is only far from the axis that we find appreciable difference as the sphere closes around in front and the paraboloid opens up. As long as we use only the central portion, we can just as well use a spherical mirror in place of a parabolic one. In fact, because it is usually cheaper to build part of a sphere than part of a paraboloid, we often use spherical mirrors of radius $2f$ when we want to obtain a mirror of focal length f.

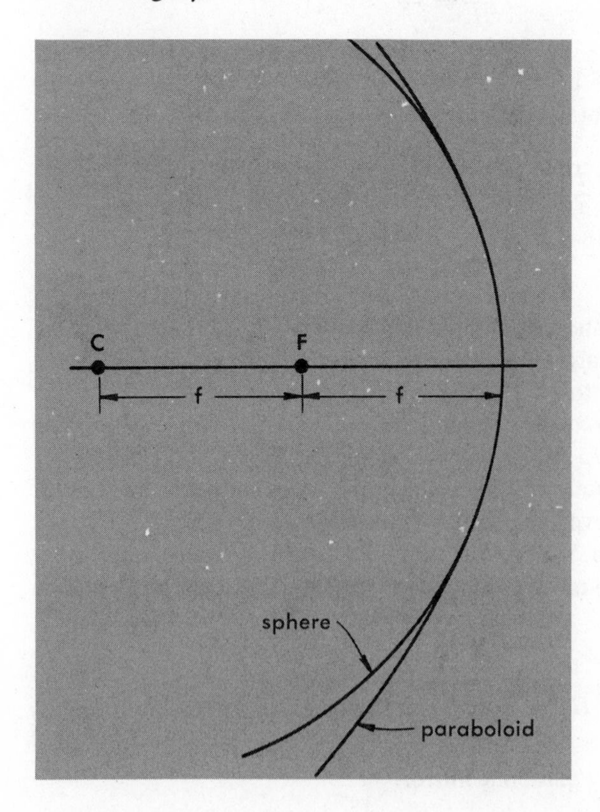

Figure 2–21
The relation of spherical and parabolic mirrors. The common principal focus is at *F*, and the center of the sphere is at *C*.

The illusory light bulb of Section 2–5 is a real image. The pencils of light that we see really converge to it and diverge from it to our eyes. It is thus distinguished from the virtual image seen in a plane mirror. There may be no light passing through a virtual image, but the light is sure to pass through a real image, and a photograph can be made by putting the film at the image as we noted in the case of the astronomical telescope. It would hardly help to put a film behind a plane mirror at an image position.

By studying parabolic mirrors (or their approximate spherical equivalents), we can learn when mirrors form real images and when the images are virtual. The clue is provided by our previous conclusions. From the relation $S_i = f^2/S_o$ we find that as we bring the object nearer to the focal point, the image runs away. When S_o gets very small, S_i, the image distance, becomes huge; and as the object passes through the principal focus, the real image disappears at infinite distance in front of the mirror.

If we continue to move the object closer to the mirror, we are not surprised to find a virtual image approaching from far behind the mirror and reaching the surface when the object does. After all, a small-enough section of the mirror is hard to tell from a plane; and when the object is close, only a small section of the mirror reflects light from the object to us. The small effective section of the mirror then acts like a plane mirror and forms a virtual image. (See Fig. 2–22.)

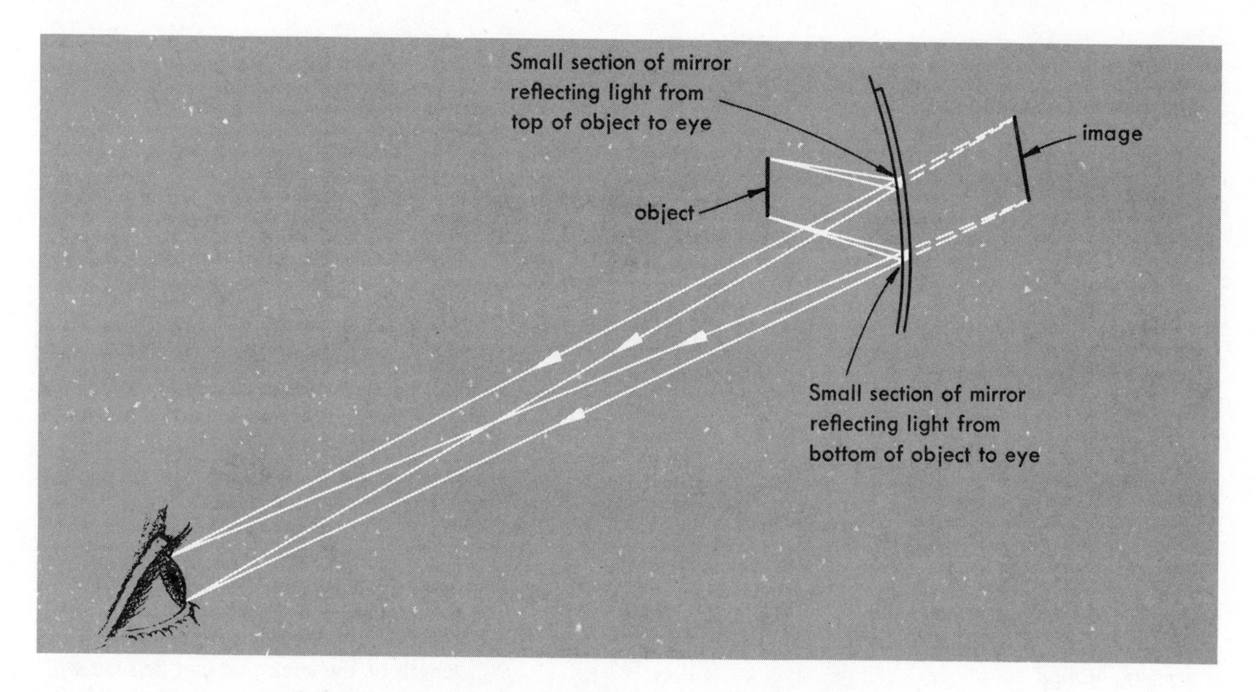

Figure 2–22
The formation of a virtual image in a concave mirror. The rays entering the eye from a point on the object are reflected by a very small section of the mirror, which can be treated approximately as a piece of plane mirror. Two such small sections are indicated. Note that they are oriented slightly differently. These differences in orientation result in a magnified virtual image.

A two-ray diagram (Fig. 2–23) now makes it easy to see that whenever the object is between the focus and the mirror we get a virtual image. We use the same two rays—the principal rays, they are called—the one parallel to the axis, which must reflect through the focus, and the one coming from

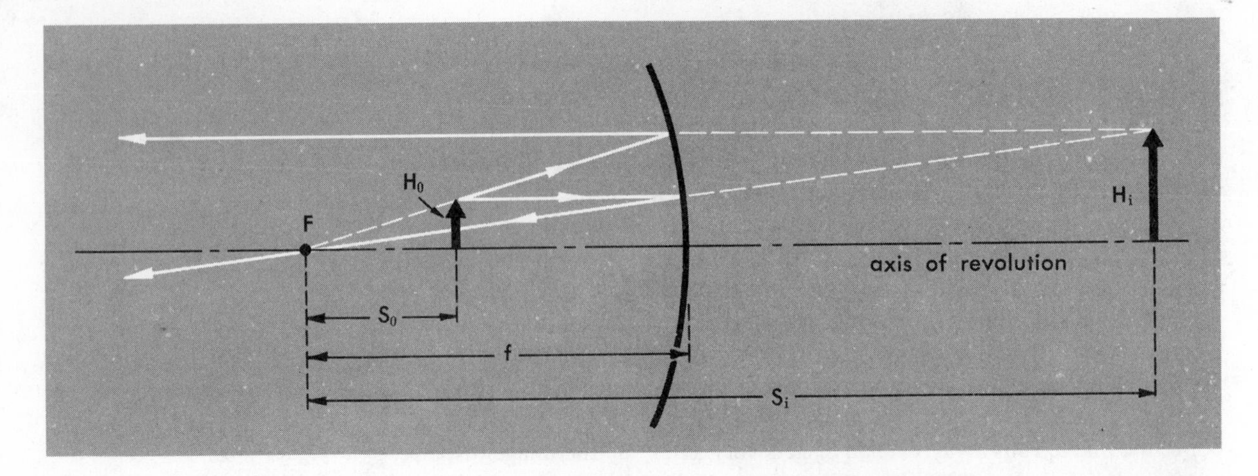

Figure 2–23
Formation of a virtual image in a concave mirror. The object lies between the principal focus F and the mirror. The principal rays are used to find the expected location of the image.

the focus to the mirror and then going out parallel to the axis. As Fig. 2–23 shows, after reflection these rays appear to come from an intersection behind the mirror, and the image is always magnified compared to the object. Ray tracing will work out the details just as before; and if you want to check experimentally, the inside of a shiny metal bowl or even a large spoon will do almost as well as a shaving mirror. An accurate sphere or parabola will help to get accurate measurements, but won't change the nature of your observations.

For Home, Desk, and Lab

1.* In Fig. 2–24, which is the angle of incidence and which is the angle of reflection? (Section 1.)

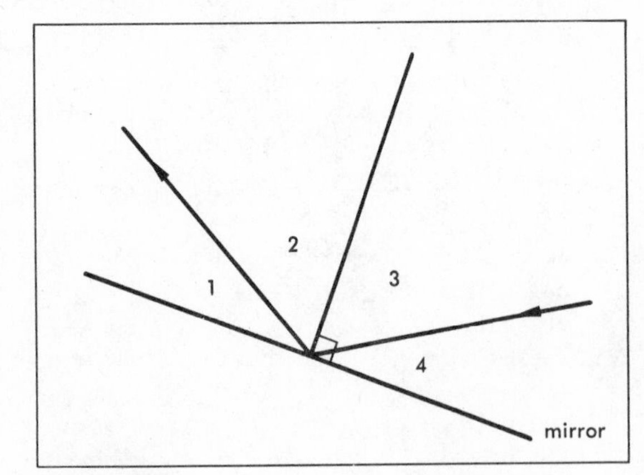

Figure 2–24
For Problem 1.

2.* How many different rays can strike a mirror at the same point and with the same angle of incidence? (Section 1.)

3.* For a given incident ray of light striking a mirror, how many possible reflected rays are there? (Section 1.)

4. If you cán see the eyes of someone in a complicated system of mirrors, is it possible for him to see your eyes?

5. Place a plane mirror (one about 30–40 cm high is convenient) with its center approximately at eye level. Hold a meter stick vertically just in front of your face, with the middle of the stick at eye level; stand in front of and facing the mirror.

   (a) Move toward and away from the mirror. Does your motion change the amount of the stick that you can see?
   (b) Formulate a general rule connecting the length of the stick that can be seen with the height of the mirror. By making a ray diagram, show that this rule holds for all distances from the mirror.
   (c) Clothing stores often have mirrors that extend all the way to the floor, designed

to allow a customer to see his or her full length. Is it necessary for this purpose for the mirror to be as long as it is?

(d) If the shortest customer has eyes at a height of 5 ft. 0 in., what is the maximum allowable height of the bottom of the mirror from the floor if the customer's feet are to be visible?

6. Two mirrors form a right angle (Fig. 2–25). A light ray *AB* is reflected from both mirrors. Prove that the reflected ray *CD* is always parallel and opposite in direction to *AB*, independent of the angle of incidence of *AB*. (This result is also true when three mirrors are at right angles like the corner of a box. This is how light from a laser is reflected from the moon.)

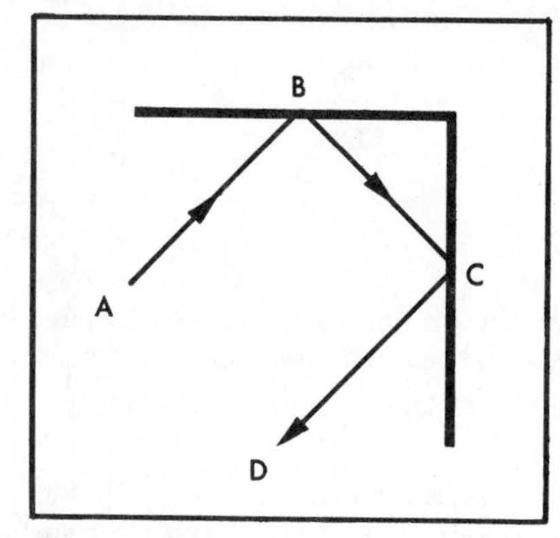

Figure 2–25
For Problem 6.

7.* Measure the focal length in millimeters on the drawing of the parabolic mirror in Fig. 2–12. (Section 3.)

8.* What would happen to the rays of light in Fig. 2–11 if a small plane mirror were placed at *F* facing the mirror? (Section 3.)

9.* Why do we need to use only two rays in Fig. 2–14 to locate the image of the star? (Section 4.)

10. (a) What will happen if, in Fig. 2–18, the real light bulb is placed at the position where the real image was previously formed?
(b) Can you state a general rule about moving an object to the position of its real image?

11.* In Fig. 2–18, if you put your eyes at the position of the letter S_i, could you see the image? (Section 5.)

12.† A nail 4.0 cm high stands in front of a concave mirror at a distance of 15 cm from the principal focus. The focal length of the mirror is 20 cm. What is the size of the image?

13. The image of a candle is 30 cm from the center of a concave mirror. The candle is 10 cm long and its image is 5.0 cm long. What is the focal length of the mirror?

14. How large an image of the sun will be formed by the Palomar telescope, whose focal length is 18 m? The sun's diameter is about 1.4×10^9 m, and it is 1.5×10^{11} m away.

15. What is the focal length of a plane mirror?

16. The distances of an object and its image in a concave mirror are often measured from the center of the mirror, instead of from the principal focus. We call these distances D_o and D_i respectively. We have $S_o = D_o - f$ and $S_i = D_i - f$ where f is the focal length. Using these relations, show that from $S_o S_i = f^2$ follows

$$\frac{1}{D_o} + \frac{1}{D_i} = \frac{1}{f}.$$

17.* Is a real image formed by a parabolic mirror ever larger than the actual object? (Section 6.)

18.* Is a virtual image formed by the parabolic mirror of Fig. 2–23 ever smaller than the actual object? (Section 2–10.)

19. In Section 2–5 the question is asked: At what distance from the principal focus must the object be placed so that the image will be at the same place? Show that this question has two answers and explain the second answer, which was not discussed in the chapter.

20. Be prepared to discuss what happens to the image as the object is moved along the principal axis of a concave mirror from infinity to the mirror.

21. Our discussion of curved mirrors has been limited to the inner, or concave, surface. The outer, or convex, surfaces of these curves will also produce images. Such mirrors are called convex mirrors and are often used as side mirrors on a car or as ornamental mirrors in a room. Using the laws of reflection and assuming the surface to be parabolic, demonstrate the following facts by suitable constructions.
(a) The area reflected to the eye by a convex circular mirror is larger than that reflected by a plane mirror of the same diameter in the same position.
(b) Light rays parallel to the axis reflect as

though they were coming from a point behind the mirror. This is the principal focus of the mirror. It is called a virtual focus. Why?

(c) Rays starting from a fixed point on the axis are reflected in such a way that they seem to come from a point on the axis behind the mirror. The image is therefore virtual.

(d) The image formed is smaller than the object and is not inverted.

(e) As the object moves in from a great distance, the image moves toward the mirror.

(f) There is a limit to the distance of the image from the mirror—that is, this distance can never be greater than a certain value. What is this value? Try drawing ray diagrams.

22 Compare Fig. 2–11, a paraboloid which reflects all rays parallel to its principal axis through its principal focus, with Fig. 2–21 and explain why all parallel rays striking the surface of the sphere would not reflect through *F*.

Further Reading

These suggestions are not exhaustive but are limited to works that have been found especially useful and at the same time generally available.

GRIFFIN, DONALD R., *Echoes of Bats and Men.* Doubleday Anchor, 1959: Science Study Series.

JENKINS, FRANCIS A., and WHITE, HARVEY E., *Fundamentals of Optics.* McGraw-Hill, 1957. Covers geometric optics on plane and spherical surfaces. (Chapters 2, 3, and 6)

MINNAERT, M., *Light and Color in the Open Air.* Dover, 1954. A very handy reference explaining many light phenomena.

Refraction

CHAPTER

3

3-1 Refraction

In order to make an experimental study of reflection we set up apparatus to observe single pencils and beams of light being reflected from surfaces. By systematic observations we arrived at two simple laws of specular reflection. These laws, in turn, make it possible to explain the formation of images in both plane and curved mirrors. We thus see that the reduction of a wide range of phenomena to two straightforward laws is a great gain in simplicity and gives us a powerful tool for prediction. It would be difficult to remember the characteristics of all the images we can produce with different mirrors and a wide variety of positions for the object. If, however, we know how to use these two laws we can arrive at a detailed description of the image formed by any combination of mirrors. Further, we can predict what kinds of images will be formed by mirrors of types that we have not studied experimentally.

A similar study will give us simple explanations of the second group of optical phenomena we discussed in Chapter 1. There we noted that light penetrates the surface of certain materials and in doing so often changes its direction. This bending of a light beam as it passes from one material into another we called *refraction.* We saw some examples of refraction in the floating-coin illusion (Fig. 1–2) and in the bent-stick illusion (Fig. 1–1). Although the effect of the bending of light is clearly shown by such observations, they tell us little about what happens to individual pencils of light. We can simplify our study by isolating a single refracted pencil of light as shown in Fig. 3–1. This photograph shows a rather broad pencil of light passing downward through air, and entering the water in an aquarium. Notice that the direction of the pencil changes abruptly at the water surface and some of the light is reflected from the surface, back into the air.

We call the incoming pencil of light the incident pencil, as we did in our study of reflection. In Fig. 3–1 the angle between this pencil and the normal to the surface is the angle of incidence. The light pencil after bending at the water surface is called the refracted pencil, and the angle that it makes with the normal is the angle of refraction.

Figure 3–1
Light passing from air into water.

By inserting a white plastic protractor into the aquarium, as in the figure, we can learn more about the direction taken by these rays. When we place the protractor perpendicular to the water surface so that the incident rays just glance along the protractor surface, the refracted rays just glance along the protractor surface also. We can repeat this experiment with different angles of incidence and different pairs of materials. For most materials we find the same result. It is summarized in the first law of refraction:

The incident ray, the refracted ray, and the normal to the surface are all in the same plane.*

Notice the similarity between the first law of refraction and the first law of reflection (Section 2–1).

3-2 Experiments on the Angles in Refraction

The second law of reflection tells us that the angles of incidence and reflection are equal. Experimental observation will help us to find a similar relationship between the angles of incidence and refraction.

If we wish to make reasonably accurate measurements of the angles of incidence and refraction, we need narrower pencils of light than those shown in Fig. 3–1, to approximate single rays. We must change the angle of incidence through a wide range and measure the angles of refraction. We must do this for various pairs of materials if we are to find a general law.

Using air and glass we carried out the experiment indicated in Fig. 3–2. We constructed a light source that can be moved easily through an angle of nearly 90 degrees. It is adjustable so that the narrow pencil of light coming from it can just glance along the surface of a piece of white paper to make the pencil visible. A semicircle of glass, placed in the path of the pencil of light, served as the second material. The back of this plate was ground until slightly rough and painted white so that the pencil is visible as it traverses the glass. The glass semicircle is mounted on a circular scale with both centers at the same point. The normal to the diameter of the semicircle passes through zero on the scale. The incoming narrow ray is directed exactly at the common center. We can therefore measure the angle of incidence and the corresponding angle of refraction. The reflected ray is also visible in the pictures.

Photographs were made with this experimental arrangement with the pencil of light adjusted to angles of incidence of 0°, 10°, 20°, 30°, 40°, 50°, 60°, 70°, and 80°. Four of the photographs are shown in Fig. 3–2, and all the data appear in Table 1. Study this series of pictures carefully. Do you agree that the following conclusions can be reached?

(1) The angle of refraction of light passing from air to glass is always less than the angle of incidence, except when the latter is 0°.

(2) The behavior of the reflected pencil is described by the ordinary laws of specular reflection.

* In some crystalline substances the incident ray, the normal to the surface, and the refracted ray are not in the same plane.

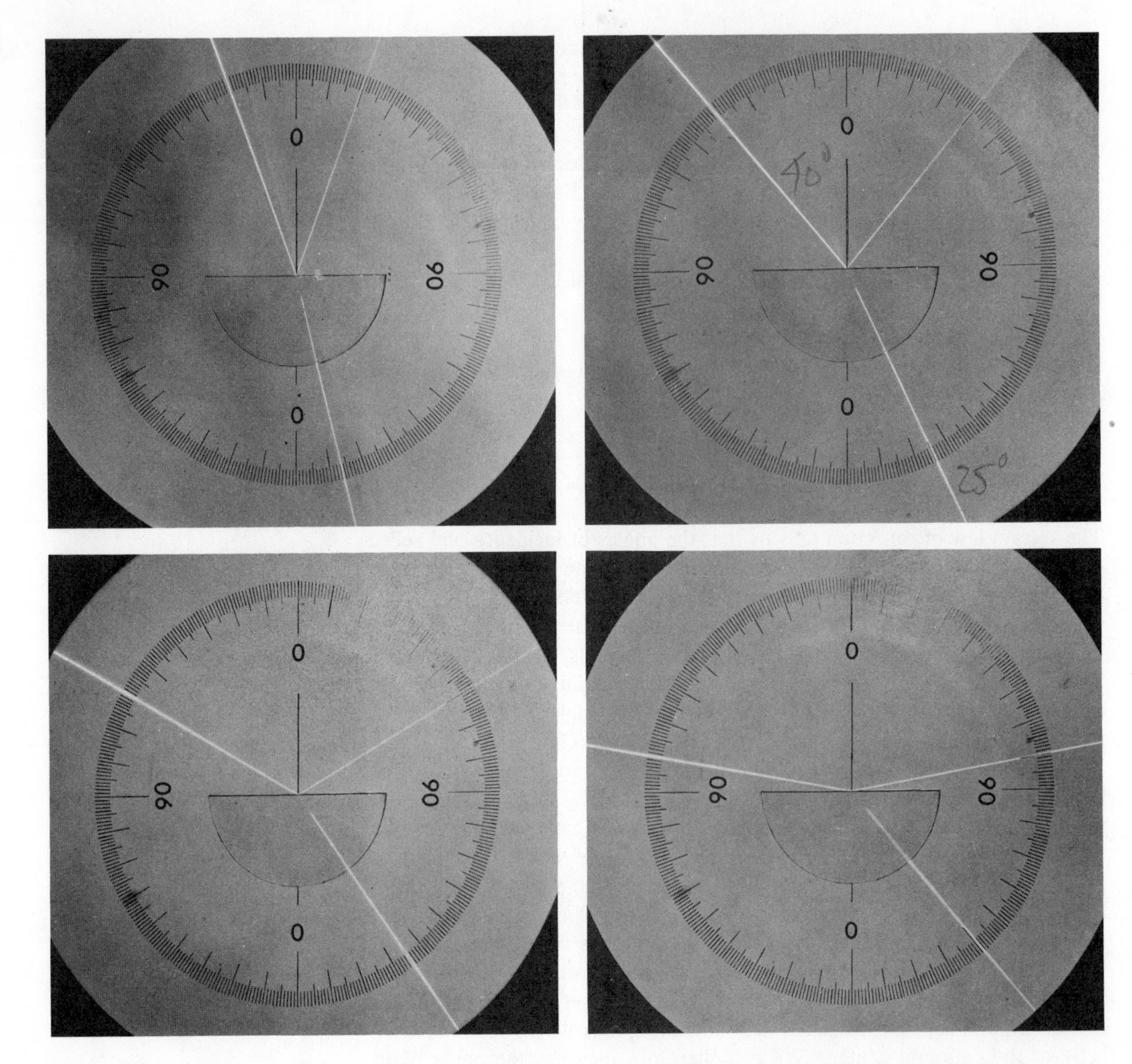

Figure 3–2
The refraction of light entering glass at different angles of incidence.

(3) The light passing through the glass travels along a radius and therefore hits the curved surface perpendicularly. It is not refracted at this boundary but goes straight out as it passes again into the air.

For the moment let us concentrate on the first conclusion, studying the relation of the angles more closely. Table 1 shows the values of the angle of incidence i and refraction r which we measured.

Certainly, angle r is always smaller than the corresponding angle i. We might try to investigate the ratio i/r to see if it is constant. We find, as shown in the last column of Table 1, that the ratios vary from about 1.5 at small angles i to more than 1.9 when $i = 80°$. The ratio is not a constant. This can also be seen in Fig. 3–3, which contains a graph of the angle r versus angle i. It was made by plotting the experimental points, which are marked, and drawing a smooth curve through the points. We can use the

Table 1

ANGLE OF INCIDENCE i IN DEGREES	ANGLE OF REFRACTION r IN DEGREES	RATIO i/r
0	0	Indeterminate
10	6.7	1.50
20	13.3	1.50
30	19.6	1.53
40	25.2	1.59
50	30.7	1.63
60	35.1	1.71
70	38.6	1.81
80	40.6	1.97

Relation between the angle of incidence i *and the angle of refraction* r *for the passage of light from air to glass.*

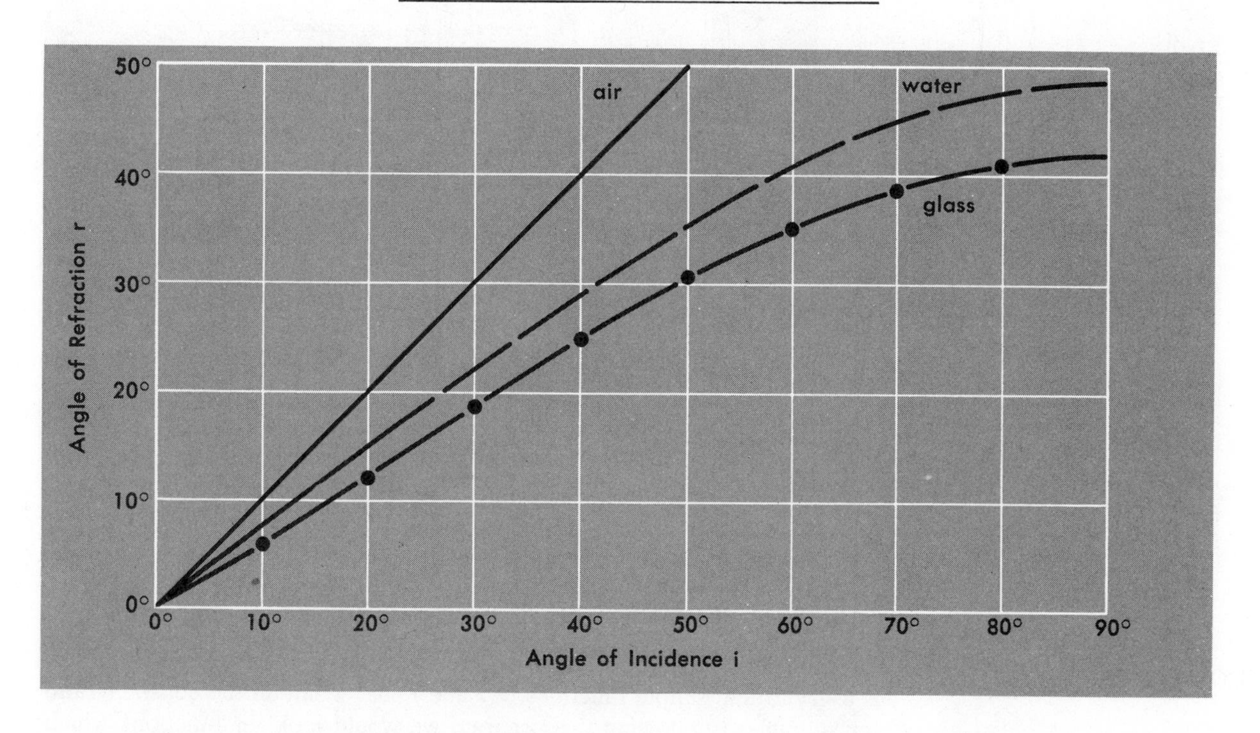

Figure 3–3
Graph of r versus i.

graph to predict what the angle of refraction would be for any given angle of incidence.

Now a question naturally arises. Is the graph we have drawn correct for all pairs of materials we might use, or is it true only for the passage of light from air into glass? To answer this question we can try the same experiment with a different pair of materials. The dashed line in Fig. 3–3 shows the graph obtained from a series of measurements made with light passing from air into water. Notice that for a given angle of incidence the angle of refraction is always greater in water than it is in glass. Furthermore, when light just passes from air into more air of the same density, the pencil of light is not bent and therefore $r = i$. This gives a straight-line graph with a 45° slope instead of a curve that bends over. Evidently, different pairs of materials lead to different graphs of r versus i.

We can perform a series of similar experiments, using many different materials, and plot a whole series of graphs of the type shown in Fig. 3–3. A book full of these graphs could be very useful in describing just how light is refracted by any pair of substances that have been studied. But it is less convenient than having a single law.

We have already tried the relation $i/r =$ *constant* and found that it does not hold for all angles. But, as Fig. 3–4 shows, for small angles of incidence the law is that i/r is constant, about 1.5 for glass and 1.33 for water. At larger angles the constancy fails, and i/r rises.

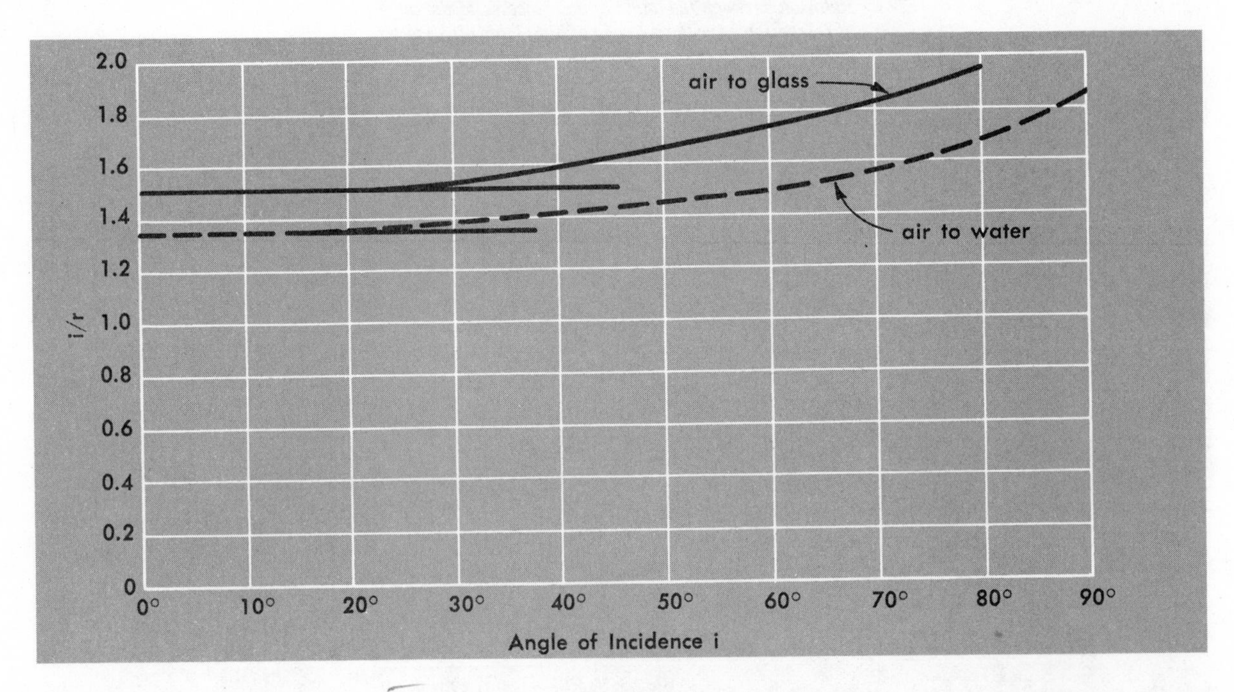

Figure 3–4
Graph of i/r versus i.

One way to look for a convenient representation of the relation of r to i is to try to find simple functions of i and r that maintain the constant ratio for all angles. Following this program we would look for functions which are proportional to the angles when they are small and which compensate for the rise in i/r when the angles get large. That we shall find such functions is not clear before we start; in fact, data on refraction were known to Ptolemy and used for a thousand years before an adequate simple law of refraction was stated. Finally in 1621 Willebrord Snell found a beautifully simple way to describe the relation of i and r. The same relation was published by Descartes in 1638 in the form we now use.

We can explain the discovery of Snell and Descartes with the aid of Fig. 3–5. In the figure a ray of light is shown coming in from the upper left. It enters the glass at the point P where it is refracted. A circle has been drawn in the plane of the rays with its center at P; and the normal to the surface of the glass is shown. The arcs AB and ED on the circumference of the circle are proportional to the angles i and r. Therefore

$$\frac{i}{r} = \frac{AB}{ED}.$$

This is the ratio plotted in Fig. 3–4, the ratio found to be nearly constant for small angles but not for large ones.

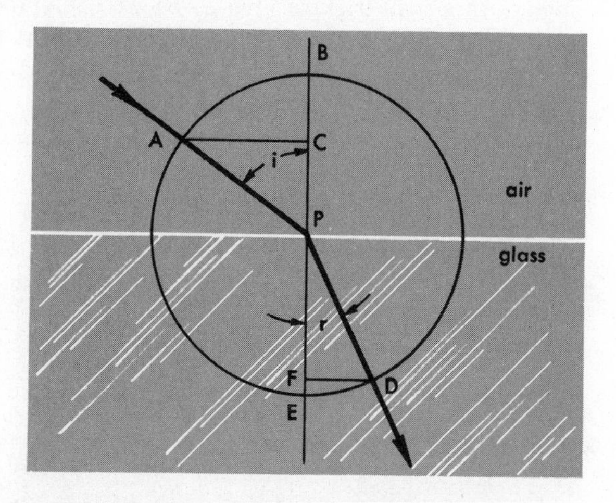

Figure 3–5
The geometry of refraction from which Snell's law is found.

In effect, what Descartes and Snell did was to try to relate i and r by examining the ratio of the semichords AC and FD. That is, in place of AB/ED they calculated AC/FD for a wide range of angles. For small angles, their ratio is the same as i/r because the chords and arcs are almost equal. But for large angles the chords and arcs are substantially different so that the chord ratio behaves differently. Let us look at this chord ratio for our experimental data taken with glass (Table 1).

We can find AC/FD by measuring the semichords in a series of figures like Fig. 3–5, and using the data in Table 1. Measuring the semichords and calculating their ratio gives the results shown in Table 2.

Table 2

i IN DEG.	AC IN CM	r IN DEG.	FD IN CM	$\frac{AC}{FD}$
10	3.47	6.7	2.30	1.50
20	6.84	13.3	4.60	1.49
30	10.00	19.6	6.71	1.49
40	12.83	25.2	8.52	1.51
50	15.32	30.7	10.20	1.50
60	17.32	35.1	11.50	1.51
70	18.79	38.6	12.50	1.50
80	19.69	40.6	13.02	1.51

The ratio of the semichords for the refraction of light in glass (see Fig. 3–5). The constancy of the ratio of the semichords (last column) illustrates Snell's law.

Our measurements of i, r, AC, and FD are not exact; our last written figures in each case are subject to question. Nevertheless the values of AC/FD are very nearly the same for all the angles of incidence. It seems very likely, therefore, that AC/FD is a constant. The value of this constant for the particular kind of glass we used is clearly between 1.49 and 1.51, probably very close to 1.50. Better experiments with smaller experimental uncertainties confirm that we have a constant whose value is close to 1.50.

We can also get the same ratio by using what are known as the sines of the angles. In a right triangle (Fig. 3–6) we define the sine of one of the acute angles as the ratio of the side opposite this angle to the hypotenuse:

$$\sin A = \frac{a}{c}.$$

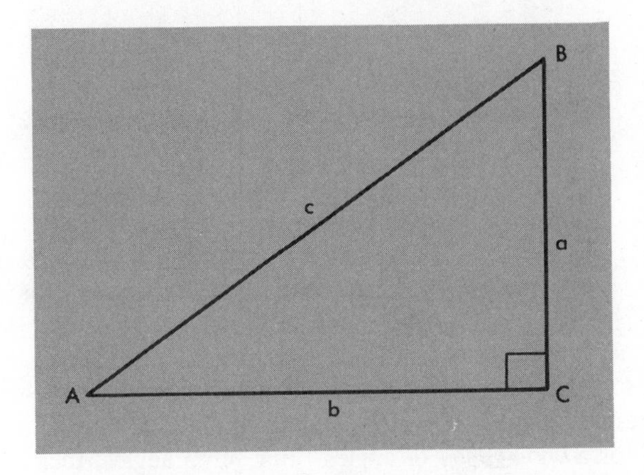

Figure 3–6
The sine of angle *A* is *a*/*c*.

The sines of all angles have been tabulated, and a table is included in the Appendix. Looking back at Fig. 3–5, we see that

$$\sin i = \frac{AC}{AP} \quad \text{and} \quad \sin r = \frac{FD}{PD}.$$

Because AP and PD are both radii of the same circle we can express $\sin r$ as FD/AP; and consequently

$$\frac{\sin i}{\sin r} = \frac{AC/AP}{FD/AP} = \frac{AC}{FD}.$$

The ratio of the sines of the angles of incidence and refraction is the same as the ratio of the semichords. We found experimentally that this ratio, in going from air to glass, is

$$\frac{\sin i}{\sin r} = 1.50.$$

Because the sines of all angles have been tabulated, this form of our result is often more convenient than the expression in terms of the ratio of the semichords. We need not construct and measure the semichords on a set of drawings; we only need to look up the values of sines in the tables. The use of the sines of i and r in place of the angles has allowed us to find a simple law of refraction describing the passage of light from air to glass at any angle.

The next question is obvious: when the light passes from air into some other substance, is the ratio $\sin i/\sin r$ also a constant—that is, independent of the angle of incidence? For many substances the experimental answer is "yes." For water, for example, $\sin i/\sin r$ is constant, and the value of the constant is 1.33. For other substances the constant ratio $\sin i/\sin r$ has different values. The value for a particular substance is a property of the substance like its boiling or melting temperature. It may even help to identify the substance. The constant is called the *index of refraction* of the substance or, if you want to be more careful, the index of refraction for light going from air into the substance. Some of these indices are listed in Table 3.

Table 3

SUBSTANCE	INDEX OF REFRACTION
Glass*	1.5–1.9
Diamond	2.42
Fused quartz	1.46
Quartz crystal	1.54
Glycerin	1.47
Carbon disulfide	1.63
Oleic acid	1.46
Water	1.33

** The index of glass depends on its composition. Most ordinary glasses have indices slightly above 1.5.*

The law that $\sin i/\sin r$ is constant for all angles of incidence is Snell's law. It tremendously simplifies our description of refraction. If we wish to know how light will be bent on entering any material from air, we need have available only the index of refraction of the material. A collection of these indices, sufficient to include nearly all important substances, can be placed on a single page of a book and can replace many books full of graphs of the type shown in Fig. 3–3. Further, when we start on the development of a theory, or model, of light in Chapter 4, we now have a simple and general law of behavior of light with which the theory must be in agreement. We can test any model that we invent by seeing whether it can account for Snell's law, the second law of refraction for which we have been searching.

3-4 The Absolute Index of Refraction

So far all our photographs of the refraction of light have been for cases of light passing from air into glass, or water, etc. What would we find for the index of refraction when the light travels from a vacuum into glass? As a matter of fact, with the limited accuracy of our experimental equipment, we would notice no difference; the same value for the index would result. This we might expect because of the very low density of air. But measurements made with great precision actually show a slightly greater index for light passing from a vacuum into glass than for air to glass. For a sample of glass with index 1.50000 in air, the index in a vacuum is 1.50044.

The index of refraction of any substance in a vacuum is called its absolute index of refraction. The value of the index from air to the substance is so close to the value of the absolute index that we rarely need to distinguish between them.

3-5 The Passage of Light from Glass (or Water) to Air: Reversibility

Thus far we have concentrated our attention on what happens to light as it enters glass or some other substance from air. What happens when light goes in the opposite direction, from glass into air? We have already found out that when light reaches the surface along the normal, the part that goes out enters the air without being bent. To find out what happens when the light approaches the glass-air surface at other angles we can make use of a glass block with parallel surfaces. As we see in Fig. 3–7, a pencil of light

Figure 3–7
Light passing through a block of glass with parallel faces.

that is incident on one side of the glass leaves the glass on the other side traveling parallel to the incident direction. The refraction in going from glass to air, therefore, is exactly the opposite of the refraction in going from air to glass.

This situation is illustrated in the ray diagram of Fig. 3–8. Here the light is traveling in the direction shown by the arrows. As it passes from air to glass, the angle of incidence is θ_{air} and the angle of refraction is θ_{glass}. The two are related according to Snell's law by the equation:

ENTER

$$\frac{\sin \theta_{\text{air}}}{\sin \theta_{\text{glass}}} = n_{\text{glass}},$$

where n_{glass} is the index of refraction of light passing from air into glass (very nearly the absolute index of glass). When the light approaches the second surface on its way from glass to air, it is incident on the surface at the angle θ_{glass} because the two surfaces are parallel. Furthermore we found experimentally that the rays entering and leaving the glass are parallel; and since the two normals are parallel, the two angles in air are equal.

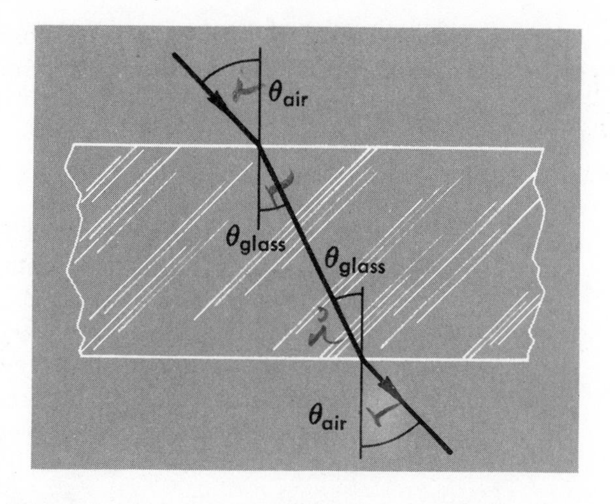

Figure 3–8
The angles and their relations for light passing through a block of glass with parallel faces.

Therefore the rays leave the glass at the angle θ_{air}. We can therefore conclude that in leaving the glass the angles of incidence and of refraction are related as

$$\frac{\sin i}{\sin r} = \frac{\sin \theta_{glass}}{\sin \theta_{air}} = \frac{1}{n_{glass}}.$$

This is Snell's law again, the ratio of sin i to sin r is a constant; and the index of refraction for light going from glass to air is just the inverse of the index for light going the other way. Because one index is the inverse of the other, we do not need two indices of refraction, one from air to glass and another from glass to air. We can use the single air-to-glass index, and write the relation of the angles in passing either from air to glass or from glass to air as

$$\sin \theta_{air} = n_{glass} \sin \theta_{glass}.$$

From experiments in which light goes from other substances into air we conclude that Snell's law applies equally well to light entering or leaving a substance. It is therefore convenient to write the law in the general form

$$\sin \theta_{air} = n_m \sin \theta_m,$$

where n_m is the index for light going from air into the material m, and θ_m is the angle between the normal to the surface and the rays in that material. The letter m stands for the name of the material. For example, for light traveling from air to diamond or diamond to air, we will have

$$\begin{aligned} \sin \theta_{air} &= n_{diamond} \sin \theta_{diamond} \\ &= 2.42 \sin \theta_{diamond}. \end{aligned}$$

The general relation, $\sin \theta_{air} = n_m \sin \theta_m$, implies that light paths described by Snell's law are reversible. In this respect they are just like light paths in which specular reflection takes place (Section 2–4). You can convince yourself of this reversibility with experiments like this: suppose one of us is under water and the other is above. We are looking each other in the eye. From the general form of Snell's law, the path of the rays of light going through the water's surface is described by

$$\sin \theta_a = 1.33 \sin \theta_w$$

regardless of whether θ_a or θ_w is the angle of incidence. Thus the light rays from your eye to mine should follow the same path as those from my eye to yours. Now comes the test. We arrange a barrier with a small hole in it so that you can just see my eye through it (Fig. 3–9). I then find that this is just the position of the barrier that also allows me to see your eye. Many experiments with many different substances confirm the prediction of Snell's law and the laws of reflection that such light paths are reversible.

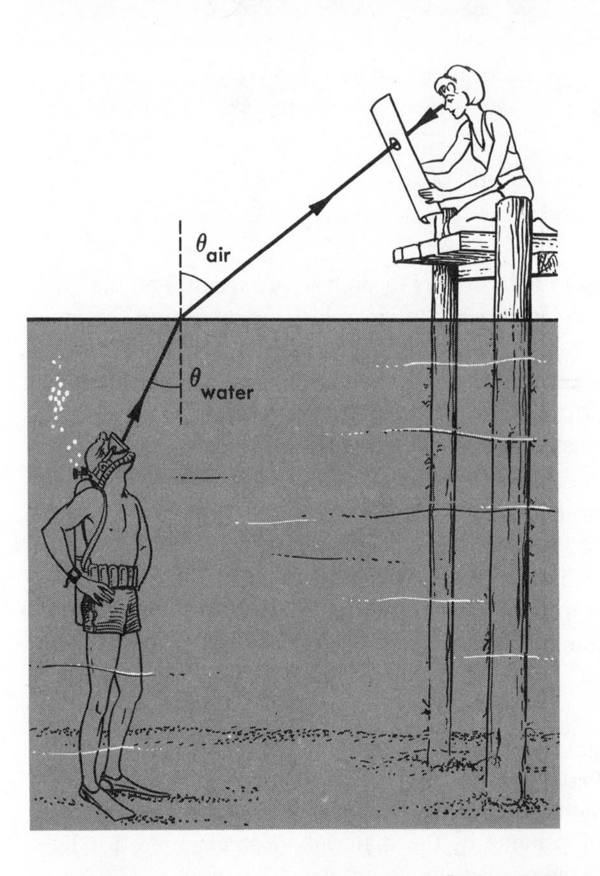

Figure 3–9
The reversibility of light paths.

3-6 The Passage of Light from Glass to Water

We can now predict what happens when light goes from glass to water. Suppose for a moment there is a layer of air between two parallel surfaces of glass and water as in Fig. 3–10. Then as the light leaves the glass, it goes into the air at angle θ_{air} such that

$$\sin \theta_{air} = n_{glass} \sin \theta_{glass}.$$

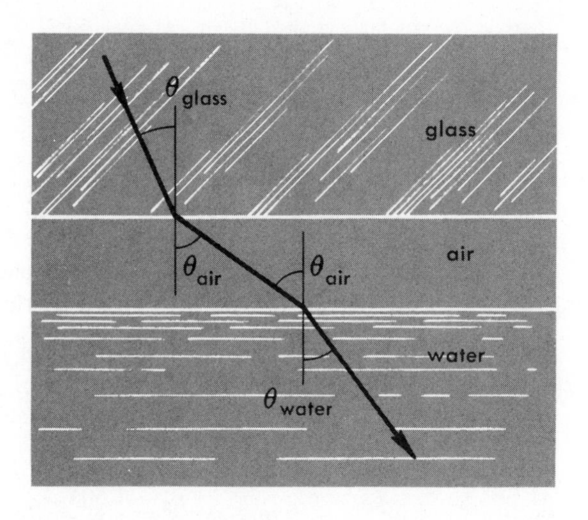

Figure 3–10
Light passing from glass to water through a layer of air.

It next enters the water and proceeds through the water at angle θ_{water} such that

$$\sin \theta_{\text{air}} = n_{\text{water}} \sin \theta_{\text{water}}.$$

Therefore

$$n_{\text{glass}} \sin \theta_{\text{glass}} = n_{\text{water}} \sin \theta_{\text{water}}.$$

This expression relates θ_{water} to θ_{glass} and is independent of the thickness of air between the glass and the water. Consequently we may hope that it relates θ_{glass} to θ_{water} in the special case when the thickness of the air layer is zero; that is, when the water and glass are touching. Indeed, if you try it, you will find that this relation holds. Further, it is not restricted to glass and water. In general, for two materials, 1 and 2, for which Snell's law holds

$$n_1 \sin \theta_1 = n_2 \sin \theta_2$$

no matter which way the light is going. The symmetry of this form of Snell's law shows the reversibility of light paths as they pass any surface for which the law is valid.

It may seem that in going from the form $\sin i/\sin r = const$ to $n_1 \sin \theta_1 = n_2 \sin \theta_2$, we have substantially modified Snell's law. That this is not the case, we can see by a slight rearrangement of the above equation. In going from material 1 to material 2, θ_1 is the angle of incidence, and θ_2 is the angle of refraction. Our general statement can then be read

$$\frac{\sin i}{\sin r} = \frac{\sin \theta_1}{\sin \theta_2} = \frac{n_2}{n_1}.$$

It still says that the ratio of the sines of the angle of incidence and refraction is a constant since n_2/n_1 is constant. This constant is called the *relative index* of the two materials or the index of refraction of light passing from material 1 into material 2. It is often written n_{12} (read $n_{\text{one two}}$).

The statement

$$\frac{\sin \theta_1}{\sin \theta_2} = n_{12},$$

therefore, takes us back to the basic statement of the law.

Moreover, since

$$n_{12} = \frac{n_2}{n_1},$$

we can find the relative indices for any pairs of materials in terms of their absolute indices. From the relatively few absolute indices, all the relative indices can be computed. For example, the relative index for light going from water to glass is

$$n_{wg} = \frac{n_g}{n_w} = \frac{1.50}{1.33} = 1.13.$$

3-7 Total Internal Reflection

If we apply Snell's law blindly to all possible cases, we soon get into trouble. A difficulty arises when we try to compute the direction of the refracted ray for light traveling from a medium with a higher absolute index of refraction to a medium with a lower index. For example, consider light passing from glass, with $n = 1.50$, to air, at the particular angle of incidence of 41.8°. The sine of this angle is 0.667. The sine of the angle of refraction is given by Snell's law as

$$\sin \theta_{\text{air}} = n_{\text{glass}} \sin \theta_{\text{glass}}.$$

Hence

$$\sin \theta_{\text{air}} = 1.50(0.667) = 1.00.$$

When $\sin \theta = 1$, θ is 90°; therefore θ_{air} is 90°. In other words, the refracted light moves parallel to the surface; it just glances along the surface. This angle of refraction is the largest one possible. What if a beam strikes the surface at an angle of incidence greater than 41.8°?

Let us try to set up an experiment that will tell us what happens. We can do this with the arrangement shown in Fig. 3–11, in which several pencils of light, diverging from one another, enter the left side of a glass prism. After refraction at the first surface, they meet the surface at the top of the prism at different angles. Notice that the two pencils in the right-hand part of the beam do not result in any refracted pencils; instead, they are totally reflected back into the glass. The phenomenon is known as *total internal reflection.* The smallest angle of incidence for which total internal reflection occurs is called the *critical angle.* If you measure the angle of incidence for the last two pencils on the right you will find that they are larger than 41.8°.

In general for light going from medium 1 into medium 2 as in Fig. 3–11 the critical angle in medium 1 occurs when $\theta_2 = 90°$; then from Snell's law,

Figure 3–11
Pencils of light entering a glass prism from the lower left. The two pencils entering nearest the bottom of the prism are totally reflected at the horizontal glass-to-air surface; the other four pencils are partially reflected at this surface.

$$n_1 \sin \theta_1 = n_2 \sin \theta_2,$$

we get

$$n_1 \sin \theta_c = n_2$$

or

$$\sin \theta_c = \frac{n_2}{n_1}.$$

Clearly, total reflection can occur only when n_1 is greater than n_2.

It may seem strange that light would suddenly stop being refracted and be totally reflected when the critical angle is reached. In fact there is no sudden change. As you see in Fig. 3–11, when the angle of incidence at the glass-to-air surface increases, more light (a brighter beam) is reflected and less light (a dimmer beam) is refracted. As the critical angle is approached, we arrive at the point where all the light is reflected and none transmitted.

Light can be piped around corners by means of total internal reflections. The "light pipe" consists of a cylindrical rod of transparent plastic. Light enters at one end of the rod nearly normal to the end surface. Any part of this light which reaches the side walls will have an incident angle greater than the critical angle and therefore will not escape into the surrounding air. Instead, a succession of total reflections will carry it along the rod and it will finally emerge at the far end.

An interesting example of total internal reflection may be seen by a fish that looks upward through the surface of a large, smooth pond or lake (Fig. 3–12). He sees the trees around the pond hanging overhead in a small

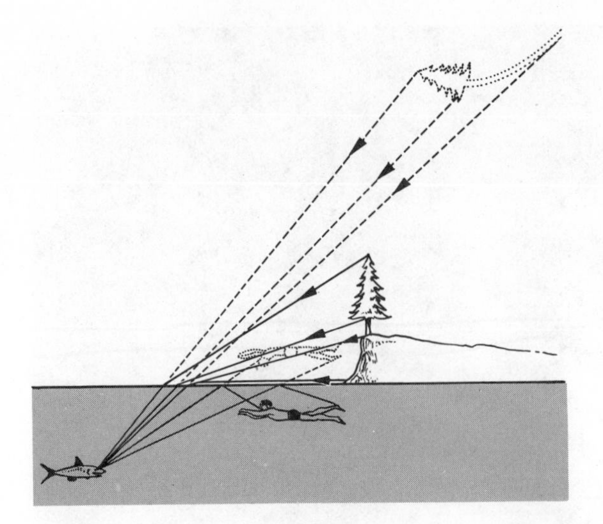

Figure 3–12
The effect of refraction and total reflection on the images seen by a fish. The rays of light from the trees appear to be coming from the virtual image up in the sky. Because the image is far away, the fish may not notice that the trunk stretches off toward infinite distance. From his point of view it is foreshortened. The fish can hardly see any of the ground, because it is compressed into such a small angle in the image. (The images of the swimmer and of the tree were drawn by following more than one ray from each of several points, such as the top of the tree, its base, and so on. Only one of the rays is shown.)

circle. They are surrounded by a mirror that reflects a man swimming near the bottom so that he appears to be swimming somewhere above the surface.

3-8 Refraction by Prisms; Dispersion

You have seen that when we send light through a slab of glass having parallel faces, the rays emerging are parallel to the incident rays. We can, however, change the direction of a light beam by using a piece of glass with two nonparallel faces (Fig. 3–13). In fact, whenever the two faces are not parallel, the light will emerge in a new direction.

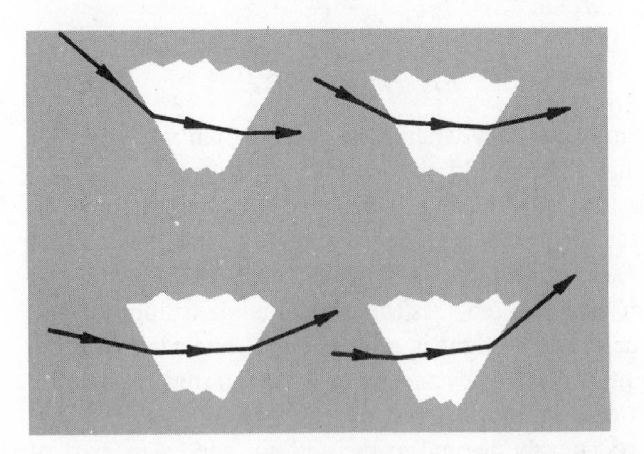

Figure 3–13
Light bent by passing through a glass block with non-parallel faces.

A careful examination shows that, even when the incident beam is made of parallel pencils of light, the beam emerging from the prism diverges, or spreads out (Fig. 3–14). To investigate this spreading, which does not seem to be quite consistent with the laws of refraction, we shall allow the light to travel a considerable distance from the prism and then examine it. We shall use a very narrow incident pencil so that the spreading will be large compared to the pencil width. A simple arrangement for doing this is

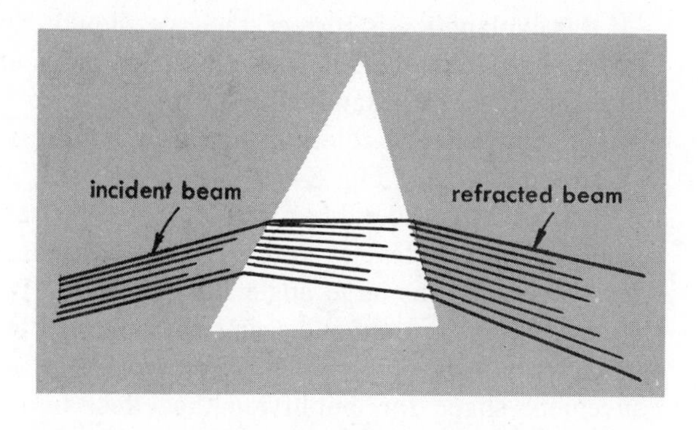

Figure 3–14
A beam of white light diverging as a result of passing through a prism. The divergence is exaggerated.

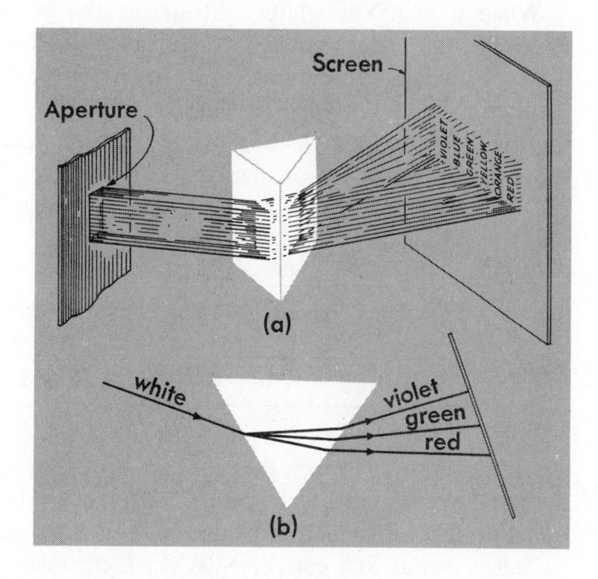

Figure 3–15
The dispersion of white light by a prism into a colored spectrum.

shown in Fig. 3–15(a). Light from a distant white source, such as the sun or an incandescent light bulb, passes through a narrow aperture. This aperture produces a narrow beam of light. It is found that the light falling on the screen is no longer "white." Instead, a brilliant spectrum of colors—violet, blue, green, yellow, orange, and red—is spread across the screen. The colors are like those that we see in a rainbow, with red at one end and violet at the other. The red deviates least from its original direction, and the violet most.

The deviation produced by a prism is determined by the angle between the surfaces through which the light passes, by the direction of incidence on the first face, and by the index of refraction of the prism. Of these, the only quantity that can differ for the different colors of light is the index of refraction. We are thus forced to give up some of the simplicity achieved in summarizing our earlier experiments. Having decided that we could describe the refractive properties of any substance in terms of a single number, its refractive index, we now find that the refractive index depends on the color of the light! Apparently white light is made up of light of different colors; and a material such as glass has different refractive indices for the different colors.

If this explanation is correct, then we should expect to find the spreading of white light into colors in some of our previous experiments on refraction. Indeed, with careful examination of the refracted beam, it is possible to see a slight separation of colors in both the experimental arrangement of Fig. 3–2 and also that of Fig. 3–7. In the first case, it is necessary to examine the light at a considerable distance from the semicircle of glass. In the second case it is necessary to use a very thick glass block. In both cases the effect is greatest for large angles of incidence. But the important thing is that the effect is present and seems to be related to the nature of the refractive index, not to some special property of a prism. The prism is simply a convenient shape for amplifying the effect into something that is easily visible.

When a beam of white light strikes the prism as illustrated in Fig. 3–15(b), the violet light is deviated most, while the red light is deviated least. Now, no matter which way light passes between air and another medium, Snell's law tells us that the bending of the light is greater the greater the index of refraction of the medium. Consequently we learn that the index of refraction of the glass is greater for violet light than it is for red. The variation of the index of refraction with color for a glass that is used in many lenses is given in Table 4.

Table 4
Index of refraction of *crown* glass

COLOR:	VIOLET	BLUE	GREEN	YELLOW	ORANGE	RED
Index:	1.532	1.528	1.519	1.517	1.514	1.513

The difference between the index of refraction for violet and for red light is seen from the table to be only 0.019, just over 1 percent of the average index of 1.52. It is not surprising that such a small variation was difficult to detect in our first experiments. In fact, unless one is particularly concerned with these color effects, it is often possible to calculate results that are in close agreement with experiment by using a single index of refraction for all colors.

The spreading out of light into a spectrum is called *dispersion*. It was first studied in the seventeenth century by René Descartes and Sir Isaac Newton. Newton performed the additional experiment of trying to break up one portion of the spectrum by inserting a prism in light of a particular color, say red. All that happens is that the red is slightly further spread out but it remains red [Fig. 3–16(a)]. Unlike the original white light, it does not split into a colored spectrum.

If we send the yellow light from sodium through a narrow slit and then a prism, only a yellow image of the slit will appear on the screen. By sufficient spreading out of the spectrum, we can see greater detail. For example, if the dispersion is great enough, this yellow image proves to be made up of two bright yellow lines. But no matter how much we spread the yellow lines we do not change their color or analyze them into anything except finer detail. The individual colors of the spectrum cannot be decomposed; they are simple and basic in a way that white light is not. Yellow light from a spectrum will always deviate the same way through a prism. It will deviate

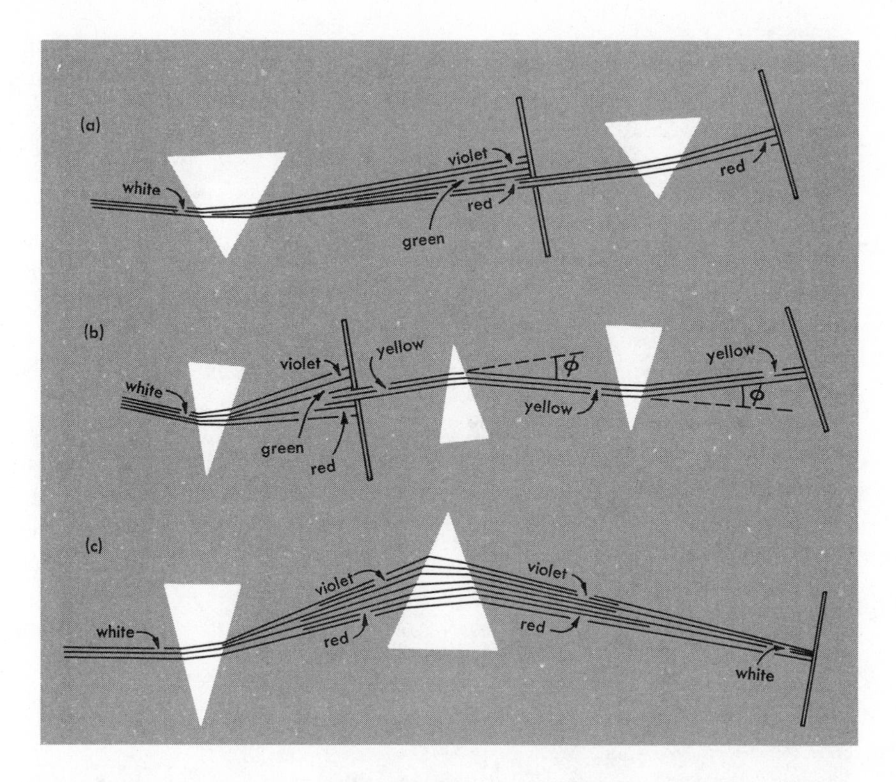

Figure 3–16
Experiments with a prism. (a) Light of one spectral color remains the same color on further analysis with a prism. (b) Light of one spectral color is always deviated through the same angle when passing on similar paths through identical prisms. (c) The spectral colors combine to make white light.

less than green and more than red [Fig. 3–16(b)], no matter how many prisms we use.

Our conclusions are supported by a further experiment of Newton's proving that white light is a combination of many colors. He recombined the spectrum to make white light. We can do the experiment by breaking up a narrow beam into a spectrum by a prism, and then placing in the spectrum a second prism with a greater angle between its faces [Fig. 3–16(c)]. Because of its greater angle this prism deviates the different colors more than the original prism and the light converges again. The pencils of light of different colors then overlap in some region beyond the second prism, and in this region the light on a screen again appears white. This is what Newton found, as you can too if you repeat his experiment with two glass or plastic prisms.

3-9 The Convergence of Light by a Set of Prisms

We found in Chapter 2 that we could control and redirect light beams by the use of curved mirrors. Devices that can accomplish similar purposes through refraction, instead of reflection, are called lenses. To understand how a lens operates, let us examine the behavior of light in passing through the combination of a plate of glass with parallel sides and the two triangular prisms shown in Fig. 3–17(a). If a parallel beam of light falls on this system from the left perpendicular to the central plate of glass, it will behave as indicated by the rays shown in the figure. The light that passes

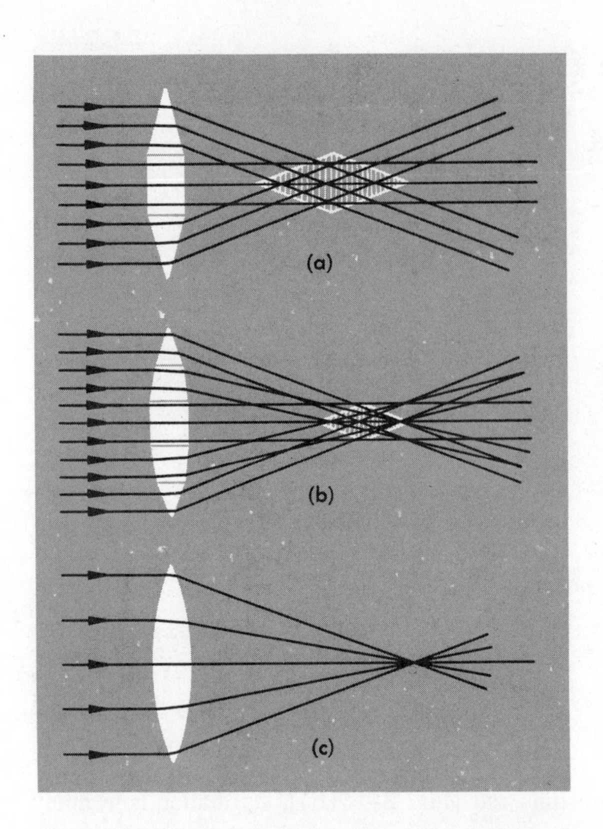

Figure 3–17
Construction of a lens by the process of subdividing prismatic sections.

through the plate in the center will continue along its original direction, since the angle of incidence is 0°. Light striking the upper prism will be deviated downward by an amount depending on the angle of the prism and on its index of refraction. Similarly, light striking the lower prism will be deviated upward. As a result, there is a region, shown shaded in the figure, through which passes almost all of the light that falls on the plate and the prisms.

The convergence of a parallel beam of light into a limited region by this system resembles the convergence of a similar beam by a set of mirrors. (See Section 2–3.) While working with mirrors, we decreased the size of the region into which the light was converged by using an increased number of mirrors, each smaller than the original one. Let us try the same scheme here. Figure 3–17 (b) shows parts of the central plate and of the two prisms cut away and replaced by pieces of new prisms. The size of the shaded region is clearly smaller than it was before.

If we continue the process of removing parts of the prisms and replacing them by sections having smaller angles, we come closer and closer to a piece of glass with the smoothly curved surface shown in Fig. 3–17 (c). This device is the limit that is approached as we increase the number of prisms indefinitely, just as the parabolic mirror of Fig. 2–11 was the limit approached as we used more and more plane mirrors to converge parallel light. In Fig. 3–18 we have actually carried out the construction indicated in Fig. 3–17. The lens produced by the process that we have outlined converges all of the parallel light that strikes it to a line, as shown in Fig. 3–19.

Figure 3–18
The experiments diagramed in Fig. 3–17.

Figure 3–19
Convergence of light by a cylindrical lens like the one shown in Fig. 3–17. Note that the light is brought to a focus along a line.

3-10 Lenses

The device we have just constructed is a cylindrical lens. Notice that we have not given any definition of the surfaces of the lens, except that they are obtained by increasing indefinitely the number of sections of prisms that are used to converge the light. It is possible to show that the resulting surface is approximated very closely by a cylinder. In other words, the lines representing the surfaces in Fig. 3–17 (c) are arcs of circles. The differences between the ideal surface which focuses parallel light to a sharp line and that of a cylinder are very slight if two conditions are satisfied: (1) the thickness of the lens is considerably smaller than its width; and (2) the width is considerably smaller than the distance from the lens to the line at which parallel light is converged.

Cylindrical lenses bring the light from a distant point source of light to a focus along a line. For most purposes we prefer that the light from a point source should be focused at a point. This focusing can be accomplished by constructing a lens whose surfaces curve equally in all directions. Such surfaces are portions of spheres. Almost all lenses are bounded by two spherical surfaces.

The line passing through the center of the lens and on which the centers of the two spheres are located is called the axis of the lens. The point on this axis at which incident parallel rays focus or converge is the principal focus, F. The distance of the principal focus from the center of the lens is known as the focal length, f.

3-11 Real Images Formed by Lenses

We have thus far concentrated our attention on the focusing of light by a lens when the light comes from a very distant object. In the practical use of lenses, we are commonly interested in the light coming from nearby objects, and we all know that lenses do form images of such objects. We can locate the images with the help of the knowledge that we have gained about the behavior of initially parallel rays.

Figure 3–20 shows a lens, an object of height H_o, and its image of height H_i. To find the location of this image, we draw the two principal rays from the top of the object, one ray parallel to the axis and the other through the principal focus F_2. The ray parallel to the axis is bent by the lens so as to pass through the principal focus F_1. We also know that rays coming from the right and parallel to the axis would be deviated to pass through the other principal focus F_2. It follows from the reversibility of light paths that the ray from the top of the object that passes through F_2 from the left must travel parallel to the axis after it has passed through the lens. All rays starting from the top of the object will converge very close to the point at which these two bent rays intersect. This point is therefore the real image of the top of the object.

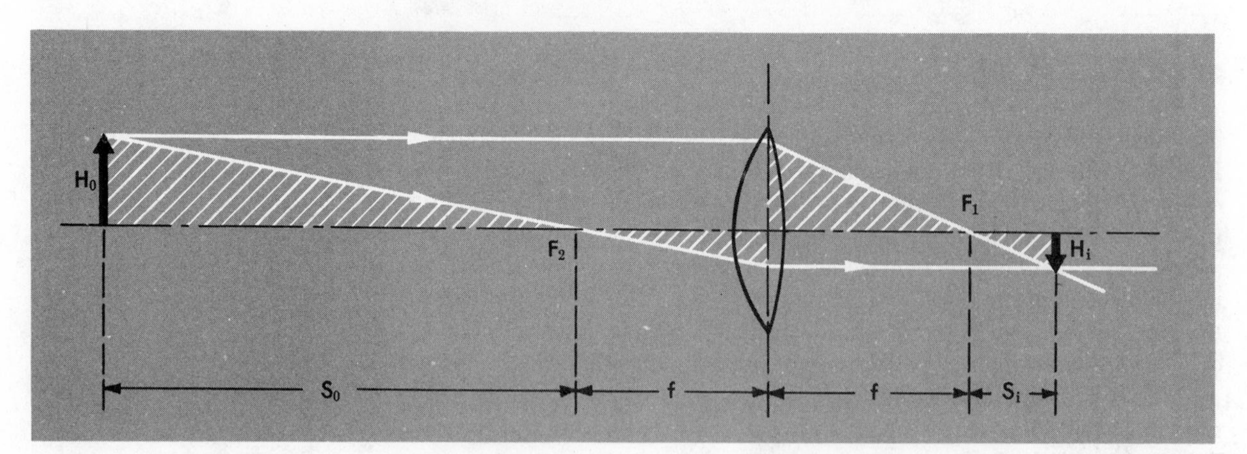

Figure 3–20
The formation of a real image by a converging lens.

We could have chosen any other point on the object and located its image in the same way. Had we done so for a number of points, we would have found that the image falls along the line that is shown in the figure.

In constructing the two principal rays, we have not considered their exact paths within the lens, but have broken them sharply. This approximate construction is good enough for our present purposes because our location of the two principal foci is accurate only if the lens thickness is small compared with the focal length. The only lenses to which our construction accurately applies are therefore thin lenses.

Convex lenses, like parabolic mirrors, focus parallel rays to a point. Lenses, therefore, obey the same equation relating image distance, focal length, and object distance as do mirrors:

$$S_i S_o = f^2.$$

The proof of this equation in the case of convex lenses is the same as for mirrors (Section 2–5). As there, we use the shaded similar triangles formed by the principal rays shown in Fig. 3–20. Considering first the shaded similar triangles to the left of the lens, we see that $H_i/H_o = f/S_o$. The shaded triangles to the right of the lens give $H_i/H_o = S_i/f$. Combining the two equations, we have

$$S_o S_i = f^2.$$

Using this equation for numerical results and Fig. 3–20 as a model for ray tracing, you can now solve a large number of situations involving lenses, either singly or in combination. You can determine object and image distances, magnification, the real or the virtual nature of the image. In short, you can do all of the things you did in the previous chapter with curved mirrors, but this time with lenses.

Optical instruments of all sorts—cameras, projectors, telescopes, microscopes—are built with combinations of these lenses. You have only barely touched on the subject, but a whole industry is devoted to the design and production of such instruments and their components. The field is tremendously complex and intricate, but is almost entirely based on one little bit of physics—Snell's law of refraction.

3-12 Light Pencils and Scaling

So far our study of the behavior of light has involved only straight lines and angles. Our ability to produce sharp shadows (Fig. 1–6) and narrow pencils of light (Fig. 1–12) suggested that light propagates in straight lines and changes direction only when it is reflected or refracted. The change in direction produced by the reflection and refraction of light seems to depend only on the angle of incidence (and the kind of glass) but not on the size of the glass. For example, we could have reduced the entire apparatus in Fig. 1–12 to one half, taken the photograph, and enlarged it by a factor of two, and it would have looked just like Fig. 1–12. Within the ranges of the sizes of objects encountered in daily life, the behavior of light seems to be independent of the size of the objects with which it interacts. Changing all sizes by the same factor, that is, changing the scale, appears to have no effect on the behavior of light.

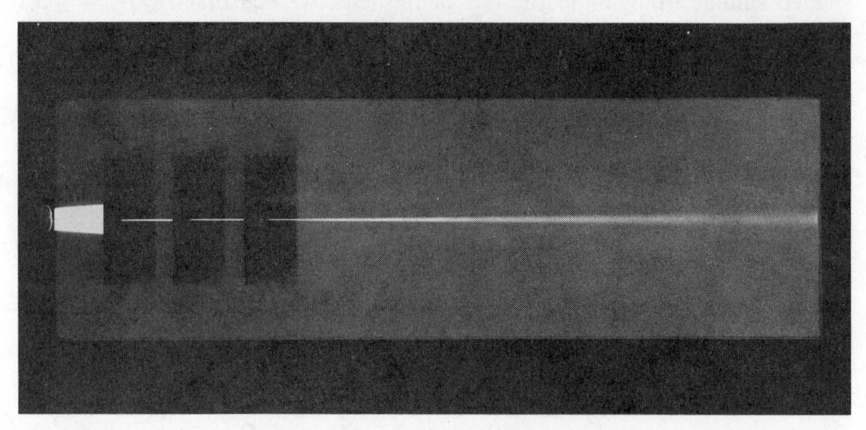

3–21
Top: A light pencil formed by light passing through two short slits 3 mm wide. The light continues in its original direction with little or no spreading. Bottom: A third slit only 0.1 mm wide is added. Now the light spreads out.

However, if we try to make the pencils of light in Fig. 1–12 very narrow by using a very narrow slit, something unexpected happens: the light spreads out beyond the slit (Fig. 3–21). Apparently, when we try to check up on the straight-line propagation of light by producing a very narrow pencil of light that will approach an ideal ray, the light changes direction in going through the slit. This bending of light by a narrow opening is called *diffraction.*

You can observe diffraction quite easily by looking at a distant street lamp, or a lamp with a thin filament, through a narrow slit produced by pressing two fingers almost together (Fig. 3–22). When you do this, the light source seems to spread out. This means that light enters the eye coming from many directions, and this can happen only if the light is bent (diffracted) by the slit. This is shown schematically in Fig. 3–23.

Diffraction is not limited to slits. Let parallel light pass through a circular hole a few millimeters in diameter and fall on a screen, and you get a sharp circular light spot. Do it with a tiny pinhole, and the curious pattern shown in Fig. 3–24 is obtained.

Light, therefore, behaves differently depending on the size of the slits or or holes it passes through. It travels through large openings in unbent straight lines. But when the openings are scaled down enough, diffraction takes place, and we must take it into account in describing the behavior of light.

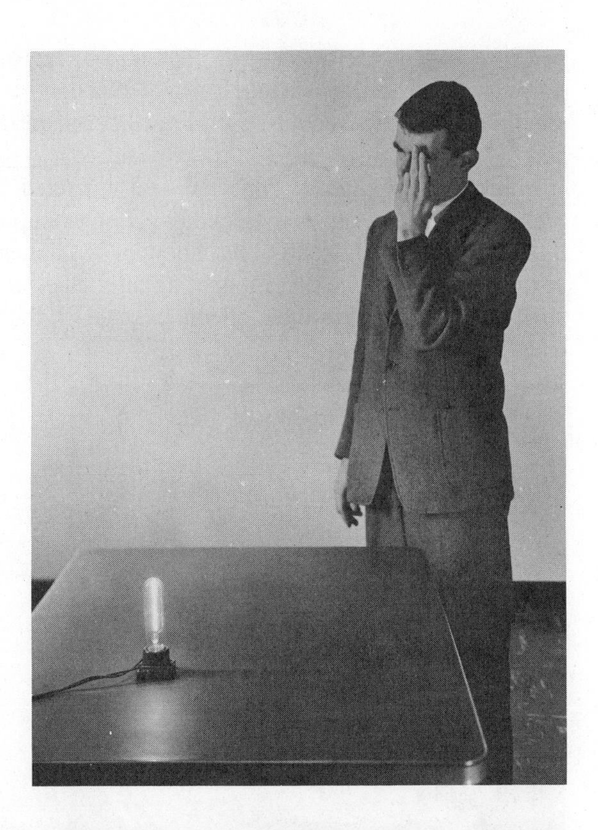

Figure 3–22
A simple way to observe diffraction.

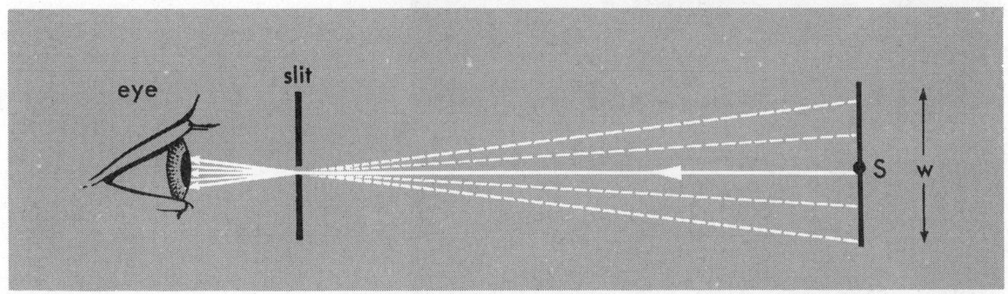

Figure 3–23
The black dot marked *s* is a vertical view of a narrow, vertical source. The diffracted light from this source appears to come from a wide source of width *w*, rather than the narrow filament *s*.

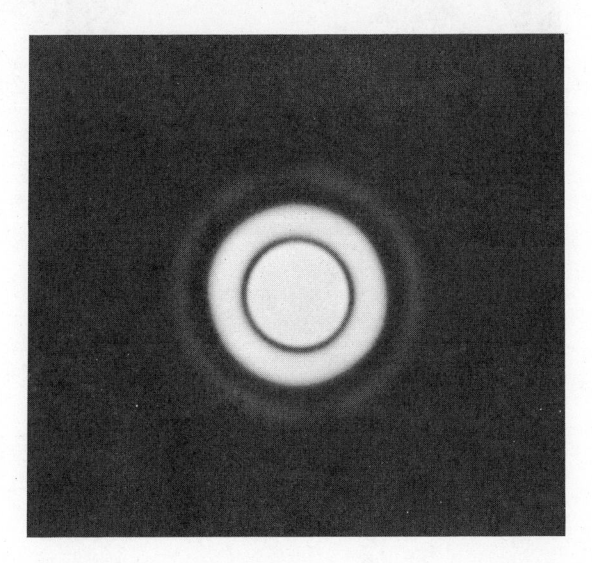

Figure 3–24
The pattern formed by a light pencil passing through a small pinhole.

Diffraction, therefore, shows that size is important in the behavior of light; a picture of the passage of light through holes or slits that are very small differs from a reduced picture of light passing through bigger holes or slits.

Similarly, how light is reflected and refracted by very thin transparent materials depends on their thickness. For example, Fig. 3–25 shows light reflected from a thin soap film. The film is thinnest at the top and thickest at the bottom. Notice that no light at all is reflected from some regions. We shall return later on to these effects of scale on the behavior of light.

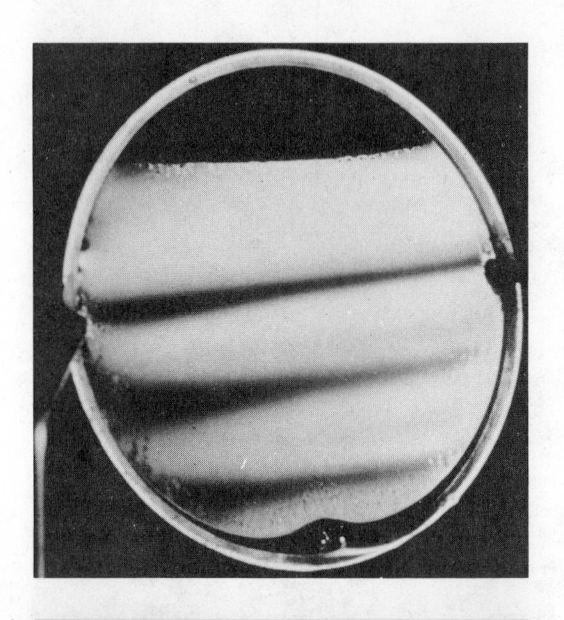

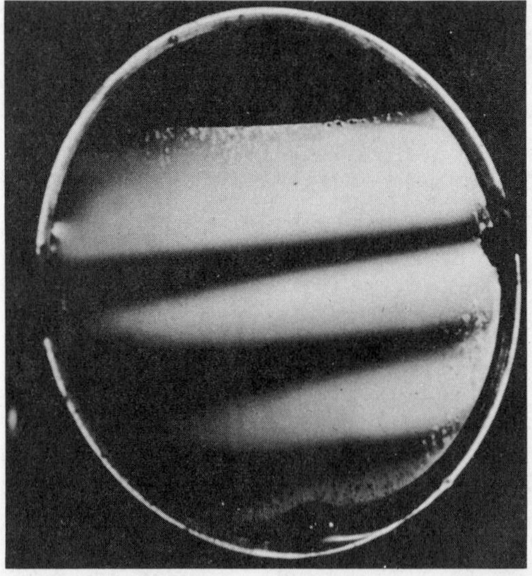

Figure 3–25
Reflection of light from a thin soap film. The film was made by dipping a metal ring in a soap solution.

For Home, Desk, and Lab

1.* Which is the angle of incidence in the drawing in Fig. 3–26? Which is the angle of refraction? (Section 1.)

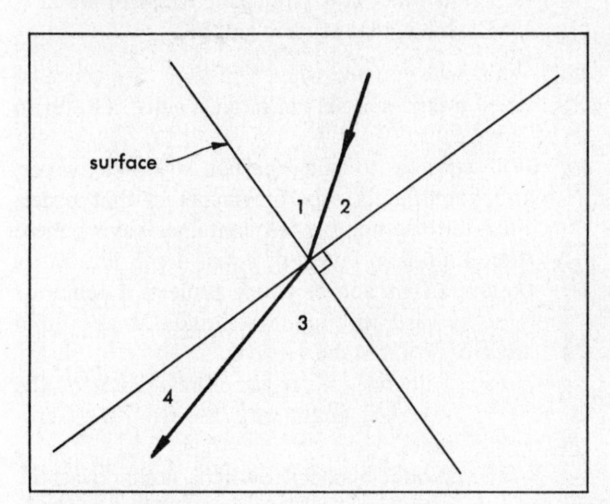

Figure 3–26
For Problem 1.

2.* Compare Fig. 3–1 with the upper right-hand photo of Fig. 3–2. In which is the light bent more? (Section 1.)

3. Can you give a reason why the incident ray, refracted ray, and the normal should all be in the same plane? Can you imagine a material in which this would not be true?

4.* Figure 3–27 shows the path of light traveling from air into glass. Is the glass on the right or the left in the drawing? (Section 2.)

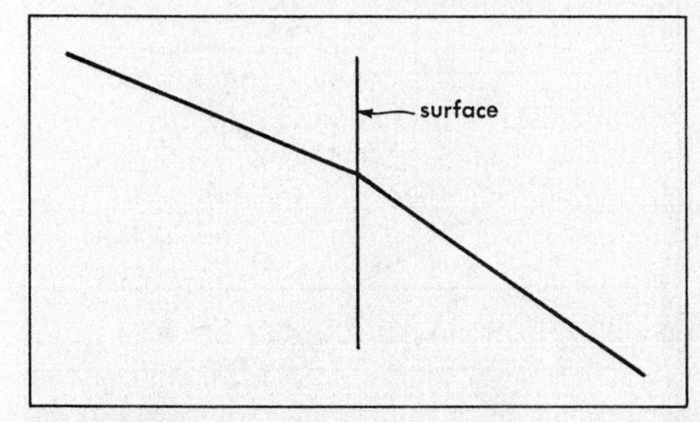

Figure 3–27
For Problem 4.

5.* In the nearly straight portion of the curve for water (Fig. 3–3), what is the approximate relation between r and i? (Section 2.)

6. (a) What are the sines of the following angles: 4°, 30°, 45°, 60°, 73°, 17.8°, 37.3°, 90°?
   (b) What are the angles that have the following sines: 0.1045, 0.0000, 0.3090, 0.8660, 1.000, 0.5000, 0.5225, 0.9636?
   (c) Plot sin i versus i from 0° to 90°.

7.* In Fig. 3–28 what is the index of refraction for light going from air into substance X? (Section 3.)

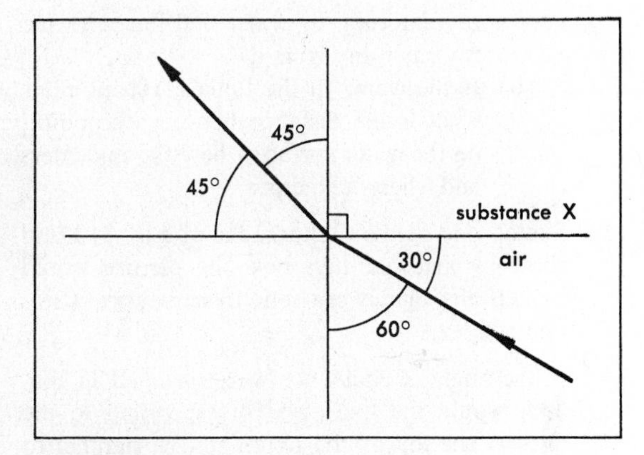

Figure 3–28
For Problem 7.

8.* Check Fig. 3–1. Do the incident and refracted pencils correspond to what they should be if the refractive material really is water? *Must* it be water? (Section 3.)

9.* For a given (nonzero) angle of incidence, which substance listed in Table 3 causes the greatest change in direction of light entering it? (Section 3.)

10. How could you predict from Fig. 3–4 that the index of refraction of water is close to 1.33?

11. A rectangular tank 8 cm deep is filled with water. A light ray enters the top surface of the water at a point just touching the side of the tank. After refraction it falls on a point on the bottom of the tank 3 cm from the same side of the tank.
   (a) What is the sine of the angle of refraction? What is the angle of refraction?
   (b) What is the sine of the angle of incidence? What is the angle of incidence of the entering ray?
   (c) Suppose that the same tank were filled with a liquid other than water and you found that in order to fall on the same point 3 cm from the side the angle of incidence of the entering ray had to be 31°. What is the index of refraction of the liquid?

12. A narrow pencil of light enters the top surface of the water in a rectangular aquarium at an angle of incidence of 40°. The refracted pencil continues to the bottom of the tank, striking a horizontally placed plane mirror which reflects it back again to the surface, and it is again refracted as it emerges into the air.
    (a) What is the angle between the incident ray entering the water and the refracted ray emerging from it?
    (b) If the water in the tank is 10 cm deep, what is the distance between the points on the water surface where the ray enters and where it emerges?

13.* Sketch one of the photographs of Fig. 3–2 and show by arrowed rays how the picture would look if the light were sent from below. (Section 5.).

14. If the angle of incidence were changed in Fig. 3–7, would the two rays corresponding to the ones at the top of the photo still be parallel to each other?

15. Light from the setting sun comes through the earth's atmosphere along a curved path to your eye, so that the sun looks higher in the sky than it really is. How do you explain this? Illustrate your answer with a diagram.

16. As the moon moves across the sky, it passes in front of stars. Compare what is seen as the moon passes in front of a star with what you would see if the moon had an atmosphere like that of the earth.

17.* In Fig. 3–29, is the relative index for light going from medium A into medium B greater or less than 1? (Section 6.)

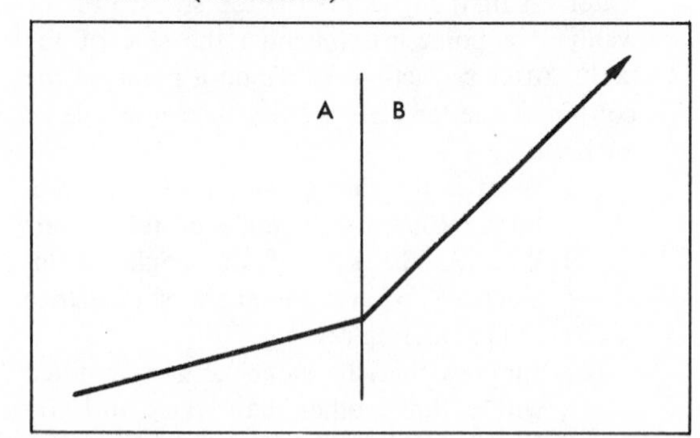

Figure 3–29
For Problem 17.

18. If the *relative* index for light going from glass into diamond is 1.61 and the *absolute* index of glass is 1.50, what is the *absolute* index of diamond?

19. If the relative index for light going from oleic acid into water is 0.91 and the index of water is 1.33, what is that of oleic acid?

20. What will you see when you look at a piece of fused quartz submerged in oleic acid? (Refer to Table 3.)

21. It is possible to place carbon disulfide, water, and kerosene in separate layers in that order, since they do not mix. A container having these three liquids in layers of equal depth has on its bottom a light source which projects a pencil of light upward through the liquids at an initial angle of 5° from the vertical.
    (a) Calculate the angle of refraction of the ray as it finally emerges from the kerosene.
    (b) Would a pencil of light travel over the same path in the reverse direction from kerosene through water into carbon disulfide?
    (c) Suppose the cylinder had been filled with water only. How would the angle of refraction of the emerging ray compare with the angle calculated above?

22.* What is the critical angle for light passing from glass to water? (Section 7.)

23. In Fig. 3–30, continue the light ray until it comes out of the glass.

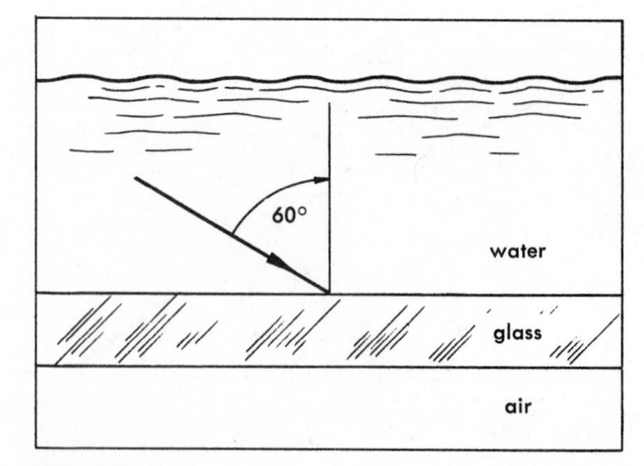

Figure 3–30
For Problem 23.

24.† Carbon disulfide (refractive index 1.63) is poured into a large jar to a depth of 10.0 cm. There is a very small light source at the center of the bottom of the jar. Calculate the area of the surface of carbon disulfide through which the light passes.

25. Why is it easier to observe dispersion in a prism rather than in the setup shown in Fig. 3–2?

26.† In Fig. 3–31, a parallel beam of monochromatic light (light of a single color) enters each box from the left. Draw what could be in each box to produce the effects shown. The single and double arrows on the emerging beams show the corresponding edges of the entering beam. The lines without arrows show the other edges of the emerging beam.

27. A crude converging lens can be constructed by placing two 30°–60°–90° glass prisms together with a glass block as shown in Fig. 3–32.
  (a) What is the focal length of this "lens" to one significant figure?
  (b) Would such a lens form a clear image? Explain.

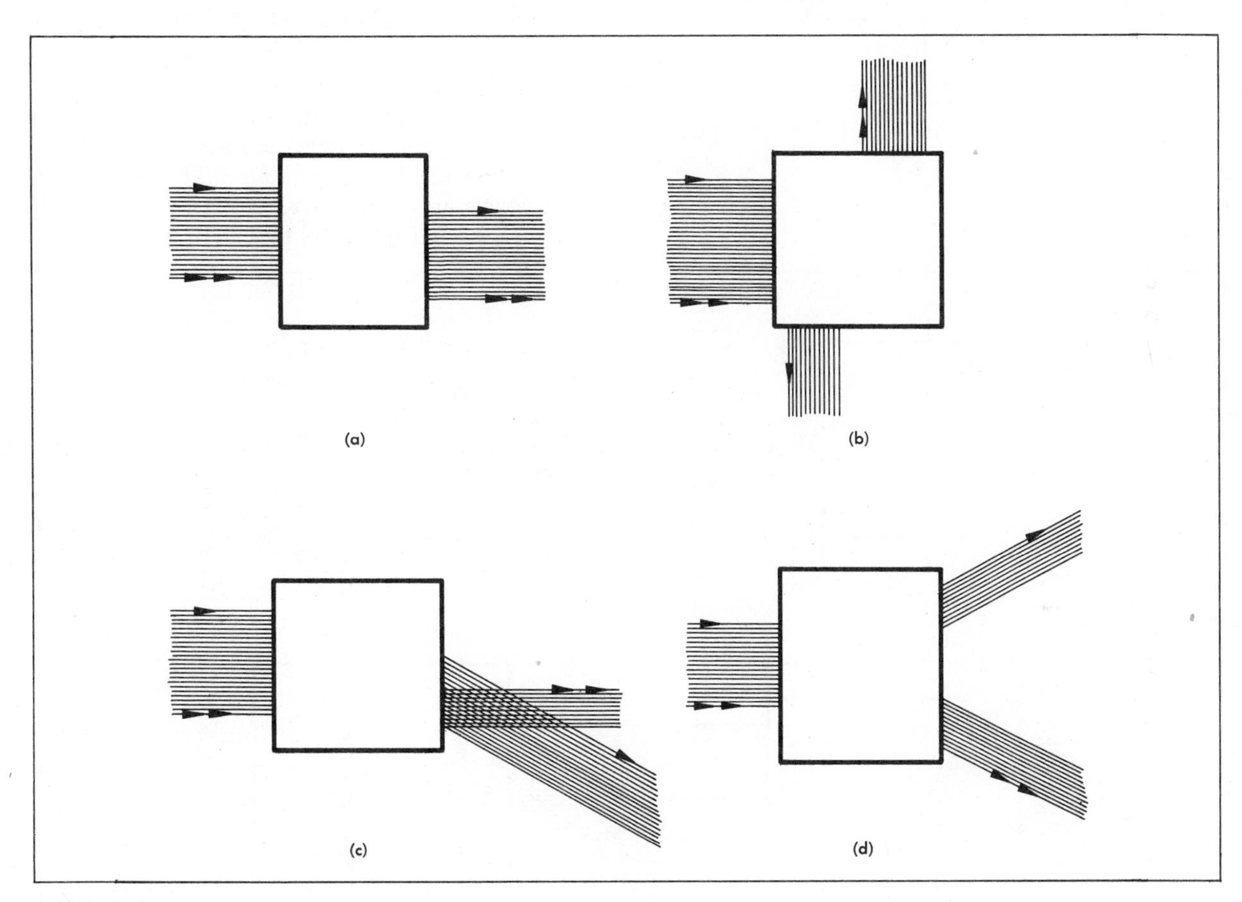

Figure 3–31
For Problem 26.

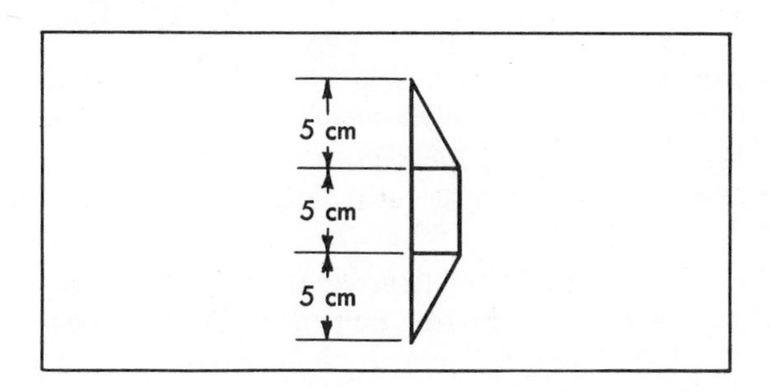

Figure 3–32
For Problem 27.

28. If two 45° prisms of glass (index = 1.50) are arranged as in Fig. 3–33, they will not converge parallel light. Why not? What will happen to the light?

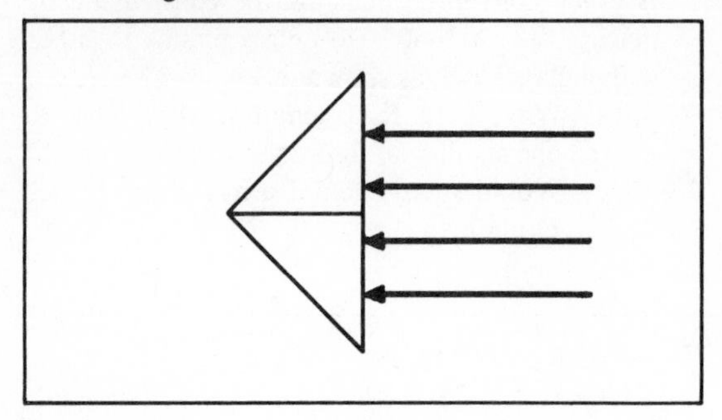

Figure 3–33
For Problem 28.

29. A lens (index = 1.50) has a focal length in air of 20.0 cm. Is its focal length in water greater or less than in air?

30.* An object is 1 meter in front of the principal focus of the lens of a camera. How far behind the other principal focus will the image be if the focal length of the lens is 5 cm? (Section 11.)

31. What is the minimum distance between an object and the real image formed by a converging lens? (Make a graph of the total distance versus the object distance. Use the focal length f of the lens as the unit of distance.)

32. In Fig. 3–34, a parallel beam of monochromatic light (light of a single color) enters each box from the left. Draw what could be in each box to produce the effects shown. The single and double arrows on the emerging beam show the corresponding edges of the entering beam.

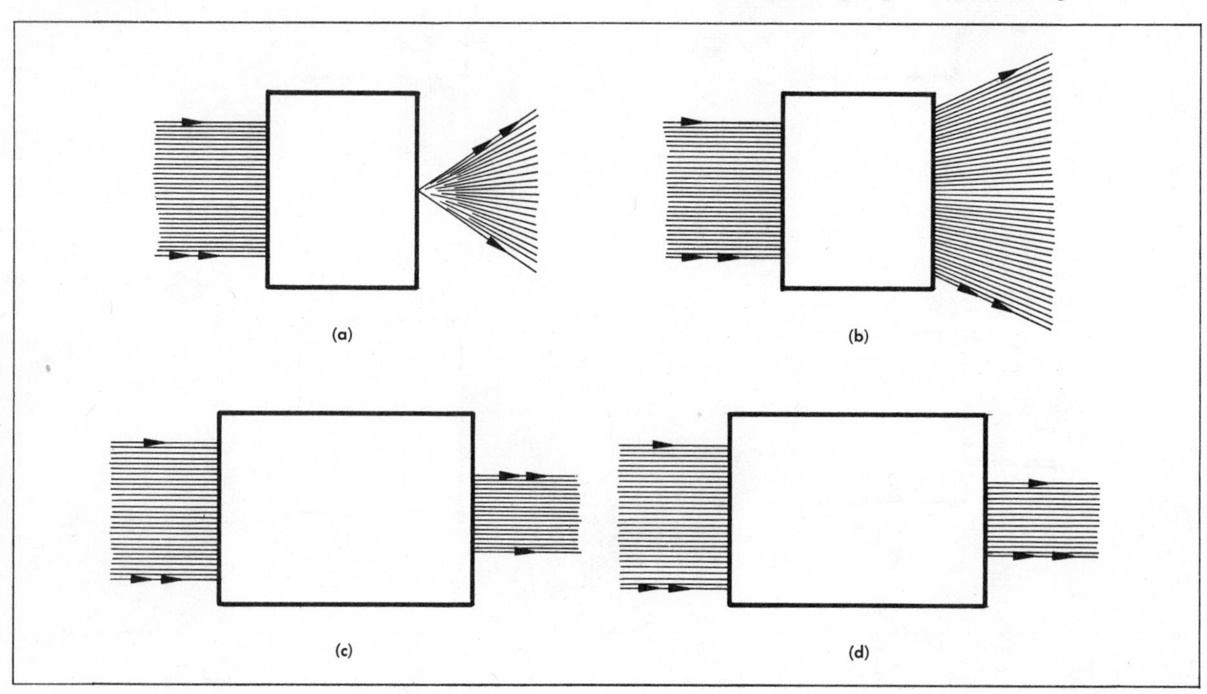

Figure 3–34
For Problem 32.

Further Reading

These suggestions are not exhaustive but are limited to works that have been found especially useful and at the same time generally available.

JENKINS, FRANCIS A., and WHITE, HARVEY E., *Fundamentals of Optics*. McGraw-Hill, 1957. Excellent reference material devoted to lenses and optical instruments. (Chapters 4, 5, 9, and 10)

NEWTON, SIR ISAAC, *Opticks, or a Treatise of the Reflections, Refractions, Inflections & Colours of Light* Based on the Fourth Edition of 1730; Dover, 1952.

The Particle Model of Light

CHAPTER

4

Let us try now to construct a model or theory of light which will account for the properties of light we have thus far studied. Such a model will allow us to draw conclusions and make predictions which in turn can be checked by further experiments. In doing so, we shall retrace ideas that were developed over several centuries. The main architect of the model we shall study in this chapter was Sir Isaac Newton.

We have seen that light always has its source in some luminous body and that it travels in essentially straight lines. Any model of light must, then, include something that starts from the luminous body and moves along a straight path. The simplest thing that we can imagine traveling in this way is a particle, such as a baseball. We might suppose that a luminous body gives off a stream of particles. You may object to this model on the grounds that baseballs or other particles do not travel in straight lines; they move in curved paths that bring them down to the earth. You do know, however, that the path of a baseball curves less and less as its speed is increased. We can expect that the paths of light particles which travel at a speed of 3×10^8 m/sec will hardly be curved by the earth's gravitational pull. With the particle model, therefore, we have no trouble in accounting for the propagation of light in straight lines.

Of course, light particles must be quite different from baseballs. In particular, unlike baseballs, they do not interact with each other, as we found out in Section 1–8. We can explain this lack of interaction by supposing that the particles are very small, so small that, even for two intense light beams, the chance that particles in one beam will collide with those of the other is extremely small. In this way we can include the lack of interaction of light with light in our model. Our assumed light particles differ from baseballs, then, by their very high speeds and their very small size. If we keep these differences in mind, we can try to predict the properties of streams of light particles by studying the behavior of particles such as baseballs or marbles.

4-1 Source Strength and Intensity of Illumination

If you illuminate this page with only a candle held a meter away, it will not be very well illuminated. An ordinary light bulb at the same distance will produce more illumination on the page—the page will appear much brighter. We say the intensity of the illumination is greater.

In order to measure the intensity of illumination, we can use a photoelectric cell or other light meter, such as those used by photographers. On these light meters higher scale readings correspond to greater illumination, and lower scale readings mean a dimmer light. Our first job is to discover a way to calibrate the meter by a method of our own that we understand.

We start by finding a set of equal sources of light. By "equal" we mean that the light meter registers the same reading for each source when placed by itself at a set distance r_0 in front of the light meter in an otherwise dark room (with dark walls to eliminate reflections). We conclude that each light source is causing the same number of light particles to fall on the light meter in one unit of time, say a second.

The particle theory of light now suggests that if we place two of these equal sources at the same distance r_0, twice as many particles of light will illuminate the light meter per second. Indeed, using any two of our standard light sources, each a distance r_0 away from the meter, we always get the same pointer reading on the meter; this reading indicates greater illumination than we get with only one standard source at the standard position.

Let us start our calibration of the scale for the light meter by marking the position of the pointer when one standard source is at the standard distance r_0. We put the number 1 right on the scale at that point. We put the number 2 on our scale at the position of the pointer when two standard sources are at the standard distance. We can do the same thing with three standard sources at the distance r_0 in front of the meter, and so on. We have now calibrated our light meter to read illumination in multiples of the illumination given by one standard source placed at the distance r_0 in front of the meter. If we have many calibrated points on our scale, we can interpolate between adjacent points and make marks on the scale that correspond to fractional values of this standard unit of intensity of illumination. When we use the meter to measure the illumination of a standard source at a distance different from r_0, our scale tells us how great that illumination is in terms of the number of our standard sources that would have to be placed the distance r_0 to give the identical effect.

We can now take one of the standard sources and move it to distances in front of the light meter which are different from r_0. Thus, we can find out how the intensity of illumination of the light meter changes as we change r, the distance from the source to the meter.

What does the particle model predict for the results of such an experiment? We suppose that illumination is directly proportional to the number of particles hitting a unit area in one second. The particles move outward in all directions from the source, traveling in straight lines, and we assume that none of them are lost as they travel. At a distance r from the point source, the particles will be spread over a sphere of radius r and of area $4\pi r^2$. (See Fig. 4–1.) At a distance $2r$ from the source, the same total

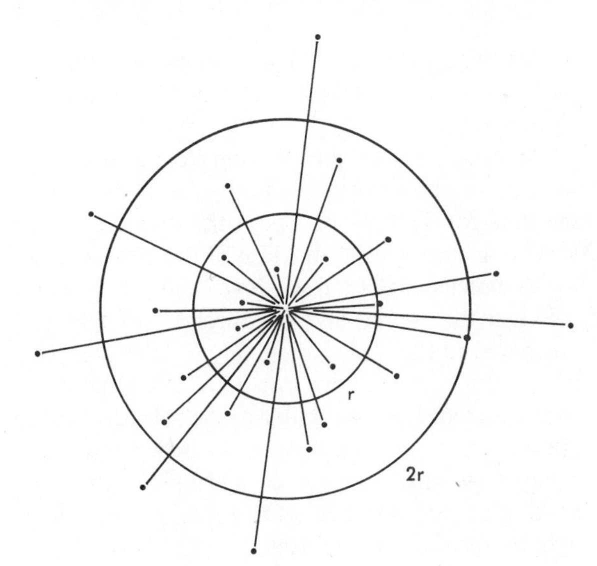

Figure 4–1
The particles which are spread over the area $4\pi r^2$ at radius r will be spread over the area $16\pi r^2$ at the radius $2r$. Thus the intensity will be only one fourth as great at distance $2r$ as it was at distance r from the point source.

number of particles must be spread over an area $4\pi(2r)^2$ or $16\pi r^2$. Thus the number of particles per unit area at $2r$, and hence the illumination at $2r$, will be only one-fourth as great as the number of particles per unit area, or illumination, at r. Similarly, at a distance $3r$, the illumination will be one-ninth as great, and at $4r$ it will have dropped to one-sixteenth. Thus, the particle model predicts that the illumination I is inversely proportional to the square of the distance r from the source. That is,

$$I = \frac{k}{r^2}$$

where k is a constant. If this prediction is correct, then the product of the intensity and the square of the distance should remain constant. Using one of our standard light sources and our calibrated light meter to check this prediction, we obtained the results shown in Table 1.

Table 1

r (METERS)	I	Ir^2
1.20	1.0	1.4
1.00	1.5	1.5
.80	2.3	1.5
.70	3.2	1.6
.60	4.3	1.5
.50	5.4	1.4
.40	8.5	1.4
.30	17.7	1.6
.25	23.5	1.4
.20	33.3	1.3

The intensity of illumination at different distances from a source.

The third column in Table 1 shows that the product of the intensity and the square of the distance is almost a constant. The data from Table 1 are plotted in Fig. 4–2.

The result that the illumination from a source of light varies inversely with the square of the distance is in agreement with the particle model. To get the data for Table 1 we chose the distance r_o in front of the meter to be very long compared to the size of our light source and also long compared to the dimensions of the sensitive element of our light meter.

When the source and the receiver of light particles are close together compared to their sizes, the light particles can pass from one to the other at odd angles to the surface instead of along paths which are nearly parallel to each other and perpendicular to the sensitive surface of the light meter. We are, therefore, not surprised if we find greater deviations from the inverse-square law where the source and light meter are close together. Also, in making the measurements for Table 1, care was taken to prevent reflected light from reaching the meter.

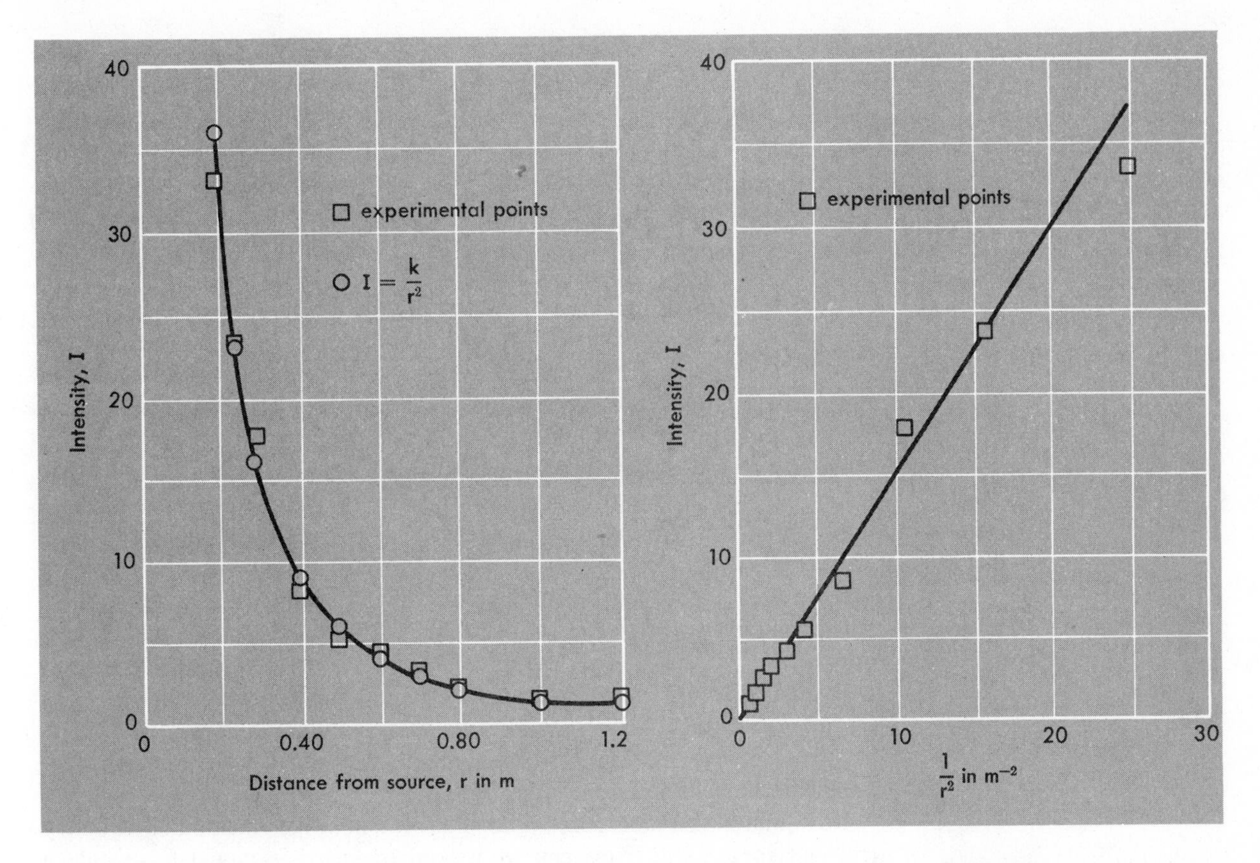

Figure 4–2
The graph on the left shows the intensity of illumination at different distances from a source. The experimental values agree closely with points computed, assuming an inverse-square relation. The graph on the right, where intensity is plotted against $1/r^2$, shows close agreement with a straight line.

4-2 Light Pressure

Whenever a baseball or a stone hits us we feel a push. The pushes exerted by the tiny molecules of a gas bumping into the walls of a container or pounding on our skin give rise to gas pressure. Do the particles of light bumping into a mirror similarly give rise to light pressure? The particle theory of light suggests that they do. Unfortunately, the particle theory we have developed so far is not specific enough to tell us how much light pressure to expect from a given stream of light particles.

But even without this detailed knowledge we do know that if light exerts a pressure at all, it must be very small for an ordinary beam of light. We do not knock over a feather by switching on a bright light. Even experiments with delicate apparatus failed to reveal the existence of light pressure until the beginning of this century. Then Peter Lebedev in Russia and Nichols and Hull in the United States succeeded in discovering and measuring this pressure. It is a very small pressure, but not all light pressure is so small. Experiments verify the prediction of the particle theory that the pressure of light increases with the intensity of illumination. At the surface of the sun, for example, light pressure is several orders of magnitude larger than the pressure of sunlight here, and at the surfaces of some stars the light pressure is enormous.

4-3 Absorption and Heating

A good reflector of light does not warm up appreciably when placed in sunlight, whereas a dark, absorbing surface can become quite warm. Dirt, shoveled up from a sidewalk along with snow after a snowstorm, melts its way down through the snow when the sun comes out. The black dirt, a good absorber, gets warm and melts the snow beneath it faster than the white reflecting snow melts by itself.

Can the particle model account for this behavior? When particles are neither reflected from a material nor transmitted by it, they must come to rest. When a piece of matter is hit by particles that neither pass through it nor bounce off it, does the material heat up? You can easily show that heating really occurs with particles by substituting a hammer for a particle and using a block of lead for the absorbing material. If you strike the block a series of rapid blows (like a series of particles), you find that the hammer does not bounce back (the particles do not reflect) and the lead gets warm. As a check to see that this heating is not an effect we get with every particle bombardment, you can repeat the experiment using a heavy steel block. The hammer now bounces back with every blow (it reflects), and the steel does not warm up. If light particles act like particles of ordinary matter, we would expect them to heat any material in which they are stopped. The particle model therefore agrees with the observed heating of materials which absorb light and with the fact that materials which reflect or transmit practically all the light do not heat up.

Most materials reflect some of the light which strikes them, and they may transmit some of it. How can we extend the particle model to include a substance which reflects some particles specularly while it absorbs other particles from the same light beam? We may imagine that the material contains small regions that stop light particles, producing heat, intermingled with regions that reflect light particles and produce no heat. Even the best reflectors, such as mirrors, warm up a little when placed in the sun, and we conclude that a mirror surface must contain a few scattered absorbing regions. A transparent substance such as glass must also contain some absorbing regions since we observe a slight decrease in the light intensity after the beam has passed through glass. Furthermore, as expected, the glass does become slightly warm.

4-4 Reflection

We know that light is reflected when it strikes a surface. Do particles such as steel balls behave in the same way? To answer this question, we merely need to throw some ball bearings onto various surfaces. The balls thrown onto a clean, smooth steel plate bounce off regularly at an angle of reflection approximately the same as their angle of incidence (Fig. 4–3). The paths of incidence and reflection are in a plane normal to the surface; also, the speed after collision with the surface is about as great as before. If either the balls or the smooth surfaces are less resilient than steel, reflec-

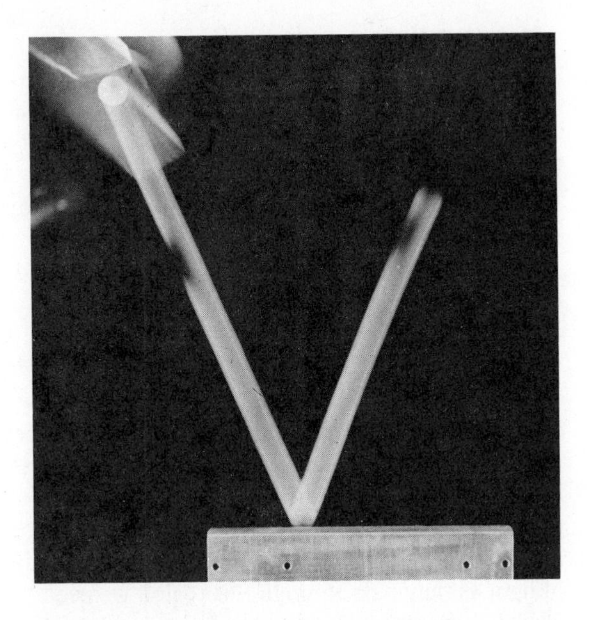

Figure 4–3
A time exposure of a ball being reflected from a steel plate. The angle of reflection equals the angle of incidence.

tion does not take place with equal angles of incidence and reflection. Also, in such cases the speed of the balls after reflection is less than before—unlike light, which does not change its speed on reflection. To account for the specular reflection of light from smooth surfaces, therefore, we have to assume that light particles reflect like ideal ball bearings bouncing off resilient surfaces. By restricting our model to such elastic particles and using only surfaces which are elastic for the particles of light, we can explain specular reflection.

The particle model can also account for the fact that light is sometimes reflected diffusely. Figure 4–4 indicates how. Individual particles bounce off with equal angles of incidence and reflection. But if the surface is

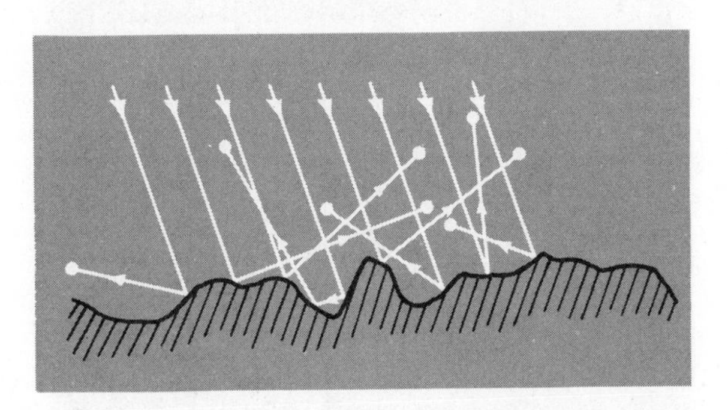

Figure 4–4
The diffuse reflection of particles. This occurs when a surface has irregularities larger than the size of the particles.

uneven, the reflected particles will travel in many directions so that the surface as a whole scatters the light particles diffusely. We might ask just how smooth a "smooth" surface must be to show specular reflection on a large scale. The answer is that the bumps in it must be very small compared with the size of the particles that strike it. A surface that appears smooth when ball bearings are bounced from it may be very rough when the small light particles are reflected. This is why metals must be highly polished before they become good mirror surfaces.

4-5 Refraction

Can our particle model of light account for refraction as well as for reflection? In refraction the light particles are deflected from their path in air to a new path as they enter the refractive material. Can we design an experiment in which moving steel balls are deflected so as to change their path in the same way?

Suppose a ball rolls along a level surface and then down a steep slope onto a lower level surface. While the ball rolls on the upper level surface, it moves in a straight line at a constant speed. This upper surface, let us assume, corresponds to the region of air through which a particle of light moves in a straight line at a constant speed. When the ball starts down the steep slope, it is pulled by gravity in a direction perpendicular to the edge of the slope. The sloping region, therefore, can model the surface region of a refractive medium such as glass. In this surface region in the glass, the particles of light are supposed to be pulled perpendicular to the boundary and toward the inside of the glass. To find out what happens to the particles of light as they pass through this region we need only look at the photograph in Fig. 4–5, which shows what happens to a ball as it rolls down the slope. You see that the ball is speeded up in the direction perpendicular to the

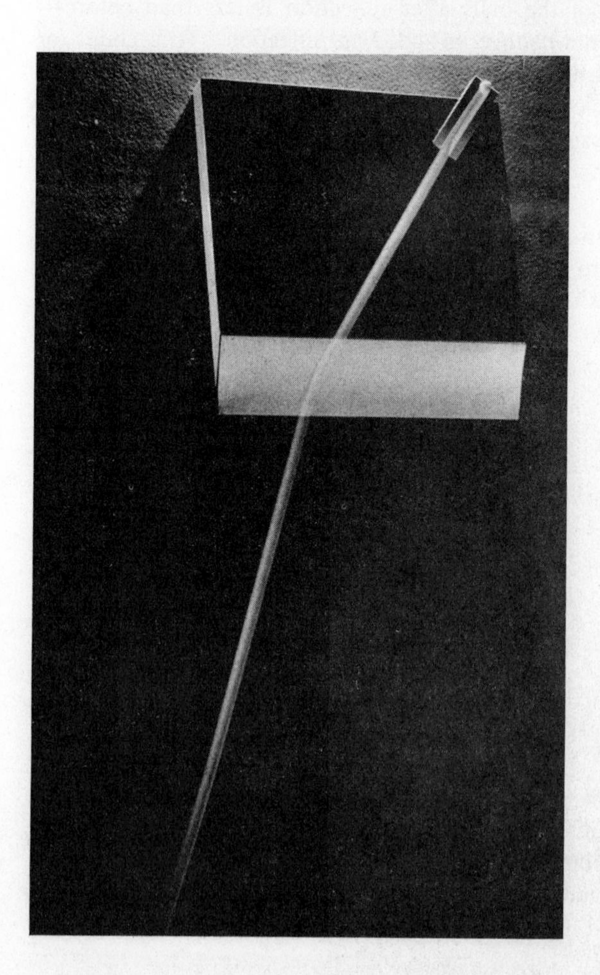

Figure 4–5
A time exposure of a ball bearing rolling from a higher to a lower surface. The change in direction of the ball shows "refraction."

edge, and the direction of its path is changed. If the ball rolling along the upper surface meets the slope at an angle, it rolls away from the slope along the lower surface in a different direction, closer to the normal. In the same way, light passing through a surface region in which it is pulled toward the inside of the refractive material should speed up and bend toward the normal as it enters the refractive material. The lower level surface on which the ball rolls corresponds to the inside of the refractive material. Here again the ball goes at a constant speed in a straight line like light inside of a piece of glass.

We can use the rolling-ball model to investigate the relation of the angle of refraction to the angle of incidence. If refraction is caused by a pull that occurs when the light passes from the outside to the inside of a refractive medium, then the relation that we find in the rolling-ball model should be the same as the relation for light. To carry out this investigation in the rolling-ball model, we always roll the balls across the upper plane at the same speed, because in a vacuum the light particles always travel at the same speed. Measurements of many pairs of angles show that the refraction of the balls agrees with Snell's law

$$\sin \theta_u / \sin \theta_l = \text{constant},$$

where θ_u is the angle between the path of the ball on the upper plane and the normal to the boundary and θ_l is the corresponding angle on the lower plane. What is more, the constant in Snell's law depends on the difference in height between the two plane regions and on the speed with which the balls move. The greater the difference in height, the greater is the index of refraction; the greater the original speed, the smaller the index. By always using the same original speed, we can model materials of different indices of refraction by planes of different heights. For example, to model the refraction of light going from air into water, the height of the sloping region must be less than the height needed to model the passage of light from air into glass. We find in each case that the rolling-ball model produces Snell's law; and by adjusting the heights, we can match any index of refraction.

In discussing the rolling-ball model we do not mean that light is made of rolling balls which fall down hills. The point of this model (which was devised by Sir Isaac Newton to explain the laws of refraction) is that refraction can be explained if at the surface of every refractive material light particles are pulled into the medium. This idea is plausible. Far inside of a piece of glass or other refractive material, a light particle is surrounded by the same material on all sides and experiences no pull. But at the surface the situation is obviously different, and a push or pull toward the inside might well occur. Then, if the pull acts like that in the rolling-ball model, refraction is successfully explained.

So far this model is successful. We therefore examine it further. We can find the relation between the speeds of a ball on the upper and lower levels by taking stroboscopic pictures (Fig. 4–6). In such pictures there are equal, short time intervals between a series of exposures. Therefore, the distance between pictures of the moving ball is a measure of its speed. When the speed is low, the pictures of the ball are closer together, and when the speed is greater they are farther apart.

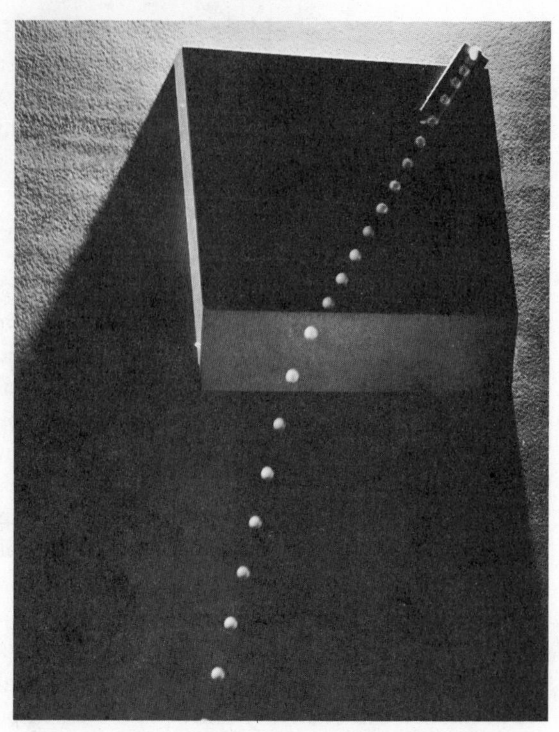

Figure 4–6
Two stroboscopic pictures of the ball bearing shown in Fig. 4–5. The speed on the upper level is the same in both pictures, since the ball bearing is released from the same height each time. Notice that the speeds on the lower level are also the same, although the angles of incidence are different in the two pictures.

Analyzing many paths of many balls hitting the slope at different angles shows that, for the same v_u on the upper plane, the ball always attains a definite speed v_l on the lower plane. The angle of incidence on the boundary does not influence v_l. Furthermore, measurements on stroboscopic photographs like those in Fig. 4–6 show that the angles and speeds are related by the equation

$$\frac{\sin \theta_u}{\sin \theta_l} = \frac{v_l}{v_u}.$$

Snell's law therefore holds, and the index of refraction must be v_l/v_u. From the speeds v_u and v_l measured with the stroboscope, and the measured angles, we then get a successful crosscheck on the model.

We can also roll balls on the lower plane toward the hill starting at the speed v_l. Then we find, as we might expect, that everything is reversed. The balls arrive on the upper surface with v_u, and Snell's law holds in reverse just as it does for light.

We now have a particle model of refraction, and—as usually happens when we have a model—it suggests many experiments. For example, our model says that the speed of light in a refractive material is independent of the angle of incidence and greater than the speed of light in vacuum. We should make measurements of the speeds of light in refractive materials and in particular find out if they agree with

$$\frac{v_m}{c} = n_m,$$

where v_m is the speed in the refractive material, c the speed of light in vacuum, and n_m the index of refraction. We shall discuss such measurements in Section 4–7. Also we should check to see that the speed of light in vacuum does not depend on the prior history of the light. According to our model, a light particle slows down in going from a refractive material into vacuum by just as much as it speeds up in going from vacuum into the material. The constancy of the speed of light in vacuum—regardless of where the light comes from—is a check on this particle model of refraction.

4-6 Some Difficulties with the Particle Theory

We have found that we can make a particle model which accounts for specular and diffuse reflection, which gives Snell's law of refraction, and which leads to the inverse-square law for the variation of the intensity of illumination with distance. The model also suggests that light should exert a pressure and that heating should be associated with light absorption; effects which we had not previously discussed are found experimentally. They serve as examples of the value of a model in leading us to new investigations. The model is successful in relating in a simple way much that we already know by giving us a structure into which our knowledge fits. It also suggests that other things should take place about which we may have had no idea. It gives us the clue to what may otherwise be unexpected, because, like the pressure of light, the unexpected may exist in the model. We can then check the model by comparing nature with what the model says.

So far with our model of light we have been very successful. Are there any respects in which the particle model is less successful? Are there aspects of the behavior of light which are poorly described by the model, or even in conflict with it?

Let us take a closer look at what happens when light is refracted. We know how to account for refraction and reflection separately in the particle model. Can we account for the fact that both take place when a light beam reaches the surface of a refractive material? To produce specular reflection the surface pushes the particles of light in one direction; to produce refraction on our model it pulls them the opposite way. How does it decide what to do? This question bothered Newton, and he attempted to explain the partial reflection and partial refraction of light by supposing that the particles of light had properties which changed periodically with time. What happens to the particle then depends upon the time at which it hits the surface.

Another attempt to explain the splitting up of light beams into reflected and refracted parts might be made by imagining that the surface is divided into little reflecting regions and little regions that refract. Either of these extensions of our model might account for a definite ratio of reflection and refraction. But they are in trouble because the ratio of refracted to reflected light decreases as the angle of incidence increases. Perhaps the difficulty can be overcome, but the model is getting complicated. When we have to fix it up too much, we look on it with suspicion.

More trouble arises when we try to account for diffraction. For example, there seems to be no simple particle explanation for the complicated diffraction pattern observed when light passes through a tiny hole (Fig. 3–24) and the reflection of light by very thin transparent films (Fig. 3–25). In addition, as we shall see in the next section, there is trouble with the speed of light.

4-7 The Speed of Light and the Theory of Refraction

According to our particle theory, the speed v_m of light in a refractive material must be greater than the speed c in vacuum. In Section 4–5, we found $v_m = cn_m$, where n_m is the index of refraction. This is a quantitative result of the model which we can test if we can measure the speed of light over a short distance—in a tank of water, for example.

In Newton's time such a measurement was not possible. The first successful measurements of the speed of light over a measured path on earth were made by Fizeau (1849). His method of measurement was based on chopping a beam of light with a rotating disc that has a hole in it. Think of directing a beam of light at the hole. Little pulses of light are let through each time the opening in the whirling disc passes the beam. Now let these pulses strike a mirror so that they are reflected back to the disc again (Fig. 4–7). On the return trip a pulse can pass through the disc only if the opening in the disc is again in the right position.

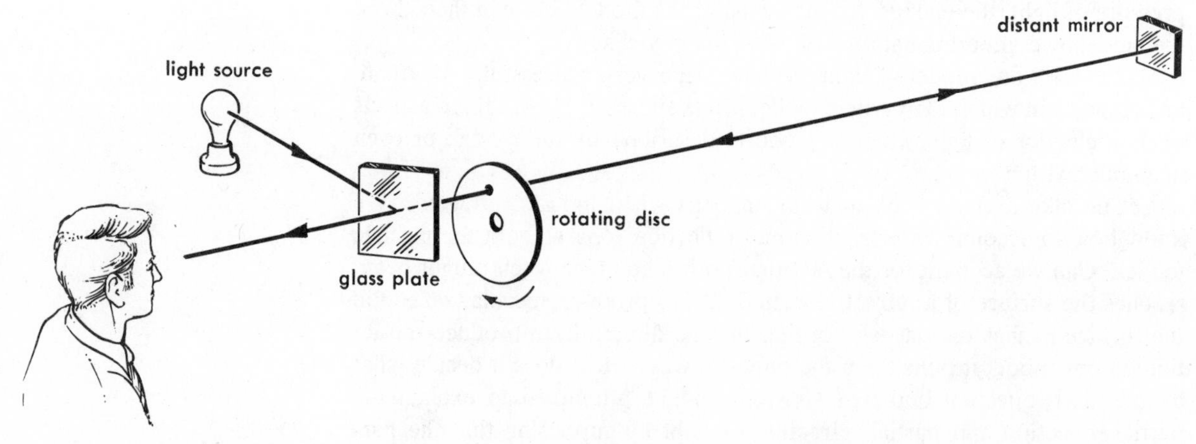

Figure 4–7
Timing a light pulse. Light from the source reflects from the glass plate and goes through the hole in the rotating disc to a distant mirror. The returning pulse passes through the disc and through the glass plate to the eye of an observer.

A disc with a single hole must rotate at least one full turn in the time that the light goes to the mirror and back. By measuring the speed of rotation at which this occurs, we can time the light as it goes the known distance from the disc to the mirror and back again. Thus we determine the speed of light.

By designing a special high-speed disc with many holes, Fizeau succeeded in measuring the speed of light in air. Other physicists improved Fizeau's method, among them the French physicist Cornu. His disc had 200 openings and could turn at 54,000 revolutions per minute. As he increased its speed from zero to the maximum he measured twenty-eight successive eclipses and brightenings of the light coming back through the disc from a mirror 23 km away. By using a rotating mirror to replace the rotating disc, Foucault in France and later Michelson in the United States increased the accuracy of this method still further (Fig. 4–8). Michelson was able to determine the speed of light to an accuracy of about 3 km/sec (1 part in 10^5).

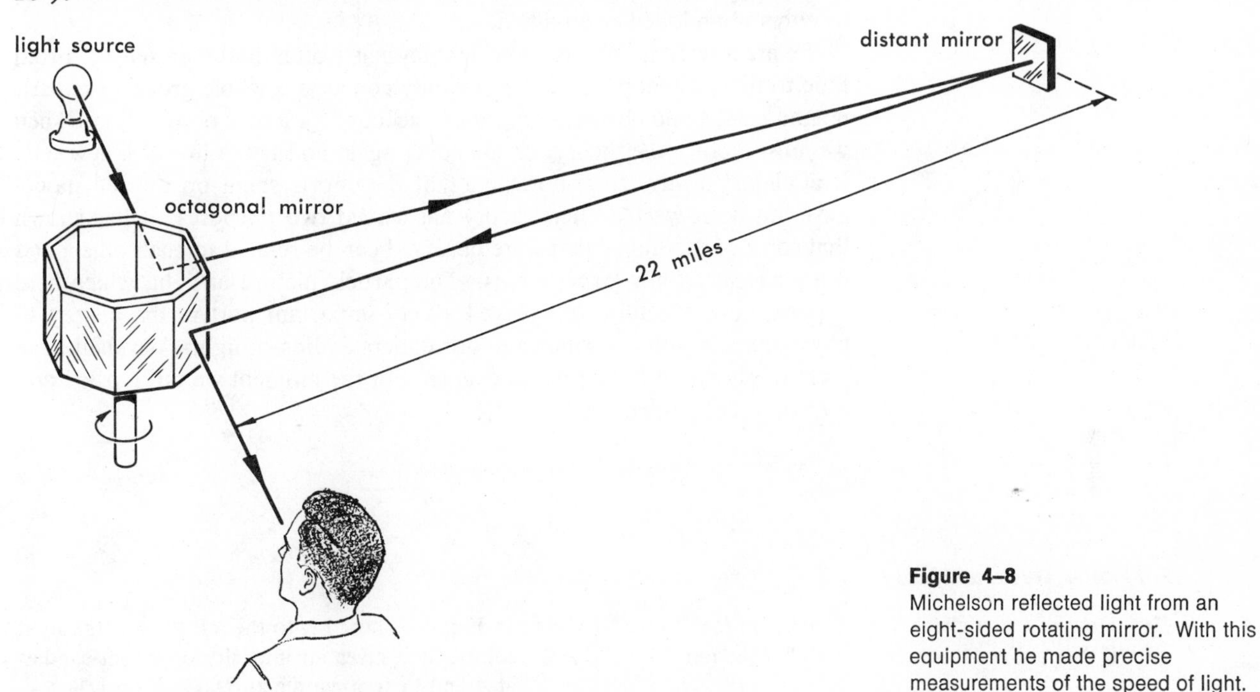

Figure 4–8
Michelson reflected light from an eight-sided rotating mirror. With this equipment he made precise measurements of the speed of light.

For this accurate measurement Michelson used a long path. But the same method can be used over short paths to find the speed of light in liquids, such as water, though with lower accuracy. In 1862, Foucault measured the speed of light in water and found it to be close to 2.23×10^8 m/sec. This is almost exactly three-fourths of its speed in air. Some years later, Michelson determined the speed of light in carbon disulfide, a liquid with an index of refraction of 1.63, and found it to be 1.71×10^8 m/sec.

When it encounters these results our particle theory of light really strikes a snag. In order to account for Snell's law, we were forced to assume that the speed of light is higher in refractive materials than in vacuum. Using Newton's particle model, we obtained $v_{\text{water}} = n_{\text{water}}c$. Consequently, according to the model the speed in water should be about 4/3 of that in air, and in carbon disulfide the speed should be about 5/3 of that in air. The observed values are almost exactly 3/4 and 3/5 of the speed in air. Not only have the experiments failed to show the ratios that we expected—they indicate that the ratios are inverted. Our particle model of light appears to fail.

4-8 The Status of the Particle Model

In this chapter we tried to develop a model of light based on the behavior of particles. This attempt was successful as long as we considered only *some* of the optical phenomena we know about, and as long as we did not always demand precise and quantitative agreement between the predictions of the theory and the results of experiments. When we included all of the phenomena and asked for quantitative checks, however, the model failed in several important respects. Since our particle theory of light failed, it must be either abandoned or modified.

We are now in a situation in which physicists often find themselves. From time to time, a theory which successfully connects a whole group of experimental results and observations fails to account for a new observation. Then we must modify the theory, or start over again and try to invent a new one. It is almost always true, however, that the efforts spent on the old model have not been wasted. The theory has served two purposes: it has shown that some of the things that were observed can be related to each other; and it has suggested new experiments. The particle picture of light, which here appears unsatisfactory, has played a very important part in the history of physics, and it still contributes to our understanding of light. We shall have more to say about it later in the course. For the moment we put it aside and try a different model.

For Home, Desk, and Lab

1.* How can you check several light sources to see if they are equal sources of light? (Section 1.)

2.* Suppose a light meter gives a reading of 1 unit when it is a distance r from a light bulb. At what distance from the same source will the meter give a reading of 2 units? (Section 1.)

3.† Can you explain the different strengths of light sources in terms of particles of different size? How about the decrease in intensity caused by inserting a partially absorbing sheet of matter in a light beam? Can different intensities at varying distance from a source be accounted for in this way?

4. What does the particle theory predict about the intensity produced by an *extremely* weak light source—so weak that only a few particles are emitted per second? How might you test this prediction?

5. It is found that a 40-candle source placed 3.0 feet from a light meter gives the same scale reading as an unknown source 1.2 feet from the meter.
   (a) What is the strength of the unknown source?
   (b) What is the light intensity in foot-candles read on the meter?

6.* In Fig. 4–9, how far to the left of A is the light source, if it gives an intensity of 8 foot-candles at A and ½ foot-candle at B? (Section 1.)

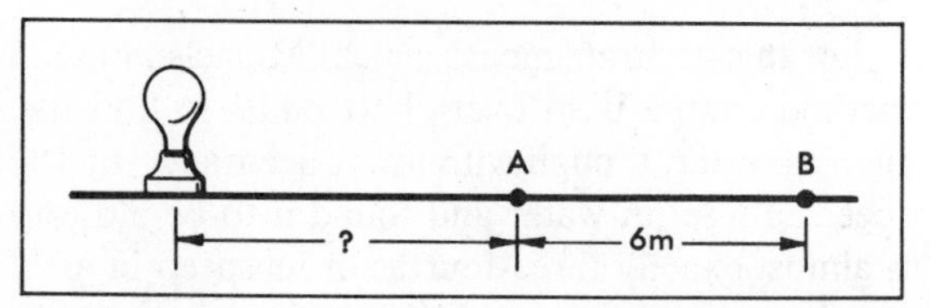

Figure 4–9
For Problem 6.

7. A light meter calibrated in foot-candles is 2.0 feet from a mirror. A very small 25-candle-power source is placed midway between them. What is the reading of the meter if the mirror is
   (a) a plane mirror?
   (b) a concave mirror with a diameter of 6 inches and a focal length of 1 foot?

8. In the use of radar, a beam of radiation is sent out from a source. Some of this radiation falls on a distant object, such as an airplane, is reflected by the object, and is detected when it returns to its starting point. The radiation used behaves like light. Assuming that the source of the radiation is equivalent to a point source of light and that the object reflecting the radiation

is a diffuse reflector, can you convince yourself that the intensity of the returning radiation varies inversely with the fourth power of the distance from the source to the reflecting object?

9. Upon which object would light exert the greater pressure, a clear piece of glass or a mirror?

10. A sensitive thermometer placed in the different parts of the spectrum from a prism will show a rise in temperature. This shows that all colors of light produce heat when absorbed. But the thermometer also shows a rise in temperature when its bulb is in either of the two dark regions beyond the two ends of the spectrum. How can the particle theory account for this?

11.* Why should a Ping-Pong table have a more even surface than the floor of a basketball court? (Section 4.)

12. Consider the box shown in Fig. 4–10, whose inside walls are perfectly reflecting. What do you predict will happen when you shine a beam of light into the box? (Use the particle model of light to arrive at your answer.)

13.* In Fig. 4–6 if you keep the speed of the ball bearing on the upper level constant by launching the ball always in the same manner, at what angle of incidence to the hill will you get the highest speed on the lower level? (Section 5.)

14.* What does the experiment of particle-model refraction suggest about the speed of light in a material, that was not apparent in our refraction experiments with light? (Section 5.)

15. The index of refraction of carbon disulfide is about 1.63. What should the speed of light be in this liquid, according to the particle model of refraction given in Section 4–5?

16. The speed of light measured in a certain medium is 2×10^8 m/sec.

    (a) According to the particle model of light, what should be the index of refraction of this medium?
    (b) Using the index of refraction from part (a), what would be the path of a light ray which passes from air into the medium at an angle of 30° from the normal to the surface of the medium?

17. Occasionally there occurs in the heavens an explosion of a star, producing what is known as a super nova. The star suddenly becomes many times brighter than before. As you know, the stars are so far away that the light from them takes many years to reach us. An explosion that we observe must have occurred a long time ago, and the light has been traveling toward us ever since. We see the explosion as a bright white light, not as a series of different colors arriving at different times.

    (a) What does this show about the speed of light of different colors in vacuum?
    (b) Try to suggest a particle model for dispersion in prisms consistent with the single speed of light of all colors in vacuum.

18.* Why is it possible to see light through the hole in the strobe disc in Fig. 4–7 when the disc is rotating very slowly? (Section 7.)

19. Using the data given in Section 4–7 about Cornu's application of the Fizeau method, compute the shortest distance from the rotating disc to the mirror that will permit the returning beam to pass through the opening immediately following the one through which it started.

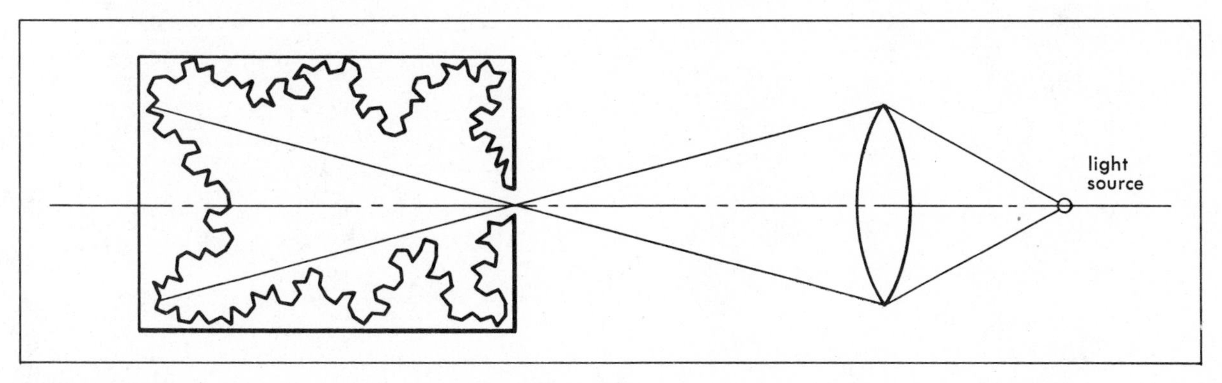

Figure 4–10
For Problem 12.

20. In the Foucault-Michelson modification of the Fizeau method, rotating mirrors are used in place of the rotating disc (Fig. 4–8). Assuming that mirrors are rotating 500 times per second and that the distance from them to the fixed mirror is 10.0 km, show that a mirror face rotates through an angle of 12° before it reflects the beam a second time and that the angle between this reflected beam and the original direction of the light is therefore 24°.

21.† It might be imagined that on each reflection the speed of light decreases slightly. Assuming that you can measure the speed of light (see Section 4–7), describe an experiment designed to test the assertion that reflection does not change the speed.

22. Can you modify your proposed experiment (Problem 21) to test the assertion that the speed of light in vacuum is independent of previous refractions?

23. Suppose that someone were to propose to you that sound consists of small, rapidly moving particles that are emitted by a source and that affect your ear when they strike it. What evidence could you use to support or to refute it?

24. We have said the light particles must be very small. Suggest an experiment to show that they must also have an extremely small mass.

25. Is it possible that the slowing down of light as it goes from a vacuum into a transparent material is some kind of friction effect?

26.* Why does the measurement of the speed of light in water cast doubt on the validity of the particle model as developed in this chapter? (Section 7.)

Further Reading

These suggestions are not exhaustive but are limited to works that have been found especially useful and at the same time generally available.

HOLTON, G., and ROLLER, D. H. D., *Foundations of Modern Physical Science.* Addison-Wesley, 1958. (Chapter 30)

JAFFE, BERNARD, *Michelson and the Speed of Light.* Doubleday Anchor, 1960: Science Study Series. (Chapter 3)

JENKINS, FRANCIS A., and WHITE, HARVEY E., *Fundamentals of Optics.* McGraw-Hill, 1957. For the student interested in the various methods of measuring velocity from Roemer to the Kerr cell. (Chapter 19)

MAGIE, WILLIAM FRANCIS, *A Source Book in Physics.* McGraw-Hill, 1935. Contains a wealth of good historical material: Roemer, Fizeau, and Foucault.

Introduction to Waves

CHAPTER

5

5-1 A Wave: Something Else That Travels

In the last chapter we considered at some length a particle model of light, in which we supposed that light consisted of a stream of particles or corpuscles. We found that this model fails to provide satisfactory explanations for some of the behavior of light that we observed. We therefore find ourselves faced with a choice: we can try to construct a better particle model that will succeed where the earlier one failed, or we can look for a new and simple model based on a completely different concept. Let us try the second approach.

The first thing to be accounted for in *any* model of light is the fact that light travels through space. In looking for a new theory, we first ask if there is anything except a particle (or a stream of particles) that can move from one point to another. The answer is "yes." Consider, for example, what happens when we drop a pebble into a quiet pond. A circular pattern spreads out from the point of impact. Such a disturbance is called a *wave.* If you watch the water closely, as such a wave moves across the surface, you will find that the water does not move forward with the wave. This is quite clear if you watch a bit of wood or a small patch of oil floating on the pond. The wood or oil moves up and down as the wave passes; it does not travel along with the wave. In other words, a wave can travel in water, but once it has passed, every drop of water is left where it was before.

If we look around us, we can find many examples of waves. For instance, we notice an American flag as it ripples in the breeze at the top of a flagpole. The ripples or waves travel out along the cloth. Individual spots on the cloth of the flag, however, hold their positions as the waves pass by. The fourth white star in the bottom line on the field of blue always remains the fourth star in the bottom line and its distances from the edges of the flag do not change. Just as the water does not travel with the water waves, so the cloth of the flag remains in place after the waves have passed through it.

Some waves are periodic or nearly so; the motion of the material repeats itself over and over. Not all waves, however, have this property. For example, when you slam the door of a room, the air in the doorway is suddenly compressed. This single short compression passes as a disturbance across the room, where it gives a sudden push to a curtain hanging over the window. Such a wave of short duration is called a *pulse.*

Here is another example of a pulse. We place half a dozen pocket-billiard balls in a straight line so that each ball is touching the next one. We then roll another ball so that it strikes one end of the row head-on. The ball at the other end of the row moves away at a speed equal to that of the ball we rolled in. A pulse has traveled through the row of balls from one end to the other. Each ball has been disturbed; this disturbance has passed along the entire line of balls; but no ball has moved from one end of the line to the other.

What is alike in these examples? In each case the disturbance travels through some medium—through the water, the cloth of a flag, the billiard balls; but the medium does not go along with the disturbance. Disturbances which travel through media are what we mean by waves. We can now answer the question we asked earlier in this section: is there anything except a particle that can move from one point to another? A wave, a thing which is not itself a particle of matter, can go from one place to another.

5-2 Waves on Coil Springs

Do waves behave like light? To find out, we must know more about them. When we know how they act, we can compare their behavior with that of light. The variety of examples we have mentioned also suggests that waves are worth studying for their own sake.

It is convenient to start our study of waves by experimenting with a coil spring. Figure 5–1 shows pictures of a pulse traveling along such a spring.

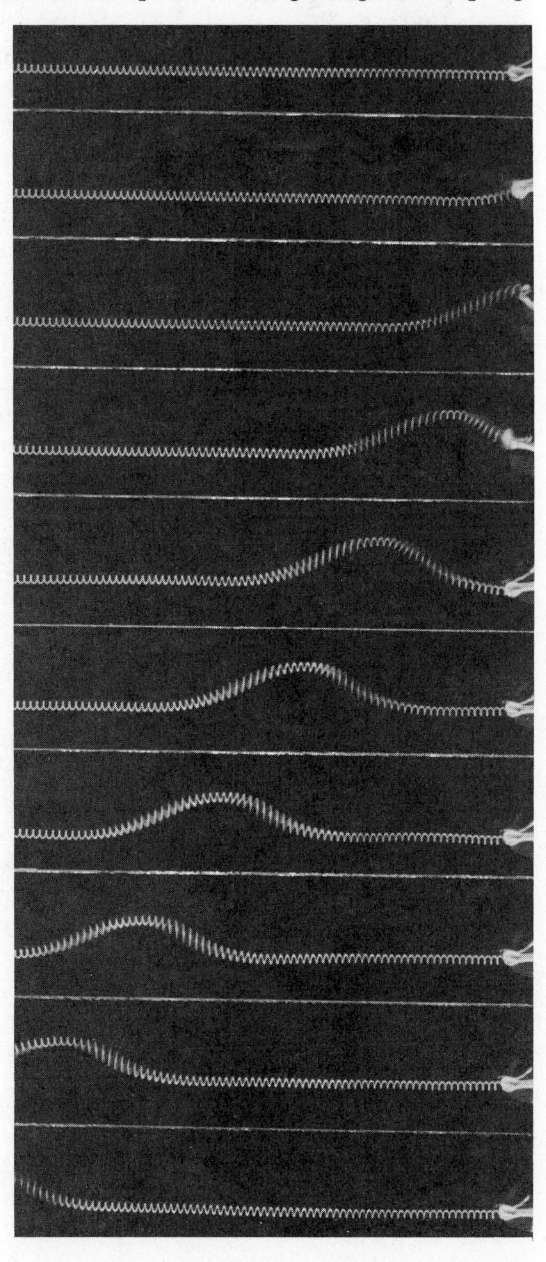

Figure 5–1
The generation and motion of a pulse along a spring shown by a series of pictures taken with a movie camera.

These pictures were taken by a movie camera at intervals of $1/24$ of a second.

We see that the shape of the pulse does not change as it moves along. Except for the fact that the pulse moves, its shape at one moment is just like its shape a short time later. Also we see that the pulse moves the same distance in each interval between pictures—it moves along the spring at constant speed.

The spring as a whole is not permanently changed by the passage of the pulse. But what happens to each small piece of spring as the pulse goes by? To help us fix our attention on one piece, we can mark it by tying a white ribbon to the spring as shown in Fig. 5–2. If we then shake the spring to start a pulse moving along it, we can see how the white ribbon is displaced. We find that it moves at right angles to the spring as the pulse passes it.

Other parts of the spring, as well as the ribbon, also move. We can see which parts are moving and which way they go if we look at two pictures, one of which is taken shortly after the other. Here we shall use two successive pictures taken from Fig. 5–2. We have printed these two pictures together in Fig. 5–3 so that we see the pulse in two successive positions just as we would see it in a rapid double exposure. Below the photo in Fig. 5–3 we have drawn the pulse in its earlier position in black, and in gray in the later position. As the arrows show, while the pulse moved from right to left, each part of the coil in the right-hand half of the pulse moved down and each part of the coil in the left-hand half moved up.

If the pulse were moving from left to right, just the reverse would be true, as we show in Fig 5–4. Here we use a schematic pulse because it is a little easier to work with and we can make the time interval between positions as short as we wish. In this way we can determine the instantaneous motion of the coil. Thus, if we know in which direction the pulse is moving, we can determine how each point of the spring moves at any particular stage in the passage of the pulse. On the other hand, if we know how the parts of the spring move, we can determine the direction in which the pulse is traveling.

We now have a good notion of how the parts of the spring move, even though there is no visible motion in any one of our pictures. Really, what we have done is to observe (1) that any pulse moves undistorted at constant speed along the spring and (2) that the spring itself moves only at right angles to the motion of the pulse. We can combine these two pieces of information to learn how each part of the spring moves at any time. Of course, we have looked only at the simplest waves, and the statement we have just made may not be true of all waves. Even in the cases we have examined, a sharp eye may detect slight deviations from our description. Nevertheless we have formed a useful first picture. With slight changes it applies to many other waves.

5-3 Superposition: Pulses Crossing

So far we have discussed the behavior of a single pulse traveling in one direction. But what happens when one pulse moves from right to left at the same time that another moves from left to right? Particularly, what happens when the two pulses meet? Do they pass through each other, or do they somehow knock each other out?

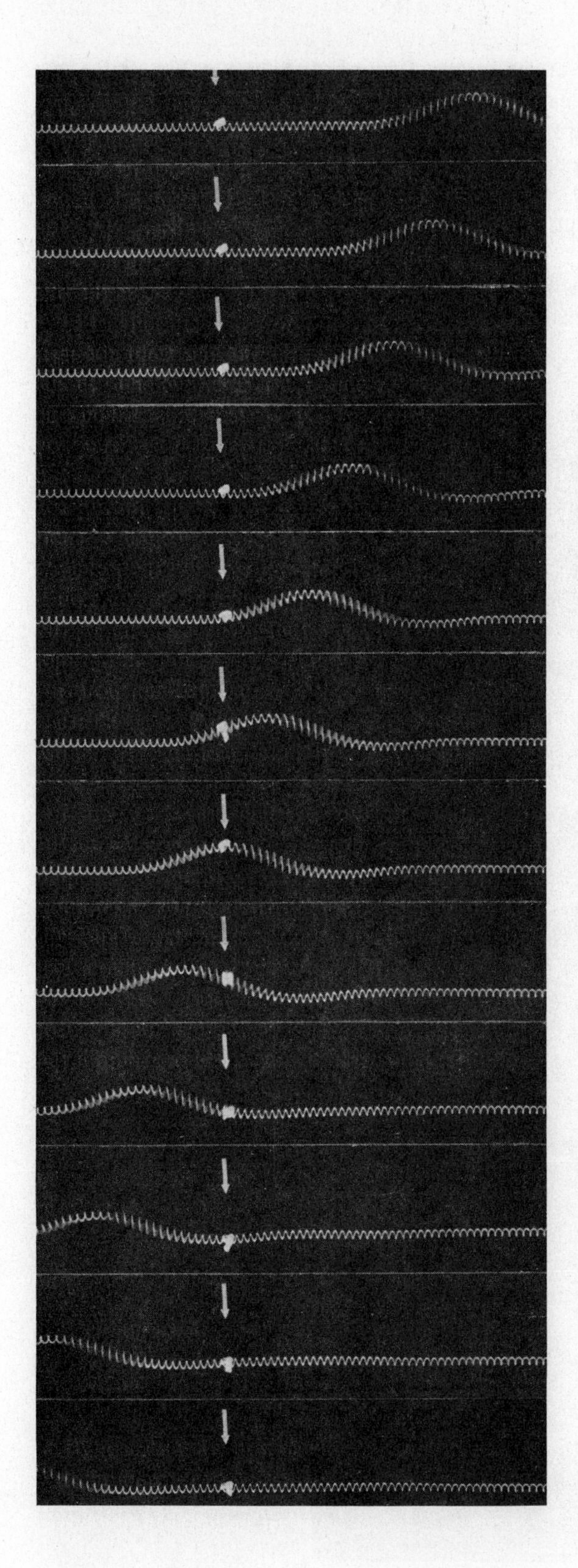

Figure 5–2
The motion of a pulse from right to left along a spring with a ribbon around one point. The ribbon moves up and down as the pulse goes by, but does not move in the direction of motion of the pulse.

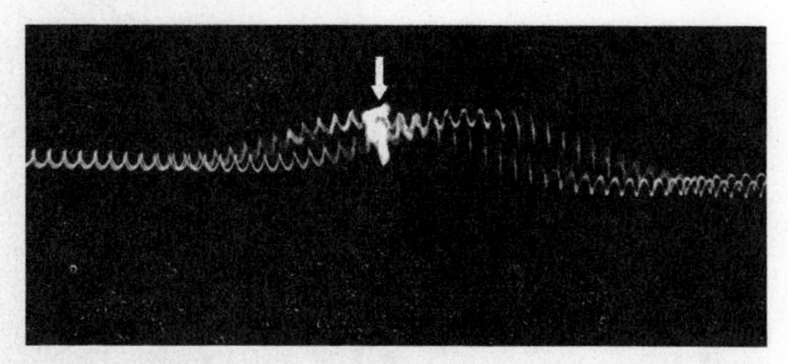

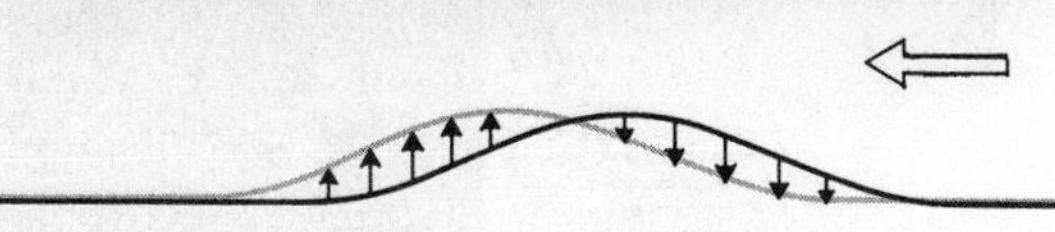

Figure 5–3
The relation between the motion of a pulse traveling from right to left and the motion of the spring. The photograph shows the pulse in two successive positions. The arrows in the diagram indicate how the spring moves as the pulse passes. The large, open arrow shows the direction of the motion of the pulse.

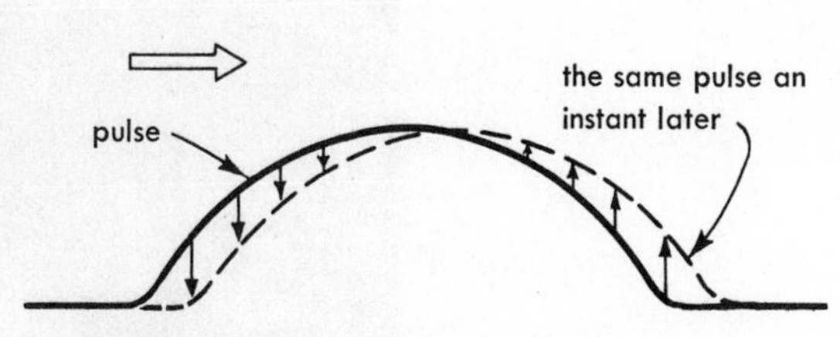

Figure 5–4
The relation between the motion of a pulse traveling from left to right and the motion of the spring.

The best way to find out is to try. The photographs in Fig. 5–5 show what happens when two pulses are started at opposite ends of a spring at the same time, one traveling from left to right and one from right to left. The top pictures show the pulses approaching each other as if each had the spring to itself. As they cross each other, the two pulses combine to form complicated shapes. But after having crossed, they again assume their

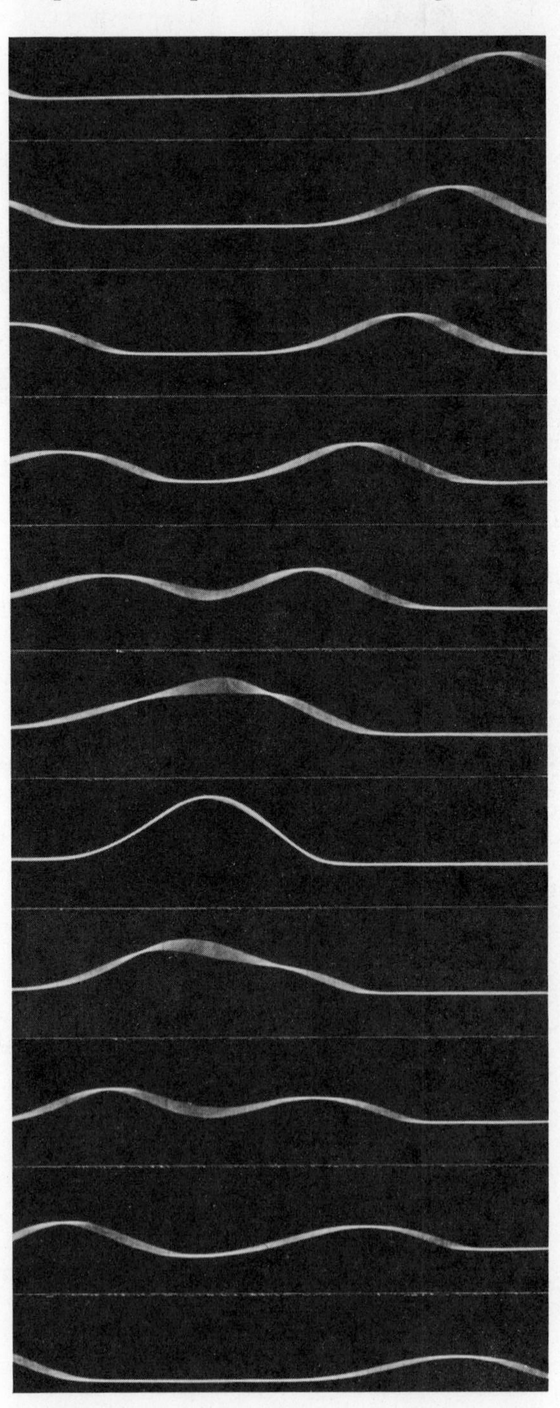

Figure 5–5
Two pulses crossing each other. Notice that the two pulses have different shapes. Thus we can see that the one which was on the left at the beginning is on the right after the crossing, and vice versa.

original shapes and travel along the spring as if nothing had happened, as is indicated by the pictures at the bottom. The left-going pulse continues to travel to the left with its original shape. The right-going pulse continues to move to the right with its earlier form. We can perform this experiment over and over with different pulses. We always get the same general result.

The fact that two pulses pass through each other without either being altered is a fundamental property of waves. If we throw two balls in opposite directions, and they hit each other, their motion is violently changed. The crossing of waves and the crossing of streams of balls made of solid matter are thus two very different processes.

Let us now take a closer look at the two pulses crossing each other (Fig. 5–5). The shape of the combined pulse does not resemble the shape of either of the original pulses. We can see its relation to them, however. We visualize each of the original pulses at the position it would occupy if alone; then we add up the displacements of the parts of the spring corresponding to the original pulses to get a new pulse. We find that the total displacement of any point on the spring at any instant is equal to the sum of the displacements of the two pulses independently. The method is illustrated in Fig. 5–6. It works for any two pulses. As a matter of fact, it also works for more than two pulses—the displacements due to any number of pulses can be added.

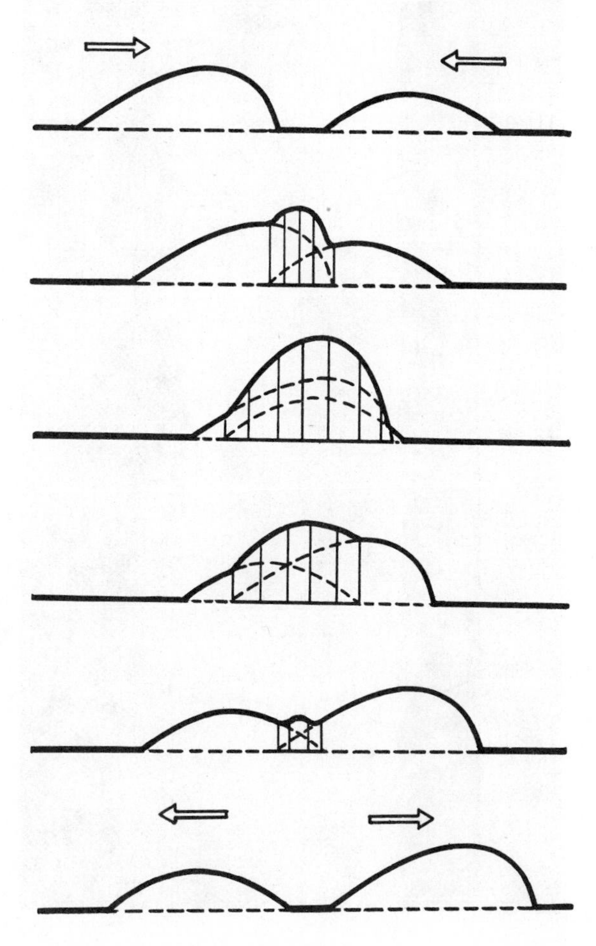

Figure 5–6
The superposition of two pulses. The displacement of the combined pulse is the sum of the separate displacements.

We can summarize the whole situation as follows. To find the form of a pulse which results from the crossing of two other pulses, we add the displacements of the individual pulses. The fact that pulses combine in this way is called the *Superposition Principle*.

Let us apply the Superposition Principle to two special cases. First, let us consider the experiment shown in the sequence of pictures of Fig. 5–7.

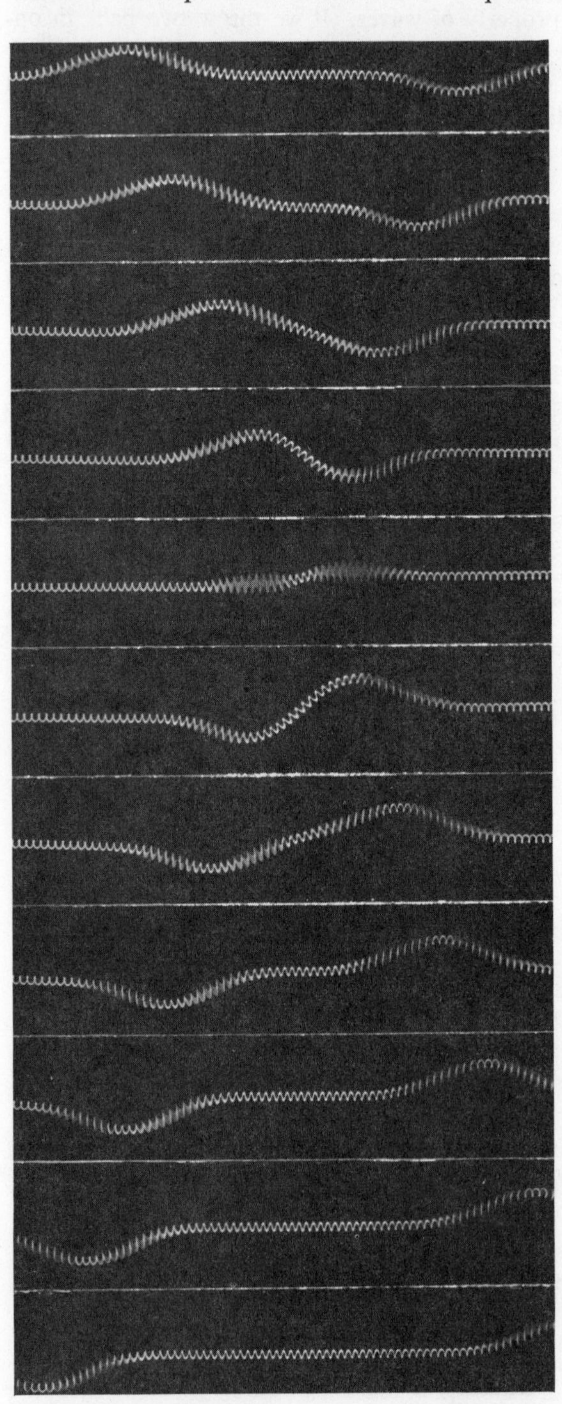

Figure 5–7
The superposition of two equal and opposite pulses on a coil spring. In the fifth picture they almost cancel each other.

Look at the combination of a pulse that displaces the spring downward and travels from the right with one that displaces the spring upward and travels from the left. Suppose that the two pulses have exactly the same shape and size and that each is symmetrical. Notice that in one picture the addition of equal displacements upward (plus) and downward (minus) leaves us with a net displacement of zero. There is clearly a moment, as the pulses pass each other, when the whole spring appears undisplaced. (See also the drawing of Fig. 5–8.) Why does the picture not look exactly like a spring at rest? Let us consider the difference between an undisplaced spring carrying two equal and opposite wave pulses and an undisplaced spring carrying no wave at all. When the spring carries no wave, all the various pieces of spring stand still at all times. On the other hand, when two equal and opposite waves are passing, there is only one instant when the spring is passing through its rest position, and at that instant the spring is moving. The motion shows up as a blur in the pictures, just as a snapshot of a rapidly moving airplane often appears blurred.

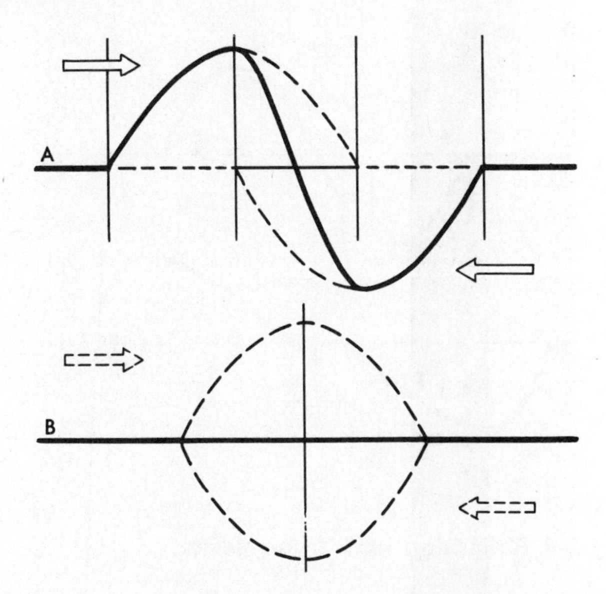

Figure 5–8
The superposition of two equal and opposite pulses. (A) Before complete cancellation. (B) At complete cancellation.

Our second special case is shown in Fig. 5–9. Here we have two similar pulses, one coming from the right and one from the left. In one the displacements are upward and in the other they are downward. These pulses differ from those of Fig. 5–7 in that neither is symmetrical, although the two are alike in shape and size.

Because neither of the pulses is symmetrically shaped, they never completely cancel each other. But there is always one point P on the spring which will stand still. That point is exactly halfway between the two pulses. As the pulses come together, they pass simultaneously through that halfway point in such a way that the highest point of one pulse and the lowest point of the other just cancel each other. The same argument applies to any other pair of corresponding points on the pulses. They always arrive at the midpoint of the spring together, one on top and one at the bottom. Consequently, the midpoint stands still.

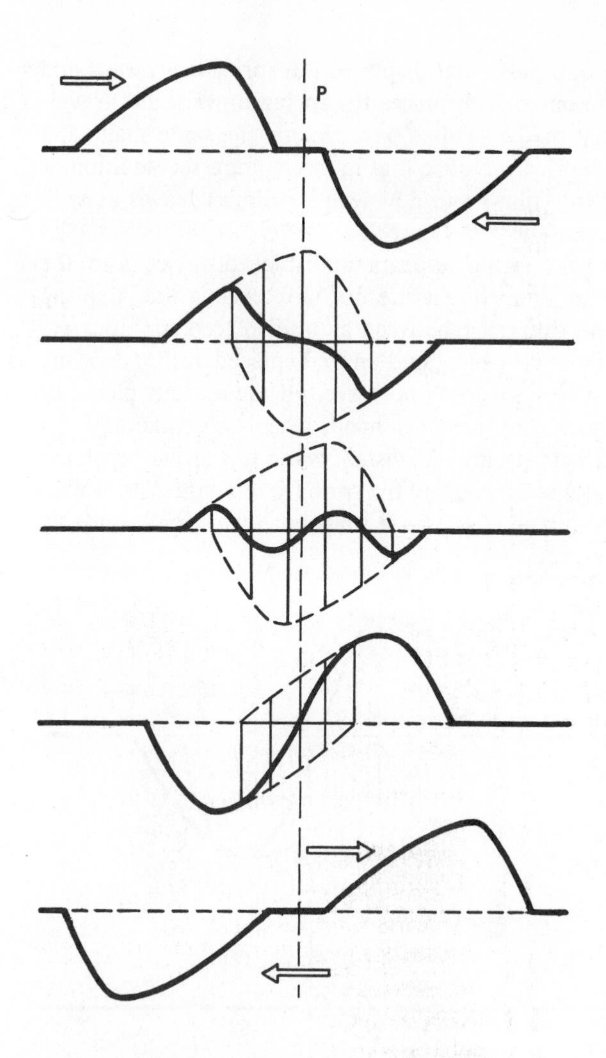

Figure 5–9
The superposition of two asymmetrical but similar pulses. Notice that the point midway between them remains at rest at all times.

5-4 Reflection and Transmission

When a pulse moving on a spring comes to an end that is held fixed, it bounces back—it is reflected. The pulse that comes back is called the reflected pulse. In Fig. 5–10 the fixed end is on the left. In the original or incident pulse, which moves to the left, the displacement is upward. The returning pulse has its displacement downward. The pulse comes back upside down, but with the same shape that it had before it was reflected.

Imagine now that instead of fixing our spring at one end, we connect it to another spring which is much heavier and therefore harder to move. Our new arrangement will be somewhere in between the two cases (a) the original spring tied down, and (b) the original spring just lengthened by an additional piece of the same material. In case (a) the whole pulse is reflected upside down; in case (b) the whole pulse goes straight on. We may, therefore, expect that under our new arrangement part of the pulse will be reflected upside down, and part of it will go on, or as we say, will be transmitted. This effect is shown in Fig. 5–11, where the original pulse comes

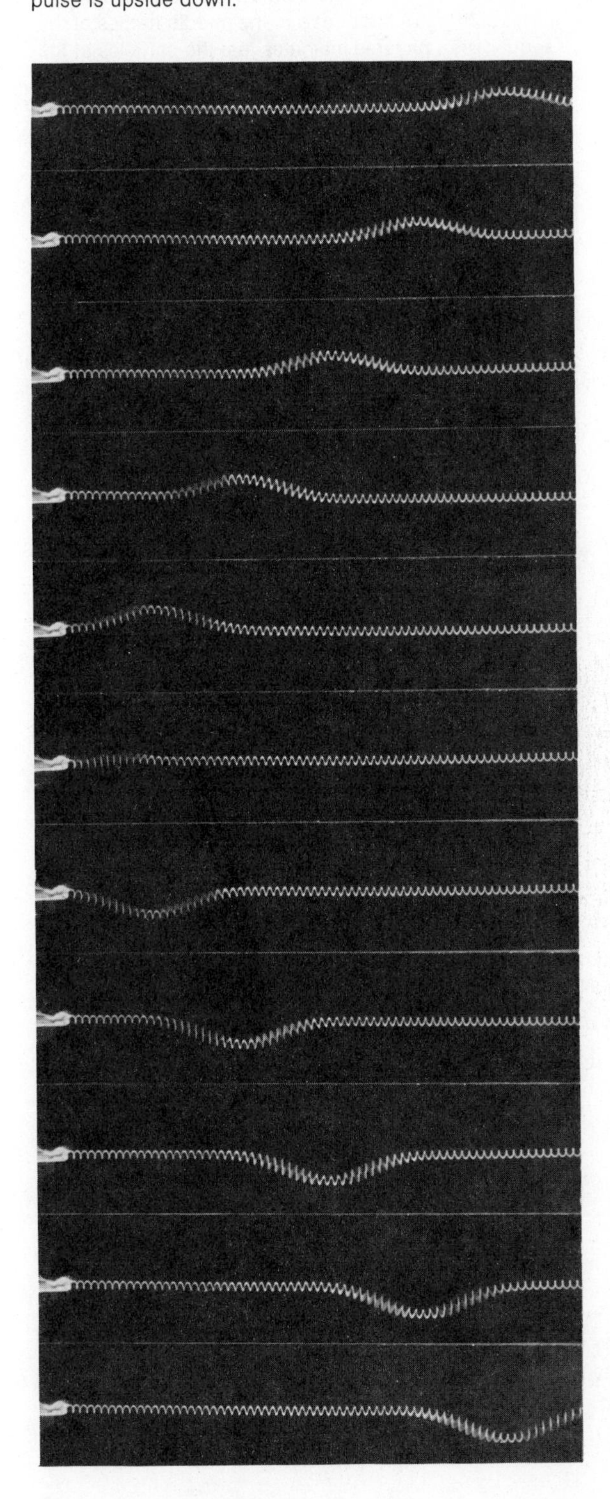

Figure 5–10
Reflection of a pulse from a fixed end. The reflected pulse is upside down.

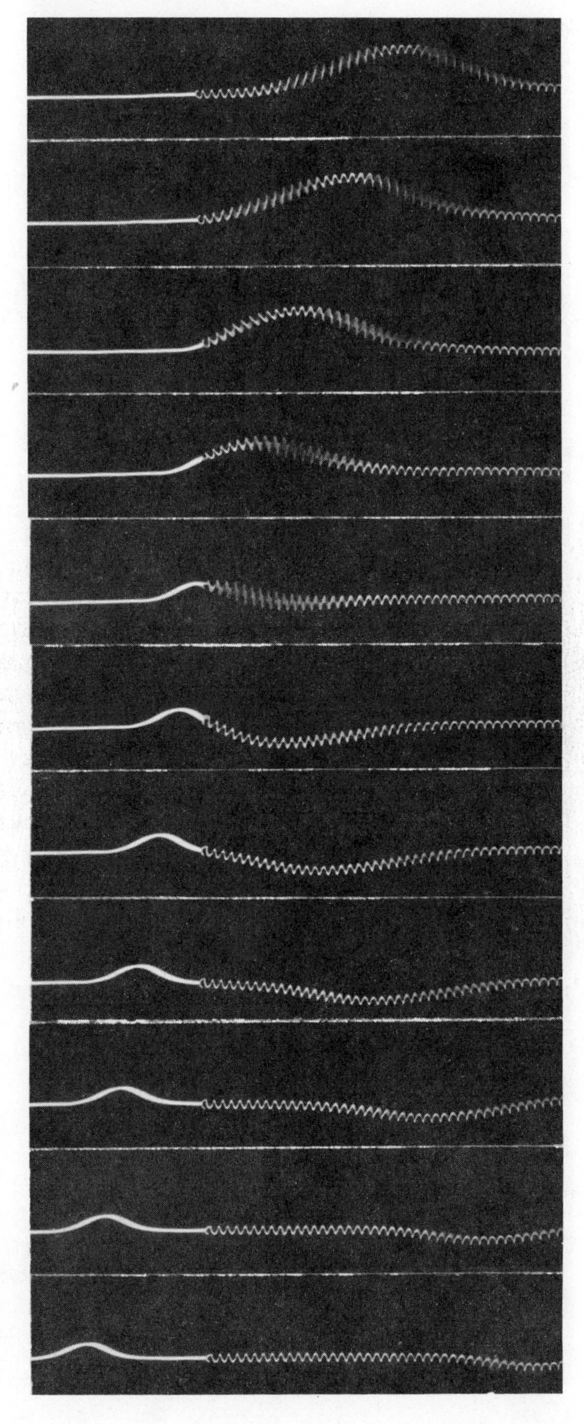

Figure 5–11
A pulse passing from a light spring (right) to a heavy spring. At the junction the pulse is partially transmitted and partially reflected. You will note that the reflected pulse is upside down.

Figure 5–12
A pulse passing from a heavy spring (left) to a light spring. At the junction the pulse is partially transmitted and partially reflected. The reflected pulse is right side up.

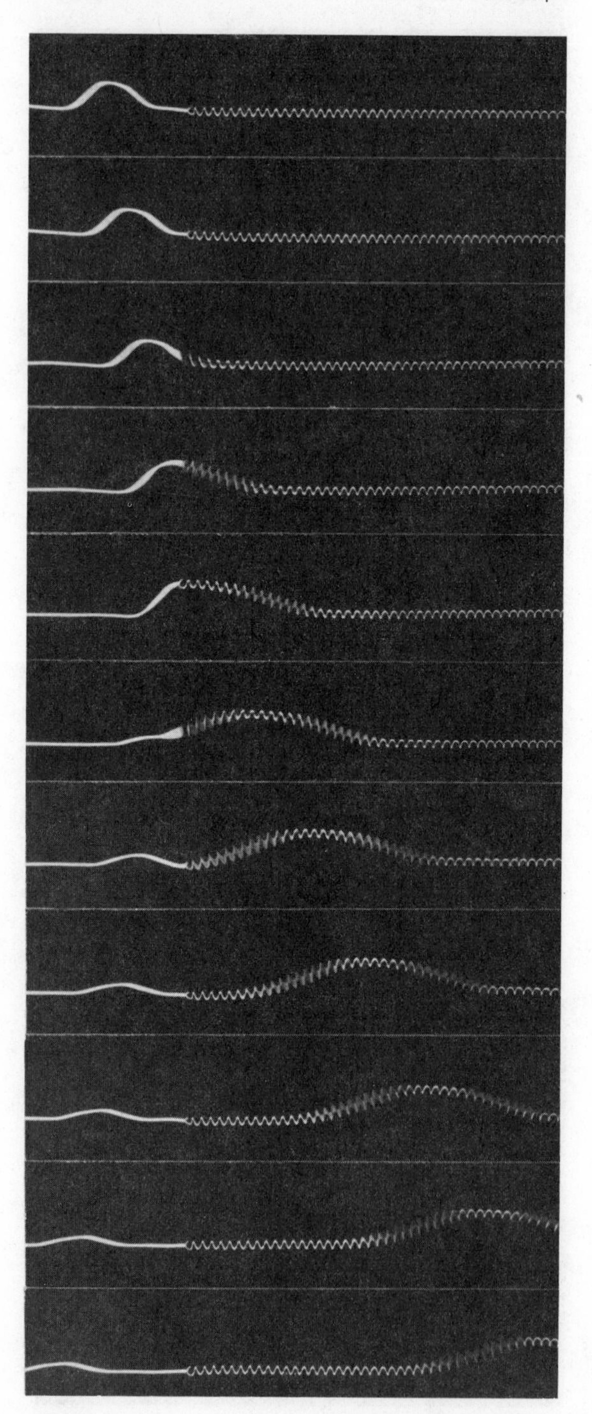

Figure 5–13
A pulse on a spring reflected from a junction with a very light thread. The whole pulse returns right side up. The blurring of the thread in the middle frames of the sequence of pictures indicates that the particles of the thread are moving at a high speed as the pulse passes. Can you determine the direction of this motion in each of the frames?

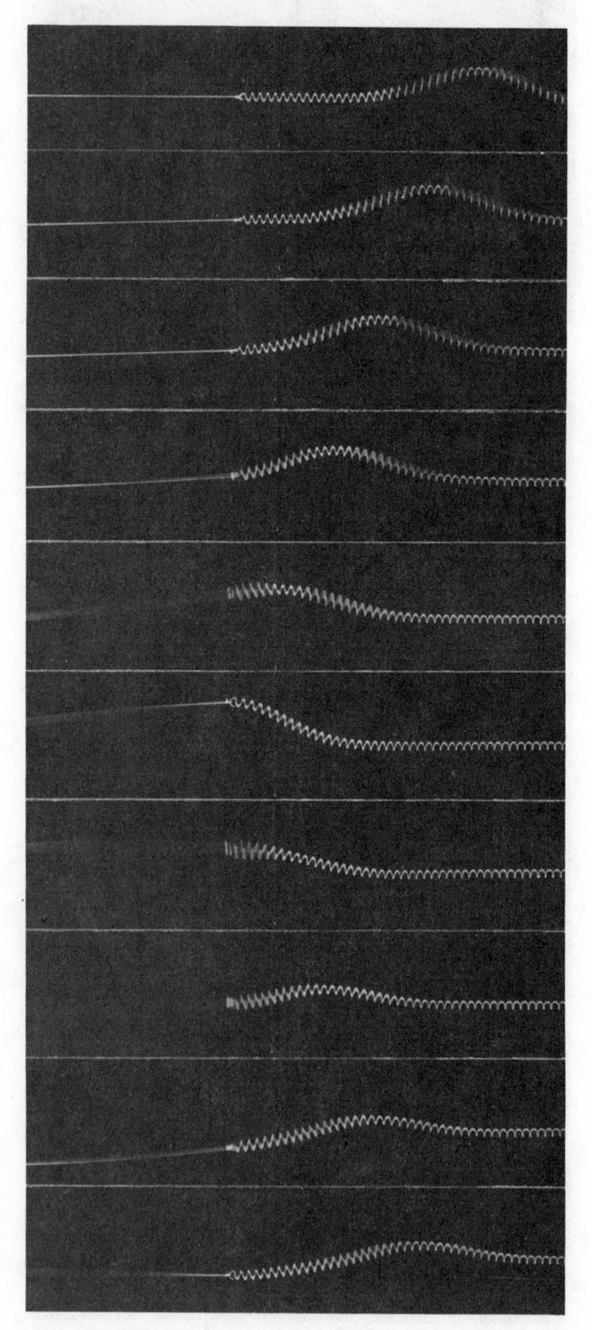

from the right and the heavier spring is on the left. We see that at the junction or boundary between the two springs—which are the media in which the wave travels—the pulse splits into two parts, a reflected and a transmitted pulse. Like superposition, the splitting into a reflected and a transmitted part is a typical wave property.

What happens when a pulse goes the other way, traveling along the heavier spring and arriving at the junction between it and the light spring? This is not so easy to foresee. We no longer can bracket the behavior between two situations in which we know the answer. But experiment tells us what takes place. In Fig. 5–12 we see a pulse moving from the left, from a heavy toward a light spring. Here, as in the opposite case, illustrated in Fig. 5–11, part of the pulse is transmitted and part is reflected, but this time the reflected pulse is right-side up.

In summary, then, when a pulse is sent along a spring toward a junction with a second spring, we observe that the whole pulse is reflected upside down whenever the second spring is very much heavier than the first. As the second spring is replaced by lighter and lighter springs, the reflected pulse becomes small and a larger and larger transmitted pulse is observed to go on beyond the junction. When the second spring is just as massive as the first, there is no reflected pulse and the original pulse is completely transmitted. Then if the second spring is made still lighter, reflection sets in again, this time with the reflected pulse right-side up. The lighter the second spring, the larger is the reflected pulse. When the second spring is negligible the reflected pulse is nearly the same size as the pulse sent in. This can be demonstrated with a heavy spring tied to a thin nylon thread (Fig. 5–13).

5-5 Idealizations and Approximations

In discussing waves along a spring, we said that their shape or size remains unchanged during the motion. Indeed, if we look again at Fig. 5–1, we can notice hardly any change in the size of the pulse as it travels along. Yet, as you have undoubtedly noticed, a pulse slowly diminishes and after several reflections it dies out completely. Is it reasonable for us to ignore the dying down of the pulse? Or has our description of wave behavior been wrong in some fundamental way?

To answer these questions, we start by observing that the time it takes a pulse to die out varies with circumstances. For example, if the spring is submerged in water, the pulse dies out more rapidly than it does in air. The water offers greater resistance to the motion of the spring than the air does. We may expect that in a vacuum the pulse will require a longer time to disappear than in air; experiments to test this, although not easy to perform, justify our expectation.

Even in a vacuum, the pulse eventually dies out because of internal resistance in the spring. The amount of this resistance depends on the material of the spring. For some materials it is very small and the pulse will keep moving for a long time.

We may imagine a spring with no internal resistance kept in vacuum. On such a spring a pulse would travel forever. By ignoring the dying out of the pulses, we have idealized our real springs and considered them as if they were free from both external and internal friction. We are entitled to do this so long as we consider the behavior of a pulse only for periods of time in which the size of the pulse changes so little that we hardly notice the change. For such times, the ideal resistance-free spring serves as a good approximation to the real one and can therefore be made the subject of our discussion of waves. There is a clear advantage in this, since the behavior of the ideal spring is much simpler than that of the real one.

A similar idealization occurred in our discussion of the superposition of two pulses. We learned that the displacement produced by the combined pulse equals the sum of the displacements due to the separate pulses. But if we make the individual pulses too big, we find that the combined displacement is less than the sum of the two displacements. Again, when we ignore this deviation from the simple superposition, we are discussing an ideal spring instead of the real one. We are making an approximation rather than a complete description of the real situation. But as long as we keep our displacements small enough so that we hardly notice these deviations, the ideal spring will be a good approximation to the real one; and it has the advantage of simplicity.

These are not the first idealizations we have made. In our discussions of the propagation of light we replaced real light pencils by ideal light rays, that is, pencils of zero width. This served us well in predicting the location of images formed by mirrors and lenses. However, when we tried to produce finer and finer pencils, we found that the pencil spread out; it did not behave like the ideal light ray.

Idealizations and approximations are very frequently made unconsciously. Consider, for example, what we mean when we say that the area of a piece of land is 1000 acres. Usually we get the area by measuring length and width and then calculating the area *as if* the land were completely flat. That is, we ignore the facts that there are little hills and valleys and that the area under consideration is really part of the surface of a sphere. We replace a surface of complicated shape by a simple plane rectangle. This procedure is useful only as long as there are no big mountains and the dimensions of the land are small compared with the radius of the earth. Under those conditions an idealized flat land serves as a good approximation.

Most of the problems we attack in science are fairly complicated, and in order to make progress toward understanding them we have to separate the essential from the inessential: that is, we have to make idealizations. In this chapter we have been studying waves. This is a very complicated matter on a real spring, but by mentally replacing the real spring by the idealized one, we separate the essential from the inessential and simplify the problem to help our understanding. Making the right idealization is one of the secrets of the successful scientist.

5-6 A Wave Model for Light?

In this chapter we learned about two important properties of waves which clearly indicate advantages of a wave model of light over the particle model.

First, we found that waves can pass through one another undisturbed. If we shine two flashlight beams across each other, each proceeds after the crossing as if the other had not been there. (See Fig. 1–13). Similarly, we can see this page despite the light crossing in all directions between us and it. This means that the crossing of light beams resembles the crossing of waves much more than it does the intersection of streams of particles.

The second important wave property is that of partial reflection and partial transmission at a boundary. Recall now what happens when light passes from one medium to another—say from air to glass. Part of the light is reflected and part of it is transmitted, as was shown by Fig. 3–2. This is just what waves do, but streams of particles do not split up this way.

These two wave properties which appear in light are good reasons to go on exploring a wave model for light; but they are far from demonstrating that a wave picture is an adequate model for light. For example, when a light beam hits a glass surface the angle of reflection equals the angle of incidence, and the direction of the refracted beam is described by Snell's law. On the basis of what we have studied so far, we cannot say whether a wave model accounts for the observed changes of direction. The waves on our spring are confined to move along one line or one dimension. Therefore, there is no way of changing the direction of propagation except to reverse it. To find if waves can really account for the behavior of light, we must have waves which move in space or at least in a plane, so that we can make a direct comparison. This we shall do in the next chapter, where we shall study waves on the surface of water.

For Home, Desk, and Lab

1.* What is similar in all examples of waves discussed in Section 1? (Section 1.)

2. Suppose you look out your window and see your neighbor across the street sitting on his porch. In how many ways could you do something to attract his attention, make him move, or otherwise influence his actions? Which ways involve mass transmission and which ways wave motion?

3.* According to Fig. 5–4, along what part of the pulse is the spring momentarily at rest? (Section 2.)

4.* In Fig. 5–2, the ribbon first moves up and then down as the wave moves past it from right to left. Would the ribbon move up first or down first if the same pulse traveled from left to right? (Section 2.)

5. Figure 5–2 shows the displacement of a point on a spring as a pulse goes by. Make a graph showing the displacement of this point as a function of time. Plot displacement vertically and time horizontally with 1⁄24-sec intervals (the interval between pictures of Fig. 5–2).

6. Sketch the motion of the spring for the pulse in Fig. 5–14.

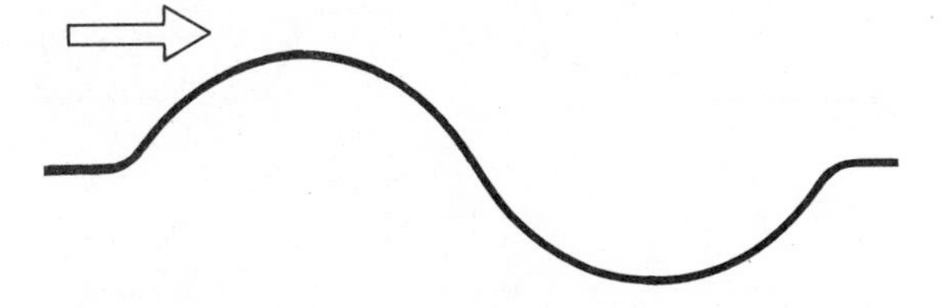

Figure 5–14 For Problem 6. A pulse moving to the right.

7.* If two pulses traveling toward each other on a coil spring have displacements in the same direction, can they cancel each other when they cross? (Section 3.)

8.* Two pulses have maximum displacements of 3 cm and 4 cm in the same direction. What will be the maximum displacement when they pass? (Section 3.)

9. Using the two pulses shown in Fig. 5–15, determine the size and shape of the combined pulse at this moment. Do the same thing for several other positions of the pulses.

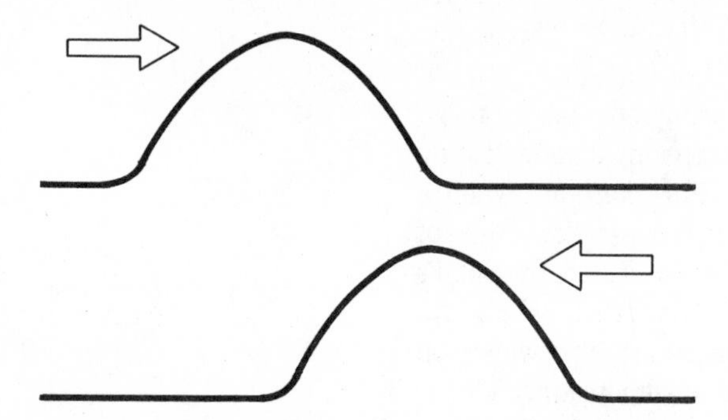

Figure 5–15 For Problem 9. Two equal pulses moving in opposite directions.

10. The seventh picture from the top in Fig. 5–5 shows two pulses at the moment of crossing. Specify the pieces of spring that are moving and their direction of motion.

11. In the fifth picture from the top of Fig. 5–7, which points are moving and in which direction do they move?

12.* A pulse, shown in Fig. 5–16, is sent along a coil spring toward the right. Draw the pulse traveling to the left which could momentarily cancel the pulse shown. (Section 3.)

Figure 5–16 For Problem 12.

13. In the sixth picture, Fig. 5–17, we see the superposition of two equal pulses, each of which is symmetrical about its center line.
    (a) The absence of blur indicates that there is no motion at this instant. Show that this is true by using the principle of superposition.
    (b) Assume that you deform the coil spring

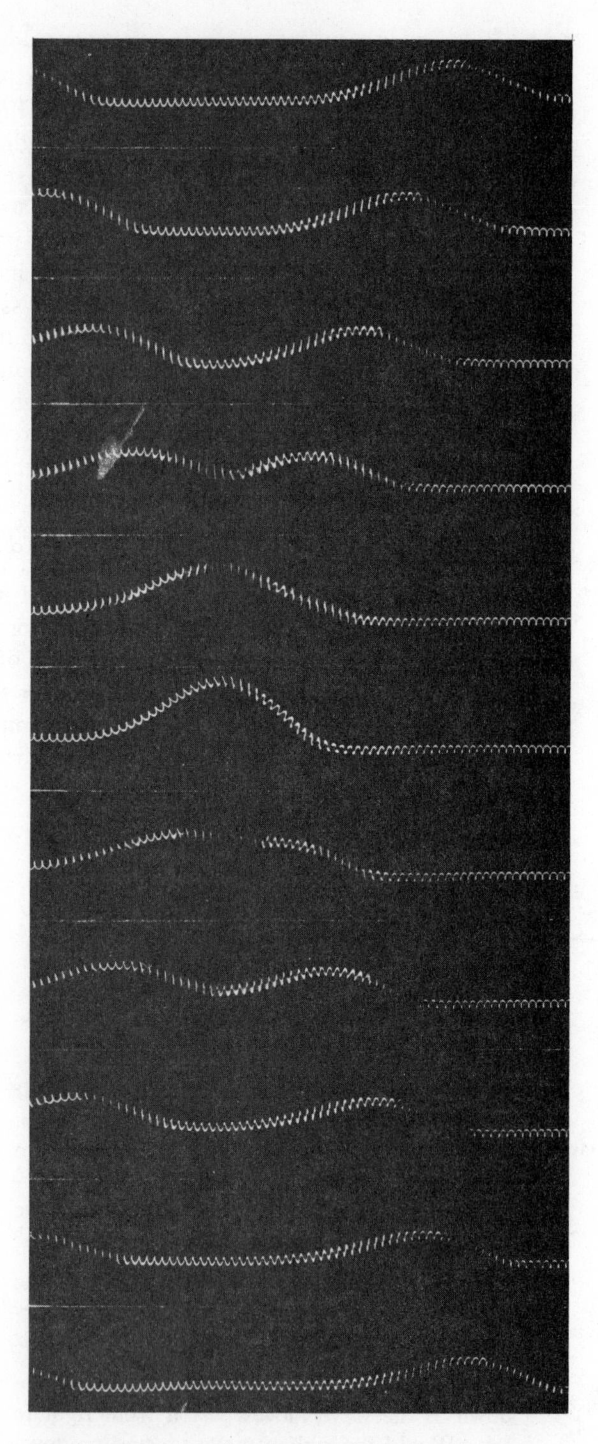

Figure 5–17 For Problem 13. Superposition of two equal and symmetric pulses.

in the same manner as shown in the sixth picture. What will happen when it is released?

14. The sixth picture of Fig. 5–10 shows the spring at an instant when the spring is almost straight. Explain why there is an instant when this happens.

15. Consider the asymmetric pulse coming from the left in Fig. 5–6. Draw the shape it will have after being reflected at a fixed end.

16.* In Fig. 5–11, which has the larger displacement, the incident or the transmitted pulse? (Section 4.)

17.* In Fig. 5–11, what is the ratio of the speed of the pulse in the light spring to the speed in the heavy spring? (Section 4.)

18. You send out a pulse at one end of a spring and it returns to you upside down and smaller in size. What can you deduce about the speed of the pulse on a second spring which is attached to the other end of your spring?

19.† Diagram (a) in Fig. 5–18 shows a pulse moving along a rope which has sections of different thicknesses. Diagrams (b) and (c) show the same rope at equal intervals of time later. Where are the junctions, and what are the relative thicknesses of the rope between them?

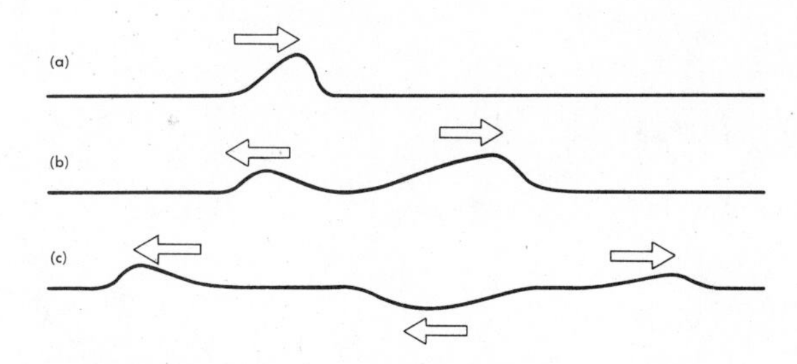

Figure 5–18 For Problem 19.

20. When light passes from air to water or vice versa, part of it is reflected. If this situation resembles that of a pulse crossing from one coil spring to another, in which case will you expect the light pulse to be reflected upside down?

21. Hold one end of a long rope with the other end tied to a rigid support. Stand looking along the rope and generate a wave by moving your hand through three fast clockwise circles.
    (a) Describe the wave generated.
    (b) Describe the reflected wave.
    (c) Describe the motion of a particle of the rope as the wave passes forward and back.

22.* How could you tell in Fig. 5–10 that the sequence of events proceeds from top to bottom and not from bottom to top? (Section 5.)

23.† Whenever a pulse travels along a spring, its size decreases as it moves along.
    (a) Under what conditions are we justified in neglecting this decrease?
    (b) What is the advantage of neglecting this decrease?

24. What do we mean by an ideal spring in the context of this chapter?

25. We can say that the surface of the sea is approximately flat. Give some examples where this is a good approximation and some examples where it is a bad approximation.

Further Reading

These suggestions are not exhaustive but are limited to works that have been found especially useful and at the same time generally available.

BASCOM, WILLARD, *Waves and Beaches.* Doubleday Anchor, 1964: Science Study Series. A good introduction to ripple-tank work.

GRIFFIN, DONALD R., *Echoes of Bats and Men.* Doubleday Anchor, 1959: Science Study Series. A helpful supplement to the treatment of waves and wave phenomena. (Chapter 2)

HOLTON, G., and ROLLER, D. H. D., *Foundations of Modern Physical Science.* Addison-Wesley, 1958. (Chapter 29)

VAN BERGEIJK, W. A., PIERCE, J. R., and DAVID, E. E., JR., *Waves and the Ear.* Doubleday Anchor, 1960: Science Study Series. Waves from a different point of view.

Waves and Light

CHAPTER

6

6-1 Water Waves

One of the first illustrations of waves mentioned in Chapter 5 was that of ripples on a pond. The spreading out of waves in the form of larger and larger circles is familiar to everyone. For example, a fish nibbling at a worm on a line gives away its presence to the angler who sees the circular waves produced by the up-and-down motion of the float attached to the line.

Because water waves move along the surface and do not extend downward to any appreciable depth, they are known as surface waves. If you have watched fish in an aquarium, perhaps you have noticed that they are undisturbed by the waves. A submarine commander does not fear a stormy sea as the captain of a surface ship might. He dives his submarine and travels along unaffected by the powerful waves above.

If we look at water waves through the side of an aquarium, we are able to see their shape. We notice that, while there are some variations, they are generally similar and look something like the illustration in Fig. 6–1. The

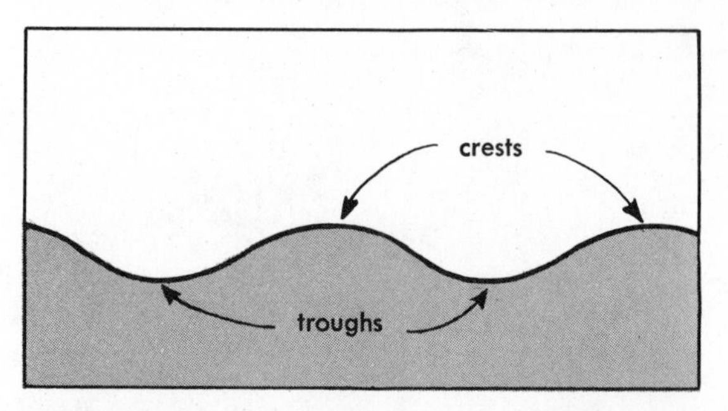

Figure 6–1
An illustration of water waves.

upper portions are called crests and the lower portions troughs. In the laboratory we use a more convenient apparatus called a ripple tank. The ripple tank (Fig. 6–2) has a glass bottom which makes it possible to project images of the waves onto a screen. These images are produced because the crests of the waves act as converging lenses and tend to focus the light from the lamp while the troughs, acting as diverging lenses, tend to spread it out. Therefore the crests appear on the screen as bright bands while the troughs appear dark.

6-2 Straight and Circular Pulses

We can generate a straight pulse, like the long bow waves caused by a passing boat, by dipping a ruler into the surface of the water of the ripple tank. The motion of the pulse is such that its crest always remains parallel to a line marking its original position. The distance between parallel lines is measured along a perpendicular; thus the direction of motion of the pulse (also called the direction of propagation) is perpendicular to the wave crest. This direction is called the *normal* to the crest. In Fig. 6–3 the crest of a straight pulse is shown as a heavy white line. Its position at a later time is shown by the dashed line. The direction of propagation is shown by the arrow.

We can also produce circular pulses in the ripple tank simply by dipping a finger into the water. Figure 6–4 is a drawing of such a pulse at two dif-

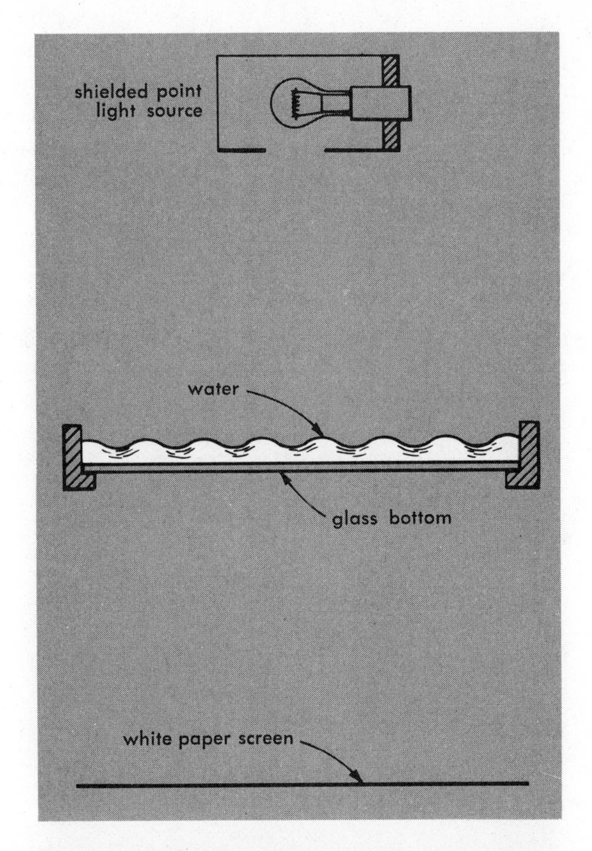

Figure 6–2
The ripple tank.

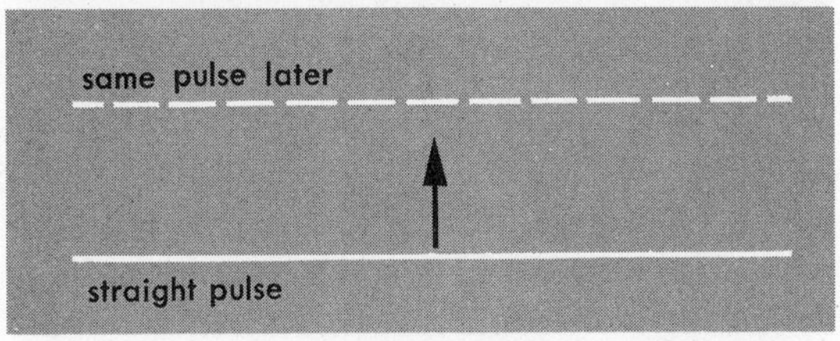

Figure 6–3
A straight pulse moves at right angles to its crest. The arrow indicates the direction.

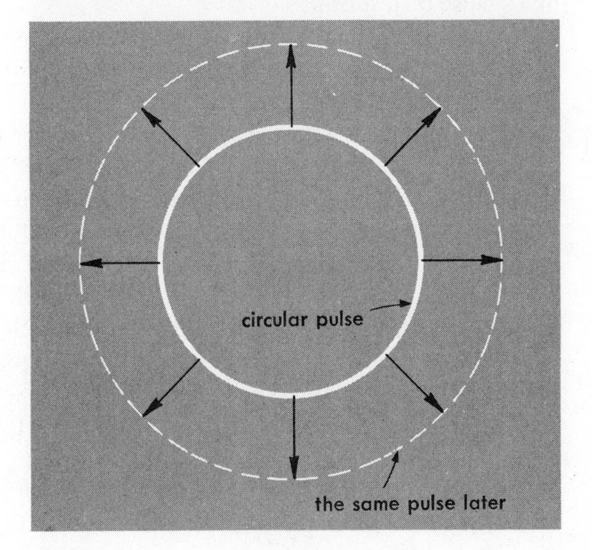

Figure 6–4
An expanding circular pulse.

ferent times. During the time interval, the pulse has expanded to form a larger circle. We cannot assign a direction to the whole circular pulse, because it moves in all directions. Instead, let us look at a segment of the circular pulse which is small enough to be considered straight (Fig. 6–5). The direction of propagation of such a segment is along the radius and away from the center of the circle. This direction is normal to the crest of the wave, just as the direction of propagation of straight waves is normal to their crests.

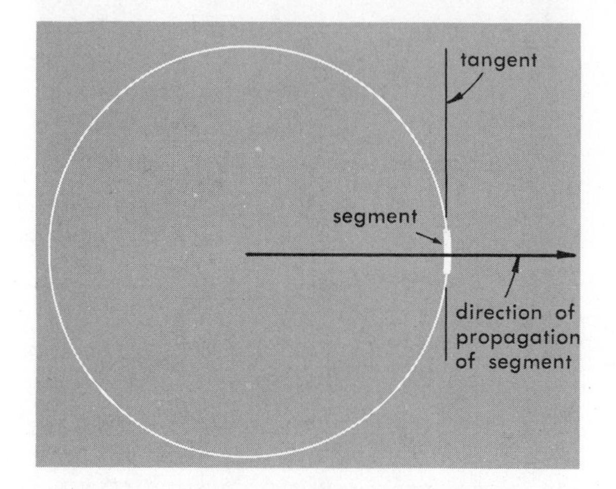

Figure 6–5
A tiny segment of a circular pulse. The segment acts as though it were straight and moves along a radius away from the center.

6-3 Reflection

We recall that a pulse on a spring can be reflected, and we may expect that water waves can also undergo reflection. Consider a straight pulse as it moves away from the ruler toward the opposite end of the tank. To reflect it we place a barrier in the middle of the tank parallel to the ruler. The pulse strikes the barrier and reflects back in the direction from which it came, just like a pulse on a spring.

Now let us change the position of the reflecting barrier so that the pulse is no longer parallel to it. In Fig. 6–6 we see two straight pulses, one approaching and one being reflected from a barrier. The angle which the incident pulse makes with the reflecting barrier is labeled i', while the angle between the reflected pulse and the barrier is labeled r'. Measure the angles r' and i' on the photograph and you will find that $r' = i'$. If we repeated the experiment with different angles i', we would always find $r' = i'$.

This result resembles the law of reflection of light from mirrors, which we obtained in Section 2–1. There we found that the angle of reflection equals the angle of incidence. But in optics each angle was measured between the direction of propagation and the normal to the reflecting surface. We can define the angles of incidence i and reflection r for waves in the same way, remembering that the direction of propagation is normal to the wave crest.

The construction in Fig. 6–7 shows that the angle of incidence i is equal to the angle i'. A similar construction shows that $r = r'$. Our observed

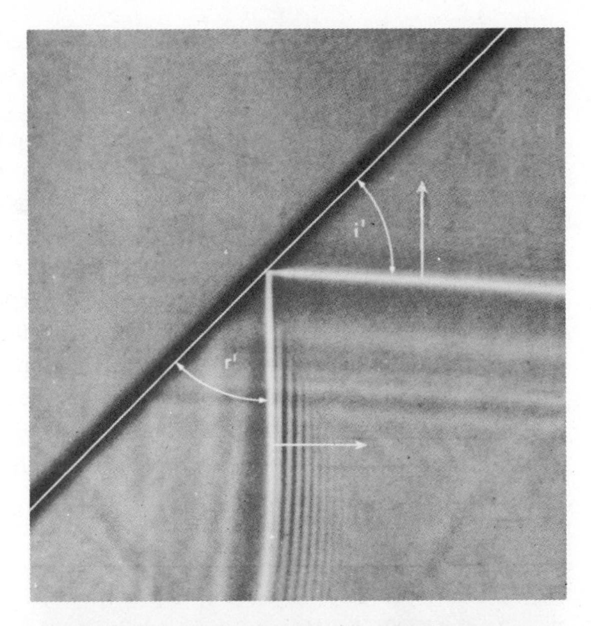

Figure 6–6
A straight pulse moving up in the picture and reflecting from a diagonal barrier. The reflected part of the pulse is moving off to the right.

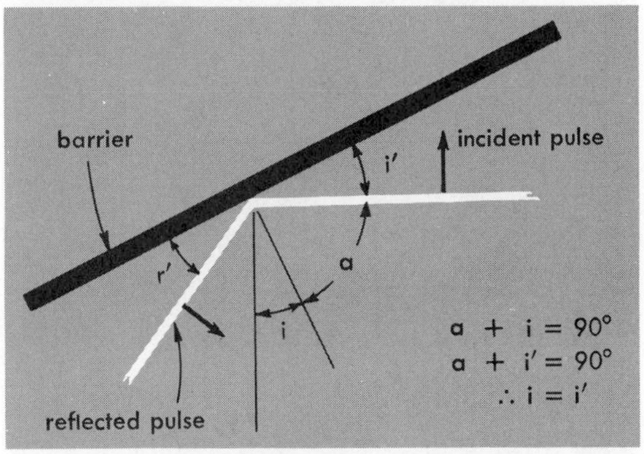

barrier
incident pulse
i′
r′
a
i
reflected pulse

$a + i = 90°$
$a + i' = 90°$
$\therefore i = i'$

Figure 6–7
A straight pulse incident on a barrier. The angle of incidence *i* is equal to the angle the pulse makes with the barrier *i′*.

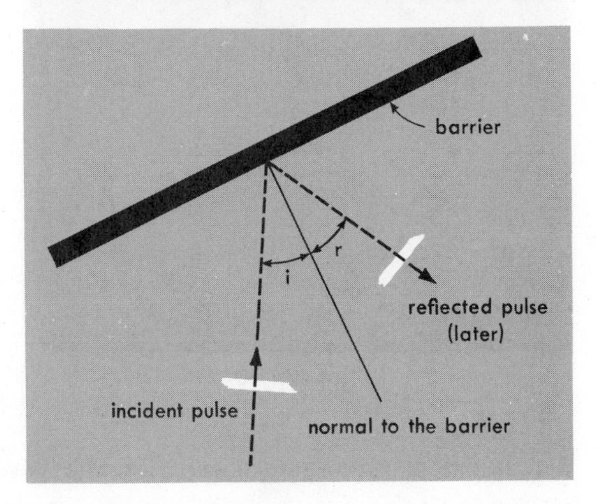

Figure 6–8
The reflection of a straight wave from a straight barrier. Just as with light, the angle of reflection *r* equals the angle of incidence *i*.

equality $r' = i'$ then becomes $r = i$, or in words, the angle of reflection equals the angle of incidence. We have shown, therefore, that waves and light follow the same law of reflection (Fig. 6–8).

An expanding circular pulse can also be reflected from a straight barrier. Figure 6–9 shows the approach and reflection of such a pulse. Notice that the reflected part of the pulse is an arc of a circle. The center of this circle is at the same distance behind the barrier as the source is in front of it. The reflected pulse appears to come from the center of the circle behind the barrier. This corresponds to the situation in optics where we placed a point source in front of a mirror. The reflected light then appeared to come from the image point behind the mirror. (See Fig. 2–6.) By using curved barriers or combinations of two or more straight barriers, we can demonstrate with the ripple tank all of the phenomena of reflection that we studied in connection with light. Just as the formation of images by mirrors followed from the laws of reflection in optics, so does the corresponding formation of images by "mirrors" in the ripple tank.

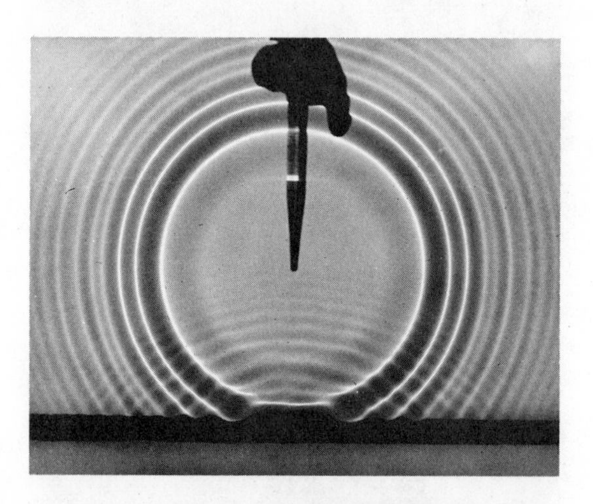

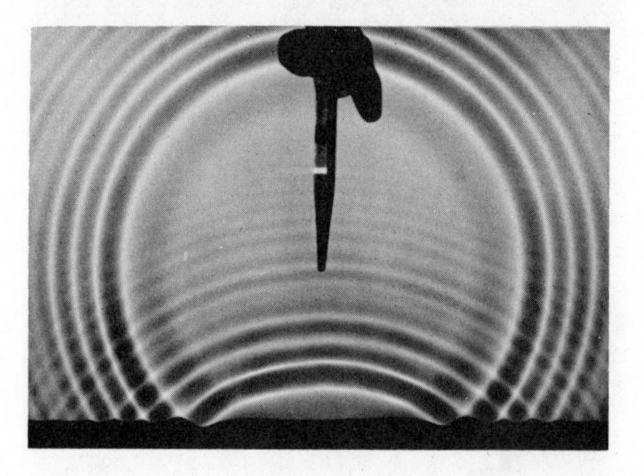

Figure 6–9
Reflection of a circular pulse from a straight barrier. In the upper photograph the pulse is approaching the barrier, while in the lower one a part of it has been reflected.

6-4 Speed of Propagation and Periodic Waves

Waves in different media propagate with different speeds. For example, we can see the waves on a coil spring speed up when we stretch the spring, and we can see the waves in a rubber hose slow down when we fill it with water. In this section we shall learn how to measure the speed of water waves in a ripple tank. There are several ways to measure this speed.

One way is to generate a straight pulse and measure with a stopwatch the time t the pulse takes to travel a specified distance l. The speed v is then equal to the distance traveled divided by the time taken:

$$v = l/t.$$

Another way is to generate two pulses, one after the other. By the time the second pulse is generated (after a time t) the first pulse has traveled a distance l. From then on both pulses travel along, while the distance l between them remains the same. We can measure this distance with a ruler, and again $v = l/t$. These methods are simple in principle, but in practice it is rather difficult to follow the pulses and measure the distances and times required.

A third method makes use of a periodic wave. To see what is meant by this term, consider pulses generated one after another at equal time intervals T. In doing this, the wave generator repeats its motion once every interval T. Such a motion is called *periodic,* and the time interval T is called the *period.* Another way of describing this periodic motion is to tell how often the motion repeats itself in a unit time interval; that is, by giving the *frequency* f of repetition. For example, if the motion repeats every 1/10 sec, the frequency is ten times per second. In general $f = 1/T$.

Let us now concentrate on some point in the tank. The pulses produced by the generator move toward this point, and they pass the point with the same frequency with which they leave the source. If ten are sent out each second, ten will pass each second. The frequency of the wave is therefore also given by $f = 1/T$, and T is the time between the passage of successive pulses. Furthermore, as the waves move, the distance between any two adjacent pulses is always the same and is called the *wavelength* λ (lambda). The wave pattern which we have been describing is called a periodic straight wave (Fig. 6–10).

We can obtain the speed of a periodic wave in a manner similar to that which we used for a pair of pulses. We know that the pulses are separated by a distance λ and that each pulse moves over this distance in a time T.

Figure 6–10
Periodic straight waves moving across a ripple tank.

Hence the speed of propagation is

$$v = \lambda/T.$$

Using the relation $f = 1/T$, we find that

$$v = f\lambda,$$

or that the speed of propagation of a periodic wave is the product of the frequency and the wavelength.

The relation that we have just obtained is by no means restricted to waves in a ripple tank. It is equally good for *any* periodic wave. Such things as the straightness of the wave, the nature of the ripple tank, and the properties of the water did not come into the argument from which we got our result. In particular, we could have followed the same procedure with circular periodic waves and would have found the relationship $v = f\lambda$ again. In this case the wavelength is measured along a radius (Fig. 6–11); and we find it is equal to the wavelength of a straight wave of the same frequency. The speed of circular waves is therefore equal to that of straight waves in the same medium. Furthermore, we could have applied the above arguments to any other kind of periodic waves—for example, periodic waves on springs—and we would have the same relation $v = f\lambda$.

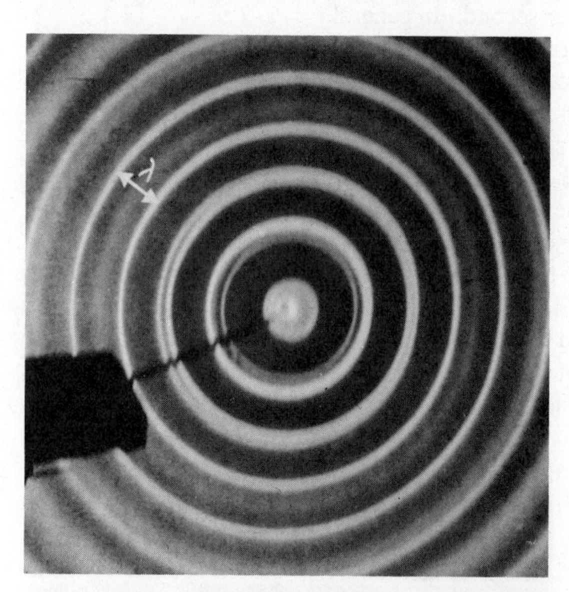

Figure 6–11
Periodic circular waves.

Now we come to the advantage of the above relation for the measurement of v. Imagine that instead of watching the wave continuously we look at it through a shutter which is closed most of the time and opens periodically for short time intervals. The stroboscope shown in Fig. 6–12 is such a device. The first time the shutter opens we will get a glimpse of the wave pattern in a certain position. During the time the shutter is closed, all of the pulses will move a distance equal to their speed times the time the shutter is closed (Fig. 6–13). If we look through the shutter while it is

Figure 6–12
Using a hand stroboscope to "stop" wave motion in the ripple tank.

periodically opening and closing, the pattern will usually appear to move. Suppose, however, that the period of the shutter is just the same as that of the wave. Then, during the time the shutter is closed, every pulse just moves up to the position of the pulse ahead of it, and we see the same pattern every time the shutter opens. That is, we see a stationary pattern from which it is easy to measure the wavelength. Moreover, since the period of the shutter is equal to the period of the wave, it can be measured by counting the number of times the shutter is opened in a given time interval; i.e., by measuring the frequency of the shutter. This gives us f. Now that we have both λ and f for the wave, we can use the general relation $v = f\lambda$ to determine the speed.

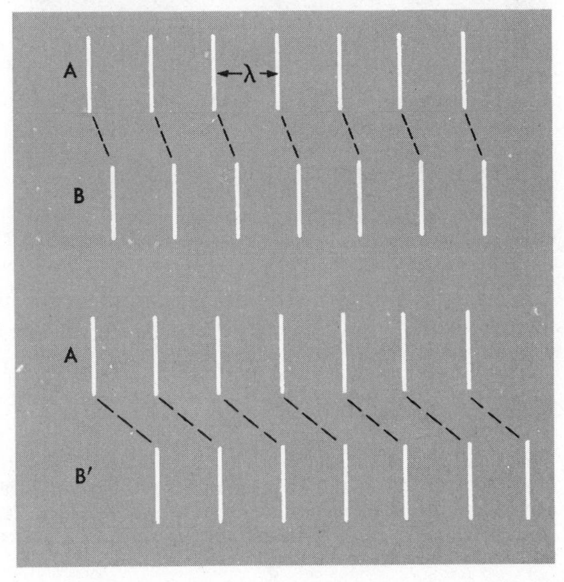

Figure 6–13
Crests of a periodic wave seen at successive openings *A* and *B* of a stroboscope shutter. In the top diagram the frequency of the stroboscope is greater than that of the waves. At the bottom, it is the same. The dashed lines, of course, are not visible.

6-5 Refraction

We mentioned earlier that the speed of propagation of waves depends on the properties of the medium through which they move. In the case of waves on the surface of water the speed depends on the depth of the water. Therefore water of two different depths can be considered to be two different media for wave propagation. This is a very useful property, because by merely changing the depth of the water in part of the ripple tank we are able to study the behavior of waves when they pass from one medium to another.

To see that the speed indeed depends on the depth, we now make half the tank shallow by placing a thick glass plate on the bottom of the back part of the tank. This divides the tank into two sections of different depths with the dividing line parallel to the waves. Let us look at the waves through the stroboscope. By turning it at the right frequency we can stop the motion of the pattern in *both* sections of the tank simultaneously. Hence the frequency is the same in both sections; it is not affected by the change in depth. But Fig. 6–14 shows that the wavelength λ_2 in the shallow part is shorter than λ_1, the wavelength in the deep part.* Since the speed of propagation is the product of the same frequency and the appropriate wavelength, we see that the speed in the shallow part (where the wavelength is smaller) is less than in the deep part—that is, $v_2 < v_1$. Furthermore, in either half of the tank, the speed of propagation is the same in all directions, as we saw in the last section.

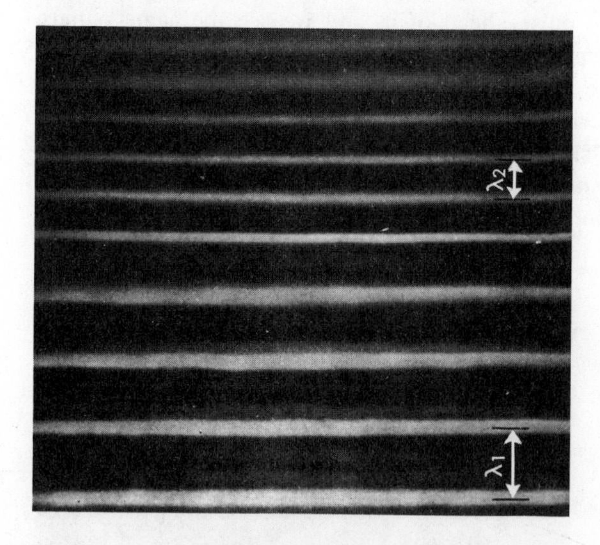

Figure 6–14
Passage of waves from deep to shallow water. The deep water is at the bottom and the shallow water at the top of the picture. Note that the wavelength is shorter in the shallow water.

Now let us repeat the experiment, but this time with the glass plate placed so that the boundary between the deep and shallow sections will form an angle with the waves. We already know that the wavelength in the shallow section is less than in the deep one. But this not the only change.

* You will notice that the waves tend to disappear toward the end of the shallow section. They die down only because we do not have an ideal state of affairs in which we can isolate the phenomenon we wish to study. Here, as in the experiments with the coil spring, there is a certain amount of resistance present which reduces the motion and finally causes it to die out.

Figure 6–15 shows that when the straight waves hit the boundary, they remain straight but change their direction of propagation. The new direction is closer to the normal to the boundary than the original direction of propagation. We remember from our study of optics that this is what happens to light when it passes from one medium to another in which its speed is less (Sections 3–2 and 4–7). For light this refraction is quantitatively described by Snell's law. The following question now suggests itself. Does Snell's law also hold true for waves such as those in the ripple tank, when they pass from one medium to another?

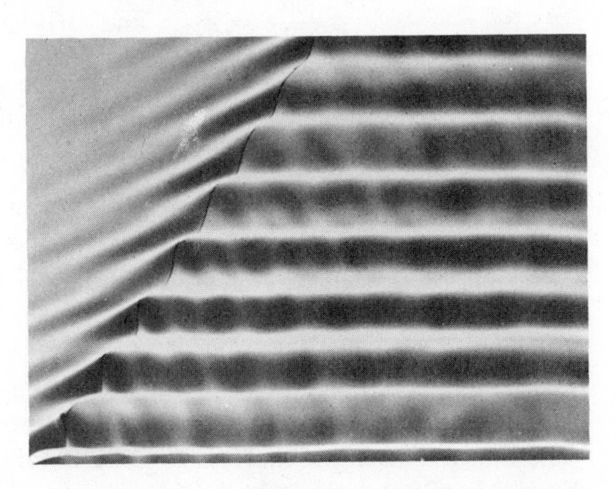

Figure 6–15
Refraction of waves at a boundary between deep and shallow sections of the ripple tank. Note the weak reflected waves.

There are two procedures open to us at this point. We can measure many angles of incidence and angles of refraction in the ripple tank, and thus find experimentally whether they are related by Snell's law; or we can find out whether Snell's law can be predicted theoretically from the properties of waves which we already know. We shall take the second course here.

In Section 6–3, we proved that the angle of incidence is equal to the angle between the incident wave crest and the barrier. Likewise, the angle of refraction is equal to the angle between the refracted wave crest and the barrier. Let us now draw two consecutive wave fronts as they are refracted at the barrier (Fig. 6–16). There is no need to add the normals, since angles equal to i and r already appear in the drawing. Then by definition

$$\sin i = \frac{\lambda_1}{AB}, \quad \sin r = \frac{\lambda_2}{AB}.$$

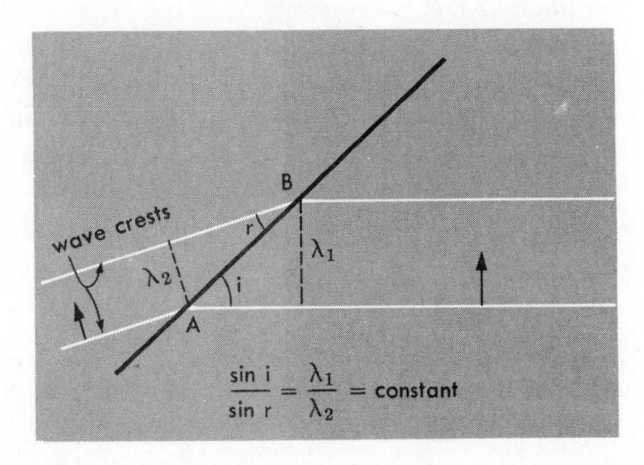

Figure 6–16
The geometry of the refraction of two consecutive wave fronts.

The values of sin i and sin r change from case to case, but their ratio is a constant independent of the angle of incidence, as we see by actually dividing sin i by sin r:

$$\frac{\sin i}{\sin r} = \frac{\lambda_1}{AB}\frac{AB}{\lambda_2} = \frac{\lambda_1}{\lambda_2} = \text{constant}.$$

The relation $\sin i/\sin r = \text{constant}$ is Snell's law, this time for waves. We shall again call the constant the index of refraction and denote it as in optics by n_{12}. Thus

$$\frac{\sin i}{\sin r} = n_{12}, \quad \text{and} \quad n_{12} = \frac{\lambda_1}{\lambda_2}.$$

We can now express the value of n_{12} in terms of the speeds of propagation of the waves in the two media. In general the speed v is related to the wavelength as $v = f\lambda$. In particular, then, $\lambda_1 = v_1/f$, and $\lambda_2 = v_2/f$. Hence

$$n_{12} = \frac{\lambda_1}{\lambda_2} = \frac{v_1/f}{v_2/f} = \frac{v_1}{v_2}.$$

This equation says that the index of refraction is equal to the ratio of the speed of propagation in the first medium to that in the second.

We arrived at Snell's law and the relation of n_{12} to the speeds by theoretical analysis of our previous results. We could get the same conclusions by direct measurements. First we could measure many pairs of angles i and r and thus establish that $\sin i/\sin r = \text{const.} = n_{12}$. Then we could measure the speeds of propagation in the two media and establish that $n_{12} = v_1/v_2$. Such measurements have often been carried out, and they do agree with our conclusions.

We now recall that the particle model of light as developed by Newton in 1669 also explained the existence of a constant index of refraction for a given pair of substances but predicted its value to be $n_{12} = v_2/v_1$ (Sections 4–5 and 4–7). Our wave model, advocated by Huygens in 1677, predicts $n_{12} = v_1/v_2$, just the inverse of Newton's result. The position of the wave model of light was strengthened by various experiments at the beginning of the nineteenth century. But this particular question was not settled for almost two hundred years. In 1862 Foucault actually measured the speed of light in air and in water and found that the speed in water was less. The exact ratio $v_1/v_2 = 1.33$ was measured by Michelson in 1883. The ratio agrees with the wave model because the index of refraction of water is 1.33 (Chapter 3).

One point still needs clarification: not all the light hitting the boundary between the two media is refracted. Part of it is reflected even if both media are transparent. The same holds for waves. In Fig. 6–15 the size of the reflected wave is rather small, but you can see if it you look closely. We conclude, therefore, that as far as refraction is concerned, waves have just the properties which we need in order to account for the behavior of light.

6-6 Dispersion

In the last section we studied the refraction of periodic waves as they pass from one medium to another. We found the index of refraction to be equal to the ratio of the speeds of propagation in the two media: $n_{12} = v_1/v_2$. We did not state the frequency of the waves, because we had previously learned that the speeds of propagation depend only on the media in which the waves travel. Accordingly, we should expect to find the same index of refraction for waves of different frequency provided we repeat the experiment with the same two media, for example, water of the same two depths.

What we actually observe is shown in Fig. 6–17 and 6–18. In Fig. 6–17 we see the refraction of a wave of low frequency (long wavelength). To indicate the direction of the refracted waves we placed a rod on the screen of the ripple tank. It is exactly parallel to the refracted wave fronts. We then increased the frequency (i.e., decreased the wavelength), leaving the rod untouched. Notice that in Fig. 6–18 the rod is no longer parallel to the refracted wave crests. The wave with the higher frequency is clearly refracted in a slightly different direction from that of the low-frequency wave, although the angle of incidence is the same in both cases. The index of refraction for the two media therefore depends somewhat on the frequency of the wave. By analogy with the dependence of the index of refraction on the color of light, which we have discussed in Section 3–8, this phenomenon is called *dispersion.* Since the index of refraction equals the ratio of the speeds of propagation in the two media, we have to conclude that the speed must depend on frequency in at least one of the two media; otherwise the ratio could not show such a dependence. A medium in which the speed of waves depends on the frequency is called a *dispersive* medium.

Figure 6–17
Refraction of low-frequency waves. The black marker is placed parallel to the refracted waves.

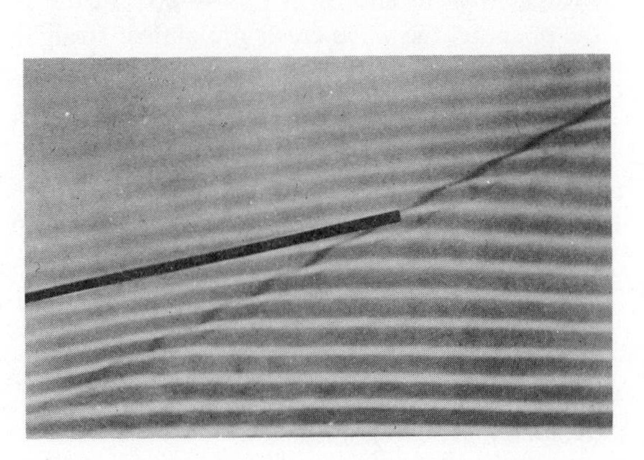

Figure 6–18
Refraction of high-frequency waves. The refracted waves are not parallel to the black marker.

In the ripple tank we can measure the speed of periodic waves of different frequencies (Section 6–4), and thus see directly that the speed changes with the frequency, provided we make our measurements accurately enough. The statement we so often make, that the speed of waves depends only on the medium, is therefore an idealization. To be sure, this idealization is a good approximation to the true state of affairs provided we are not concerned with small changes in the speed (Section 5–5).

We made a similar idealization in our study of the refraction of light in Chapter 3. You will recall that the index of refraction of light at first appeared to be dependent only on the two media through which the light was passing—for example, air and glass. Then a closer examination of refraction, using prisms, showed that the index changes slightly with color; it is a little larger for violet than for red.

The index of refraction of waves depends slightly on the frequency. That of light depends slightly on the color. Is there perhaps a relation between the dependence of the index of refraction of waves on the frequency and that of light on color? It is tempting to assume that light is a periodic wave and that different colors correspond to waves of different frequencies. At this point we cannot prove that our assumption is correct, but in the next chapter we shall learn how to measure the wavelength of light of different colors. We can then find the corresponding frequencies, and we shall indeed establish that light waves of various frequencies appear to our eyes as light of various colors.*

6-7 Diffraction

Our study of refraction and dispersion clearly shows that the wave model of light succeeds where the particle model fails. Yet the particle model predicts correctly that light should propagate in straight lines and cast sharp shadows. Can a wave model also account for these properties of light? Again, a good way to investigate these questions is to experiment with waves in the ripple tank. We use a straight wave generator and two barriers parallel to it placed in line with an opening between them (Fig. 6–19). These barriers would cast sharp shadows if light were incident upon them from the direction of the generator. What happens when we send in a periodic straight wave is shown in Fig. 6–20. In the middle of the pattern beyond the opening, the wave crests are almost straight, but at the sides they curve, giving the impression of circular waves originating from the edges of the opening. This means that after passing through the opening the wave spreads out; it is diffracted.

The diffraction of waves makes it hard to understand how we can explain the straight-line propagation of light with a wave model. If light is a wave,

* By comparing the bending of waves in Fig. 6–17 and 6–18 with the dispersion of light by glass (see Fig. 3–15), we might come to the conclusion that the wavelength of violet light is longer than that of red light. However, blue light does not refract more than red in all materials. In some materials it is the other way around. Therefore, studying the dispersion of water waves does not establish the relation between color and wavelength of light.

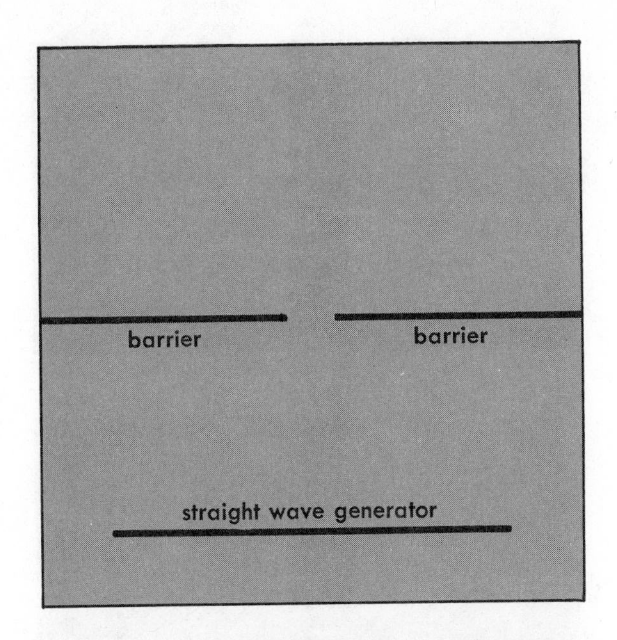

Figure 6–19
Ripple-tank arrangement for the experiment shown in Fig. 6–20.

Figure 6–20
Straight waves passing through an opening. Note the curving of the waves around the ends of the barriers.

when it passes through a small hole some of the light should bend instead of traveling straight ahead. In a wave model of light there must be at least one important difference between light waves and water waves.

What could this difference be? We know the wavelength of the water waves. We do not know the wavelength of light waves; it may be very different. Let us, therefore, examine the diffraction of water waves of different wavelengths. In Fig. 6–21 we see three pictures of periodic waves with different wavelengths, each passing through the same opening. We notice a definite trend. In the first picture the wavelength λ is seven-tenths of the width w of the opening. There the part of the straight wave which gets through the opening is almost entirely converted into a circular wave. Or, in other words, the opening acts like a source of circular waves when straight waves fall on it. In the second picture λ is five-tenths of w. In this case the wave which gets through is not so curved as in the first picture. It has a straight section in the middle but part of it still bends at the sides. In the third picture λ is three-tenths of w and here the bending fades away close

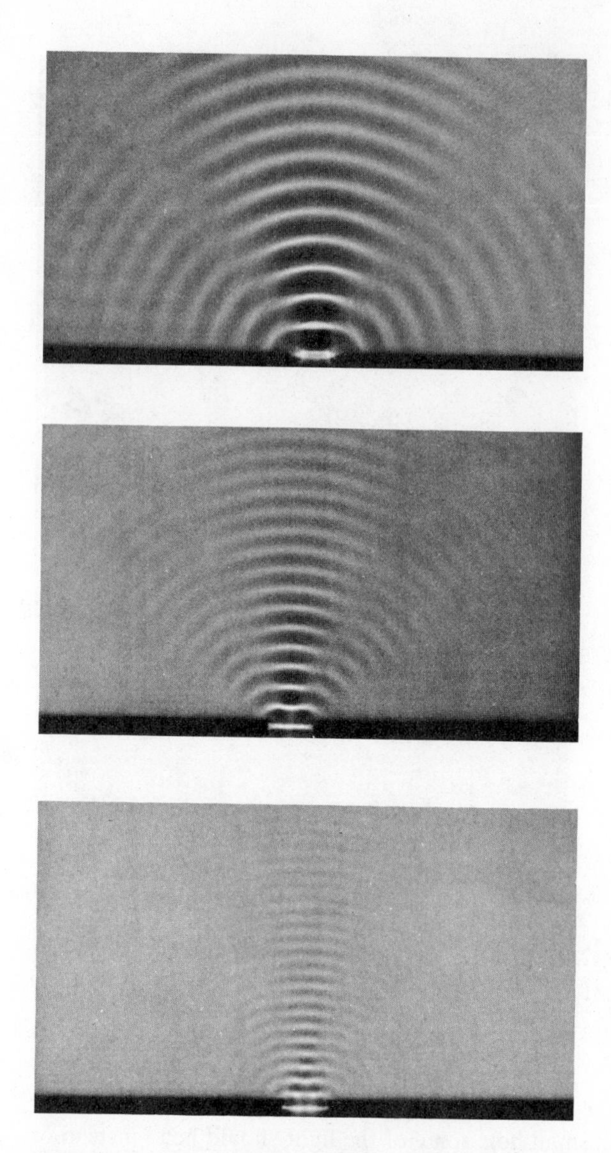

Figure 6–21
Three views of waves passing through the same opening. Note the decrease in bending of the shorter wavelengths.

to the forward direction and we obtain a nearly sharp shadow. We can also keep λ fixed and change w; then we find that the amount of bending does not depend on λ and w separately but only on the fraction λ/w. To sum up, waves are strongly diffracted when they pass through an opening of size comparable to their wavelength, and there is hardly any diffraction if the wavelength is very small compared to the width of the opening.

We all know that light passing through a keyhole is not bent but seems to continue in its original direction. Therefore, if light is a wave, its wavelength must be much less than the size of a keyhole. But to be certain that light is a periodic wave we should be able to do an experiment which will show diffraction. Two such experiments were described in Section 3–12. Suppose that we repeat the experiment in which we examine a light source through a narrow slit between two fingers. When the slit is half a centimeter across, a long, thin light source viewed through it appears normal. As the slit is narrowed to a width of about a tenth of a millimeter, however, the source appears spread out in a direction perpendicular to your fingers. Moreover, you can hold your fingers in any direction and the spreading out always appears perpendicular to them. This is a clear indication that the light is diffracted by the opening.

Briefly then, from our discussion in this chapter, light can be described as waves of very small wavelengths.

For Home, Desk, and Lab

1.* If a stick 10 cm long were placed horizontally in the ripple tank shown in Fig. 6–2, about how long would its shadow be? (Section 1.)

2.* In Fig. 6–7, if $i = 25°$, what is the value of r'? (Section 3.)

3.* In Fig. 6–22, the heavy black lines are crests; the arrows represent the direction of propagation of the pulses.
   (a) Which is the incident pulse and which is the reflected pulse?
   (b) What is the angle of incidence? (Section 3.)

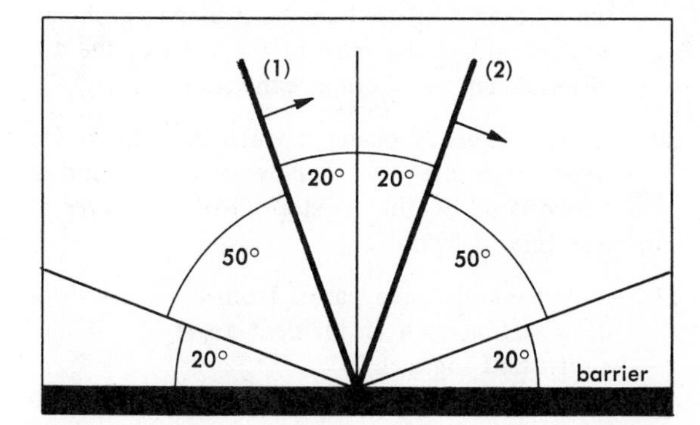

Figure 6–22 For Problem 3.

4. A straight pulse approaches a barrier at an angle of 30°. What is the direction of motion of the pulse after reflection? Indicate it on a diagram.

5. In Fig. 6–9, where is the object source? Where is the image source?

6. Describe the wave motion that results when you dip your finger into the center of a circular tank of water. What would be the motion under ideal conditions?

7.† Suppose we place in a ripple tank a barrier in the shape of an ellipse as in Fig. 6–23. When a circular pulse is generated at point A, it reflects from the barrier and converges at point B.
   (a) From this experiment what can you say about the geometry of an ellipse? (Hint: consider tiny segments of the circular pulse originating from A and see how the ellipse must be shaped so that all segments reach B at the same time.)
   (b) What will happen if we generate a pulse at point B?
   (c) Will such a convergence also happen when you dip your finger in at some point other than A or B?

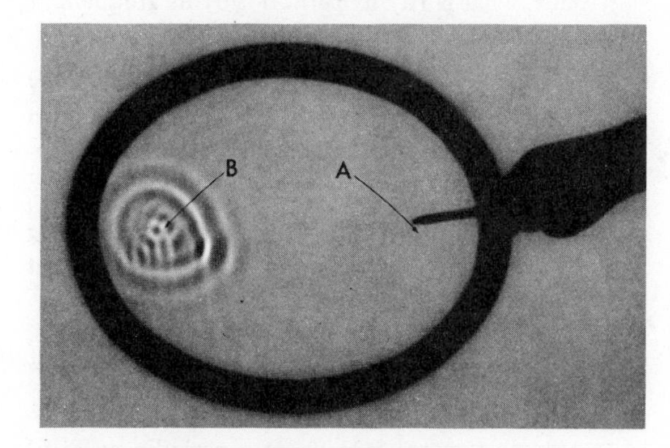

Figure 6–23 Reflection from an elliptical barrier. A pulse has been generated at *A* and is photographed as it converges on *B*.

8. A circular pulse generated at the focus of a parabolic reflector is reflected as a straight pulse. What does this tell you about the geometric properties of a parabola? (Hint: consider tiny segments of the circular pulse originating at the focal point and see what must be the shape of the parabola to produce a straight pulse.)

9. In Fig. 6–24, a straight pulse approaches a right-angled barrier at an angle of 45°.
   (a) How does it reflect?
   (b) What happens if the wave is incident at some other angle?

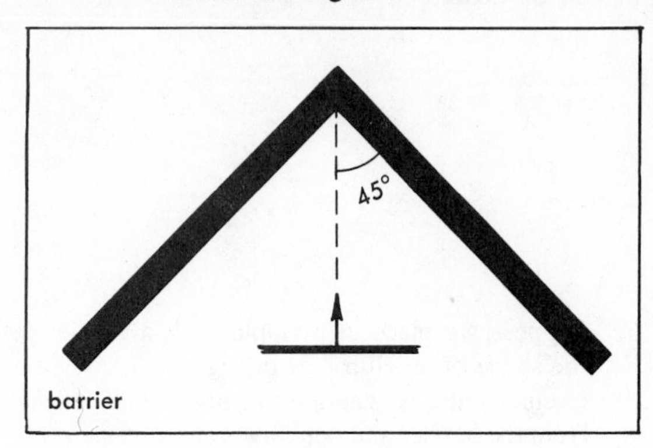

Figure 6–24 A straight pulse approaching a right-angled barrier.

10.* A machine gun fires 10 rounds per second. The speed of the bullets is 850 m/sec.
   (a) What is the distance in the air between the flying bullets?
   (b) What happens to the distance between the bullets if you step up the rate of fire? (Section 4.)

11.* If a wave generator produces 12 pulses in 3 seconds, what is (a) its period? (b) its frequency? (Section 4.)

12.* Would you increase or decrease the frequency of the generator in a ripple tank in order to produce waves of greater wavelength? (Section 4.)

13. (a) In a ripple tank, when one pulse is sent every $\frac{1}{10}$ sec, we find that λ is 3 cm. What is the speed of propagation?
   (b) In the same medium we send two pulses, the second one $\frac{1}{2}$ sec after the first. How far apart are they?

14. Assume you are looking at a periodic wave of frequency $f = 4$ per sec through a two-slit stroboscope. What do you expect to see if you rotate the stroboscope with a frequency of 1, 2, and 4 revolutions per sec?

15. A point source in the ripple tank produces circular periodic waves. By using a stroboscope to stop the motion, we measure the difference in radius between the first and sixth circular crests and find it to be 10 cm.
   (a) What is the wavelength?
   (b) Why did we not calculate the wavelength by using the radius, say, of the fifth pulse only?
   (c) Why do we use this method of measurement rather than take the difference between neighboring crests?

16. Circular waves are produced in the ripple tank by drops of water which fall at a constant rate. The wavelength is observed to be 1.2 cm. The experiment is repeated with the source moving at a uniform rate from one end of the tank to the other, along the center line. An observer near the end of the tank toward which the source is moving measures the wavelength and finds it to be 0.8 cm.
   (a) What is the ratio of the wave velocity to the source velocity?
   (b) What wavelength would an observer at the starting point measure after the source has moved some distance away from him?

17.* In Fig. 6–14, where is the boundary between the area of deep and shallow water? (Section 5.)

18.* What is the index of refraction in passing from the deep to the shallow water in Fig. 6–14? (Section 5.)

19. Measure the index of refraction in Fig. 6–15 by the method you used in the previous problem, and by finding the ratio of the sines of the appropriate angles. Compare the results.

20.* If the frequency changed when periodic waves went from deep to shallow water, would a stroboscope be able to stop all of the waves at one time? (Section 5.)

21. A ripple-tank wave passes from a shallow to a deep section with an incident angle of 45° and a refracted angle of 60°.
   (a) What is the ratio of speeds in the two sections?
   (b) If the wave speed is 25 cm per second in the deep section, what is it in the shallow one?

22. (a) A tire on an automobile wheel has a circumference of 7.0 feet. When the wheel is turning 200 times per minute, what is the speed of the automobile in feet per min?
   (b) A light wave whose frequency is 6.0×10^{14} per sec is passed through a liquid. Within the liquid the wavelength is measured and found to be 3.0×10^{-5} centi-

meters. What is the speed of light in this liquid?

(c) What is the wavelength in vacuum (from which the frequency was calculated)?

(d) What is the index of refraction of the liquid for light of this frequency?

23. The ripple tank is arranged so that the water gradually becomes shallow from one side to the other. Because of this, on one side of the tank the speed of a wave crest is different from that on the other side. As a result, straight waves become curved (Fig. 6–25). In the picture the pulses are moving toward the top of the page.

(a) Which is the shallow side?

(b) Does a similar phenomenon occur with light? Be prepared to discuss this in class.

24.† Water waves traveling in the deep section of a ripple tank at 34 cm/sec meet a shallow part at an angle of 60°. In the shallow part all waves travel at 24 cm/sec. When the frequency is increased slightly, the waves are found to travel at 32 cm/sec in the deep section.

(a) Compute the angle of refraction for each case.

(b) Considering the ripple-tank conditions, is it easier to measure the two speeds and find their difference directly, or to measure it indirectly by the angular difference found in (a)?

(c) How can we detect small differences in the speed of light?

25. We set up the ripple tank as shown in Fig. 6–26 and generate a periodic straight wave. The resulting wave pattern is shown in the photograph in Fig. 6–27.

(a) Explain what is taking place.

(b) Of what optical arrangement is this a model?

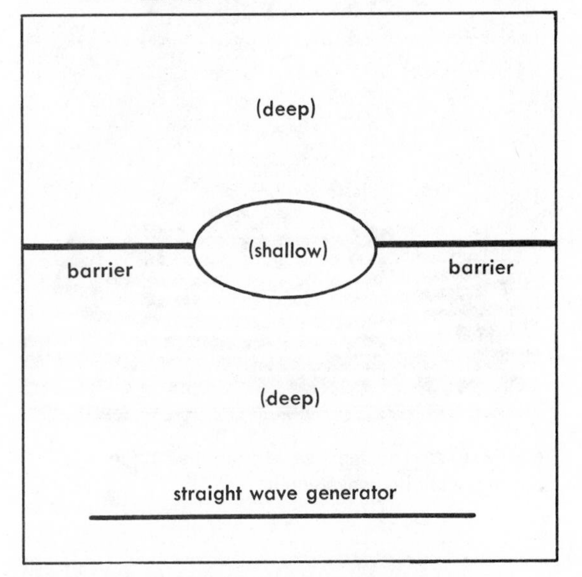

Figure 6–26 The ripple-tank arrangement used in making the picture in Fig. 6–27. The oval region between the barriers has shallow water, while water in the rest of the tank is deep.

Figure 6–25 Curving of a straight wave when the water becomes more and more shallow from one side to the other.

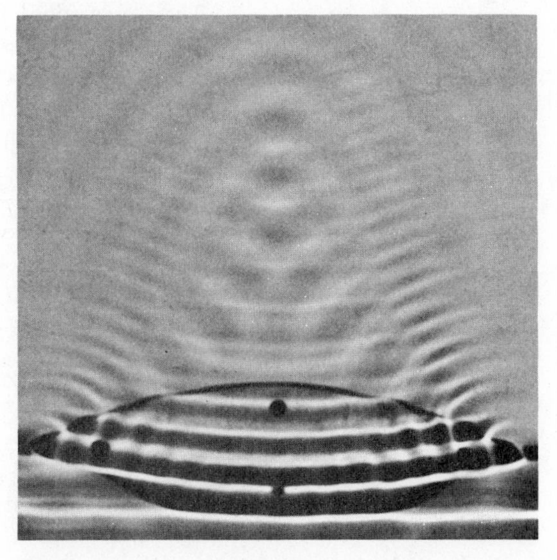

Figure 6–27 The wave pattern with periodic waves and an arrangement as in Fig. 6–26.

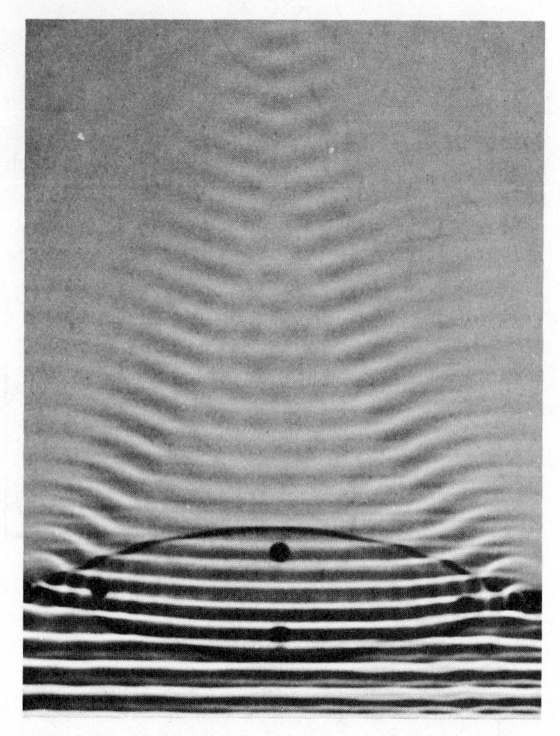

Figure 6–28 The same wave pattern as in Fig. 6–27 but with a shorter wavelength.

26. In Fig. 6–28 we see a photograph taken under the same conditions as those in Problem 25 except that here the waves have a shorter wavelength.
    (a) How do the two photographs differ?
    (b) What can you conclude from this difference?
27. Suppose the oval region in Fig. 6–26 is deep and the surrounding region is shallow.
    (a) What will happen to the straight waves?
    (b) Of what optical arrangement is this a model?
28. How do we know that the wavelength of light must be very much less than a centimeter?
29. Imagine our eyes were sensitive only to light of wavelength 0.1 mm. How would this affect our ability to see? Could you thread a needle?
30. If sound is a wave phenomenon, how would you explain the common experience of hearing sounds from around corners?
31. Sound waves in air usually travel at about 330 meters per second. Audible sounds have a frequency range of about 30 to 15,000 cycles per second. What is the range of wavelengths of these sound waves?

Further Reading

These suggestions are not exhaustive but are limited to works that have been found especially useful and at the same time generally available.

ANDRADE, E. N. DA C., *An Approach to Modern Physics.* Doubleday Anchor, 1956. For more information about sound and how well it conforms to the wave model. (Chapter III)

BENADE, ARTHUR H., *Horns, Strings, and Harmony.* Doubleday Anchor, 1960: Science Study Series. An excellent general reference for the student interested in the physics of sound and musical instruments. (Chapters 2 and 4)

BONNER, FRANCIS T., and PHILLIPS, MELBA, *Principles of Physical Science.* Addison-Wesley, 1957. (Chapter 16)

JAFFE, BERNARD, *Michelson and the Speed of Light.* Doubleday Anchor, 1960: Science Study Series. (Chapters 4, 5, 8, and 11)

Interference

CHAPTER
7

In Chapter 6 we studied the properties of waves to see if they can account for the common properties of light. We had already learned how light is reflected from a mirror, and we found that waves obey the same law of reflection. We had also learned how light is refracted when it passes from one medium to another, and we found that the refraction of waves follows Snell's law just as that of light does.

In the last section of the preceding chapter we recalled that light propagates in straight lines and produces sharp shadows. We then experimented with waves in the ripple tank and found they were diffracted when they passed through an opening. We could reduce the amount of diffraction by decreasing the wavelength but could not completely eliminate it. Therefore we changed our approach; we went back to light, and found that light is also diffracted when it passes through a very narrow slit. This was a rather convincing demonstration of the wave nature of light. In this chapter we shall continue to study in detail the properties of waves and in the next chapter do the corresponding experiments with light.

7-1 Interference on a Spring

One of the most striking results of the experiments with waves on a spring described in Chapter 5 was that two pulses traveling in opposite directions passed right through each other. The shape of the displacement of the spring could be explained by adding the displacements of the individual pulses (the principle of superposition). For example, Fig. 5–9 shows the successive shapes of a spring as two opposite pulses pass through each other. The point P halfway between the two pulses remains undisturbed because adding the displacements of the individual pulses at this point produces a cancellation at every instant during the crossing. This behavior of two opposite pulses is somewhat difficult to see because you have to look at the pulses exactly at the moment they cross. However, if we use periodic waves, we can observe the cancellation more easily.

When a periodic wave travels along a spring which is tied down at one end, every individual pulse is reflected upside down. Now we know that a reflected pulse superposes with every oncoming pulse it meets. Thus, suppose we first consider only two of the pulses, a and b, a wavelength λ apart as they travel toward the reflecting end (Fig. 7–1). Some time after the first pulse is reflected it will meet the second pulse and there will be a cancellation at the midpoint P between them (Fig. 7–2). In Fig. 7–3 the pulses are shown as they meet. Because they were originally a distance λ apart we can see that the point P is a distance $\lambda/2$ from the reflecting end. The next pulse c, which reaches P later, will be superposed with the reflected pulse b so that the same cancellation occurs again. Because the wave is periodic, this will happen every time a pulse passes P, and, although the motion of the spring as a whole is complicated, the point P always remains at rest. We call such a point a *node*. There are other nodes spaced $\lambda/2$ apart, as you can see by working out where the next few must be (for example, the cancellation of a and c).

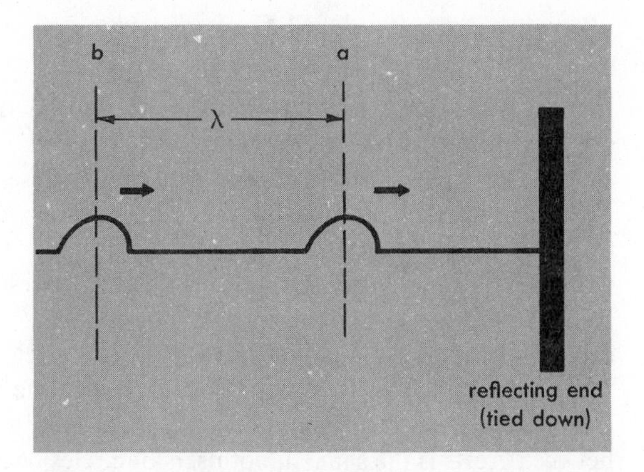

Figure 7–1
Two pulses travel toward the reflecting end of a spring.

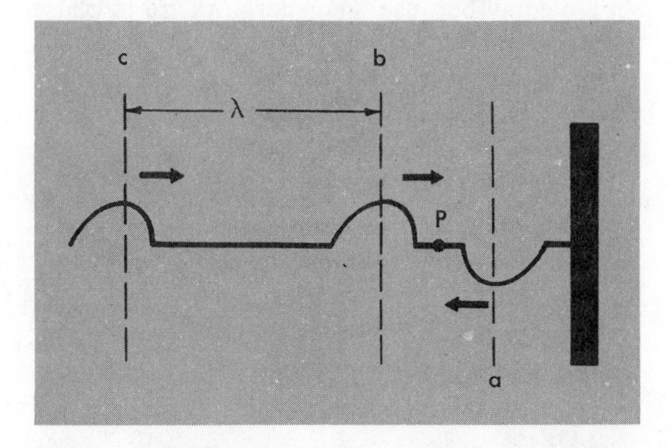

Figure 7–2
The spring after one of the pulses has been reflected. The reflected pulse *a* is upside down and traveling toward the incident pulse *b*. A third pulse *c* is approaching at a distance λ behind *b*.

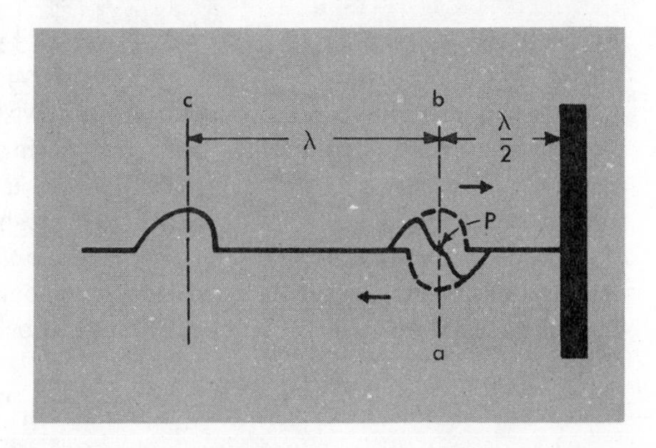

Figure 7–3
The pulses *a* and *b* of Fig. 7–1 and 7–2 meeting at point *P*.

It is clear that we could have obtained nodes just as easily by sending appropriate periodic waves from opposite ends of a long spring. The use of the fixed end as a means of producing a wave moving in the opposite direction is only a convenience.

The phenomenon we have just described—the superposition of two periodic waves which produces a series of nodes—is called *interference*. Rather than try to find the corresponding effect in light at once, we shall first go on to make a systematic study of interference between water waves in a ripple tank. Then we shall look for interference in light.

7-2 Interference from Two Point Sources

For the purpose of studying interference in the ripple tank, we shall use point sources generating circular waves. Imagine two point sources side by side a distance d apart, each one generating pulses at the same frequency. Furthermore let them move so they dip into the water together—that is, so each source produces a crest at the same instant. When this is the case, we say that the sources are *in phase.* We can represent the waves which the sources produce by drawing two sets of concentric circles side by side with centers a distance d apart (Fig. 7–4). The circles represent the crests of the waves expanding from each source. Since the sources are periodic, the crests are always the same distance apart—one wavelength. The distance between crests is the same in both sets of circles because the wavelengths are the same for both sources. The radii of corresponding circular crests in each set are equal because the generators are in phase.

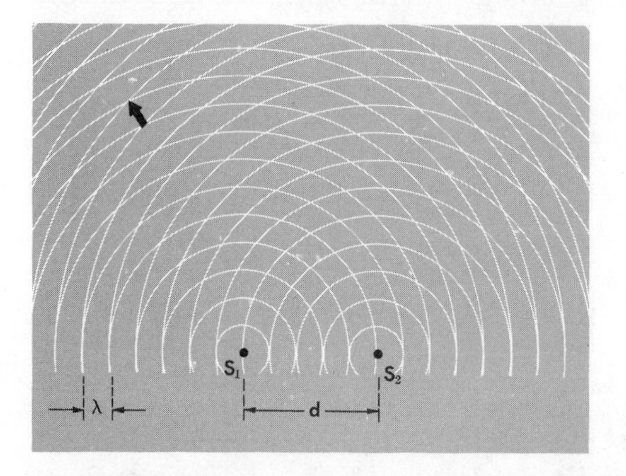

Figure 7–4
The circles represent the crests of waves from two sources S_1 and S_2 a distance d apart. The sources are periodic and are in phase. The black arrow points out a region we shall examine more closely later (Fig. 7–7).

Figure 7–5
The pattern we predict by applying the principle of superposition to the waves from the two point sources shown in Fig. 7–4. Lighter areas show where crest meets crest; zigzag area is undisturbed water, where crest meets trough; places where troughs meet are left dark.

What will happen when the waves from the two sources overlap? Let us try to predict the resulting wave pattern by using the principle of superposition. Where two crests cross each other, a "double crest" will be formed. Such "double crests" will produce bright regions on the screen of a ripple tank. In Fig. 7–5 (a) we have emphasized these regions by making them white. Where a crest from one source meets a trough from the other, the water will be practically undisturbed and these regions will

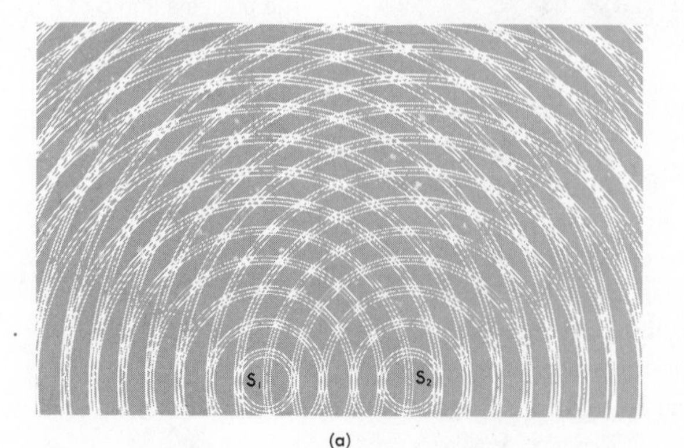

(a)

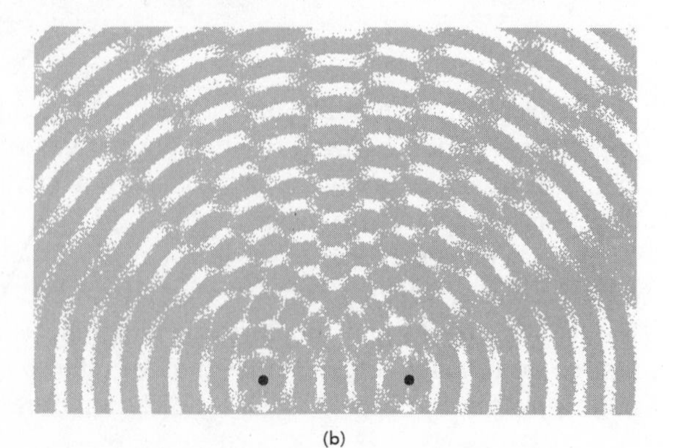

(b)

appear less bright on the screen. In these regions we have used some dots in Fig. 7–5 (a) to give a light gray appearance. Finally, where two troughs meet, a very dark image will be formed on the screen. In Fig. 7–5 (a) we have left these regions gray. In Fig. 7–5 (b) we have suppressed the construction lines, leaving only the pattern we expect to see. Therefore, the superposition of the waves in Fig. 7–4 should result in the pattern in Fig. 7–5 (b). An actual photograph of the waves from two point sources shows that this prediction is correct (Fig. 7–6).

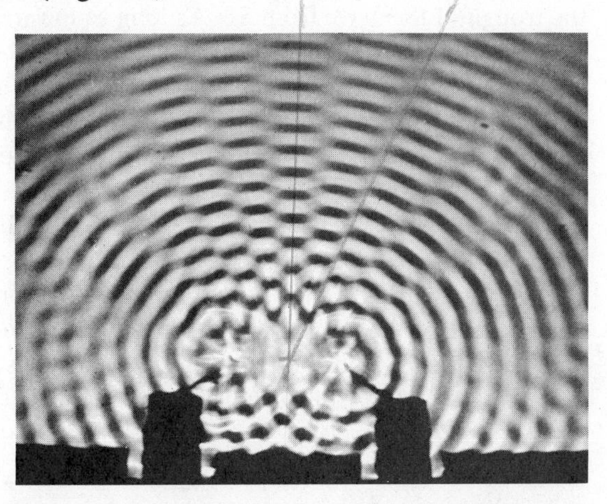

Figure 7–6
A photograph of the interference pattern from two point sources in phase. Notice the nodal lines radiating outward.

We have just constructed an interference pattern by superposing the waves from two sources. We have found that at one moment this predicted pattern agrees with what we see. Now let us consider how the waves in the pattern move. We shall start by finding out how one "double wave crest" moves. In Fig. 7–7 we show the two wave crests that cross to form the double crest at the tail of the arrow in Fig. 7–4. The dashed arcs in Fig. 7–7 represent the same two crests a short time later. Each crest has expanded away from its source; and as a result the double crest moves away from the region of the sources in the direction of the arrow. In one whole period T, the crest from each source will have moved out a whole wavelength λ and the double crest will have moved from the tail to the head of the arrow in Fig. 7–4. The double crests and double troughs all over the ripple tank have the same kind of outward motion. Consequently, each row of double crests and double troughs moves away from the source region while new double crests and troughs are formed near the sources. Each row is a moving train of waves.

Figure 7–7
The solid white lines show the double crests at the tail of the black arrow in Fig. 7–4. The dashed circles show two crests a short time later. The two black dots represent the sources.

What happens in the dotted regions in Fig. 7–5(b), the regions between the moving wave trains? Here crests lie on troughs at all times and the water surface is not wavy. To see why, let us examine the line between two wave trains at any given instant. In Fig. 7–5 (a) pick one of the dotted regions. Start at the crest of a wave from S_1. Because it lies on a trough from S_2, there is no net displacement of the water surface. It is practically undisturbed. Now let us move out from the sources along this region. As we do, we descend from the crest of the wave from S_1 and we go up from the trough of the wave from S_2. As long as the waves are nearly symmetrical —so that a trough looks like an inverted crest—the addition of the displacements of the two waves continues to give zero displacement. As we look farther out, we come to a trough from S_1 at the same place that we reach a crest from S_2. The upward and downward displacements still cancel; and we can see that by continuing out along the line through the intersections of crests and troughs, we shall find practically undisturbed water. Looking at Fig. 7–6 (or, even better, at the surface of a ripple tank), we can see these lines of undisturbed water extending outward from the region of the sources and separating the moving wave trains of reinforced crests and troughs. By analogy with the nodes on a spring, we call these lines of zero disturbance nodal lines. In Fig. 7–8 we have drawn them as thick black lines.

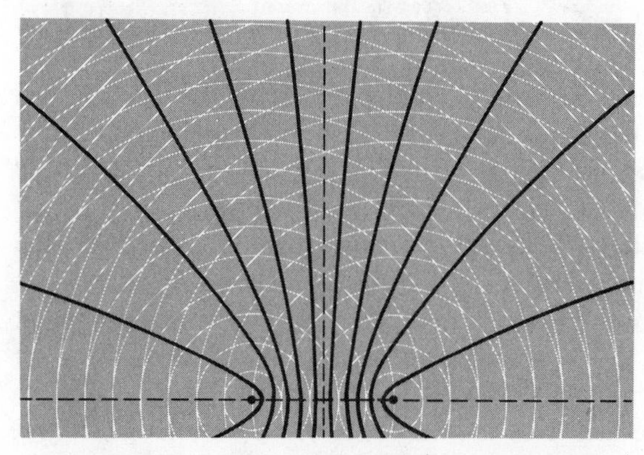

Figure 7–8
The nodal lines from two sources. Between the nodal lines are moving double crests and moving double troughs.

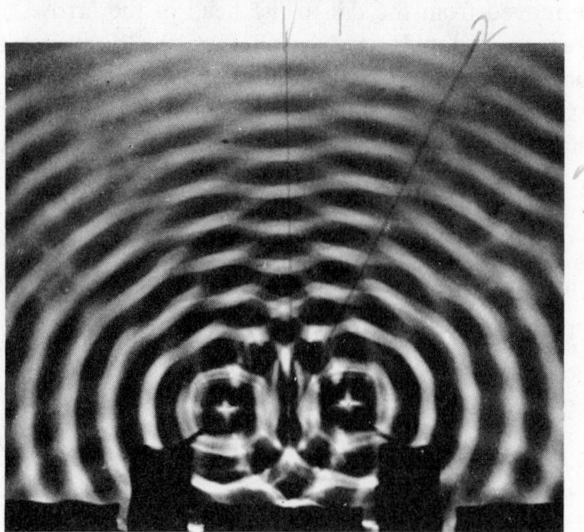

Figure 7–9
A two-source interference pattern like that in Fig. 7–6, but with a longer wavelength.

With a different wavelength or a different separation of the sources, the interference pattern changes in detail; but the general structure of such patterns is the same. Another pattern with a longer wavelength is shown in Fig. 7–9. Again you can see the nodal lines, and the waves moving out between them. The waves are slightly blurred because the photograph was exposed 1/50 of a second, and in that time the waves moved an appreciable fraction of a wavelength.

Although we considered water waves in our investigation, we really have not used any special property of water waves to obtain our result. We used only the principle of superposition, which is common to all waves. The results of this entire chapter apply equally well to all waves.

7-3 The Shape of Nodal Lines

If we examine Fig. 7–6 and 7–9 we notice that, although the nodal lines are slightly curved near the sources, they soon become quite straight. Another striking fact is that the number of nodal lines decreases as the wavelength increases. Let us number the nodal lines so we will be able to refer to them conveniently. To do this we reexamine Fig. 7–8 and notice that the pattern is symmetrical—that is, it looks exactly the same to the left of the central line as it does to the right. This should not be too surprising, as our sources also look exactly the same left and right. For this reason we need count only half the nodal lines, say those to the right. Thus we call the first one to the right of the central (dotted) line the first nodal line. The next one is called the second nodal line, and so on. When we want to talk about a nodal line and we don't care which one, we usually say the *n*th nodal line, where *n* is an integer (1st, 2nd, 3rd, etc.).

Let us call a point on the first nodal line *P*, and connect it to the two sources by drawing the lines PS_1 and PS_2 (Fig. 7–10). We call these lines the path lengths from *P* to S_1 and from *P* to S_2. By counting the crests on the diagram you can see that $PS_1 = 3\lambda$ and $PS_2 = 2\frac{1}{2}\lambda$, so that the difference in path length is

$$PS_1 - PS_2 = \tfrac{1}{2}\lambda.$$

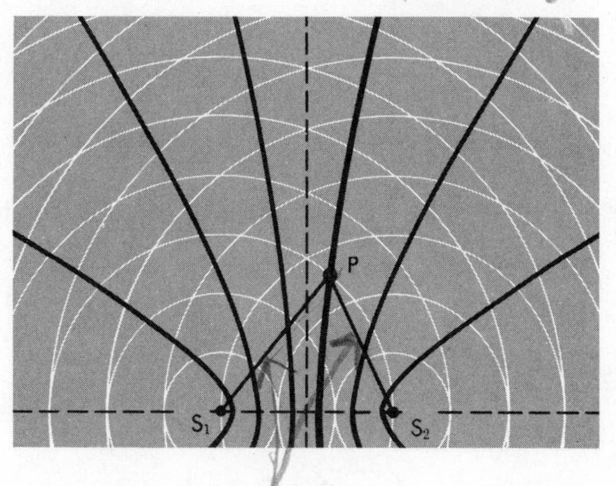

Figure 7–10
The first nodal line. For any point *P* on the line the difference in path length from *P* to S_1 and *P* to S_2 is half a wavelength.

If we had taken any other point on the first nodal line, we would have found the same difference, ½λ, between the path lengths. Indeed, we may say that the first nodal line is composed of those points P for which the difference in the path length is ½λ, so that one crest and one trough always arrive there at the same time.

The second nodal line can be characterized in a similar way. In this case, if P is any point on the second nodal line, the difference in path lengths is

$$PS_1 - PS_2 = \tfrac{3}{2}\lambda,$$

which can also be seen from Fig. 7–10. Continuing this procedure, we arrive at an equation describing the nth nodal line:

$$PS_1 - PS_2 = (n - \tfrac{1}{2})\lambda.$$

Using this equation we can construct the nodal lines by finding the intersections of circles of radii r centered at S_2 with circles of radii $r + (n - ½)\lambda$ centered at S_1, where r has a different value at each point along the nodal line, while n is constant along any one line (Fig. 7–11).

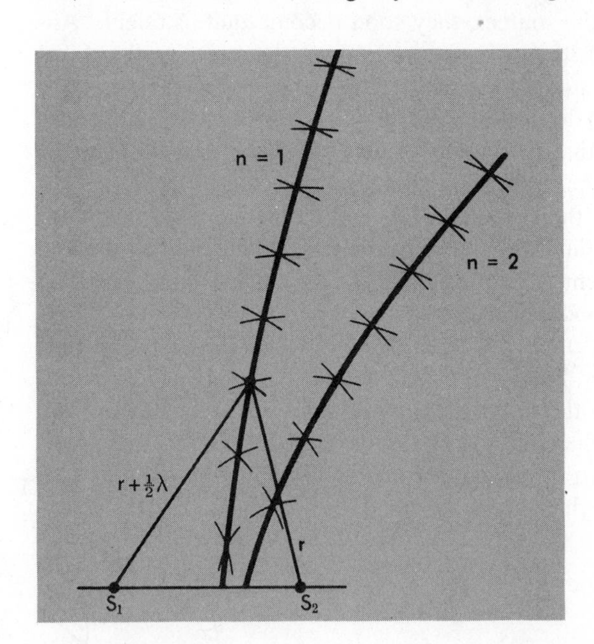

Figure 7–11
Construction of a nodal line. Arcs are swung around the two sources with radii r and $r + (n - ½)\lambda$ respectively. The nodal curve runs through the intersections. The curves shown are for the first and second nodal lines: $n = 1$ and $n = 2$.

7-4 Wavelengths, Source Separation, and Angles

In a ripple tank we can measure the path lengths to any point on a nodal line; and, using

$$PS_1 - PS_2 = (n - \tfrac{1}{2})\lambda,$$

we can find the wavelength λ. We do not need to stop the waves to make such a measurement. The nodal lines stand still while we take our time measuring PS_1 and PS_2.

It is often necessary to make our measurements at a point P which is far away from S_1 and S_2. But if we then measure the two large lengths PS_1 and PS_2 directly and subtract to find the small difference between them, we shall have a hard time getting sufficient accuracy. We may subtract away most of our measurement, leaving only our errors. We therefore look for a more accurate way to measure the path difference.

For any point P the difference in the path lengths $PS_1 - PS_2$ depends on the angle between PS_1 and d. Consider Fig. 7–12 (a), which shows the two sources S_1 and S_2, and a point P very far away compared to the source separation d. The distance PA is made the same as PS_2 so that the angles 1 and 2 are equal and $PS_1 - PS_2 = AS_1$. The farther away P is, the more nearly parallel the lines PS_1 and PS_2 become. We shall consider only points P that are far enough from S_1 and S_2 so that for all practical purposes PS_1 and PS_2 are parallel. Then we can draw Fig. 7–12 (b). Because angles 1 and 2 have become right angles, the triangle AS_1S_2 is a right triangle. Therefore from the definition of the sine of an angle (Section 3–3), the sine of θ in Fig. 7–12 (b) is

$$\sin \theta = \frac{AS_1}{d}.$$

Since AS_1 is the path difference, we find

$$PS_1 - PS_2 = d \sin \theta.$$

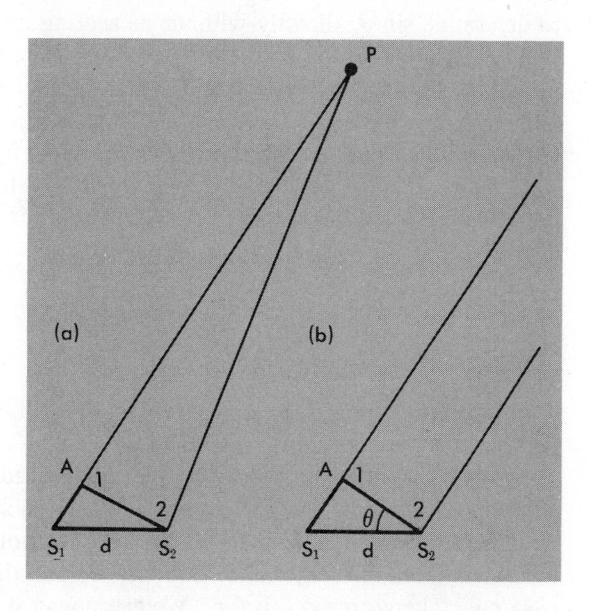

Figure 7–12
(a) The path difference AS_1 can be determined in terms of the source separation d and an angle. (b) When P is far from the sources, $AS_1 = d \sin \theta$.

This equation expresses the path difference in terms of source separation and angle. The angle θ tells us the direction of P with respect to the sources. When $\theta = 90°$, for example, P is to the right on the horizontal line passing through the sources; when $\theta = 0$, P is in the direction of the top of Fig. 7–12 (b).

Now when P is on the nth nodal line

$$PS_1 - PS_2 = (n - \tfrac{1}{2})\lambda.$$

Consequently,

$$(n - \tfrac{1}{2})\lambda = d \sin \theta_n$$

or

$$\sin \theta_n = (n - \tfrac{1}{2})\lambda/d$$

as long as P is far away from S_1 and S_2. Incidentally, this result tells us that far from the sources the direction of a nodal line does not change. It is given by the angle θ_n. Far from the sources, therefore, the nodal lines must be straight as we noticed in the last section. Actually, if these straight portions of the nodal lines are extended back toward the sources, they pass through the midpoint of the line between the sources.

In the last section we also noticed that the number of nodal lines increases as the wavelength decreases. We can relate this observation to our equation for the direction of the nodal lines far from the sources. Since $\sin \theta_n$ cannot be greater than 1, then $(n - \tfrac{1}{2})\lambda/d$ cannot be greater than 1. The largest value of n which satisfies this condition is the number of nodal lines on each side of the center line. This number depends only on λ/d, and increases as λ decreases. We can make an approximate measurement of the wavelength, merely by counting the number of nodal lines.

To make an accurate determination of λ, we can find the direction of the nth nodal line, that is, the angle θ_n, and calculate λ from the equation $\sin \theta_n = (n - \tfrac{1}{2})\lambda/d$. In the ripple tank, θ_n is easily found; but this is not always the case with other waves. Therefore we shall look for a way to determine $\sin \theta_n$ directly without measuring the angle θ_n itself. Let the point P in Fig. 7–13 be on the nth nodal line far away from the two sources S_1 and S_2, so that the lines CP and S_1P are practically parallel to each other and both are practically perpendicular to AS_2. Since the center line is perpendicular to d, we see that $\theta_n' = \theta_n$. But, from the figure, $\sin \theta_n' = x_n/L$ where L is the distance PC and x_n is the distance from P to the center line. Therefore we have

$$(n - \tfrac{1}{2})\lambda/d = \sin \theta = \sin \theta' = x/L$$

or

$$\lambda = \frac{d(x/L)}{n - \frac{1}{2}},$$

where we have omitted the subscript n on θ, θ', and x, but you must remember that θ, θ', and x refer to the nth nodal line.

An example will show how simple the procedure is. Suppose we are working with sources 10 cm apart. We may pick a point P on the third nodal line, measure its distance L from the midpoint of d, and measure the distance x from the center line. Suppose we find that L is 80 cm long and x is 48 cm. Therefore, x/L is 0.6. We can now test the accuracy of our ratio by measuring values of L and the corresponding values of x for other points on the third nodal line. If all the points are far enough away, x/L remains 0.6.

Now because we are working with the third nodal line we must use $n - \tfrac{1}{2} = 5/2$ and with $d = 10$ cm we find

$$\lambda = \frac{d(x/L)}{n - \frac{1}{2}} = \frac{10(0.6)}{\frac{5}{2}} = 2.4 \text{ cm}.$$

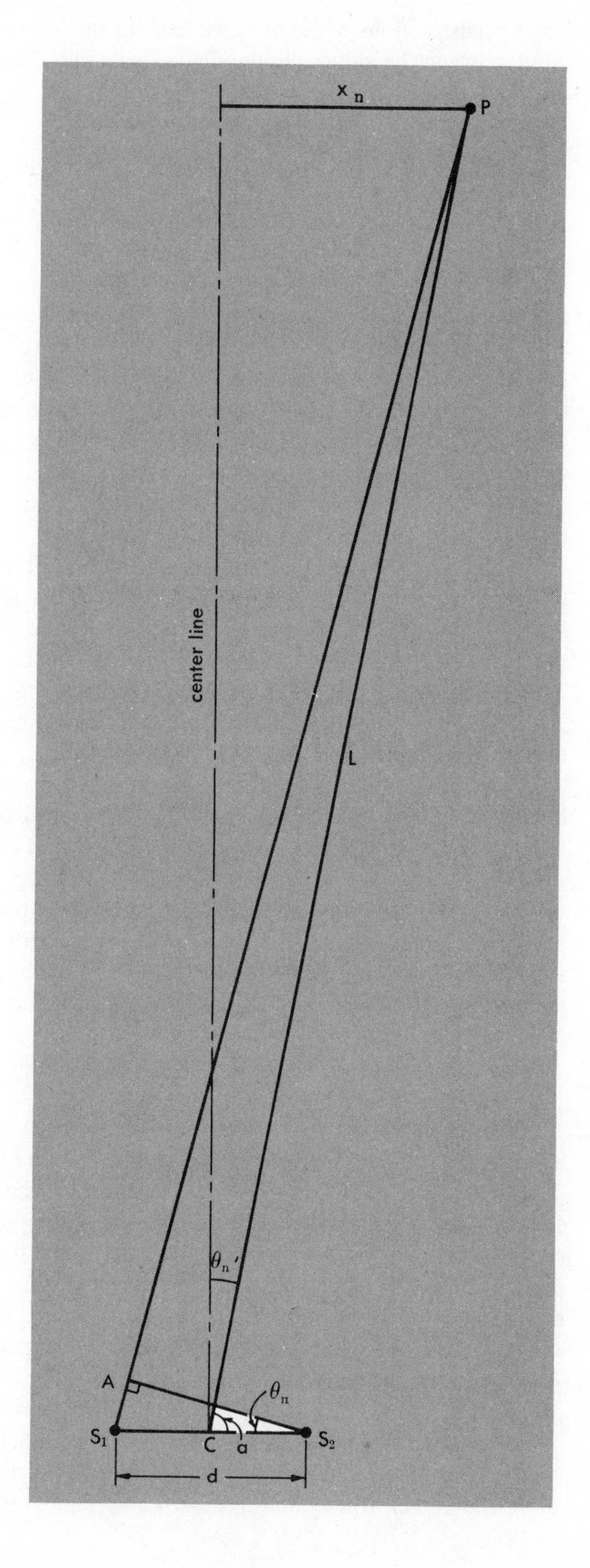

Figure 7–13
When P is far from the sources, $\theta_n + a = 90°$. Since $\theta_n' + a = 90°$, $\theta_n = \theta_n'$. Also, $\sin \theta_n' = x_n/L$, and so $\sin \theta_n = x_n/L$.

Using the various nodal lines we can get several evaluations of λ. Agreement between the values obtained gives us a check on our reasoning and on our measurements.

7-5 Phase

Two generators with the same period are *in phase* when they always dip into the water together, producing crests at the same instant. However, it is not necessary for two sources with the same period to be in phase. For example, one of the sources may always dip into the water somewhat later than the other, after a time delay t. Since the natural time unit for a particular periodic motion is its period T, it is convenient to measure the time delay between the dips of S_1 and S_2 in fractions p of the period. Thus we use $p = t/T$ to measure the delay. For example, if each source dips every $\frac{1}{6}$ second and S_2 always dips $\frac{1}{18}$ second after S_1, then the fraction p is one third.

When two sources of the same frequency do not dip together we say that they are *out of phase*. The fraction p describes the *phase* delay of one source with respect to the other. There are no delays which are longer than the period T because we always measure the delay of the second source from the most recent dip of the first source, and its dips come a time interval T apart. Consequently there are no phase delays greater than 1. The value of p is always between 0 and 1.

Let us now use the two point generators of waves operating so that S_2 has a phase delay p with respect to S_1. What will the interference pattern look like? We can find out by drawing two sets of concentric circles representing the wave crests from each source. As in Fig. 7–4, the crests in each set are always one wavelength λ apart; however, this time the sources are not in phase, and the radii r_1 and r_2 of corresponding crests from the two sources are not equal (Fig. 7–14). The radii of the delayed crests from S_2 are smaller than those of the corresponding crests from S_1 by a distance l equal to the fraction p of a wavelength λ:

$$l = p\lambda.$$

As an example let us see what happens when one of the sources is half a period behind the other. Then the distance l is half a wavelength, and the phase delay p is $\frac{1}{2}$. In Fig. 7–15 we have drawn the wave crests for such a situation and constructed the nodal lines by joining the points where a crest crosses a trough. We see that the pattern of nodal lines is different from the pattern for two sources in phase. For the same ratio λ/d the nodal lines are at the places where the reinforced crests used to be, and the reinforced crests are now where the nodal lines used to be. Compare Fig. 7–15 with Fig. 7–10. In each of these figures $\lambda = \frac{1}{3}d$, but in Fig. 7–10 the sources dip in phase while in Fig. 7–15 there is a phase delay $p = \frac{1}{2}$.

The photograph in Fig. 7–16 was taken with a phase delay $p = \frac{1}{2}$ between the sources. We can see, for instance, that there is now a nodal line along the center where in Fig. 7–6 (and Fig. 7–9) there are reinforced crests.

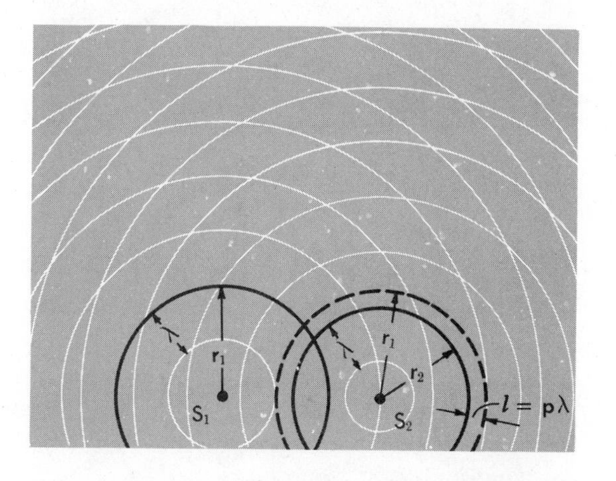

Figure 7–14
The waves from two point sources out of phase. Source S_2 has a phase delay p with respect to S_1. The difference between the radii of corresponding crests is the distance $l = r_1 - r_2 = p\lambda$.

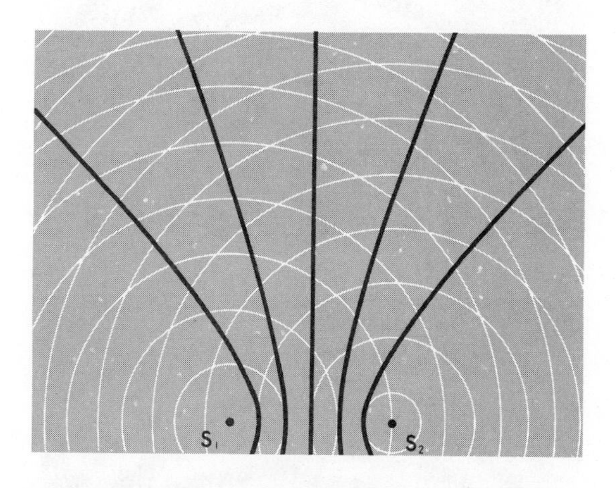

Figure 7–15
The pattern of nodal lines when there is a phase delay $p = \frac{1}{2}$.

Figure 7–16
A photograph of the ripple-tank screen when two sources are operating with a phase delay $p = \frac{1}{2}$.

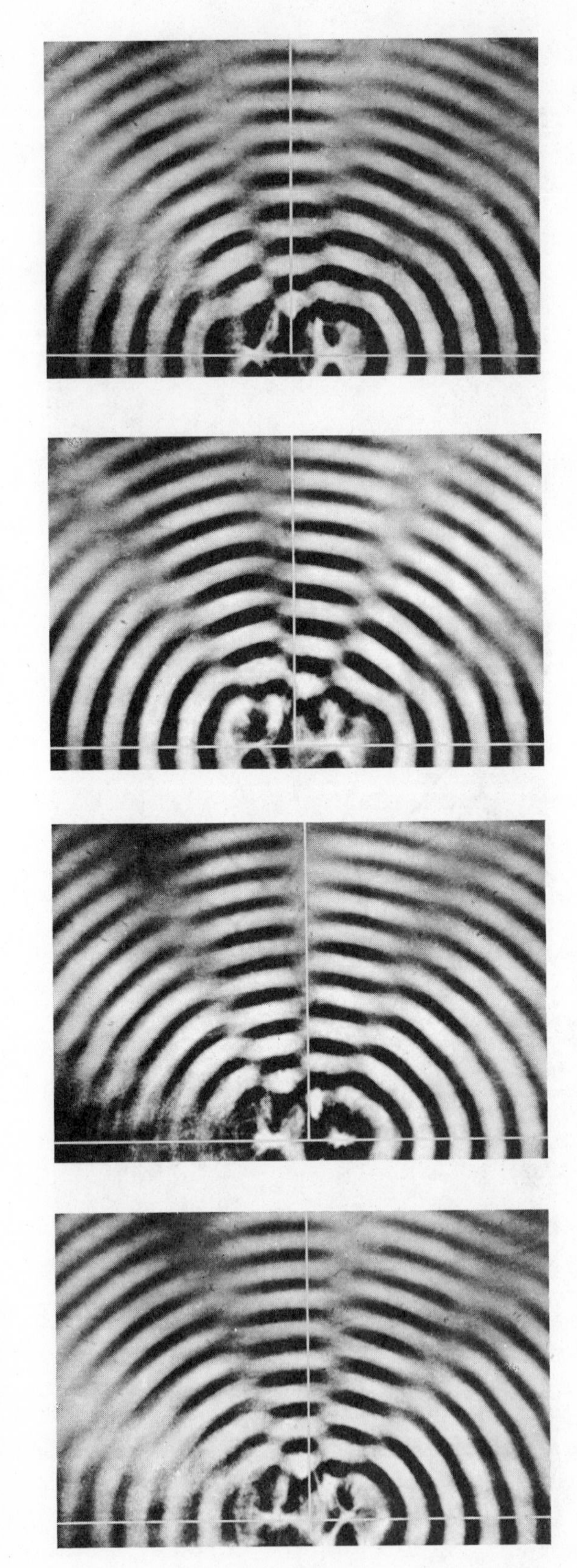

Figure 7–17
The interference from two point sources with different phase delays p of the right-hand source. In the top picture the sources are in phase; $p = 0$. In the subsequent pictures p increases, passing through $p = \frac{1}{2}$ in the third photograph.

We have given examples of the interference pattern for two particular choices of the phase delay, $p = 0$ and $p = \frac{1}{2}$. Actually we could have chosen any phase delay from 0 to 1, and in each case the interference pattern would have been different. The series of pictures in Fig. 7–17 shows the interference patterns in the ripple tank for different phase delays of S_2. Changing the phase delay causes the whole pattern of nodal lines to shift in a definite direction. As the phase delay of S_2 increases, the radii of the crests from S_2 fall behind those from S_1 by an increasing distance $l = p\lambda$. Consequently the nodal lines bend more sharply around S_2, away from the center line. This is just what we see in the photographs.

7-6 Summary and Conclusion

The interference pattern from two point sources in the ripple tank is characterized by a set of lines where the water surface remains undisturbed—the nodal lines. When the sources are in phase these nodal lines are distributed symmetrically about the center line, the perpendicular bisector of the line S_1S_2 joining the sources. In the immediate neighborhood of the sources the nodal lines are curved, but not very far away from the sources they become nearly straight. If the straight portions are extended back toward the sources, they pass through the center point between the sources. The number of nodal lines and the angle between any of these lines and S_1S_2 is determined by λ/d, the ratio of wavelength to source separation. This means that we can determine either d or λ if we know the other.

Even for the same wavelength and source spacing, different interference patterns are obtained for different phase delays p between the two sources. When the sources are in phase there is an even number of nodal lines symmetrically arranged about the center line. When the source S_2 produces crests later than S_1 with a time delay $t = pT$, crests from S_2 are all at smaller radii than the corresponding crests from S_1. The corresponding radii differ by the distance $l = p\lambda$. As a result the points where crests from S_2 cross troughs from S_1 are closer to S_2. This means that the nodal lines are bent away from S_1 and swing in closer to S_2; and the pattern becomes asymmetrical. Only for $p = \frac{1}{2}$ do we again get a symmetrical pattern. Then one nodal line covers the center line and there is an odd number of nodal lines.

When the distance between the sources is fixed and we generate waves of one particular wavelength, the interference pattern depends on the phase delay between the sources. Consequently, to maintain a fixed interference pattern, the phase delay must also remain constant. For two sources which run continuously at the same frequency, the phase delay will remain constant. But if each of the two sources is turned on and off in an irregular fashion, the phase delay will vary and with it the interference pattern. This shifting of the interference pattern will be of great importance in the next chapter, where we shall discuss the interference of light waves.

1.* If a periodic wave whose pulses each had the shape shown in Fig. 7–18 were traveling along a spring, and you wanted to produce nodes by sending appropriate periodic pulses in the opposite direction, what shape pulses would you use? (Section 1.)

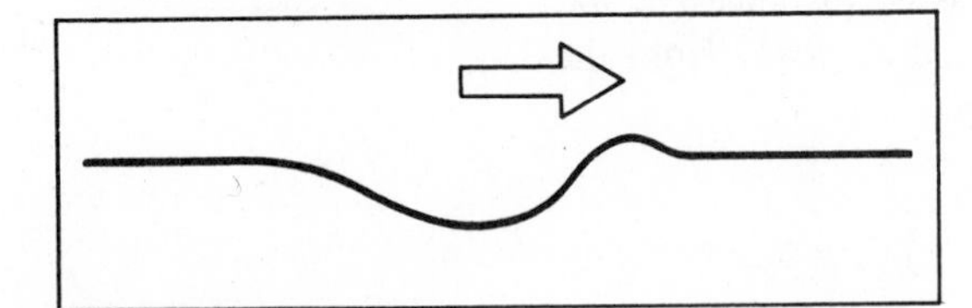

Figure 7–18 For Problem 1.

2.* In Fig 7–3 how far from the wall will pulse *c* meet pulse *a* to make the next node? (Section 1.)

3.† We showed in the text (Fig. 7–1, 7–2, and 7–3) that when pulses are incident periodically on the fixed end of a spring, the point *P*, a distance $\lambda/2$ from the end, never moves and is, therefore, a node. Extend the argument to show that the point P_2, a distance $3\lambda/2$ in front of the end, is a node.

4.* In Fig. 7–19 the circles represent the crests of waves produced by sources S_1 and S_2. At which of the points *A*, *B*, and *C* is there a "double crest," a "double trough," or a nodal point? (Section 2.)

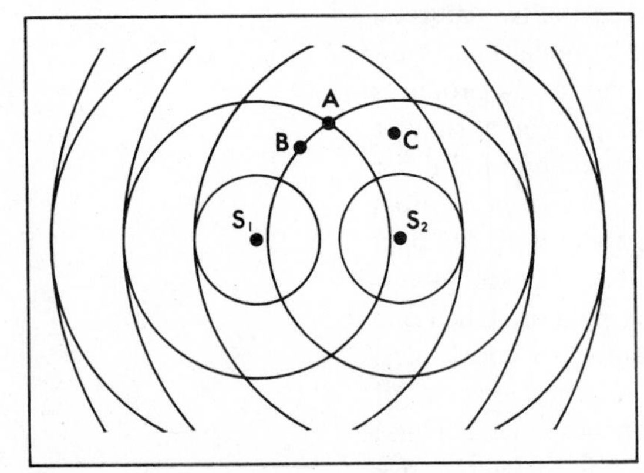

Figure 7–19 For Problem 4.

5. Draw the sets of concentric circles and the interference pattern from two sources with $d = 5\lambda$ at the time:
   (a) when the generators have just produced crests,
   (b) when they have just produced the following troughs.

   How have the reinforced crests moved during the time interval between these drawings?

6. Draw the interference pattern for the case $d = 5\lambda$ on a piece of paper large enough so that you can see the nodal lines become straight at a great distance from the sources. Continue these straight lines back toward the sources and show that they all pass close to the midpoint of the line joining the sources.

7.* In Fig. 7–4, is there a nodal line along the perpendicular bisector of the line joining S_1 and S_2? (Section 2.)

8. Fold two pieces of ruled paper into long strips about 2 cm wide and hold them as in Fig. 7–20. Imagine that the lines are wave crests. Your fingers then represent the sources of the waves. Notice how the crests from both sources add together. Now by sliding the free ends sidewise, locate nodal lines and moving wave regions.

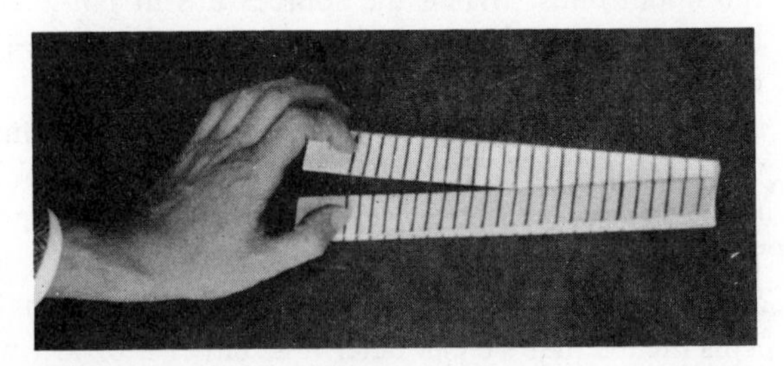

Figure 7–20 For Problem 8.

9.* Which nodal line in Fig. 7–10 is made up of all points *P* such that $PS_2 - PS_1 = 3\lambda/2$? (Section. 3.)

10.* You know the distances from a point on a nodal line to the two point sources in a ripple tank. What else do you have to know to calculate the wavelength of the waves? (Section 3.)

11. Construct the nodal lines for two point sources with $\lambda/d = 1/3$ by the method of Fig. 7–11. Is this really a different method from that used in Problem 6?

12. What would happen to the nodal lines if one of the two sources made weaker and weaker waves and then quit?

13.* For two point sources separated by a distance *d* is there a range of wavelengths λ which will produce an interference pattern with no nodal lines? (Section 4.)

14. Two sources 6.0 cm apart operating in phase produce water waves with a wavelength of 1.5 cm. Draw the nodal lines far from the sources. Determine the position of each line by means of intersecting arcs of circles drawn from the two sources. Measure the angle between the second nodal line and the center line of the pattern. Compare the sine of this angle with $(n - \frac{1}{2})\lambda/d$.

15. Suggest an interference experiment to prove that sound is a wave phenomenon. How could you use such an experiment to determine the wavelength of sound?

16. (a) From Fig. 7–6 and 7–9 find the ratio λ/d by using the equation $\sin \theta_n = (n - \frac{1}{2})\lambda/d$.
(b) The dimensions of the photographs are one fourth the actual size. By measuring d on the photographs estimate the actual value of λ.

17. Look up the definition of "hyperbola" and show that nodal lines are hyperbolas.

18. In Fig. 7–21, $L = 50$ cm, $d = 10$ cm, $\alpha = 30°$. What are γ and β? Find γ and β when $L = 500$ cm. Does this convince you that it is a good approximation to set $\gamma \approx \beta \approx \alpha$ when L is much greater than d?

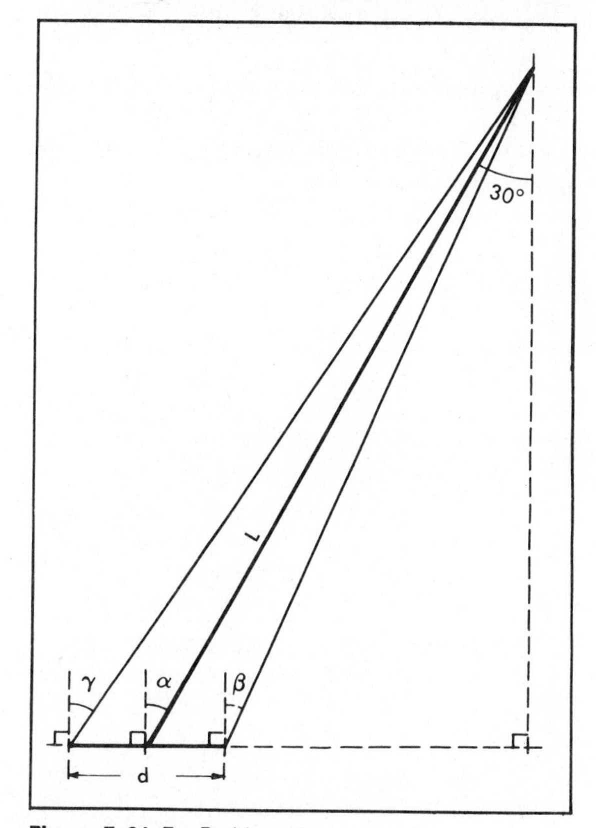

Figure 7–21 For Problem 18.

19. One red and one blue car are going around a circular racetrack 5.0 km in circumference. They move at constant speed. Each car takes 2.5 minutes for each lap. The blue car always comes around 0.50 minute behind the red.
(a) What is the phase delay p of the blue car with respect to the red car?
(b) What is the speed of each car?
(c) If the track were only 4.0 km long, would this change the answers to (a) and (b)?

20.* In Fig. 7–14, $l = 0.25$ cm and $\lambda = 0.70$ cm. What is the phase delay? (Section 5.)

21.* Is a phase delay of 0 the same as a phase delay of 1? (Section 5.)

22. In radioastronomy, one method for "looking" at different directions in the heavens is rather like using an adjustable-phase two point source rippler in reverse.

Antennas A and B, tuned to the same wavelength, are set up a certain distance apart, and the signals they receive are mixed in an amplifier after the signal from one of them has been fed through a phase adjuster (Fig. 7–22). Maximum signal is obtained if the signals entering the receiver at C and D are in phase.

The phase adjuster changes the phase by changing the length of the transmission cable

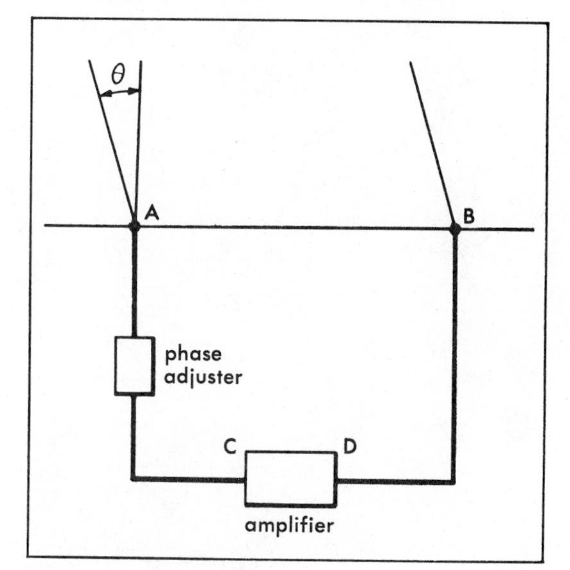

Figure 7–22 For Problem 23.

between A and C. When the cable AC is equal in length to the cable BD, the signals from a point directly overhead will be in phase at the receiver. To "look" in the direction θ, the cable AC must be lengthened. If the distance AB is 100 m, and if the frequency to which the receiver is tuned is 20 megacycles/second, what phase difference must be introduced to make the system look in a direction 5° away from the vertical?

23. Standard radio broadcasting stations operate on wavelengths of between 200 and 600 meters. The usual antennas are vertical towers that behave like point sources of waves in the plane of the earth's surface.
    (a) How could you arrange two point source radio antennas to give good coverage of a city built in a strip along the seacoast and waste a minimum of power over the water or in unpopulated areas?
    (b) If the city expanded inland, how would the station be able to provide better signal strength to the newly populated areas?

24.† Two sources in a ripple tank are operating at frequencies of 15 cycles per second and 16 cycles per second. Describe the resulting pattern of nodal lines.

25. What will be the appearance of the interference pattern of two point sources if the phase delay between them is altered suddenly?

Further Reading

These suggestions are not exhaustive but are limited to works that have been found especially useful and at the same time generally available.

BONNER, FRANCIS T., and PHILLIPS, MELBA, *Principles of Physical Science.* Addison-Wesley, 1957. (Chapter 16)

GRIFFIN, DONALD R., *Echoes of Bats and Men.* Doubleday Anchor, 1959: Science Study Series.

JAFFE, BERNARD, *Michelson and the Speed of Light.* Doubleday Anchor, 1960: Science Study Series. (Chapters 4, 5, 8, and 11)

Light Waves

CHAPTER

8

8-1 Can We See Interference in Light?

In the last chapter we studied the interference patterns produced in a ripple tank by two point generators. Now we wish to do similar experiments with light, in order to see whether it has all the properties of periodic waves.

In designing an interference experiment with light we must keep in mind some important differences in the way we observe water waves and light waves. We can see water waves by standing at any place where light reflected from the waves reaches our eyes. We easily recognize a nodal line in the ripple tank because we can see where the water is undisturbed. Imagine now that we replace the two point generators by two light sources. How can we find out whether there are places where the light waves from the two sources cancel each other, places corresponding to the nodal lines in the ripple tank? In the ripple tank we could see the waves, but we cannot look across a light beam and see the light waves. (Recall Fig. 1–10.) To see this light, we must get our eyes directly in the path of the light, or else place a reflecting screen in the path and observe the light that is reflected.

You can get an idea of the problem involved in detecting the interference of light if you imagine that you have to study the interference of water waves in a completely dark room, so that you cannot see the water. In this case you could locate the nodal lines by putting your finger in the water and moving it slowly across the tank. Most of the time you would feel the waves moving up and down; but when you came to a nodal line, you would not feel any motion. Similarly, when we observe light waves, we can either move our eyes or place a reflecting screen in the path of the light. Where light waves are reinforced on the screen, we can see light. Where a nodal line intersects the screen, we see a dark region.

At what angles do we expect to see these bright and dark bands? We learned from the ripple tank that, in the interference pattern generated by two point sources, the angles between the nodal lines depend on λ/d. For a given wavelength λ, these angles all increase as the source separation d decreases. We already know that the wavelength of light waves is much less than a tenth of a millimeter (Section 6–7); therefore, to achieve an observable separation of the nodal lines at a reasonable distance from the sources, we must place the sources close together. This means that they must be very small.

Also, to observe clear nodal lines we need sources which emit waves of a definite wavelength or at most a small range of wavelengths; otherwise, the nodal lines corresponding to one wavelength will be buried in the crests resulting from the others. Imagine, for instance, that each source emits waves of the two different wavelengths which give the patterns shown in Fig. 7–6 and 7–9. The total pattern you would then observe would be a combination of the patterns shown in the two photographs, and there would be no places which remain undisturbed in the total pattern.

In the following discussion we shall assume that the light we observe contains a narrow-enough range of wavelengths so that the nodal lines, although they may become somewhat fuzzy, are not wiped out.

Imagine that we have two very small light sources placed a small distance d apart and that a screen is placed at a large distance L away from them, as

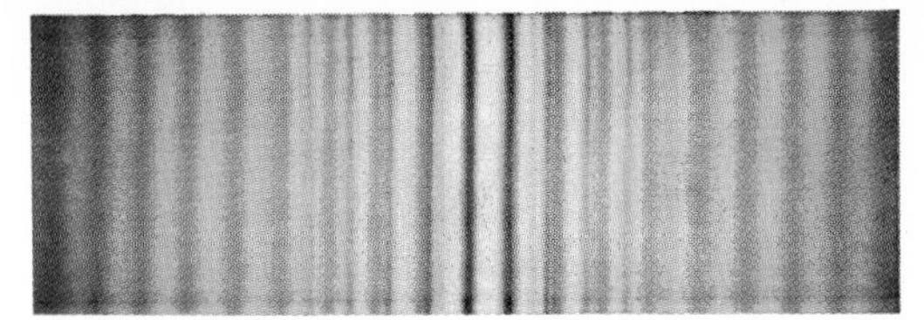

Figure 8–A
Interference pattern produced by white light passing through two narrow slits.

Figure 8–B
Interference patterns of red light and blue-violet light made with exactly the same setup used to make the white-light interference pattern in Fig. 8–A.

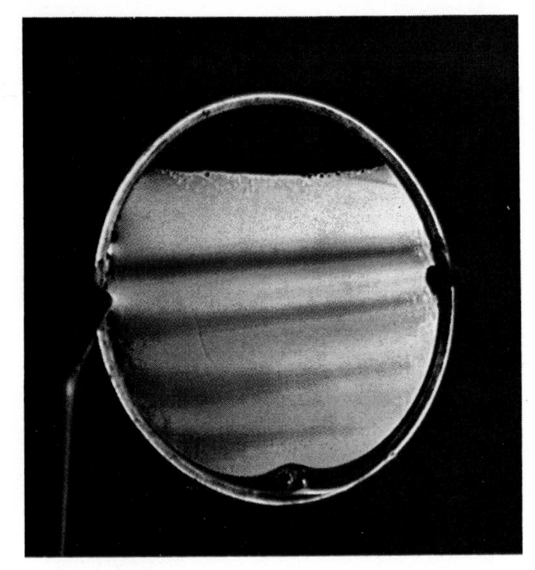

Figure 8–E
Interference seen in white light reflected by a thin soap film. The film fills a circular aperture. The picture was taken shortly after the film had drained enough so that the upper region is less than $\lambda/4$ thick for all wavelengths of visible light.

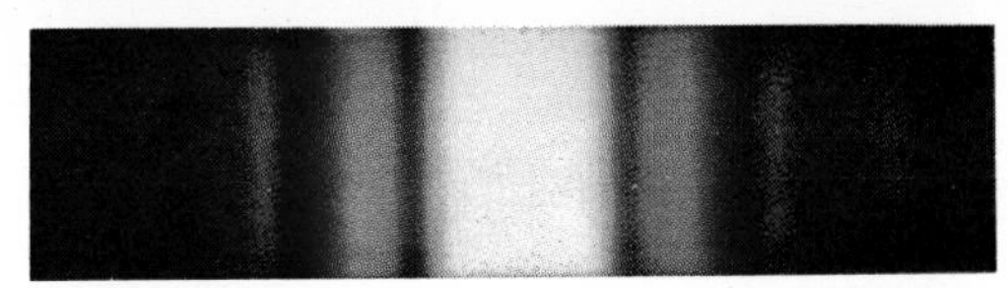

Figure 8–C
White light passing through a single slit produced this pattern.

Figure 8–D
With the same setup as in Fig. 8–C, red light produced this pattern. Both patterns are called diffraction patterns.

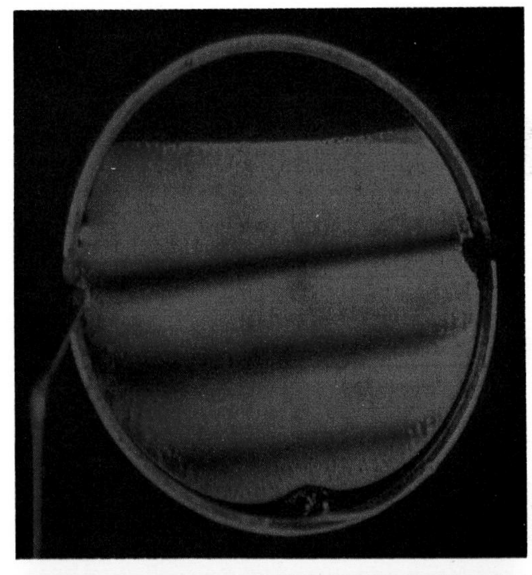

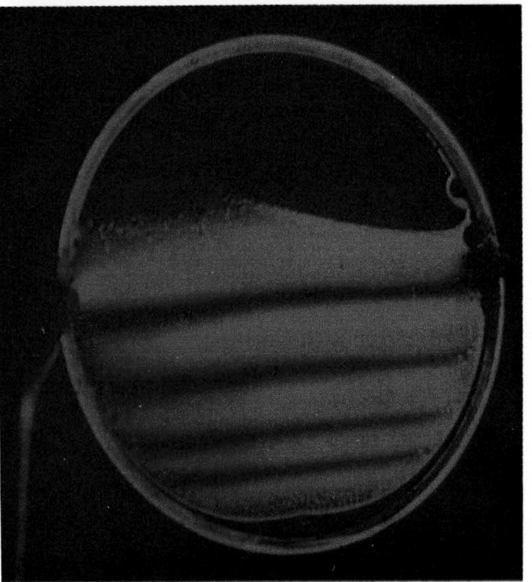

Figure 8–F
Right: interference seen in red light reflected from the same soap film. Top: at almost the same time that Fig. 8–E was photographed. Below: at a later time. Because the film has drained more, the dark region extends farther down at the top.

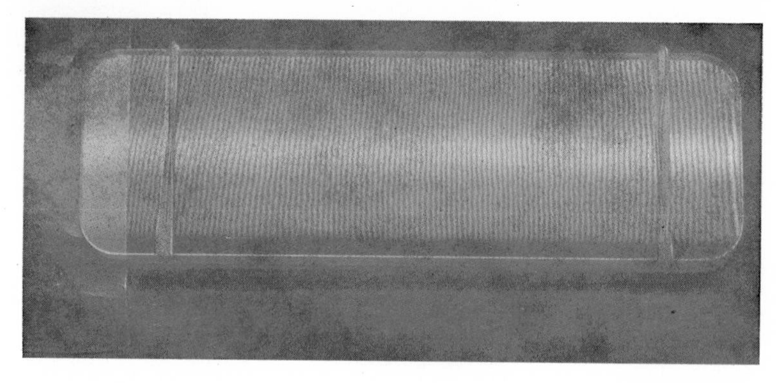

Figure 8–G
An interference pattern produced when light of one spectral color is reflected from the two sides of an air wedge made by separating two pieces of plate glass by a thin sheet inserted at the left end.

How a Colored Diffraction Pattern is Formed by White Light

From top to bottom in Fig. 8-H you see the diffraction patterns of the same slit in red light, green light, blue light, and white light. If you look down the center line, you see that light of every spectral color is at a maximum. Their combination in the pattern at the bottom looks white. Now look down the line AA. The red light is moderately intense; the green almost absent; and the blue completely gone. In the white light pattern we see red. A little bit to the left of the line AA in addition to red, there will be more yellow and green and there you see yellow. To the right of the line the red disappears but the intensity of blue rises. What do you see in the white light pattern?

Try the line BB, a little farther out on the same side of the center. Here the red light is almost absent, the green light is bright and the blue light has almost disappeared. The resulting yellowish color in the white light pattern arises from the yellow-green region of the spectrum. Run down any line of your own choosing and see how the white light pattern is formed.

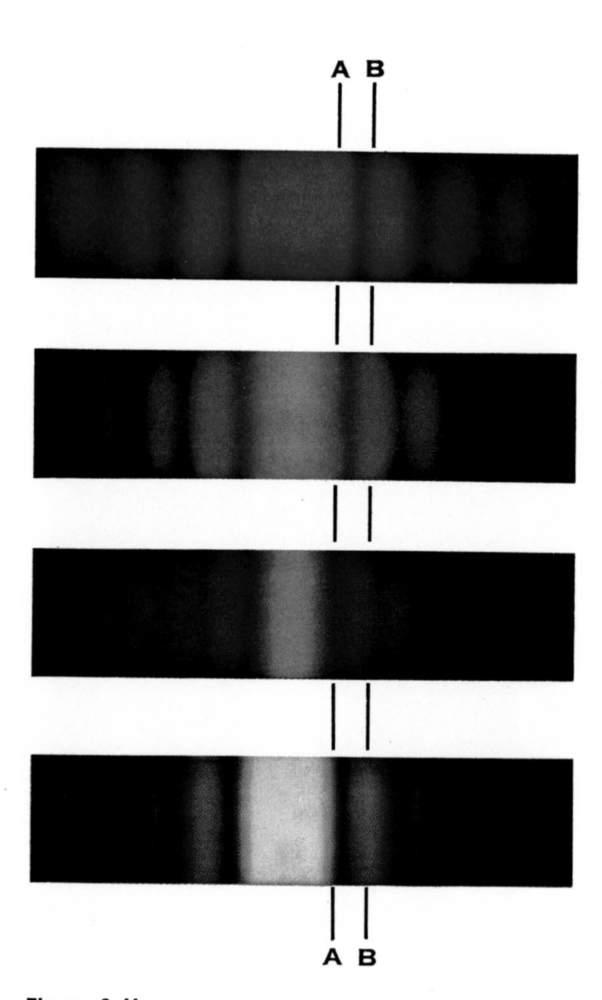

Figure 8–H
Formation of a white-light diffraction pattern from the diffraction patterns of the spectral colors.

shown in Fig. 8–1. This arrangement seems similar to that of a ripple tank in which there are two neighboring sources of circular waves. Consequently, from the results of Sections 7–4 and 7–5, we might expect to see a pattern of bright and dark areas on the screen. The bright areas should be where light waves from the two sources reinforce, and the dark areas should occur where the waves cancel. Just as in the ripple tank, if the two sources emit waves in phase there should be a bright area in the middle of the screen. On each side of this central bright area, there should be dark areas where the first nodal lines meet the screen; and as we go away from the center, bright and dark areas should alternate.

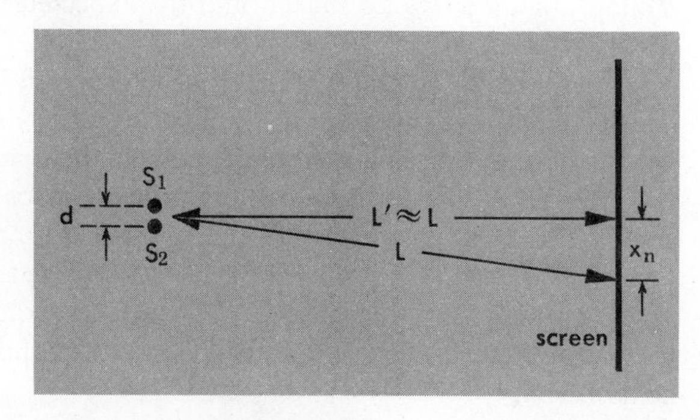

Figure 8–1
This apparatus resembles the setup for producing interference in a ripple tank. But with two separate, ordinary sources of light S_1 and S_2, we cannot find interference. No pattern appears on the screen. Using the special, phase-locked sources of light called lasers, this arrangement can be made to show a pattern (Section 8–3).

In this pattern of alternating bright and dark areas, the center of the *n*th dark area on each side should be a distance $x_n = (n - \frac{1}{2})L\left(\frac{\lambda}{d}\right)$ from the center of the pattern, if x_n is much smaller than L, so that $L' \approx L$. Consequently, the spacing between the centers of neighboring dark areas should be $\Delta x = L\left(\frac{\lambda}{d}\right)$. Also, as in the ripple tank, if one source emits its waves with a phase delay p compared with the other, the pattern should shift off center. It should shift sidewise on the screen by the fraction p of the distance between the dark areas.

Can we observe these interference patterns? We already know that if light is a wave, its wavelength must be much smaller than a tenth of a millimeter. Consequently, $\left(\frac{\lambda}{d}\right)$ for any practical sources will be a very small fraction. However, we can choose a very large distance L to the screen; therefore, it should be possible to see the bright and dark areas.

In reality, when we try the experiment we have just described we do not see an interference pattern. No matter how we vary d or L, the screen is always uniformly illuminated.

Does our failure to observe an interference pattern prove that the wave model of light is a failure? Not necessarily. As we have just mentioned, the interference pattern produced by two sources depends on λ, on d, and also on the phase delay p. If the phase delay between the sources changes rapidly, the nodal lines and the dark areas on the screen must shift position rapidly. Our eyes cannot follow these rapid shifts, so the screen will appear uniformly bright. In other words, the interference patterns may possibly be produced, but we fail to see them because they shift too rapidly.

8-2 Interference of Light Waves: Young's Experiment

We can show that our inability to observe interference in the experiment just described is due to the rapid shifting of phases. To do this, we resort to a trick introduced in 1801 by Thomas Young. He found a simple way to lock together the phases of two light sources so that any interference pattern that is formed will not shift rapidly. The trick is to start with a single light source and split the light from it into two parts in phase with each other. These two parts, which act as if they came from two sources locked in phase, are then allowed to come back together to interfere with each other. With Young's method, we do indeed find the expected interference pattern on a distant screen.

Young's experimental arrangement is shown in Fig. 8–2. He used sunlight passing through a pinhole as a single source of light. The light spreading out from this pinhole source fell upon an opaque barrier which contained two pinholes, placed very close together and located equidistant from the source. Light originating from the pinhole source at any moment passed through the other two pinholes at the same time; the light at these two pinholes was then always in phase. With such an arrangement, the interference pattern of the light emerging from the pair of pinholes did not shift; it could be observed.

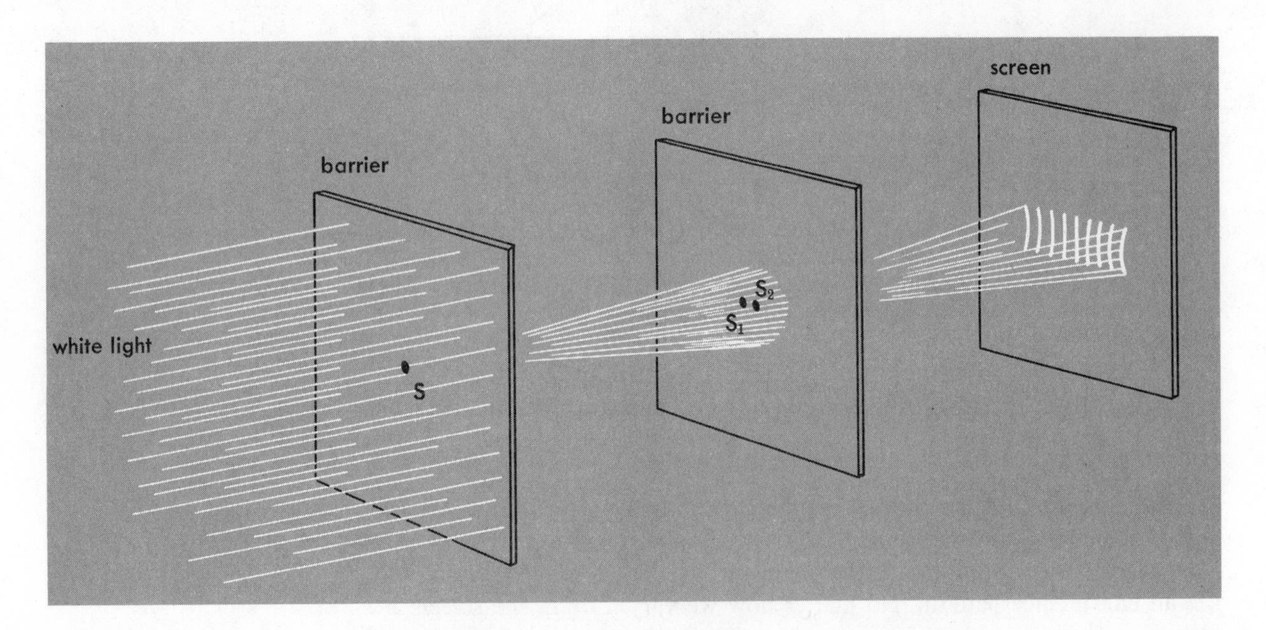

Figure 8–2
Young's experiment. The sunlight reaching the screen from the pinholes S_1 and S_2 all comes through the pinhole S. An interference pattern is visible on the screen.

Today we can do this experiment more easily. We replace the sunlight and pinhole source by an incandescent light bulb having a long, straight filament; and we replace the two pinholes in the opaque barrier by two long, narrow slits. The slits must be very close together (about 1/10 mm apart) and placed with their long dimension parallel to the line filament. Figure 8–3 is a diagram of the wave pattern producing the interference, looking at the line source and slits from above. The similarity to a ripple-tank pattern is very clear. If you hold the page at the level of your eye and sight from the right edge of the figure toward the source, the nodal lines can be seen distinctly. These are areas of no wave disturbance and hence of no light; they have been marked "dark" where they intersect the screen. On the screen, we would see an interference pattern consisting of alternate bright and dark areas of various colors. The bars are parallel to the slits, with a bright one at the center, just as predicted by the wave model.

A permanent record of the interference pattern produced by this system can be made by putting a camera in place of the white screen to record the result on film. Such a picture is reproduced in Fig. 8–A opposite page 146.

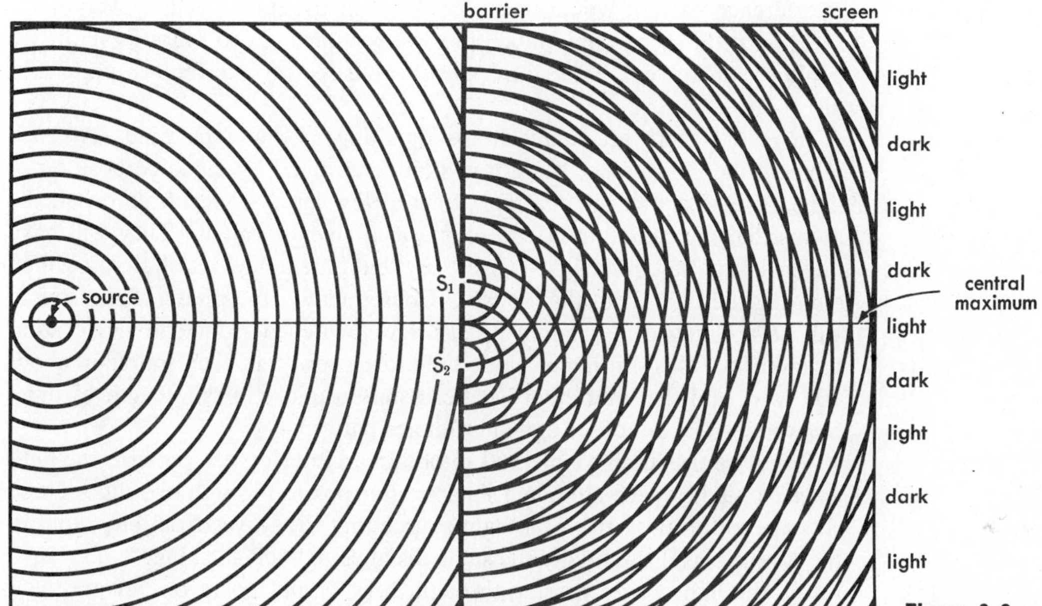

Figure 8–3
Waves from a line source of light passing through the slits S_1 and S_2 interfere to give alternate bright and dark bands on a screen. In this idealized diagram, the slit width is less than the wavelength.

8-3 The Phase of Light Sources: Atoms

We can couple two generators in a ripple tank with any desired phase delay, start them, and let them run as long as we wish. We do not have such control over light sources. That is shown by our inability to observe interference between light waves unless Young's method (or some similar method) is used to lock the phase of the two sources. To understand why the phase delay between two sources usually shifts rapidly, consider two separate sources of light. The light from each source comes from a large number of individual atoms; each atom sends out a burst of light waves only during a very short time. When we turn on the sources we start the overall process of the emission of light, but we do not control the individual atoms; they emit bursts of light waves at random intervals.

To see the significance of this situation for our interference experiments, let us consider a pair of atoms, one in each source. In all probability these atoms will emit light out of phase by some fraction p_1. The light waves from these two atoms will produce an interference pattern that depends on the value of p_1. A short time later these particular atoms will have stopped radiating, so we shift our attention to another pair of atoms which happen to be radiating at that moment. Since we have no way of influencing their behavior, these atoms will also be out of phase, but this time probably by a different fraction p_2. They will, therefore, produce a different interference pattern. Still later we consider a third pair of atoms, and so on. A typical time during which an atom emits light has been found to be about 10^{-9} sec. If we consider only one atom radiating at any moment in each source, we must change atoms every 10^{-9} sec to have the radiation continue. Then the interference pattern would also change in an irregular fashion every 10^{-9} sec. Since this is certainly much too fast for our eyes to follow, even if we could see such weak light sources, we would see no interference pattern at all.

In reality there are many atoms radiating at the same time. At each moment the interference pattern will be determined by the superposition of light waves from all the emitting atoms. Now let us concentrate on all the atoms that are radiating in one source at a given moment. They will have finished radiating in about 10^{-9} sec. By that time a new set of atoms will be emitting the light from that source. Consequently, the phase of the source will have changed at random just as if we switched from one single atom to another. This shifting of phase therefore goes on in each source in about 10^{-9} sec, and the interference pattern shifts rapidly in this length of time. No wonder we cannot see any interference pattern.

When we use Young's method to obtain an interference pattern, the light wave emitted by each atom in a single source passes through two slits. These slits are so far away from the light source that the waves from any atom in the source travel essentially the same distance to each slit (Fig. 8–4). As a result, the light waves passing through the slits and going on beyond them leave the slits with no measurable phase delay. The slits, therefore, act as sources of waves, and these sources are locked in phase. On the screen beyond them we see the interference pattern that we expect.

Figure 8–4
Because the source is far from the slits, the paths from any atom in the source to the two slits are almost parallel and the distances almost equal.

Within the last few years, a family of remarkable new light sources has been developed. In these sources, atoms within solids or gases are made to emit light, not at random but in almost locked phase, one atom tending to control another. The fixed phase condition can be maintained thousands of times longer than the 10^{-9} second during which an individual atom in a hot filament emits light. These new devices are called lasers (for Light Amplification by Stimulated Emission of Radiation).

In an experiment done in 1963, two independent lasers were made to work so well as phase-locked sources that light interference fringes were seen, in exact analogy to the interference pattern we see produced in a ripple tank by two independent wave generators. A technical difficulty prevented this from being an easy experiment, however. Although the phase-lock was good, the wavelengths could not be made precisely equal, which caused the position of the central fringe to drift rapidly across the screen, just as the ripple-tank pattern drifts when the two generators run at slightly different rates. The laser fringes were so fleeting that it was necessary to use high-speed television techniques to make a "snapshot" of the fringes at one position.

8-4 Color and Wavelength of Light

The photograph in Fig. 8–A (opposite page 146) was taken by letting white light from a small source pass through two closely spaced parallel slits; the light spreading out from the two slits was brought back together again by the lens of a camera to form the interference pattern on the photographic film. Notice that the bright area in the center (called the central maximum) is white, but the edges of the bright areas on the sides are colored.

We know that white light is a mixture of all colors, and the colored edges of the areas suggest that different colors are deflected at different angles in the interference pattern. The idea that different colors interfere at different angles implies that the wavelength of light is associated with its color. We already had some reason to suspect an association of color and wavelength when we studied dispersion. Now we can settle the question by further interference experiments. Let us, therefore, look at the interference pattern of light of one color. In Fig. 8–5 we see the interference pattern for red light alone. It is a series of alternating bright and dark areas. The spacing of the areas in the photograph depends on the separation d of the two slits,

Figure 8–5
Interference pattern produced with red light. This is shown in color in the upper part of Fig. 18-B, opposite page 146.

the distance L from the slits to the photographic plate, and the wavelength of the light. Consequently, when we know the distances d and L, the separation of the areas in the photograph tells us the wavelength of the light. As we saw in Section 8–1, the separation Δx of neighboring dark areas is given by

$$\Delta x = L\left(\frac{\lambda}{d}\right).$$

On rearranging this equation we get the wavelength as

$$\lambda = \left(\frac{d}{L}\right) \Delta x.$$

By measuring across a large number of areas we can get an accurate value of Δx to use in computing the wavelength λ. For red light this turns out to be

$$\lambda_{\text{red}} = 6.5 \times 10^{-7}\ \text{m}.$$

We can go through the same procedure with light of any other color. If, for example, we use a blue-violet filter to select the light, we obtain a wavelength

$$\lambda_{\text{b-v}} = 4.5 \times 10^{-7}\ \text{m}.$$

Other typical wavelengths, similarly measured, are given in Table 1. Our early conjecture that color is associated with wavelength is borne out by experiment.

Table 1
Wavelengths of Light in Vacuum

A. The wavelengths in angstrom units of the colors of the spectrum. The visible spectrum ranges from about 4000 A in the deep violet to about 7500 A in the deep red. (1 A = 10^{-10} m)

Color	Wavelength
Violet	$<$ 4500 A
Blue	4500–5000
Green	5000–5700
Yellow	5700–5900
Orange	5900–6100
Red	$>$ 6100

B. The wavelengths of some of the strongest visible lines in the spectra of a few common gases.

Lithium	Neon	Sodium
6103.6	5400.6	5890
6707.8	5832.5	5896
	5852.5	
Mercury	6402.2	**Potassium**
4358.4		4044
5460.7		4047

The colors associated with definite wavelengths are known as spectral colors. They are the same ones seen in the rainbow. Not all colors we see are of this kind. For example, purple is not a spectral color. If we use an interference experiment to analyze the light that comes through a purple filter, we find that it is made of blue light and red light. Color vision is in fact very complicated. We sometimes see what appears to be a spectral color when only light of other wavelengths enters our eyes. The analysis of light by locating maxima and minima in an interference pattern makes it possible to characterize the light accurately in terms of wavelength. Thus, interference patterns extend our knowledge far beyond what we can learn with our unaided eyes.

The experiments we have been describing are ones that you can do yourself. Indeed, you can check up on our evaluation of the wavelengths of red and blue-violet light just by making some measurements of Fig. 8–B. This photograph was taken with a single source of white light and two slits. We made the different patterns that you see, one above the other, by placing a red-colored filter above a blue-violet-colored filter in the path of the light between the source and the photographic film. The ratio of the spacing of the areas in the two halves of the picture should correspond to the ratio of the wavelengths of red and of blue-violet light. By measuring the photograph we find that the separation of the areas in the red pattern is greater than that in the blue-violet by a factor of about 1.4. We can compare this to the ratio of the wavelengths measured independently. That ratio is

$$\frac{\lambda_{\text{red}}}{\lambda_{\text{b-v}}} = \frac{6.5 \times 10^{-7}\text{ m}}{4.5 \times 10^{-7}\text{ m}} \approx 1.4.$$

We have now succeeded in showing that the spectral colors of light are related directly to its wavelength in vacuum. We have measured the wavelength of light of one color by measuring the spacing of the areas in the interference pattern from two slits. We therefore have a primitive spectroscope whose operation we understand.

A much more efficient spectroscope that sorts out the colors of light by interference can be made by using many equally spaced slits. The effect of the many slits is to intensify the amount of light emerging from the slits in any direction and to make the maxima for any wavelength sharper. These are known as grating spectroscopes and are commonly used to analyze light into its component wavelengths. For some purposes these spectroscopes are superior to those that use a prism to spread the light out into a spectrum.

8-5 Diffraction: An Interference Effect in Single Slits?

If you scratch only one slit with a razor blade (or a needle to make a slightly wider opening) and look through it at a white light source, you will see something like the photograph in Fig. 8–C. There is a bright, broad central region bordered by colored areas of lower intensity. In light of one color the pattern is similar. Figure 8–D shows what you find if you use a red light source—a bright center, then dark regions alternating with progressively less intense areas of light.

Although the pattern of light and dark differs from Young's pattern, it looks suspiciously like an interference pattern. There are nodes and maxima in light of one color, and color effects when white light is employed. Because this interference effect takes place with only one slit, it may seem to make our interpretation of Young's pattern questionable. To clarify the situation we shall study more carefully the behavior of light passing through a single slit.

If we look back at our earlier discussion of the diffraction of waves by a slit (Section 6-7), we notice that the single slit does not act like a point source of waves unless its width is less than a wavelength. Normally, the slits we use in observing diffraction or interference in light are much wider than the wavelength of light. (Both the narrower slits we scratch with a razor blade and the wider ones made with a needle are many wavelengths wide.) Perhaps they are 10^{-1} mm wide while the wavelength of light is between 4 and 7×10^{-4} mm. The explanation of the interference patterns we see in light passing through single slits must be found by carefully examining the behavior of light as it passes through a narrow slit which is, nevertheless, many wavelengths wide.

In Fig. 6–20 we saw how straight waves traveling through a narrow slit in a ripple tank spread out or diffract. The waves on the far side of the barrier arise from the propagation of the parts of the crests and troughs that enter the slit. Instead of producing the crests and troughs in the slit by letting a straight wave fall on it, it seems possible that we might produce them by a line of point sources moving up and down together at the position of the slit. As a test of this idea let us replace the slit by a line of very small point sources, separated by a small fraction of a wavelength. We shall use enough sources so that they just fill the space of the original slit. If we now run the sources in phase and at the same frequency as that of the original wave, we see (Fig. 8–6) that the pattern produced does look like the diffraction pattern of the slit.

It seems reasonable that the wave patterns formed when straight waves are diffracted by a slit are the same as the patterns formed by the waves from a large number of point sources evenly spread through the width of the slit. We shall now assume this equivalence, and try to explain the interference patterns we observe through single slits by superposing the waves from the effective point sources in the slit. This general procedure of predicting the further propagation of waves by replacing wave fronts with sources was applied by Huygens to a number of problems including reflection and refraction.

8-6 A Theory of Diffraction by a Slit

In our study of the interference pattern from a line of close-spaced point sources occupying the width of a slit, we shall consider the pattern only at distances very large compared to the slit width. Waves reaching a distant point have traveled away from the sources along almost parallel paths (Fig. 8–7). We start with the total wave produced by the sources at a distant point directly in front of the slit. The path lengths from all the sources to that point are almost equal, and crests from all the sources arrive there together. This reinforcement of all the individual waves means that strong waves go out from the slit along the center line.

Next we examine a point P some distance off the center line so that the angle θ between BP and the center line in Fig. 8–7 is no longer zero. (Point B is at the center of the slit.) Because PB is essentially parallel to PA, the angle between PA and the center line is also θ. But PA is perpendicular to the wave front; the center line is perpendicular to the line of sources; hence, the angle between the wave front and the line of sources is θ, also. The path lengths from the various sources to point P are no longer equal. In particular, PC is shorter than PA. The individual crests do not arrive at P at the same time, and the total wave is weaker than it is along the center line.

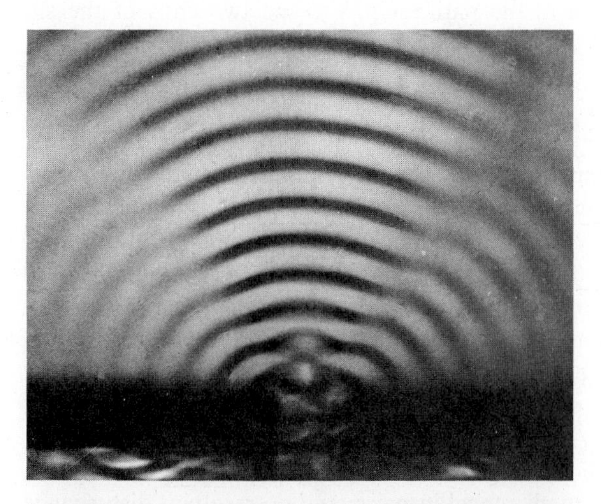

Figure 8–6
A diffraction pattern of straight waves passing through a slit and, below, an interference pattern of a line of equally spaced point sources extending across the slit. Near the sources the effect of source separation leads to some difference in the patterns. Far away the two patterns are the same.

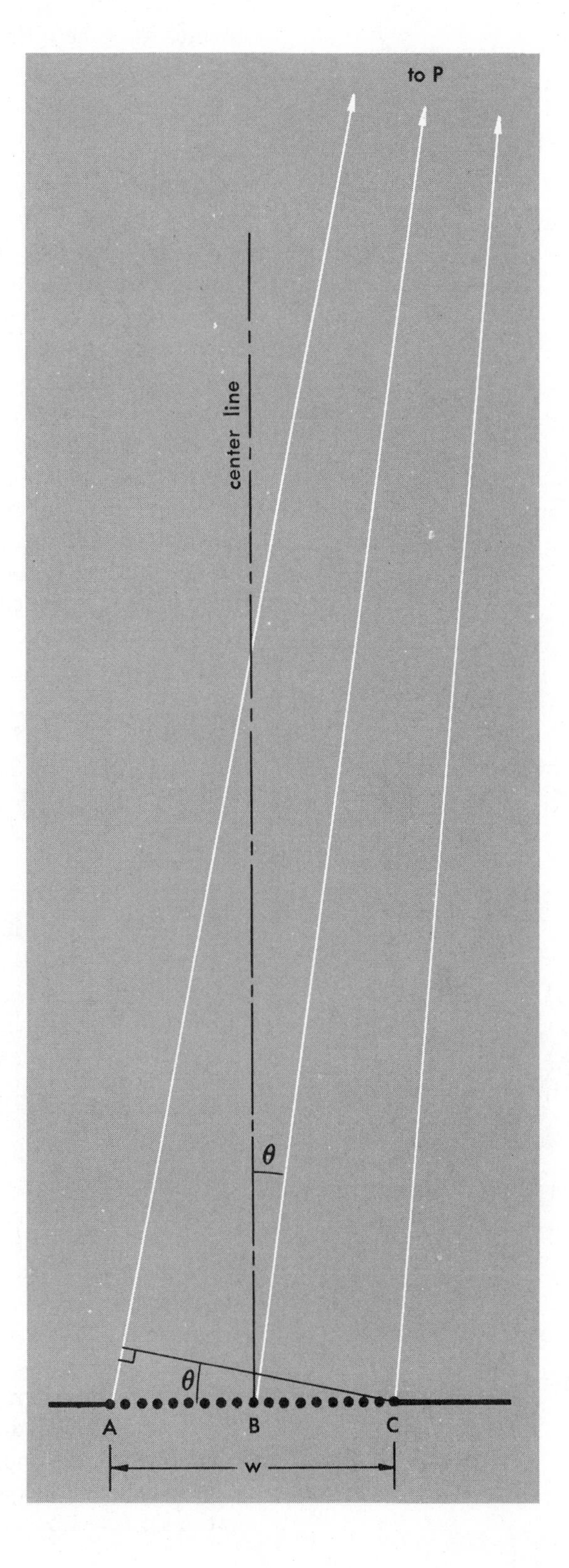

Figure 8–7
Light rays from the various sources to a distant point *P* (off the top of the page) are nearly parallel. However, if the distant point is not on the center line, the paths are of different lengths.

We now move P farther away from the center line so that the angle increases until the difference in the path lengths to the two ends of the slit becomes

$$PA - PC = \lambda.$$

At this point $\sin\theta = \lambda/w$ as we see in Fig. 8–8. Also, as the figure shows, $PB - PC = \lambda/2$. We can now show that the waves from the individual sources reaching any faraway point P at this angle cancel each other. Thus the total intensity far away in this direction is zero. To understand this cancellation we match the sources in pairs which cancel. Consider the sources just to the left of C and of B. They give crests at P which come ½ wavelength apart. In other words, at P a trough from one of these sources arrives with a crest from the other, and they cancel. Moving to the next pair of sources to the left, the second points to the left of C and B respectively, the same thing happens. Again we get the same result for the third pair, the fourth pair, and for all succeeding pairs. We have thus superposed the effects of all the sources; they all cancel in pairs; and no resultant disturbance takes place at P. It does not matter in what order we add up the effects of all the sources; we must always get the same result. The trick of adding them in pairs is just an easy way of seeing what the result is. Thus we have shown that there is complete cancellation of waves at the angle θ given by $\sin\theta = \lambda/w$.

FIG 8.8

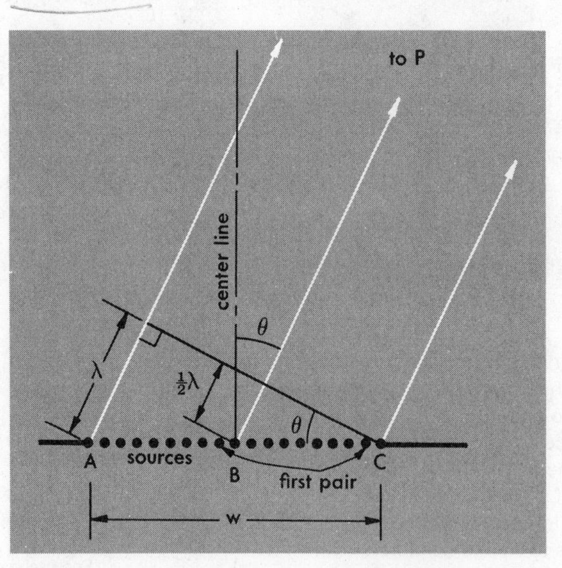

Figure 8–8
When $\sin\theta = \dfrac{\lambda}{w}$, waves from the sources cancel.

As the angle θ becomes larger the cancellation is no longer perfect, and the intensity rises. It goes through a maximum and then falls to zero again when $\sin\theta = 2\lambda/w$. We can understand the cancellation at this angle by reference to Fig. 8–9. All the sources in interval 1 on the figure can be paired with those in interval 2 so that all pairs cancel; and the sources in interval 3 can be paired with those in interval 4 to produce complete cancellation. The intensity in this direction is therefore zero.

It is natural to expect a maximum intensity approximately halfway between $\sin\theta = \lambda/w$ and $\sin\theta = 2\,\lambda/w$, near $\sin\theta = \frac{3}{2}\,\lambda/w$. At this angle

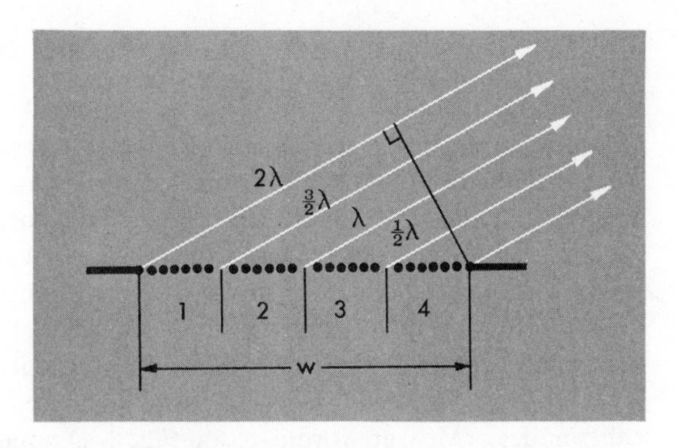

Figure 8–9
When $\sin \theta = \frac{2\lambda}{w}$, there is cancellation.

(Fig. 8–10) we can divide the sources into three intervals with path differences of ½ λ between their ends. Pairing the sources in the two intervals on the right we find that they all cancel out as before. Only the third interval is left over. The effects of the sources at each end of the third interval cancel. But there are no other sources which can be paired off this way. Consequently, there will be at least partial reinforcement at $\sin \theta = \frac{3}{2} \lambda/w$; but the intensity is less than that which would be observed along the center line from one third of the sources in the slit. The light intensity in this maximum is therefore considerably less than it is in the maximum along the center line where all the sources in the slit contribute to a complete reinforcement.

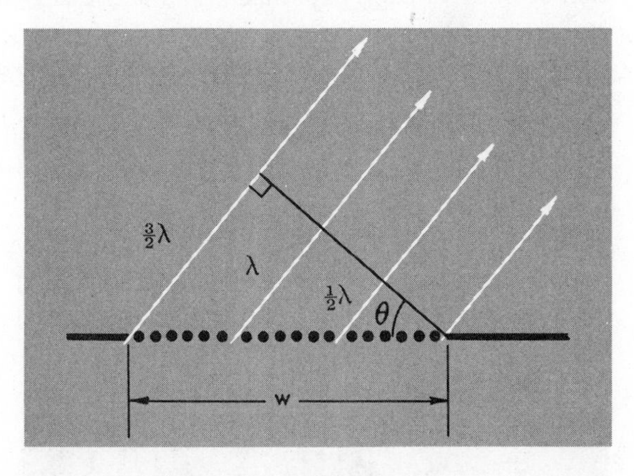

Figure 8–10
When $\sin \theta = \frac{3\lambda}{2w}$, partial reinforcement occurs.

Beyond the node at $\sin \theta = 2 \lambda/w$, as $\sin \theta$ increases, the intensity rises to a still weaker maximum near $\frac{5}{2} \lambda/w$, and falls to zero again at $3 \lambda/w$. It continues its rise and fall as θ increases, while the maxima get progressively weaker, as shown in Fig. 8–11.

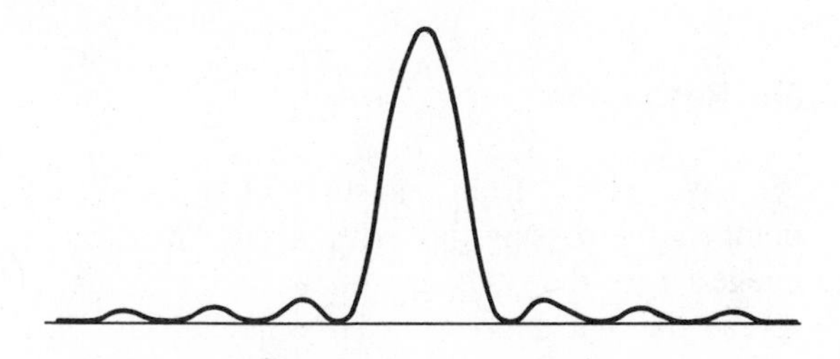

Figure 8–11
A single-slit diffraction pattern for light of one definite wavelength. The intensity of the light is plotted vertically as a function of the distance from the center line.

We now have a theory which connects the appearance of the diffraction pattern of a slit with the width of the slit and the wavelength of light. Does the theory account for actual observations? To be sure, we must construct slits of accurately known width and measure the distances from the center line to the dark regions which occur in the diffraction pattern of light of a single color. When we make such measurements, we find that the distance between the nodes at the sides of the pattern is constant. Also, as expected, the distance across the central bright band is twice the distance between the other nodes. (Measure it yourself on Fig. 8–D opposite page 146.) Furthermore, the intensities of the maxima decrease with the distance from the center, as predicted by the theory. We therefore have reason to believe that the theory is good.

Finally, we can find the wavelength of the light by measuring the distance between the center of the pattern and the first node. Using

$$x/L = \sin\theta = \lambda/w,$$

we find values of λ for light of various colors. These values are the same as those found by using two slits or a many-slit spectroscope. Our interpretation of both Young's experiment and single-slit diffraction is probably correct.

We can now see why we did not need to worry about the diffraction pattern of single slits when we discussed Young's experiment. We performed Young's experiment with very narrow slits (cut with two razor blades). For such slits the central bright area of the diffraction pattern of each slit covers a rather large angle. Far enough away from the slits the central bright regions will largely overlap, and in this overlapping region we see the simple double-slit pattern. The other maxima of the single-slit diffraction pattern occur far out to the sides, and they are too weak to be seen easily. For this, the slits have to be wider. (See Fig. 8–12.)

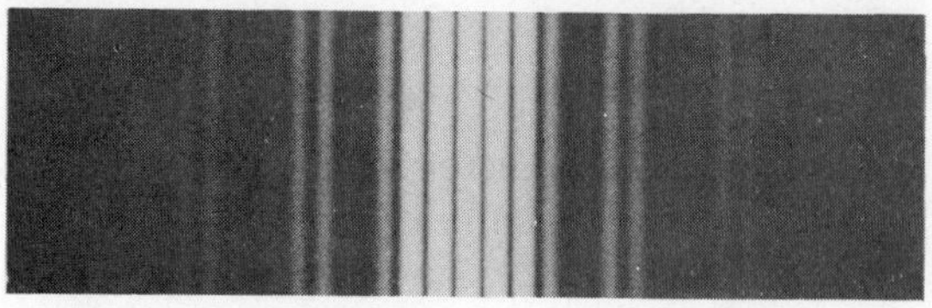

Figure 8–12
When two identical, fairly wide slits are used to make an interference pattern, and when they are separated by a distance comparable to their width, the resulting pattern combines the features of a single-slit diffraction pattern and the two-slit interference pattern. (Courtesy: Bruno Rossi, "Optics," Addison-Wesley Publishing Co., 1957.)

8-8 Resolution

We have learned from our study of single-slit diffraction that light from a point source passing through a small slit or pinhole spreads out to give an image larger than the size of the hole. We also learned that small holes spread the light more than large holes. The diffraction caused by a small hole is of great importance in designing microscopes and telescopes. It determines the ultimate limit of their magnification.

To see why diffraction limits magnification, let us consider what happens when two point sources, close together, send light through a pinhole onto a screen. In Fig. 8–13 (a), I_1 is the image of source S_1, and I_2 is the image of source S_2. These images are, in fact, diffraction patterns produced by the tiny pinhole, and they are large and fuzzy. In Fig. 8–13 (b) everything is the same but the hole is smaller. This spreads the light still more, and the images are now so large that they overlap. It is difficult to decide, from looking at the screen, whether the pattern is that of two separate sources or a single odd-shaped source. When the hole is so small, or the sources are so close that the images cannot be distinguished, we say the sources are unresolved. When we can separate them, we say they are resolved. The *resolution* of an optical instrument is a measure of its ability to give separated images of objects that are close together.

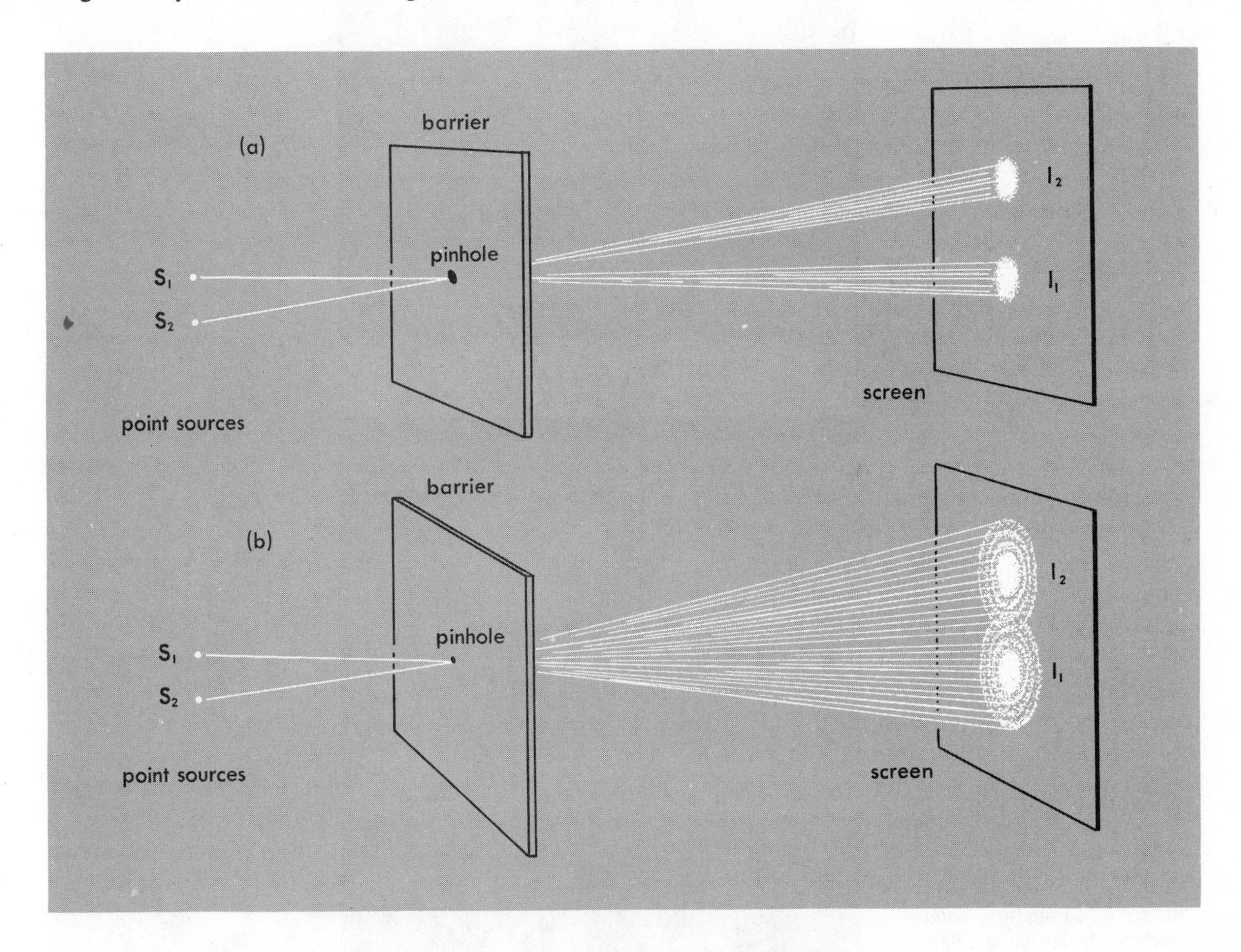

Figure 8–13
(a) When light from two sources S_1 and S_2 passes through a small hole, the images are fuzzy but resolved. (b) When the hole is very small, the images overlap and are unresolved. (See Fig. 8–14.)

If we substitute a lens for the pinhole, we can focus the light from two point sources and produce what appear to be sharp images. Careful examination of these images, however, shows that a lens cannot eliminate the spreading of light by diffraction because light in passing through a lens is passing through a hole of limited size. Figure 8–14 shows a series of photographs of three point sources seen through lenses of the same focal length but different diameter. In the top picture the light passes through a small-diameter lens; in the second picture it passes through a larger lens; and in the last, through a very large lens. In the top picture, the images are unresolved. As the lens becomes larger, diffraction decreases and resolution improves, until with a large lens the images are so clearly separated that even if the sources were closer together, we could still resolve them.

In a microscope, where the light passes through a small objective lens, we expect that diffraction will be important. We can usefully increase the magnification of a microscope until diffraction stops us from resolving nearby

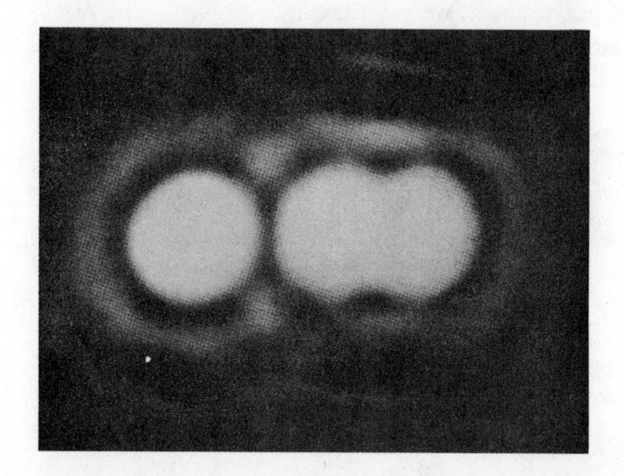

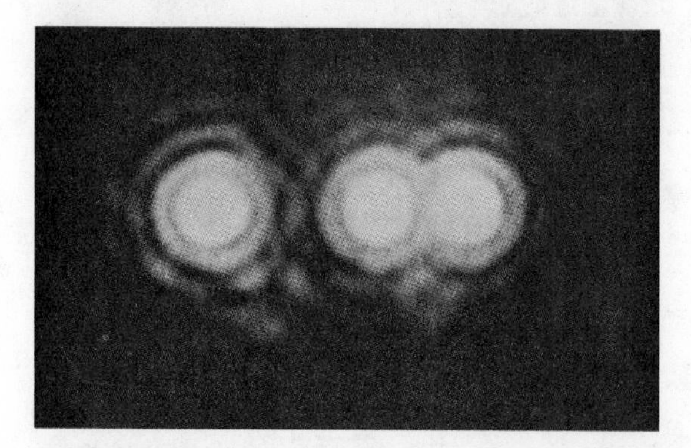

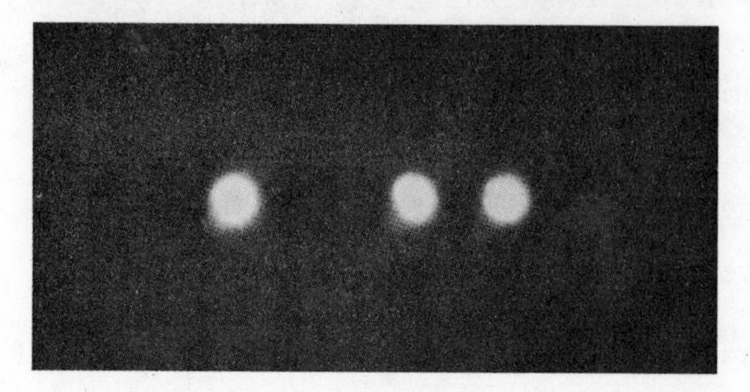

Figure 8–14
Diffraction patterns from three point sources produced by lenses of different diameters. As the diameter is increased (top to bottom), the resolution improves. (Courtesy: Francis Sears, "Principles of Physics, III," Addison-Wesley Publishing Co.)

objects. A further increase in magnification makes everything look larger, but it does not enable us to see objects which are closer together. Increasing the magnifying power of microscopes even with the best of lenses is futile beyond the point where detail fails to be resolved because of diffraction.

Telescopes also are subject to diffraction. The greater the size of the objective, whether mirror or lens, the better is the resolution. The large Hale telescope on Mt. Palomar gives sufficiently good resolution to resolve individual stars in the Andromeda Nebula, a feat that smaller telescopes are unable to accomplish.

8-9 Interference in Thin Films

Of the many interference patterns we can see, among the most common are those we observe when light is incident on a thin soap film or an oil film on a puddle of water. The result is a striking colored interference pattern (Fig. 8–E opposite page 146).

To study this effect we shall examine the pattern obtained with a single color (Fig. 8–F opposite page 146). In the reflected light from the film there is a wide dark region at the top. In this region the soap film is very thin, because the water has drained down toward the bottom. If we look farther down the film, the thickness increases, until finally we come to a thickness where reflected light is clearly visible. Still lower down, at greater thickness, a dark area appears in the photograph; the reflection is absent. Then continuing farther, bright and dark areas succeed one another as the thickness gets greater.

That the observation of a bright or a dark area really depends on the thickness of the film is borne out by watching the areas while a newly formed film is draining. When the top has drained to a certain thickness, a bright area forms there. As the draining continues, the areas moves down the film, and a dark area forms in the thinning region above. Area after area moves down the film, each staying with its appropriate thickness. Finally a dark region spreads down from the top when the film is very thin.

How can we explain these alternating bright and dark areas? When light passes through a sheet of refractive material, some of the light is reflected at the first surface it meets, and some is reflected at the second surface (Fig. 3–7). A small part of the light that has been reflected at the bottom of the glass is reflected again at the top instead of passing out. Some light may be partially reflected several times inside the glass, but for a weakly reflecting material like glass or a soap film, only the reflection as the light enters and the first reflection inside are usually appreciable. It is the interference between these two reflected light waves that produces the bright and dark areas in the reflected light of a single color and the color effects with white light.

To understand these interference patterns we shall start with a very thin film like one at the top of the soap film shown in Fig. 8–F opposite page 146. Let us assume that the film in this region is very thin compared to the wavelength of visible light. Our thin film is illustrated in Fig. 8–15. Because its thickness is much less than the wavelength of the light, the path difference between the two reflected crests is very small compared to a wavelength, and we might expect that the two reflected waves add together so that a reflected crest in one and a reflected crest in the other are superposed almost in phase. The reflected light should then be appreciable. Instead we see that no reflection occurs. The whole top of the soap film is dark. Apparently the waves in the two reflected beams cancel each other instead of reinforcing.

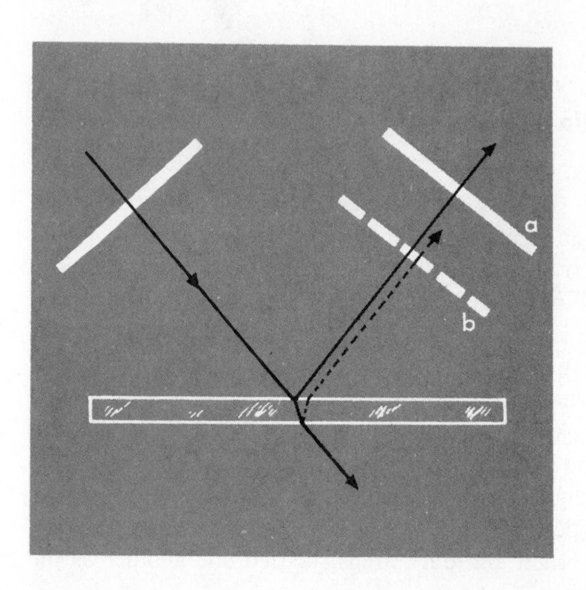

Figure 8–15
A crest of a light wave striking a sheet of transparent material is partially reflected and refracted. (The lines marked by arrows are normal to the crests and show the path of the light.) The crest b in the part of the light that has been reflected from the bottom surface lags behind the crest a in the light reflected from the top surface. For a given angle of incidence the distance it lags behind depends on the thickness of the transparent sheet and the speed of the wave in the transparent material.

This cancellation may seem puzzling; but when we recall the behavior of pulses reflected on coil springs, it is less surprising. When a pulse traveling along a spring is reflected from a heavier spring in which the speed of waves is smaller, the pulse turns upside down. When a pulse is reflected from a lighter spring where the speed is greater, it stays right side up (Section 5–4). Here with light we have an analogous situation. The light reflected from the outside of a soap film is reflected from a boundary with a more refractive material in which the wave moves more slowly. The light beam reflected inside the thin film is reflected from a boundary with air, a less refractive medium in which the wave moves faster. Perhaps we can picture the more refractive medium as acting like a heavy spring. Then light waves reflected from the top of the film in Fig. 8–15 should turn upside down. An incident crest turns into a reflected trough. On the other hand, an incident crest reflected inside from the bottom of the film will be reflected as a crest. Because one wave is turned upside down and the other is not, the two waves cancel and we see no reflected light.

Let us now consider what happens as the soap film becomes thicker. The light wave which penetrates the top surface has a longer path than the light wave reflected from the top surface. In the light reflected from the bottom, therefore, the crests fall farther and farther behind as the thickness increases. Where the path difference becomes one half of a wavelength (the wavelength in the soap film), the crests have dropped behind by a half wavelength. Instead of coinciding with troughs in the other reflected wave as they do for a very thin film, they have now been delayed until they coincide with crests. The crests in the two beams add together, and the total reflected light intensity is at a maximum. If the incident light waves are traveling perpendicular to the film, the path difference is $\frac{1}{2}\lambda$ when the thickness of the film is exactly $\frac{1}{4}\lambda$ thick, and the two reflected waves reinforce. In Fig. 8–15 the incident light wave is not perpendicular so that if the thickness of the film is $\frac{1}{4}\lambda$ the reflected light will be close to maximum intensity.

As the thickness of the soap film increases still further, the crests in the internally reflected wave fall still farther behind. When the length of the path back and forth inside the film is a full wavelength the crests in the internally reflected wave have fallen behind by a whole wavelength. They now coincide with troughs in the other reflected wave, just as they did when the film was extremely thin. The crests and troughs in the two reflected waves cancel again, and we see a dark bar in the soap film in this region.

As the thickness continues to increase every quarter wavelength, we shift from a region in which there is no reflection to a region of maximum reflection, or from a region of maximum reflection to a region of no reflection again. There is no reflection from the film when its thickness is $\frac{1}{2}\lambda$, λ, $\frac{3}{2}\lambda$, etc. Maximum reflection occurs when the film is $\frac{1}{4}\lambda$ thick, $\frac{3}{4}\lambda$, $\frac{5}{4}\lambda$, and so on. It is for this reason that the bright and dark areas seen in a thin wedge are equally spaced. The thickness of the wedge increases at a constant rate, so the areas are equally spaced.

Thin wedges, such as air between glass plates, also give interference patterns. Like soap-film patterns they consist of parallel light and dark areas but the areas are equally spaced (Fig. 8–G opposite page 147), because the difference in path length between the plates varies uniformly with the distance along the plates. Furthermore, if the angle between the plates is reduced the separation of the areas is increased.

8-10 Interference in Light Transmitted Through Thin Films

When the intensity of reflected light changes as a function of thickness, what happens to the light transmitted by the film? When there is no reflected light we expect that the intensity of the transmitted light will be the same as that of the incident light. Experiments agree with our expectation. At thickness $\frac{1}{2}\lambda$, λ, $\frac{3}{2}\lambda$, etc., all the light is transmitted. Furthermore, at those thicknesses for which reflection takes place, the intensity of the transmitted light is less by just the intensity of the reflected light.

We can understand in terms of interference why the transmitted light decreases when the reflected light increases. Like the reflected light, the transmitted light is made of two interfering waves. They are the wave that passes straight through the film without any reflections and a wave which is internally reflected twice (Fig. 8–16). Other internal reflections give waves too weak to be important.

We now examine how the two waves *a* and *b* in Fig. 8–16 superpose for two different thicknesses of the film. Let us start with a film ½λ thick for which there is no reflection. In this case the twice-reflected wave *b* is delayed by a whole wavelength with respect to the unreflected wave *a* because of its extra journey back and forth across the film. A crest in the twice-reflected wave, therefore, comes out of the bottom of the film along with a crest in the unreflected wave. The two crests add together to produce crests in the transmitted light which are larger than those in the unreflected beam.

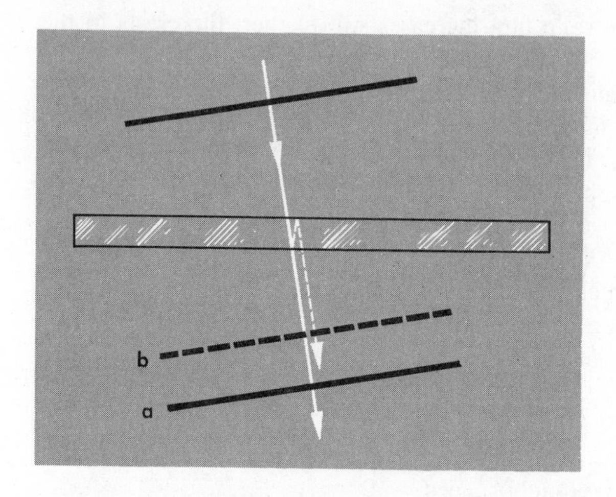

Figure 8–16
The crest *b* in the part of the light that is internally reflected twice lags behind the crest *a* in the wave that passes directly through the film.

On the other hand, when there is a maximum reflection, at the thickness ¼λ for example, the twice-reflected wave *b* travels an extra distance ½λ compared to the unreflected beam *a*. The crest in *b* therefore coincides with a trough in *a* and the superposition of *a* and *b* results in smaller troughs and smaller crests in the transmitted light. We thus see that when there is no reflected light from the film, the transmitted beam is more intense; and when there is maximum reflection from the film, the transmitted light is less intense. If we work through the details quantitatively, we find that the sum of the intensity transmitted and the intensity reflected is, indeed, equal to the intensity of the incident light.

8-11 Color Effects in Interference

In explaining interference patterns we have concentrated on those produced by light of a single spectral color. We can now go back to the interference patterns of white light, shown opposite pages 146 and 147 and explain the many bright colors in those patterns. We consider the interference pattern of each spectral color separately; then, by noting which spectral colors are present and which are absent at a given place, we can predict the color that we shall see at that place in the white light pattern. For example, if at one place blue light is absent but red light is intense, we shall see a color belonging to the red end of the spectrum—yellow or red. If at another place only blue and red light are present, we shall see purple. The procedure is illustrated in Fig. 8–H and in the accompanying text opposite page 147. There it is applied to the white light diffraction pattern of a single slit. You can apply the same procedure to Young's interference pattern, to thin film interference patterns, and in fact to any other interference pattern.

8-12 The Wavelength of Light and Scaling

Our study of light has shown that as long as we deal with openings larger than a few hundredths of a centimeter or films thicker than this, size has hardly any effect on the behavior of light. But when we deal with very small openings (Fig. 3–24) or very thin layers (Fig. 3–25) the behavior of light depends on the size. An enlarged photograph made by light passing through a small pinhole is not the same as a picture of light passing through a larger hole. This is evidence that light has some characteristic length of its own. Only through the wave model of light can we find what this length is. It is the wavelength of the light. Whenever the dimensions of openings, lenses, etc., are large compared to the wavelength of light, the wave nature of light is almost undetectable and the sizes of the openings, etc., are not important. But scaling them down to a size comparable to the wavelength of light affects the behavior of light very drastically. However, the behavior of light does remain unchanged when we also scale down the wavelength of the light by the same factor as we scale down the size. For example, we can reduce the width w of a slit and get the same diffraction pattern if we also reduce the wavelength of the light by the same factor so that λ/w remains unchanged.

The wavelength of light is not the only characteristic length in physics which is not immediately apparent. We shall see other examples where the existence of a characteristic length has an effect on scaling up or down in size.

For Home, Desk, and Lab

1.* In a ripple tank the size of a lake, with long waves on it, how would you search for nodal lines with a boat in the dark? (Section 1.)

2. Why can't we see interference from two incandescent filaments?

3. In an interference pattern produced by white light passing through two narrow slits, the distance between the black bars is 0.32 cm. The distance between slits is about 0.02 cm, and the distance to the screen on which the bars are observed is 130 cm. Find the average wavelength of white light.

4. A source of red light produces interference through two narrow slits spaced a distance $d = 0.01$ cm apart. At what distance from the slits should we place a screen so that the first few interference bars are spaced one centimeter apart? What will be the spacing of the bars if we then use violet light?

5. When a source of light of wavelength λ is used in a two-slit experiment with narrow slits separated a distance d, at what angles do you expect to find the *maxima* in the light intensity in the interference pattern?

6.† A double-slit source of constant phase delay can be produced, using the direct light from a single slit and the reflected light from a glass plate, as shown in Fig. 8–17. The interference fringes can be observed by means of a simple magnifier focused on the edge of the glass plate. When observed in this manner, the dark lines are found to be equally spaced, starting at the reflecting surface of the plate.
   (a) What does this indicate about the phase delay between the two sources?
   (b) If the wavelength is 5400 angstroms and the black bars are 0.9 mm apart, how far is the source slit above the plane of the surface of the glass plate when the slit is 60 cm from the edge of the glass plate?

7.* How far would light travel in air in 3×10^{-9} sec? (Section 3.)

8. Calculate the period of yellow light. About how many wavelengths are included in a light wave during emission of a light burst by a single atom?

9.* A double-slit interference pattern was produced with the yellow light of sodium, and a spacing of 1.0 mm between nodes was obtained. What would be the spacing if violet light of potassium were used in the same system? (Section 4.)

10.* What is the ratio of the spacing of the nodes in Fig. 8–H (facing page 147) between red and blue light? (Section 5.)

11. Suggest an optical method for measuring the width of a narrow slit.

12.* Why is the partial reinforcement referred to in Fig. 8–10 for a single-slit interference pattern so much less than the central maximum for the same slit? (Section 6.)

13. (a) When yellow light passes through a slit 1 mm wide, at what angles are the first three nodes in the diffraction pattern?
   (b) If the slit is 10 times as wide?
   (c) If it is 1/10 as wide?

14. In Fig. 8–7, how will the diffraction pattern be affected
   (a) if you cover the right-hand half of the sources?
   (b) if you cover the extreme left quarter and the extreme right quarter?

15. When a source of wavelength λ is used with three slits, each separated from its neighbor by the distance d, show that you get maximum intensity at the same angles as for two slits only.

16. Two 3-inch loudspeakers emit a steady pitch whose frequency is 1000 vibrations per second. These sources are in phase and are 2 meters apart.
   (a) At what angles would you expect to hear no sound? (The speed of sound is about 300 m/sec)
   (b) What do you think would happen if you tried this experiment in a room with hard-surfaced walls?

17. In Fig. 8–12, identify the dark regions arising from the diffraction pattern of each slit and the dark bars arising from interference between the slits.

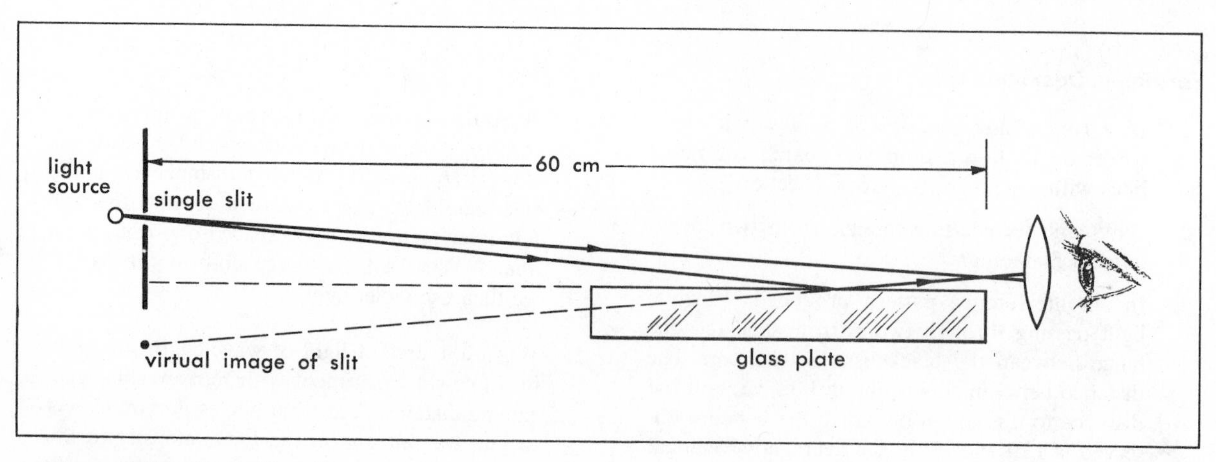

Figure 8–17 For Problem 6.

18.* How does the magnification in the top picture in Fig. 8–14 compare with that in the bottom picture? (Section 8.)

19. Microscopes in which the object is illuminated by ultraviolet light can give higher magnifications than microscopes that use visible light.
    (a) How do you explain this?
    (b) How are the images seen if no visible light is used?
    (c) Since glass is opaque to ultraviolet light, how can such a microscope be made?

20. Two images can just be resolved when the central maximum of one falls on the first node of the other.
    (a) Show that the resolution of a narrow slit depends upon λ/w where w is the slit width.
    (b) About how close together can two line sources be placed and still be resolved if they are viewed through a 0.01-cm slit (about the smallest size you can make easily) 3 meters away from the sources?

21. Stars are often photographed through a blue filter. What is the advantage of this?

22.* What would happen to the distance between the lines in Fig. 8–G (facing page 147) if red light were used instead of green? (Section 9.)

23.* What is the wavelength in water of green light whose wavelength in vacuum is 5600 A? (The speed of light in water is ¾ c.) (Section 9.)

24.* Several different materials are used to make $\lambda/4$ thick films for red light at 6100 A. Will these films all have the same thickness? (Section 9.)

25.* In Fig. 8–G (facing page 147), if at the left-hand side you put 2 sheets between the glass plates instead of one, how will it affect the pattern? (Section 9.)

26. From the photograph of a soap film at the top in Fig. 8–F, plot a graph of film thickness against vertical distance down the film.

27. Two pieces of plate glass 10 cm long make an air wedge like that shown in Fig. 8–G. The plates are separated at one end by a human hair with a diameter of 0.09 mm. The reflected interference pattern is observed by looking in a direction perpendicular to the surfaces of the plates.
    (a) What is the spacing of the bars if the incident light is blue?
    (b) How many bright bars are seen per centimeter if the incident light is red?
    (c) Is the light reflected from the end where the plates are in contact a maximum or a minimum?
    (d) Can you tell from this experiment which reflected waves are turned upside down?

28. Suppose that in Fig. 8–16 the thickness of the film is ½λ.
    (a) What is the path difference between rays a and b?
    (b) How many times are the waves inverted?
    (c) What is the overall phase delay between rays a and b?

29. Lenses are often coated with a thin film to reduce the intensity of reflected light.
    (a) If the index of refraction of the coating is 1.3, what is the smallest thickness that will give a minimum reflection of yellow light?
    (b) Such lenses often show a faint purple color by reflected light. Why?

Further Reading

These suggestions are not exhaustive but are limited to works that have been found especially useful and at the same time generally available.

FINK, DONALD G., and LUTYENS, DAVID M., *The Physics of Television.* Doubleday Anchor, 1960: Science Study Series. A careful description of physical and psychological color and its relation to the production of color television. (Chapter 6)

MAGIE, WILLIAM FRANCIS, *A Source Book in Physics.* McGraw-Hill, 1935. (See especially Thomas Young.)

MURCHIE, GUY, *Music of the Spheres.* Houghton Mifflin, 1960. While music, light, and color are emphasized, a much broader area of wave phenomena is covered. (Chapters 11 and 12)

VAN BERGEIJK, W. A., PIERCE, J. R., and DAVID, E. E. JR., *Waves and the Ear.* Doubleday Anchor, 1959: Science Study Series.

WEISSKOPF, VICTOR F., *Knowledge and Wonder.* Doubleday Anchor, 1963: Science Study Series. (Chapters 3 and 9)

WIGHTMAN, W. P. D., *The Growth of Scientific Ideas.* Yale University Press, 1953. Provides resource material on the part Young, Huygens, and others played in the development of wave theory. (Chapters XII and XIV)

Motion Along a Straight-Line Path

CHAPTER
9

A freight train is rolling down the track at 40 miles per hour. Out of the fog a mile behind, a fast express appears, going at 70 miles per hour on the same track. The express engineer slams on his brakes. With the brakes set he needs two miles to stop. Will there be a crash? What we are called upon to do here is to predict where the two trains will be at subsequent times, and to find in particular whether they are ever at the same place at the same time. In a more general sense, we are asking about the connections between speeds, positions, and times.

The general subject of such relationships is called kinematics. In studying kinematics we do not concern ourselves with questions such as "Why does the express train need two miles to stop?" To answer such a question we would need to study in detail how the brakes slow down the train. Such questions as these will be considered in later chapters. Here we just consider the description of motion. We shall start with the discussion of motion along a straight-line path. Then in the next chapter we shall extend the discussion to describe more general motions.

In both of these chapters we shall draw on our ability to measure time and position, for all motion is the changing of position as time goes on. Usually, we shall not think consciously of the time and position measurements, but without them we would in fact be talking words without meaning.

9-1 Position and Displacement Along a Straight Line

The first step in the study of motion is to describe the position of a moving object. Consider a car on an east-west stretch of straight highway. To answer the question "Where is the car?" we have to specify its position relative to some particular point. Any well-known landmark can serve as our reference point, or origin for measuring position. The choice of landmark is arbitrary so long as it is clearly stated and understood by everyone. We then state how far the car is from the landmark and in which direction, east or west, and the description of position is complete. Thus, for example, we say that the car is 5 miles west of the center of town, or it is 3 miles east of the Sandy River bridge. It is not enough to say only, "five miles from the center of town." You would not know whether this means 5 miles east or 5 miles west.

Similarly, if you wish to describe the position of a point on a straight line that you have drawn, you must specify some origin and state a distance and direction from that origin. But this time the direction cannot be given as east or west, for the line may not run that way. You might try "right and left," but how would someone standing on the other side of the line interpret these directions? To get a description of direction along the line about which we can all agree, we shall call the line on one side of the origin positive, on the other side negative; we can then specify position on the line by a positive or negative number which gives both the distance (in some convenient units) and the direction of that point from the origin. We shall refer to such a number, with its sign and units, as the coordinate of the point. If we call the line the x coordinate line, we shall label these coordinates as x_1, x_2, x_3, etc. (Fig. 9–1).

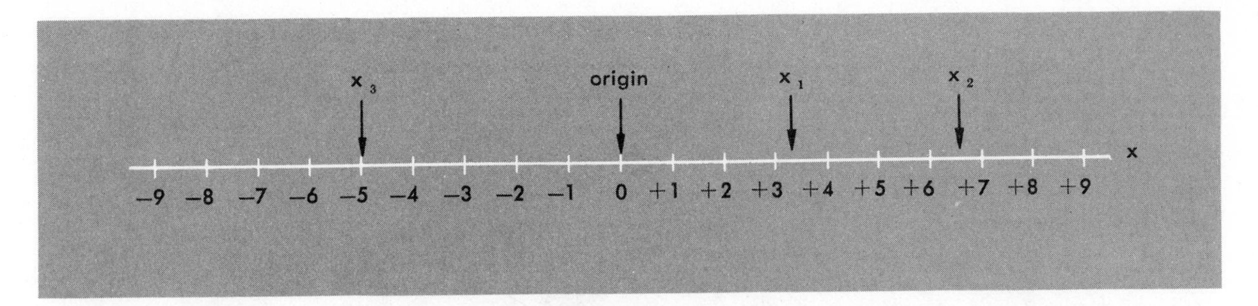

Figure 9–1
The *x* coordinate line.

We shall often want to refer to the change of position in our study of motion and we shall give it a special name, the *displacement*. If an object moves from position x_1 to position x_2, the displacement is given by the difference $x_2 - x_1$, that is, the later position coordinate minus the earlier one. Displacement can be either positive or negative (positive when x_2 is greater than x_1, negative when x_2 is less than x_1). Whether the displacement is positive or negative depends only on the direction of motion; it does not depend on where on the x coordinate line the displacement takes place. The two displacements in Fig. 9–2 (a) are positive and equal to each other. The displacements in Fig. 9–2 (b) are negative and also equal to each other.

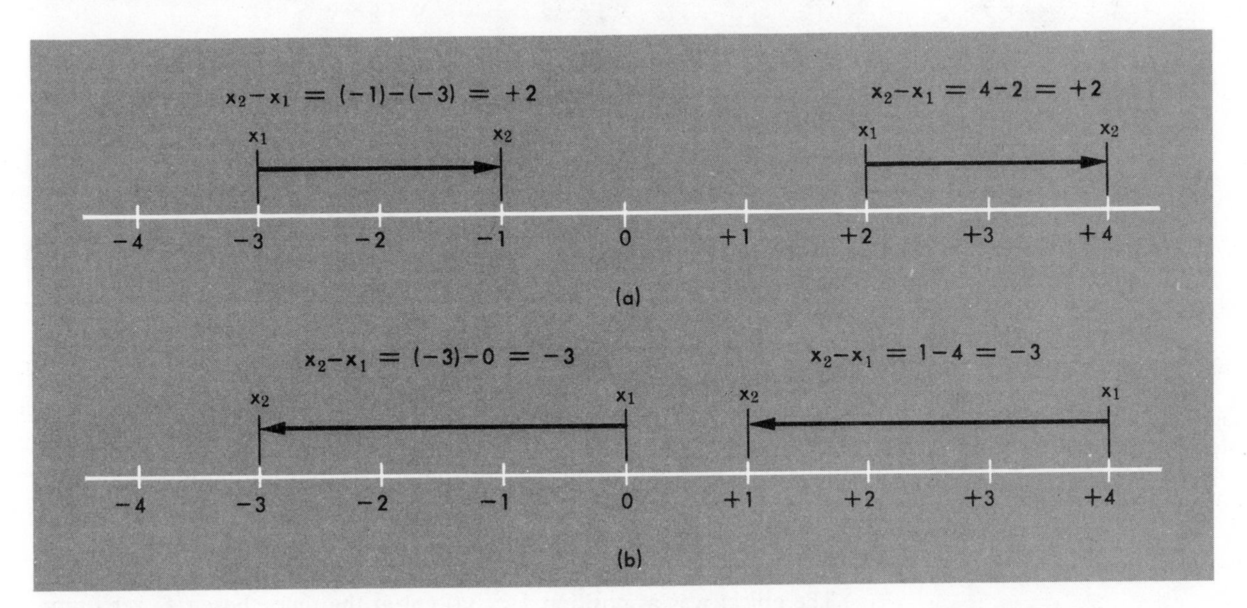

Figure 9–2
Two equal displacements, (a) positive and (b) negative.

Displacements are also independent of the point chosen for the origin of the coordinate line. Figure 9–3 shows the position coordinates of the same points as those in Fig. 9–2 but on a coordinate line whose origin is at a different place. The position coordinates are different, but the displacements, being differences, are the same.

Differences, or changes, occur so often in science and mathematics that a special notation is used to express them. The Greek letter delta, written as Δ (Greek capital D), is usually chosen to stand for "difference" or "interval" or "change of," or "increase of." Thus Δa means "change in a" or "increase in a" and is read as "delta a." It makes no sense to separate the

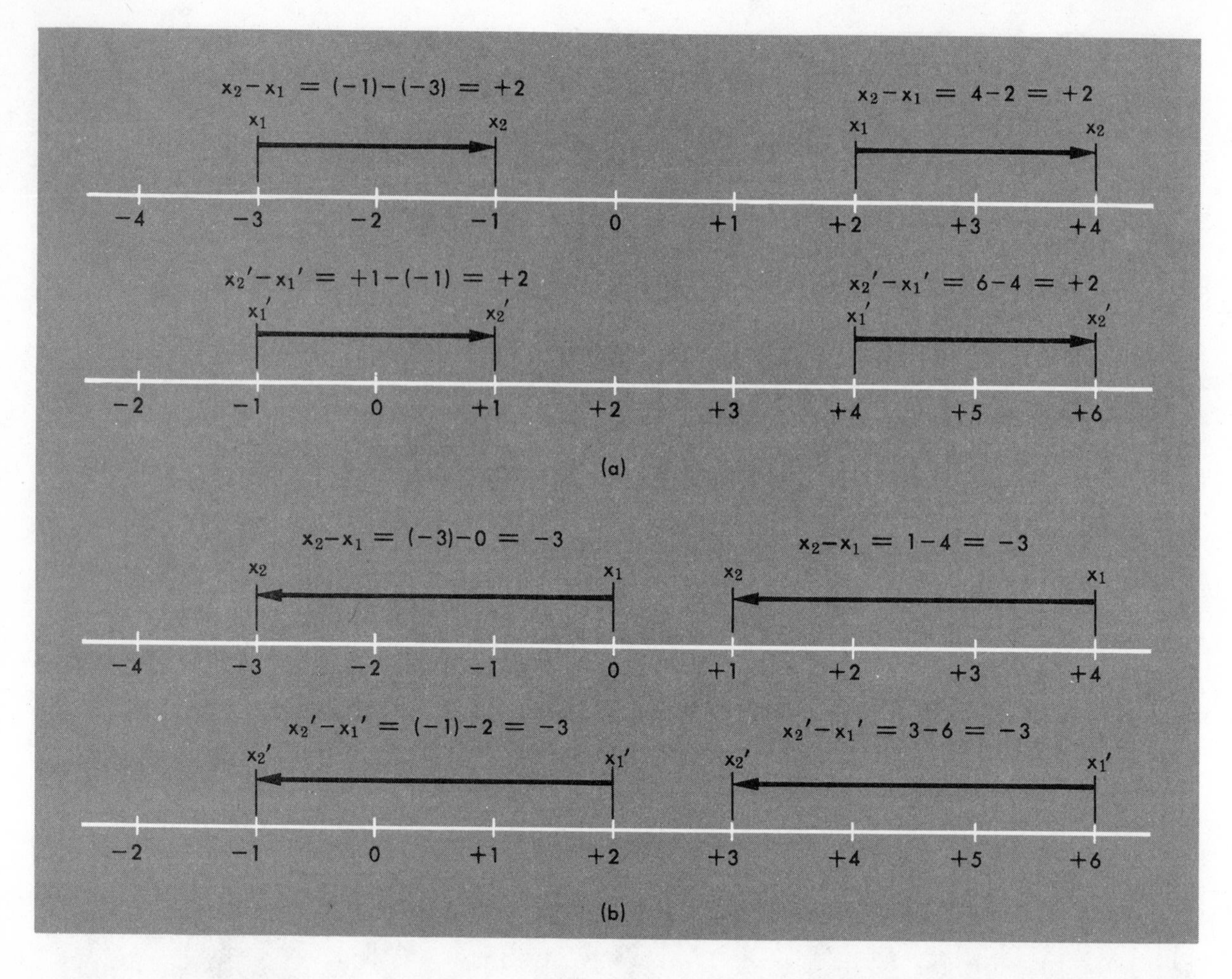

Figure 9–3
In (a) the same displacements as in Fig. 9–2 (a) are shown on top; underneath they are referred to a coordinate axis with a different origin. In (b) the displacements of Fig. 9–2 (b) are referred to a different origin.

Δ from the *a*. The whole symbol Δa has a special meaning: the change in *a* or an interval of *a*. It does not mean Δ multiplied by *a*.

To describe the motion of an object along a coordinate line, it is often convenient to make a graph of position against time. In such graphs, we usually plot the time along the horizontal axis and the position along the vertical axis. Figure 9–4 is an example of such a graph. There are many (qualitative) features about the motion which you can learn immediately from the graph.

The object was at position $x = 3.0$ cm at the time chosen as zero time. It stayed there till $t = 3.0$ sec. At that instant it started moving away from the origin. Its farthest position was $x = 6.5$ cm and it arrived there at $t = 10.2$ sec. It then reversed its direction, crossed the origin, and stopped again at $x = -2.0$ cm, etc.

9-2 Steady Motion: Constant Velocity

Moving fast and moving slowly are very familiar terms to everyone, but you may not have noticed that there are two different (although related) ways

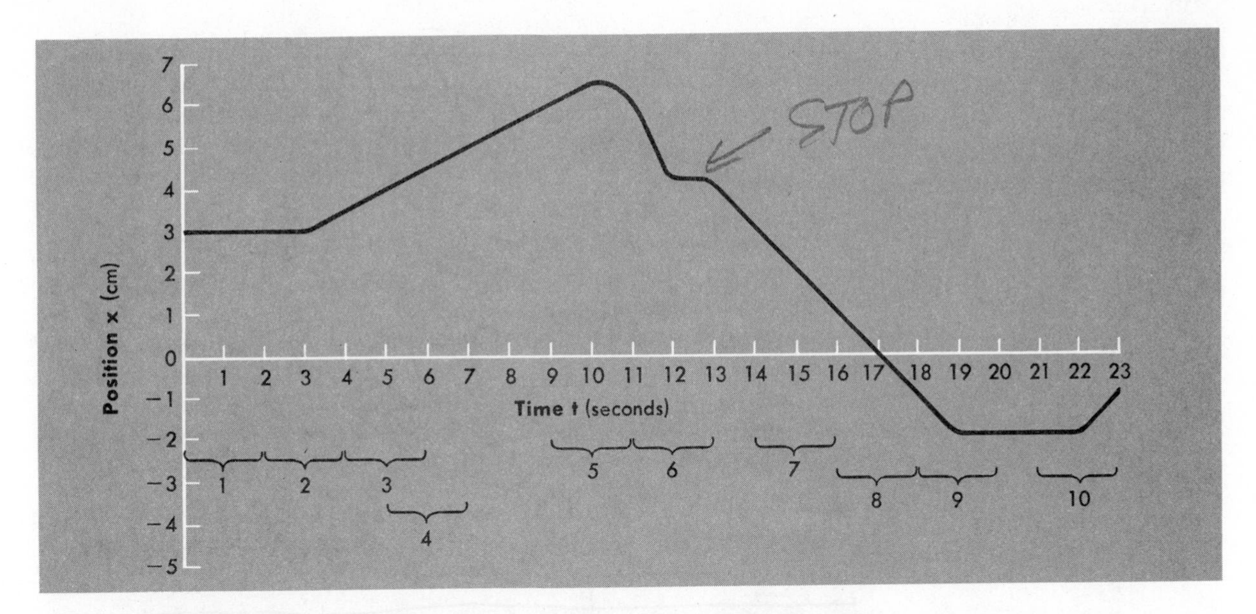

Figure 9–4
A position-time graph.

of expressing this distinction quantitatively. In sports we say that a runner *a* is faster than runner *b*, if *a* covered the same distance as *b* in less time. When it comes to highways, we say that on the expressway you can drive faster because you are allowed to cover more miles in one hour than on a side road.

We can use the graph showing position as a function of time (Fig. 9–4) to find when the object was moving fast or slow. We shall do so by the second method, that is, by comparing displacements made in equal time intervals. Since a time interval is the difference between two time coordinates t_1 and t_2, it is appropriate to designate the difference $t_2 - t_1$ by Δt.

As an example let us compare the displacements of the object whose motion is described in Fig. 9–4 for time intervals $\Delta t = 2$ sec, beginning at various times (Table 1).

Table 1

NUMBER OF INTERVAL	t_1 (SEC)	x_1 (CM)	t_2 (SEC)	x_2 (CM)	Δt (SEC)	Δx (CM)
1	0	3.0	2.0	3.0	2.0	0
2	2.0	3.0	4.0	3.5	2.0	0.5
3	4.0	3.5	6.0	4.5	2.0	1.0
4	5.0	4.0	7.0	5.0	2.0	1.0
5	9.0	6.0	11.0	6.0	2.0	0
6	11.0	6.0	13.0	4.0	2.0	−2.0
7	14.0	3.0	16.0	1.0	2.0	−2.0
8	16.0	1.0	18	−1.0	2.0	−2.0
9	18.0	−1.0	20.0	−2.0	2.0	−1.0
10	21.0	−2.0	23.0	−1.0	2.0	1.0

Table 1 tells us that the object moved fastest during intervals 6, 7, and 8, and that it was moving to the left. (In these intervals Δx is largest in magnitude, and negative.)

In intervals 1 and 5 the displacement was zero. Does this mean that the object was at rest during those time intervals? The table alone is not enough to settle the question. Going back to the graph in Fig. 9–4, you see that at any instant during interval 1—that is, between $t = 0$ and $t = 2.0$ sec—the object was at rest at $x = 3.0$ cm. However, during interval 5—between $t = 9.0$ sec and $t = 11.0$ sec—the object was first moving to the right (upward on the x scale) and then to the left (downward on the x scale). It just happened that at the end of the time interval it was at the same position as at the beginning.

Now let us examine the motion during intervals 6, 7, and 8; in all three the displacement was -2.0 cm. Was the motion the same in these intervals? To answer this question, we shall redraw Fig. 9–4 on a larger scale and divide each time interval into two equal parts, and find the corresponding displacements (Fig. 9–5). The results are shown in Table 2.

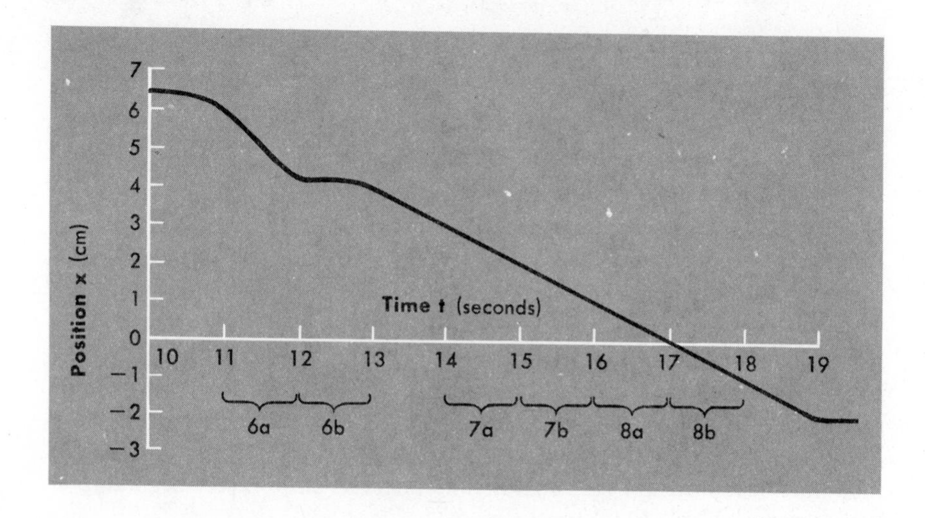

Figure 9–5
A part of Fig. 9–4 redrawn on a magnified scale.

Table 2

NUMBER OF INTERVAL	t_1 (SEC)	x_1 (CM)	t_2 (SEC)	x_2 (CM)	Δt (SEC)	Δx (CM)
6a	11.0	6.0	12.0	4.2	1.0	−1.8
6b	12.0	4.2	13.0	4.0	1.0	−0.2
7a	14.0	3.0	15.0	2.0	1.0	−1.0
7b	15.0	2.0	16.0	1.0	1.0	−1.0
8a	16.0	1.0	17.0	0	1.0	−1.0
8b	17.0	0	18.0	−1.0	1.0	−1.0

You see from the table that the subdivision of interval 6 shows that there were unequal displacements, whereas the subdivision of intervals 7 and 8 showed equal displacements to within the accuracy of the reading of the graph. Further subdivisions of intervals 7 and 8 show that for any equal time intervals these smaller displacements are also equal. A motion for which this is the case is called *steady motion*. On a position-versus-time graph, portions corresponding to steady motion must be straight-line segments, since only for a straight line do equal changes along one axis correspond to equal changes along the other.

We can look at the steady motion in intervals 7 and 8 in another way. In each second the displacement is 1.0 cm; in 2.0 sec it is twice as much or 2.0 cm; in 3.0 sec it is three times as much or 3.0 cm, and so on, as long as the motion is steady. We can generalize this result as follows: if for any equal time intervals the displacements are equal, then the displacement is proportional to the time interval:

$$\Delta x = v \, \Delta t$$

where v, the proportionality constant, is the *velocity.* Since Δx has the dimension of length and Δt the dimension of time, v has the dimension of length divided by time. Its units depend on the units in which the displacement and time are expressed. For example, if Δx is expressed in cm and Δt in seconds, then v is given in cm/sec. This is seen best by writing

$$v = \frac{\Delta x}{\Delta t}.$$

For the straight section of the graph in Fig. 9–4 which we have just discussed, $v = \dfrac{-2.0 \text{ cm}}{2.0 \text{ sec}} = -1.0$ cm/sec. The sign of the velocity is always the same as the sign of the displacement Δx, because Δt is always positive. The magnitude of the ratio $\dfrac{\Delta x}{\Delta t}$ is a measure of the steepness of the straight part of the x vs t graph; it is called the *slope* of the line.

When the ratio of two changes is involved, as it is in determining a velocity, it is understood that the change in the numerator "takes place during" the interval of the denominator. Thus $v = \dfrac{\Delta x}{\Delta t}$ (which is read "v equals delta x over delta t,") means "to find the velocity, take the change in position Δx and divide it by that time interval Δt during which it took place." In general, when we write $\Delta a/\Delta b$, we mean that we shall use the change in a that corresponds to a given change in b.

9-3 Instantaneous Velocity

We have seen that for steady motion the change in position, i.e., the displacement, is proportional to the change in time. But most motions are not steady, and their position-versus-time graphs will not have straight segments. Is there a measure for how fast an object moves when its motion is not steady?

Consider the position-versus-time graph shown in Fig. 9–6. How fast is the object moving at $t = 50$ sec? The motion around that time is not steady, as you see from the fact that the line is curved. Now let us look with a magnifying glass at only the part of the graph between $t = 45$ sec and $t = 55$ sec (Fig. 9–7). The magnified part of the graph looks straighter than the whole graph, because it is only a small portion of it. A still greater magnification shows us the interval which covers only 0.5 sec before and after the 50-sec mark (Fig. 9–8). In this small interval the line is almost straight, and we can find the velocity by measuring the slope of the "straight" line. We choose two points 1 and 2 in Fig. 9.8 near 50 sec; then, reading from this graph, we find

$$t_1 = 49.86 \text{ sec}, \qquad x_1 = 38.42 \text{ m}.$$
$$t_2 = 50.16 \text{ sec}, \qquad x_2 = 38.58 \text{ m}.$$

Consequently, the slope is given by

$\dfrac{\Delta x}{\Delta t} = \dfrac{x_2 - x_1}{t_2 - t_1} = \dfrac{+0.16 \text{ m}}{0.30 \text{ sec}} \approx +0.53$ m/sec, and the velocity at the point 50 sec from the start is very close to $+0.53$ m/sec. Thus we can say that the velocity of the object at $t = 50$ sec is very nearly 0.53 m/sec in the positive direction.

The magnified part of a graph looks straighter than the whole graph because in the magnified picture we look at only a small portion of the unmagnified graph. When we magnify sufficiently, we look at only a small interval of x and t. In effect, therefore, we find the slope of a small portion of the curve by taking the ratio

$$\frac{\Delta x}{\Delta t} = \frac{x_2 - x_1}{t_2 - t_1}$$

Figure 9–6
Position-time graph for an object with continually changing velocity.

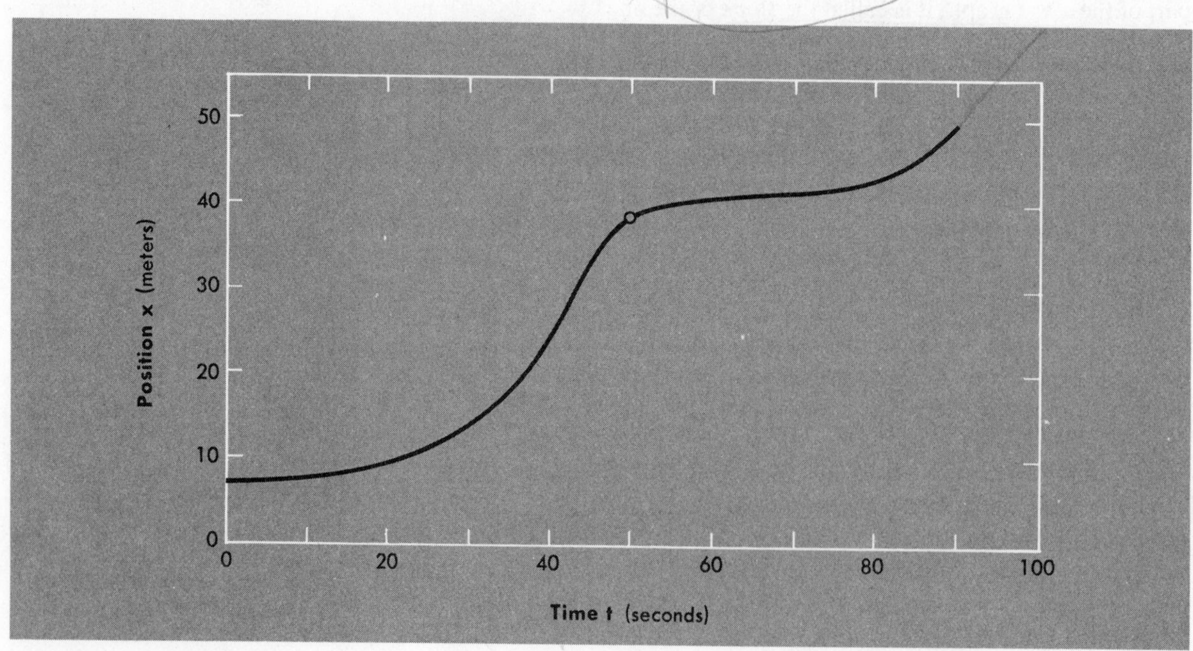

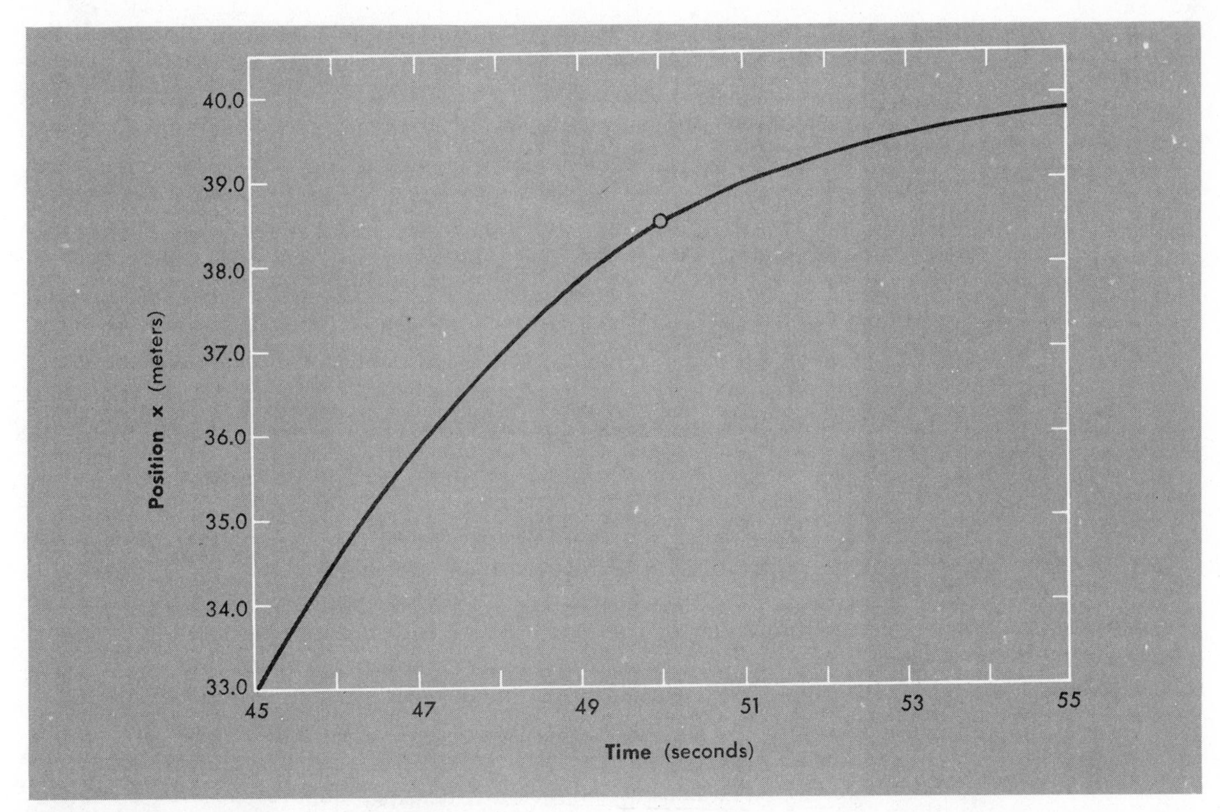

Figure 9–7
In this figure, part of the graph is enlarged.

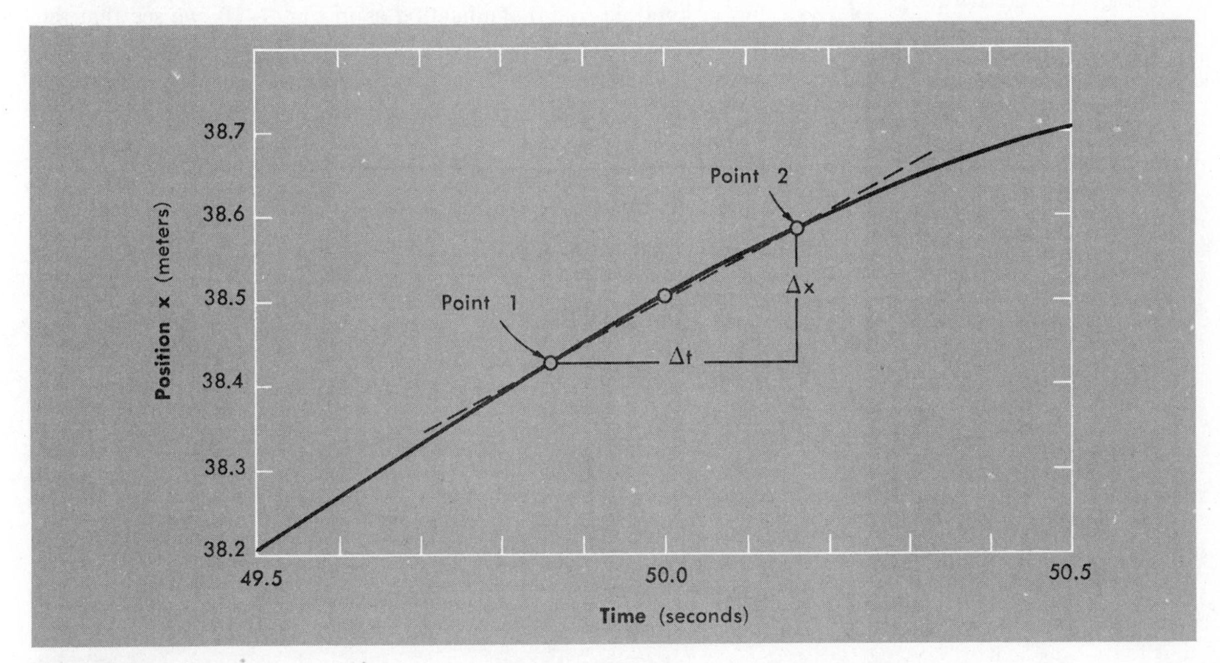

Figure 9–8
At a magnification of 100, a very small portion of the graph appears to be almost straight.

for a pair of points 1 and 2 which are very close together. The points we use must be close enough together so that the graph is essentially a straight line in between.

When the part of the graph we use is nearly straight, we find nearly the same slope, and thus nearly the same velocity, no matter which points we take to evaluate it. In Fig. 9–8, for example, if we let points 1 and 2 move along the curve until they are closer to the 50-sec mark, we get almost the same result as before. Try it yourself. Furthermore, the points we use to find $\Delta x/\Delta t$ will be close together when Δt is small; and if Δt is small enough, $\Delta x/\Delta t$ will not change much for a smooth position-time curve around a given point. We therefore say that the velocity at a particular instant or the *instantaneous velocity* is given by the limit of $\Delta x/\Delta t$ as Δt "approaches zero," that is, as Δt gets smaller and smaller. Symbolically this statement is written

$$v = \lim_{\Delta t \to 0} \frac{\Delta x}{\Delta t}.$$

(We read "$\lim\limits_{\Delta t \to 0}$" as "the limit as Δt approaches 0 of . . .") The word "limit" here stands for the result obtained when we take t_1 and t_2 close enough together so that there is no noticeable change produced in the value of the ratio by using a still smaller interval between t_1 and t_2.

In developing the idea of instantaneous velocity we used a magnified picture of the position-time graph. Is there any simple way of getting the velocity without going through the steps of magnifying? The answer is "yes": draw a tangent and measure its slope. Suppose we drew a straight line tangent to the curve of Fig. 9–6 at the point for which $t = 50$ sec, as in Fig. 9–9. In this figure we can easily distinguish the tangent from the graph of x vs t; but at hundred-fold magnification as in Fig. 9–10, we see that the tangent line and the curve are hardly distinguishable over a small enough interval around the point of tangency. In this region they have the same slope. We can therefore use the slope of the tangent line to determine the

Figure 9–9
To find the velocity at any instant, draw a line tangent to the curve at the point being considered. Then, by taking any two points on the tangent line, find the slope. The value of the slope is the velocity.

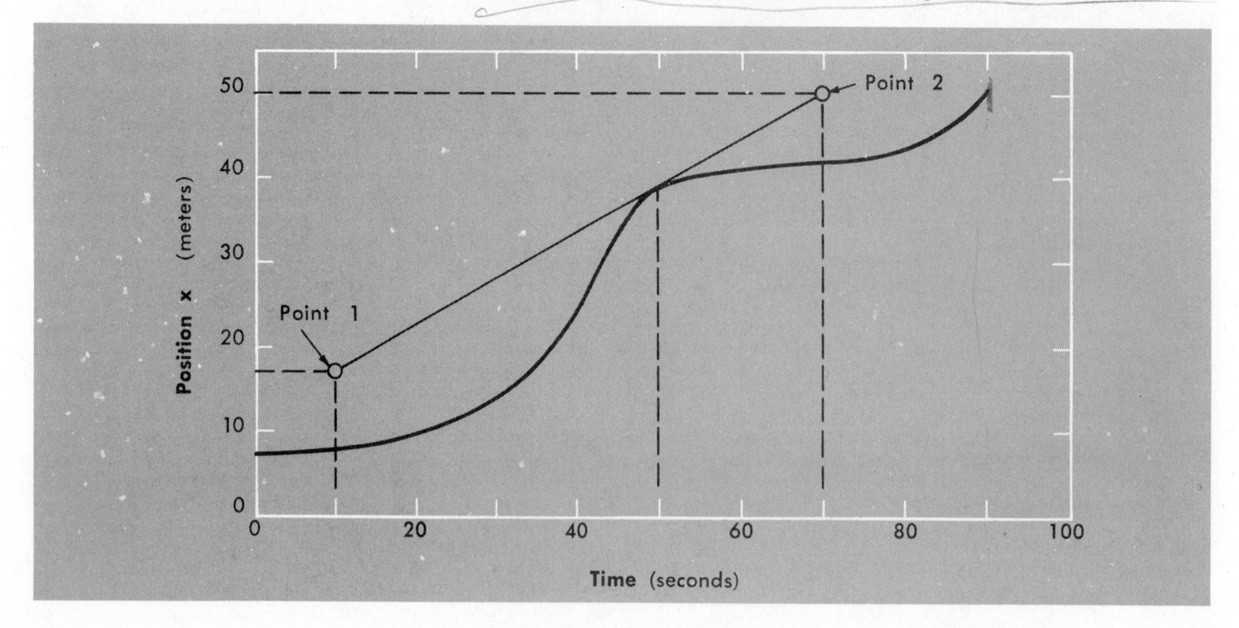

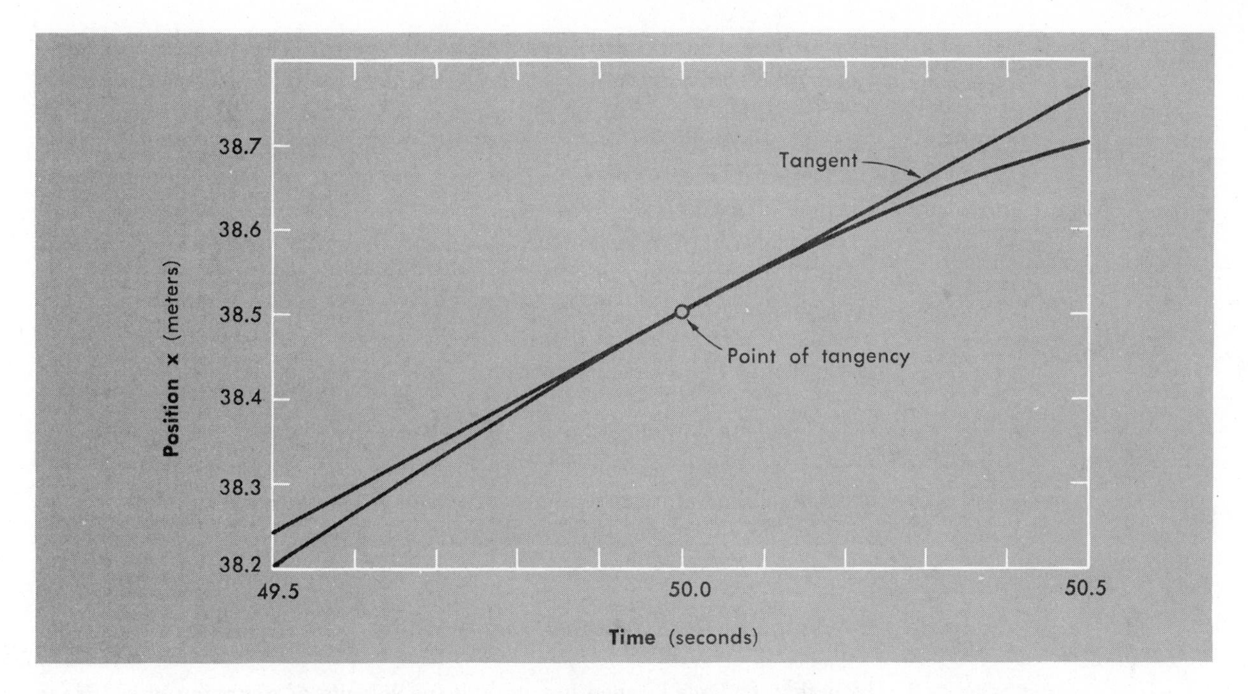

Figure 9–10
With 100-times magnification of Fig. 9–9, the curve and its tangent are indistinguishable near the point of tangency.

slope of the graph at any particular point. Figure 9–9 shows us how we can use this method to get the speed of the object at $t = 50$ sec. First we draw the tangent line to the curve at the point chosen. Then we choose two convenient points (labeled 1 and 2) on the tangent line. We read the values of x_1, x_2 and of t_1, t_2 to find $x_2 - x_1 = +32$ m and $t_2 - t_1 = 60$ sec, so that the slope is

$$\frac{\Delta x}{\Delta t} = \frac{x_2 - x_1}{t_2 - t_1} = \frac{+32 \text{ m}}{60 \text{ sec}} \approx +0.53 \text{ m/sec}.$$

Therefore, at $t = 50$ sec the object has an instantaneous velocity of 0.53 m/sec.

We use the word instantaneous to distinguish the velocity at one moment from the *average velocity* over an interval of time. The average velocity in the time interval Δt is defined as

$$v_{\text{av}} = \frac{\Delta x}{\Delta t}.$$

AVE VELOCITY

Graphically, it is the slope of the straight line connecting the two points on the position-time graph at the ends of the time interval Δt (see Fig. 9–11).

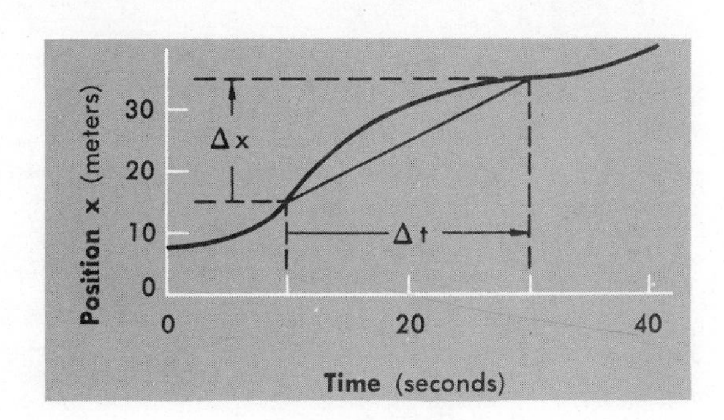

Figure 9–11
The slope of the straight line $= v_{\text{av}} = \Delta x/\Delta t$.

The average velocity is that constant velocity which would change the position by the actual amount Δx in the actual time interval Δt.

In general, the value of the average velocity depends on which time interval Δt we are examining. In Fig. 9–9, the average velocity during the time interval 0 sec to 60 sec is not the same as the average velocity during the time interval 40 sec to 60 sec.

If, during Δt, the instantaneous velocity is changing noticeably, it will differ at almost every moment from the average velocity for that interval. At some moments it may be greater, at other moments smaller, than the average velocity. Only at times when the tangent to the position-time curve is parallel to the average-velocity straight line are they the same. For example, on the graph in Fig. 9–11 they are the same only at $t = 18$ sec. Only for steady motion does the instantaneous velocity equal the average velocity at every moment. In this case, no matter whether we choose a long or a short time interval, the average velocity will be the same.

9-4 Velocity-Time Graphs from Position-Time Graphs

Now that we have learned how to find the velocity of a moving object from a position-versus-time graph, we can use this knowledge to make graphs of velocity from position-time graphs. As an example, consider the graph shown in Fig. 9–12. Between $t = 0$ and about $t = 3$ sec the motion is steady and we can measure the slope directly anywhere. Between $t = 3$ sec and $t = 5$ sec the velocity is changing so we shall have to draw tangents at several points and measure their slopes. After $t = 5$ the graph is straight again and we can again measure the slope directly anywhere. The smallest value of the velocity is at the beginning and is $\frac{\Delta x}{\Delta t} = \frac{-0.30 \text{ m}}{2.0 \text{ sec}} = -0.15 \frac{\text{m}}{\text{sec}}$. The largest is at the end: $\frac{\Delta x}{\Delta t} = \frac{0.40 \text{ m}}{1.0 \text{ sec}} = +0.40$ m/sec.

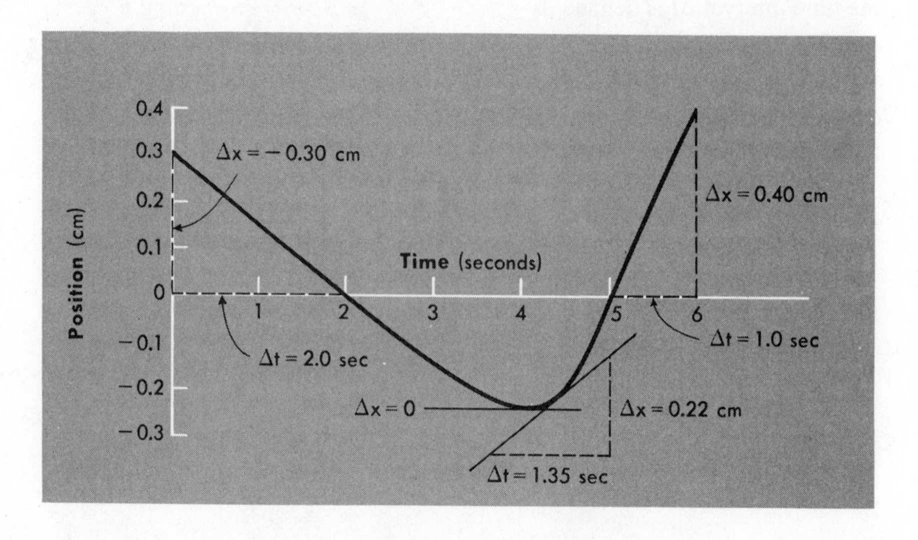

Figure 9–12
Finding the slope at various points on a position-time graph.

Choosing the scale for the velocity axis in such a way that it will contain the values above, we can start the v vs. t graph as shown in Fig. 9–13. The next step depends on how accurate we wish the graph to be. If high accuracy is required, we have to go back to Fig. 9–12 and measure the slope of the curved part at several points. If a rough sketch is sufficient, we can complete the graph by hand as shown in Fig. 9–14.

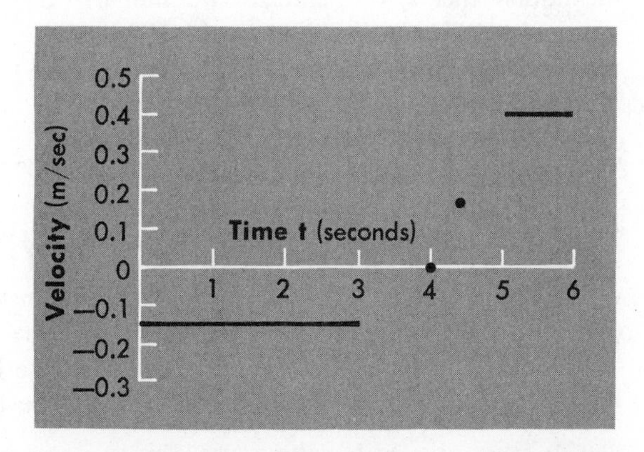

Figure 9–13
Beginning the construction of the velocity-time graph from the slopes in Fig. 9–12.

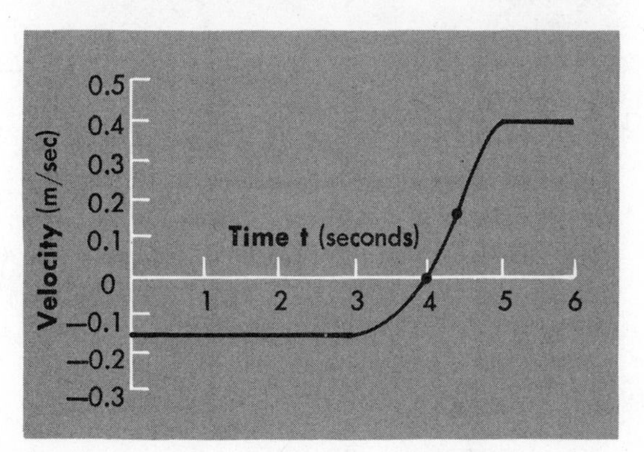

Figure 9–14
A rough sketch of the completed graph started in Fig. 9–13.

We can generalize the method which we used in the preceding example for constructing a velocity-time graph from a position-time graph. Find first the largest and smallest values of the velocity to determine the scale for the v axis. Then identify sketches of steady motion and plot the corresponding values of the constant velocities. Enter as many values of the instantaneous velocity of the curved part of the position-time graph as needed and draw the velocity-time curve.

9-5 Displacement from Velocity-Time Graphs

A car speedometer and a clock can be used to measure velocity as a function of time for a moving car (provided it only moves forward). We can plot such data on a velocity-time graph. How can we use such a graph to obtain a position-time graph? We shall answer this question in two steps, first for motion with constant velocity and then for motion with variable velocity. For motion with constant velocity the displacement, or change in position, is proportional to the change in time:

$$\Delta x = v\Delta t.$$

Here the time interval Δt is simply the time passed from the beginning of the motion: $\Delta t = t - t_0$. We can have our clock read zero at the beginning of the motion; then $t_0 = 0$, and $\Delta t = t$. Similarly, $\Delta x = x - x_0$, where x_0 is the position of the object at time $t = 0$. If we choose the origin of the coordinate line to be at the position of the object at time $t = 0$ then $x_0 = 0$ and the relation $\Delta x = v\Delta t$ becomes

$$x = vt.$$

In words, for steady motion starting from the origin of the coordinate axis, the position at any time is given by the product of the velocity and the time.

The velocity-time graph corresponding to constant v is a straight line parallel to the t axis (Fig. 9–15). At any time the product vt describes the "area" of the rectangle between the graph and the horizontal axis. We put the word "area" in quotation marks because the base of the rectangle is

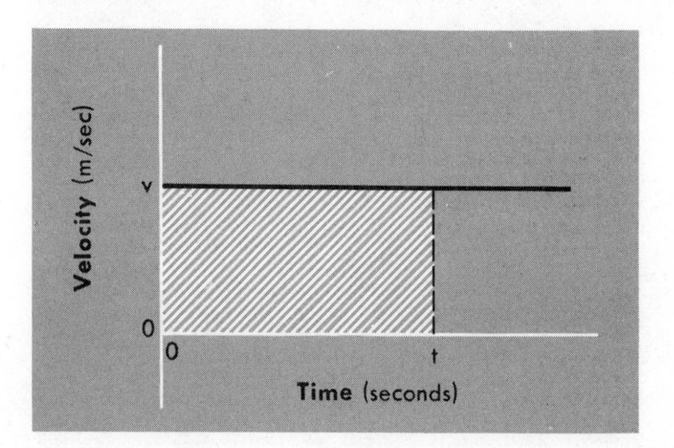

Figure 9–15
The shaded area is given by the product vt.

given in seconds and the height in m/sec; thus the "area" has units of (m/sec) × sec = m and not in m^2 as is the case in geometry. Had we expressed the velocity in miles/hour and the time in hours, the "area" would be given in miles and not in $(\text{miles})^2$. By finding the area of the rectangle for various values of t, we obtain the corresponding values of x, and from these values can plot a position-time graph. For the special case of constant velocity this would be a rather unnecessary trouble because we can plot the graph for $x = vt$ directly (Fig. 9–16).

The importance of the fact that the area between a velocity-time graph and the x axis is the displacement is that it holds for any velocity-time graph. To see that this is the case, consider the velocity-time graph for a real car speeding up (Fig. 9–17). We can approximate the graph in Fig. 9–17 closely by the "stepped graph" in Fig. 9–18. This graph represents the motion of an imaginary car which starts at the same position as the real car, but which has a velocity that changes in a series of steps. At the beginning of each step, its velocity is equal to the velocity of the real car. While the

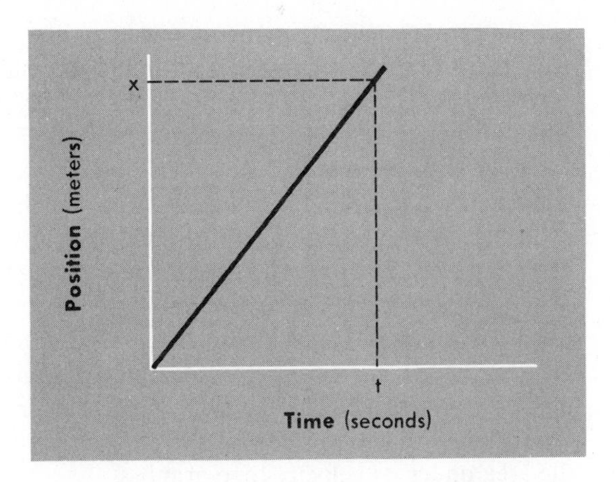

Figure 9–16
The graph of $x = vt$.

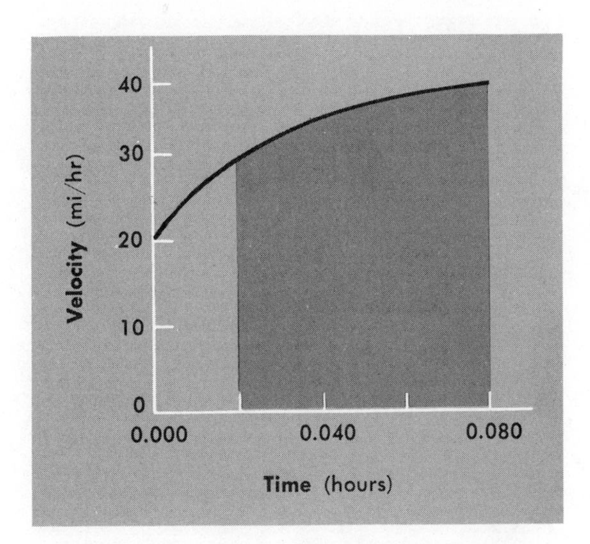

Figure 9–17
The graph of velocity as a function of time for an accelerating car.

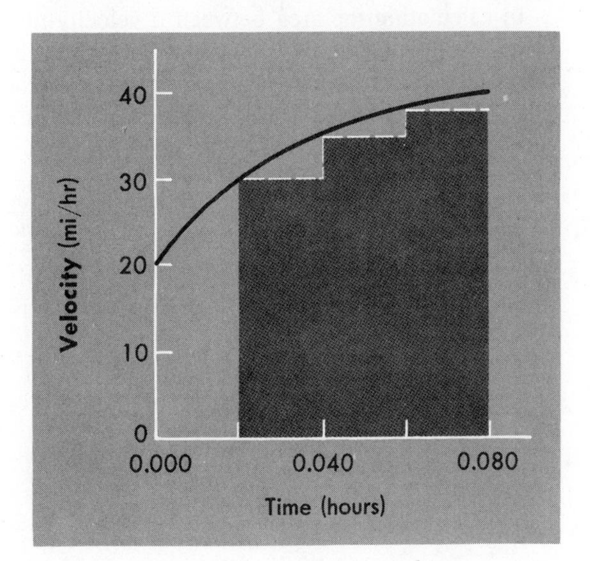

Figure 9–18
We can approximate the displacement of the car in Fig. 9–17 by that of an imaginary car traveling with velocities different from those of the real car as shown in this figure.

imaginary car's velocity remains constant, the velocity of the real car increases, so the real car gets farther and farther ahead of the imaginary car. At the end of the step the imaginary car's velocity suddenly increases to the value of the real car's velocity at that instant. The imaginary car can never catch up with the real car, however, because the real car is going faster at almost every instant. If we make the steps much smaller and more frequent, as in Fig. 9–19, the two cars will differ less in velocity, and hence the real car will never get far ahead. The shaded area, which gives the displacement of the imaginary car during the time interval, because it contains only rectangles, sections of constant velocity, will also very nearly give the displacement of the real car during this interval. And this shaded area, when we use many steps, is practically the shaded area under the graph of Fig. 9–17 for the real car. Thus, we see that the displacement of the real car is given by the area under its velocity-time graph.

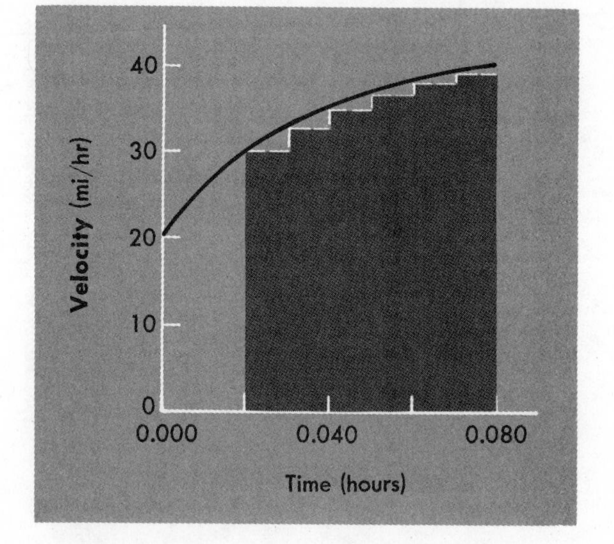

Figure 9–19
If the imaginary car in Fig. 9–18 changes its velocity more frequently, it approximates the motion of the real car more closely.

In calculating the area between a velocity-time curve and the t axis we have to remember that when v is negative, $v\Delta t$ is also negative. Therefore, for portions of a velocity-time curve which are below the t axis, the area is negative. To find the total displacement the portions must be added with their proper sign (Fig. 9–20).

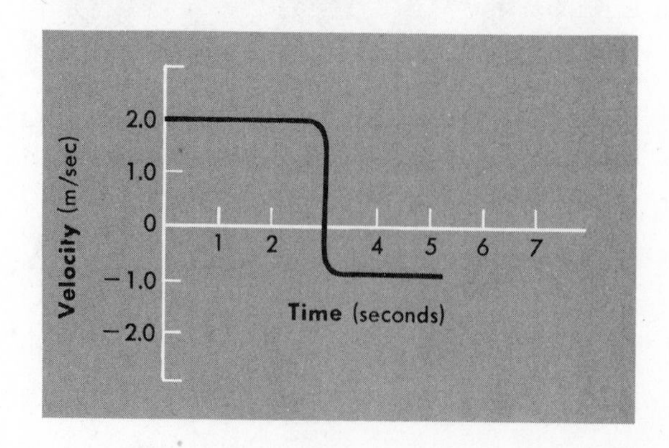

Figure 9–20
Between $t = 3.0$ sec and $t = 5.0$ sec the velocity is negative. Thus the area between the curve and the time axis in this region describes a negative displacement.

Finally, the area between a velocity-time graph and the t axis gives only the displacement, or change of position, of the moving object as a function of time. To be able to find position itself as a function of time we have to know where the object was at $t = 0$, that is, its initial position. For example, all four position-time graphs shown in Fig. 9–21 correspond to the velocity-time graph shown in Fig. 9–20. The displacements during any time interval are the same for all four graphs, but the position at any instant of time is different in each case.

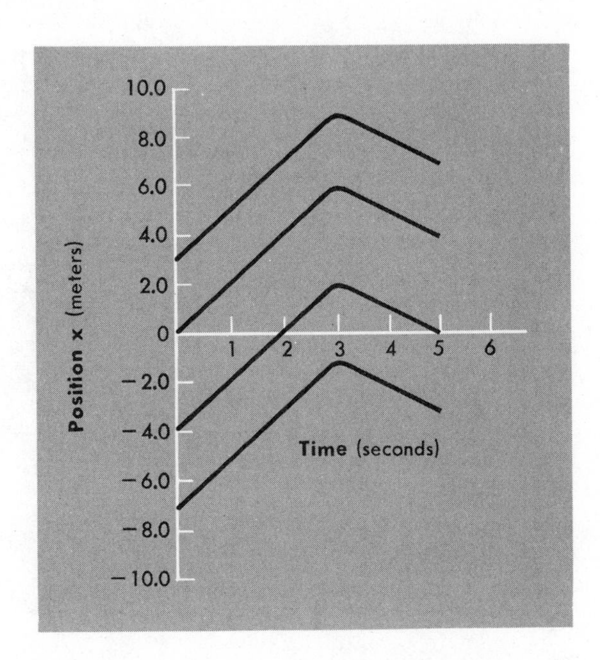

Figure 9–21
The motions described by the four graphs differ only in the initial position of the object. All four graphs yield the velocity-time graph shown in Fig. 9–20.

9-6 Acceleration

A driver might say that his car can reach 50 mi/hr in 10 seconds starting "from a standstill." He would be talking about the *acceleration* of his car, how fast its velocity changes. We usually express the acceleration, the rate of change of velocity, as the change of velocity in one second. In these terms a car that reaches 50 mi/hr in 10 seconds from rest has an average acceleration of 5 mi/hr per second.

Just as the slope of a displacement-time graph is a measure for the rate of change of the displacement, so is the slope of a velocity-time graph a measure of rate of change of the velocity or acceleration.

Consider the velocity-time graph for a car shown in Fig. 9–22. Since the graph is a straight line, the change in velocity is proportional to the length of the time interval

$$\Delta v = a\Delta t.$$

The constant of proportionality a is the acceleration. It is given by the slope $a = \dfrac{\Delta v}{\Delta t}$. The unit of acceleration depends on the units used to express the change in velocity and the time interval during which this change takes place. For example, in Fig. 9–22 the velocity is measured in miles per hour, and the time interval in seconds. Therefore

$$a = \frac{\Delta v}{\Delta t} = \frac{20 \text{ mi/hour}}{5.0 \text{ sec}} = \frac{4 \text{ mi/hr}}{\text{sec}} = 4 \frac{\text{mi}}{\text{hour-sec}}$$

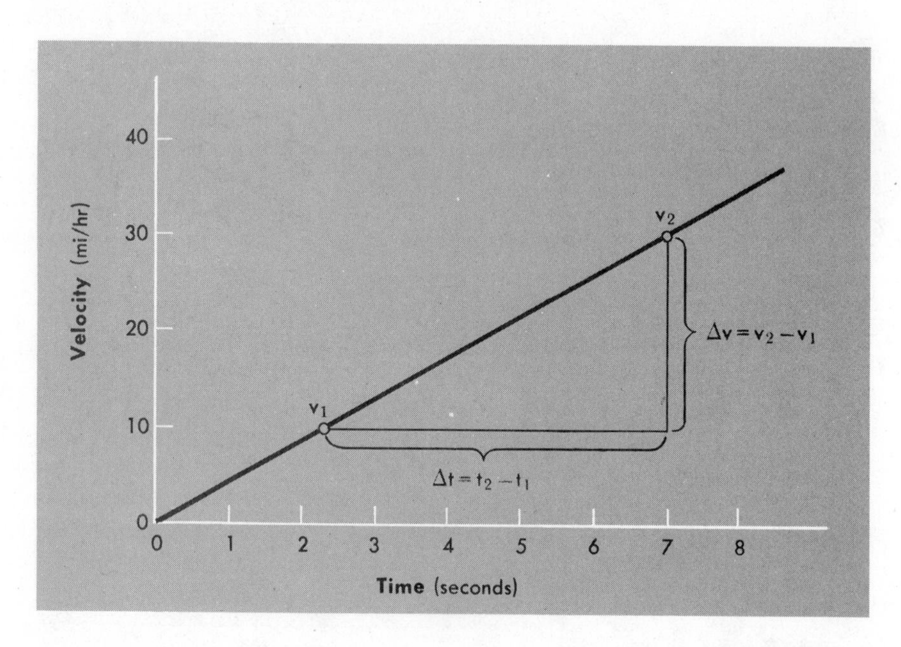

Figure 9–22
A velocity-time graph with constant slope.

There are 1.6×10^3 m in one mile and 3.6×10^3 sec in one hour. If speedometers were made to measure velocities in m/sec, 20 mi/hour would correspond to $\dfrac{20 \times 1.6 \times 10^3}{3.6 \times 10^3} = 8.9$ m/sec, and the acceleration would be $\dfrac{8.9 \text{ m/sec}}{5.0 \text{ sec}} = \dfrac{1.8 \text{ m/sec}}{\text{sec}} = 1.8 \dfrac{\text{m}}{\text{sec-sec}}$. Whatever the units, an acceleration is always given in terms of $\dfrac{\text{length}}{(\text{time}) \times (\text{time})}$.

Since Δt is always positive, the sign of a is the same as the sign of Δv. In Fig. 9–23, Δv is always positive; hence the acceleration is positive, although at $t = 0$ the object moved faster than at $t = 2.0$ sec, and at $t = 4.0$ sec it was momentarily at rest! This will not seem so strange if you consider the fact that before $t = 4.0$ the object was moving to the left and the change in its velocity between $t_1 = 0$ and $t_2 = 2$ sec was $v_2 - v_1 = -4 - (-8) = +4 \dfrac{\text{m}}{\text{sec}}$. A positive acceleration can either mean speeding up, if the velocity is positive, or slowing down if the velocity is negative. Conversely, a negative acceleration can mean slowing down if the velocity is positive or speeding up if the velocity is negative.

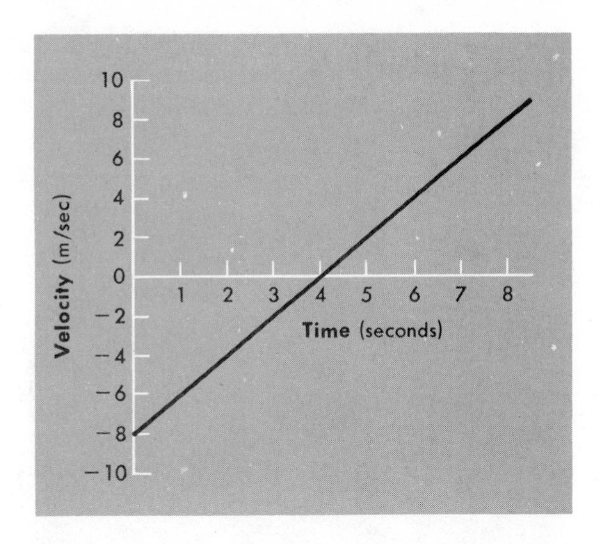

Figure 9–23
A constant acceleration can mean slowing down, coming to rest, and then speeding up in the opposite direction.

To illustrate this, consider what happens when you throw a ball straight up. As soon as it leaves your hand, it begins to slow down; after it reaches its maximum height, it speeds up on the way down. If you choose up as the positive direction, then the ball has a negative acceleration throughout its flight.

What is the acceleration of an object when its velocity-time graph is not a straight line (Fig. 9–24)? We can define the average acceleration between t_1 and t_2 as

$$a_{av} = \frac{v_2 - v_1}{t_2 - t_1} = \frac{\Delta v}{\Delta t}$$

in analogy with the definition of the average velocity $v_{av} = \dfrac{x_2 - x_1}{t_2 - t_1} = \dfrac{\Delta v}{\Delta t}$.

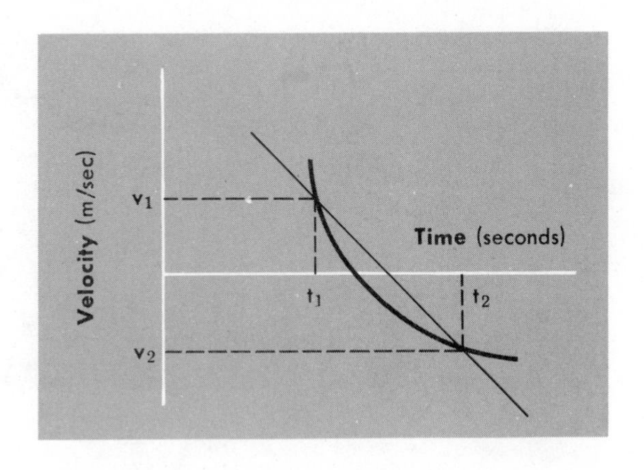

Figure 9–24
The average acceleration between t_1 and t_2 is $a_{av} = \dfrac{v_2 - v_1}{t_2 - t_1}$.

To continue this analogy we define the *instantaneous acceleration* at time t by

$$a = \lim_{\Delta t \to 0} \frac{\Delta v}{\Delta t}.$$

It is given by the slope of the tangent to the velocity-time graph at time t (Fig. 9–25). By finding the instantaneous acceleration at different times, we can plot an acceleration-time graph.

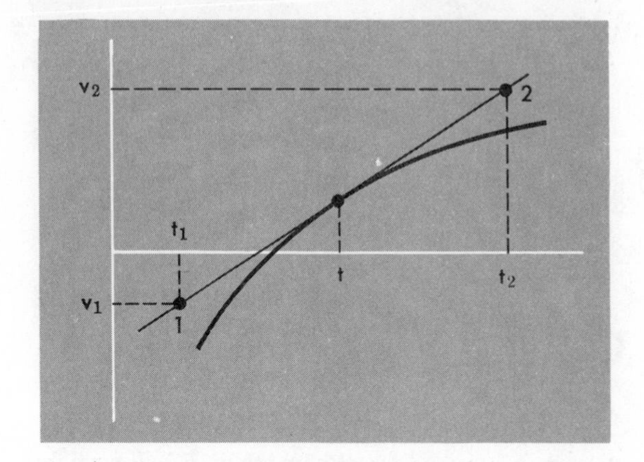

Figure 9–25
To find the instantaneous acceleration at time t, draw the tangent to the graph and use two convenient points to calculate its slope.

Recall that the slope of a position-time graph gives the velocity, and the slope of a velocity-time graph gives the acceleration. Similarly, since the area between a velocity-time curve and the t axis gives the change in position, so the area between an acceleration-time curve and the t axis gives the change in velocity.

To find the position of an object from a velocity-time graph, we must know its initial position; the change in position obtained from the area in the velocity-time graph is not enough. Completing the analogy between velocity and acceleration, we must know the initial velocity if we are to find the velocity from an acceleration-time graph. These relations are summarized in Fig. 9–26.

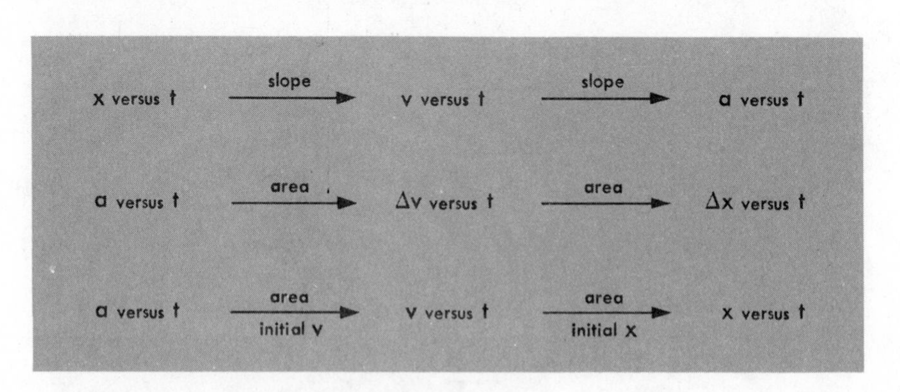

Figure 9–26
The relationships between position, velocity, and acceleration as a function of time.

9-7 Constant Acceleration: Some Useful Relations

Motion with constant acceleration is not easily produced. However, it can serve as an approximate description of motions with almost constant acceleration, which occur in many situations. You will find it useful, therefore, to assemble some basic relations between constant acceleration, velocity, position, and time for ready reference.

Let us start with an object moving with a constant acceleration a which at $t = 0$ is at rest at the origin of the coordinate axis. The area between the acceleration–time curve and the t axis will give the velocity as a function of time [Fig. 9–27(a) and (b)].

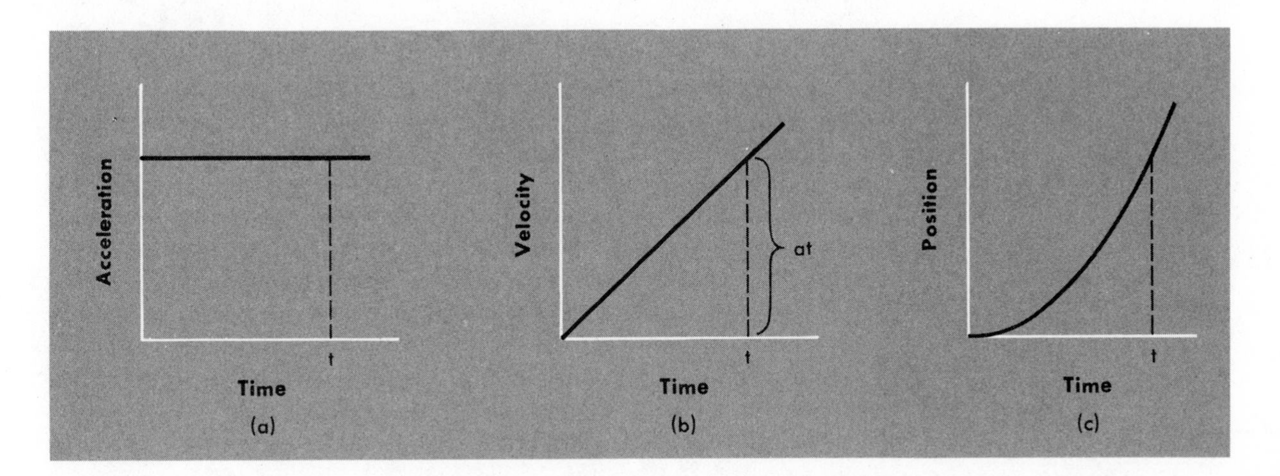

Figure 9–27 The acceleration-time, velocity-time, and position-time graphs for constant acceleration.

$$v = at. \tag{1}$$

The area between the velocity-time curve and the t axis yields the position as a function of time, since the object was at the origin at $t = 0$.

The area under the graph is a triangle. Its area is given by

$$x = \tfrac{1}{2}\,(\text{altitude})\,(\text{base}) = \tfrac{1}{2}\,(at)\,t = \tfrac{1}{2}\,at^2. \tag{2}$$

This is the equation of a parabola shown in Fig. 9–27 (c). Eq. (1) expresses the velocity as a function of time. For the generalization of equations (1) and (2) to motion starting at any point with any initial velocity, see the special section on the next page.

Often it is useful to express the velocity in terms of position. We can do that by substituting the value of $t = \frac{v}{a}$ from Eq. (1) into Eq. (2).

$$x = \tfrac{1}{2}a\left(\frac{v}{a}\right)^2 = \frac{v^2}{2a}$$

$$\text{or} \quad v^2 = 2ax. \tag{3}$$

To relate the velocities at the points to the distance traveled between them, we have, from Eq. (3): $v'^2 = 2ax'$ and $v^2 = 2ax$. Therefore,

$$v'^2 - v^2 = 2a(x' - x) = 2a\Delta x. \tag{4}$$

We shall use this relation in the forthcoming chapters.

More General Forms of Equations 1 and 2

In the derivations given below, we shall call the initial velocity, position, and time v, x, and t respectively. The final velocity, position, and time we shall call v', x', and t'.

Figure 9–28 is an acceleration-time graph for a constant a. The area from t to t' between the straight line representing the constant acceleration and the t axis is the change in velocity. This area is just

$$v' - v = a(t' - t) \quad (1a)$$

or

$$v' = v + a(t' - t) \quad (1b)$$

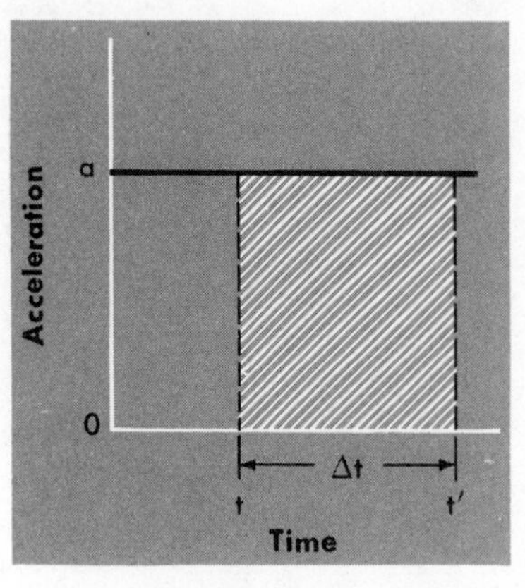

Figure 9–28

Figure 9–29 shows a graph of this function.

To get the displacement, the change in position, from Fig. 9–29 we have to find the sum of the areas of the shaded rectangle and triangle. This is

$$x' - x = v(t' - t) + \tfrac{1}{2}(v' - v)(t' - t).$$

Substituting the expression for $v' - v$ from Eq. (1a), we get

$$x' - x = v(t' - t) + \tfrac{1}{2}a(t' - t)^2 \quad (2a)$$

In terms of the initial position and velocity, and the initial and final times we have for the final position x',

$$x' = x + v(t' - t) + \tfrac{1}{2}a(t' - t)^2 \quad (2b)$$

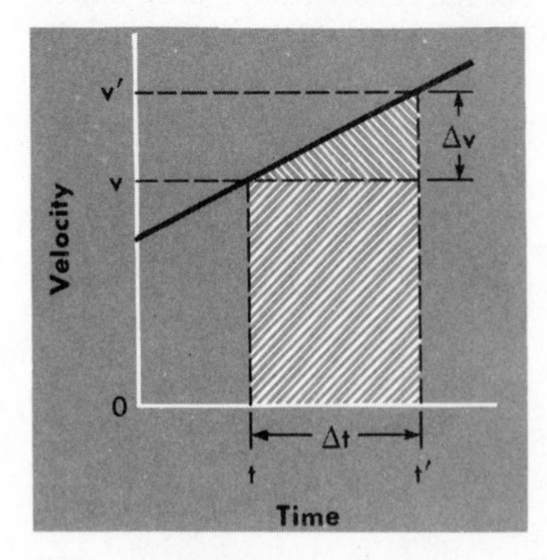

Figure 9–29

Equations (1b) and (2b), like equation (4) in the text, apply to any case of constant acceleration, no matter what the initial position, velocity, or time.

For Home, Desk, and Lab

1.* In the table below, which displacements are equal? (Section 1.)

Table

	x_1(m)	x_2(m)
(1)	5	8
(2)	7	−2
(3)	−5	−2
(4)	15	12
(5)	0	2
(6)	−5	−8
(7)	−5	0

2.* Does the graph of a car trip in Fig. 9–30 represent a real situation? Explain. (Section 1.)

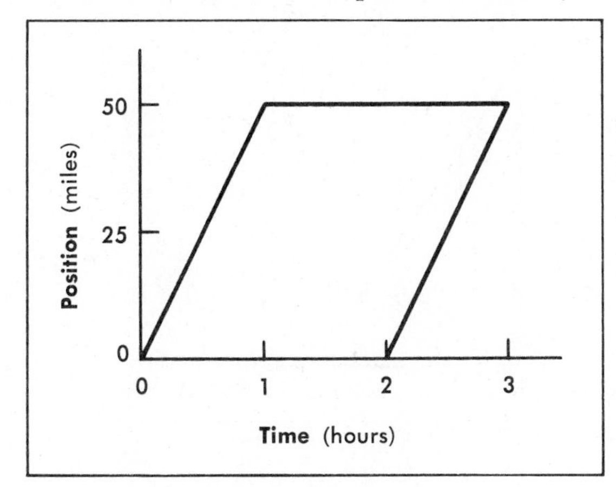

Figure 9–30 For Probiem 2.

3. Express the following velocities in mi/hour. Give examples of objects that move with such velocities.
   (a) 1 m/sec
   (b) 10 m/sec
   (c) 30 m/sec
   (d) 250 m/sec
   (e) 8000 m/sec

4. Identify the parts of the graph in Fig. 9–4 where the motion is steady, and determine the velocity of the object in those regions.

5. Find the slopes of the graphs in Fig. 9–31. State the units in each case.

6. Identify all points on the graph of Fig. 9–4 at which the instantaneous velocity is zero.

7. The position-time graph of a car is shown in Fig. 9–32.
   (a) At what time is the car going at the greatest velocity?
   (b) How fast is it traveling at that time?
   (c) How fast is the car going at 0.70 hour?
   (d) What is the average velocity for the first 0.70 hour?

8.* What is the average velocity over the first 90 sec in the graph of Fig. 9–9? (Section 3.)

9. A train travels 60 mi/hr for 0.52 hr, 30 mi/hr for the next 0.24 hr, and then 70 mi/hr for the next 0.71 hr. What is its average velocity?

10. Sketch a velocity-time graph for a shuttle train which runs between cities *A* and *C* with an intermediate stop at city *B*. All cities are on a straight line.

11.* What is the displacement of a car which travels at a steady velocity of 40 mi/hr (a) for 3 hr, (b) for ½ hr? (Section 4.)

12. A man walks to the corner to mail a letter and comes back. Sketch graphs showing his velocity and position plotted against time.

13. The position-time graph of a car traveling along a road is shown in Fig. 9–33. Make a graph of its velocity versus time.

14. John rode his bicycle as fast as he could from his house to Tom's house. After a short time he rode back as fast as he could. Figure 9–34 shows a position-time graph of his trip. Plot the velocity-time graph of John's trip. From the information given and your graph, what would you give as a plausible description of the road between John's and Tom's houses?

15. Sketch a velocity-time graph for the object whose displacement-time graph is given in Fig. 9–4.

16.† Car *A* is stopped at a traffic light. The light turns green and *A* starts up. Just as it does so, car *B* passes it, going at a steady velocity. Their velocity-time curves are shown in Fig. 9–35.
   (a) How long does it take car *A* to be going as fast as car *B*?
   (b) At that time, how much is car *B* ahead of car *A*?
   (c) Which car is ahead, and by how much, at the end of 0.010 hour?
   (d) At what time does car *A* catch up with car *B*?
   (e) How far have they traveled from the traffic light by the time car *A* catches up?

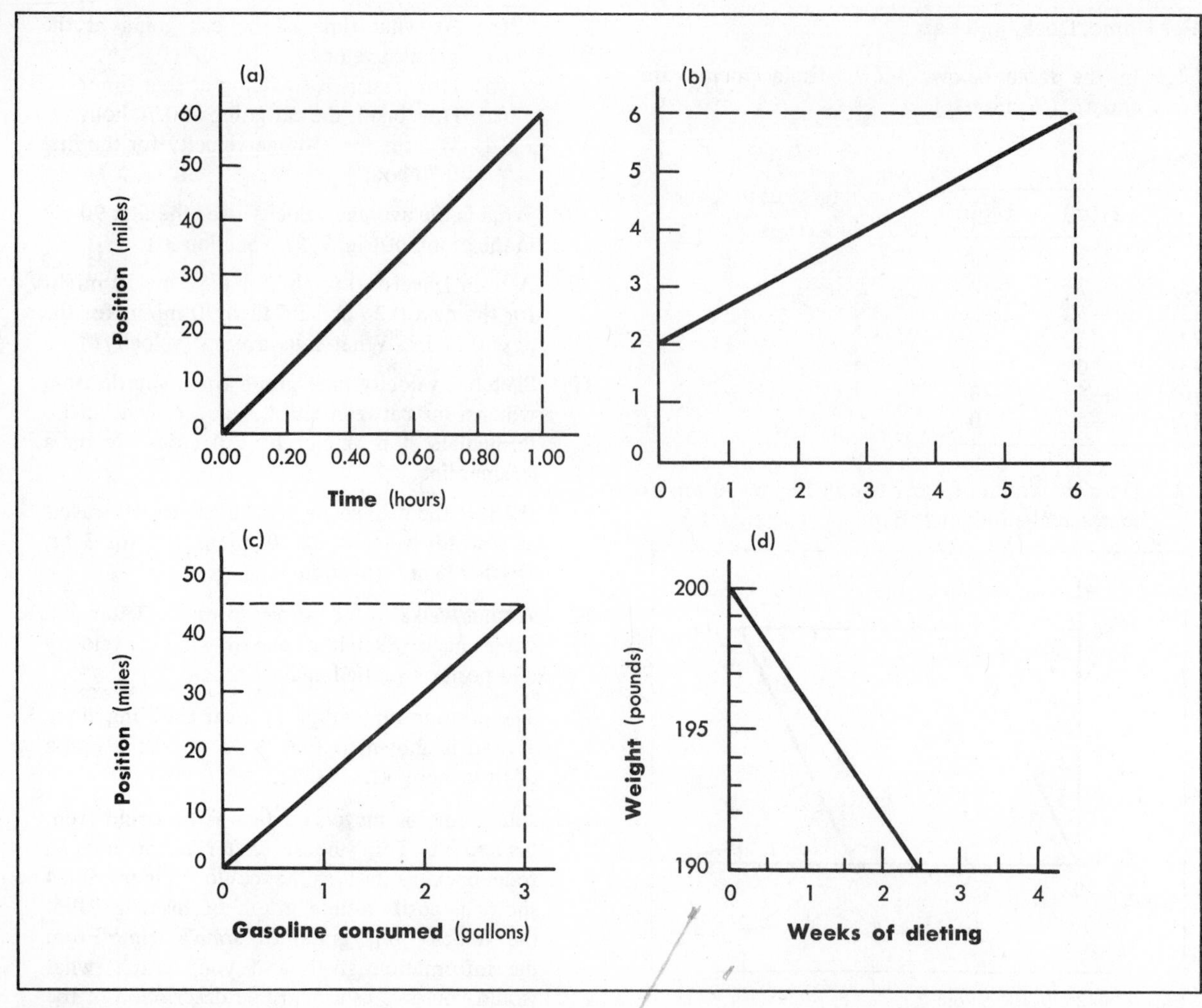

Figure 9–31 For Problem 5.

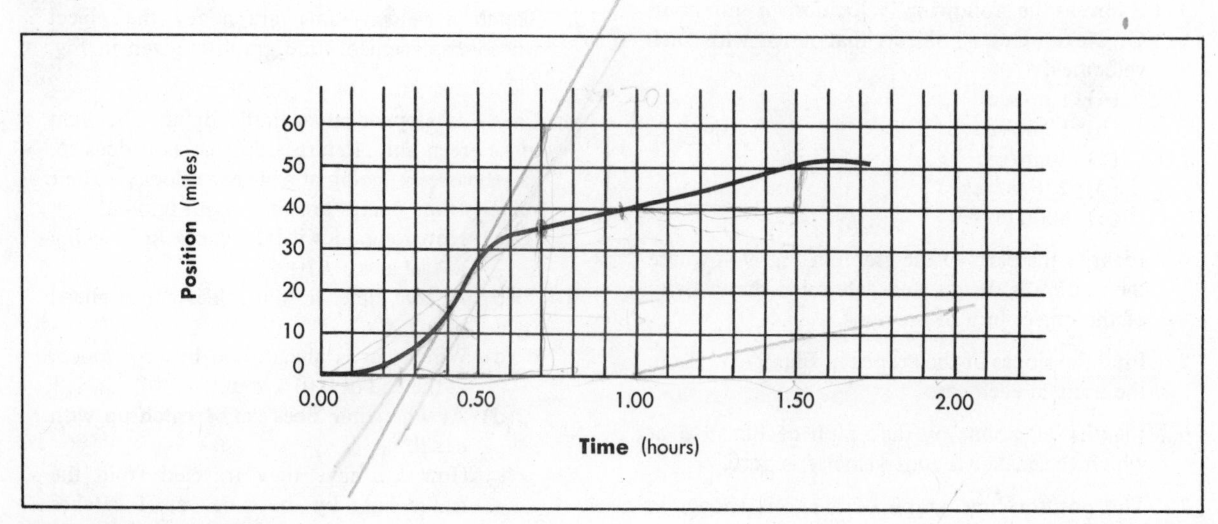

Figure 9–32 For Problem 7.

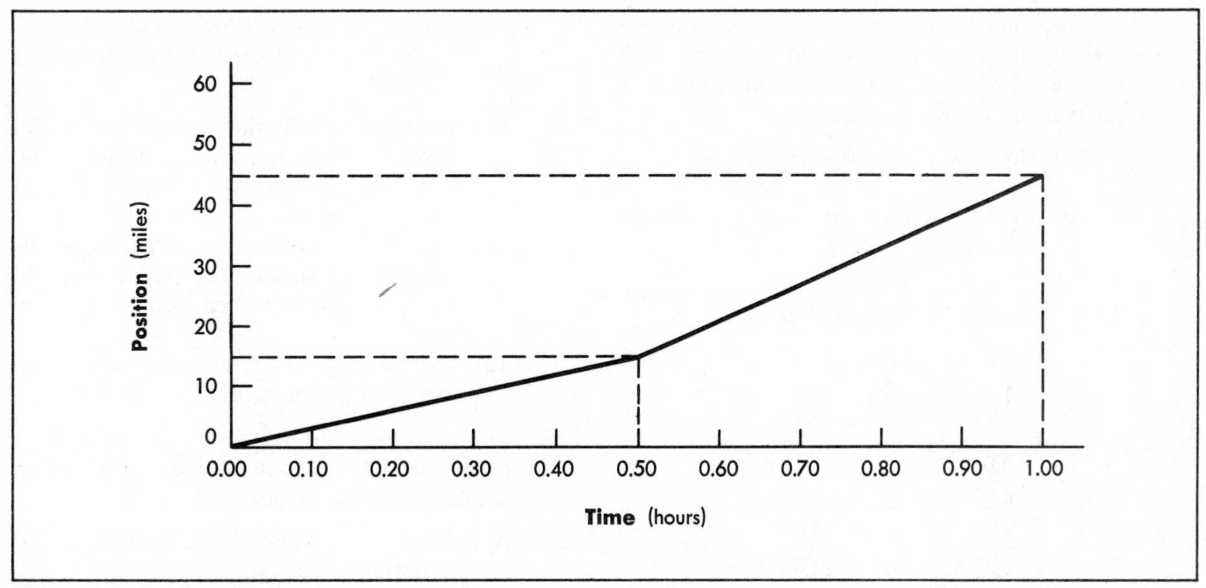

Figure 9–33 For Problem 13.

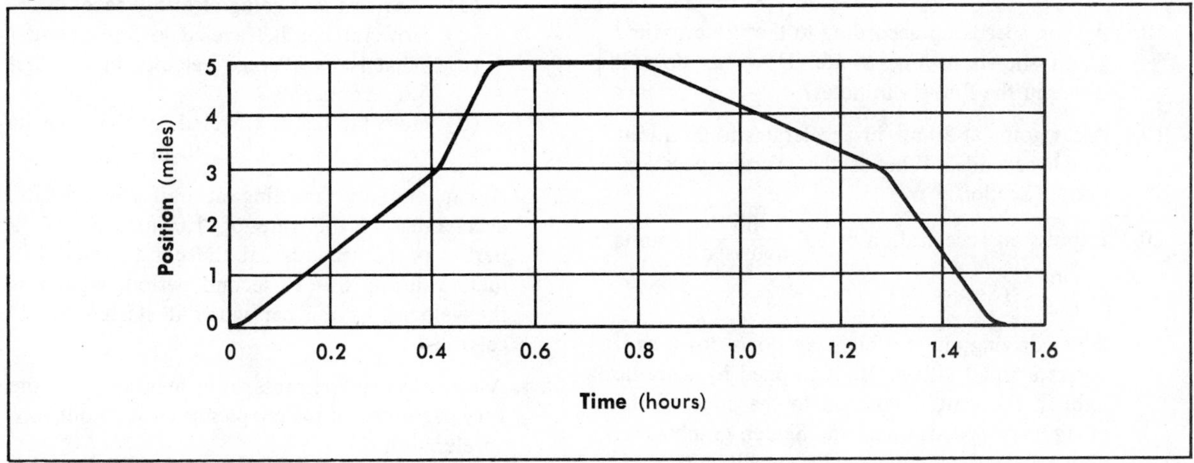

Figure 9–34 For Problem 14.

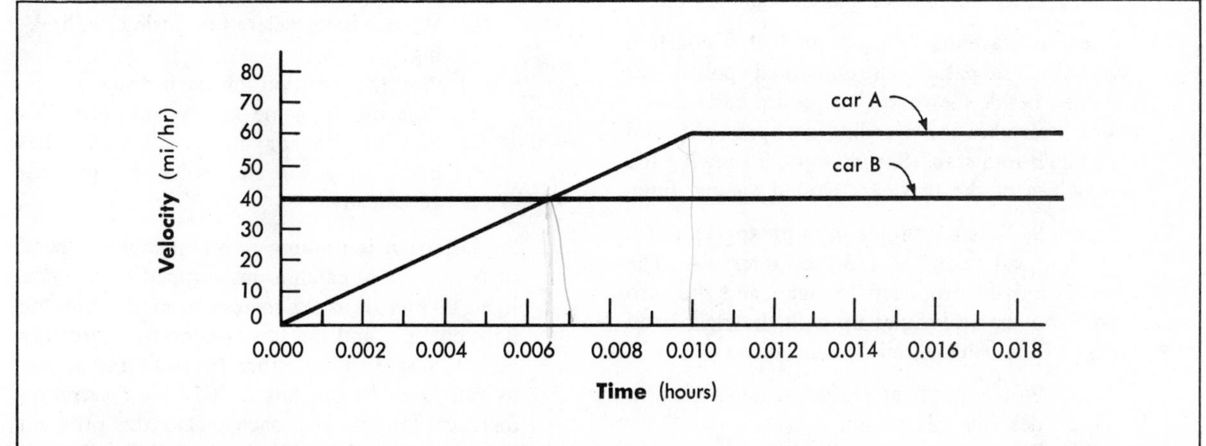

Figure 9–35 For Problem 16.

17. The accompanying data show the instantaneous velocity of a car at intervals of 1 second. Plot the velocity versus time, and use your graph to answer the following questions.

    (a) How fast is the car going at 2.6 sec? At 4.8 sec?
    (b) How far did the car travel between the two instants in part (a)?

TIME (SEC)	VELOCITY (M/SEC)
0.0	10.0
1.0	12.4
2.0	14.8
3.0	17.2
4.0	19.6
5.0	22.0
6.0	24.4

18. A train speeds up according to the velocity-time graph shown in Fig. 9–36. How far does it travel in the first six minutes?

19.* A car going at 20 mi/hr accelerates to 60 mi/hr in 6.0 seconds. What is the average acceleration? (Section 6.)

20.* Express an acceleration of $1.0 \frac{\text{mi}}{\text{hour-sec}}$ in units of $\frac{\text{m}}{\text{sec}^2}$. (Section 6.)

21. A car driving along a highway slows down as it enters a small village. It is stopped by a traffic light in the center, goes on to the edge of the village, and speeds up again. Sketch graphs, one above the other, showing its velocity, its position, and its acceleration plotted against time.

22. A car is traveling along a road at a constant velocity. It passes an unmarked police car parked beside the road. The police car accelerates, overtakes the speeding car, passes it, and signals it to a stop. Sketch a graph showing the velocities of the two cars plotted against time.

23. Figure 9–37 is a multiple-flash photograph of a moving ball taken at 1/30-sec intervals. The ball is moving from left to right, and the zero point on the scale lines up with the right-hand edge of the ball's initial position.

    (a) Plot a graph of position against time to describe this motion.
    (b) From your graph in (a), construct a velocity-time graph.
    (c) What does your velocity-time graph tell you about the acceleration of the moving ball?

24. (a) Sketch a velocity-time graph for a ball thrown straight up from the time it is still at rest in the hand till it is momentarily at rest on the ground.
    (b) Should the area between the curve and the time axis above the time axis be equal to the corresponding area below the time axis?
    (c) Use the graph in (a) to sketch the acceleration-time graph for the same motion.

25. From the graph of velocity versus time for a car shown in Fig. 9–38, deduce the graph of acceleration versus time.

26. A bobsled has a constant acceleration of 2.0 m/sec^2 starting from rest.

    (a) How fast is it going after 5.0 seconds?
    (b) How far has it traveled in 5.0 seconds?
    (c) What is its average velocity in the first 5.0 seconds?
    (d) How far has it traveled by the time its velocity has reached 40 m/sec?

27. A car, initially traveling at uniform velocity, accelerates at the rate of 1.0 m/sec^2 for a period of 12 seconds. If the car traveled 190 meters during this 12-second period, what was the velocity of the car when it started to accelerate?

28.† Assume that the express train mentioned at the very beginning of the chapter stops with uniform acceleration.

    (a) How long does it take to stop?
    (b) What is the acceleration during the braking?
    (c) Plot the position of each train versus time on the same graph and from this find whether or not a collision takes place. If you prefer, solve this part algebraically.

29. A pedestrian is running at his maximum speed of 6.0 m/sec to catch a bus stopped by a traffic light. When he is 25 meters from the bus the light changes and the bus accelerates uniformly at 1.0 m/sec^2. Find either (a) how far he has to run to catch the bus or (b) his frustration distance (closest approach). Do the problem either by use of a graph or by solving the appropriate equations.

30. A rocket which placed a satellite in orbit attained a velocity of 2.90×10^4 km/hr in 2.05 minutes.

(a) What was the average acceleration, in km/hr-sec? In m/sec^2?

(b) If the rocket had enough fuel to maintain this same acceleration for an hour, what velocity would it have at the end of the hour if it started from rest?

(c) How far would it travel during this hour?

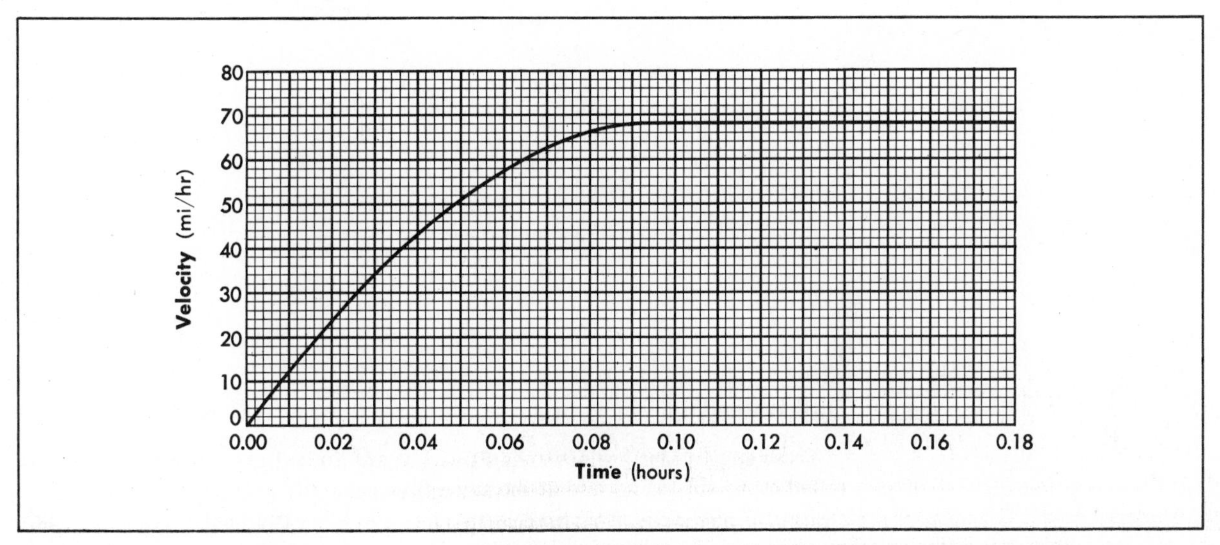

Figure 9–36 For Problem 18.

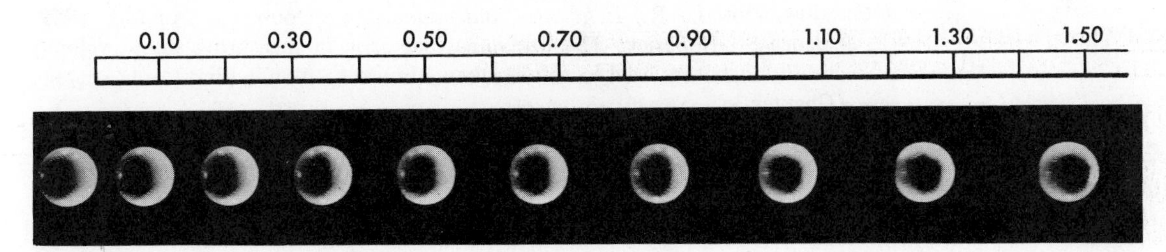

Figure 9–37 For Problem 23. The scale is measured in meters.

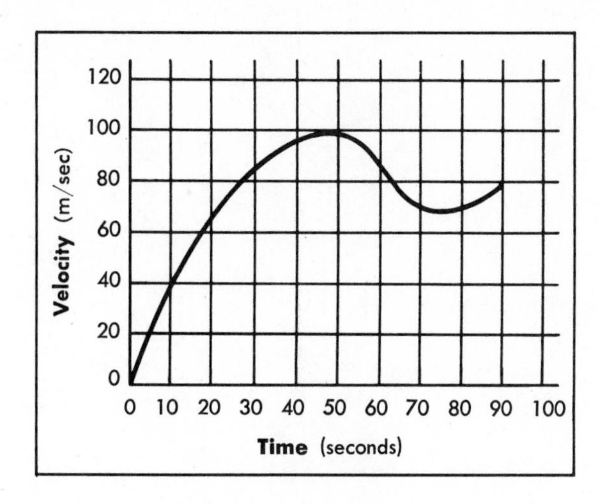

Figure 9–38 For Problem 25.

Further Reading

These suggestions are not exhaustive but are limited to works that have been found especially useful and at the same time generally available.

COHEN, I. BERNARD, *The Birth of a New Physics*. Doubleday Anchor, 1960: Science Study Series. Aristotelian concepts of motion are contrasted with the inertial and accelerated motions of modern physics. (Chapters 2 and 5)

GRIFFIN, DONALD R., *Echoes of Bats and Men*. Doubleday Anchor, 1959: Science Study Series. The problems that arise in measuring sound velocity and the ingenuity used in solving them should be illuminating to the student. (Chapter 3)

HOLTON, G., and ROLLER, D. H. D., *Foundations of Modern Physical Science*. Addison-Wesley, 1958. A good parallel discussion. (Chapter 1)

Vectors

CHAPTER

10

In the last chapter we considered only motion along a straight line—motion in only one dimension. Displacement, velocity, and acceleration all had to point along the same straight line; they could take on positive or negative values as the situation demanded, but were allowed no other freedom of direction. Clearly, most situations in nature cannot be fitted into such a strict set of requirements; the motion of a swimmer going across a river, the course of an airplane on a windy day, the path of a satellite orbiting the earth. We cannot describe such motions with the methods of Chapter 9. We must extend those methods to include the possibility of displacements, velocities, and accelerations in several different directions, perhaps all occurring at the same time; we must learn to treat motion in a three-dimensional world. However, much of the extension required to go from one to three dimensions is already present in a description of two-dimensional motion. We shall concentrate, then, on motion on a plane surface.

10-1 Position and Displacement

For motion along a straight line, we described the position of an object in terms of its distance and direction from a fixed landmark. Thus we said, "The car is 5 miles west of the center of town," and we had only two directions to consider, east and west. If we now allow the car to turn off the straight highway and wander about on the back roads, we can describe its position in a similar way, but we need to give more information. We might say, for example, "The car is 5.5 miles west and 1.2 miles north of the center of town." We are describing its *two* coordinates, measured along two perpendicular reference directions in the plane. In more general terms, we can represent the two reference directions by two perpendicular coordinate lines, x and y, each with positive and negative directions and each with appropriate units. A position on the plane is then given by two coordinate numbers; the x coordinate is stated first, the y coordinate second. Several points are located by their coordinates in Fig. 10–1.

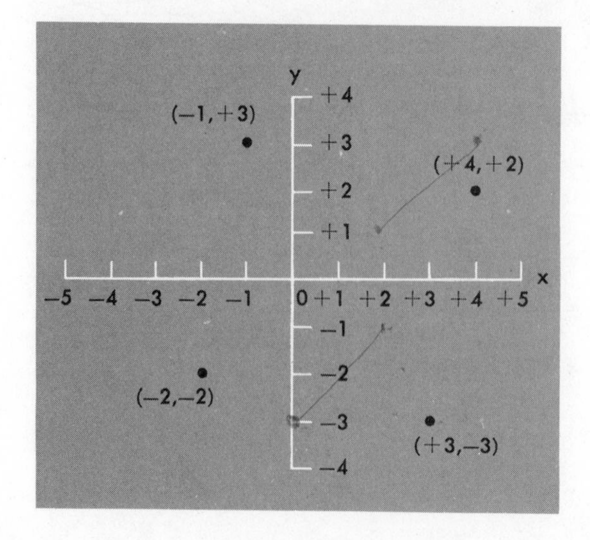

Figure 10–1
Position on a plane can be specified by the two coordinate numbers of the point: first, the distance out from the origin along the x axis; second, the distance up, parallel to the y axis.

Suppose that in Fig. 10–2 the position (+ 4, + 2) is the position of a ship relative to a lighthouse at the origin and the x and y scales represent distances in miles. The ship is moving, and some time later it is at the position (+ 5, + 4) as shown by Fig. 10–2. The arrow drawn between these two points is the displacement of the ship. Since displacements depend on direction as well as on magnitude, we shall indicate them by a boldface "d" with a half arrow over it ($\vec{\mathbf{d}}$). In this case, its length gives the distance between the ship's initial and final position, and the direction of the displacement is just the direction of the large arrow in the figure.

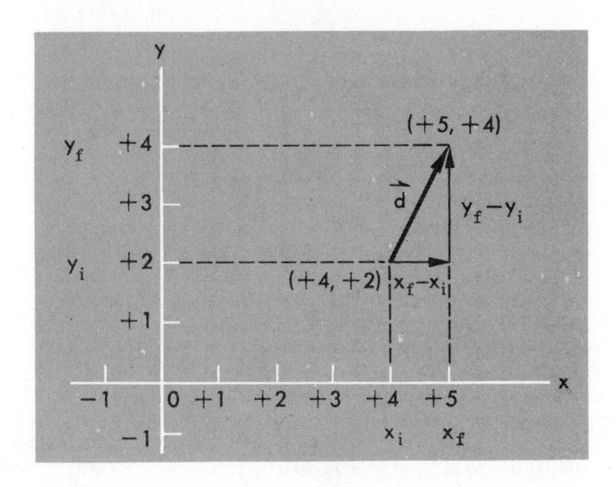

Figure 10–2
The arrow between the positions (+4, +2) and (+5, +4) is a displacement in two dimensions.

We can also have displacements that do not lie in the plane of the x and y axes, as is shown in Fig. 10–3. To describe such a displacement we construct a third axis, the z axis perpendicular to the x, y plane. In this case, to describe the displacement we need six coordinates instead of four. They are (x_i, y_i, z_i) and (x_f, y_f, z_f). In Fig. 10–3 they are (4, 1, 3) and (6, 2, 5) respectively.

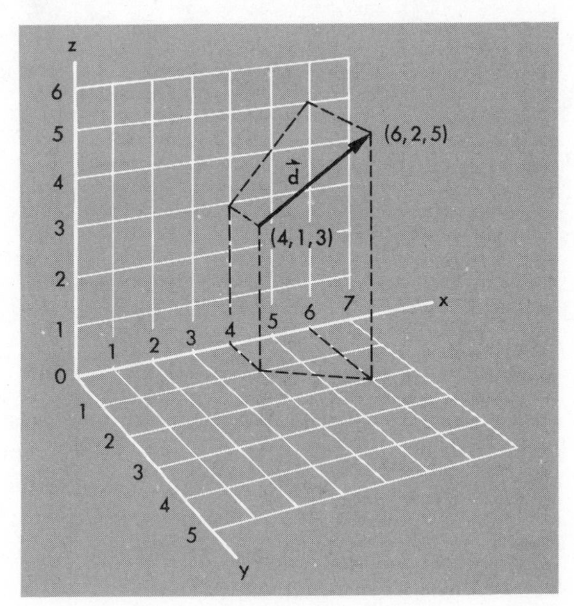

Figure 10–3
The x and y axes lie in a horizontal plane, and the z axis is in a vertical plane. The displacement $\vec{\mathbf{d}}$ is not in either plane. Its tail is 3 units above the x-y plane and its head is 5 units above the x-y plane.

Returning to displacements in the x, y plane, Fig. 10–4 shows two displacements in one plane with both positive and negative coordinates. Notice that although the two displacements have different coordinates, they have the same magnitude and direction. The two are therefore equal displacements but in different locations. Thus we see that a displacement can be moved to any location and remain unchanged.

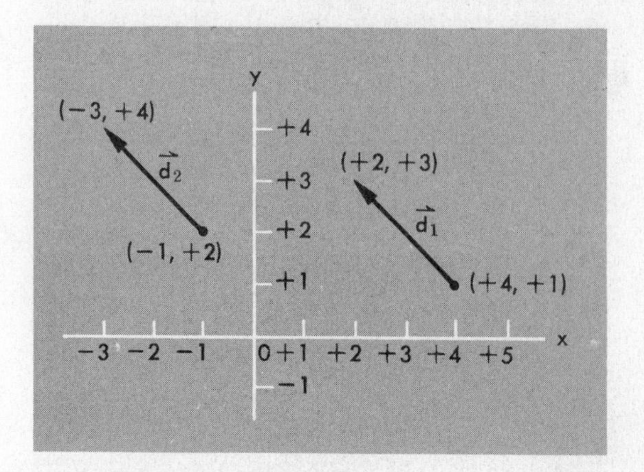

Figure 10–4
Two equal displacements lying in different quadrants.

A displacement is also independent of the position of the origin of the coordinate system, as Fig. 10–5 shows. Again, although the coordinates describing the positions change, the displacements themselves remain unchanged. In this respect the situation in two or three dimensions is the same as in one dimension.

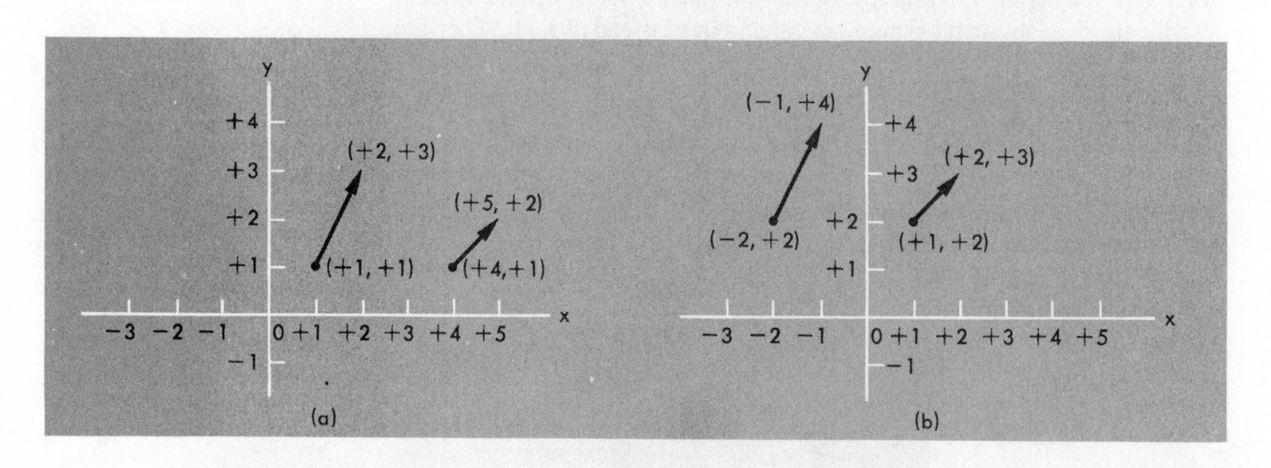

Figure 10–5
In (b) the origin in (a) has been moved down one unit on the y axis and 3 units to the right on the x axis.

Although it is often convenient to describe displacements by specifying their x and y displacements in some coordinate system, it is not necessary to do this because, as we have just seen, displacements are independent of any coordinate system. We are often interested only in the relation of one displacement to another, and not in their coordinates; in such cases we dispense with a coordinate system, and use a geometric representation in terms of arrows only.

In Chapter 9, displacements along a line were given by positive and negative numbers, and when such displacements were added they simply obeyed the rules of ordinary arithmetic. Displacements in two or three dimensions do not add according to these rules. In Fig. 10–6 we have added $\vec{s}$ to $\vec{r}$ to

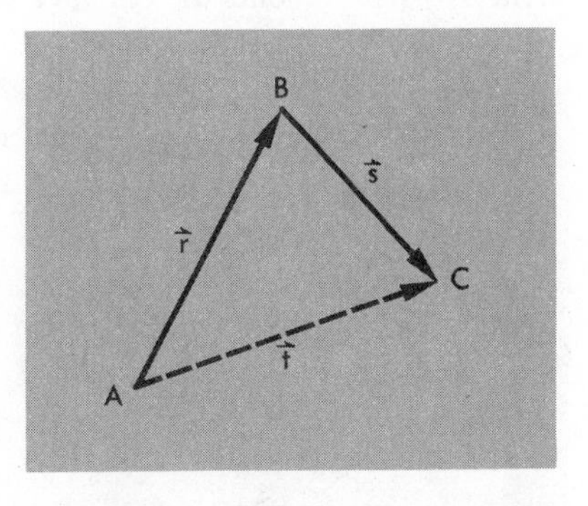

Figure 10–6
The displacement $\vec{\mathbf{s}}$ can be added to the displacement $\vec{\mathbf{r}}$ by the use of a scale drawing. In this case 0.75 cm = 1 m. The displacement $\vec{\mathbf{r}}$ therefore represents 4.0 m, $\vec{\mathbf{s}}$ represents 3.0 m, and the sum, $\vec{\mathbf{t}}$, has a magnitude of 4.2 m.

get the displacement $\vec{\mathbf{t}}$, by drawing the line between the tail of one arrow and the head of the other. The dashed arrow is the sum of $\vec{\mathbf{s}}$ and $\vec{\mathbf{r}}$. In symbols we write

$$\vec{\mathbf{s}} + \vec{\mathbf{r}} = \vec{\mathbf{t}}.$$

As you can see, if $r = 4.0$ m and $s = 3.0$ m, t is less than 7.0 m. In fact, it is only 4.2 m. The equation $\vec{\mathbf{s}} + \vec{\mathbf{r}} = \vec{\mathbf{t}}$ in this case means that moving an object first from A to B and then from B to C is exactly equivalent to moving the object directly from A to C; the object ends up at the same position in either case.

We shall often want to add two displacements even though they may not share any common point. As we have seen in Section 10–1, displacements are independent of location, so we shall add displacements by placing them "head to tail" as in Fig. 10–6. Thus if we want to find the sum of the two displacements shown in Fig. 10–7(a), we can move one of them "parallel to itself" until its "tail" coincides with the "head" of the other, as shown in Fig. 10–7(b).

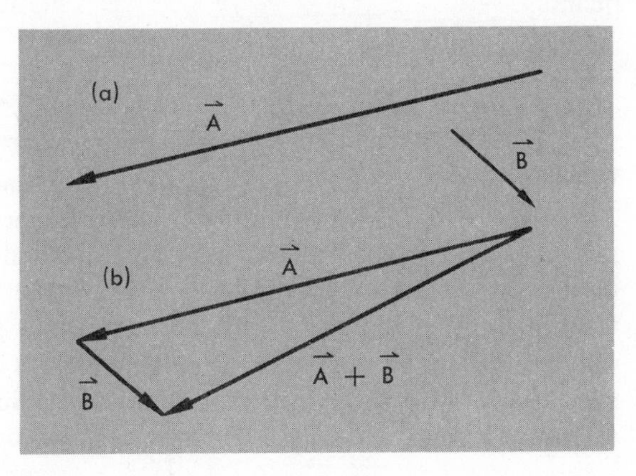

Figure 10–7
Addition of two vectors after moving one parallel to itself.

To subtract one displacement from another we simply place the two displacements tail-to-tail and draw the line connecting their tips. The direction of this displacement is from the displacement being subtracted to the other displacement (Fig. 10–8). If, for example, $\vec{\mathbf{a}}$ represents your distance and direction from your house (4.5 miles) and $\vec{\mathbf{b}}$ represents your distance and direction from your house some time later (6.3 miles), then the displacement $\vec{\mathbf{b}} - \vec{\mathbf{a}}$ represents the distance (3 miles) between P_1 and P_2.

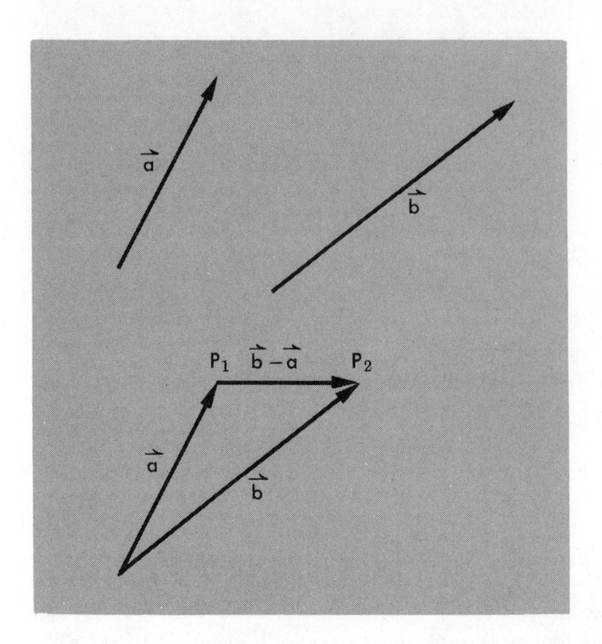

Figure 10–8
To subtract the two vectors at the top, we place them tail-to-tail and draw the arrow connecting their tips.

There are other quantities of importance in physics which add and subtract like displacements. All such quantities are called *vectors.* We shall always represent vectors in diagrams by heavy arrows having lengths proportional to the magnitudes and having directions indicated by the arrowheads. Notice that vectors are quantities which cannot be described by single numbers. *Scalars,* on the other hand, are physical quantities which can be specified by single numbers (and the appropriate physical units). The temperature of a room is a good example of a scalar physical quantity, just as a displacement of 3 kilometers northeast is a good example of a vector quantity.

10-3 Vector Components

If a jet plane flies from Los Angeles to San Francisco, its displacement is 350 miles. How far west and how far north has it traveled? The answer to this question depends both on the magnitude and the direction of the displacement. To get the westward displacement of the plane we draw a line from the head of the plane's displacement arrow perpendicular to an east-west line passing through the tail (Fig. 10–9). The displacement $\vec{\mathbf{d}}_w$ is the displacement we are looking for; it is 230 miles. Similarly, to find the northward displacement we draw a line from the head of the plane's displacement

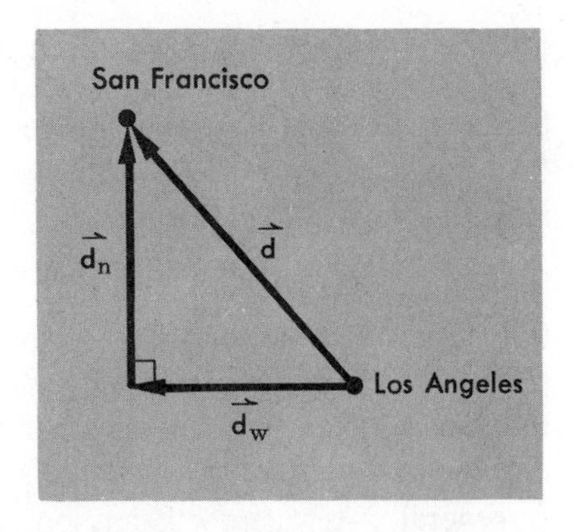

Figure 10–9
$\vec{\mathbf{d}}_w$ is the westward displacement of $\vec{\mathbf{d}}$, and $\vec{\mathbf{d}}_n$ is the northward displacement.

arrow perpendicular to a north-south line as shown in the figure. The displacements $\vec{\mathbf{d}}_w$ and $\vec{\mathbf{d}}_n$ in Fig. 10–9 are called *vector components* of $\vec{\mathbf{d}}$.

Whenever we construct a pair of vector components that are perpendicular to each other, their vector sum is always equal to the original vector. This you can see from Fig. 10–9.

We can resolve any vector in a plane into two vector components along two perpendicular axes (Fig. 10–10). Note that while the vectors $\vec{\mathbf{A}}$, $\vec{\mathbf{B}}$, and $\vec{\mathbf{C}}$ can have any direction, $\vec{\mathbf{A}}_x$, $\vec{\mathbf{B}}_x$, and $\vec{\mathbf{C}}_x$ necessarily point along the x axis in either the positive or negative direction. Since they all lie along the same line, we can represent their length by a number in suitable units and indicate which way they point by a "+" or "−" sign. Thus, when we deal only with vector components along the x axis, we can dispense with the vector notation by simply writing A_x, B_x, C_x. The situation is similar with respect to the y axis. $\vec{\mathbf{A}}_y$, $\vec{\mathbf{B}}_y$, and $\vec{\mathbf{C}}_y$ all point up or down along the y axis. They, too, can be represented by positive or negative numbers A_y, B_y, C_y. The vector sum $\vec{\mathbf{A}}_x + \vec{\mathbf{B}}_x$ or $\vec{\mathbf{A}}_y + \vec{\mathbf{B}}_y$ is represented by $A_x + B_x$ and $A_y + B_y$ respectively. However, $\vec{\mathbf{A}}_x + \vec{\mathbf{A}}_y$ *cannot* be represented by $A_x + A_y$, since the $\vec{\mathbf{A}}_x$ and $\vec{\mathbf{A}}_y$ do not point along the same line. To distinguish the vector $\vec{\mathbf{A}}_x$ from the number A_x, we shall refer to the first as a *vector component* and to the second simply as a *component*. Describing a vector $\vec{\mathbf{A}}$ in terms of its components A_x and A_y is the same as describing a point in a plane by its two coordinates A_x and A_y.

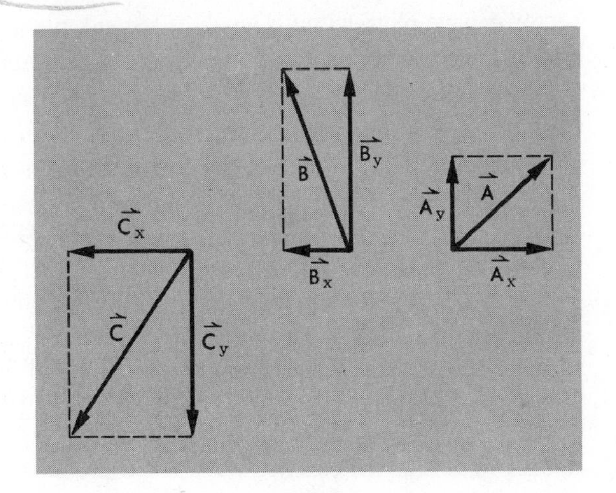

Figure 10–10
Resolving vectors into vector components along the x and y axes.

10-4 Multiplying Vectors by Scalars

Suppose that you are given any vector $\vec{\mathbf{A}}$. What do you think is meant by $2\vec{\mathbf{A}}$? Because multiplying a quantity by two means adding the quantity to itself, it is logical to define

$$2\vec{\mathbf{A}} = \vec{\mathbf{A}} + \vec{\mathbf{A}};$$

that is, $2\vec{\mathbf{A}}$ is a vector twice as long as $\vec{\mathbf{A}}$ pointing in the same direction [Fig. 10–11(a)]. Similarly, since division by a number is equivalent to multiplying by the reciprocal of the number [Fig. 10–11(b)] we can expect division by a number to follow the same rules as multiplication by a number. In general, $k\vec{\mathbf{A}}$, where k is any positive number, means a vector parallel to $\vec{\mathbf{A}}$ and k times as long.

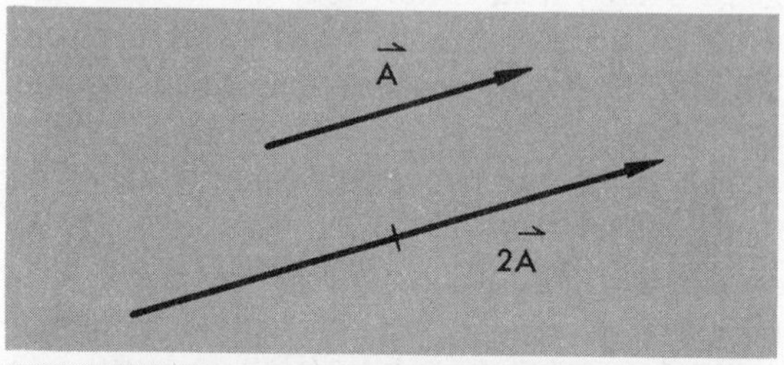

Figure 10–11a
Multiplication of a vector $\vec{\mathbf{A}}$ by the scalar $k = 2$.

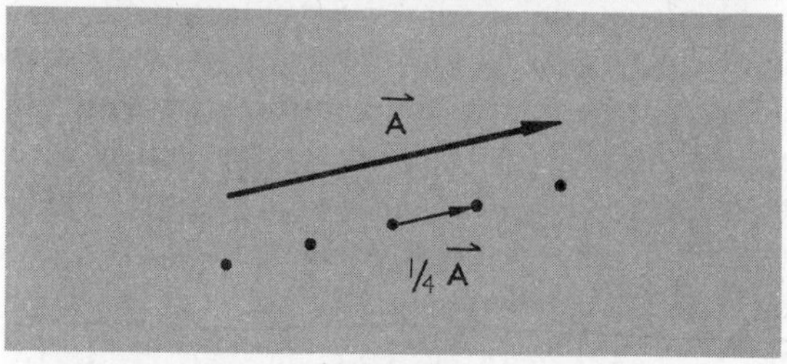

Figure 10–11b
Dividing a vector by 4 is equivalent to multiplying it by ¼.

Briefly, then, multiplying or dividing a vector by an ordinary number means multiplying or dividing the magnitude by the absolute value* of that number. The direction remains unchanged if the number is positive, and is reversed if the number is negative.

Now we extend this rule to cases in which both the scalar k and the vector $\vec{\mathbf{A}}$ have physical units, such as second, mile, 1/second, miles/second, etc. If k were a pure number, with no physical units, then $k\vec{\mathbf{A}}$ would have the same units as $\vec{\mathbf{A}}$. But when k has the units of a physical quantity, the units of $k\vec{\mathbf{A}}$ are found by multiplying the units of k by the units of $\vec{\mathbf{A}}$. In other words, $k\vec{\mathbf{A}}$ is a different physical quantity from $\vec{\mathbf{A}}$. It would make no sense, for example, to add $k\vec{\mathbf{A}}$ to $\vec{\mathbf{A}}$ unless k were a pure number. To avoid confusion, it is best to represent the new vector on a new diagram.

What we have just said about division of vectors by scalars applies directly to acceleration and velocity. To get both of these we divide vector displacements by time. This gives just another vector, as we have seen, and so both velocity and acceleration are vectors and have all the properties of vectors.

* The *absolute value* of a number is the number without its algebraic sign. Thus, the absolute value is always positive. For example, the absolute value of -3.5 is 3.5.

We are sometimes concerned only with the magnitude of a velocity vector and not its direction. In such cases we use the word *speed* when we refer only to the magnitude of the velocity. Since we have shown that velocity is a vector quantity, we should therefore write a general definition of velocity in vector notation. For a displacement $\Delta\vec{\mathbf{d}}$ occurring in a time interval Δt, the average velocity during the interval is

$$\vec{\mathbf{v}}_{\text{av}} = \frac{\Delta\vec{\mathbf{d}}}{\Delta t}.$$

Clearly $\vec{\mathbf{v}}_{\text{av}}$ has the direction of $\Delta\vec{\mathbf{d}}$.

If the time interval is taken ever smaller, in the limit the instantaneous velocity is

$$\vec{\mathbf{v}} = \lim_{\Delta t \to 0} \frac{\Delta\vec{\mathbf{d}}}{\Delta t}.$$

The direction of $\vec{\mathbf{v}}$ is just along the tangent to the path at the point under consideration.

In order for an instantaneous velocity to be called "constant," two things must be constant: its magnitude *and* its direction. A body moving with constant velocity would, therefore, move in one direction along a straight line at constant speed.

10-5 Velocity Changes and Constant Vector Acceleration

Figure 10–12 is a multiple-flash photograph of a ball projected horizontally with an initial velocity of 2.0 m/sec. Let us analyze its motion by finding its instantaneous velocity vectors at successive intervals as it falls.

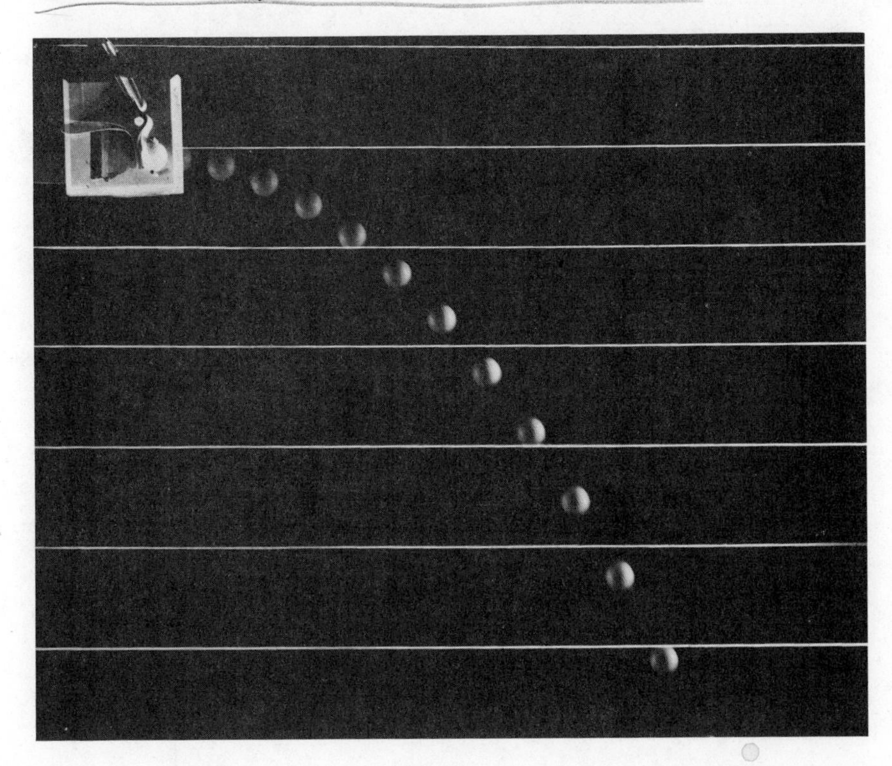

Figure 10–12
A flash photograph of a golf ball projected horizontally from the mechanism shown with an initial velocity of 2.00 m/sec. The light flashes were 1/30 second apart. The white lines in the figure are a series of parallel strings placed behind the golf ball, 15 cm apart. Why do the strings appear to be in the foreground?

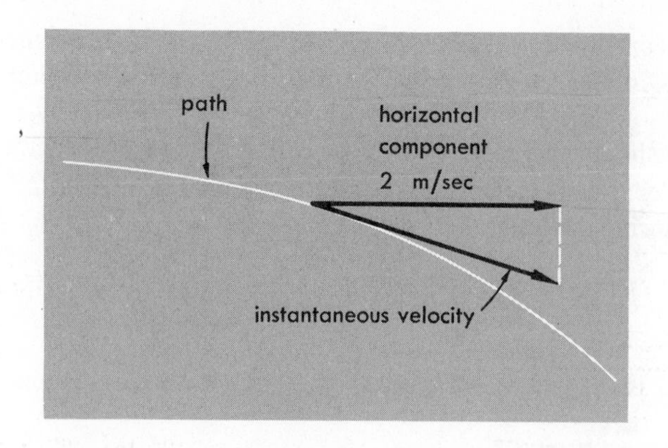

Figure 10–13
How to find the instantaneous velocity vector of the ball in Fig. 10–12. It is tangent to the path and of such length that its horizontal component is equal to the initial horizontal velocity of the projectile.

One way of finding the instantaneous velocity in this case is to note that the horizontal component of the velocity is constant. This follows from the fact that the horizontal displacement is the same in each time interval. We then get $\vec{\mathbf{v}}$ from this fact and the fact that the instantaneous velocity vector always points in the direction of the path (Fig. 10–13). Using this method we plotted the instantaneous velocity vectors at equal time intervals (Fig. 10–14).

Figure 10–14
The position and velocity of the golf ball in Fig. 10–12 are shown here on a single graph.

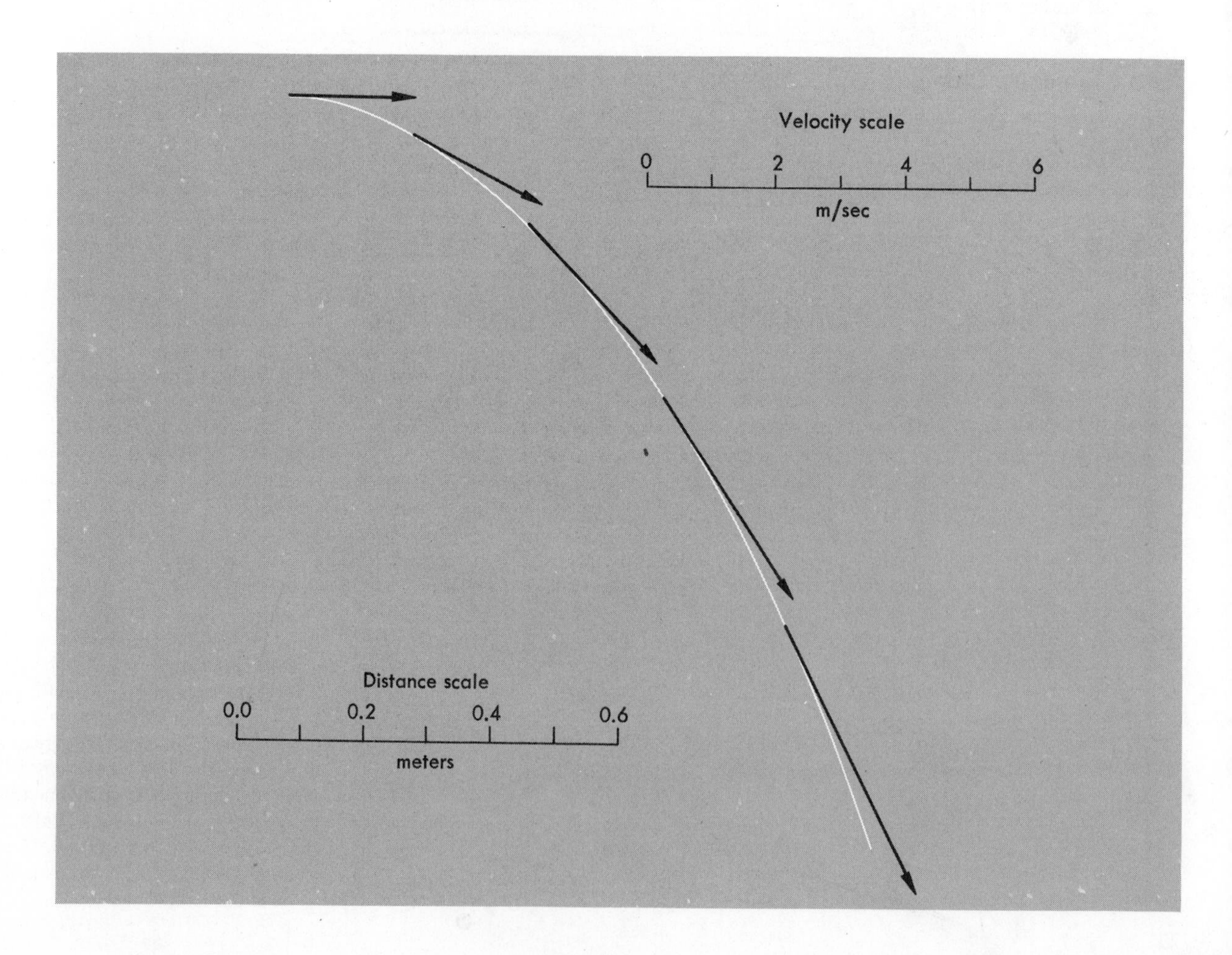

Figure 10–15 shows that the instantaneous velocity vector of the ball changes steadily. In each successive interval it increases by the same amount, $\Delta\vec{\mathbf{v}}$. The average acceleration is therefore constant. Because the average acceleration is constant throughout the motion, we can be almost certain that the instantaneous acceleration is also constant throughout the motion. To confirm this, of course, we might want to take additional strobe photographs of the falling ball, using shorter and shorter time intervals to approach the limit as $\Delta t \to 0$. But even without this we can be pretty certain of our conclusion. Consequently, for the instant of the nth flash, we can write the instantaneous velocity as $\vec{\mathbf{v}}_n = \vec{\mathbf{v}}_i + n\Delta\vec{\mathbf{v}}$. Here $\vec{\mathbf{v}}_i$ is the instantaneous velocity vector with which we started, and $\Delta\vec{\mathbf{v}}$ is the constant change that occurs in each interval. By adding n of these changes to the original velocity, we get the velocity n intervals further along.

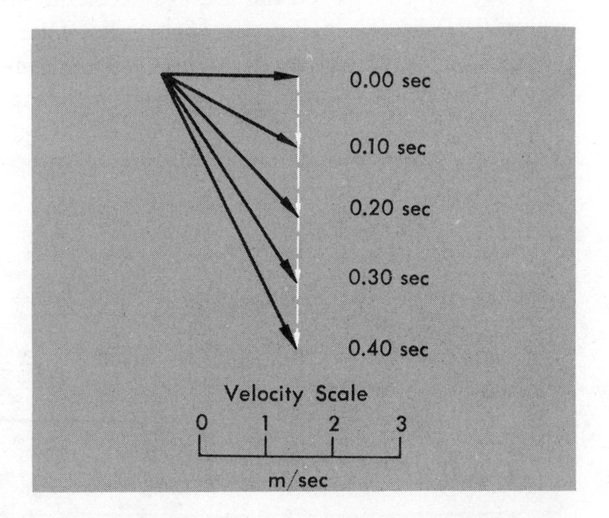

Figure 10–15
A sequence that shows only the velocity vectors of Fig. 10–14. Successive vectors are found by adding a constant vector directed vertically downward.

We can rewrite the last equation so that it more closely resembles the equations we developed for the description of motion along a straight-line path. (See Chapter 9, especially Section 9–7 and box.) There we defined the acceleration along the path as $a = \Delta v/\Delta t$. Here by dividing $\Delta\vec{\mathbf{v}}$ by Δt we shall introduce the vector acceleration $\vec{\mathbf{a}} = \Delta\vec{\mathbf{v}}/\Delta t$. Using it, our last equation becomes

$$\vec{\mathbf{v}}_n = \vec{\mathbf{v}}_i + n\Delta t\,\frac{\Delta\vec{\mathbf{v}}}{\Delta t}.$$

Replacing $n\Delta t$ by t, the time during which the velocity has changed from its initial value $\vec{\mathbf{v}}_i$ to its value when the nth flash occurs, we can express the velocity at time t as $\vec{\mathbf{v}} = \vec{\mathbf{v}}_i + \vec{\mathbf{a}}t$. Note that $\vec{\mathbf{v}}_i$ is in the horizontal direction, while $\vec{\mathbf{a}}$ is downward. Therefore, as time progresses, $\vec{\mathbf{v}}$ will point more and more downward.

10-6 Changing Acceleration

In the last section we described motion with constant vector acceleration. We introduced the vector acceleration to describe how the velocity vector changes. When the acceleration vector is not itself constant, we can define the average acceleration $\vec{a}_{av}$ by $\vec{a}_{av} = \Delta\vec{v}/\Delta t$, where $\Delta\vec{v}$ is the vector change in $\vec{v}$ during the time interval Δt. This vector acceleration has the same direction as the change $\Delta\vec{v}$ of the velocity. Since this change need not be in the same direction as $\vec{v}$, the acceleration $\vec{a}$ may point in any direction with respect to the motion, as we saw in the case of the ball in the last section.

If the acceleration is itself changing as time goes on, the average $\vec{a}$ will depend on the time interval we choose. Let us take an example. Suppose a speedboat moves along the path shown in Fig. 10–16(a). At time t_1, it is moving with the vector velocity $\vec{v}_1$, and at the later time t_2 it is moving with vector velocity $\vec{v}_2$. What is the average acceleration in the time interval $\Delta t = t_2 - t_1$? To find out we must determine the vector change $\Delta\vec{v} = \vec{v}_2 - \vec{v}_1$ in the velocity. Then the average acceleration is $\vec{a} = \Delta\vec{v}/\Delta t$. The procedure is indicated in Fig. 10–16(b). We take the vector difference between $\vec{v}_2$ and $\vec{v}_1$, divide $\Delta\vec{v}$ by Δt, and plot the average acceleration vector $\vec{a}$ for the time Δt on a new diagram with an appropriate scale.

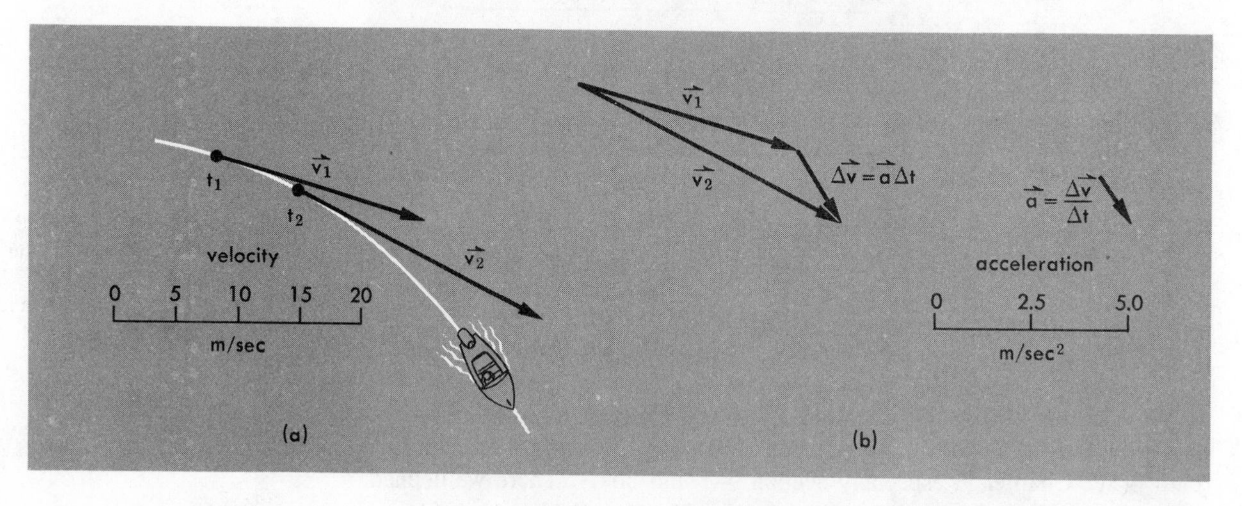

Figure 10–16
The white line shows the path of a speedboat. To find the average acceleration in the interval $\Delta t = t_2 - t_1$, first find the vector difference $\Delta\vec{v} = \vec{v}_2 - \vec{v}_1$ and divide this by Δt. The result is the average acceleration vector $\vec{a}$, which can now be plotted with an appropriate scale as shown above.

As the speedboat moves on, its acceleration may change. In fact, we do not expect that it has remained constant even over the time interval from t_1 to t_2. In such situations we need to know the instantaneous vector acceleration at various times t rather than only the average acceleration vector over various intervals.

To find the instantaneous acceleration vector at any particular time t, consider Fig. 10–17, which shows the successive velocity vectors of our speedboat at time intervals of 10 seconds. If we make a composite picture in which the tails of these vectors coincide, we get Fig. 10–18. The changes in velocity during each successive time interval are also shown in this figure. We see that the velocity changes $\Delta\vec{v}$ in each of the four intervals differ both in magnitude and in direction. Therefore, since the time intervals are the same, the average accelerations are different.

If, instead of using 10-second intervals, we had used 2-second intervals, we would have gotten the pictures shown in Fig. 10–19. Since it is difficult to see the details, a magnified portion of the first 10 seconds is shown in Fig.

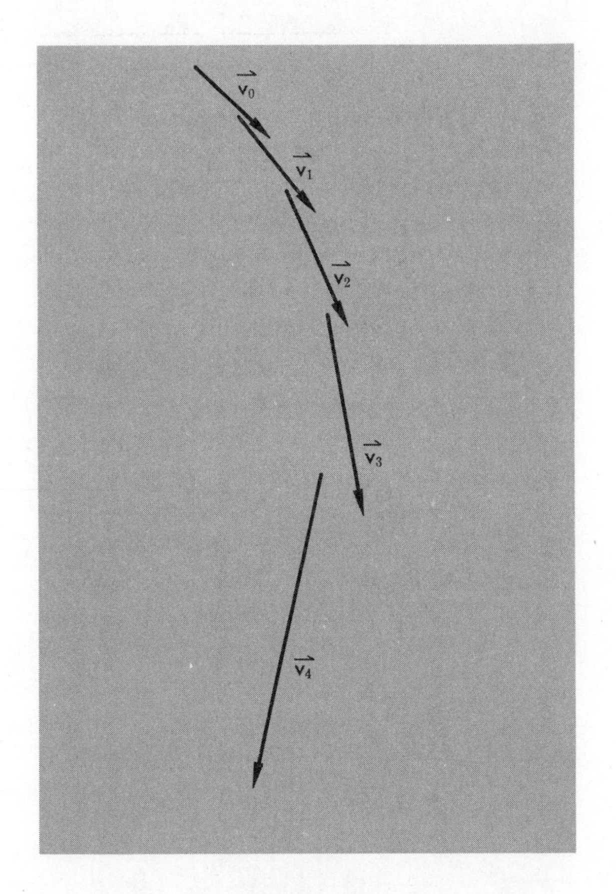

Figure 10–17
Successive velocity vectors of the speedboat at 10-second time intervals.

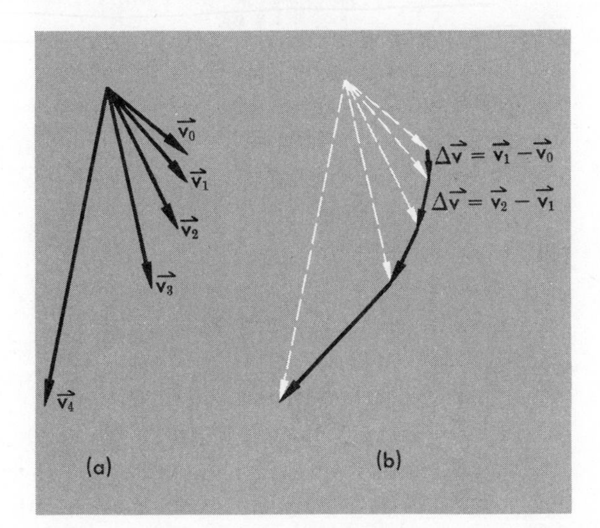

Figure 10–18
The vectors of Fig. 10–17 are shown here drawn from the same origin. The acceleration is not constant in direction or magnitude, as shown in (b).

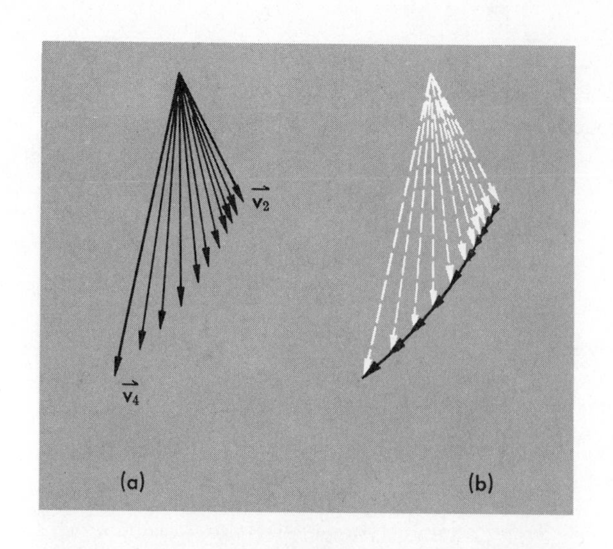

Figure 10–19
(a) If we take 2-second time intervals, the motion described in Fig. 10–17 looks like this when the vectors are drawn from the same origin. (b) The changes of velocity become more nearly equal.

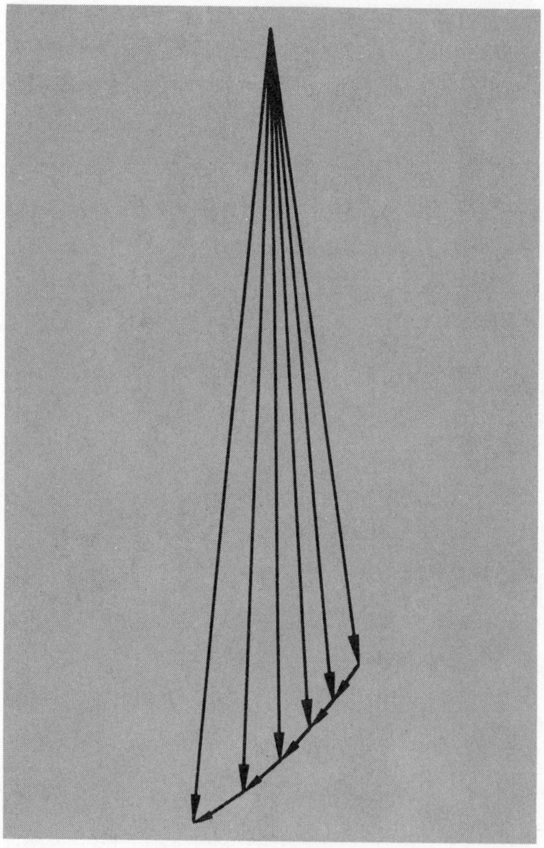

Figure 10–20
When the time intervals become very short, the successive velocity changes are nearly equal in magnitude and direction. Here are the first 10 seconds from Fig. 10–19. The vectors are magnified 5 times.

10–20. Notice that all five of the velocity changes have more nearly the same magnitude and direction than those of Fig. 10–18(b).

If we had chosen a time interval shorter than 2 seconds, the successive velocity changes would have been even more alike. Usually we can choose time intervals so short that the acceleration does not change appreciably either in magnitude or direction in going from one interval to the next. The instantaneous acceleration at time t is the limit of the average acceleration as the time interval approaches zero.

$$\vec{\mathbf{a}} = \lim_{\Delta t \to 0} \frac{\Delta \vec{\mathbf{v}}}{\Delta t}.$$

10-7 Centripetal Acceleration

The really new thing that we learn by considering vector acceleration as compared with acceleration along a straight line is that there is an acceleration even if the magnitude of the velocity stays constant and only its direction changes. The simplest and most important example of this is a body moving in a circular path at constant speed. The acceleration of such a body is called centripetal acceleration. In Fig. 10–21(a) a circular path is shown and on it a sequence of velocity vectors at equal time intervals. These vectors are all of the same length, but each points in a direction different from any other. If we construct a velocity diagram, as in Fig. 10–21(b), in which the velocity vectors are plotted from a common origin, we see that the successive changes $\Delta \vec{\mathbf{v}}$ in velocity are all of the same length but different in direction one from the other. Because the successive changes in velocity are not parallel to each other, the vector acceleration, which is the rate of change of velocity, cannot be constant. By taking smaller and smaller time intervals, as in Fig. 10–21(c), we can see that the instantaneous acceleration vector is directed perpendicular to the velocity vector at each instant of time. The fact that the magnitude of the centripetal acceleration vector is constant *and* the angle between the acceleration and velocity vectors is always 90° means that they rotate at the same uniform rate.

Figure 10–21
Part (a) shows the velocity of a body moving at constant speed in a circular path. In (b) the velocity vectors are drawn from a common origin, showing that the changes in velocity are in different directions. When shorter time intervals are taken, as in (c), the instantaneous acceleration vector is seen to be perpendicular to the velocity vector.

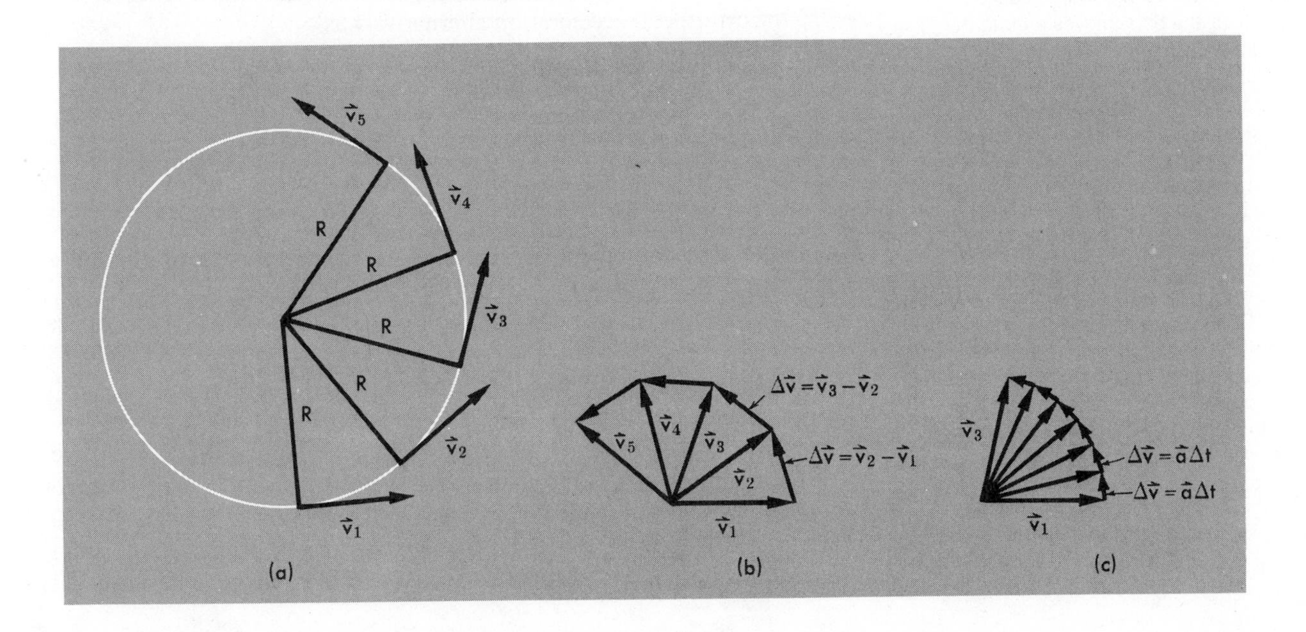

We shall now obtain two useful mathematical formulas for centripetal acceleration. Consider an object moving around a circle of radius R in a time T (period of revolution). Then its speed v is given by the circumference divided by the period:

$$v = \frac{2\pi R}{T}.$$

In the same time T the velocity vector has also turned completely around and now points the way it did originally. Consequently the velocity vector turns steadily, with the same period T as the object itself (Fig. 10–22).

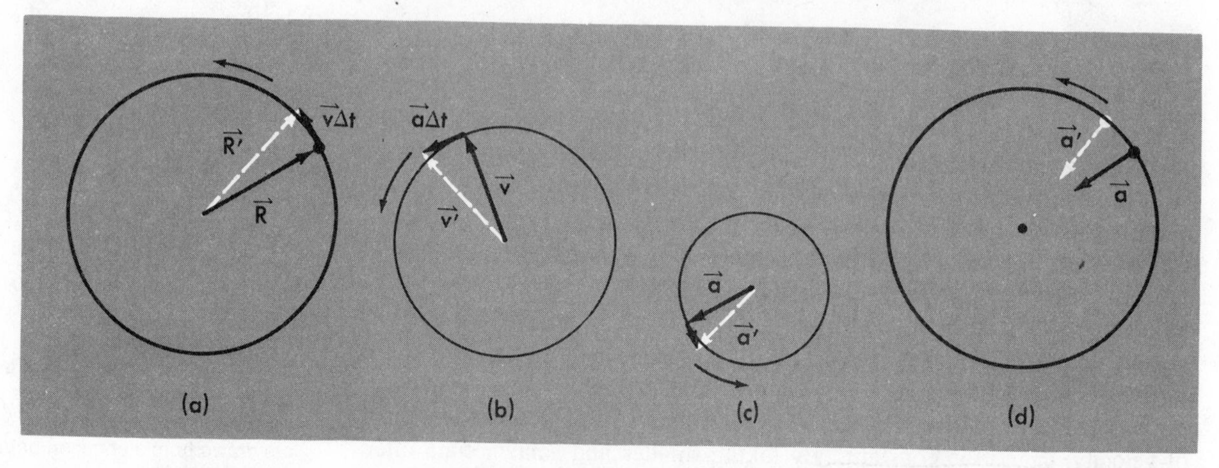

Figure 10–22
(a) The path of a body moving in a circle at constant speed. The vector $\vec{\mathbf{R}}$ drawn from the center of the circle to the body rotates uniformly. In a short time interval Δt the body's displacement is $\vec{\mathbf{v}}\Delta t$. During a complete period T, the body moves around the circle a distance $vT = 2\pi R$. (b) The changing velocity vector of the body. $\vec{\mathbf{v}}$ rotates at the same speed as $\vec{\mathbf{R}}$, but at an angle of 90° to it. In a short time interval Δt, the velocity changes an amount $a\Delta t$. In one revolution, the tip of $\vec{\mathbf{v}}$ moves once around the circle, so $aT = 2\pi v$. Part (c) shows the changing acceleration vector, which is always perpendicular to the velocity. Part (d) again shows the circular path of the body. Here we have drawn the acceleration vectors in the same direction as in (c), but with their tails on the body. The acceleration vector always points from the body toward the center of the circle in which it moves.

The tip of the velocity vector therefore goes around a circle of radius v and circumference $2\pi v$ in the time T; and the magnitude of the acceleration is

$$a = \frac{2\pi v}{T}.$$

By combining the last two equations, we can eliminate either T or v, expressing the centripetal acceleration alternatively as

$$a = \frac{v^2}{R} = \frac{4\pi^2 R}{T^2}$$

Since the acceleration is toward the center, we can write

$$\vec{\mathbf{a}} = -\frac{4\pi^2 \vec{\mathbf{R}}}{T^2}.$$

These expressions for centripetal acceleration will be valuable in our study of satellite motion, the planetary system, and atomic physics.

10-8 The Description of Motion; Frames of Reference

We have studied vectors largely in order to describe motion, to describe the successive positions of an object in space and how fast it moves through them. The simplest and most fundamental motion is that of a single object. A planet, such as Jupiter, moves in the sky. We can represent its position at each instant of time by the tip of a vector which we imagine drawn from the earth to the planet. Such a vector is called a position vector. (The radius R in Fig. 10–22(a) is a position vector rotating about its tail.) An airplane drones on its course toward some destination; its positions can also be represented by vectors (Fig. 10–23). Neither the planet nor the airplane is a tiny point; and, if we wish to know what happens within them, we need much more information than is given by these position vectors. But for many of the needs of astronomy or of navigation, it is sufficient to think of each as a point at the tip of a vector. As the planet or airplane moves, the vector moves, changing in magnitude and direction.

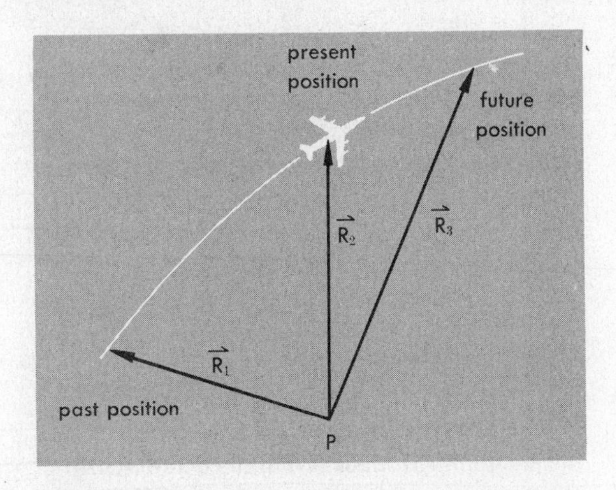

Figure 10–23
The position of an airplane with respect to a reference point may be shown by position vectors giving distance and direction.

In each of our examples the position vector extends from some identifiable point called the origin to the moving object, from the earth to Jupiter, from the control tower to the airplane. Furthermore, motion is always measured with respect to some frame of reference, as are positions. The airplane's motion is described with respect to the earth's surface, for example, and so is the motion of the falling ball in Fig. 10–12.

The motion of the airplane with respect to the air is different from its motion with respect to the ground if there is a wind blowing (Fig. 10–24).

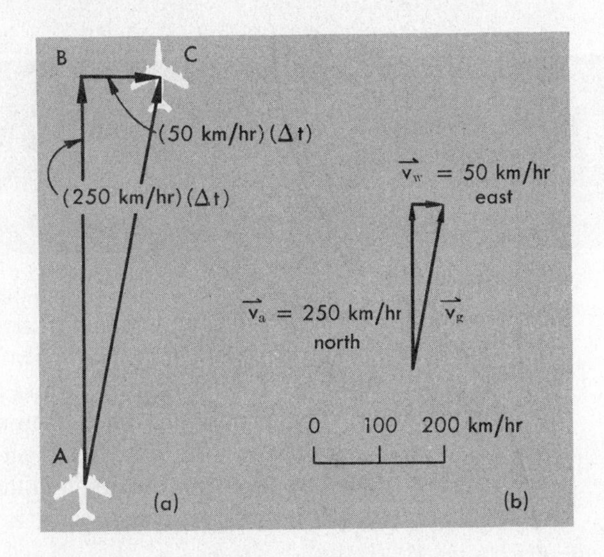

Figure 10–24
Part (a) shows the addition of the displacement vector $\overrightarrow{\mathbf{AB}}$, which represents the airplane's motion relative to the air, and $\overrightarrow{\mathbf{BC}}$, which represents the displacement of the air relative to the ground. Their sum is the vector $\overrightarrow{\mathbf{AC}}$, the displacement of the airplane relative to the ground. When the displacements are divided by the elapsed time, the velocity vectors are found as shown in part (b). The velocity of the airplane relative to the ground is the vector sum of its velocity relative to the air and the velocity of the wind relative to the ground.

Similarly, the motion of a swimmer swimming across a river with respect to the river is different from his motion with respect to a point on the river bank (Fig. 10–25). The motion of Jupiter with respect to the sun is simpler than its motion with respect to the earth. To the driver of a moving car a raindrop falling vertically with respect to the earth rushes almost horizontally toward him. Motion is described differently depending on the frame of reference.

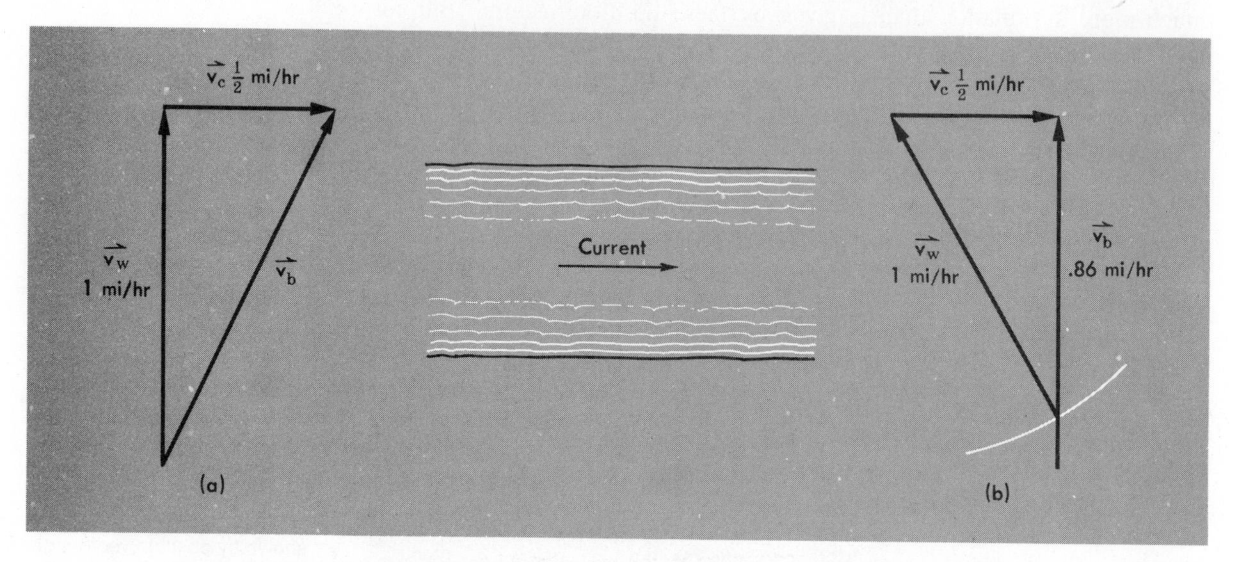

Figure 10–25
(a) If you swim across a river at a velocity $\vec{v}_w$ through the water at right angles to the current $\vec{v}_c$ your velocity relative to the bank will be $\vec{v}_b$. Does the current help you to reach the other side? (b) If you swim so that the vector component of $\vec{v}_w$ parallel to the banks is equal in magnitude but opposite in direction to $\vec{v}_c$ then $\vec{v}_b$ will be at right angles to the bank.

In general, we wish to view motion in such a way as to make it appear simple. We therefore place ourselves mentally in a frame of reference in which the motions are easy to describe; we draw a coordinate system with its origin at the most convenient point, and draw the position vectors locating the object from that point. Another example will show what we mean. Suppose we stand on the earth again and look at the motion of a point on the edge of a wheel of a slowly moving car. The point moves through the complicated curve illustrated in Fig. 10–26. A position vector from us as origin to a point on the wheel performs an extremely complicated motion. But a point on the earth is not always the most convenient origin for position vectors. We are far better off if we get in the car and hang out of the window to look at the wheel. The point on the wheel then moves steadily around a circle, and the motion of the position vector looks far simpler.

Figure 10–26
Path of a point on the rim of a rolling wheel. This curve, a cycloid, shows the path as it appears to an observer standing alongside. One small light was mounted on the rim of a wheel, and another light at the center. The camera shutter was held open while the wheel rolled.

In describing motion we try to put the origin in the most convenient place even though we are sometimes physically unable to go there. The motions of the planets, for example, look very complicated when we describe them with position vectors whose origin is in the earth. Copernicus pointed out that the planetary motions are much simpler to describe when we assume that the sun is the center of the solar system, and move the origin of the position vectors to the sun. Ever since Copernicus pointed out the importance of placing the origin in a convenient place, the proper choice of origin has been an important technique used by physicists to describe motion in simple terms.

10-9 Kinematics and Dynamics

What we have been discussing is the branch of physics called *kinematics* (after the Greek word *kinema,* meaning "motion"). It is that part of our subject which treats of the description of motion, without taking into account what it is that moves or what causes the motion. The science of mechanics, indeed much of physics, is dominated by the study of motion; but that study is complete only when we extend it to what is called *dynamics* (after the Greek word *dynamis,* meaning "power"). In dynamics one discusses the causes of motion, what is moving, and how its nature affects the motion. In our study of kinematics, we needed to measure only positions and times; in dynamics, the pushes and pulls which cause and resist and determine motions must be taken into account as well.

Motion can be simple or complex. The method of physics is to analyze first the simpler cases, to extract what we can from them, and to go forward to more and more complicated cases. It would be a mistake to think that a quick elementary study of physics can explain the breaking of the surf on a shore, the path of a jet plane in the sky, or the intricate rhythms of a diesel engine. But these motions and many more can be handled by pursuing the same methods of analysis that we have outlined here in simple cases. When, instead of a single separate object like a pebble or a car, a whole fluid like air or water is in motion, even the kinematics may become difficult. The eddies behind a spinning propeller and a flying baseball (Fig. 10–27) are complex indeed. Their study is a specialized kind of unraveling, requiring enormous "bookkeeping" operations to keep track of every small bit of the moving air. Powerful mathematical methods are needed, but the basic ideas used in solving such problems flow out of the discussion we have given; the underlying laws are not different. One may well be wary, though, of too simple an explanation of what is obviously kinematically complex. You cannot design airplanes or understand the weather without checking theories with experiments or without interpreting experimental results in terms of theories. We must go back and forth between theory and the experimental study of the real complex motions. Nevertheless, great progress has been and will be made with such a process. Nowhere in the whole domain of the motions of the natural world, here on earth or in the skies, in our large-scale machines or in our means of transport, have we yet found any case to which the analysis of mechanics does not apply. Where the kinematics is hard, progress is slow; where the kinematics is less demanding, progress is fast. In this course, where we are trying to learn fundamentals, we do not want

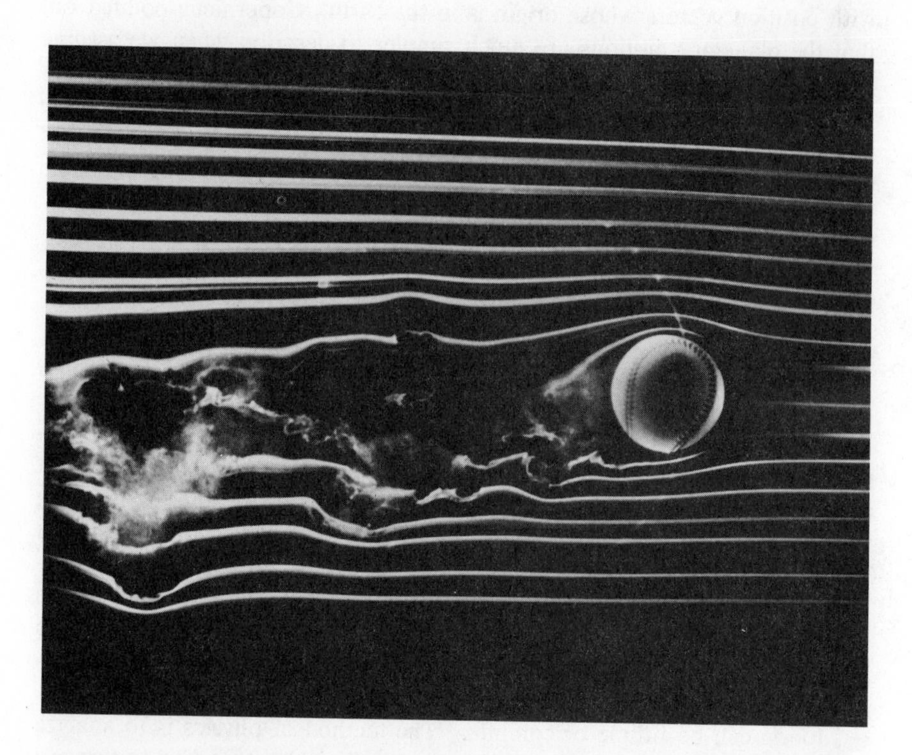

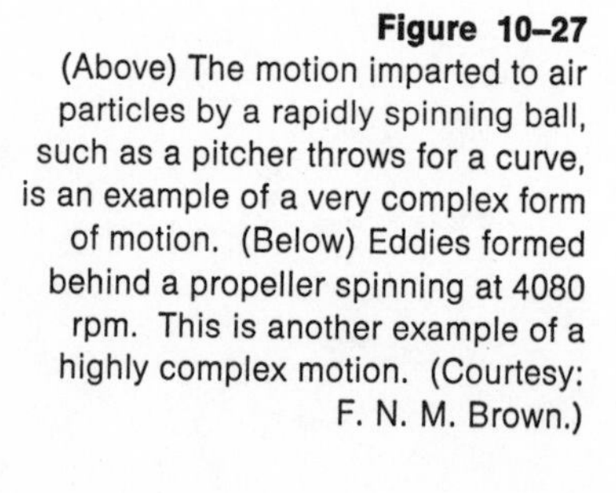

Figure 10–27
(Above) The motion imparted to air particles by a rapidly spinning ball, such as a pitcher throws for a curve, is an example of a very complex form of motion. (Below) Eddies formed behind a propeller spinning at 4080 rpm. This is another example of a highly complex motion. (Courtesy: F. N. M. Brown.)

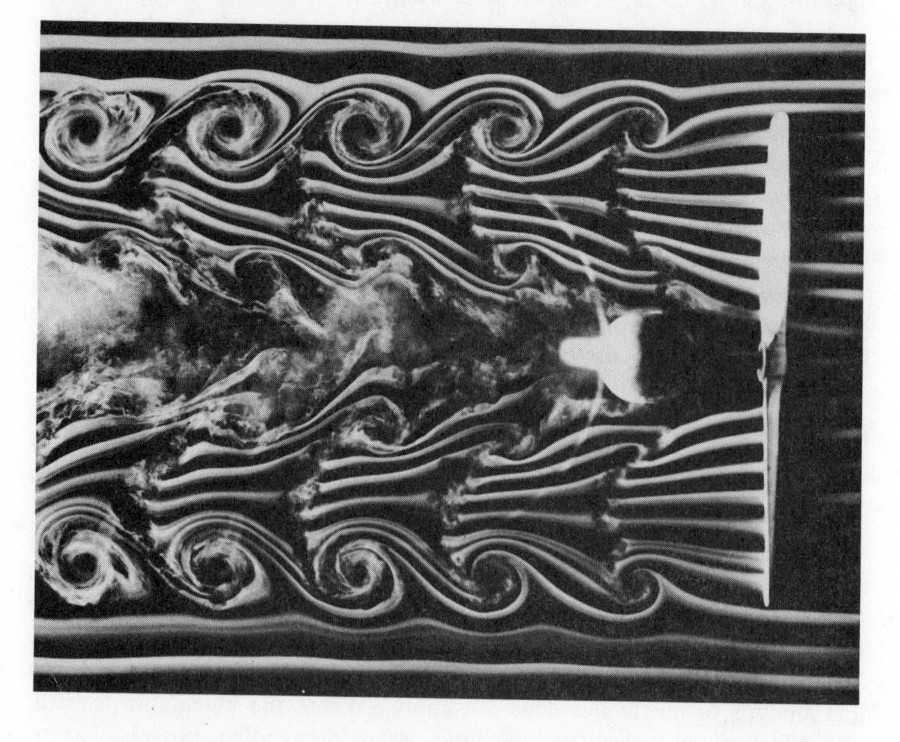

to produce the false impression that there are no complex cases of interest; there are, especially in technology. In engineering and technology, the aim is to get something done for people to use; in a science such as physics, the aim is to understand and hence to predict and to control. In physics we seek out the simple and examine it, even if it is costly and difficult to attain. In engineering the orderly methods developed in this research are painstakingly and often brilliantly applied to more and more complex cases, which have clear usefulness for men. The complicated, trial-and-error extension always rests on the foundation of the simple and the well understood.

10-10 The Speed of Light

We have pointed out earlier that from everyday experiences we can develop a wide extension or extrapolation. This is the kind of thing we have just been doing, basing our study of motion on the ideas of space and time we have built up from experience. But if such schemes are pushed too far, they may go wrong. They need to be tested. For fifty years now, we have been finding that our usual notions of space and time, extremely reliable for most of the motions that we notice daily, do not work for extremely fast motions. For speeds which begin to be of the order of magnitude of that of light, the kinematics we have just discussed begins to go wrong. It turns out that the speed scale cannot be extended indefinitely; if you add velocities that are too big, you reach a region where the rules of addition go wrong. There arises a natural limit to speed, which cannot be surpassed. This universal speed limit is the speed of light in free space, which is known to be very close to 3×10^8 m/sec. Speeds up to 10^5 m/sec relative to the sun span the whole range of speeds of every large object in the solar system, from planets and meteors to the spacecraft of man. Only small particles, electrons and their kin, move appreciably faster. These closely approach the speed of light. For their study, the ideas of relativity kinematics are needed. But for everything else —solar-system astronomy, engineering, or any large-scale laboratory physics, the kinematics we have studied, that of Newton, is accurate enough. This whole topic is an example of how you can begin with familiar ideas, which hold well over a wide range, and reform them entirely when you reach another order of magnitude. The fact that familiar ideas may be modified to fit extreme conditions does not end their meaning and usefulness in the domain for which they were originally built up and in which they have been amply tested. Newtonian kinematics is a good approximation to relativity kinematics whenever the speeds are small compared with the speed of light.

For Home, Desk, and Lab

1.* What are the length and the direction of a displacement from (1, 2) to (3, 4)? (Section 1.)

2.† You may walk the following distances, one after another, in any order you choose: (a) go 3 meters due east; (b) go 2 meters due north; (c) go 3 meters due west. What is the farthest you can be from your starting point at the end of your walk?

3. A man follows this route: From his house he travels four blocks east, three blocks north, three blocks east, six blocks south, three blocks west, three blocks south, two blocks east, two blocks south, eight blocks west, six blocks north, and two blocks east. How far and in what direction will he be from home?

4. Suppose you have three sticks 4.00, 5.00, and 6.00 ft long which can be joined by connecting the ends at a 90° angle.
   (a) What is the largest displacement possible between the unconnected ends of the sticks?
   (b) Displacement vectors can be added in any order and the result is the same. Why do you get different lengths between the ends of the sticks in this problem when you change the order of the sticks?

5. What is the magnitude of a displacement whose components along the perpendicular x, y, and z axes are respectively 4.00 m, 2.50 m, and 8.50 m?

6. In Fig. 10.5, what are the x and y vector components of the short displacement in both (a) and (b)?

7. A ship sails 10 miles in a northeast direction. What are the components of its displacement (a) to the north? (b) to the east?

8.* What are the north and east components of a velocity vector of magnitude 100 mi/hr and direction 30° west of south? (Section 3.)

9.* What is the component in the y direction of a vector of length 2 cm in the x direction? (Section 3.)

10. (a) Find the result of adding a vector 2 cm east to one 3 cm northwest.
   (b) Find the result of adding a vector 8 cm east to one 12 cm northwest.
   (c) Compare the results of parts (a) and (b), and state a theorem about adding a pair of vectors which are multiples of another pair. Can you prove the theorem in general?

11. Suppose the ball in Fig. 10–12 had an initial horizontal velocity of 1.5 m/sec. instead of 2.00 m/sec. When it crosses the lowest white line, what is (a) its displacement along the x axis? (b) along the y axis (c) its velocity?

12.* Suppose you were told that Fig. 10–12 represented a ball thrown *upward.* In what way would Fig. 10–14 have to be changed in order to correspond to the motion of the ball? (Section 5.)

13.* How can a motion with constant speed be an accelerated motion? (Section 7.)

14.† An object moving in a circular path with a constant speed of 2.0 m/sec changes direction by 30° in 3.0 seconds.
   (a) What is its change in velocity?
   (b) What is its average acceleration during the 3.0 seconds?

15. A watch has a second hand 2.0 cm long.
   (a) Compute the speed of the tip of the second hand.
   (b) What is the velocity of the tip of the second hand at 0.0 seconds? at 15 seconds?
   (c) Compute its change in velocity between 0.0 and 15 seconds.
   (d) Compute its average vector acceleration between 0.0 and 15 seconds.

16. Suppose an airplane flies in a circle of circumference 10 miles at a constant speed of 100 mph.
    (a) What is the change in velocity in one fourth of a revolution?
    (b) What is the change in velocity in one half of a revolution?

17. Show that the expression $\frac{v^2}{R}$ for centripetal acceleration has the units of acceleration.

18. When you ride a bicycle at 10 mi/hr your speed is about 5 m/sec.
    (a) At this speed what is the centripetal acceleration of a point on the rim of a wheel (radius, 0.30 m)?
    (b) How does this acceleration compare with the acceleration of the ball in Fig. 10–12?

19. The outermost tip of the eccentric screw on a motor used to generate waves in a ripple tank is 0.75 cm from the center of the motor's shaft. The period of the rotating shaft is about 0.025 sec when the motor is running at its top speed.
    (a) What is the acceleration of the tip of the screw?
    (b) What would the acceleration be at half the above speed?

20. The planet mercury has a period of 7.6×10^6 sec and it is 5.8×10^{10} m from the sun. What is its acceleration toward the center of the sun?

21. The astronauts in an Apollo spacecraft circled the moon once every 6.5×10^3 sec. They were about 1.7×10^6 m from the center of the moon. What was their centripetal acceleration?

22.* If you are in a free balloon being carried along by the wind at a constant velocity, and you hold a lightweight cloth flag in your hand, which way will the flag wave? (Section 8.)

23.* Describe the motion of a swing as it is seen from another swing exactly like it, when both swings started together in the same direction with the same push. (Section 8.)

24. A bug placed at the center of a phonograph turntable crawls at a constant speed directly toward a point on the edge of the turntable. Describe the motion of the bug as seen by you from above when the turntable has a speed of 16 revolutions per minute and it takes the bug 15 seconds to crawl from the center to the edge of the turntable.

25. A plane flying north at 320 km/hr passes directly under another plane flying east at 260 km/hr.
    (a) What is the horizontal component of the displacement of the second plane relative to the first, 20 minutes after they pass each other? 50 minutes after they pass?
    (b) What is the horizontal component of the velocity of the plane flying east relative to the plane flying north?
    (c) Does the direction of this velocity vector (relative to the earth) change?

26. An ocean liner is traveling at 18 km/hr. A passenger on deck walks toward the rear of the ship at a rate of 4.0 m/sec. After walking 30 meters he turns right and walks at the same rate to the rail, which is 12 meters from his turning point.
    (a) What is his velocity relative to the water surface while walking to the rear? While walking toward the rail?
    (b) Draw the displacement vectors relative to the water surface for his stroll. What was the total displacement from his starting point?

27. An airplane is flying toward a destination 200 miles due east of its starting point, and the wind is from the northwest at 30 mi/hr. The pilot wishes to make the trip in 40 min.
    (a) What should be the heading?
    (b) At what airspeed should he fly?

28. A man rows a boat "across" a river at 4.0 mi/hr (i.e., the boat is always kept headed at right angles to the stream). The river is flowing at 6.0 mi/hr and is 0.20 mi across.
    (a) In what direction does his boat actually go relative to the shore?
    (b) How long does it take him to cross the river?
    (c) How far is his landing point downstream from his starting point?
    (d) How long would it take him to cross the river if there were no current?

29. An airplane maintains a heading of due south at an airspeed of 540 mi hr. It is flying through a jet stream which is moving east at 250 mi/hr.
    (a) In what direction is the plane moving with respect to the ground?
    (b) What is the plane's speed with respect to the ground?
    (c) What distance over the ground does the plane travel in 15 min?

Further Reading

These suggestions are not exhaustive but are limited to works that have been found especially useful and at the same time generally available.

GALILEI, GALILEO *Dialogues Concerning Two New Sciences,* translated by H. Crew and A. de Salvio. Macmillan, 1914. (Reprinted by Dover.) The section entitled "The Fourth Day."

HOLTON, G., and ROLLER, D. H. D., *Foundations of Modern Physical Science.* Addison-Wesley, 1958. (Chapter 3)

ROGER, ERIC M., *Physics for the Inquiring Mind.* Princeton University Press, 1960. Many sample problems in vector math. (Chapter 2)

Vectors—A Programmed Text for Introductory Physics. Appleton-Century-Crofts, 1962: Basic Systems. An excellent book for the student who wants additional explanation and practice in vector additions and subtractions.

Newton's Law of Motion

CHAPTER
11

Automobiles move down highways or wind in and out of traffic; passenger planes fly high above us; jet planes and artificial satellites streak across the sky; the stars perform their regular progression. What makes each of them go? What makes anything move? Is there a single cause common to all motion? Is any cause necessary?

So far in our study of motion we have been concerned entirely with the description of motions without regard to their causes—in other words, we have been dealing with kinematics. But description alone can never enable us to satisfy our desire to do something new, to control motions, to go beyond the mere description of what occurs. Now we shall take the next step: we shall examine the causes of motions or of changes in motion—that is, dynamics.

Newton's law of motion, on which we now base our understanding of dynamics, does go beyond kinematic description. For example, we make use of it when we design rockets and launch man-made satellites and space vehicles to the moon and planets. In the next few chapters, once we have understood Newton's law, we shall apply it to the motions of the moon and planets. Like Newton, we shall find the connection between the time it takes a planet to move around the sun and the gravitational attraction between chunks of matter. Finally, we shall use the same law of motion to study electrical forces and to enter the submicroscopic world. With this single law we shall investigate motion throughout all this range.

11-1 Ideas About Force and Motion

Questions about the causes of motion arose in the mind of man more than twenty-five centuries ago, but our present answers were not found until the time of Galileo (1564–1642) and Newton (1642–1727).

Let us start in terms of our own personal experience. What sort of thing do we associate with the "cause of motion"? The answer is muscular pulls or pushes (Fig. 11–1). To move a piano across a room you have to push hard. To move a sheet of paper off your desk takes very little push. These pulls or pushes we call *forces*. The notion of force as used in physics certainly started this way. Later, as understanding grew, the idea of force was extended to include all causes of motion. The pull of a magnet on a nail is a force; it can change the nail's motion in the same way that a muscular force can.

More specifically, what is the relation between force and motion? Suppose we move a desk across the floor. We must apply a force all the time to keep it moving steadily from one side of the room to the other. Similarly, a horse must keep pulling on a wagon to keep it rolling at constant speed. Everyday experience seems to indicate that it is necessary to exert a force constantly in order to maintain a steady motion, such as motion in a straight line with constant speed (Fig. 11–2). Aristotle (384–322 B.C.) had noted this fact. He concluded that a constant force was required to produce a constant velocity. It then follows that bodies would come to rest in the absence of force.

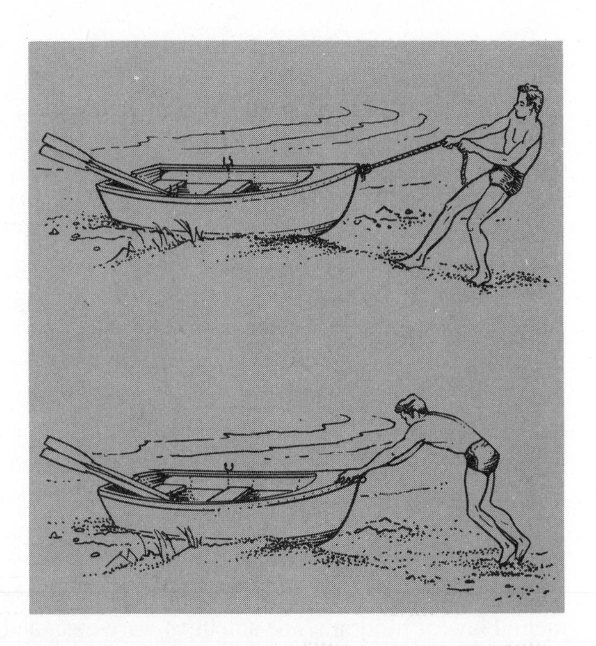

Figure 11–1
All pushes and pulls are called forces.

Figure 11–2
Steady motion seems to require a steady force.

The hypothesis that in the absence of outside forces bodies would come to rest and stay at rest helps us to understand a great many observed motions, but it does not explain *all* the motions which occur in nature. For example, the Greeks were aware that bodies fall with increasing speed without the application of any evident outside force. They were also acquainted with the motions of the sun, moon, and stars, which seem to occur without pushes or pulls to maintain them. To the Greeks there seemed to be three kinds of motion. We must explain not only the motion of things pushed around on the surface of the earth, but also the motion of bodies falling to earth and the unceasing motions of heavenly bodies. Aristotle explained that ordinary matter falls toward the earth because the earth is the center of the universe to which matter naturally moves. He proposed that celestial matter was fundamentally different in nature from matter on earth and that it obeyed different laws. To Aristotle, celestial matter had the built-in property of supplying from within itself the force necessary to maintain the observed motions.

We should not think that separate models for three different classes of observed motions are foolish. We often do the same thing. When we see a piece of metal that attracts iron nails, we say it is a magnet—a different kind of matter from wood; and we may investigate its magnetic behavior using a model that is different from the model we use to describe its nonmagnetic behavior. However, explaining as much as we can with as few assumptions as possible by use of a single model is preferable to making a separate model for each new observation. As far as we can, we describe wood, magnets, etc. in a single model, as simple as we can make it. Likewise, we try to explain all motion on one theory rather than three.

A modern Aristotle would hardly explain the unceasing nature of celestial motion by invoking a distinct kind of matter. We can send our own earthly matter into the celestial realm. The world of motion on earth and the unceasing motions of the planets are now united. The artificial satellites offer us an excellent demonstration that we need assume no difference between earthly matter and celestial. Our understanding of the motion of falling bodies, of heavenly bodies, and of bodies which we ourselves push and pull along the surface of the earth is now described in a single fundamental law of motion. The satellites were designed, built, and fired according to this law. Their behavior is one of many pieces of evidence that Newton's law of motion encompasses the three types of motions described by Aristotle.

11-2 Motion Without Force

For two thousand years after the time of Aristotle, the apparent difference between celestial motion and motion on earth halted significant progress in dynamics. Then, in the seventeenth century, Galileo took the first big step toward creating a single explanation of both these types of motion. He asserted that "... any velocity once imparted to a body will be rigidly maintained as long as there are no causes of acceleration or retardation, a condition which is approached only on horizontal planes where the force of friction has been minimized." This statement embodies Galileo's law of inertia. Briefly it says: When no force is exerted on a body, it stays at rest or it moves with constant velocity.

How did Galileo reach the startling conclusion, so different from everyday experience, that constant motion requires no force? He was studying the motions of various objects on an inclined plane. He noted that "in the case of planes that slope downward there is already present a cause of acceleration, while on planes sloping upward there is retardation." (See Fig. 11–3.) From this experience he reasoned that when the plane slopes neither upward nor downward there should be neither acceleration nor retardation: "... motion along a horizontal plane should be constant." Of course Galileo knew that such horizontal motions were not in fact constant, but he saw that when there was less friction, bodies moved for a longer time with nearly constant velocity. Because of his arguments he was convinced that friction provided the forces which stop bodies in horizontal motion, and that in the absence of all forces the bodies would continue to move

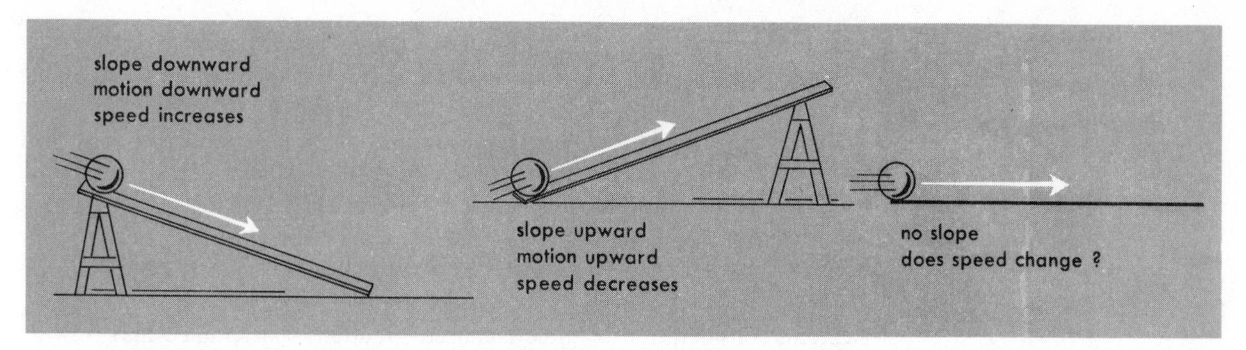

forever. He therefore stated his result for the idealized situation in which *no* forces act.

In a second series of experiments, Galileo showed that if he placed two of his inclined planes facing each other (as in Fig. 11–4, top), an object starting from rest would roll down one and up the other until it almost reached its original height. Friction prevented it from attaining this height, but Galileo saw that this height was the limit to the motion. He reasoned that if the slope of the upward plane is decreased, as in the middle of Fig. 11–4, the distance that the object will travel to reach its original height will increase. If, as in the bottom portion of Fig. 11–4, the slope is finally reduced to zero, so that the second plane is a horizontal surface, the object will never attain its initial height. It should travel on forever. "From this," Galileo again concluded, "it follows that motion along a horizontal plane is perpetual."

Galileo's experiments are not difficult, nor is there any evidence that he performed them with exceptional skill. Some, like the extension of the experiment at the bottom of Fig. 11–4 to the idealized case of perpetual

Figure 11–3
From observing motion on inclined planes, Galileo reasoned that motion along a horizontal plane is steady.

Figure 11–4
Galileo observed that a ball tends to rise to its original height regardless of the slope of the incline. With zero slope, the original height can never be reached. Therefore, motion along a horizontal plane should be perpetual.

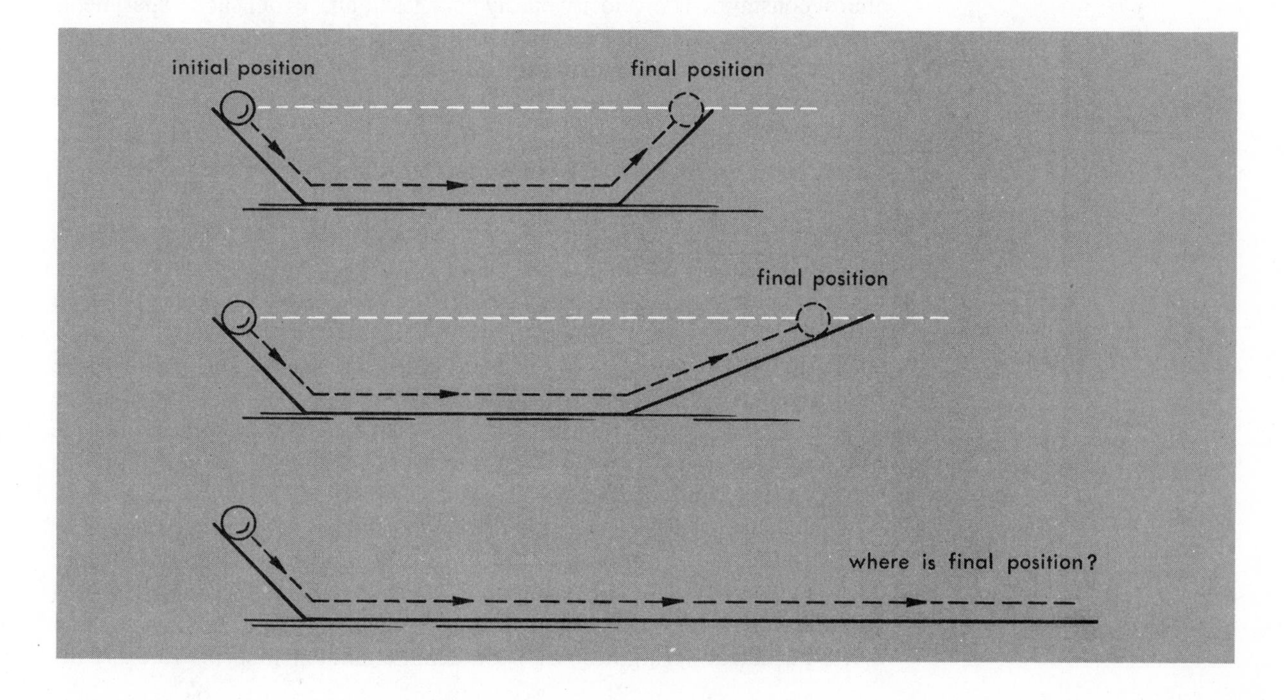

motion, were not "real" experiments. They were experiments in thought alone. But they were based on solid fact. It is this combination of thinking and fact that distinguishes Galileo's work. It was this combination which allowed him to pick out the useful idealization despite the great variety of observed motions. His principle of inertia was the great breakthrough which enabled Newton to build up our present understanding of dynamics.

Many of the motions Galileo analyzed, and those which Newton later studied, were so highly idealized that they seem to have very little to do with motions of real systems as we observe them. But it was only by careful consideration of these idealized situations that Galileo and Newton made their great contributions to dynamics. In the same way, we must look hard at very simple and idealized motions to obtain a real understanding of the basis of dynamics. Then, and only then, will we be ready to apply dynamics to the ordinary complex world.

With modern equipment we can do experiments that almost realize Galileo's idealized experiment on motion without force. To get the flash photograph of Fig. 11–5, we used a Dry Ice puck sliding across a glass surface. In Fig. 11–6 you can see the puck we used. It is a heavy metal disc, highly polished on the bottom surface and carrying a container full of pieces of Dry Ice. This frozen carbon dioxide changes slowly into a gas which escapes through a small hole in the center of the bottom of the metal disc and flows between the bottom and the glass surface (Fig. 11–7). Thus a thin layer of gas is provided continuously between the metal disc and the glass surface; the puck floats on the gas layer, and friction is almost completely eliminated.

The photograph of Fig. 11–5 was taken by flashing a light 24 times in 10 seconds, or every $^{10}/_{24}$ sec, as the puck coasted over the smooth glass surface. As you can see, the displacement of the puck between flashes is almost constant. The velocity hardly changes at all. From such experiments we can calculate that if the puck were set going at about 10 miles per hour on a long enough horizontal surface, it would coast for about ½ mile!

11-3 Changes in Velocity When a Constant Force Acts

Galileo's law of inertia tells us that an object on which no force acts moves with unchanging velocity. If the velocity changes, we conclude that some force is acting on the object. What is the relation between force and change in velocity?

We shall begin our study of this question with the simplest experiment we can imagine. We shall apply a single force to a single object. To minimize friction we use one of the same Dry Ice pucks on the same glass surface with which we examine motion with no force. The force we now apply is therefore the only force we need consider.

We need some way to recognize when the applied force is constant. For this purpose we use a spring in the form of a rubber loop attached to the puck [Fig. 11–8 (a)]. It is common experience that the force exerted by a spring increases in some manner with the stretch of the spring. Also, whenever a particular spring is stretched a definite amount it seems to exert the

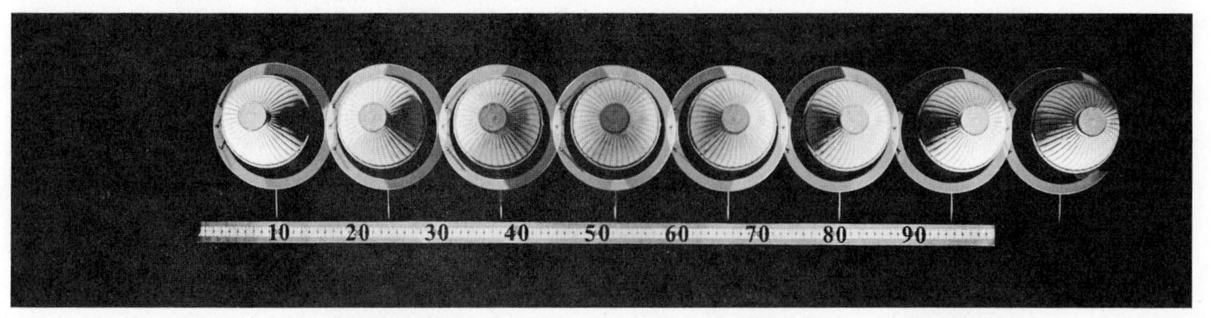

Figure 11–5
The motion of a Dry Ice puck. It moved from left to right while the light was flashed 24 times in 10 sec, or every 10/24 sec. The scale at the bottom is in centimeters. Here is a close approach to the ideal situation of motion without force. The puck has a nearly constant displacement in equal time intervals.

Figure 11–6
A Dry Ice puck, as described in the text, rests on a smooth glass surface. With such apparatus we can study almost frictionless motion.

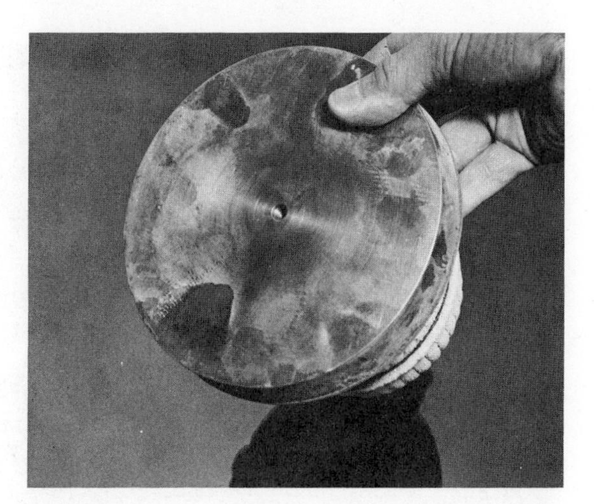

Figure 11–7
Dry Ice inside the puck changes into a gas which escapes through the hole in the bottom of the puck.

same force. We shall assume, therefore, that the force exerted by our loop is the same whenever it is stretched the same amount [Fig. 11–8 (b)].

Now we pull our puck in such a way that the loop is always stretched the same amount, and record the motion of the puck by flash photography. The result of such an experiment is shown in Fig. 11–9. In this experiment the light flashed at intervals of $10/24$ second. We can see the successive positions of the puck, and we can see that the extension of the loop remained constant. The puck started from rest and moved on a straight-line path in the direction of the force we applied. Therefore, we can analyze this motion using the methods that we learned in Chapter 9.

Let us call the direction of motion of the puck the positive direction and label it as an x axis. Clearly, in each time interval the displacement, Δx, of the puck increased. Therefore, the average velocity $v = \frac{\Delta x}{\Delta t}$ increased

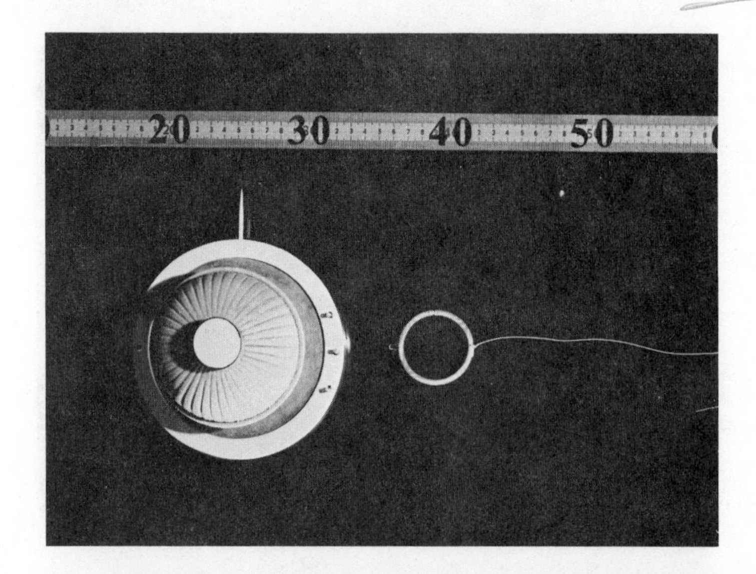

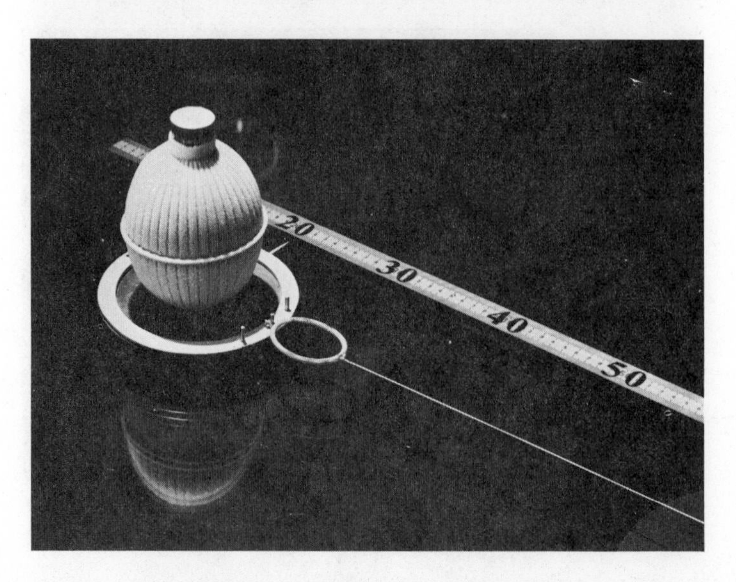

Figure 11–8
(a) Unextended rubber loop before being mounted on puck. (b) The loop extended. Whenever this loop is stretched the same definite amount, we seem to get the same force.

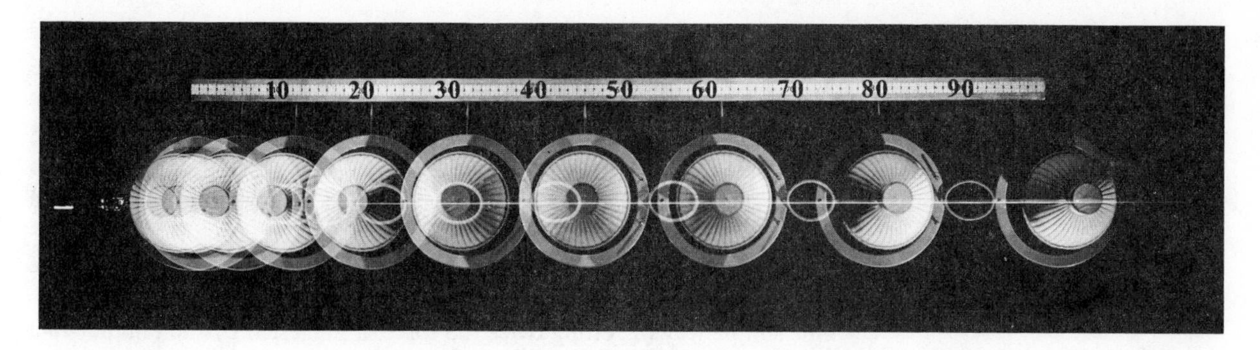

(to the right) in each time interval. By measuring the successive displacements, we can calculate how the average velocity changed. As the caption under Table 1 shows, in each 10/24 second after the motion started, the average velocity increased by about 5.8 cm/sec, a constant amount within the limits of accuracy of our experiment. Dividing the change in velocity of 5.8 cm/sec by the time interval of 10/24 sec, we see that the average velocity changed at the constant rate of 14 cm/sec^2. Since the average velocity changed at a constant rate throughout the motion, we can safely assume that the instantaneous velocity also changed at the same constant rate. That is, the acceleration is constant and equals 14 cm/sec^2.

Figure 11–9
The flash photograph shows the puck being pulled to the right. The light flashes were separated by 10/24 sec. A constant force was applied by keeping the loop extension constant. The displacement of the puck in each interval marked on the photograph has been measured and appears in Table 1.

Table 1
Data from Experiment Shown in Fig. 11–9

INTERVAL NO.	POSITION x(CM)	AVERAGE VELOCITY IN INTERVAL $\Delta x/\Delta t = v$(CM/FLASH)	CHANGE IN AVERAGE VELOCITY Δv(CM/FLASH)
1	4.1	4.1	
2	10.4	6.3	2.2
3	19.2	8.8	2.5
4	30.4	11.2	2.4
5	44.0	13.6	2.4
6	60.1	16.1	2.5
7	78.6	18.5	2.4

The zero of position was taken as the location of the puck at the instant of the first flash. The value of x *is the position of the puck at the end of the given interval. The average velocity, measured in cm/flash, is numerically equal to the displacement during the given interval. The last column simply gives the amount by which the average velocity changed between successive intervals. We find that* Δv *was constant within the limits of accuracy of our experiment and was 2.4 cm/flash. Since we had 24 flashes in 10 seconds, or 2.4 flashes/sec, the change in average velocity was (2.4 cm/flash) (2.4 flashes/sec) = 5.8 cm/sec between successive time intervals of duration 10/24 sec.*

The particular value 14 cm/sec^2 occurs in this experiment because we pulled with a particular force on a particular object. When we pull with other forces or pull on other objects, we usually obtain other values of the acceleration. But all experiments like the one just described show that under influence of a constant force the acceleration is constant.

11-4 Dependence of Acceleration on Magnitude of Force

What happens when we apply a different constant force to the same body? Let us apply twice the force and see what happens. This simple suggestion raises a new problem. We have seen that we could use a spring loop to remove the human element from the operation of applying a force of a particular size, giving us confidence that we applied the same force throughout our last experiment. But how can we use the loop to apply twice the force?

One simple way of getting twice the force is suggested by the familiar fact that two men can push harder than one. For example, two men may be needed to push a stalled car which one alone could not move. We can arrange two loops to give twice the force that one gives. Let us construct a second loop as nearly identical with the first as we can make it. When we stretch the second loop the same amount as the first, it should exert an equal force. We can assure ourselves that the forces are equal by doing our last experiment over again with the new loop. When the force is the same as before, the puck speeds up in just the same way as before. We thus show that the new loop gives the same force as the old one.

Now we can apply twice the original force to the puck. We hook both loops to the puck side by side and pull each of them in the same direction (Fig. 11–10). We make sure that each loop is extended the same amount as in the original experiment with one loop, and we observe the motion in the same way as before. In this way we can double the force on the puck, leaving everything else the same.

We have used this procedure to put a doubled force on the same puck as before. The result is shown in Fig. 11–11. What do we find? The data in Table 2, taken with exactly the same arrangement as before but with the force doubled, shows that the velocity increases twice as fast as before. Instead of $\Delta v/\Delta t = 14 \text{ cm/sec}^2$ we obtain $\Delta v/\Delta t = 28 \text{ cm/sec}^2$.

Further experiments show that this result is general. Whenever we double the force on a given object, we double the acceleration. Moreover, if we make the force three times as large by putting three identical loops side by side, we triple the acceleration. From many measurements of this kind we conclude that the acceleration of a body is proportional to the force that acts on it.

11-5 Inertial Mass

The acceleration a produced by a given force F depends on the object on which the force acts. Applying equal forces to a baseball and to an elephant produces less acceleration of the elephant.

Because larger bodies are less easily accelerated by forces, it is convenient to write the proportionality between F and a in the form:

$$F = ma$$

The proportionality constant m depends on the object. Its value increases with the size of the body, at least for objects made of one uniform substance.

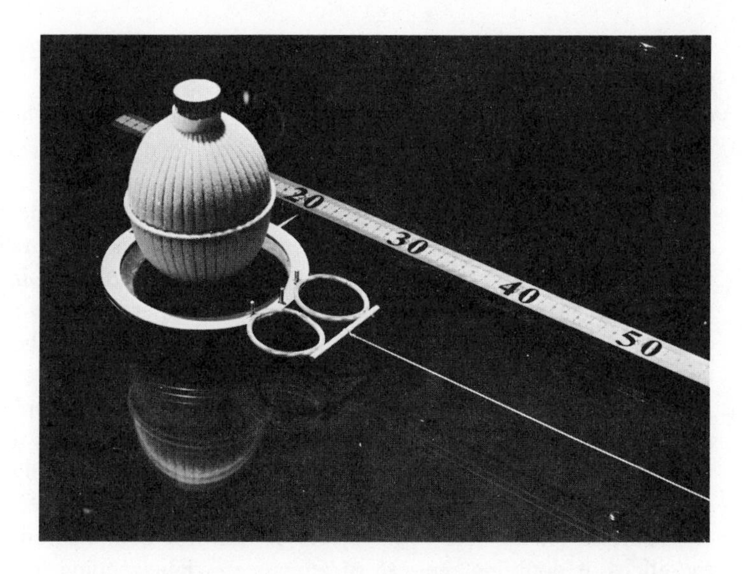

Figure 11–10
To apply twice the original force, we attach two identical loops to the puck.

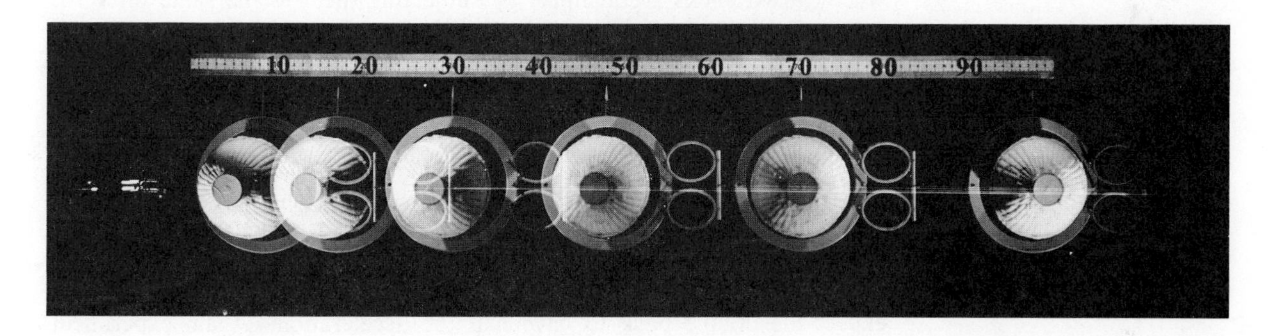

Figure 11–11
The puck used in the first experiment is accelerated by twice the force used before. The interval between flashes is again 10/24 second.

Table 2
Data from Experiment Shown in Fig. 11–11

INTERVAL NO.	POSITION x(CM)	AVERAGE VELOCITY IN INTERVAL $\Delta x/\Delta t =$ v(CM/FLASH)	CHANGE IN AVERAGE VELOCITY Δv(CM/FLASH)
1	8.4	8.4	
2	21.5	13.1	4.7
3	39.3	17.8	4.7
4	61.9	22.6	4.8
5	89.3	27.4	4.8

These are the results of an experiment in which the applied force was twice that used in the first experiment (Table 1). The flash rate was again 2.4 flashes per second. Note that the change of average velocity and hence of instantaneous velocity is just twice as great as it was before.

The constant m is called the *inertial mass* of the body. By rearranging the above equation, we define m as F/a for a given object. This ratio is constant for a given body and it tells how difficult it is to accelerate the body. The greater the force needed to produce a given acceleration, the greater the inertial mass of the object.

Naturally we wish to know whether inertial mass, this measure of the difficulty of accelerating a body, is a new property of a body. Is it unrelated to anything we know already, is it some familiar property in a different guise, or is it some combination of familiar properties? To answer these questions, we investigate the relation of the inertial mass to the shape, size, composition—in fact to any other known property of an object.

Of course, we do not look about at random. As the example of the elephant and baseball shows, our experience already indicates that size is a reasonable place to start. But volume alone will not do. A hollow elephant, on a float in the Mardi Gras parade, is not so hard to accelerate as a real elephant; so let us start with objects of uniform composition. Let us try the effect of doubling the size of an object while keeping its composition uniform. One easy method is to use two identical pucks. We can decide whether two pucks are identical by subjecting them to the same force for equal times. If each gains the same speed, they have identical inertial masses.

Let us pull the two identical pucks side by side, applying the same force to each with identical loops. The two pucks travel along together, each gaining the same Δv as they progress. Next we join the pucks rigidly together and pull them with both loops yoked side by side. Again, we would expect them to gain speed in the same way. Experiments show that this is so. Now let us try pulling the connected pucks with the force of only one loop. This is half the previous force; and from what we know already, we expect the double puck to have half the acceleration. Again experiment confirms our expectation. In other words, when we apply one standard force to two standard pucks, the double puck has just half the acceleration of a single puck.

Our results for identical bodies are not surprising. The real questions about inertial mass arise when we consider objects made of different materials. We cannot have a piece of silver identical with a piece of gold. But we can find pieces that have the same inertial mass, pieces for which F/a is the same. Such pieces are certainly not identical in size or composition. Inertial mass is, then, not entirely a matter of size.

What happens when we join together a piece of silver and a piece of gold of the same inertial mass? To find the inertial mass of the new body we pull it with various forces F for certain times Δt. We find that F/a, the inertial mass of the new body, is just twice the inertial mass of each original piece. Thus when we add the pieces together, we also add their inertial masses. In fact, we can take any piece of gold and any piece of silver and measure their individual inertial masses, m_1 and m_2. When we join these bodies together we find that the inertial mass, m, of the combination, again measured as F/a, is now equal to the sum of the two original inertial masses: $m = m_1 + m_2$. The same is true for bodies made of any other substances. Inertial masses are additive.

In concluding that inertial masses add, we have concluded that the inertial mass does not depend on the shape or the chemical nature of an object. We did not specify how we combined silver and gold pieces, nor do we need to. If we melt down any object and cast it again in any shape we choose, the ratio F/a will not be affected.

Let us go even further. Suppose we measure the inertial mass of a flashbulb, then flash it so that the magnesium and oxygen inside combine to form magnesium oxide. We find that the chemical combination has not affected the inertial mass. Or, we can place separate solutions of sodium carbonate (washing soda) and calcium chloride in a closed vessel (Fig. 11–12). We measure the inertial mass of this system, then upset the container so that the substances react to form calcium carbonate (an insoluble white solid) and a solution of table salt. When we measure the inertial mass again, we find no change.

Figure 11–12
Apparatus used in an experiment to show that inertial mass does not change in chemical reactions.

What, then, do we know about inertial mass? It increases in proportion to the amount of a single substance in an object. When various pieces of matter are put together, the inertial masses add irrespective of the nature of the materials involved. Finally, inertial mass is conserved in chemical reactions.

11-6 Inertial and Gravitational Mass

The properties of inertial mass remind us of the properties of mass as measured on a balance. When a balance is in equilibrium we say that we have equal masses on each side. The masses thus measured are called *gravitational* masses because we are comparing the gravitational pull of the earth on them. Actually the earth is not important; the property we measure is a property of the body alone. The balance works just as well at the top of a mountain, where the pull of the earth on each object is weaker. It would work as well on the moon, where the masses we compare would be pulled still more weakly. The important thing in measuring gravitational mass is that we compare gravitational pulls on the objects when they are in the same place in relation to the other pieces of matter in the universe.

In terms of the measurements by which we determine them the gravitational mass and the inertial mass have no connection. To measure inertial mass we apply a force to an object and find its acceleration. Gravity is irrelevant. On the other hand, when we measure gravitational mass by using a balance in equilibrium under gravitational forces, we have no acceleration but we must have gravitational forces. Two measurements could hardly differ more completely. Nevertheless the properties of the gravitational mass are strikingly similar to those we have just found for inertial mass. Gravitational masses of any substances add. Gravitational mass is conserved in chemical reactions.

The additivity of masses of each kind and the conservation of each kind of mass in chemical reactions suggest that the gravitational and inertial mass may be proportional for any object. This proportionality can be tested by measuring the inertial and gravitational masses of many different objects of many different compositions. In fact, such experiments have been done many times over. To within our best experimental accuracy (one part in 10^8), the inertial masses of all objects are proportional to their gravitational masses.

The equivalence of inertial and gravitational mass—their experimentally revealed proportionality—makes it convenient to use the same standard of mass for both. The standard kilogram—a carefully protected cylinder of platinum alloy at Sèvres in France—is the international standard unit of both gravitational and inertial mass. To find the inertial mass m of an object in kilograms, we accelerate the object and accelerate a standard kilogram mass m_s with the same force. Then we know that $m = F/a$ while $m_s = F/a_s$; therefore

$$\frac{m}{m_s} = \frac{a_s}{a}.$$

Because $m_s = 1$ kg, the mass m in kg is given by the ratio a_s/a. For example, if a certain force accelerates a kilogram mass at ½ m/sec^2 and another object at 2 m/sec^2, the mass of the other object is ¼ kg.

It is often difficult to establish the ideal conditions in which we can apply a known force to an object and be sure that no unknown forces influence the resultant motion. But we need not measure the inertial mass of every object in which we are interested in this direct way. Because of the equivalence of inertial and gravitational mass, a measurement of gravitational mass tells us the inertial mass. We need not normally worry about the distinction between them, and we usually use the word "mass" alone to refer to either one.

11-7 Newton's Law: Dynamical Measurement of Force; Units

The relation $F = ma$ embodies Newton's law of motion.* When we know the mass of an object, we can use Newton's law in either of two ways. We

* What we call Newton's law is often called his second law, and Galileo's principle of inertia, which is a special case of the general law, is sometimes called Newton's first law. The names of Newton's first and second laws don't change the content, but it may be important to know them so that you can understand what someone means when he says, "According to Newton's first law. . . ."

can predict the acceleration a of the object if we know the force F; or we can determine the force by observing the acceleration it imparts to the object.

Suppose that we observe the same object in two different experiments. Perhaps we see that it accelerates three times as fast in the second experiment as in the first. We can then conclude that the force acting in the second experiment is three times as great as the force in the first experiment. In other words, we can use the acceleration $a = \Delta v/\Delta t$ of a given object as a measure of the force, just as we used it earlier to establish the equality of the forces exerted by two identical spring loops when they produced equal acceleration on the same object.

We now have a dynamical method to determine forces. We start with a kilogram mass. The force which accelerates it at the rate of 1 m/sec^2 will be our unit of force. It is called a *newton*.

From $F = ma$ we can determine any force which acts on a known mass by measuring the acceleration it produces. The acceleration in meters/sec^2 times the mass in kg gives the force in newtons.

11-8 Newton's Law and Moving Bodies

We have discussed Newton's law of motion under what appear to be rather special conditions. For simplicity, we have accelerated masses from rest using a constant force. Does the connection that we have found between force and acceleration hold equally well if we change the magnitude of the force while a body is in motion?

Suppose we push some object, initially at rest, with a steady force for a definite time. It will accelerate as long as we push it. If we stop pushing, acceleration ceases; the body moves on at constant velocity. If we begin pushing again, we again get acceleration. Suppose we apply a force in the direction opposite to the motion. Then we expect the acceleration to be in the direction of the force. Because it is opposite to the motion, the object should slow down instead of speeding up. Experiments show that it does indeed slow down, and the deceleration is F/m.

Whether a body is standing still or sailing through outer space at 10^5 m/sec, a force will accelerate it: we always find $a = F/m$. We need not consider what velocity the body now has, nor what process produced this velocity. No matter what forces acted in the past or what the present velocity of a body is, a given force applied along the line of motion will produce the same acceleration.

11-9 How Forces Add; the Net Force

So far we have studied the motion of an object acted on by a single force. What happens when two or more forces act on the same object? Recall, for example, the Dry Ice puck pulled by two identical loops yoked one next to the other. With this arrangement the force on the puck is twice that of the single loop; and the acceleration is twice the acceleration imparted by one loop. The acceleration is proportional to the sum of the two individual forces.

We can also use two identical loops, each stretched the same amount, to pull in opposite directions (Fig. 11–13). Then no acceleration takes place. For example, if you and a friend pull equally hard on a book, but in opposite directions, the book does not accelerate. Apparently the net force, the force which changes the motion, is obtained by adding the forces in the same way as displacements or vectors in Chapter 10. As far as their effect on the motion is concerned, two forces of equal magnitude and opposite direction just cancel, and one of them can be considered to be the negative of the other.

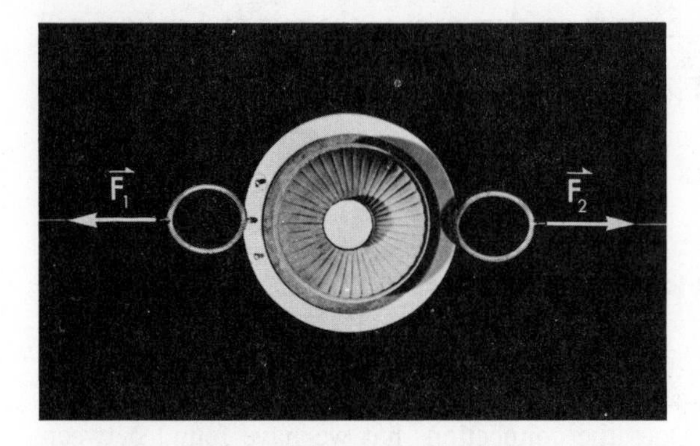

Figure 11–13
Two forces, equal in magnitude but opposite in direction, are acting on a Dry Ice puck. The net force is zero, and the observed acceleration is zero.

In general, when we apply forces on an object in opposite directions, the acceleration of the object is found to be proportional to the algebraic sum of the forces. When a force of 1 newton acts to the left, and a force of 3 newtons pulls to the right on a Dry Ice puck, the puck accelerates to the right in just the way it does when a single force of 2 newtons acts on it. The net force of 2 newtons is the sum of the individual forces taken as indicated in Fig. 11–14. Furthermore, when any number of individual forces act on an object, we find that Newton's law of motion holds and the observed acceleration arises from the net force.

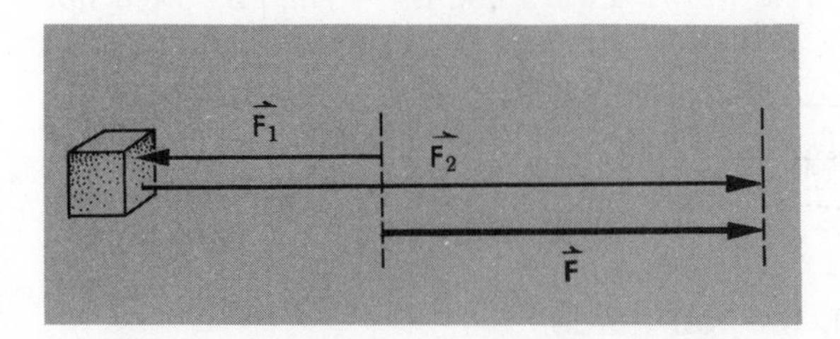

Figure 11–14
A force $\vec{F}_2$ of 3 newtons is acting to the right on an object, and a force $\vec{F}_1$ of 1 Newton is acting to the left. Adding them like displacements, we obtain the sum $\vec{F}$ of 2 newtons acting to the right. This is the net force that appears in Newton's law.

Two forces need not pull in the same or in opposite directions; they may be pulling at an angle to each other. What is then the direction and magnitude of the net force? Suppose we pull equally hard on an object with each of two identical loops, as in Fig. 11–15. We find that the object accelerates along the line bisecting the angle between the directions of the two forces.

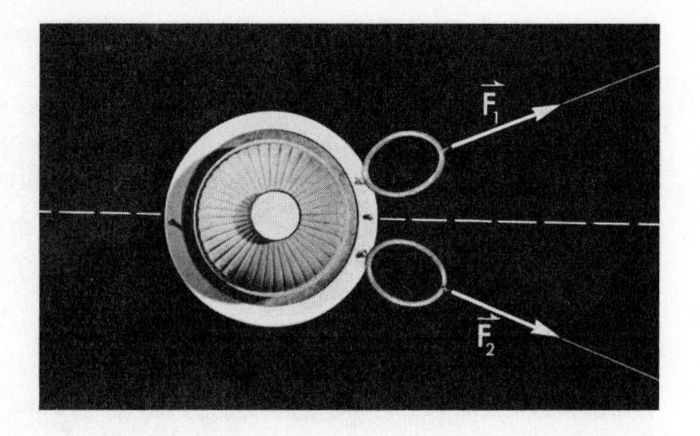

Figure 11–15
Two forces of equal magnitude act at an angle to each other. The object accelerates along the dotted line bisecting the angle between the directions of the two forces. We conclude that the net force acted along this line.

Apparently the net force is the vector sum of the two separate forces. Experiments show that this is indeed the case. The acceleration imparted by the two loops in Fig. 11–15 is given by $F = ma$, where F is the magnitude of the vector that we obtain by treating the individual forces as vectors and adding them together to get the net force vector (Fig. 11–16). When the two forces are not equal or when there are more than two forces, the magnitude and direction of the net force is given by the vector sum of the individual forces. This net force determines the acceleration in accord with $a = F/m$.

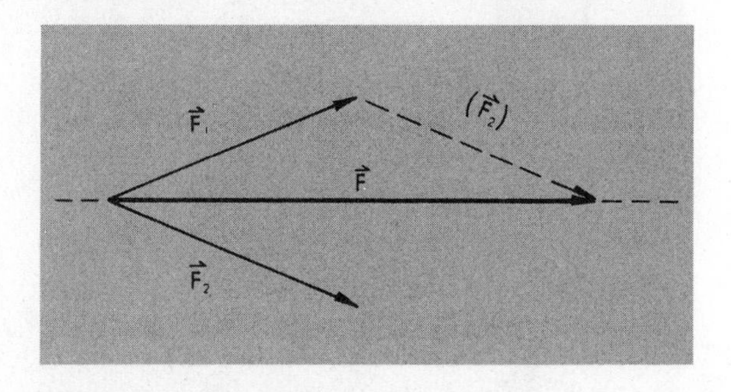

Figure 11–16
The vector addition of the two forces that are shown in Fig. 11–15. The sum is the net force which determines both direction and magnitude of the acceleration of a given mass.

Let us sum up what we have learned so far. We started out by studying the acceleration of bodies from rest under the influence of a single force. This led us to Newton's law of motion. We then investigated what happens to a body which is already moving when a force acts on it in the direction of motion or opposite to it. We found that Newton's law still holds. We then asked what happens if several forces act on a body. Again, Newton's law holds as if a single force, the net force, acts on the body. The net force is the vector sum of all the forces.

11-10 The Vector Nature of Newton's Law

Newton's law is even more general than we have yet indicated. So far we have applied a net force only along the direction of motion of an object or opposite to it. As a result of these net forces we change the speed of the object but not the direction of its motion. Forces, however, can be applied in other directions. We can change the direction of a moving ball by pushing across its direction of motion. In general, then, we see that forces change the velocity vector of a moving object either in magnitude or direction, or both. The force is itself a vector and Newton's law relates it to the rate of change of the vector velocity. We should, therefore, write the law as

$$\vec{\mathbf{F}} = m\frac{\Delta\vec{\mathbf{v}}}{\Delta t} = m\vec{\mathbf{a}},$$

where $\vec{\mathbf{a}}$ is the vector acceleration. In the next chapter we shall see some of the evidence for the vector nature of Newton's law and discuss some of its implications.

11-11 Forces in Nature

When we pull an object, we cannot usually be sure that the force we exert is the only force acting on the object (Fig. 11–17).

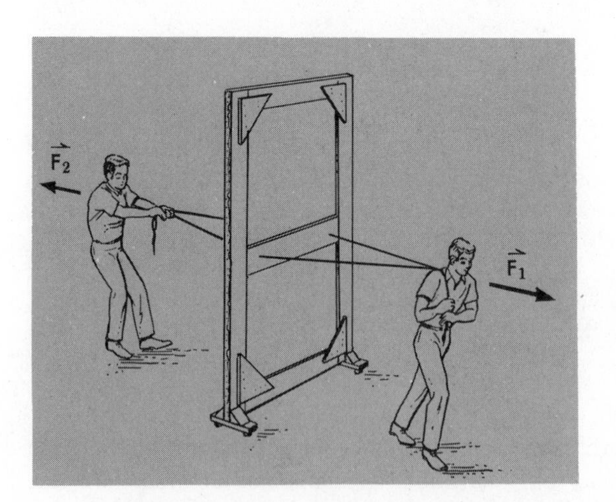

Figure 11–17
The force we exert may not be the only force that is acting.

Sometimes the origin of forces on an object is not immediately apparent, or forces may arise as a result of the motion of the object. For example, the wind pushing on the surface of a balloon exerts a force, and we must pull on the balloon in the opposite direction to prevent it from blowing away. Even if the air is still, when we move the balloon through it, the air gives rise to a force opposing the motion. When the balloon is in motion its acceleration is not given by the force we apply. It is given by the net force. If the net force is zero the acceleration is zero and the balloon moves at constant velocity.

We can measure the force exerted by a steady wind going past the balloon by measuring the extension of a rubber loop in a string holding the balloon still. This force increases with the speed of the wind. As we all know, when the speed of the wind increases, it exerts a greater frictional force.

Frictional forces also become evident when we try to move an object along a solid surface. Unlike the retarding force which arises when we pull a balloon through the air, frictional forces between solids are often nearly independent of the speed of the object. It was undoubtedly the common occurrence of friction that led the Greeks to conclude that force was needed to maintain a constant motion. The force needed is equal and opposite to the forces of friction.

The idea of force as a cause of motion is valuable because it enables us to predict what motion will occur in a given situation. Some forces are independent of motion; for example, the force of gravitational attraction, the weight of a body. As we shall see in the next chapter, this force is the same, whether the body moves or stands still. If we know our geographical position, we know what gravitational force to expect, and we can make predictions about the motions of a falling object. Other forces depend on the relative motion of one body with respect to another. One of the essential problems we face is to learn about the forces in nature. Then we can use the observed forces to make predictions about motion and to design mechanical apparatus.

For Home, Desk, and Lab

1.* What is the difference between *kinematics* and *dynamics?* (Section 1.)

2.* Was it necessary for Galileo to measure time in order to get the extrapolation which led him to the law of inertia? (Section 2.)

3. A ball is released from rest on the left-hand incline of Fig. 11–4 at a height of 10 cm above the lowest point.
   (a) If there is no friction, how high vertically will it rise on the right-hand incline?
   (b) If the right-hand incline rises 1 cm for every 10 cm of horizontal distance, how far will the ball travel horizontally on it?
   (c) If the incline rises only ½ cm for every 10 cm of horizontal distance, how far will the ball go?

4.* If you roll a putty ball around the inside of a bowl, it soon comes to rest in the center. A steel ball will orbit quite a few times before coming to rest in the center. What would be the "ideal" motion of a ball in a bowl? (Section 2.)

5. Why is it particularly dangerous to drive on an icy highway?

6.* If a ball is rolling with a velocity of 20 cm/sec and no force acts on it, what will be its velocity after 5 sec? (Section 2.)

7.* In Table 1, what would you predict for the average velocity during the eighth time interval if the puck continued to be pulled by the stretched loop? (Section 3.)

8.* Suppose a Dry Ice puck increases its velocity so that it travels first at 10 cm/sec, at 12 cm/sec at the end of the next second, 14 cm/sec at the end of the second after that, and so on. What can you tell about the force acting on the puck? (Section 3.)

9. A body is pulled across a smooth horizontal surface by a spring loop that is kept stretched by a constant amount. It is found that the body is accelerated at 15 cm/sec^2. What will be the acceleration of the body if it is pulled by two loops, each just like the first loop, side by side and stretched by that same amount? (See Fig. 11–10.)

10. An object sliding on a low-friction bearing is pulled with a constant force. In a time interval of 0.3 sec the velocity changes from 0.2 m/sec

to 0.4 m/sec. In a second trial, the object is pulled with another force. In the same length of time the velocity now changes from 0.5 m/sec to 0.8 m/sec.

(a) What is the ratio of the second force to the first?

(b) If the body is pulled with the second force for 0.9 sec, what change in velocity results?

11. A block is pulled along a horizontal surface by 2, 4, 6, and 8 parallel bands of rubber. All bands are alike and each is stretched the same length in each experiment. The graph (Fig. 11–18) shows the resulting accelerations plotted versus the number of bands.

(a) What can you conclude from the fact that the plot yields a straight line?

(b) What does the intercept of the graph with the horizontal axis measure?

(c) Can you use the graph to predict the acceleration of the block produced by stretching one rubber band by the standard amount?

(d) Suppose you repeat the experiment, changing only the surface on which the block is pulled. How will the new graph relate to the old one?

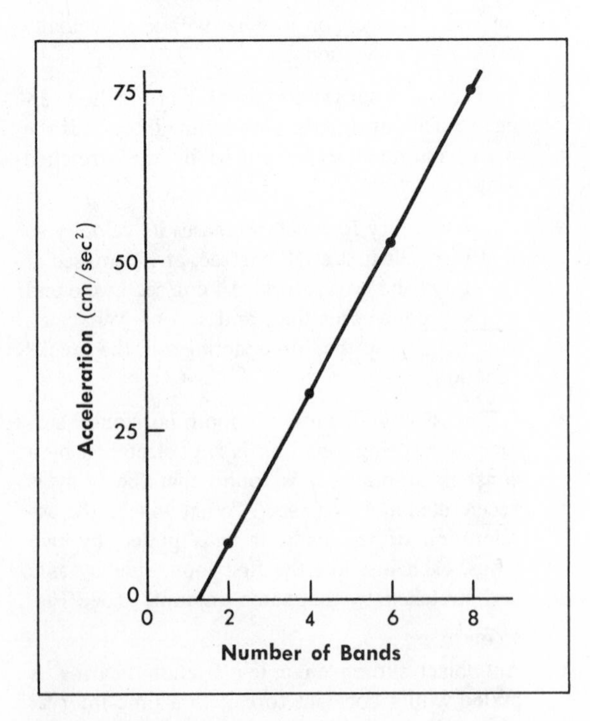

Figure 11–18 For Problem 11.

12.* In Fig. 11–9, suppose the photograph represented a body moving from right to left. In which direction would the force be acting in this case? (Section 4.)

13.* A car can accelerate by 3 m/sec². What is its acceleration if it is towing another car like itself? (Section 5.)

14. Why is the flask in Fig. 11–12 closed? Be prepared to explain in class.

15. Why do we make such careful distinctions between gravitational and inertial mass, rather than talking about one mass only, since they are equivalent?

16.* Two objects are accelerated separately by the same force. Object A accelerates at 20 cm/sec² and B at 60 cm/sec². What is the ratio of (a) their inertial masses, (b) their gravitational masses? (Section 6.)

17.* A carton is balanced by two identical cans on a balance. A certain force accelerates the carton by 2 m/sec². How much will the same force accelerate one of the cans? (Section 6.)

18.* A body with a mass of 0.5 kg is accelerated at 4 m/sec². How large a force is acting? (Section 7.)

19.* A boy gives a 0.5-kg ball a speed of 6 m/sec from rest in 0.2 sec. What average force did he apply to the ball? (Section 7.)

20. A force of 5 newtons gives a mass m_1 an acceleration of 8 m/sec², and a mass m_2 an acceleration of 24 m/sec². What acceleration would it give the two when they are fastened together?

21. The graph in Fig. 11–19 shows the velocity, along a straight line, of an object of mass 2 kg,

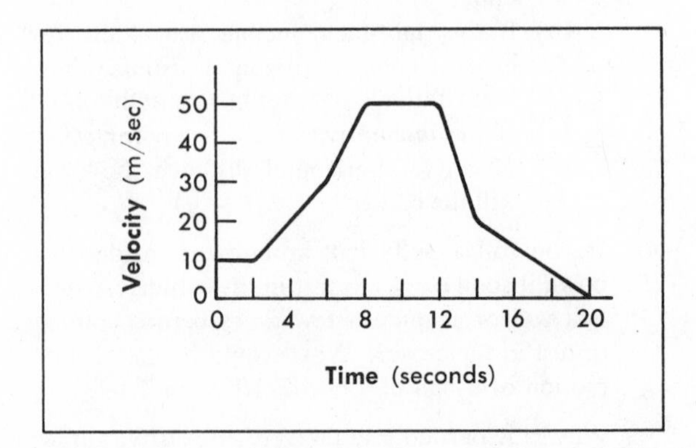

Figure 11–19 For Problem 21.

as a function of time. Plot a graph of the force as a function of time.

22.† A block of mass 3.0 kg is moving along a smooth horizontal surface with a velocity v_0 at an instant of time $t = 0$. A force of 18 newtons is applied to this body opposite to the direction of its motion. The force slows the block down to half its original velocity while it moves 9.0 m.
(a) How long does it take for this to occur?
(b) What is v_0?

23.† Consider the two masses m_1 and m_2 in Fig. 11–20. At time $t = 0$, m_1 is at rest and m_2 is moving with velocity v_0. Is it possible to apply the same constant force (in magnitude and direction) for the same time to both masses to bring them to equal velocity at a given moment? Try to solve this problem by a qualitative argument and then check your conclusion by writing down the necessary equations. Consider all cases: $m_1 < m_2$, $m_1 = m_2$, $m_1 > m_2$.

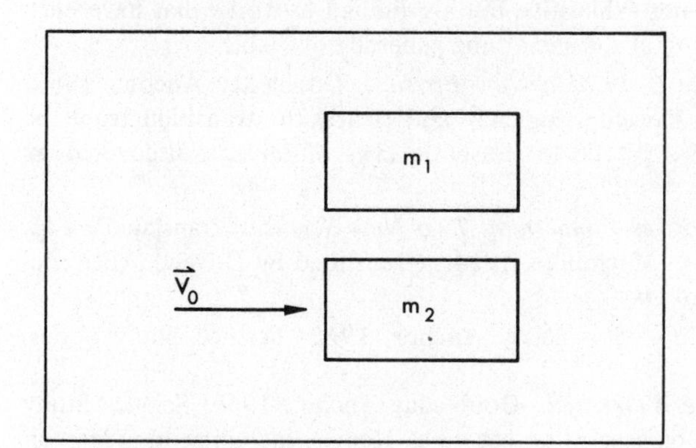

Figure 11–20 For Problem 23

24. You observe an object covering distance in direct proportion to t^3, where t is the time elapsed.
(a) What conclusion might you draw about the acceleration? Is it constant? Increasing? Decreasing? Zero?
(b) What might you conclude about the forces? Be prepared to discuss in class.

25.* If several forces of different magnitudes and directions act on an object, in what direction will the object accelerate? (Section 9.)

26.* In Fig. 11–13, what would be the acceleration of the puck if its mass were 0.50 kg, $F_2 = 10$ nt, and $F_1 = 6$ nt? (Section 9.)

27.* In Fig. 11–15, suppose $F_1 = F_2 = 2.0$ nt. What would be the net force on the puck? (Section 9.)

28. A block of mass 8.0 kg, starting from rest, is pulled along a horizontal tabletop by a constant force of 2.0 newtons. It is found that this body moves a distance of 3.0 m in 6.0 sec.
(a) What is the acceleration of the body?
(b) What is the ratio of the applied force to the mass?
(c) Since your answer to part (b) is not equal to that to part (a) (at least, it shouldn't be), what conclusions can you draw about this motion? Give numerical results, if possible.

29. Two men wish to pull down a tree by means of a rope fastened near the top. If they use only one rope, the tree will come down on top of them. To prevent this, they tie two ropes 10.0 meters long to the same point and stand on the ground 10.0 meters apart when they pull. If each pulls with a force of 300 newtons, what is the force exerted by the ropes on the tree?

30. Two men and a boy are pulling a boat along a canal. The two men pull with forces $\vec{F}_1$ and $\vec{F}_2$ whose magnitudes and directions are indicated in Fig. 11–21. Find the magnitude and direction of the smallest force which the boy could exert to keep the boat in the middle of the canal.

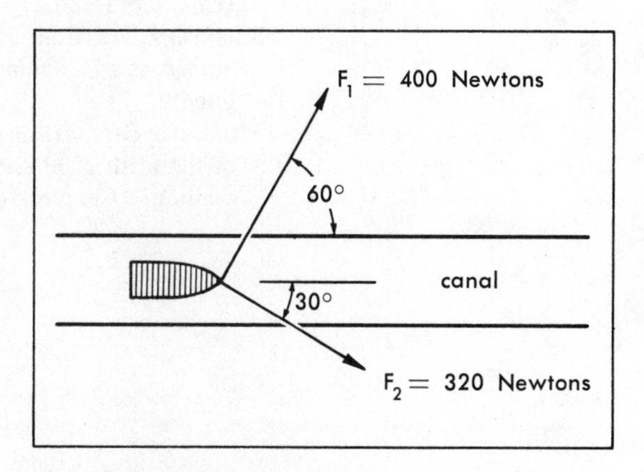

Figure 11–21 For Problem 30

31.* You have to push with a force of 200 nt to slide a refrigerator across a floor at constant velocity. What is the force of friction acting on the refrigerator? (Section 9.)

32. The retarding force of air resistance on a balloon is proportional to the square of the velocity. For a certain balloon, inflated a certain amount, this force is given in newtons by $F_R = .2v^2$ where v is the velocity in m/sec. The balloon and the air inside have a combined mass of 10 gm.
    (a) Draw graphs of the balloon's acceleration as a function of velocity when you pull it with a 1.8-newton force and with a 7.2-newton force.
    (b) What is the maximum velocity that the balloon will reach in each case?
    (c) If the mass were 5.0 gm, how would this affect the maximum velocity?
    (d) What do you think would be the effect on the maximum velocity if you inflated the balloon to a larger volume?

33. Aristotle thought that a constant force was required to produce a constant velocity and from this he concluded that, in the absence of force, bodies would come to rest.
    (a) Name several situations where a constant force seems to produce a constant velocity.
    (b) How do you explain each of the situations in (a) in the light of Newton's law of motion?

34. How would you define a unit of mass if people had placed a standard spring at Sèvres instead of a standard mass?

Further Reading

These suggestions are not exhaustive but are limited to works that have been found especially useful and at the same time generally available.

COHEN, I. BERNARD, *The Birth of a New Physics.* Doubleday Anchor, 1960: Science Study Series. Develops logically and clearly the transition from the natural philosophy of Aristotle to the basic laws of physics discovered by Galileo and Newton.

GALILEI, GALILEO, *Dialogues Concerning Two New Sciences,* translated by H. Crew and A. de Salvio. Macmillan, 1914. (Reprinted by Dover.) (See section entitled "The Third Day.")

GAMOW, GEORGE, *Gravity.* Doubleday Anchor, 1962: Science Study Series. (Chapters 1 and 2)

KOESTLER, ARTHUR, *The Watershed.* Doubleday Anchor, 1960: Science Study Series. An intriguing biography of Johannes Kepler, including his ideas on gravity.

MURCHIE, GUY, *Music of the Spheres:* Houghton Mifflin, 1960. A broad picture of the birth of physics from Galileo to Newton. Interesting and inspirational reading. (Chapter 10)

Motion at the Earth's Surface

CHAPTER

12

In this chapter we shall start our study of the forces occurring in nature. We shall determine them from the motions of the bodies on which they act, or, when possible, by balancing the unknown forces against known forces that we can apply. Here on the surface of the earth the force of gravity which pulls objects toward the center of the earth is familiar to us all. We shall first study this gravitational attraction.

12-1 Weight and the Gravitational Field of the Earth

Objects on the surface of the earth are pulled toward its center. Different objects are pulled down by gravitational forces of different magnitude. The magnitude of this force on an object is called the weight of the object. It may be measured by hanging the object on a spring calibrated in newtons.

If we take a standard kilogram mass and measure its weight anywhere on the surface of the earth, we find it to be very nearly 9.8 newtons. Actually the weight of an object differs slightly from place to place on the earth's surface. However, the fractional variation in weight is the same for all objects. For example, at the North Pole a 1-kg mass weighs 9.83 newtons and at the equator it weighs 9.78 newtons. At the pole a mass of 2 kg weighs 19.66 newtons, and at the equator it weighs 19.56 newtons. Both masses change weight by ½ percent, when moved from pole to equator. The weight of the 2-kg mass is always exactly twice that of a 1-kg mass at the same place. Therefore, the gravitational force $\vec{\mathbf{F}}$ acting on an object of gravitational mass m_g can be written

$$\vec{\mathbf{F}} = m_g\vec{\mathbf{g}},$$

where $\vec{\mathbf{g}}$ is the proportionality factor between the gravitational force and the gravitational mass. The mass is independent of position on the earth's surface—if two masses balance (on an equal-arm balance) at one place on the earth, they balance at all other places. Therefore, since the force of gravity on an object varies slightly from one place to another, the proportionality factor $\vec{\mathbf{g}}$ must also vary slightly from place to place near the surface of the earth, but it is the same for all masses at any one place.

The factor $\vec{\mathbf{g}}$ is the gravitational force per unit mass. It is a vector with the dimensions of newtons per kilogram. We can measure and tabulate it for a

Figure 12–1
The gravitational field of the earth. The inner vectors represent the magnitude and direction of the field at the earth's surface. The middle set gives values for a height of 3×10^6 meters. The outer vectors are for a height of 6×10^6 meters. (The radius of the earth is 6×10^6 meters.)

large number of locations on the surface of the earth. Such a collection of the values of a physical quantity depending on position is called a field. In particular, the gravitational force per unit gravitational mass at different locations on the earth is called the gravitational field of the earth. Figure 12–1 represents part of the gravitational field around the earth, and Table 1 gives its magnitude at a number of different places.

Table 1
Gravitational Field of the Earth

PLACE	LATITUDE	ALTITUDE, METERS	MAGNITUDE OF THE FIELD, NEWTONS/KG
North Pole	90°	0	9.832
Greenland	70°	20	9.825
Stockholm	59°	45	9.818
Brussels	51°	102	9.811
Banff	51°	1376	9.808
New York	41°	38	9.803
Chicago	42°	182	9.803
Denver	40°	1638	9.796
San Francisco	38°	114	9.800
Canal Zone	9°	6	9.782
Java	6° South	7	9.782
New Zealand	37° South	3	9.800

12-2 Free Fall

How does an object move in the gravitational field $\vec{g}$ of the earth? Suppose we let it fall. Its motion depends on the net force acting on it and on its *inertial* mass. The net force $\vec{F}$ acting on the falling body is given by the vector sum of the gravitational attraction $\vec{F}_g = m_g\vec{g}$ (the weight of the body) and the air resistance $\vec{F}_a$. It is

$$\vec{F} = \vec{F}_a + m_g\vec{g}.$$

According to Newton's law of motion, the acceleration of the object is proportional to this net force and inversely proportional to the inertial mass m_i; that is,

$$\vec{a} = \frac{\vec{F}}{m_i} = \frac{\vec{F}_a + m_g\vec{g}}{m_i}.$$

Now for objects which are relatively dense, air resistance is very small at low speeds and can be neglected compared to the weight. In such circumstances, the gravitational attraction alone is important, and we expect the acceleration of the object to be

$$\vec{a} = \frac{m_g}{m_i}\vec{g}.$$

Here something remarkable appears. As we found out in Chapter 11, the ratio $\frac{m_g}{m_i}$ is the same for all objects. Consequently the acceleration does not depend on the mass of the objects. In fact, by measuring m_i and m_g in terms of the same standard kilogram, we have made m_i equal to m_g; that is, $\frac{m_g}{m_i} = 1$. It then follows that for a freely falling body

$$\vec{a} = \vec{g}.$$

In other words, we predict that all objects moving under the influence of gravitation alone are accelerated in just the same way when located at the same place in the gravitational field. In addition, we should be able to measure the gravitational field strength g at any given location by observing the acceleration of any object as it moves through this position.

What do we find in fact? All bodies of sufficiently high density (large mass per unit volume) do fall with the same acceleration at the same place. The magnitude of the acceleration with which they fall is very nearly 9.8 m/sec^2 anywhere near the surface of the earth. It makes no difference whether we measure it in a laboratory on the first floor of a building or on the top floor, whether the body starts from rest or has been projected with any reasonable vertical velocity.

Careful measurements of the multiple-exposure photograph of a falling object (Fig. 12–2) lead to the results shown in Table 2. The exposures are at $\frac{1}{30}$-sec intervals. These results illustrate our statement that the gravitational field near the earth's surface results in an acceleration of falling objects of about 9.8 m/sec^2 directed down.

We have limited our attention to compact, dense objects carefully selected to minimize the frictional resistance of the air. But if you drop a Ping-Pong ball it falls only a short distance before the force of air resistance balances the force of gravity, and the ball moves at constant velocity. In general, air resistance becomes greater with higher velocity. Therefore, if an object falls far enough, it will gain so much velocity that the air resistance becomes equal in magnitude to the weight. The body continues to fall, but with constant velocity. This final constant velocity is called the terminal velocity of the falling body.

We can check that these deviations from free fall are indeed the result of air resistance by performing experiments in a vacuum. When we remove the air, we find that all objects, regardless of shape or density, fall with the same acceleration at a particular position near the earth's surface. Furthermore, because $\vec{g}$ does not change direction or magnitude appreciably unless we move through distances comparable with the size of the earth, the acceleration is closely the same for falling objects anywhere within a room, within a building, a city, or even a state. In the region within which the gravitational field $\vec{g}$ is effectively constant and where gravitation alone is important, all objects fall with constant acceleration equal to $\vec{g}$. Starting from rest, in time t they pick up *downward* velocity

$$v = gt$$

and move down through the distance

$$d = \tfrac{1}{2}gt^2$$

given by the area under the graph of velocity versus time (Fig. 12–3). By using the flash photo of Fig. 12–2 again, you can show that the distance d moved from rest increases as $\frac{1}{2}gt^2$.

Figure 12–2
A flash photograph of a falling billiard ball. The position scale is in centimeters, and the time interval between successive positions of the ball is 1/30 second. This motion is analyzed in Table 2.

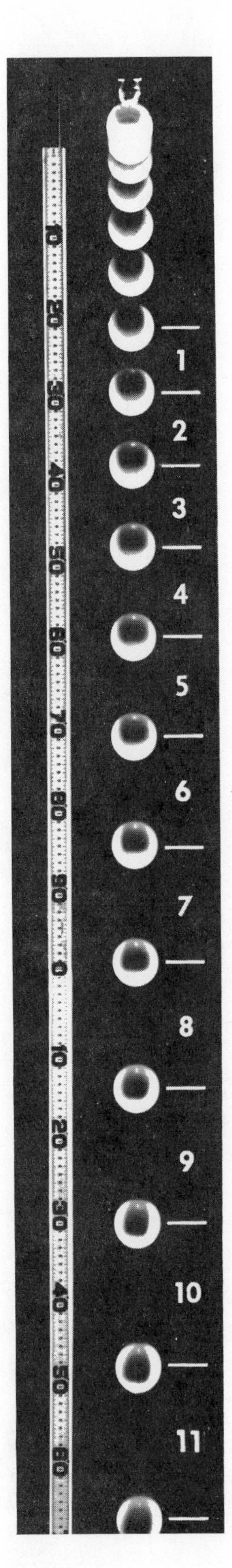

Figure 12–3
A velocity-time graph of a freely falling body. The vertical displacement, or distance fallen, is given by the area under the curve—the area of the triangle with base t and height gt. This area is $\frac{1}{2} gt^2$.

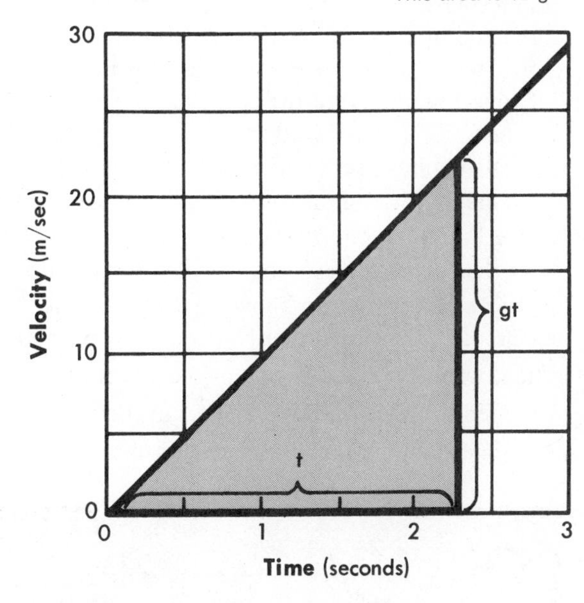

Table 2

INTERVAL NUMBER	DISPLACEMENT Δx(CM)	AVERAGE VELOCITY $\Delta x/\Delta t = v$(CM/SEC)	CHANGE IN AVERAGE VELOCITY Δv(CM/SEC)	ACCELERATION $\Delta v/\Delta t$(M/SEC^2)
1	7.70	231		
2	8.75	263	32	9.6
3	9.80	294	31	9.3
4	10.85	326	32	9.6
5	11.99	360	34	10.2
6	13.09	393	33	9.9
7	14.18	425	32	9.6
8	15.22	457	32	9.6
9	16.31	489	32	9.6
10	17.45	524	35	10.5
11	18.52	556	32	9.6
			Average acceleration	9.8

The analysis of the motion shown in Fig 12–2. The calculated values of the acceleration are constant within the limits of accuracy of our measurements. Even though an accurate ruler was brought right up to the picture of the ball, the last significant figure in the Δx column is quite uncertain and is just a reasonable estimate of a fraction of a millimeter. It has been retained, however, to reduce successive errors that would accumulate if we rounded off early. Notice that we have kept only three significant figures in the velocity column.

12-3 Projectile Motion: the Vector Nature of Newton's Law of Motion

Objects that drop straight down under the influence of gravitational attraction alone all accelerate at the same rate. Do all objects also accelerate at that same rate when they move in other directions in the gravitational field? Figure 12–4 shows a multiple-flash photograph of two balls. The first ball started falling from rest at the moment that we projected the second one horizontally. You see that the vertical motions of the two balls are identical despite the fact that the horizontal motions differ. Also you see that the horizontal motion is at constant horizontal velocity, like motion when there is no force. The presence of the downward force does not change the horizontal motion; and the existence of horizontal motion does not change the effect of the downward force on the vertical motion. The horizontal and vertical motions are independent. Each one follows from the appropriate component vector of the force: $\vec{\mathbf{F}}_v = m\vec{\mathbf{g}}$, and the vertical motion is the same as free fall; $\vec{\mathbf{F}}_h = 0$, and the horizontal motion is without acceleration.

At the end of the last chapter we found that the net force $\vec{\mathbf{F}}$ acting on a body is a vector. As we know from Chapter 10, the acceleration $\vec{\mathbf{a}}$ is also a vector. This suggested that Newton's law of motion is a vector law. In addition, when the force $\vec{\mathbf{F}}$ acts in the direction of motion, the acceleration $\vec{\mathbf{a}}$ is related to the force by $\vec{\mathbf{F}} = m\vec{\mathbf{a}}$. In the last chapter, however, we did not investigate any motion in which the net force $\vec{\mathbf{F}}$ acted at an angle to the

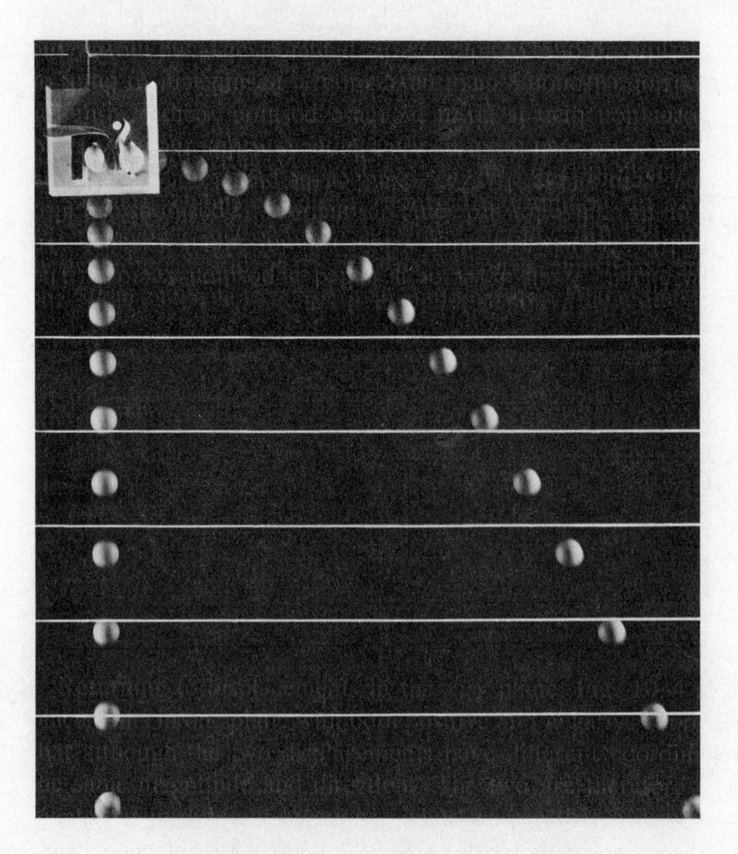

Figure 12–4
A flash photograph of two golf balls, one projected horizontally at the same time that the other was dropped. The strings are 15 cm apart, and the interval between flashes was 1/30 second.

velocity $\vec{\mathbf{v}}$. We might, therefore, ask the questions: Is Newton's law of motion valid when $\vec{\mathbf{F}}$ and $\vec{\mathbf{v}}$ are in different directions? Is the acceleration still in the direction of the force? Is the magnitude of $\vec{\mathbf{a}}$ the same for the same size force?

Projectile motion is the first case we have examined in which $\vec{\mathbf{F}}$ and $\vec{\mathbf{v}}$ are in different directions. The observed fact is that the vertical gravitational force $\vec{\mathbf{F}}$ produces the same vertical acceleration whether there is horizontal motion or not. Although this observation is not enough to prove that $\vec{\mathbf{F}} = m\vec{\mathbf{a}}$, it provides supporting evidence for the idea that Newton's law in this simple form holds regardless of the direction of motion. After discussing the path of a projectile, we shall examine experiments with other forces acting across the direction of motion. We shall see that the vector law $\vec{\mathbf{F}} = m\vec{\mathbf{a}}$ also holds for those experiments.

12-4 Projectile Motion: Determination of the Path

In studying the motion of a projectile we encounter a new problem. Previously in our study of dynamics, we have considered objects moving in a straight line under the action of a net force along the line of motion. The second ball in Fig. 12–4, however, traveled in a curved path. One of the important problems in dynamics is the determination of the path in such a situation.

If we know the position and velocity of an object at one moment, the path it follows as a function of time can be found from the force on the object and Newton's law of motion. For the projectile, however, we do not need to go back to Newton's law and the gravitational force. We can find the path by combining the known vertical and horizontal motions. As we saw in the last section, they go on independently. For this purpose we choose a horizontal axis of reference (the x axis) and a vertical reference axis (the y axis) so placed that the origin ($x = 0, y = 0$) is at the point where the body is projected (Fig. 12–5). When the ball is projected with horizontal velocity v_o, we know that it continues to move along the x direction at this velocity. After a time t, the x coordinate of the position of the ball is therefore $x = v_o t$. We also know that the vertical motion is the same as free fall. The y coordinate at the same time t is therefore $y = -\frac{1}{2}gt^2$. (The minus sign just tells us that the ball goes down rather than up.) These equations contain all the information about bodies projected horizontally with an initial velocity equal to v_o. The common value of the time t in the equations relates them to the motion of one single body rather than to the motion of two different bodies.

The path which the body follows is a curve, and we can express this curve by an equation relating the vertical position y to the horizontal position x at the same instant of time. To find this equation we eliminate the time t from the two equations $y = -\frac{1}{2}gt^2$ and $x = v_o t$. From the second equation we see that $t = x/v_o$; and putting this expression for t into the first equation, we get

$$y = -\tfrac{1}{2}gt^2 = -\tfrac{1}{2}g\left(\frac{x}{v_0}\right)^2 = -\frac{g}{2v_0^2}x^2.$$

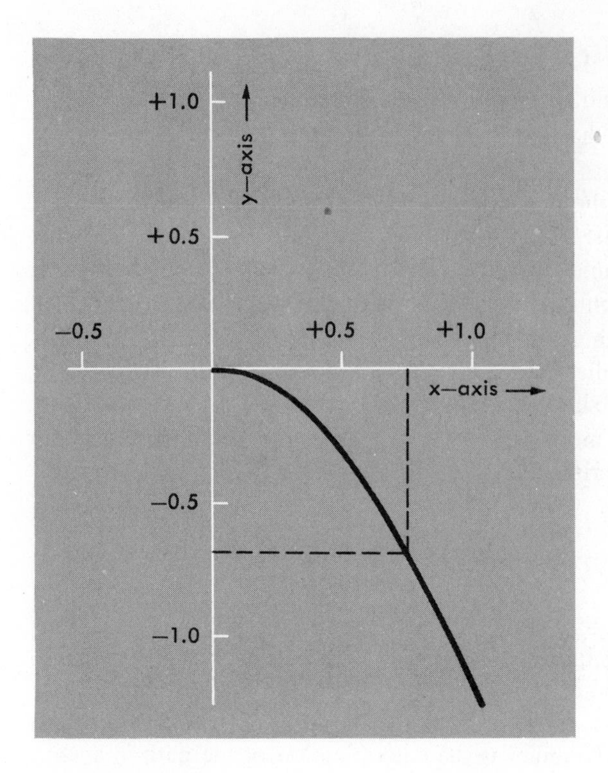

Figure 12–5
The path of the ball in Fig. 12–4 is plotted on a pair of coordinate axes. The scales on the figure measure the distance along the coordinate axes in meters. The x coordinate of the ball at any time t is v_0t (where $v_0 = 2$ m/sec), and its y coordinate is $\frac{1}{2}gt^2$. For example, when $t = 0.38$ sec, then $x = 0.75$ m and $y = -0.7$m.

The equation

$$y = -\frac{gx^2}{2v_0^2}$$

is the equation of the path of the object. As shown in Fig. 12–5, the path is a parabola with its vertex at the place where the object is moving horizontally.

In Fig. 12–6 we have plotted several possible paths which correspond to different values of the initial horizontal velocity v_o. You can see that when the horizontal velocity is large, the parabola is rather flat. The projectile moves a long way sidewise before falling any great distance. On the other hand, small values of v_o give sharply curved parabolas. The projectile moves a shorter distance horizontally in the time it takes to fall a given amount.

Here we have analyzed the problem of a projectile which is fired horizontally. The more general case of a projectile fired with an initial velocity $\vec{\mathbf{v}}_o$ at any angle with the horizontal can be handled in the same way. We again use the fact that the vertical and horizontal motions are independent. From the initial velocity vector $\vec{\mathbf{v}}_o$, we find the initial horizontal and vertical components. The horizontal component of velocity never changes, and the vertical component undergoes a uniform change at the rate

$$\frac{\Delta v_y}{\Delta t} = -g.$$

When we work through the details the result is again a parabolic path with a nearly flat region whose extent is determined by the horizontal velocity. The only difference is that the top of the parabola is not at the starting point of the motion (Fig. 12–7).

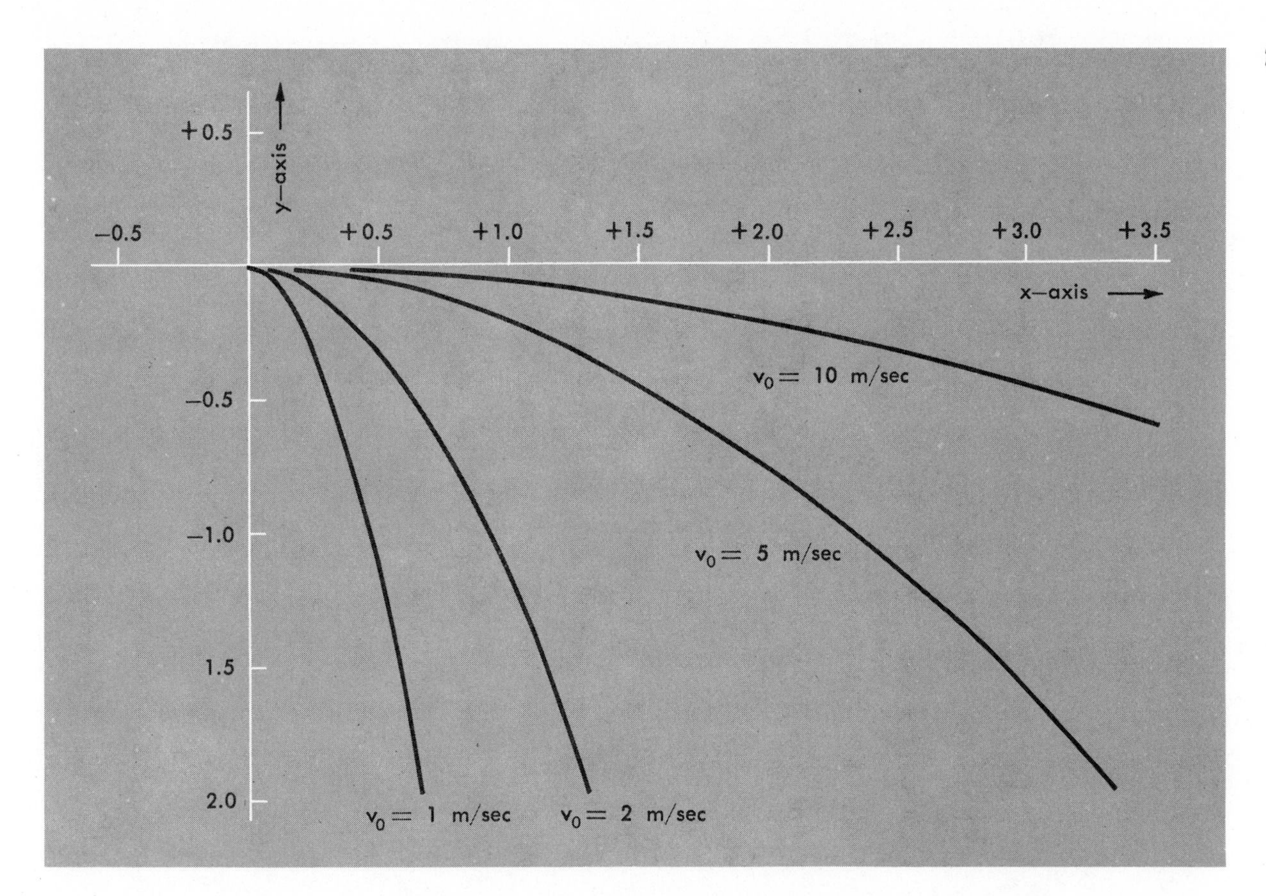

Figure 12–6
Several possible paths for a body projected horizontally. Note that the parabola's shape depends on the horizontal velocity v_0.

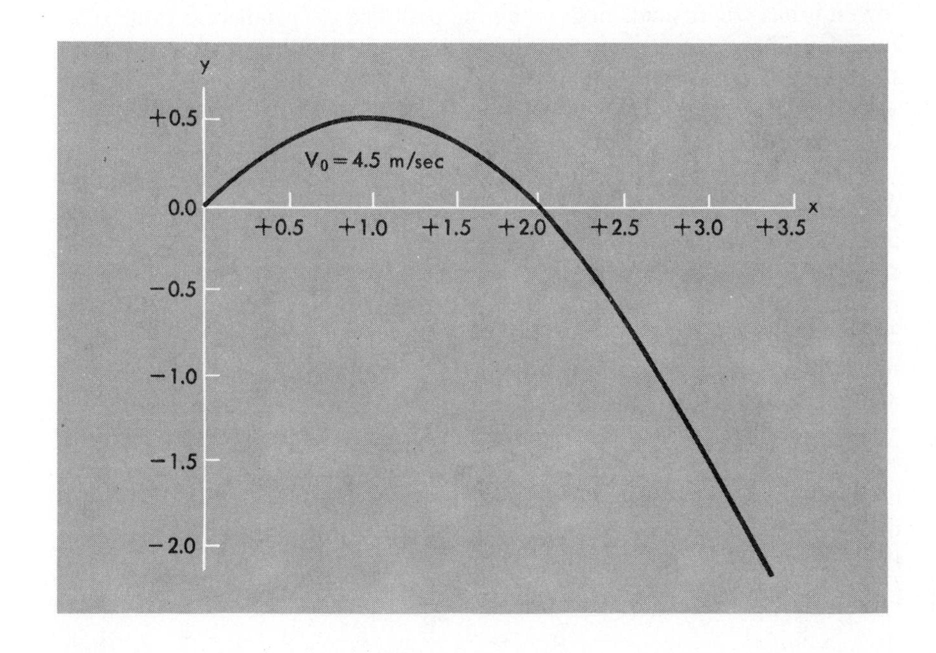

Figure 12–7
The path of a body projected with an initial velocity v_0 of 4.5 m/sec at a 45° angle with the horizontal.

What we learned in this section on projectile motion resembles what we learned earlier in Chapter 10. Yet there is an important difference. In Chapter 10 we dealt only with the description of motion (kinematics), whereas here we considered the force which caused the motion (dynamics). The gravitational force $m\vec{g}$ and Newton's law in vector form allow us to predict as well as describe what happens. In our study of the dynamics of other motions we shall lean heavily on kinematic descriptions given in Chapter 10.

12-5 Deflecting Forces and Circular Motion

An object moves on a curved path because the force that pushes it has a component perpendicular to the direction of motion. The component of force along the path changes the speed, but it does not change the direction. On the other hand, a force perpendicular to the motion shoves the object sidewise so that the path curves. If the total force is perpendicular to the velocity, it leads to a perpendicular acceleration which changes the direction without changing the magnitude of the velocity vector as shown in Fig. 12–8.

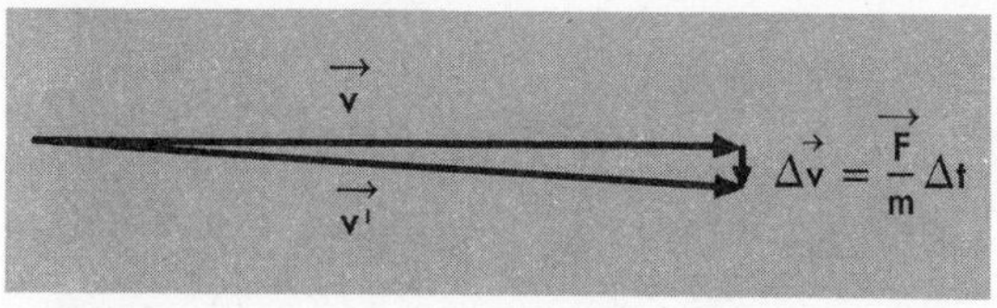

$\vec{v}$

$\vec{v}'$

$\Delta\vec{v} = \frac{\vec{F}}{m}\Delta t$

Figure 12–8
The change in the velocity of a body when a force acts perpendicular to its path. Such a force changes the direction of the motion, but not the speed. The magnitude of $\vec{v}'$, the velocity after the force $\vec{F}$ has acted for the very short time Δt, is the same as that of $\vec{v}$.

While a projectile is moving on its parabolic path, the force of gravity acting on it has components both along the path and perpendicular to it. The direction of the velocity changes because the perpendicular component of force produces a perpendicular acceleration, and the magnitude of the velocity changes because the component of the force along the path produces an acceleration along the path.

In this section we shall concentrate on change in direction alone, that is, circular motion at constant speed. How is such uniform rotation produced? Suppose we push steadily on a body, always pushing at right angles to its motion. Because there is never a component of force in the direction of motion, there is no acceleration along the path; the speed of the body stays constant while the direction of motion changes. Now if we keep the magnitude of the force constant, the direction of the path will change the same amount in every equal time interval, and the path must be a circle.

Thus we have reached the important conclusion that a deflecting force of constant magnitude perpendicular to the motion makes a body move in a circle with constant speed. The force and the acceleration of the body are directed toward the center of the circle, and the velocity vector is tangent to the circle at every point. In Section 10–7 we have shown that the acceleration for such motion is given by

$$\vec{a} = -\frac{4\pi^2\vec{R}}{T^2},$$

where $\vec{R}$ is the position vector and T is the period.

Now, Newton's law $\vec{\mathbf{F}} = m\vec{\mathbf{a}}$ tells us how the force and acceleration are related. Therefore, with our mathematical expression for the acceleration in terms of the radius R of the motion and its period T, we can relate the force to the circular motion it produces. The result is

$$\vec{\mathbf{F}} = m\vec{\mathbf{a}} = -\frac{m4\pi^2\vec{\mathbf{R}}}{T^2}.$$

The magnitude of the force can also be related to the speed v and radius R as

$$F = \frac{mv^2}{R}.$$

Remember that the direction of $\vec{\mathbf{F}}$ is inward, toward the center of the circle, and the force is called *centripetal* force.

We can check our equations experimentally. Figure 12–9 shows an apparatus designed to do this with a Dry Ice puck. The puck rests on a horizontal glass table. One end of the string is attached to a bearing at the center of the table, and the other end is connected to a spring loop fastened to the puck.

In the experiment the puck was given just the right push to make it move around a circle. Figure 12–10 is a stroboscopic flash photograph of the motion. The light flashed 2.4 times a second and was stopped before one complete revolution was registered. The spacing between successive pictures of the puck shows that the speed was essentially constant. The loop on the puck is stretched, and the stretch stays the same throughout the motion. From the stretch we can find the force which makes the puck run around in a circle.

Figure 12–9
An arrangement for measuring the centripetal force acting on a body moving in a circular path. The puck is at rest; no force is acting; the loop on the puck is not extended.

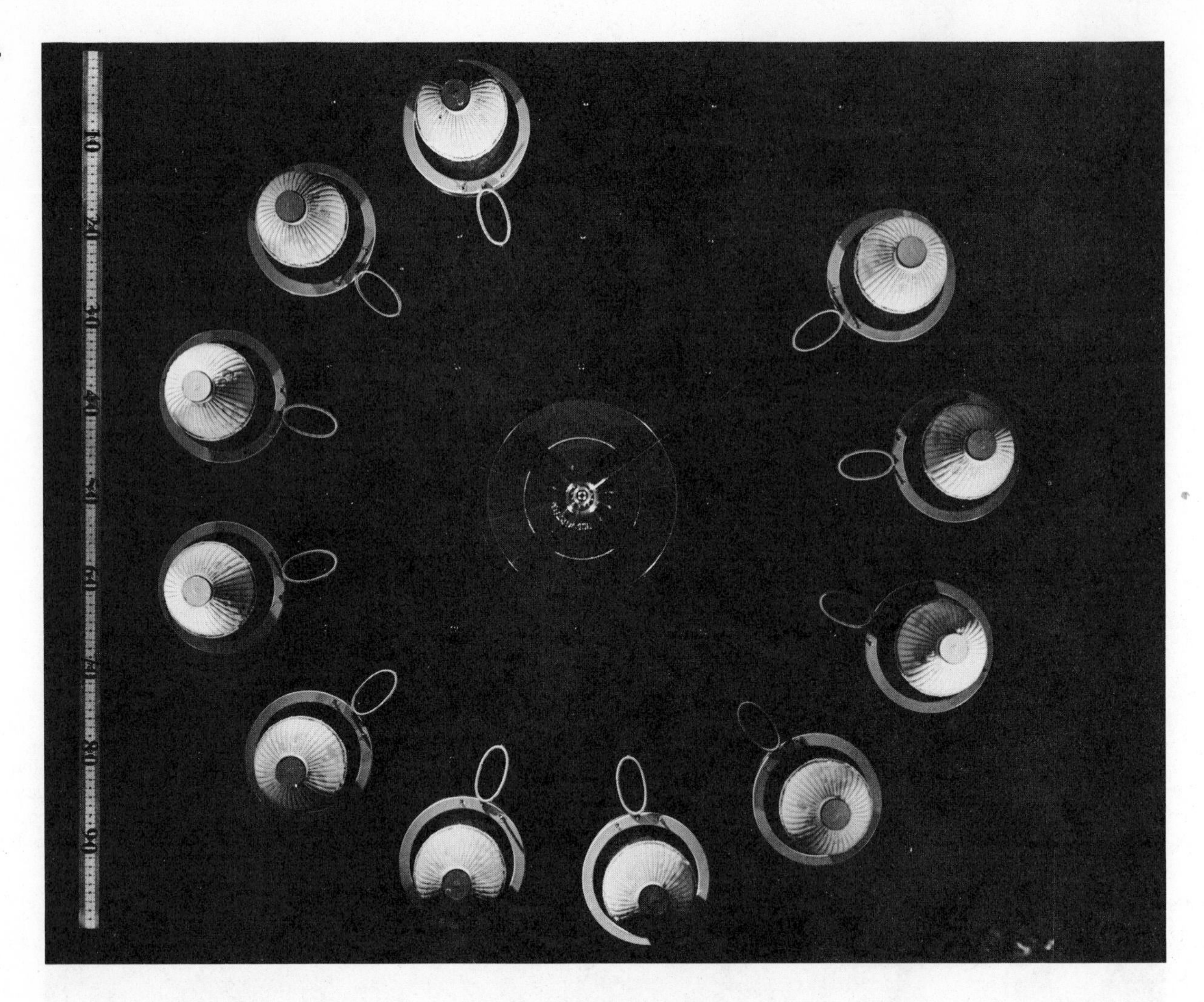

Figure 12–10
A puck moving in a circle at uniform speed. Note that the loop is extended the same amount at each position, indicating that constant force is acting.

The measured extension of the loop tells us the force is 2.4 newtons. The time taken to move through 10 intervals from the first position to the last is 10 intervals $\times \frac{1}{2.4}$ sec/interval. In this time the puck moved through an angle of 286°. Therefore the time T for one complete revolution is $10 \times \frac{1}{2.4} \times \frac{360°}{286°} = 5.2$ sec. The radius was 0.44 meter, and the mass of the puck and loop was 3.9 kg. Using these values in our expression for the magnitude of the centripetal force F which should make the puck go around the circle, we get

$$F = \frac{m4\pi^2 R}{T^2} = \frac{(3.9 \text{ kg})4\pi^2(.44 \text{ m})}{(5.2 \text{ sec})^2} = 2.5 \text{ nt.}$$

We see that this experiment confirms our expression for the centripetal force to within the accuracy of our measurements.

Suppose the string breaks. The puck no longer moves in a circle, since the centripetal force has been removed. Since no external force now acts on

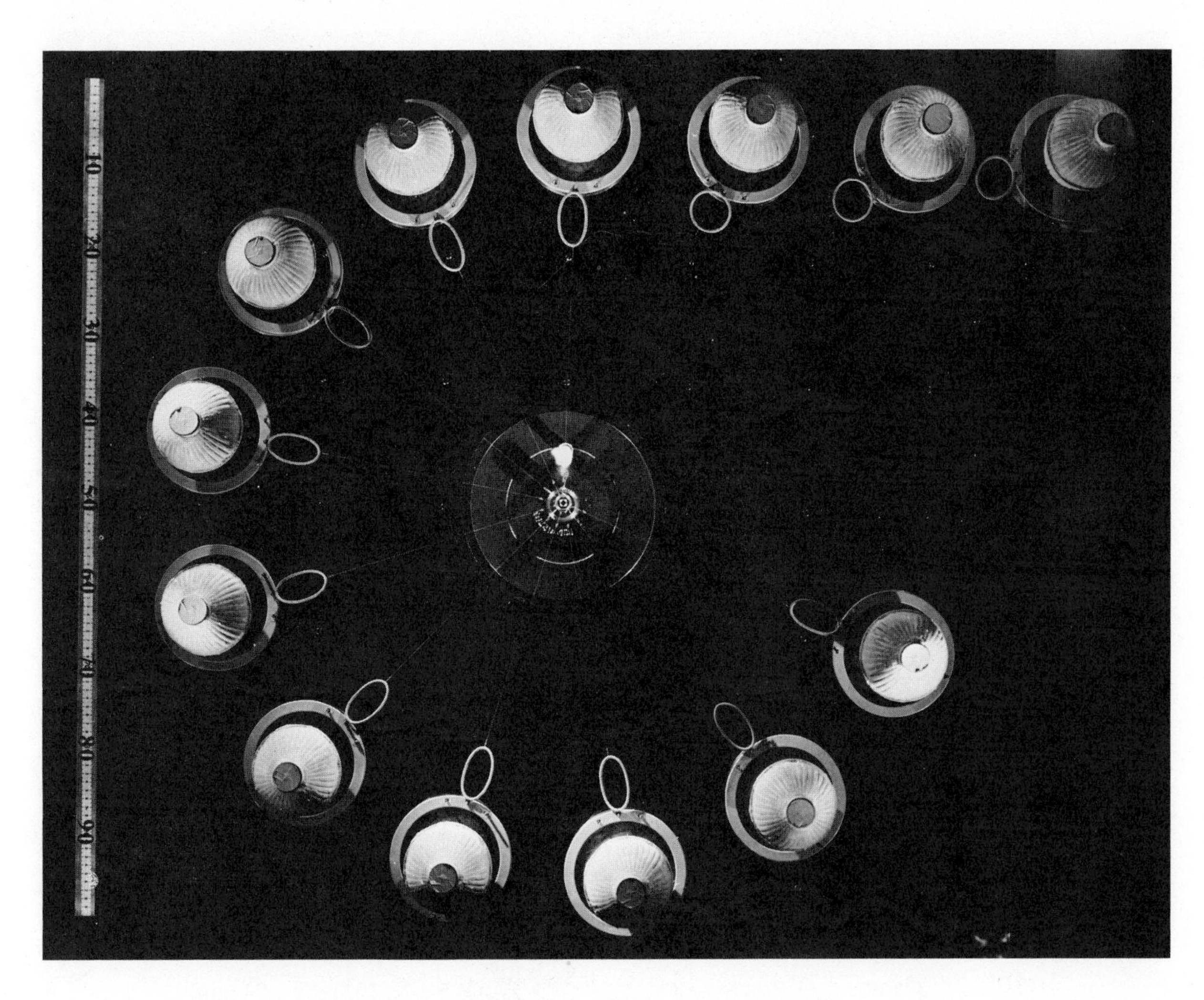

the puck, it moves in a straight line at constant speed. This motion is shown in Fig. 12–11, where the string was burned through by a gas torch. Notice that after the string breaks, the puck moves along a tangent to the original circle; it does not move out along a radius.

Figure 12–11
A puck in circular motion photographed at a flash rate of 2.4 per second. When the puck reached the top of the photograph, the string attached to it was burned with a torch. The puck continued to move in a straight line with the velocity it had when the string broke.

12-6 Earth Satellites

Only for distances small compared with the earth's radius is the weight of a body constant in direction and magnitude. Rockets that propel satellites into orbit travel so far that we cannot neglect the change in direction of the force of gravity. Because the earth is almost spherical, we conclude that the force of gravity always points toward its center. If we look from outer space, we see that the direction of the force is completely different for different points around the earth (Fig. 12–1).

If a satellite is given the right velocity it will move at constant speed in a circular orbit around the earth. However, since it is very difficult to project an object in exactly the right direction, and with just the right speed, artificial satellites actually follow slightly elliptical orbits (see next chapter). We shall analyze only the special case of circular motion of a satellite, since it is relatively simple mathematically.

Let us compute the period of revolution of a satellite moving around the earth. For motion in a circle the magnitude of the acceleration is $a = v^2/R$, and since the centripetal force is the gravitational attraction of the earth, the magnitude of this acceleration must be equal to g, the gravitational field strength. Hence the speed of the satellite is given by

$$\frac{v^2}{R} = g \quad \text{or} \quad v^2 = gR,$$

where R is the radius of the circle and g is the value of the acceleration of gravity where the satellite is. Suppose the satellite is 400 km above the earth. Then R = radius of earth + 400 km = 6.8×10^6 meters; and g is about 8.6 m/sec^2 out there. Therefore

$$v^2 = gR = 8.6 \times 6.8 \times 10^6 \text{ m}^2/\text{sec}^2,$$

and

$$v = 7.6 \times 10^3 \text{ m/sec}.$$

This is a speed of about 5 mi/sec or 18,000 miles per hour.

To get the period T, we note that the circumference of the circle ($2\pi R$) is the distance covered in one revolution at constant speed v. Therefore

$$2\pi R = vT \quad \text{or} \quad T = \frac{2\pi R}{v}.$$

We can calculate the period of the satellite by substituting $R = 6.8 \times 10^6$ m and the value $v = 7.6 \times 10^3$ m/sec that we have just computed. The result is

$$T = 2\pi \frac{6.8 \times 10^6}{7.6 \times 10^3} = 5.6 \times 10^3 \text{ sec} = 93 \text{ min}.$$

The calculations we have made for the period and velocity of an artificial satellite were needed to put the first artificial satellite in orbit. Newton calculated that a projectile fired from a high mountain with a velocity of about 5 miles per second would go into orbit around the earth (Fig. 12–12). The existence of artificial satellites is one of the triumphs of the development of the dynamics of Galileo, Newton, and others.

Why, then, was the first artificial satellite not placed in orbit in the seventeenth century? You can guess the answer. There were no guns or rockets powerful enough. Man's understanding of broad scientific laws often runs ahead of technology. The detailed application of scientific knowledge requires much time and labor.

Sometimes it is the other way round—technology runs ahead of science. Now that our technology has enabled us to establish artificial satellites, collect samples of lunar materials, and send probes to other planets, we shall gain new observations of the universe. Some of these observations have previously been denied us by the curtain of the earth's atmosphere. The new data will give us new knowledge of the history of the solar system, of

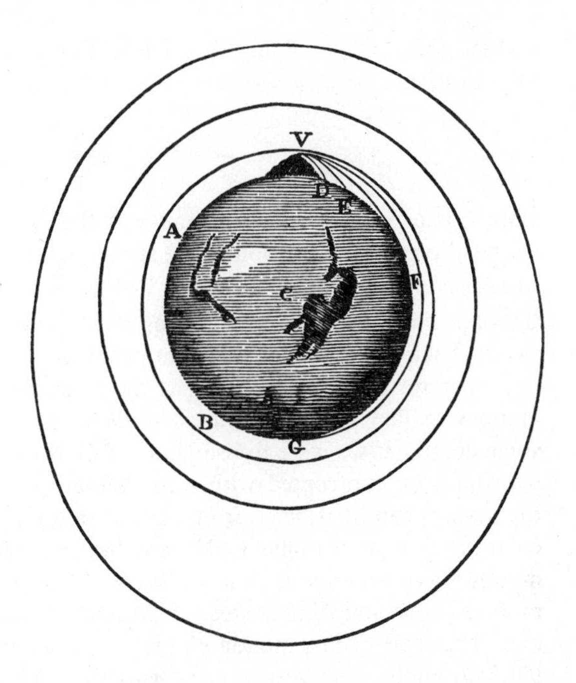

Figure 12–12
A drawing from Newton's "System of the World" (attached to later editions of the "Principia") showing the paths that a body would follow if projected with various speeds from a high mountain. As you can see, Newton was aware that a body would go into an orbit around the earth if its speed was great enough. The orbits from *V* ending at *D*, *E*, *F*, and *G* are for projectiles thrown with larger and larger horizontal velocities. Newton recognized that air resistance would prevent the motion of satellites close to the earth from following his ideal paths or from continuing for a long time. He pointed out that satellites could move permanently in the outer orbits.

cosmic rays, and of the density of matter in interplanetary space, and will help shape our ideas of the universe. In the long run, basic knowledge and technological applications go hand in hand—one helps the other.

12-7 The Moon's Motion

The moon is an earth satellite. We can compute its centripetal acceleration from the following observations. The period of the moon's motion is 27.3 days, or 2.3×10^6 sec; and the distance from the earth to the moon is about 3.8×10^8 meters (about 240,000 miles). The magnitude of the moon's acceleration toward the earth is

$$a = \frac{4\pi^2 R}{T^2} = \frac{4 \times \pi^2 \times 3.8 \times 10^8 \text{ m}}{(2.3 \times 10^6 \text{ sec})^2}$$
$$= 2.7 \times 10^{-3} \text{ m/sec}^2.$$

This acceleration is much smaller than the acceleration of a satellite near the earth's surface. Comparing it with $g = 9.8$ m/sec^2, we see that the gravitational attraction has fallen off by a factor of about 2.7×10^{-4}. This evidence of the weakening of gravitational attraction as the separation increases was one of the things that led Newton to his law of gravitation, as we shall see in the next chapter.

12-8 Simple Harmonic Motion

When we stretch a spring, it pulls back with a force proportional to the stretch. Also, when we compress a spring, it pushes against us with a force proportional to the compression. If we attach a mass to a stretched spring and let go, it will oscillate to and fro. For a small stretch or compression this force can be described by

$$\vec{\mathbf{F}} = -k\vec{\mathbf{x}}.$$

Here k is a positive constant, and $\vec{\mathbf{x}}$ is the displacement of the mass from the position of no distortion of the spring (chosen positive for stretch, negative for compression). The minus sign in the equation indicates that $\vec{\mathbf{F}}$ is in the opposite direction from $\vec{\mathbf{x}}$; thus it is a restoring force which pulls the system back toward its equilibrium position.

Linear restoring forces such as these always lead to similar to-and-fro motions, called *simple harmonic motion.* Let us examine this motion in some detail. If we were to use Newton's law directly to predict the motion, we would be confronted with a mathematically complicated problem; but the motion can be derived from circular motion, as we shall now see. Circular motion in a plane looks just like the simple harmonic to-and-fro motion if we examine it from the edge of the plane. To get an idea of the motion, move your thumb steadily around a horizontal circle level with your eye. The straight-line motion you see is simple harmonic motion, as we can tell by matching such a motion with the motion of a mass on a spring (Fig. 12–13).

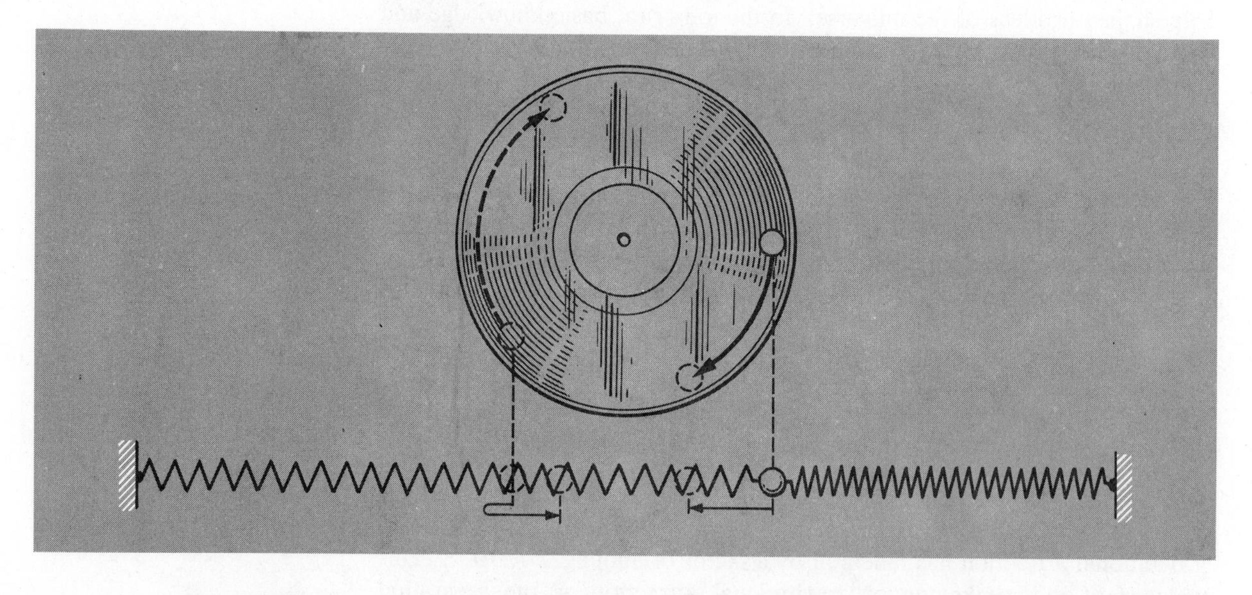

Figure 12–13
The to-and-fro motion of a mass on a spring can be matched by one component of the motion of a mass on a circular turntable. We must choose an appropriate radius for the turntable and turn it at the right speed.

To study this motion, we once again return to the fundamental property of vectors which allows us to represent any motion in a plane as the combination of two independent motions in fixed directions at right angles to each other. We shall describe circular motion at constant speed in terms of these rectangular components. We shall then examine one of these components in detail and also find the force associated with the motion described by that component.

Figure 12–14 shows a circle around which a body of mass m is moving with constant speed; let us examine the horizontal component of this mo-

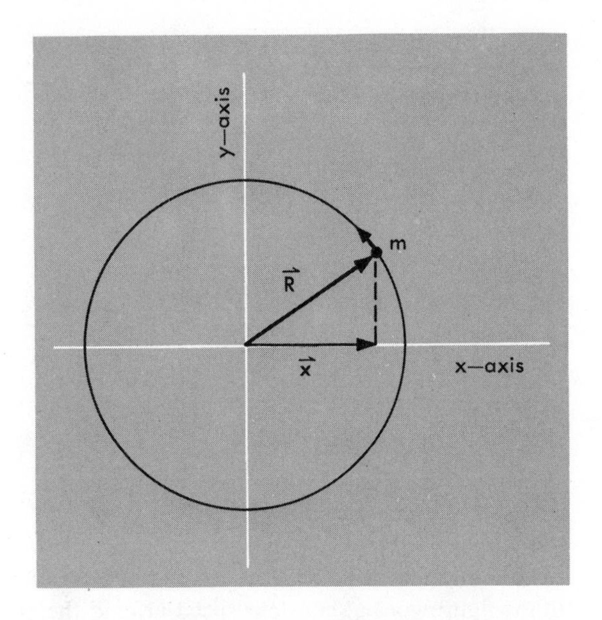

Figure 12–14
A mass m is moving around a circle at constant speed. This motion can be thought of as consisting of two components at right angles—a horizontal and a vertical component. We shall look closely at the horizontal x component.

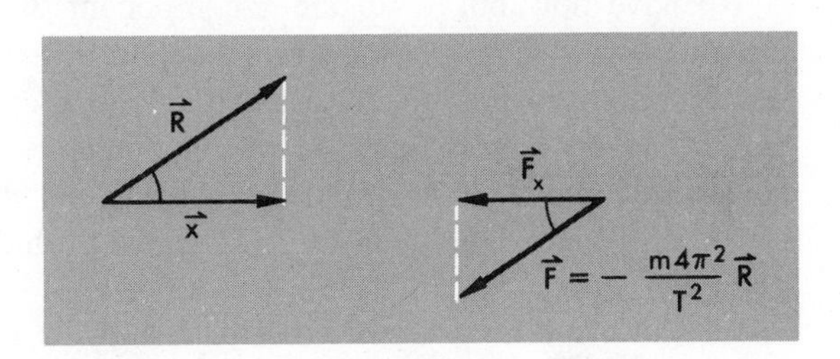

Figure 12–15
A right triangle is formed by the position vector $\vec{\mathbf{R}}$ and its horizontal vector component $\vec{\mathbf{x}}$. The centripetal force $\vec{\mathbf{F}}$ and its horizontal component $\vec{\mathbf{F}}_x$ form another right triangle. The triangles are similar because the angle between $\vec{\mathbf{R}}$ and $\vec{\mathbf{x}}$ is equal to the angle between $\vec{\mathbf{F}}$ and $\vec{\mathbf{F}}_x$ (since $\vec{\mathbf{R}}$ is parallel to $\vec{\mathbf{F}}$, and $\vec{\mathbf{x}}$ is parallel to $\vec{\mathbf{F}}_x$).

tion. In Fig. 12–15 we have drawn a right triangle with the position vector $\vec{\mathbf{R}}$ and its vector component $\vec{\mathbf{x}}$ in the horizontal direction. We have also drawn the similar right triangle with the centripetal force $\vec{\mathbf{F}}$ and its vector component $\vec{\mathbf{F}}_x$. From these similar triangles we see that the magnitudes of $\vec{\mathbf{F}}_x$ and $\vec{\mathbf{x}}$ are in the same ratio as those of $\vec{\mathbf{F}}$ and $\vec{\mathbf{R}}$ and that $\vec{\mathbf{F}}_x$ and $\vec{\mathbf{x}}$ point in opposite directions.

Since $\vec{\mathbf{F}} = -\frac{m4\pi^2}{T^2}\vec{\mathbf{R}}$ (see Section 12–5), the relation between $\vec{\mathbf{x}}$ and $\vec{\mathbf{F}}_x$ is therefore

$$\vec{\mathbf{F}}_x = -\frac{m4\pi^2}{T^2}\vec{\mathbf{x}}.$$

(This equation is simply the x component of the equation $\vec{\mathbf{F}} = -\frac{m4\pi^2}{T^2}\vec{\mathbf{R}}$.)

Since the mass m and the period T are constant for any particular motion, $m4\pi^2/T^2$ is a constant, and the equation can be more simply written

$$\vec{\mathbf{F}}_x = -k\vec{\mathbf{x}},$$

where the factor $k = m4\pi^2/T^2$ is the constant of proportionality between $\vec{\mathbf{F}}_x$ and $\vec{\mathbf{x}}$. This shows that the force $\vec{\mathbf{F}}_x = -k\vec{\mathbf{x}}$ is the force that produces the to-and-fro motion of $\vec{\mathbf{x}}$. Because the motion along the x axis depends only on the force along the x direction, any such force produces a motion exactly

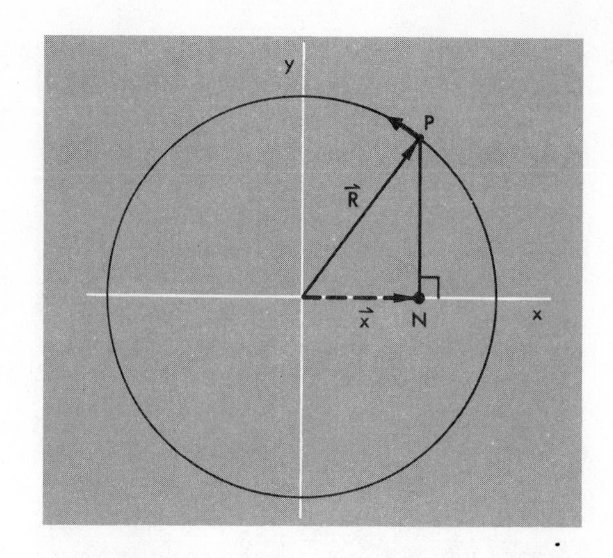

Figure 12–16
The mass at N was originally displaced from the origin by a distance R along the x axis. If it is acted on by a force such that $\vec{\mathbf{F}} = -k\vec{\mathbf{x}}$, then it will move back and forth along the x axis; its motion will be the same as the x component of the motion of a point P moving uniformly in a circle of radius R.

like the motion of point N at the end of $\vec{\mathbf{x}}$ in Fig. 12–16 when the point P in the figure goes at constant speed around the circle.

We have now linked up the x component of circular motion with the motion of any mass m when acted on by a restoring force $\vec{\mathbf{F}} = -k\vec{\mathbf{x}}$ no matter what provides the force, and this link enables us to calculate the period. If we have a mass m moving under a force $\vec{\mathbf{F}} = -k\vec{\mathbf{x}}$, we can always imagine a circular motion that will match the motion of m. The period of the actual motion of m is the same as that of this matching circular motion.

We can use our earlier discussion to find the relation between the force constant k in $\vec{\mathbf{F}} = -k\vec{\mathbf{x}}$ and the period in which the mass moves to and fro. The period depends only on the mass m and on the force constant k. By rearranging $k = \frac{m4\pi^2}{T^2}$, we find

$$T = 2\pi\sqrt{\frac{m}{k}}.$$

This expression is reasonable. If the restoring force increases rapidly with distance (that is, if k is large), the mass is pushed back and forth rapidly: T becomes small. On the other hand, if the mass is large it responds more sluggishly to the force; and so the period is greater the larger the mass. One result of our reasoning may seem astonishing: the period does *not* depend on the amplitude R of the motion. This is a result you can check experimentally.

A simple pendulum is just a body of mass m on the end of a string of length l. (Fig. 12–17.) We shall describe the displacement of the mass from its equilibrium position by the vector $\vec{\mathbf{s}}$. The vector component of $\vec{\mathbf{s}}$ in the horizontal direction is $\vec{\mathbf{d}}$ (Fig. 12–18); this is the horizontal displacement of m from equilibrium.

The motion of the mass m is very nearly simple harmonic motion. To show this, we must show that there is a linear restoring force, a force which is proportional to the negative of the displacement. As we see in Fig. 12–19, the weight of the body $m\vec{\mathbf{g}}$ can be broken into rectangular com-

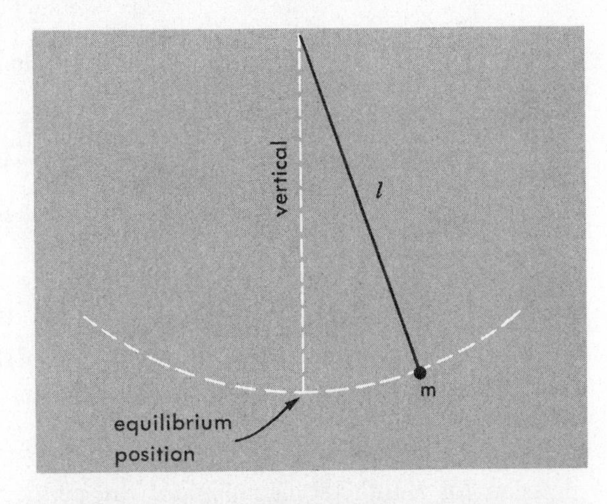

Figure 12–17
A simple pendulum: a mass *m* on a string of length *l* moves along the arc of a circle.

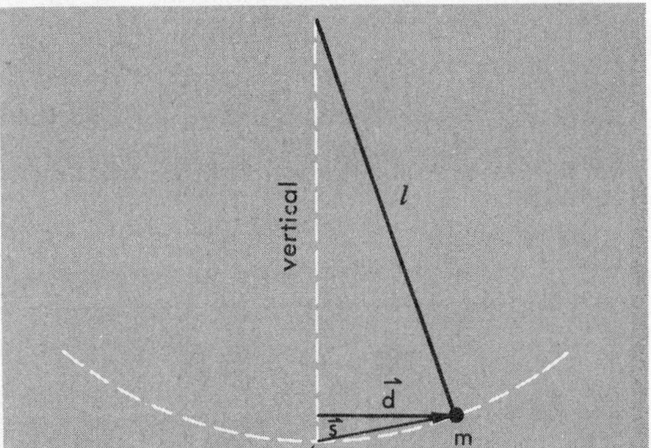

Figure 12–18
When the mass is displaced only a little way from its equilibrium position, there is very little difference between the actual displacement *s* and the horizontal component of the displacement *d*. Imagine that the displacement is much smaller than it is shown here, compared with the length of the string.

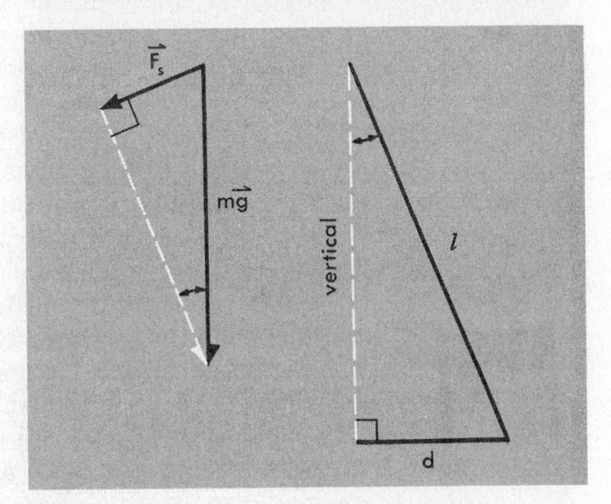

Figure 12–19
In the left-hand triangle the weight $m\vec{\mathbf{g}}$ has been broken into components along the string (dashed line) and perpendicular to the string ($\vec{\mathbf{F}}_s$). The other triangle shows the string of length *l*, the horizontal displacement *d* of the mass from the vertical, and the vertical center line. The two right triangles are similar, since $\vec{\mathbf{F}}_s$ is perpendicular to *l* and $m\vec{\mathbf{g}}$ is perpendicular to *d*.

ponents. One component, shown dotted in the figure, is taken along the string and simply keeps the string pulled tight. The other component is perpendicular to the string, points along the circular path, and pulls the mass back toward the equilibrium position. From the similar triangles of Fig. 12–19, the magnitude F_s of this last component is given as

$$\frac{F_s}{mg} = \frac{d}{l} \quad \text{or} \quad F_s = \frac{mg}{l} d.$$

Now, as long as d and s are small compared with l, there is very little difference between the lengths of d and s. Thus, we can write

$$F_s = \frac{mg}{l} s.$$

The direction of $\vec{s}$ is very nearly perpendicular to the string if s is small, or is very nearly parallel to $\vec{F}_s$ but in the opposite direction. Therefore $\vec{F}_s = -\frac{mg}{l}\vec{s}$ and we do have a linear restoring force of the form $\vec{F}_s = -k\vec{s}$ necessary to give simple harmonic motion. From the proportionality constant, $k = \frac{mg}{l}$, we can find the period of the pendulum. We substitute this expression for k into the equation for the period, $T = 2\pi\sqrt{\frac{m}{k}}$, and find $T = 2\pi\sqrt{\frac{l}{g}}$.

The period depends only on the length of the pendulum and the gravitational field strength. As always when the force is basically gravitational, the mass does not matter. The amplitude or size of the swing also does not matter as long as it is so small that d and s are nearly the same. A pendulum of accurately known length and a good clock can thus be used to measure g accurately.

12-9 Experimental Frames of Reference

Turn around. Apparently the room rotates the other way. The corner of the room goes around a circle; so does a marble on the table. But no force mv^2/r on the marble is required to accelerate the marble while it moves in its apparently circular path. Under the action of no net force the marble is simply at rest when we describe it with respect to the room, but it appears to be accelerating when you describe its motion with respect to yourself.

Suppose you are in a car that is accelerating in a straight line. You look out and see a ball flying through the air perpendicular to the direction of your motion. It appears to curve violently in the horizontal plane, although you realize that it moves on a straight path over the ground. If you describe the motion of the ball with respect to the car, the ball undergoes an acceleration which does not correspond to any sensible force.

In Chapter 10 frames of reference were discussed in connection with the description of motion—that is, in connection with kinematics. The essential point made there is that any frame of reference can be used to describe a motion truly. So the sensible thing to do is to use the frame of reference that provides the simplest description. In other words, we have complete freedom of choice and use this freedom to make life as easy as possible.

But how about dynamics? Have we the same freedom? It is sad but true that here we no longer can use every reference frame and keep Newton's law of motion. Newton's law of motion is *not* correct in all frames of reference. To relate force and acceleration in some frames requires more complicated laws. How then do we select reference frames in which Newton's law of motion is correct? The answer is: from experimental observations. Remember how we were led to Galileo's principle of inertia and to the statement of Newton's law of motion; and remember that any conclusion which is drawn from experimental observation is true only within the accuracy of the measurements. Since we arrived at Newton's law of motion from laboratory experiments on the earth, we know that a frame of reference rigidly attached to the earth is a good one—at least within the accuracy of our measurements. On the other hand, an accelerating car is not a frame of reference in which Newton's law holds.

12-10 Fictitious Forces in Accelerated Frames

We are all aware of the strange forces which appear when motion is observed in some frames of reference. If you ride in a closed car at constant speed on a straight, smooth road, no forces act on you except those which act when you sit in a chair at rest. But when you drive around a curve, especially at high speed, you are conscious of a force: the door of the car pushes on your body. If you think of this experience in the frame of reference of the car, it appears to be a cause for concern. A force acts on you, but you do not move. You are likely to say that another force pushes you against the door and holds you there. This force acting away from the center of the curve just balances the force exerted by the door. In this way you may explain why you don't move with respect to the car. But here is a real dilemma: how can a change in the motion of a reference system (the car) actually create new forces? Is this centrifugal force real?

To answer these questions we start by looking at a simple experiment. We take a Dry Ice puck, for example, hold it at rest on a flat, frictionless surface, and let it go. What happens? Exactly nothing. The puck stays at rest, and we conclude that no net force acts on it. In particular, no horizontal force acts on the ice puck.

Now we perform the same experiment on a frictionless table located on a large, rotating merry-go-round. Our whole merry-go-round rotates uniformly with respect to the earth. Now what do we observe when we let the puck go? The puck does *not* stay at rest but moves *relative to us*—that is, relative to the merry-go-round—in a curve of the sort seen in Fig. 12–20. Now we are really in trouble! Here our observations show that, in the absence of forces, a body moves in a curve, not in a straight line with constant speed. What a contradiction of Galileo's principle of inertia! How can we properly reconcile our observations with Galileo's law? We might say: we are mistaken to think that no force acts on the puck; such a motion shows that there *is* a complicated force. But what exerts this force? Nothing of which we are aware. So we are right back in the quandary from which we started.

Now look at the whole situation from the standpoint of an observer on the ground. He sees us moving in a circle with constant speed. We are holding an ice puck which moves in the circle with us. When we let go of

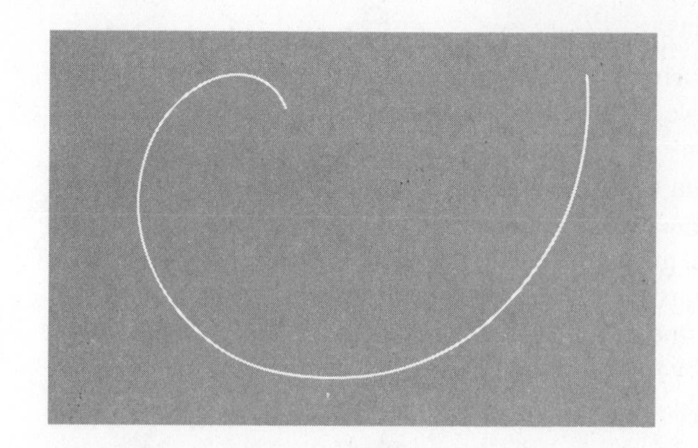

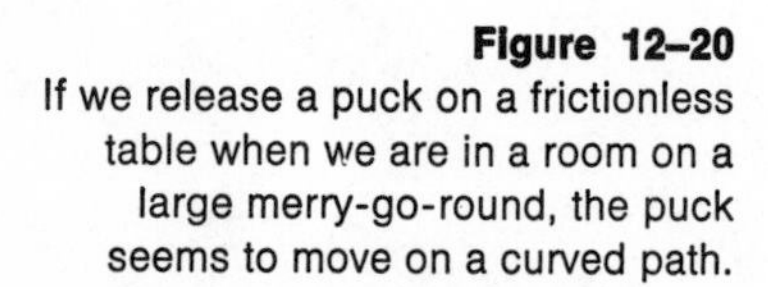

Figure 12–20
If we release a puck on a frictionless table when we are in a room on a large merry-go-round, the puck seems to move on a curved path.

the puck, it moves in a straight line, tangent to the circle, at constant speed, as Galileo's principle says it must. The man on the ground understands our dilemma and points out to us that our troubles have come, not from a breakdown of the laws of dynamics, but from the fact that we made our observations in a rotating system of reference. This rotating system is an accelerated system, and we can solve all our problems by allowing for the motion of our rotating reference system. Now we know the cause of our trouble. The apparent violation of Galileo's principle and Newton's law of motion is due to the *acceleration* of the rotating frame of reference relative to the ground.

Thus if the observer on the merry-go-round holds the puck at rest relative to him, he finds that he must pull on it with a force which is constant in magnitude and directed inward toward a fixed point. He reasons that, since the puck is at rest (relative to him) there must be an equal and opposite force acting outward *on the puck*. This somewhat mysterious force he calls *centrifugal* force. On the other hand, the observer on the ground sees that the puck is accelerated toward the center of the merry-go-round and that the rotating observer is applying the needed *centripetal* force. The observer on the ground has no need for the centrifugal force.

Now what about the force that we actually feel when we sit in a car going around a corner? This is the force from the door on us. This force is real enough; but it may be mysterious when we forget that we are moving around a curve. In the reference system of the earth everything is clear. The door of the car must exert a real *centripetal* force on us to cause us to go around the corner instead of straight ahead. In this example the motion in the frame of reference where Newton's law holds is accelerated, and the forces producing the acceleration are real. The car, including the door and us, will not turn the corner unless it is pushed sidewise by the road. The centrifugal force, on the other hand, is *not* real. It is not a force in the sense of a push or a pull exerted by one object on another; rather it is a fictitious or "phony" force that we introduce to correct for the acceleration of our rotating frame of reference. We invent it to make it appear that Newton's law of motion applies in the frame of reference of the car.

Centrifugal force is a term which is widely used and poorly understood. If we use it at all, we must remember that it is a device used to make corrections for the dynamical description of motions in a rotating frame of reference. In a frame in which Newton's law is valid, it simply does *not* exist.

Now we have raised another problem. We know that the earth spins about its axis once a day relative to the sun or to the fixed stars. Are we in a rotating frame of reference, or are the sun and the fixed stars themselves rotating about the earth? In other words, is Newton's law "really" valid when we use the earth as the reference frame, or is it more accurate in some other frame of reference with respect to which the earth turns? If Newton's law really applies to some frame of reference other than the earth, we should be able to find out by noticing that to explain the observed motions of bodies with respect to the earth we must introduce fictitious forces like centrifugal force. These fictitious forces must be small or they would long since have been apparent in our experiments. Also, we expect them to be small if they correspond to the daily rotation of the earth. To force a mass m to go around daily at the equator requires a centripetal force (Section 12–5)

$$4\pi^2 mR/T^2 = (3 \times 10^{-2}\ \text{newton/kg})m$$

or about 1/300 of the force of gravity. An experiment to make sure that 1/3 percent of mg is used to produce this motion must be more precise than the experiments we performed in establishing Newton's law of motion.

The French physicist Foucault performed a famous experiment to demonstrate the fact that the earth rotates on its axis in a frame of reference where Newton's law holds. To perform this experiment we construct a pendulum by hanging a bob on a string, the top end of which is fastened to the ceiling. When set in motion, this pendulum will swing back and forth; and, if the amplitude of the motion is not too large, the bob will perform the simple harmonic motion which we have met in Section 12–8.

To start such a bob in motion we might, for example, pull the bob aside from its equilibrium position and release it from rest. It will then oscillate in very nearly straight-line motion. If we are in a proper frame of reference for Newton's law, the motion of the pendulum should continue indefinitely in the plane containing the line of oscillation and the point of suspension of the pendulum (see Fig. 12–17). There is no component of force perpendicular to this plane, nor has the pendulum bob an initial velocity component perpendicular to the plane. Therefore Newton's law of motion predicts that, if such a pendulum starts swinging in this plane, it will continue to swing in the same plane indefinitely. This is what will happen in an inertial frame of reference, one in which Newton's law is valid.

When we perform this experiment in the laboratory we find that a pendulum does pretty well; it seems to stay in the plane in which it started and obeys Newton's law of motion in fine shape. But if we wait long enough—hours instead of minutes—we find that slowly but inexorably the plane of the motion rotates relative to its initial orientation. In fact, if we performed this Foucault pendulum experiment at the North Pole we would find that in just 24 hours the plane in which the pendulum swings has rotated completely around through 360 degrees, as seen by us (Fig. 12–21). This provides the experimental evidence, obtained from an experiment done on

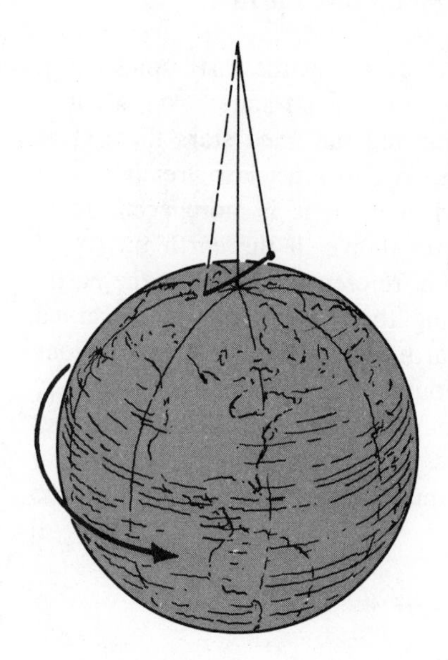

Figure 12–21
A Foucault pendulum at the North Pole. The pendulum swings in nearly straight-line motion while the earth rotates beneath it. To a man standing on the earth, the plane of the pendulum's motion would seem to rotate.

the earth, that the earth is indeed in rotation and that frames of reference rigidly attached to it are rotating frames of reference. It also shows again how little we are likely to notice the effects of the earth's rotation in laboratory experiments. A pendulum with a period of one second oscillates about 10^5 times in a day. If the frictional forces on such a pendulum were so small that it could really swing completely free in the inertial frame, the plane of oscillation would appear to us to rotate through an angle of about 10^{-3} degree in each swing at the North Pole.

These experiments show that the earth is indeed a rotating frame of reference in which Newton's law is not exactly valid. But for all except the most precise experiments we may ignore effects of the earth's rotation. If we want to be as accurate as possible, however, we must use Newton's law in the frame of reference of the fixed stars rather than in that provided by the earth.

12-12 Newton's Law and a "Coasting" Spacecraft

When a spacecraft is moving outside the earth's atmosphere with its engines off, gravity is the only force acting on it. The acceleration of the spacecraft is the same whether it is approaching the earth, leaving it, or in orbit around it. A frame of reference attached to the spacecraft is, therefore, an accelerated frame of reference. Would an astronaut doing dynamical experiments inside the spacecraft have to invent fictitious forces in order to make Newton's law hold? In other words, is the law of motion in an accelerating spacecraft the same as in an accelerating car?

As you know, a ball on the floor of a car accelerating in the forward direction rolls to the back of the car. If you have seen the inside of a spacecraft on television, you probably noticed that objects left in mid-air stay there and do not change their position relative to the spacecraft (provided the spacecraft does not tumble or rotate). If an astronaut gives an object a little push, the object moves at constant velocity. We can predict that a brick pulled with a constant force (like the puck in Fig. 11–9) will move with constant acceleration. Newton's law of motion seems to hold in a "coasting" spacecraft without the introduction of fictitious forces. What, then, are the reasons for the difference in behavior between an accelerating car and a spacecraft accelerated by gravity?

The force which accelerates a car acts only on those objects which interact with the body of the car in some way. A passenger accelerates with the car because the back of the seat exerts a force on him. A brick on the floor of the car accelerates because of the frictional force. However, there is no force acting on a frictionless puck inside an accelerating car. To an observer on the road it will remain at rest (or move with constant velocity); to an observer in the car the puck will accelerate toward the back end, thus giving rise to the idea of a fictitious force. To keep it at rest requires a real force.

In a spacecraft the force of gravity acts on all objects whether they interact with the spacecraft or not. Moreover, the gravitational force is proportional to the mass of the objects in the spacecraft; thus to an observer on earth they all have the same acceleration as the spacecraft. To the astronauts inside, the objects remain at rest unless nongravitational forces act on them. Then they will accelerate relative to the spacecraft according to Newton's law, taking into account only the nongravitational forces.

To put it differently, in any accelerated frame of reference, a fictitious force is needed to make Newton's law valid. If the accelerating force is the force of gravity, the fictitious force exactly cancels the gravitational forces, and the frame of reference behaves just like an inertial frame of reference with no gravitational forces. No force is needed to keep objects at rest, and in this case no effect of gravity is felt inside the spacecraft. This is why astronauts are said to be weightless. To observers on earth their weight is not zero but is their mass times the gravitational field at their position in space.

For Home, Desk, and Lab

1.* If the gravitational force on an object is 49.0 newtons where the gravitational field is 9.80 nt/kg, what is the object's mass? (Section 1.)

2.* The gravitational field at the surface of the moon is one sixth of that at the surface of the earth.
   (a) How much would a 70-kg man weigh on the moon? On the earth?
   (b) What would be his mass on the earth? On the moon? (Section 1.)

3. Which two locations would you pick from Table 1 to investigate the change in g with altitude?

4.* What is the magnitude of the gravitational field of the earth at a point where the acceleration of gravity is 9.81 m/sec^2? (Section 2.)

5. What happens if you throw a Ping-Pong ball down faster than its terminal velocity?

6. If different objects had different ratios of their inertial and gravitational masses, would they all accelerate at the same rate in the gravitational field of the earth? Explain.

7.* In Fig. 12–2, how long would you make your scale in order to photograph the ball over 30 time intervals, if the ball started from rest at the beginning of the first interval? (Section 2.)

8. What is the acceleration of a falling 0.2-kg ball if the gravitational field is 9.80 nt/kg and the air resistance acting on the ball is 0.5 nt?

9. A ball is thrown straight up with a velocity of 15 m/sec. (Neglect air resistance.)
   (a) How fast will it be going after 1.2 sec?
   (b) How far above the ground will it be at that time?
   (c) How fast will it be going after 2.3 sec?
   (d) How far above the ground will it be at that time?
   (e) What is the ball's acceleration at the top of its rise?

10.† A stone is dropped off a cliff of height h. At the same instant a ball is thrown straight up from the base of the cliff with initial velocity v_i. Assuming the ball is thrown hard enough, at what time t will stone and ball meet? (Neglect air resistance.)

11. Two 0.5-kg Dry Ice pucks on a flat table are tied together by a string.
    (a) What is the total gravitational force on these pucks?
    (b) At what rate will they speed up if pulled across the table by a horizontal force equal to the gravitational force found in (a)?
    (c) If one puck remains on the table and the other hangs over the edge, what will be the acceleration of the pucks?

12. An object of mass m is moving on a frictionless inclined plane that makes an angle of 30 degrees with the horizontal.
    (a) What is the net force on the object?
    (b) What is its acceleration down the plane?

13.* How can you tell that the horizontal component of velocity of the ball projected horizontally in Fig. 12–4 is constant and its vertical component is not? (Section 3.)

14. A baseball is thrown from center field to home plate. If the ball is in the air 3.00 seconds and we neglect air resistance, to what vertical height must it have risen?

15. Plot the trajectory of an object that is thrown into the air with an initial velocity of 10 m/sec at an angle of 45 degrees with the horizontal.

16.* What is the centripetal force needed to keep a 3.0-kg object moving in a circle of 2.0 m radius at a speed of 4.0 m/sec? (Section 5.)

17.* What is the effect on the speed and direction of motion of an object when a force acts on it in a direction perpendicular to its path? (Section 5.)

18.* What is the velocity of the puck in Fig. 12–10 when it is at the bottom of the photograph? (Section 5.)

19.* What is the acceleration of the puck in Fig. 12–10 when it is at the bottom of the photograph? (Section 5.)

20. How fast must a plane fly in a loop-the-loop of radius 1.00 km if the pilot experiences no force from either the seat or the safety belt when he is at the top of the loop? In such circumstances the pilot is often said to be "weightless."

21. Some people speak of an astronaut re-entering the atmosphere as weighing several times his own weight. We sometimes say that an astronaut, before re-entering the atmosphere, is weightless.
    (a) Do these statements make sense in the light of our definition of weight as the force of gravity on an object?
    (b) What is meant by "several times his own weight" and "weightless" above? Be prepared to discuss these questions in class.

22. A string 5.0 m long of diameter 2.0 mm just supports a hanging ball without breaking.
    (a) If the ball is set to swinging, the string will break. Why?
    (b) What diameter string of the same material should be used if the ball travels

7.0 m/sec at the bottom of its swing? (The breaking strength is proportional to the cross section of the string.)

23. The gravitational field at the surface of the moon is 1/6 of that at the surface of the earth. The radius of the moon is about 1/4 of the radius of the earth. Calculate the period of a lunar landing module circling the moon close to the surface.

24. An electron (mass $= 0.91 \times 10^{-30}$ kg) under the action of a magnetic force moves in a circle of 2.0 cm radius at a speed of 3.0×10^6 m/sec. At what speed will a proton (mass $= 1.6 \times 10^{-27}$ kg) move in a circle of the same radius if it is acted upon by the same force?

25. A body moving with a speed v is acted upon by a force that always acts perpendicular to the motion of the body but increases steadily in magnitude.
    (a) Draw a sketch of the trajectory.
    (b) Does the speed of the object increase, decrease, or remain unchanged?

26.* How must the force on an object vary with the displacement if simple harmonic motion is to take place? (Section 8.)

27.* At what points on the circle in Fig. 12–16 will *P* have the same velocity as *N*? (Section 8.)

28.* If the force constant of a spring is 8 newtons/meter, what suspended mass will give a period of 1 sec? (Section 8.)

29.* What must be the length of a simple pendulum if it is to have a period of one second? (Section 8.)

30. A block of mass m rests on a horizontal platform. The platform is driven vertically in simple harmonic motion with an amplitude of 0.098 meter. When at the top of its path, the block just leaves the surface of the platform. (This means that at this point its acceleration is 9.8 m/sec² downward.)
    (a) What is the period of the simple harmonic motion?
    (b) When the block is at the bottom point of its path, what is its acceleration?
    (c) What is the force exerted by the platform on the block at this bottom point?

31. (a) What is the period of a pendulum consisting of a mass of 2.0 kg suspended on a light cord 2.4 meters long if $g = 9.8$ newtons/kg?
    (b) Notice that we can use a pendulum to measure g. If the period of this pendulum is 3.0 sec, what is g?

32. A 2-kg ball is suspended from a spring. When disturbed in a vertical direction, the ball moves up and down in simple harmonic motion at a frequency of 4 cycles per second.
    (a) What is the period of the motion?
    (b) What is the upward force exerted on the ball by the spring when the ball is at the midpoint of its up-and-down path?
    (c) How much did the spring stretch when the ball was first attached to its end (before the oscillatory motion was started)?

33. A ball is thrown vertically upward with a velocity of 24 meters/sec from a railroad flatcar moving horizontally with a velocity of 4 meters/sec. Describe the path of the ball as seen by an observer (a) on the flatcar, and (b) by the observer on the ground nearby.

34. While a bus is in motion along a level, straight road, we roll a marble from one side to the other across the floor of the bus. Its path is a straight line relative to the bus. Later we roll it again, and this time the path is a parabola which bends toward the front of the bus (concave toward the front of the bus). Describe the motion of the bus in each case. Be prepared to discuss in class.

35. A long-range gun is located in the Northern Hemisphere. You wish to hit a point due north of it. In what direction should you aim the gun? (Section 11.)

36. When a spacecraft is moving through space on its way from the earth to the moon, the gravitational force of the earth will at first exceed the gravitational force of the moon. Later on, when the spacecraft is nearer to the moon, the gravitational force of the moon will exceed the gravitational force of the earth. In the science-fiction story "From the Earth to the Moon," written about 100 years ago, Jules Verne has his astronauts falling from the rear end to the front end of the spacecraft when it passed the point in space where the gravitational force from the earth was equal to the gravitational force from the moon. Do you think Verne had a correct understanding of the physics of the case? Explain how you think the men in the spacecraft would feel when they passed that point in space.

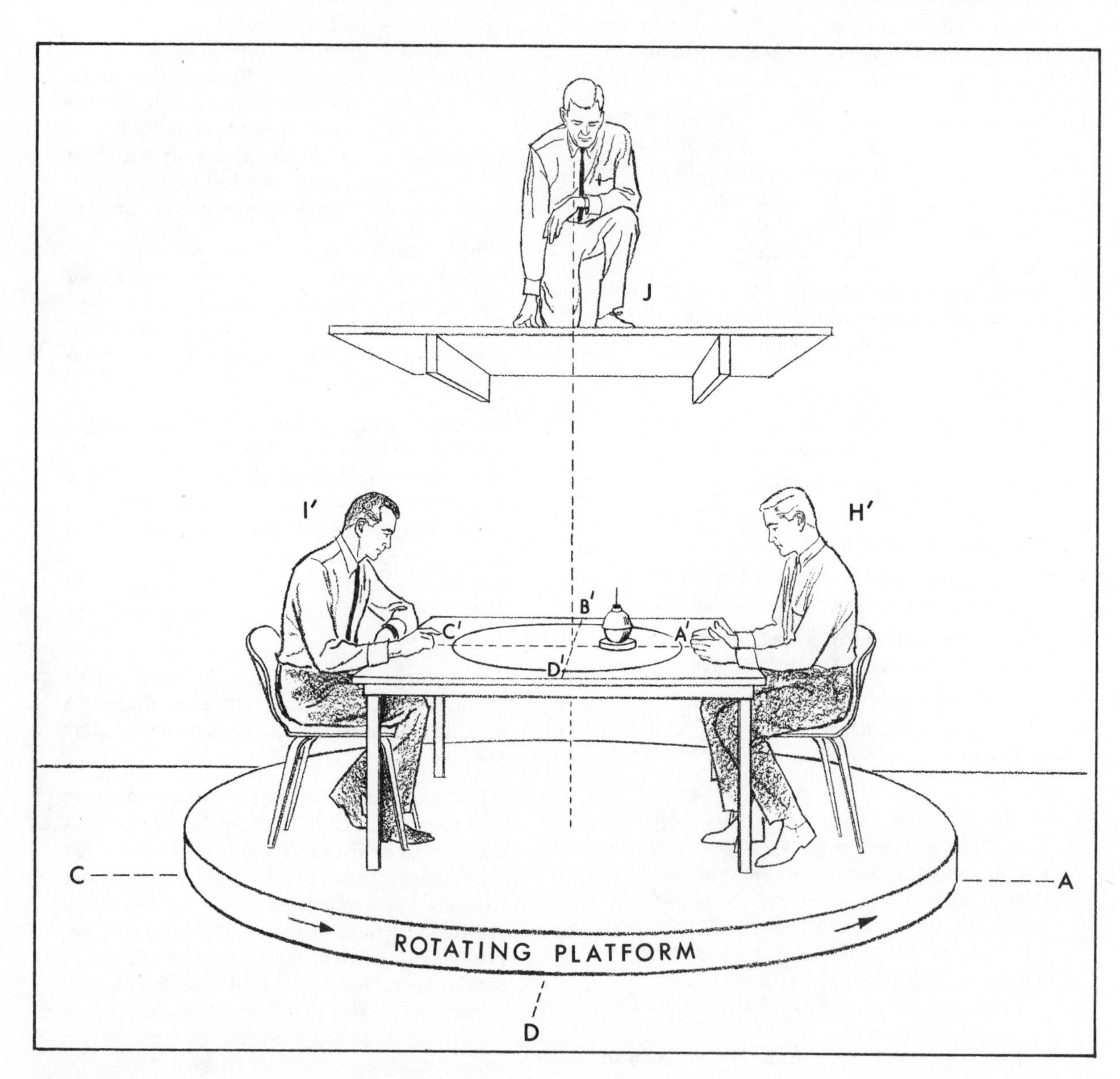

Figure 12–22
For Problem 37.

37. For some experiments in relative motion, a smooth, level table, shown in Fig. 12–22, is centered on a platform which rotates uniformly at a rate of one revolution in 12 seconds. Two perpendicular lines are drawn through the center of the table, intersecting a circle of 1.20-meter radius at the points A', C', B', and D'. Two men, H' and I', sit on the platform at opposite ends of the line $A'C'$. A third man, J, is above the table so that he can observe the motion of a Dry Ice puck in a stationary frame of reference. He has four marks on the floor, forming two perpendicular reference lines AC and BD through the center of the table. (In the diagram, B is behind the table.)
   (a) H' holds the puck in his hand at A'. What is its velocity in J's frame of reference at the instant A' crosses the line AC?
   (b) As H' passes A he gives the puck a sudden push so that it travels along line AC with a speed of 0.40 m/sec. Construct a vector diagram to show the velocity that H' gave the puck in J's frame of reference.
   (c) Construct a diagram which shows the positions of the puck and of A' at 1, 2, 3, 4, 5, and 6 seconds in J's frame of reference.
   (d) Make use of the diagram constructed for (c) to construct a diagram which will show the motion of the puck on the table, as seen by H' and I'. (Hint: Construct this moving-frame diagram on tracing paper which can be laid over the diagram of the fixed-frame motion. Plot the position of the puck for each 15° rotation of the moving frame. These points can be connected to give a picture of the motion in the moving frame.)

38. With the same situation as in Problem 37:
   (a) Suppose that H' launches the puck as he passes A so that I' will catch the puck as he passes D. Construct a diagram which shows the motion of the puck in J's frame of reference. What is the speed of the puck in this frame of reference?
   (b) Make use of the diagram constructed for (a) to construct a diagram which will show the motion of the puck on the table as seen by H' and I'.
   (c) Construct a vector diagram to show the velocity that H' gave the puck so that it could travel as in (a).

39. With the same situation as in Problem 37:
   (a) With what speed and in what direction can H' launch the puck as he passes A so that, as J sees it, the puck remains at A?
   (b) What is the motion of the puck as seen by H' and I'?
   (c) The puck has a mass of 0.50 kg. After H' has let go of it, what force acts on it according to J? What force acts on it according to H' and I' to make it move as it does in their frame of reference?

Further Reading

These suggestions are not exhaustive but are limited to works that have been found especially useful and at the same time generally available.

BONNER, FRANCIS T., and PHILLIPS, MELBA, *Principles of Physical Science.* Addison-Wesley, 1957. Combination of history and physics from Galileo to Newton. (Chapters 2 and 3)

COHEN, I. BERNARD, *The Birth of a New Physics.* Doubleday Anchor, 1960: Science Study Series. (Chapters 1, 2, 4, 5, and 7)

GAMOW, GEORGE, *Gravity.* Doubleday Anchor, 1962: Science Study Series. A very complete treatment of the subject. (Chapters 1, 2, 3, and 4)

HOLTON, G., and ROLLER, D. H. D., *Foundations of Modern Physical Science.* Addison-Wesley, 1958. Projectile motion, circular and simple harmonic motion are described at length. (Chapters 3 and 5)

Universal Gravitation and the Solar System

CHAPTER

13

There are few people who do not recognize the Big Dipper in the night sky. The most striking fact about the heavens is that the constellations, groups of stars such as the Big Dipper, maintain unaltered shapes. They move just as though they were tacked on the inside of a large rotating sphere, and we were at the center of this sphere watching (Fig. 13–1). Against this background of "fixed stars" the sun and moon move smoothly, as if they were attached to other spheres rotating at different rates about the earth. According to this view, the earth, large and immobile, is located at the center of a universe made of celestial matter which revolves about it. Such a universe is called geocentric (earth-centered).

Figure 13–1
A one-hour time exposure taken with the camera pointed at the North Star. The circular arcs show the apparent motion of the stars. It was this circular motion that led the Greeks to visualize the stars as attached to a sphere which turned about the earth. (Yerkes Observatory Photograph.)

Seven heavenly bodies which appeared to move among the fixed stars were known to ancient man. The sun and moon, Mercury, Venus, Mars, Jupiter, and Saturn were called planets from the Greek word meaning "wanderer." With the exception of the motion of the sun and moon, the motions of these bodies appear irregular when viewed over long periods of time (Fig. 13–2). Their erratic motion focused the attention of ancient men on the planets. They were brighter than the stars; and because their brightness changed, their distances from the earth seemed to change. They became associated with various human endeavors and emotions (Venus with love, Mars with war, etc.), as though they formed an intermediary between the immutable perfection of the stars and the restless imperfection of the earth. Later, the astrologers saw in the planetary positions indications of the future course of the lives of human beings.

Finding a reasonable explanation for the peculiar motion of the planets was a major concern of the ancient astronomers. It is related that Plato, the Greek philosopher (427–347 B.C.), set the following problem for his students: The stars seem to move in perfect circular paths around the earth,

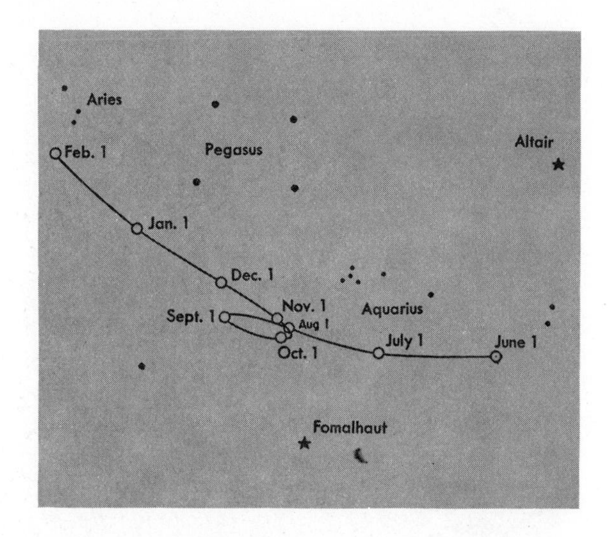

Figure 13–2
The peculiar apparent motion of the planet Mars with respect to the fixed stars. Mars, at various intervals, appears to reverse its direction of motion. (From: R. H. Baker, "Astronomy," D. Van Nostrand Co., Inc.)

but the planets seem to trace irregular paths. What are the combinations of perfect circular paths along which the planets really move? The form of this question reveals the fact that the circle was believed to be the most perfect of all curves and hence alone worthy of describing celestial motions. The efforts of all astronomers for many centuries were devoted at least partly to answering this question.

13-1 Early Planetary Systems

Plato's pupil Eudoxus tried to represent planetary motions by a collection of moving spheres, each with its center at the earth. Each planet was attached to the surface of a sphere which rotated uniformly about an axis attached to two opposite points on the surface of a larger sphere (Fig. 13–3 and 13–4). While the inner sphere spun uniformly on its axis, the axis itself was carried around by a uniform motion of the outer sphere. In fact, the axis of the outer sphere might itself be attached to the surface of a still larger sphere; in this way the number of spheres could be extended to represent more complex motions. Finally, the whole system spun around inside the celestial sphere which carried the fixed stars. With a sufficient number of spheres moving around inside of other spheres, Eudoxus obtained a good approximation to the motion of a planet. His successors improved the accuracy of his model by using still more spheres. In the course of history many planetary systems depending on the motions of the spheres were developed, and large numbers of spheres were eventually employed. In one system, there were thirteen spheres for Mercury alone.

Other Greek astronomers tried to solve Plato's problem in a different way. For example, Apollonius and Hipparchus (third and second centuries B.C.) developed a system in which a planet moves on a circle the center of which moves on another circle. The work of these early Greek astronomers led to the system of Claudius Ptolemy of Alexandria in the second century A.D. His system of circles moving on other circles repro-

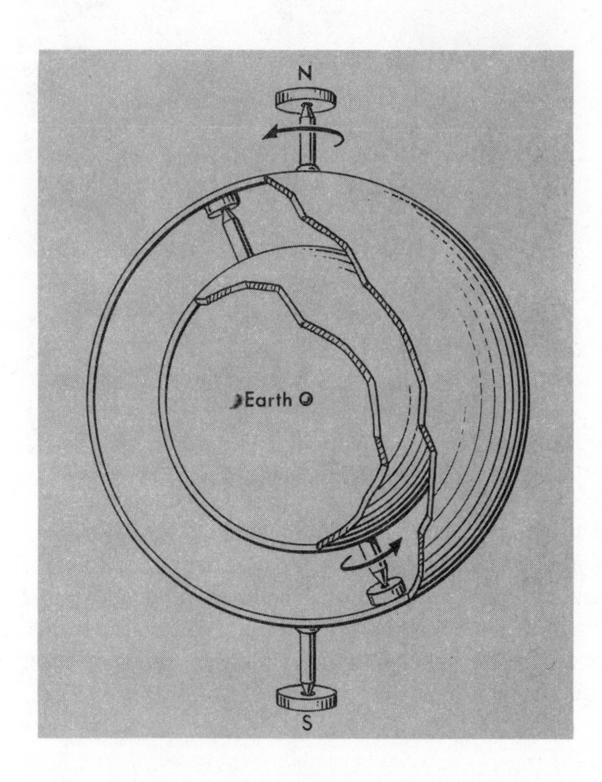

Figure 13–3
Eudoxus's planetary system. The motions of the sun, moon, and planets can be approximated by an arrangement of spheres turning within spheres. Here the outer sphere is the sphere of the fixed stars. The sphere rotates every 24 hours from east to west on an axis through the earth's north and south poles. The inner sphere has the sun fixed to a point on its surface. The axis of the inner sphere is connected to the outer sphere at two points. The inner sphere completes a rotation once a year.

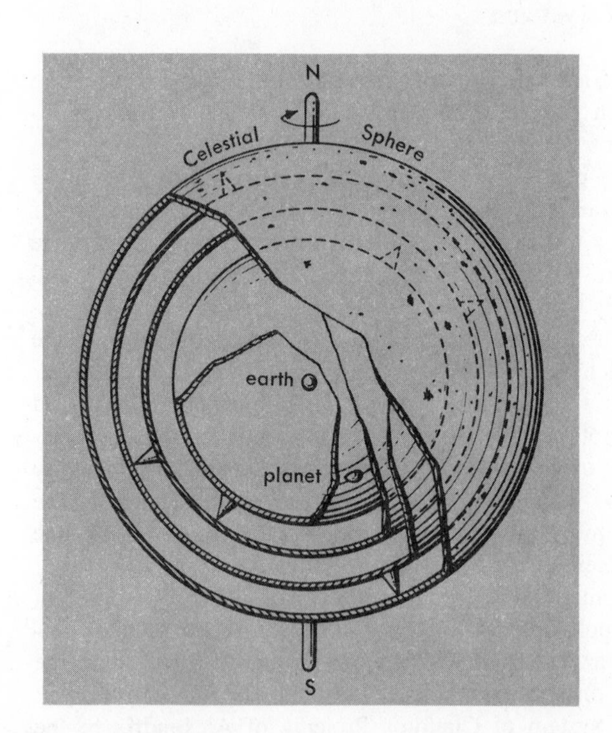

Figure 13–4
Here a planet is mounted on the surface of the innermost sphere. By placing the axes of the spheres at the proper angles, and choosing suitable speeds and directions of rotation, one can reproduce the motion of a planet against the background of the fixed stars as observed from the earth to a good approximation.

duced the observed motions of the planets reasonably accurately (Fig. 13–5). But his curves describing the planetary orbits were so complicated that there were many complaints from those who studied them. Alphonso X, King of Castile in 1200, was prompted to say that, had he been consulted at the creation, he would have made the world on a simpler and better plan. A simplified Ptolemaic orbit of a planet is shown by the heavy line in Fig. 13–6; the circular motions from which this orbit was supposed to have originated are also shown.

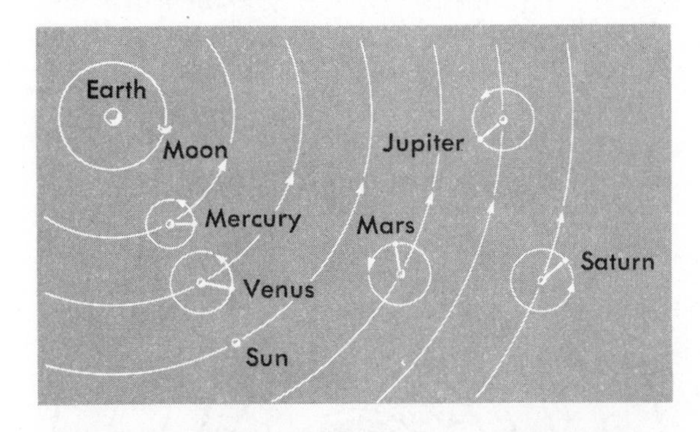

Figure 13–5
A simplified diagram of the Ptolemaic system of planetary motions.

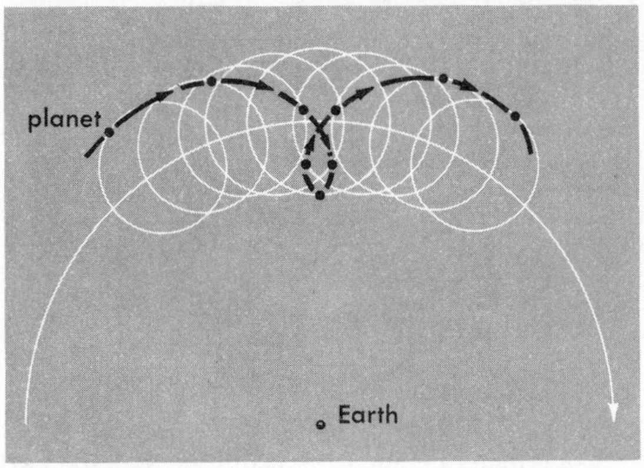

Figure 13–6
The motion of a single planet in the Ptolemaic system. The planet was considered to revolve in a small circle whose center moved in a circular orbit about the earth.

13-2 Copernicus' Planetary System

The Polish astronomer Nicolaus Copernicus (born in 1473) felt that the Ptolemaic system was too complex. The simplicity of uniform circular motion desired by Plato had been buried in complicated constructions. The truth, Copernicus thought, must be simpler. He therefore set out to give a simpler answer to Plato's problem by choosing a different center for the system of circles.

Like others before him, Copernicus realized that the motion of the fixed stars might be explained by assuming that the earth rotates.* Our situation

* Although the main current of Greek and medieval thought embodied a geocentric theory, Heracleides (about 370 B.C.) believed that the earth rotated on its axis, and Aristarchus (third century B.C.) thought that the earth moved around the sun.

with respect to the heavens is then much like that of a passenger in an airplane. As the airplane turns in flight above a big city, the streets and avenues seem to go around. So as the earth rotates, the stars appear to move.

Once he imagined that the earth rotates daily, Copernicus found that the orbits of the planets could be greatly simplified by choosing the sun rather than the earth as the center of the planetary system. Then the earth would be neither at the center of the universe nor at rest. Perhaps it was a planet, and revolved along with the other planets about the sun. In Fig. 13–7 we show the simple orbits followed by the earth and the planets as they move around the sun according to Copernicus.

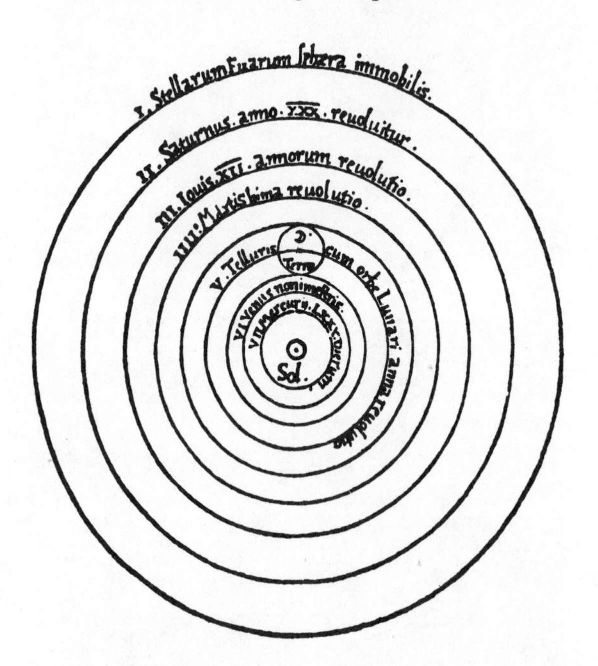

Figure 13–7
The orbits of the planets according to the planetary system of Copernicus.

Copernicus justified his proposal that the earth moved by saying, "And although it seemed to me an absurd opinion, yet, because I knew that others before me had been granted the liberty of supposing whatever circles they chose in order to demonstrate the observations concerning the celestial bodies, I considered that I too might well be allowed to try whether sounder demonstrations of the revolutions of the heavenly bodies might be discovered by supposing some motion of the earth. . . . I found after much and long observation, that if the motions of the other planets were added to the motions of the earth, . . . not only did the apparent behavior of the others follow from this, but the system so connects the orders and sizes of the planets and their orbits, and of the whole heaven, that no single feature can be altered without confusion among the other parts and in all the universe. For this reason, therefore, . . . have I followed this system."

Copernicus' system also included a large, immobile sphere on which the fixed stars were located. About it he said: "The first and highest of all the spheres is the sphere of the fixed stars. It encloses all the other spheres and is itself self-contained; it is immobile; it is certainly the portion of the universe with reference to which the movement and positions of all the other heavenly bodies must be considered. If some people are yet of the opinion that this sphere moves, we are of a contrary mind; . . ." Then, describing the planetary spheres and their periods of rotation in which the earth appears as one of six planets while the moon is clearly designated as a satellite of the earth, he concluded: "In the midst of all, the sun reposes unmoving. Who, indeed, in this most beautiful temple would place the light-giver in any other part than that whence it can illumine all the other parts? . . ."

13-3 Objections to Copernicus' Theory

It must be understood that in order to achieve his simplified orbits Copernicus was obliged to discard the entire picture of the universe that had been developed from the time of Aristotle. The question as to whether or not the earth moved was a very serious one. All of medieval cosmology and physics was based on the idea that the earth was at rest at the center of the universe. In part this belief was based on man's inner conviction that his earth must be at the center of things. But in addition there seemed to be good evidence of the earth's special position. For one thing, if the earth moves, what is pushing it and why is this motion not felt? Or, to take another example, why do stones fall toward the earth if it is not the center of the universe?

Copernicus expected a great deal of criticism, and he delayed publication of his book so long that he first saw a printed copy of it on the day he died. Anticipating many of the objections, he attempted to answer them beforehand. To the argument that his earth rotating so rapidly about its own axis would surely burst like a wheel driven too fast, he countered: "Why does the developer of the geocentric theory not fear the same fate for his rotating celestial sphere—so much faster because so much larger?" To the argument that birds in flight would be left behind by the rapidly moving earth, Copernicus answered that the atmosphere is dragged along with the earth.

In fact, there were many arguments and counterarguments. The Copernican theory was denounced as "false and altogether opposed to the Holy Scriptures." Martin Luther branded Copernicus a fool and a heretic. The argument over this bold new concept of the universe raged for more than 100 years before the notion that the earth could move was generally accepted.

13-4 Tycho Brahe

Tycho Brahe, the Danish astronomer born in 1546, could not accept the Copernican system, despite its simplicity. He contributed instead an improved geocentric system in which the sun goes around the earth and the other planets go around the sun (Fig. 13–8). Also, to test astronomical models, he set out to make a really accurate map of the positions of the fixed stars and to determine the apparent positions of the planets as seen from the earth over a long period of time. He began his observations with an instrument consisting of a pair of joined sticks, one leg to be pointed at

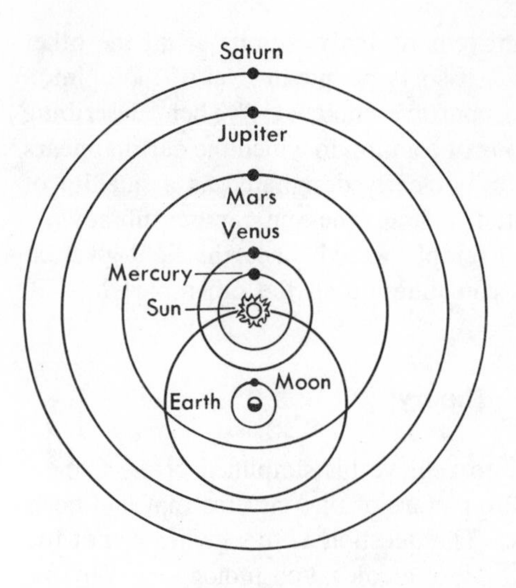

Figure 13–8
Tycho Brahe's geocentric system.

a fixed star, the other at a planet. In this way he could measure the angle between them. Later he constructed great sextants and compasses with which he made wonderfully careful observations (Fig. 13–9). He catalogued the positions of 777 stars so accurately that his observations are still used, and his measurements of the planetary angular positions over a period of twenty years contain no error larger than 0.067 of a degree. This angle is about the same as the angle subtended by the head of a pin held at arm's length from your eye.

Figure 13–9
Tycho Brahe's mural quadrant. A brass arc of over 6-foot radius was mounted on a wall and equipped with movable sights. An observer (upper right) sighted on a star through the window in the wall at the left. Other assistants noted the time and recorded the angle of the observation. The sector of wall enclosed by the arc is covered with a mural painting of Tycho and other aspects of his work.

Tycho's observations of planetary positions were far more accurate than those available to Copernicus. They soon showed that the Copernican orbits were only roughly correct. A search then began for a more accurate description of the orbits. This objective was reached, after Tycho's death, by the German astronomer Kepler, who had worked in Tycho's laboratory during the last eighteen months of Tycho's life.

13-5 Kepler

Johannes Kepler, born in 1571, was a striking contrast to Tycho Brahe. Tycho posessed tremendous mechanical ability and skill but relatively little interest in mathematics. Kepler was clumsy as an experimenter but was fascinated by the power of mathematics. He was akin to the ancient Greeks in his reverence for the power of numbers and was intrigued by mathematical puzzles.

After Kepler had learned the elements of astronomy, he became obsessed with the problem of finding a numerical scheme underlying the planetary system. He wrote, "I brooded with the whole energy of my mind on this subject." He devoted his life to the analysis of the tables of planetary positions which Tycho had left him. In tackling the problem of translating the observations of Tycho Brahe into mathematical descriptions of planetary motions, Kepler acted like any scientist today who attempts to explain experimental findings in terms of simple mathematical laws rather than mere tables of numbers. With mathematical laws we can not only reproduce observed data, but we can predict the results of observations not yet made. Furthermore, mathematical laws are easier to remember and to communicate than mere tables of numbers.

In his first book Kepler described his attempts to understand why there were precisely six planets in the solar system. He established a connection between the six orbits and the five regular geometrical solids* (Fig. 13–10).

* By a regular solid body we mean a symmetrical body with identical flat faces. Only five kinds of regular solid bodies can be constructed.

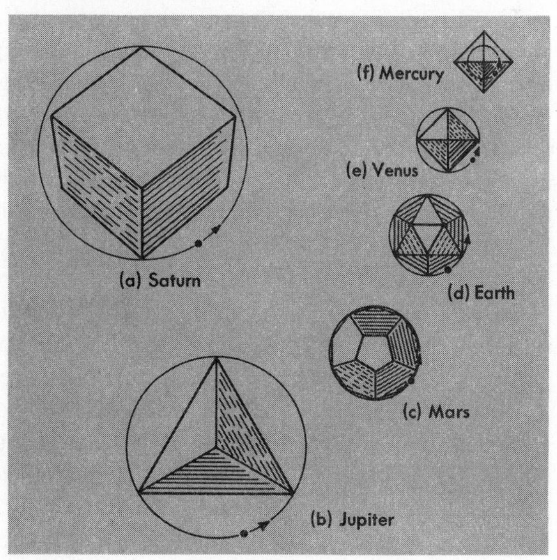

Figure 13–10
Kepler's law of planetary orbits was based on the five regular solids. According to the law, a sphere, with a radius equal to that of the orbit of Saturn, is circumscribed about a cube (a). A sphere inscribed within this cube has a radius equal to the radius of the orbit of Jupiter.

In (b) the sphere of the orbit of Jupiter is shown with a tetrahedron inscribed. A sphere inscribed within the tetrahedron gives the radius of the orbit of Mars.

In (c) the sphere for Mars has a dodecahedron inscribed. A sphere inscribed within that gives the orbit of the earth (d).

We may continue this process of alternate inscribing of spheres and regular solids, using the icosahedron (20-sided) and the octahedron. These will give us the orbits of Venus (e) and Mercury (f), which is on a sphere inscribed inside the octahedron. Kepler considered the five solids as bridging the intervals between planetary orbits. Since there are only five regular solids, Kepler believed there could be only six planets.

From this construction he obtained ratios of radii agreeing fairly well with the values then known for the planetary orbits.

Kepler was ecstatic. He wrote, "The intense pleasure I have received from this discovery can never be told in words. I regard no longer the time wasted; I tired of no labor; I shunned no toil of reckoning, days and nights spent in calculation, until I could see whether my hypothesis would agree with the orbits of Copernicus, or whether my joy was to vanish into air."

The relation between the radii of planetary orbits is typical of the kind of result Kepler set out to achieve with Tycho's data. It often happens, however, that even the prettiest correlation of the data turns out to have no deep meaning in explaining the nature of things. Today this discovery of Kepler's is all but forgotten. His system is destroyed by the fact that there are more than six planets. But a seventh planet was not discovered until many years after his death.

Kepler discovered other mathematical relations, relations which have survived the test of later observations. He began his great analysis of Tycho's data with an exhaustive study of the motion of Mars. On what sort of curve had Mars moved during the twenty years of Tycho's observations? Would Mars move in a simple curve if one imagined the earth to be at rest, or if one imagined the earth to be in motion as Copernicus believed? Kepler adopted the Copernican idea that the earth spun about its axis while moving in an orbit about the sun. Following tradition, he first tried a system of circles moving on circles to obtain possible orbits. He made innumerable attempts, each involving long and laborious calculations. He had to translate each of Tycho's measurements of the angle between Mars and the fixed stars into a position of the planet in space with respect to a fixed sun about which the earth itself was moving.

After about seventy trials using the "eccentric circle" type of orbit, Kepler found one scheme which agreed fairly well with the facts. Then, to his dismay, he found that this curve, when continued beyond the range of the data he had used, disagreed with other observations that Tycho had made of Mars' position.

The disagreement between Tycho's data and Kepler's calculations was about 0.133 of a degree. (This is the angle through which the second hand on a watch moves in about 0.02 second.) Might not Tycho have been wrong by this small amount? Could not the cold of a winter's night have numbed his fingers or blurred his observations? Kepler knew Tycho's methods and the painstaking care he took. Tycho could never have been wrong even by such a small amount. Thus, on the basis of Tycho's data, Kepler rejected the curves he had constructed. What a tribute this was to the experimental skill of Tycho Brahe!

Saying that "upon this eight minutes [he] would yet build a theory of the universe," Kepler began again. Discarding the ancient and cherished belief in uniform motion, he considered possible changes of the speed of a planet as it moved in its orbit about the sun. He now made his first great discovery. He found that a line from the sun to a planet sweeps out equal areas in equal times. This has come to be known as Kepler's second law (Fig. 13–11).

After the discovery of his second law, Kepler finally abandoned his attempts to construct planetary motions out of combinations of uniform circu-

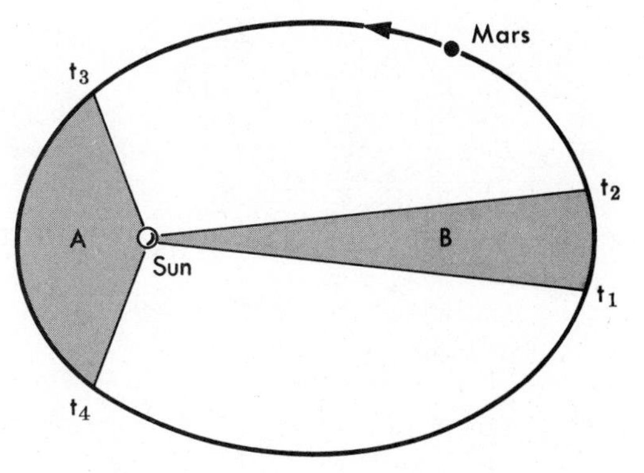

Figure 13–11
Kepler's law of equal areas. Mars travels along its orbit with varying speed, moving fastest when it is closest to the sun. Kepler found that for equal time intervals, $t_2 - t_1 = t_4 - t_3$, the areas swept out by a line between the sun and the planet were equal (area B = area A). In the drawing, the elongation of the ellipse has been exaggerated to illustrate the equal-area law more clearly.

lar motions and began to try various ovals as possible orbits. After many more laborious calculations, he finally achieved one of his most important results—his so-called first law. Each planet, he found, moves in an elliptical orbit with the sun at one focus. Imagine Kepler's delight. After years of effort, he had finally found a simple curve which described the motions of planets.

Kepler then set to work to find a connection between the size of a planet's orbit and its period, the time of one revolution of the planet about the sun. After many trials he found the precise relation for which he was searching: for all the planets the ratio of the cube of the radius of the orbit to the square of the period is the same.* Once he happened upon this ratio, the regularity was striking. (See Table 1.) The constancy of the ratio R^3/T^2 is called Kepler's third law.

With this triumph Kepler wrote, ". . . what sixteen years ago I urged as a thing to be sought . . . that for which I joined Tycho Brahe . . . at last I have

Table 1
Kepler's Third Law

PLANET	RADIUS R OF ORBIT OF PLANET IN A.U.	PERIOD T IN DAYS	R^3/T^2 IN $(\text{A.U.})^3/(\text{DAY})^2$	MODERN VALUES R^3/T^2 IN M^3/SEC^2
Mercury	0.389	87.77	7.64×10^{-6}	3.354×10^{18}
Venus	0.724	224.70	7.52	3.352
Earth	1.000	365.25	7.50	3.354
Mars	1.524	689.98	7.50	3.354
Jupiter	5.200	4,332.62	7.490	3.355
Saturn	9.510	10,759.20	7.430	3.353

The values of the orbits and periods in this table are those used by Kepler. In Kepler's day, the radii were known only in terms of the radius of the earth's orbit. The radius of the earth's orbit is called an astronomical unit (A.U.) of length. The nearly constant values of R^3/T^2 *illustrate Kepler's third law. The last column is based on accurate modern measurements of the orbits and periods.*

* The radius R of an orbit is defined by taking one half the sum of the shortest and longest distances between the sun and the planet. Because the planetary orbits are not very different from circles, the distance from the sun to any point on a planetary orbit will do for most purposes.

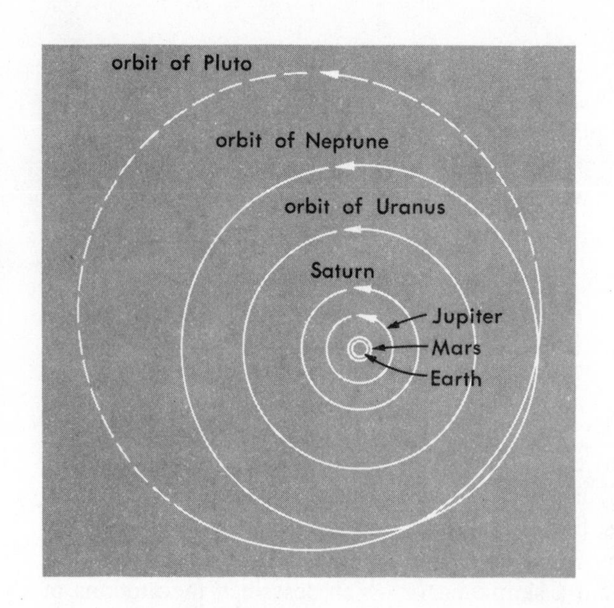

Figure 13–12
Approximate orbits of the principal planets. (The orbits of Mercury and Venus are too small to show on this drawing.) They are very nearly circles, except for Pluto, and are in almost the same plane. Only careful measurements show them to be ellipses. The orbits of Pluto and Mercury are the most elliptical, and the plane of Pluto's orbit makes an angle of about 17° with the planes of the other orbits. In the figure, the portion of Pluto's orbit below the plane of the paper is shown by a dashed line.

Table 2
The Solar System

OBJECT	MASS (KILOGRAMS)	RADIUS (METERS)	PERIOD OF ROTATION (SECONDS)	MEAN RADIUS OF ORBIT (METERS)	PERIOD OF REVOLUTION (SECONDS)
Sun	1.98×10^{30}	6.95×10^{8}	2.14×10^{6}	—	—
Mercury	3.28×10^{23}	2.57×10^{6}	5.05×10^{6}	5.79×10^{10}	7.60×10^{6}
Venus	4.83×10^{24}	6.31×10^{6}	2.1×10^{7}	1.08×10^{11}	1.94×10^{7}
Earth	5.98×10^{24}	6.38×10^{6}	8.61×10^{4}	1.49×10^{11}	3.16×10^{7}
Mars	6.37×10^{23}	3.43×10^{6}	8.85×10^{4}	2.28×10^{11}	5.94×10^{7}
Jupiter	1.90×10^{27}	7.18×10^{7}	3.54×10^{4}	7.78×10^{11}	3.74×10^{8}
Saturn	5.67×10^{26}	6.03×10^{7}	3.60×10^{4}	1.43×10^{12}	9.30×10^{8}
Uranus	8.80×10^{25}	2.67×10^{7}	3.88×10^{4}	2.87×10^{12}	2.66×10^{9}
Neptune	1.03×10^{26}	2.48×10^{7}	5.69×10^{4}	4.50×10^{12}	5.20×10^{9}
Pluto	6×10^{23}	3×10^{6}	5.51×10^{5}	5.9×10^{12}	7.82×10^{9}
Moon	7.34×10^{22}	1.74×10^{6}	2.36×10^{6}	3.8×10^{8}	2.36×10^{6}

brought to light and recognize its truth beyond my fondest expectations. . . . The die is cast, the book is written to be read now or by posterity. I care not which—it may well wait a century for a reader as God has waited six thousand years for an observer."

Kepler had carried astronomy through a momentous advance. He had translated the magnificent tables of data of Tycho Brahe into a simple and comprehensive system of curves and rules. Kepler's system earned him the title "Legislator of the Heavens."

Here are statements of Kepler's three laws:

I. Each planet moves in an elliptical path with the sun at one focus.

II. The line joining the sun and the planet sweeps out equal areas in equal times.

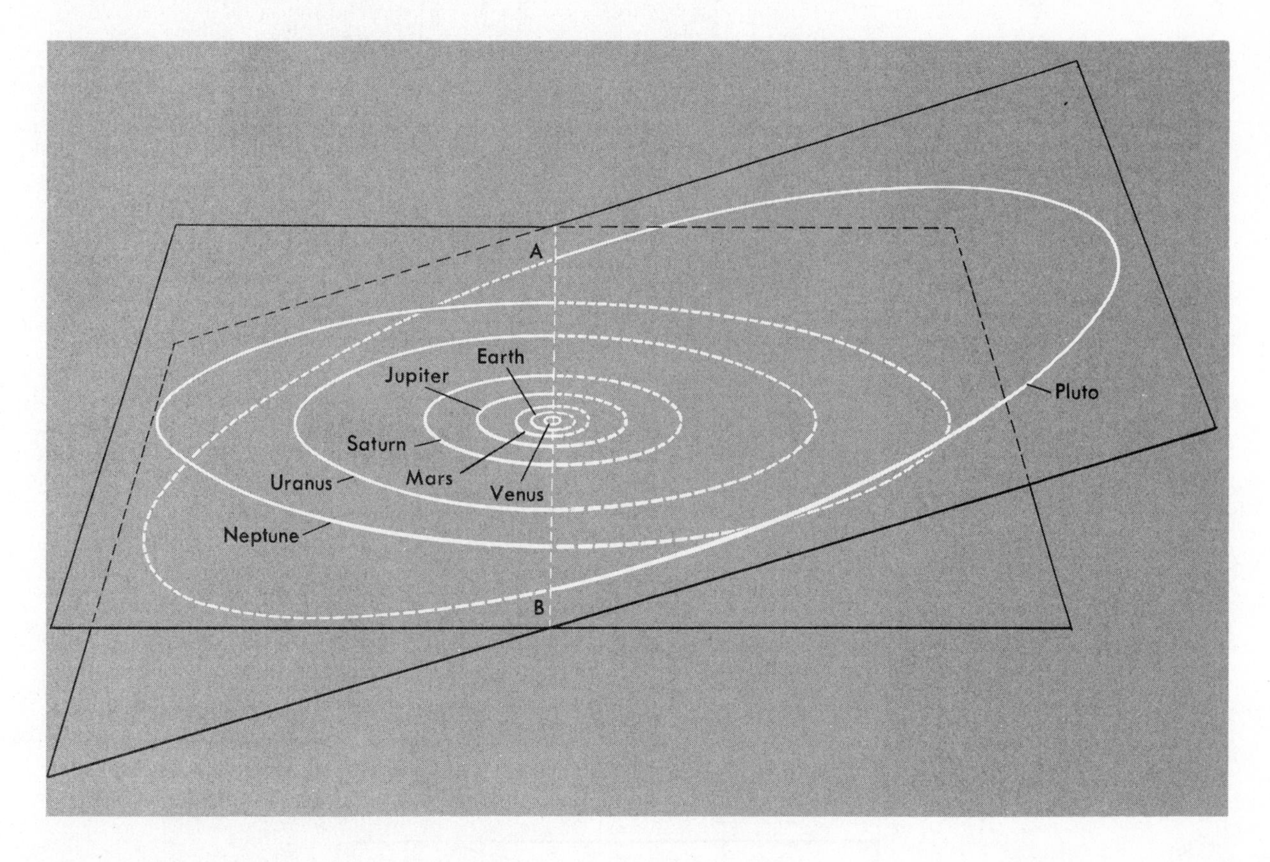

Figure 13–13
The orbits of the known planets as seen from an angle. Notice that all but the outermost planet, Pluto, revolve about the sun in approximately the same plane. In this figure the orbit of Mercury is too small to show.

III. The ratio $\frac{R^3}{T^2}$ is the same for all planets.

If this constant ratio is called K, the third law may be written

$$\frac{R^3}{T^2} = K.$$

Kepler's three laws give a more accurate representation of the planetary orbits than had been obtained from either the Ptolemaic or the Copernican system with all their complexity of circles moving on circles. They were based on observations made before the invention of the telescope!

13-6 Kinematic Description and the Dynamical Problem

Kepler's laws are the kinematics of the planetary system. They give a simple, accurate description of the motions of the planets, but they do not provide an explanation of the motions in terms of forces. Ptolemy's description of planetary motion is also kinematics. What is the essential difference in the two kinematic descriptions?

Both descriptions are reasonably accurate. Each allows us to predict where we shall see a particular planet at a given time. The difference lies in the point of view. Kepler's description appears simpler to us. He followed

Copernicus in choosing the fixed stars as a framework with respect to which the planetary motions are specified, and like Copernicus he used the sun as the origin from which to measure the planetary positions. It is simpler to describe planetary motions using the sun as the origin than it is to describe the motions as seen from the earth, and we can conclude from this simple description what the planetary motions must look like when viewed from earth.

On the other hand, there is nothing wrong in describing the motion directly, using the earth as the origin. It is merely difficult and confusing because the motions we see look complicated and irregular. Convenience—not principle—determines the frame of reference and origin we use in kinematics.

The situation is like that of a man on the ground watching the motion of a point on the edge of the wheel of a car (Fig. 10–26). He sees the point move on a cycloidal path with a speed that varies periodically from zero to a maximum value. On the other hand, a man who describes the motion with respect to the axle of the wheel finds that the point moves at constant speed around a circle. Both men are right; and if we take account of the motion of the axle with respect to the ground, the two descriptions are equivalent. In just the same way, geocentric and heliocentric descriptions of the planetary system can be equivalent, as are the systems of Tycho Brahe and of Copernicus (Fig. 13–7 and 13–8); and which we adopt is a matter of convenience. For navigation, for example, the earth-centered view is preferable

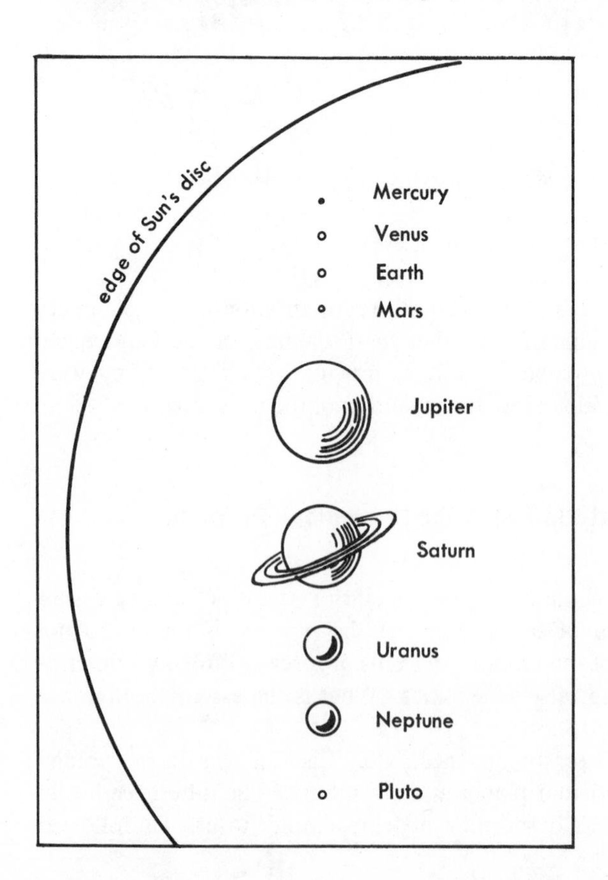

Figure 13–14
The approximate sizes of the planets compared to the sun. The total mass of all the planets equals only 1.34×10^{-3} of the sun's mass.

—we don't care how simple the motions would appear from the sun. We want to know where we are when we see the planets in certain directions at a certain time. Consequently, the geocentric language is used in celestial navigation of ships and airplanes today.

When we wish to explain planetary motions, however, the situation is different. In the first place we should certainly suspect that the dynamical explanation will be easier to arrive at and to understand, the simpler the description of the motion. In the second place, we already know that Newton's law of motion applies only in certain frames of reference. Even the uniform motion that goes on in the absence of a net force on an object will not look uniform to an observer who spins around like a top. In this case we know that complicated fictitious forces will appear, and then we cannot find a simple dynamical explanation. Consequently, in making the transition from kinematics to dynamics, it is important to find a preferred frame of reference where fictitious forces do not confuse us.

To explain planetary motion dynamically, we need to choose an appropriate frame of reference. Can we choose that frame so that the earth is at rest in it? To this question the answer seems to be "no." In any such frame, the motions of the planets imply irregular forces and no dynamical explanation has been found. We can only retire to the Aristotelian view that planets are different from other matter and behave according to their own special laws of motions.

The Ptolemaic system was appropriate to this Aristotelian view. It also fitted with a geocentric dynamics for earthly objects. It led to the idea that the orbits of the planets should appear simplest when viewed with the earth as center (and not around some other point). The greater simplicity of the planetary orbits in Kepler's heliocentric description thus undermines the whole Aristotelian picture.

In the heliocentric system, on the other hand, the earth becomes a planet like other planets. Then there is no reason left to have a special geocentric

Table 3
A Scale Model of the Solar System

OBJECT IN THE SOLAR SYSTEM	OBJECT IN THE MODEL	DISTANCE FROM "SUN"
Sun	basketball	
Mercury	half a pinhead	13 m
Venus	appleseed	25 m
Earth	appleseed	34 m
Mars	small appleseed	52 m
Jupiter	golf ball	180 m
Saturn	Ping-Pong ball	320 m
Uranus	a marble	0.65 km
Neptune	a marble	1.0 km
Pluto	small appleseed	1.3 km
Nearest star	basketball	8×10^3 km

It is impossible to show in a small drawing both the relative sizes and the distances of the planets on the same scale. The above table gives some idea of the relative sizes and distances in terms of common objects. To get the true dimensions of the solar system, every dimension should be multiplied by 4.4×10^9.

dynamics. Instead we can look again for a single dynamics that includes the motions of objects on the earth and of all the planets including our own. Indeed the heliocentric view provided the clue from which we have constructed a dynamical explanation of planetary motion.

In the rest of this chapter we shall see how the heliocentric system fits in with the new dynamics of Galileo and Newton. There we shall follow in the footsteps of Newton. With the heliocentric description and a frame of reference tied to the fixed stars, we shall find that a simple law of force between chunks of matter leads to the observed planetary motions. It explains them on the basis of the same dynamics that applies here on earth. Since the time when this law of gravitational force was postulated by Newton, it has also been experimentally tested for small pieces of matter here on earth. It is valid here as well as through astronomical distances. Thus the search for a dynamics in which motions here and in the heavens are of just the same kind was successful. In a geocentric picture no such unified explanation is available.

13-7 Newton

Isaac Newton was born in 1642, the year Galileo died. He brought together the discoveries of Copernicus, Kepler, Galileo, and others in astronomy and in dynamics. To these he added his own findings and fused them into a structure that still stands today, one of the greatest achievements of science. So profound and clear was his understanding, that he was able to apply the laws of motion successfully to an astonishing number of phenomena, from the movement of the planets to the rise and fall of the tides.

Between the time of Kepler and that of Newton a great change had taken place in scientific thought. After Galileo's work the feeling grew that there were universal laws governing the motion of bodies and that these laws might apply to motion in the heavens as well as on earth. The scientific discussions in the Royal Society of London often centered about the question, "What sort of force does the sun exert on the planets which causes the planets to move according to the laws which Kepler has discovered?" With Kepler's laws as a guide Newton answered this question. He created a planetary dynamics which was so successful that for many years scientists complained that nothing was left to be done.

Newton's first effort to understand the motion of celestial bodies was directed toward a study of the motion of the moon. He knew that if no force acted on the moon it would move in a straight line with constant speed. However, as viewed from the earth, the moon follows a nearly circular path. Consequently, there must be an acceleration toward the earth and a force which produces it. He stated:

"Nor could the moon without some such force be retained in its orbit. If this force was too small, it would not sufficiently turn the moon out of a rectilinear course; if it was too great it would turn it too much and draw the moon from its orbit toward the earth."

What is the force that causes the moon to move about the earth? Newton said that the answer came to him while he was sitting in a garden. He was

Figure 13–15
Sir Isaac Newton.

thinking about this problem when an apple fell to the ground; the force which the earth exerted on the apple might, he thought, also be exerted on the moon. The moon might be a falling body.

In Chapter 12 we calculated the acceleration of the moon toward the earth and found it to be about 2.7×10^{-3} m/sec^2, nowhere near the value of 9.8 m/sec^2 which is the acceleration of a falling body at the earth's surface. Newton carried out essentially the same calculation. At first he did not have available a very accurate value for the radius of the moon's orbit. He did know, however, that it was about sixty times the radius of the earth; and, using a rough value for the earth's radius, he could obtain the radius of the moon's orbit and calculate the moon's acceleration. When he found out how small the acceleration of the moon is, Newton must have asked himself questions like these: Why is the acceleration of a falling body so much greater than that of the moon? Does the force with which the earth attracts a body decrease as the body gets farther and farther away? If so, what is the exact relation between the force and the distance of separation?

Newton had assumed that the earth pulled on the moon in the same general way that it pulled on the falling apple. If this assumption was to be retained, any postulated law of force would have to explain the acceleration g of a body at the earth's surface and the much smaller value of the acceleration of the moon. Newton explained many years later that he was led to the correct force law by working backward from Kepler's third law. He temporarily left the forces that were exerted by the earth and considered instead the forces exerted by the sun on the planets, the centripetal forces which kept the planets moving in their orbits. Newton wanted to know how the force on a planet varied with the radius of the planet's orbit. We shall now see how this force may be calculated.

One of Kepler's triumphs was his description of planetary orbits as ellipses. But the planetary orbits are nearly circular, and for simplicity we shall approximate them as circles with the sun at their common center. Let us consider a planet moving around the sun with a period T in a circular orbit of radius R. As we learned in Section 12–5, the centripetal acceleration of a planet or any object moving uniformly around a circle is $a = \frac{4\pi^2 R}{T^2}$. Therefore the centripetal force on the planet must be

$$F = ma = \frac{m4\pi^2 R}{T^2},$$

where m is the mass of the planet. This is the force that acts on the planet.

In order to eliminate the period T and express the force as a function of R and m alone, Newton used Kepler's third law $\frac{R^3}{T^2} = K$, or $T^2 = \frac{R^3}{K}$. On substituting $\frac{R^3}{K}$ for T^2 in the equation $F = \frac{m4\pi^2 R}{T^2}$, we find the force on the planet is

$$F = 4\pi^2 K \frac{m}{R^2}.$$

The force is proportional to the mass of the planet and inversely proportional to the square of the distance from the sun (Fig. 13–16).

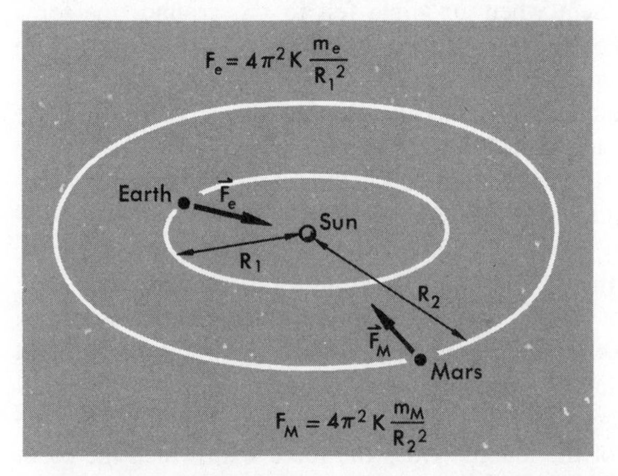

Figure 13–16
The gravitational force exerted by the sun on a planet is proportional to the mass of the planet and inversely proportional to the square of its distance from the sun.

Eventually Newton was able to show that any body moving under the action of this force must move in an elliptical orbit with the sun at one focus and that the line connecting the sun with the body will sweep out equal areas in equal times. Furthermore, we know from the method by which we found the force that Kepler's third law will follow from it. The whole system of planetary motion described by Kepler's laws, therefore, follows from this law of force and Newton's law of motion.*

13-8 Universal Gravitation

Notice that the factor $(4\pi^2 K)$ in the law of force between the sun and a planet enters the equation from the law of periods. It applies to any planet of any mass on any orbit around the sun. Therefore, $(4\pi^2 K)$ depends only

* Huygens and Hooke also used Kepler's third law and Newton's law of motion to infer that F is proportional to $1/R^2$, but they did not show that Kepler's other laws then followed. Newton provided the law of motion, found the law of force, and also showed that it resulted in Kepler's description of planetary motions.

on the properties of the sun; it measures the strength of the sun as the source of the force of attraction.

The force between the sun and a mass m is

$$F = \frac{(4\pi^2 K_s)m}{R^2},$$

where $4\pi^2 K_s$ refers to the sun and R is the distance between the sun and the mass m. Perhaps the force between the earth and a mass m is

$$F = \frac{(4\pi^2 K_e)m}{R^2},$$

where $(4\pi^2 K_e)$ is the strength of the earth as a source of gravitational attraction and R is now the distance between the earth and the mass m. With such ideas Newton returned to the problem of the moon's motion about the earth. Measuring R from the center of the earth, the value of the gravitational field strength g (the acceleration of a mass falling at the earth's surface) is then

$$g = \frac{F}{m} = \frac{(4\pi^2 K_e)}{R_e^2},$$

where R_e is the radius of the earth, the distance between the center of the earth and a mass m at its surface. Also, the value of the gravitational field strength at the distance of the moon, which is the acceleration of the moon toward the earth, is

$$a_m = \frac{(4\pi^2 K_e)}{R_m^2},$$

where R_m is the distance from the center of the earth to the center of the moon. On dividing this equation by the last one we get

$$\frac{a_m}{g} = \frac{R_e^2}{R_m^2} \quad \text{or} \quad a_m = g\frac{R_e^2}{R_m^2}.$$

Since Newton knew that $\frac{R_e}{R_m}$ is about 1/60 and g is 9.8 m/sec^2, he found that $a_m \approx 2.7 \times 10^{-3}$ m/sec^2. This is about the same value of a_m that he obtained from the radius and period of the moon (Section 13–7).

Now Newton had obtained the acceleration of the moon in two different ways: from R_m and the period of the moon's motion without any reference to the inverse-square law of force, and from g at the earth's surface by using the inverse-square law in the ratio $(R_e/R_m)^2$. The approximate agreement of the values he obtained gave him support for the idea that the force between earth and moon was of the same kind as that between the sun and the planets. They were both gravitational forces like the force on the falling apple.

Newton probably did not show that all of Kepler's laws follow from the law of gravitational force until several years after his original discoveries. But he discovered the law of force and applied it to the problem of the moon when he was twenty-four years old. Of this period Newton later wrote: "And the same year I began to think of gravity extending to ye orb of the Moon, and . . . from Kepler's Rule (third law) . . . I deduced that the forces which keep the Planets in their Orbs must (vary) reciprocally as the square of their distances from the centers about which they revolve: and thereby compared the force requisite to keep the Moon in her Orb with the force of gravity at the surface of the earth, and found them to answer pretty nearly. All this was in the two plague years of 1665 and 1666, for in those days I was in the prime of my age for invention, and minded Mathematicks and Philosophy more than at any time since."

Newton certainly suspected that the inverse-square law of attraction applied not only to the sun and planets, not only to the earth and moon, but also to any two chunks of matter. This suspicion leads immediately to the question: What property of a body determines its gravitational attraction for other masses? What property of the earth determines how large $4\pi^2K_e$ is for the earth? What determines $4\pi^2K_s$ for the sun? Perhaps $4\pi^2K$ depends on a new property of a body; but if gravitational attraction is a property of all bodies, it is reasonable to suppose that $4\pi^2K$ depends on the quantity of matter in the body. The simplest assumption is that $4\pi^2K$ is proportional to the mass of the body. Then $4\pi^2K_e = Gm_e$ for the earth, and $4\pi^2K_s = Gm_s$ for the sun, where G is the proportionality factor between $4\pi^2K$ and m for any body.

Newton made this assumption. With it, the magnitude of the gravitational force of attraction that a body of mass m_1 exerts on a body of mass m_2 at a distance R becomes

$$F_{1-2} = (4\pi^2K_1)\frac{m_2}{R^2} = Gm_1\frac{m_2}{R^2}.$$

Furthermore, because every mass attracts every other piece of matter gravitationally, the mass m_2 also exerts a gravitational force on m_1. Since $4\pi^2K^2 = Gm_2$, the magnitude of this force of attraction exerted by m_2 on m_1 is

$$F_{2-1} = (4\pi^2K_2)\frac{m_1}{R^2} = Gm_2\frac{m_1}{R^2}.$$

Therefore, the two forces are equal in magnitude, although they are opposite in direction (Fig. 13–17). The expression $F = G\frac{m_1m_2}{R^2}$ for the

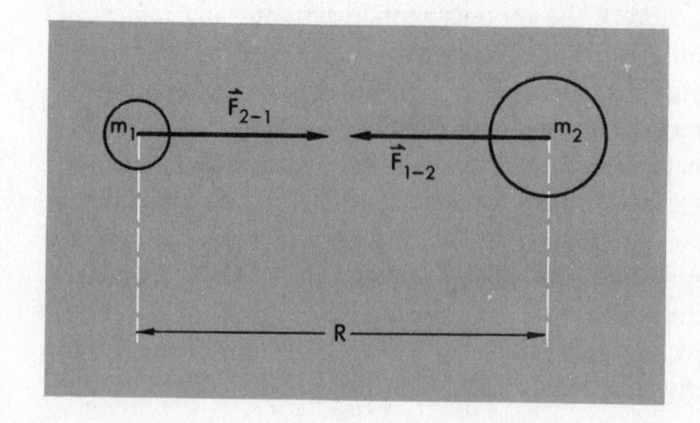

Figure 13–17
The force of gravitation exerted by m_1 on m_2 is equal and opposite to the force exerted by m_2 on m_1.

magnitude of the attraction summarizes Newton's law of universal gravitation: any two bodies attract each other with forces proportional to the mass of each and inversely proportional to the square of the distance between them. The universal gravitational constant, G, does not depend on what objects are considered, where they are, or what their state of motion may be.

We do not know in detail how Newton arrived at the law of universal gravitation. In addition to the reason that we have given to make the law plausible, a number of other considerations may have suggested the same result. For example, as we shall see in the next chapter, Newton eventually assumed that the forces of interaction between two bodies are always equal and opposite, and he may already have had this idea in mind when he formulated the law of universal gravitation. Whatever the steps by which Newton was led to the law of universal gravitation, the law eventually stands or falls by the consistency between the predictions we can make from it and the actual behavior of the universe. On the basis of this law Newton was able to take great strides in building up a theoretical system of the universe.

13-9 Some of Newton's Later Accomplishments

Newton applied the law of universal gravitation to a wide variety of problems. As we mentioned previously, he derived all three of Kepler's empirical laws from the law of universal gravitation. He then examined the tides and explained them on the basis of the gravitational force exerted by the moon, both on the earth and on the oceans. He began to analyze the small irregularities (perturbations) of planetary orbits. These small deviations of the planets from their predicted elliptical paths can be explained by the small gravitational interactions among the planets themselves. Not only is the earth attracted by the sun, but also in varying degree it is attracted by each of the other planets. These attractions are relatively small because of the small masses of the planets in comparison with the mass of the sun; but their effects can be observed and they are predicted correctly by Newton's theory.

Later this perturbation theory led to the discovery of a new planet. In the nineteenth century seven planets were known. Of these, the original six behaved well; but the seventh, Uranus, which had been discovered by Herschel in 1781, did not act quite as expected. When the perturbations of its orbit by the other planets had been computed, the result did not agree with the details of the observed motion. The astronomers Adams and Leverrier arrived independently at the conclusion that there must be another yet undiscovered planet (still farther away from the sun but close enough to influence the motion of Uranus); and on September 23, 1846, the astronomer Galle found the new planet where Leverrier had told him to look. This new planet was named Neptune.

Among the many other problems to which Newton applied the law of universal gravitation, one is of particular interest to us. It concerns the calculation of the acceleration of the moon from the inverse-square law of force and the value of g at the surface of the earth (Section 13–8). When he originally performed this calculation, Newton used the distances R_e and

Satellitum tempora periodica.

1d. 18h. 28′ $\frac{3}{5}$. 3d. 13h. 17′ $\frac{9}{10}$. 7d. 3h. 59′ $\frac{3}{5}$. 16d. 18h. 5′ $\frac{1}{5}$.

Diſtantiæ Satellitum à centro Jovis.

Ex Obſervationibus	1.	2	3	4	
Caſſini	5.	8.	13.	23.	Semidiam. Jovis.
Borelli	5 $\frac{2}{3}$.	8 $\frac{2}{3}$.	14.	24 $\frac{2}{3}$.	
Tounlei *per Micromet-*	5,51.	8,78.	13,47.	24,72.	
Flamſtedii *per Microm.*	5,31.	8,85.	13,98.	24,23.	
Flamſt. *per Eclipſ. Satel.*	5,578.	8,876.	14,159.	24,903.	
Ex temporibus periodicis.	5,578.	8,878.	14,168.	24,968.	

Hypoth. VI. *Planetas quinque primarios Mercurium, Venerem, Martem, Jovem & Saturnum Orbibus ſuis Solem cingere.*

Mercurium & Venerem circa Solem revolvi ex eorum phaſibus lunaribus demonſtratur. Plenâ facie lucentes ultra Solem ſiti ſunt, dimidiatâ è regione Solis, falcatâ cis Solem; per diſcum ejus ad modum macularum nonnunquam tranſeuntes. Ex Martis quoque plena facie prope Solis conjunctionem, & gibboſa in quadraturis, certum eſt quod is Solem ambit. De Jove etiam & Saturno idem ex eorum phaſibus ſemper plenis demonſtratur.

Hypoth. VII. *Planetarum quinque primariorum, & (vel Solis circa Terram vel) Terræ circa Solem tempora periodica eſſe in ratione ſeſquialtera mediocrium diſtantiarum à Sole.*

Hæc à *Keplero* inventa ratio in confeſſo eſt apud omnes. Eadem utique ſunt tempora periodica, eædemq; orbium dimenſiones, ſive Planetæ circa Terram, ſive iidem circa Solem revolvantur. Ac de menſura quidem temporum periodicorum convenit inter Aſtronomos univerſos. Magnitudines autem Orbium *Keplerus* & *Bullialdus* omnium diligentiſſimè ex Obſervationibus determinaverunt: & diſtantiæ mediocres, quæ temporibus periodicis reſpondent, non

Aa a 2 diffe-

Figure 13–18
A page from Newton's "Philosophiae Naturalis Principia Mathematica" (London, 1687). The periods of revolution of the four largest moons of Jupiter are given at the top of the page. The table gives the radii of the orbits of the moons as measured by different observers. The bottom line of the table gives the radii as calculated from the periods using Kepler's third law.

R_m from the center of the earth. Although the center of the earth is a natural place from which to measure R, Newton was not sure that it was the right place. Because he suspected that the inverse-square law of attraction applied to any two pieces of matter, Newton thought that the gravitational attraction of the earth for an object should be the resultant of the forces attracting it toward each piece of matter in the earth.

The different pieces of matter in the earth are located at various distances from an object at the earth's surface. Do they, all acting together, produce the same force on this object that they would produce if all were concentrated at the center of the earth? Does the force fall off as $1/R^2$ even close to the surface of the earth? Before he would be satisfied, Newton had to solve the mathematical problem of adding up all the forces arising from all the pieces of matter of which the earth is made. He had to prove that this vector sum gave the inverse-square law of force outside the surface of the earth.

Now we can solve this problem by applying an elegant mathematical theorem; but in Newton's time this theorem and its basis were unknown. Newton himself invented the mathematics (now called the calculus) necessary to solve this and similar problems. When he obtained the answer it turned out that his original assumption was correct. When the forces from each piece of matter fall off as the square of the distance, spherical bodies do attract one another as though the entire mass of each were concentrated at its center. Newton was delighted.

13-10 Laboratory Tests of the Law of Universal Gravitation

The direct way to see if Newton's law of universal gravitation is consistent with the behavior of all matter is to measure the gravitational forces between chunks of matter in the laboratory. We should measure the gravitational attraction between two masses, using objects of various materials to see that the mass alone determines the attraction. If the law were found to be consistent with our measurements, we would then naturally evaluate the universal constant of proportionality G.

Such experiments are difficult. Even when two stones are placed close together, they do not attract each other noticeably. With a crude estimate of G we can see why. According to the law of universal gravitation, the gravitational force on a mass m at the surface of the earth is

$$F = G\frac{m_e m}{R_e^2}.$$

The gravitational field strength g is therefore

$$g = \frac{F}{m} = G\frac{m_e}{R_e^2}.$$

In this equation we know g; it is 9.8 m/sec^2. We know the radius of the earth $R_e = 6.38 \times 10^6$ meters; and Newton knew an approximate value for it. To determine G, therefore, we need only to estimate the mass of the earth.

Newton made such an estimate. He guessed a reasonable value for the average density of the earth, about five times the density of water, and multiplied it by the volume of the earth. The mass of the earth is then estimated as about 6×10^{24} kg; and the order of magnitude of G is $10^{-10}\,\frac{\text{m}^3}{\text{kg-sec}^2}$.

We now know G accurately. It is $G = 0.667 \times 10^{-10}\,\frac{\text{m}^3}{\text{kg-sec}^2}$. Applying this result to two stones 0.1 m apart and each of 1 kg mass, we find that their gravitational attraction should be about 10^{-8} newtons, about one billionth of the force pulling them toward the earth. Newton also concluded that "the gravitation [between such stones] must be far less than to fall under the observation of our senses." He therefore directed his attention to the calculations about the gravitational interactions of the planets and their satellites which we mentioned in the last section.

One hundred years later, in 1798, Henry Cavendish succeeded in measuring the gravitational interaction of laboratory-sized objects. The apparatus he used is diagramed in Fig. 13–19. Two small spheres were mounted on opposite ends of a light rod 2 meters long. This rod hung horizontally with its center below a fine vertical wire from which it was suspended. At the ends of the rod and on the side of the case which contained the apparatus, Cavendish mounted ivory rulers with which to measure the position of the rod. When Cavendish placed two large masses near the small spheres at the ends of the rod, the small spheres were attracted toward the big masses and the wire suspension twisted.

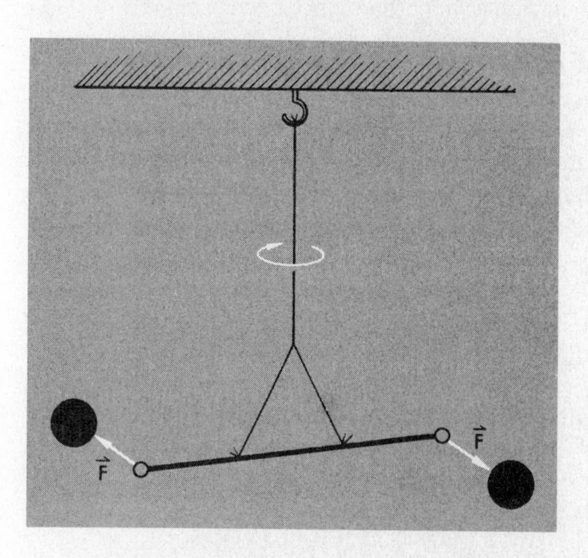

Figure 13–19
A simplified diagram of the apparatus used by Cavendish to verify the law of universal gravitation for small objects and to measure the gravitational constant G.

Cavendish recorded the position of the rod with the suspension twisted when the big masses were located as shown in Fig. 13–19. Then he moved each of the large masses to the symmetrical position on the other side of the little spheres. The gravitational attraction twisted the suspension in the opposite direction, and he measured the new position. By measuring the changes in position when known forces were applied to the small balls, Cavendish could find the strength of the gravitational forces between the small balls and the large masses by the following reasoning: The force at various twists of the suspension can be found dynamically by taking away the large masses and letting the rod carrying the small balls swing horizontally, twisting and untwisting the wire suspension. The motion of the rod then depends on the known masses of the small balls and on the forces exerted on them by the twisted suspension. The size of these forces can therefore be determined from the motion. Now consider the whole system at rest with the large masses in position. The net force on each ball must be zero; but the suspension is twisted because of the gravitational forces between the large masses and the balls on the rod. The net force, zero, is the resultant of the gravitational force and the force exerted by the twisted suspension.

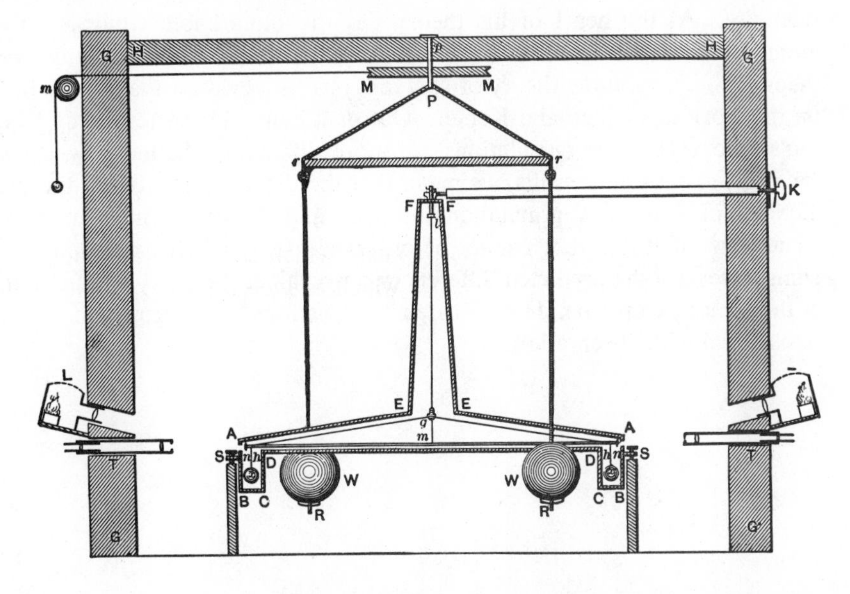

Figure 13–20
The sketch of the Cavendish apparatus which appeared in Cavendish's original paper. The two large masses are labeled *W* and the small masses *x*. Notice that the whole device is mounted in a large case *G* with outside controls to move the weights and adjust the horizontal rod. Scales at *A* near the ends of the rod were illuminated by the lamps *L* and observed through the telescopes *T*.

The gravitational force is therefore equal in magnitude and opposite in direction to the force exerted by the suspension. Since we now know the force exerted by the suspension, we also know the gravitational force.

Actually Cavendish performed numerous experiments. He had to evaluate possible extraneous effects, such as convection currents in the air due to slight temperature differences. He wanted to make sure that he was not measuring magnetic forces by mistake. He needed many determinations to make sure that his results were reproducible and to determine their accuracy. With these experiments Cavendish determined G. He expressed his answer in terms of the mean density of the earth, which he found to be close to 5½ times that of water, very close to Newton's estimate.

By using different substances for the various objects, we can modify the Cavendish experiment to show that the masses alone determine the gravitational attraction. By changing the relative positions of the small and large masses, we can check up on the inverse-square law at laboratory rather than planetary distances. Many modifications of the Cavendish experiments have been carried out, and so far Newton's law of universal gravitation is consistent with all of them.

13-11 A Small Discrepancy

Three centuries have passed since Newton's work on gravitation. During that time the law of gravitation has been tested through the most detailed calculations of the motion of the planets and their moons. In almost every case, the calculations have predicted orbits which are in agreement with actual observation. There is, however, one exception: an extremely small irregularity in the orbit of the planet Mercury which is not predicted by Newton's law of gravitation. Even though the discrepancy is minute, an improved theory is needed to explain it.

Such a theory was provided by Albert Einstein in his general theory of relativity. At the heart of his theory lies the remarkable equivalence of gravitational and inertial mass. Einstein welded them into one entity. His theory is built upon the theory of Newton, just as Newton's theory was built on the work of Galileo and Kepler. Out of it come all the results of Newton's theory (but the calculations are more difficult). Actually, when we say it gives the same results, we mean that the differences between the predictions of Einstein's gravitational theory and Newton's mechanics are usually so small that they cannot be observed. Only under exceptional circumstances are the predicted differences observable. Mercury's orbit is one of these rare exceptions. Here Einstein's predictions of the orbit bring theory into accord with observation.

For Home, Desk, and Lab

1. Three fireflies X, Y, and Z are on a moving bicycle at night. X is at the very center of one of the axles, which turns with the wheel. Y is on the rim of the wheel. Z is on the frame of the bike outside the circumference of the rim. Draw sketches and use a few words to describe the motions
   (a) of X and Y as viewed by Z.
   (b) of Y and Z as viewed by X.
   (c) of X and Z as viewed by Y.
   (d) of all three as seen by an observer standing near the bicycle.

2.* (a) In Fig. 13–5, what is the ratio of the greatest distance to the shortest distance between the earth and Mars?
   (b) What about Fig. 13–8? (Section 4.)

3. The fixed stars go around in about one day as seen by an observer on the earth. About how long do the following revolutions appear to take?
   (a) The sphere of the fixed stars as seen by someone on the moon. (From the earth, we always see the same side of the moon.)
   (b) The earth about the moon as seen from the moon.
   (c) The sun as seen from the moon. Does the sun appear to rotate as if it were one of the fixed stars? Does it appear to rotate faster or slower? Remember that the earth and moon go around the sun once every year.

4. Does Tycho Brahe's geocentric system account for the fact that there is a maximum angle between Venus and the sun as seen from the earth?

5. The earth's orbit around the sun is nearly circular, and the moon moves in a nearly circular orbit around the earth. The radius of the earth's orbit is 1.5×10^{11} meters, and that of the moon's orbit is 4×10^{8} meters.
   (a) How often is the moon between the earth and the sun?
   (b) How far has the moon moved around the sun in the interval between two successive times when the moon is between the earth and the sun?
   (c) Sketch the orbit of the earth going around the sun and on the same figure sketch the orbit of the moon going around the sun.
   (d) Does the moon show retrograde motion as seen by an observer at the sun? (Retrograde motion is motion like that shown in Fig. 13–2.)

6.* What area is swept out by Jupiter in making one fourth of a revolution around the sun? (See Table 2.) (Section 5.)

7. How fast, in m^2/sec, is area swept out by the radius from sun to earth?

8.* Figure 13–12 shows the orbits of the planets about the sun. When does Pluto move (a) fastest? (b) slowest? (Section 5.)

9.* Table 1 does not include Uranus, Neptune, and Pluto. What value do you predict for the ratio $\frac{R^3}{T^2}$ for each of these planets? (Section 5.)

10.* If a small planet were discovered whose distance from the sun was eight times that of the earth, how many times longer would it take to circle the sun? (Section 5.)

11. Between September 21 and March 21 there are three days fewer than between March 21 and September 21. These are the dates when day and night are of equal length, and between them the earth moves 180° around its orbit with respect to the sun. From this and Kepler's law of equal areas, explain how you can determine the part of the year during which the earth is closest to the sun.

12. Astronomers have observed that Halley's comet has a period of 75 years and that its smallest distance from the sun is 8.9×10^{10} meters, but its greatest distance from the sun cannot be measured because it cannot be seen. Use this information together with the second footnote in Section 13–5 to compute its greatest distance from the sun. (It was Newton who told Halley how to compute the orbit of a comet. Halley found and calculated the orbit and period of the comet that bears his name in the course of a general analysis he made of comets' orbits.)

13.* The radius of the moon's orbit is 60 times greater than the radius of the earth. How many times greater is the acceleration of a falling body on the earth than the acceleration of the moon toward the earth? (Section 7.)

14.* (a) Calculate K_e from Table 2.
(b) Using the value of K_s from Table 1 and your answer to (a), determine the ratio of the mass of the earth to the mass of the sun. (Section 8.)

15.* At what height above the earth's surface will a rocket have half the force of gravitation on it that it would have at sea level? Express your answer in earth radii. (Section 8.)

16. (a) At what height will a satellite moving in the plane of the equator stay over one place on the equator of the earth? One way to get the answer is to compare this satellite with the moon, which is at 59.5 earth radii from the center of the earth and takes 27 days to go around the earth.
(b) How fast is the satellite accelerating toward the earth?
(c) Using the inverse-square law and g at the surface of the earth, find the gravitational field strength at the height of the satellite. Compare with your answer to (b).

17. Find the weight of a 100-kg man on Jupiter.

18. A 70-kg boy stands 1 meter away from a 60-kg girl. Calculate the force of attraction (gravitational) between them.

19. A satellite circles the earth once every 98 minutes at a mean altitude of 500 km. Calculate the mass of the earth. The masses of planets are actually calculated from satellite motions, and one reason for establishing artificial earth satellites is to get a better value for the mass of the earth.

20. (a) If T is the period of a satellite circling around just above the surface of a planet whose average density is ρ, show that ρT^2 is a universal constant.
(b) What is the value of the constant?

21. The earth is acted upon by the gravitational attraction of the sun. Why doesn't the earth fall into the sun? Be prepared to discuss your answer.

22. (a) What is the speed of the moon around the sun compared to that of the earth around the sun?
(b) If the earth could be removed suddenly without disturbing the motion of the moon, what would be the subsequent path of the moon?
(c) Calculate the ratio of the force of attraction exerted by the sun on the moon to the force exerted by the earth on the moon.
(d) Why does the sun not capture the moon, taking it away from the earth?

23. Assume the earth is perfectly round and has a radius of 6400 km.
(a) How much less does a man with a mass of 100 kg apparently weigh at the equator than at the poles because of the rotation of the earth?
(b) How fast would the earth have to spin in order that he would exert no force on a scale at the equator?
(c) How many times larger is the speed of rotation in (b) than the actual speed?

24. A 10,000-kilogram spaceship is drifting on a long mission toward the outer edge of the solar system. It has put out a small experimental satellite which revolves around it at a distance of 120 meters under their mutual gravitational attraction.
    (a) What is the period of revolution of the satellite?
    (b) What is the speed of the satellite?

25. An astronomer observes a planet with a small satellite revolving around it in a circular orbit of radius r with a period of time T.
    (a) What is the mass of the planet?
    (b) What is the acceleration of the satellite toward the planet?
    (c) What is the gravitational force on the satellite?
    (d) The astronomer measures the radius of the planet and finds it to be 1/10 of the radius of the orbit of the satellite. What is the gravitational field strength of the planet at its surface?

26.† Two unequal masses m_1 and m_2, which are isolated in free space, attract each other with a gravitational force $F = G\frac{m_1 m_2}{R^2}$. What are the accelerations of m_1 and m_2? Is your answer in conflict with the justified expectation that an observer on m_1 will see that m_2 is coming toward him with the same acceleration as that with which an observer on m_2 is seeing m_1 coming toward m_2?

27. Astronomical observations indicate that the sun is describing a circular orbit around the center of our galaxy. The radius of the orbit is about 30,000 light-years ($= 2.7 \times 10^{20}$ m) and the period of one complete revolution is about 200 million years. In this motion the sun is acted on by the gravitational pull of the great quantity of stars lying inside its orbit.
    (a) Calculate the total mass of these stars from the data given.
    (b) How many stars of mass equal to the sun (2×10^{30} kg) does this represent?

Further Reading

These suggestions are not exhaustive but are limited to works that have been found especially useful and at the same time generally available.

ANDRADE, E. N. DA C., *Sir Isaac Newton*. Doubleday Anchor, 1958. Science Study Series, 1965.

BONDI, HERMANN, *The Universe at Large*. Doubleday Anchor, 1960: Science Study Series. Explains the part universal gravitation plays in determining the orbits of satellites from the astronomer's viewpoint. (Chapter 9 and 10)

COHEN, I. BERNARD, *The Birth of a New Physics*. Doubleday Anchor, 1960: Science Study Series. (Chapters 1, 2, 3, 4, 6, and 7)

GAMOW, GEORGE, *Gravity*. Doubleday Anchor, 1962: Science Study Series. (Chapters 2, 4, and 8)

HOLTON, G., and ROLLER, D. H. D., *Foundations of Modern Physical Science*. Addison-Wesley, 1958. (Chapter 11)

KOESTLER, ARTHUR, *The Watershed*. Doubleday Anchor, 1962: Science Study Series. An excellent detailed study of Kepler.

WEISSKOPF, VICTOR F., *Knowledge and Wonder*. Doubleday Anchor, 1963: Science Study Series. (Chapters 1, 3, and 9)

Momentum and the Conservation of Momentum

CHAPTER

14

14-1 Impulse

Try to make a baseball and a 16-lb shot go at the same speed. As you know, it is a lot harder to get the shot going. If you apply a constant force $\vec{\mathbf{F}}$ for a time Δt, the change in velocity is given by $m\Delta\vec{\mathbf{v}} = \vec{\mathbf{F}}\Delta t$. So, in order to get the same $\Delta\vec{\mathbf{v}}$, the product $\vec{\mathbf{F}}\Delta t$ must be greater the greater the mass m you are trying to accelerate.

To start a 16-lb shot from rest and give it the same final velocity as a baseball (also started from rest), we must push either harder or longer. What counts is the product $\vec{\mathbf{F}}\Delta t$. This product $\vec{\mathbf{F}}\Delta t$ is the natural measure of how hard and how long we push to change a motion. It is called the *impulse* of the force.

We can exert a given impulse in many different ways: a big force for a short time, a smaller force for a longer time, or even a force that changes while it acts. In Fig. 14–1 we have plotted a constant force F against the time it acts. The plot is just a horizontal straight line of height F above the time axis and of length $\Delta t = t_2 - t_1$, equal to the time during which the force acts. The area of the rectangle under this line is $F\Delta t$, the size of the impulse in the time interval. For any constant force acting for any time interval, we can always get the magnitude of the impulse as the area under the force-time curve for this time interval, and the direction is the direction of the force.

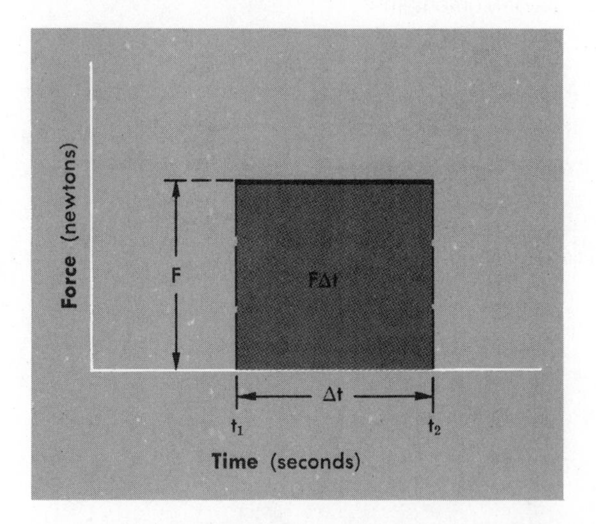

Figure 14–1
A constant force F is plotted against time for the interval $t_2 - t_1 = \Delta t$. The area $F\Delta t$ under the curve gives the impulse of this force in time Δt.

Now suppose the force changes from the constant value $\vec{\mathbf{F}}_1$ to the constant value $\vec{\mathbf{F}}_2$ along the same direction. For a time Δt_1 the force is constant and the impulse is $\vec{\mathbf{F}}_1\Delta t_1$. This impulse makes the change $m(\Delta\vec{\mathbf{v}})_1$ in the motion. In the next time interval, Δt_2, the force is again constant, this time $\vec{\mathbf{F}}_2$, and there is an impulse $\vec{\mathbf{F}}_2\Delta t_2$ which results in the change $m(\Delta\vec{\mathbf{v}})_2$ in the motion. That is, $\vec{\mathbf{F}}_1\Delta t_1 = m(\Delta\vec{\mathbf{v}})_1$ and $\vec{\mathbf{F}}_2\Delta t_2 = m(\Delta\vec{\mathbf{v}})_2$.

If we add these two statements vectorially we get

$$\vec{F}_1\Delta t_1 + \vec{F}_2\Delta t_2 = m(\Delta\vec{v})_1 + m(\Delta\vec{v})_2 = m[(\Delta\vec{v})_1 + (\Delta\vec{v})_2].$$

Thus, the total impulse during $\Delta t = \Delta t_1 + \Delta t_2$ just equals the mass times the total change in vector velocity that occurs during Δt.

When we deal with a continuously varying force we can still obtain the total impulse by adding up all the impulses in short time intervals. We can make each time interval so short that $\vec{F}$ is essentially constant during it. Then the total impulse, which is the sum of all the $\vec{F}\Delta t$'s, gives the total change $m\Delta\vec{v}$ in the motion.

When the force changes magnitude but always points in the same direction, we can add up the impulses by finding the area under the F–t curve. A simple case is shown in Fig. 14–2. When the force changes continuously as in Fig. 14–3 we can break up the area under the F-t curve into a large number of small areas that are nearly rectangles. (See shaded area.) The total impulse is the sum of all these impulses as long as the force points in the same direction. If the force changes direction, however, this graphical method is not enough. We must add the little impulses $\vec{F}\Delta t$ in their correct directions, using vector addition to find the total impulse. If the first little impulse occurs at time t when the moving object has velocity $\vec{v}$, and the final impulse occurs at time t' when the object has velocity $\vec{v}'$, then the total impulse $= m(\vec{v}' - \vec{v}) = m\vec{v}' - m\vec{v}$.

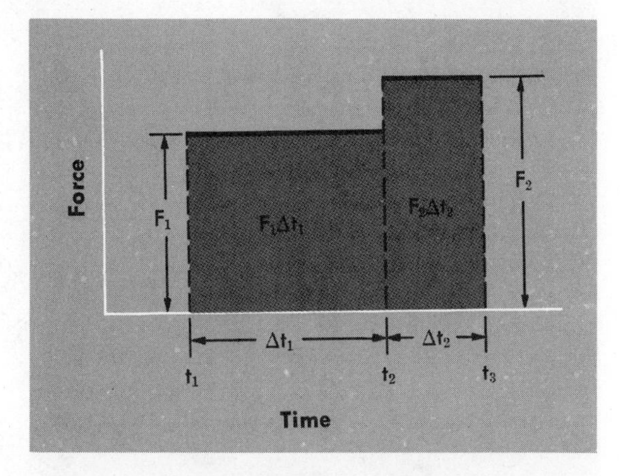

Figure 14–2
The force is F_1 in the time interval Δt_1, and changes to F_2 in time interval Δt_2. If the direction does not change, the total impulse in time $\Delta t_1 + \Delta t_2$ is $F_1\Delta t_1 + F_2\Delta t_2$, the shaded area under the curve.

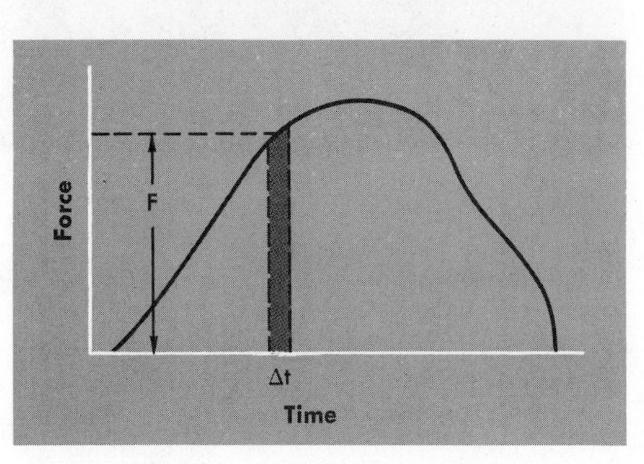

Figure 14–3
A force-time plot of a force that is continuously changing.

(We shall often use the symbol "′," which is read "prime," to distinguish a quantity at a later time from that same quantity at an earlier time.)

Often a large force acts only for a very short time interval. Think what happens when you hit a tennis ball with a racket (Fig. 14–4). Without the most elaborate equipment it is very hard to tell the size of the forces which act during the collision. But the total impulse of the force is easily found by observing the net change in $m\vec{\mathbf{v}}$.

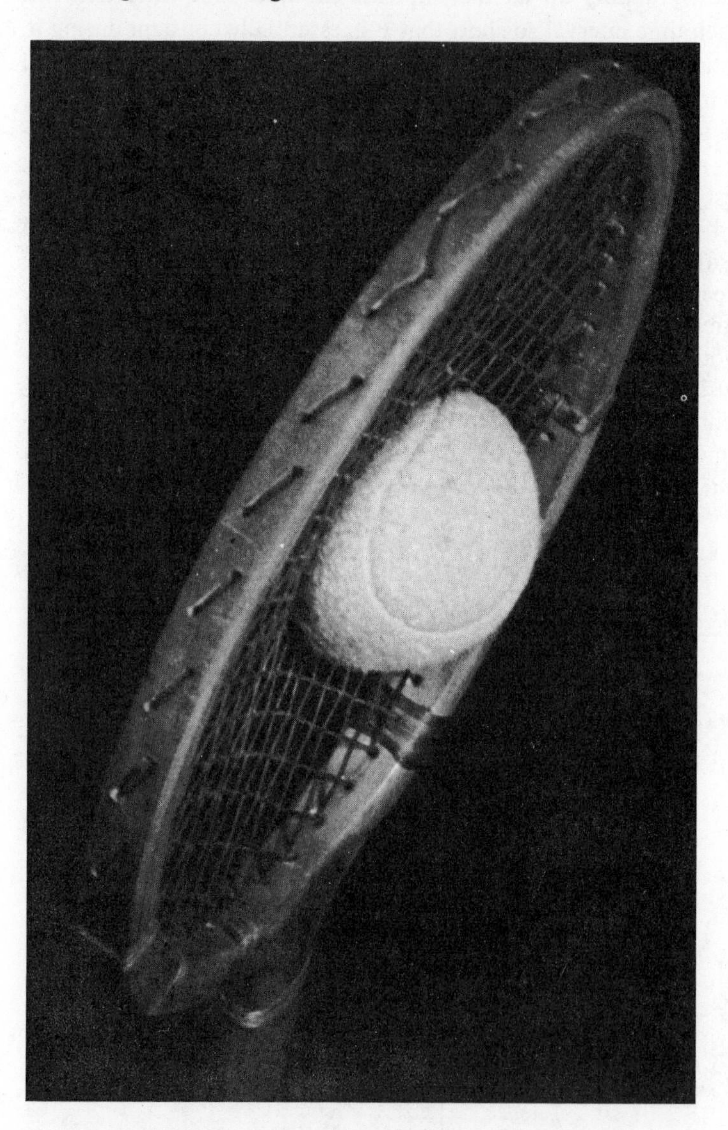

Figure 14–4
This high-speed photograph shows a familiar example of a force that acts for only a very short time. Notice the large distortion of the racket and the ball. (Courtesy: Harold E. Edgerton.)

14-2 Momentum

Suppose we apply the same impulse to two quite different objects, say a baseball and a 16-lb shot, both initially at rest. Since the initial value of the quantity $m\vec{\mathbf{v}}$ is zero in each case, and since equal impulses are applied, the final values $m\vec{\mathbf{v}}'$ will be equal for the baseball and the shot. Yet, because the mass of the shot is so much greater than the mass of the baseball, the velocity of the shot will be very much less than the velocity of the baseball. The product $m\vec{\mathbf{v}}$, then, is quite a different measure of the motion than simply $\vec{\mathbf{v}}$ alone, but it will prove to be of utmost importance in physics. We call it the *momentum* of the body, and measure it in kilogram-meters per second. Because we shall use momentum often, we introduce the symbol $\vec{\mathbf{p}}$ to represent the momentum of a body:

$$\vec{\mathbf{p}} = m\vec{\mathbf{v}}.$$

Velocity and momentum, although related, tell us different things. The knowledge of the velocity only tells how fast (and in which direction) an object moves. It tells us nothing about the effort required to get it moving, or, for that matter, to stop it. The momentum, on the other hand, does not tell the speed of the object (although it tells us the direction in which it moves), but it determines the impulse necessary to get it going and the impulse necessary to stop it. Briefly, the velocity is a kinematical quantity—one which enters the geometrical description of where and when—whereas the momentum is a dynamical one, connected with the impulses and therefore with the causes of changes in the motion of masses.

Notice that the momentum does not depend on the way in which the body acquired its present motion. The equation $\vec{\mathbf{p}} = m\vec{\mathbf{v}}$ contains nothing but the mass and its motion now. The impulse which set it going may have been delivered in any of an infinite number of ways (as we know from the last section), or the mass may always have been moving with momentum $\vec{\mathbf{p}}$. Similarly, the impulse required to bring the mass to rest may be applied in infinitely many ways or not at all; we still know from the present momentum how big it would be. It would require a reverse impulse equal to $-m\vec{\mathbf{v}}$ to stop the body whether we use a big force and short time or a little force and long time.

Because of its connection with the impulse which occurs naturally in Newton's law $\vec{\mathbf{F}}\Delta t = m\Delta\vec{\mathbf{v}}$, we expect momentum to fit naturally into Newtonian dynamics. In fact, Newton expressed his law of motion in terms of the momentum $m\vec{\mathbf{v}}$, which he called the quantity of motion. We can easily express Newton's law in terms of the change in momentum instead of change in velocity:

$$\vec{\mathbf{F}}\Delta t = m\Delta\vec{\mathbf{v}} = m(\vec{\mathbf{v}}' - \vec{\mathbf{v}})$$

where $\vec{\mathbf{v}}$ *and* $\vec{\mathbf{v}}'$ are the velocities before and after the impulse $\vec{\mathbf{F}}\Delta t$. But the right-hand side of the last equation can be writtten as:

$$m(\vec{\mathbf{v}}' - \vec{\mathbf{v}}) = m\vec{\mathbf{v}}' - m\vec{\mathbf{v}} = \vec{\mathbf{p}}' - \vec{\mathbf{p}} = \Delta\vec{\mathbf{p}},$$

the change in the momentum. Therefore

$$\vec{\mathbf{F}}\Delta t = \Delta\vec{\mathbf{p}};$$

or, in words, the impulse equals the change in the momentum. Then the average force which would produce the required impulse during Δt is just

$$\vec{\mathbf{F}}_{\text{av}} = \frac{\Delta\vec{\mathbf{p}}}{\Delta t}.$$

If the actual force is changing, we can find its instantaneous value at any time provided we know how the momentum varies with time. We simply take Δt smaller and smaller until we can find a value for $\frac{\Delta \vec{\mathbf{p}}}{\Delta t}$ which does not change much with further decreases in Δt. In other words, the instantaneous force is

$$\vec{\mathbf{F}} = \lim_{\Delta t \to 0} \frac{\Delta \vec{\mathbf{p}}}{\Delta t}.$$

The force at any time equals the rate of change of the momentum. It was in this form rather than in terms of $\vec{\mathbf{F}} = m\vec{\mathbf{a}}$ that Newton originally formulated his law.

For the discussion of the motion of a single object, both forms of Newton's law are equally convenient. The great importance of momentum will become apparent in the next sections where we deal with the motions of two objects which exert forces on each other.

14-3 Changes in Momentum when Two Bodies Interact

A boy and a man are standing next to each other on a smooth sheet of ice. The boy shoves the man and they both start to move, sliding away in opposite directions with the boy moving away somewhat faster than the man. If we perform such experiments, we find that whenever two people are at rest and one pushes the other, they go off in opposite directions. We also find that their speeds are inversely proportional to their masses. For example, if a 50-kg boy shoves an 80-kg man hard enough so that the man goes off at .25 m/sec, we find that the boy goes off in the opposite direction at .40 m/sec (Fig. 14–5).

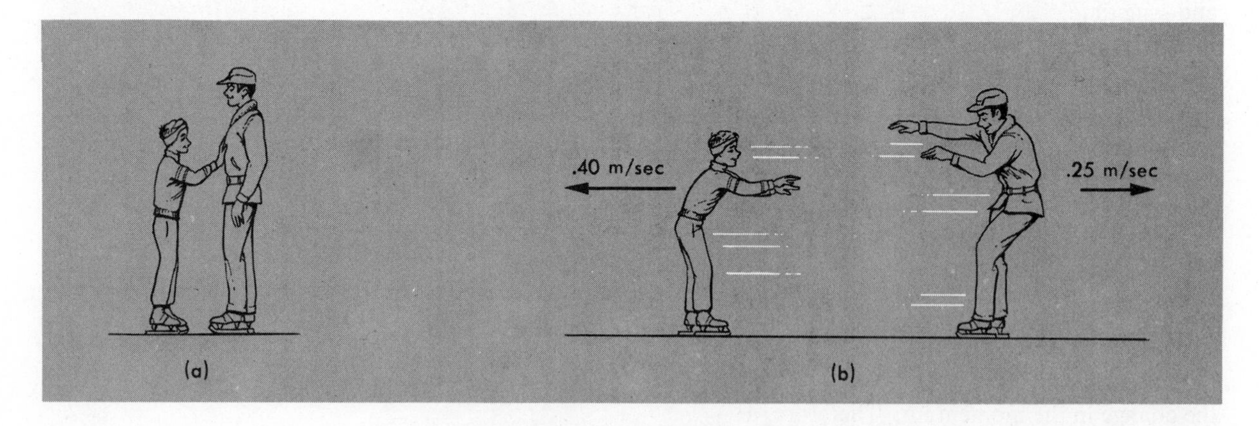

Figure 14–5
On a frozen pond a 50-kg boy pushes an 80-kg man hard enough to give him a speed of 0.25 m/sec. The boy moves off at 0.40 m/sec.

These experimental facts are most easily expressed in terms of the momentum of the man and the momentum of the boy. The momentum of the man is $mv = 80 \text{ kg} \times 0.25 \text{ m/sec} = 20 \text{ kg-m/sec}$, and the momentum of the boy in the opposite direction is $mv = 50 \text{ kg} \times 0.40 \text{ m/sec} = 20 \text{ kg-m/sec}$. After the shove their momenta are equal in magnitude and opposite in direction.

With Dry Ice pucks we can perform an accurate experiment of the kind that we have been describing. We fasten one end of a spring to the edge of one puck; we bend the spring and tie the ends together with a thread (Fig. 14–6). Then we put a second puck close to the first. When the two pucks are at rest on a level glass plate, we burn the thread and see what happens when the spring "explodes." The subsequent motion of the pucks is shown in Fig. 14–7.

Figure 14–6
The apparatus used to show a simple kind of explosion. A compressed spring is held between two Dry Ice pucks. The spring is fastened to the larger puck and held by a thread.

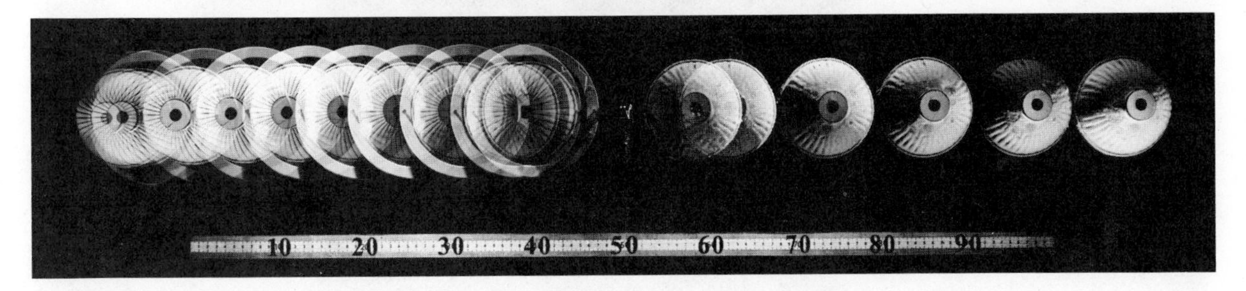

Figure 14–7
A multiple-flash picture of the motion of the two pucks after being pushed apart by the spring. The flash rate was 4 per second, and the scale is in centimeters. The photograph was taken from above.

In this experiment the mass of the large puck, including the spring mounted on it, was 3.9 kg, and the mass of the small puck was 2.0 kg. A glance at the picture shows that the two pucks went off in opposite directions. In each interval between flashes of the light the small puck moves farther than the large one. The small puck must have been given a higher velocity than the large one. By making measurements on the photograph you can tell that the small puck was moving at 0.48 m/sec after the shove, and the large puck was moving at 0.24 m/sec. The momentum of the small puck was $mv = 2.0 \text{ kg} \times 0.48 \text{ m/sec} = 0.96 \text{ kg-m/sec}$ to the right, while the momentum of the large puck was $mv = 3.9 \text{ kg} \times 0.24 \text{ m/sec} = 0.94 \text{ kg-m/sec}$ to the left. Within experimental accuracy these momenta are equal and opposite.

By observing the motions of two bodies starting from rest and interacting with each other, we find time after time that the resulting changes in momentum are equal and opposite. It does not matter what produces the forces of interaction. They may arise from our muscles, from a distorted spring, or from a chemical explosion.

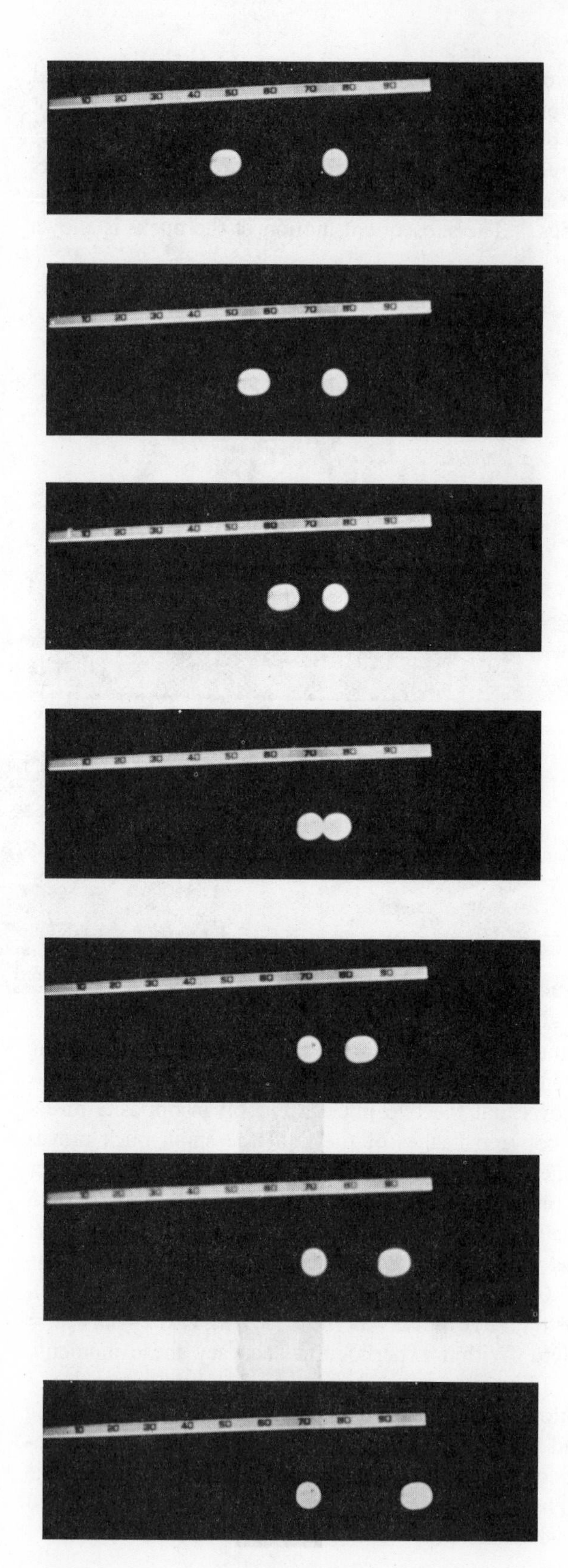

Figure 14–8
Enlarged movie frames showing a head-on collision between two billiard balls of equal mass. The upper pictures show the ball on the left moving toward the stationary ball. The moment of collision is shown in the fourth picture from the top. The time interval between pictures is 1/48 second; the scale is in centimeters. Notice that all of the momentum of the incident ball is transferred to the ball initially at rest.

Take a common chemical explosion as another example. When a man fires a rifle, the gases from the explosives exert violent forces inside the barrel. The bullet is pushed out in one direction and the rifle is pushed back in the other. We can show that the rifle's momentum and the bullet's momentum are equal and opposite: we suspend the rifle on long strings, take suitable high-speed flash photographs of the bullet, and time the very much slower motion of the recoiling rifle with a stop watch. In ordinary use the recoil velocity of the gun is rapidly decreased by an impulse from the man's shoulder, and a later measurement of its momentum does not show the equality and opposition so directly.

So far we have examined only cases in which both of the interacting bodies were originally at rest. What can we say about the changes in the momenta of two interacting bodies when one or both are originally in motion?

In Fig. 14–8 a moving billiard ball collides with a billiard ball at rest. The incident ball stops and the ball it hits goes off with the same velocity with which the incident ball came in. The two billiard balls have the same mass. Therefore, the momentum of the struck ball after the collision is the same as that of the incident ball before collision. The incident ball has lost all its momentum, and the ball it struck has gained exactly the momentum which the incident ball lost. The changes in momentum are again equal and opposite.

The collision illustrated in Fig. 14–8 is a very special one: the incident billiard ball hit the ball at rest head on. Usually one billiard ball hits another somewhere on the side, and the two balls go off in different directions. Such a collision is illustrated in Fig. 14–9. In the picture the ball originally in motion came in at the bottom. After the collision, as we should expect, one ball goes off to the right; the other goes off to the left.

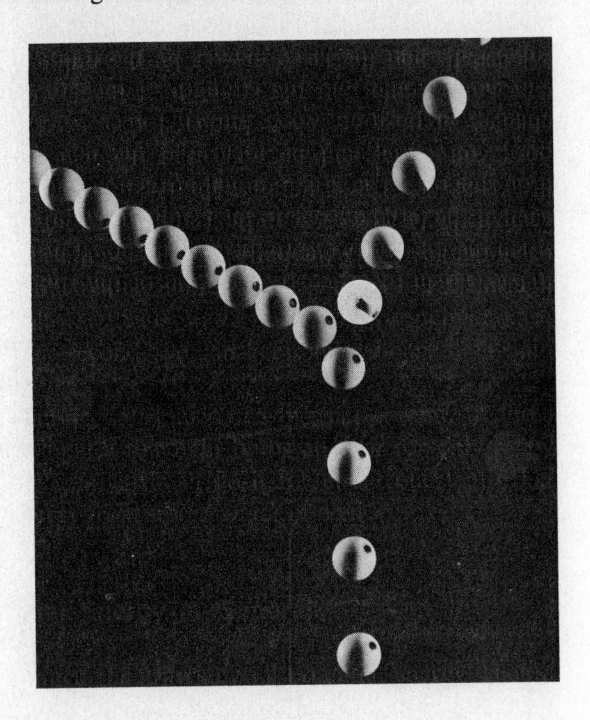

Figure 14–9
A multiple-flash photograph (flash rate 30 per second) of an off-center collision between two balls, each of mass 173 grams. The dotted ball entered at the bottom of the picture and struck the striped ball at rest. (The camera was pointed straight down, and the balls were moving horizontally.) To get the velocities and momenta diagramed in Fig. 14–10, we used the fact that the photograph is about 12% actual size. Could this approximation affect the conclusion that $\Delta\vec{\mathbf{p}}_1 = -\Delta\vec{\mathbf{p}}_2$?

The flash photograph shows the speeds and directions of the balls both before and after collision. Using these measured speeds and the observed directions of motion, we find the velocity vectors, $\vec{v}_1$, $\vec{v}_1'$, and $\vec{v}_2'$, representing the velocity of the first ball before and after the collision and that of the second ball after the collision. Then we obtain the momenta of the balls by multiplying the vector velocities by the masses: $\vec{p} = m\vec{v}$ for each ball. (Because the masses of the balls here are equal, using $\vec{p}$ instead of $\vec{v}$ just changes the scale in this case.) In Fig. 14–10 (a) we have plotted the momentum vectors $\vec{p}_1$, $\vec{p}_1'$ and $\vec{p}_2'$, representing the momentum of the incident ball before the collision, and the momentum of each ball after the collision.

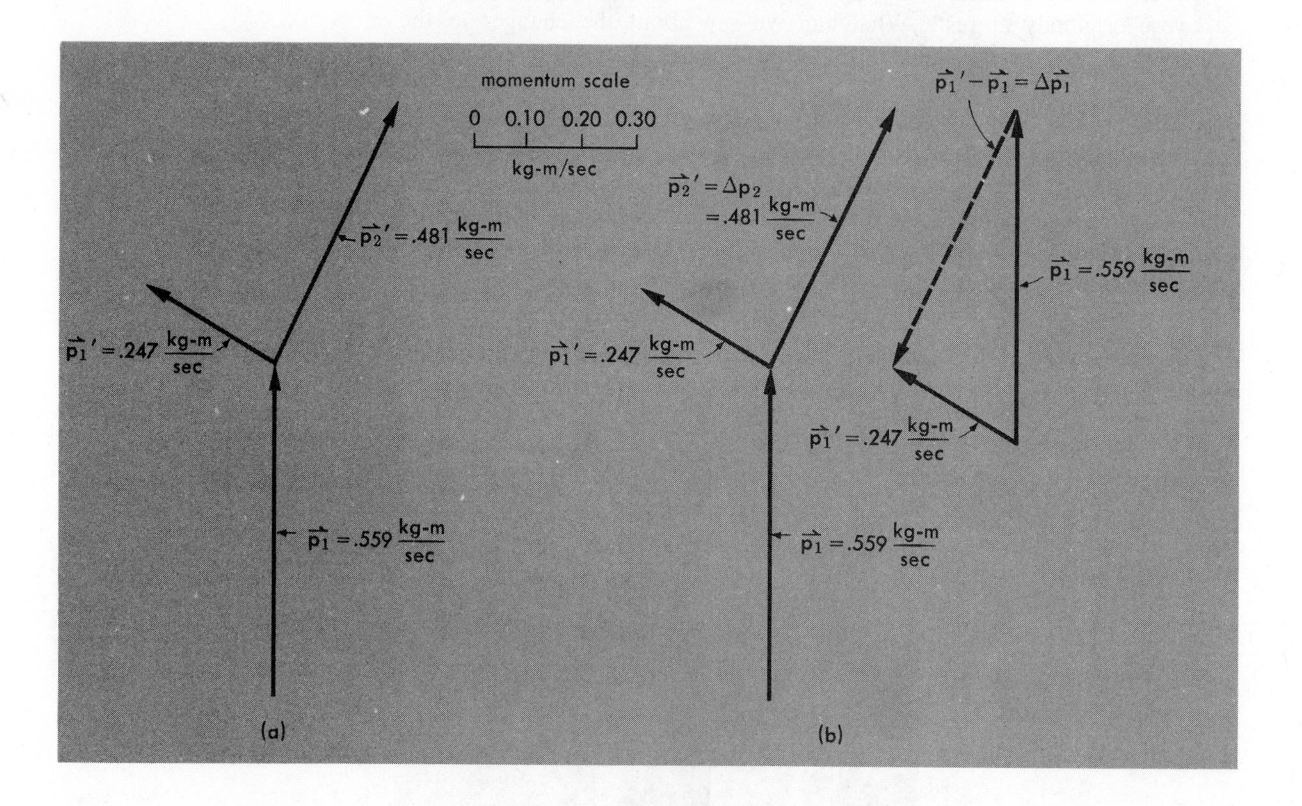

Figure 14–10
(a) The vectors representing the momenta of the balls in Fig. 14–9. The value $\vec{p}_1$ is the original momentum of the dotted ball; $\vec{p}_1'$, its momentum after the collision; and $\vec{p}_2'$ the momentum acquired by the ball originally at rest. In (b) we obtain $\Delta\vec{p}_1 = \vec{p}_1' - \vec{p}_1$ graphically. It is shown dashed in the figure. Note that $\Delta\vec{p}_1$ is closely equal and opposite to $\Delta\vec{p}_2 = \vec{p}_2'$.

Now we shall find out whether the momenta of the two balls change by equal amounts in opposite directions in this experiment.

In Fig. 14–10 (b), $\Delta\vec{p}_1$ is obtained graphically. We draw $\vec{p}_1$ so that its tail coincides with the tail of $\vec{p}_1'$. The vector $\Delta\vec{p}_1$ is the dashed line between the heads of $\vec{p}_1'$ and $\vec{p}_1$. We see that it is equal and opposite to $\Delta\vec{p}_2 = \vec{p}_2'$, the change in the momentum of the struck ball. The change in momentum of the incident ball is compensated by an equal and opposite change in the momentum of the other ball, or

$$\Delta\vec{p}_1 = -\Delta\vec{p}_2 .$$

In the last section we have seen examples in which two bodies interact. The motion of each body changes, and the change in momentum of one body is equal and opposite to the change in momentum of the other. As we go further we shall come across more and more examples. No single example can yield a general proof that the momenta of two interacting bodies always change equally and oppositely. But our experience with the interactions of two bodies, in all circumstances, strongly suggests that equal and opposite changes of the momenta are the invariable rule in nature.

This rule of nature can be stated in a different form. We introduce the total momentum $\vec{\mathbf{p}} = \vec{\mathbf{p}}_1 + \vec{\mathbf{p}}_2$ of the two bodies. Because any changes in the momenta $\vec{\mathbf{p}}_1$ and $\vec{\mathbf{p}}_2$ are just opposite, the total momentum $\vec{\mathbf{p}}$ never changes. That is, $\Delta\vec{\mathbf{p}} = 0$, or $\vec{\mathbf{p}}$ is constant. We call the statement that the total momentum is constant the law of conservation of momentum.

Let us look at the conservation of momentum when two balls of different masses, both initially in motion, collide. Such a collision is illustrated in Fig. 14–11. By measuring the distances that the balls have moved between

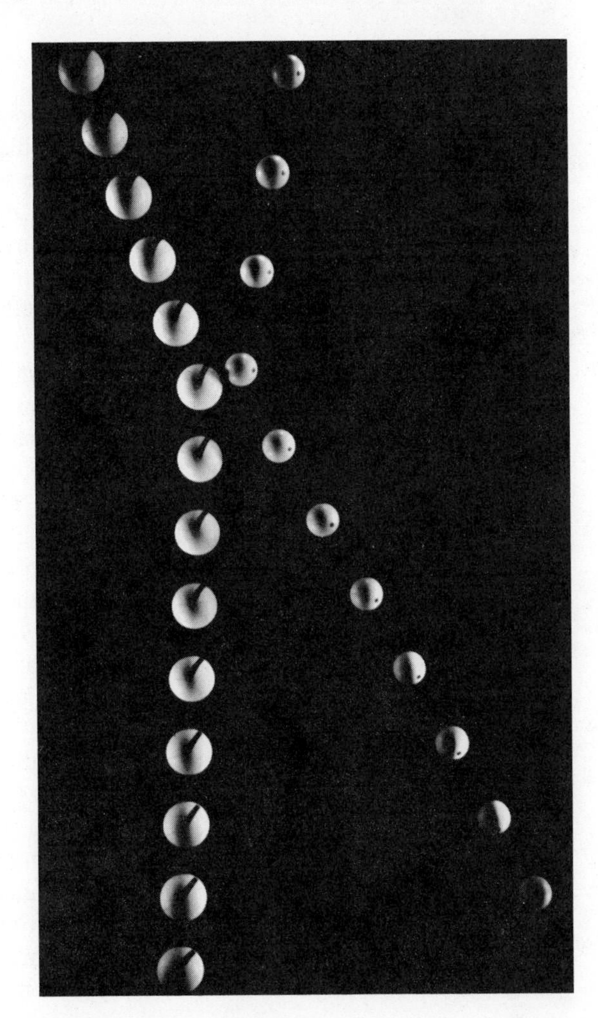

Figure 14–11
A multiple-flash photograph of a collision between two balls—one of mass 201.1 gm and the other 85.4 gm. Again the flash rate was 30 per second and the setup was the same as that described in Fig. 14–9. Both balls were initially in motion and entered from the top of the picture.

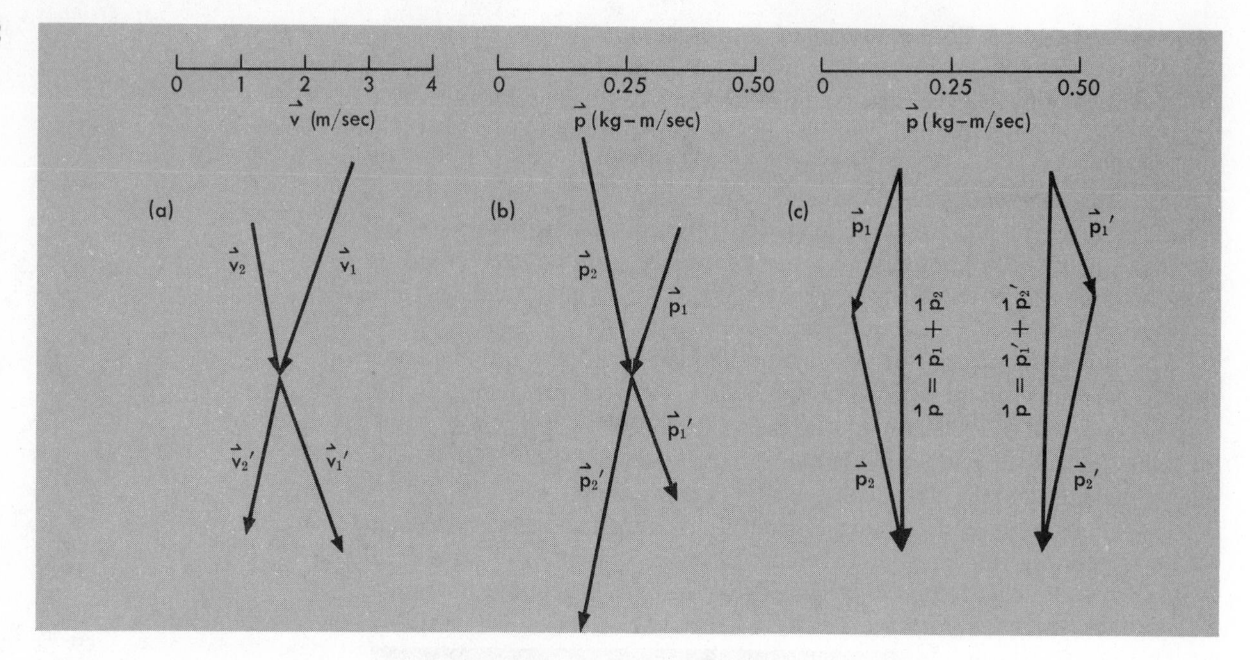

Figure 14–12
(a) The velocity vectors of the balls shown in Fig. 14–11. The subscript "$_1$" refers to the smaller ball, and "$_2$" to the larger. The primes indicate velocities and momenta after the collision. (b) The momentum vectors before and after the collision. (c) On the left we have added the momenta before the collision. On the right we have done the same for the momenta after. As you can see, $\vec{\mathbf{p}}_1 + \vec{\mathbf{p}}_2 = \vec{\mathbf{p}}_1{}' + \vec{\mathbf{p}}_2{}'$. In fact, when we made the original drawing, they were equal within about 1 part in 200.

flashes, we determine their velocities. The velocity vectors are indicated in Fig. 14–12 (a). Multiplying by the masses, which were measured as 85.4 grams for the first ball and 201.1 grams for the second, we obtain the momentum vectors for the individual balls before and after the collision [Fig. 14–12 (b)]. Then, on the left in Fig. 14–12 (a), we have added together the momentum vectors of both balls before the collision to obtain the total momentum; and on the right we have added together the momentum vectors of the two balls after the collision. You see that the two measured values of the total momentum are almost precisely the same. This experiment is consistent with the law of conservation of momentum.

Let us try one more kind of collision: one where the two colliding masses stick together and continue as a unit after the collision. Figure 14–13 shows a collision between a golf ball and a putty ball. The golf ball had a mass of 45.7 gm, and the putty ball, originally at rest, had a mass of 69.7 gm. The combined mass of the two balls stuck together after the collision was therefore 115.4 gm. This total mass was 2.53 times the mass of the incident ball. By measuring the ratio of the incident and final velocities in the flash photograph, you will find that the incident velocity was about 2.53 times the final velocity. Therefore, the final momentum was just equal to the initial momentum.

From now on we shall assume the conservation of momentum for two interacting bodies. We consider it to be a general law based on experience. Indeed, the kind of experiments which we have summarized in our discussion here is the kind that originally led to the conservation law. Because the laws of collisions were of such obvious importance for the development of mechanics, in 1668 the Royal Society of London (then only eight years old) initiated an investigation into the problem. Three able men produced almost similar solutions: John Wallis, an English mathematician; Sir Christo-

pher Wren, the English scientist and architect who designed St. Paul's Cathedral in London; and Christian Huygens, a Dutch physicist. Wallis was first, and so he gets the credit for the general principle of conservation of momentum.

Figure 14–13
A flash photograph of a collision between a golf ball and a putty ball. The golf ball moved in from the left and struck the putty ball. The two balls moved off to the right, stuck together. Measure the ratio of the incident and final speeds yourself, and check them against the calculation in the text.

In a paper published in 1669, and later in his book "Mechanica" (1670), Wallis developed fairly clear ideas of impulse and momentum and the connection between them. He believed that an impulse which started one body toward another would give the same momentum to the single body or to the two bodies combined after collision. He thus arrived at the conservation of momentum for one kind of collision. He also argued that conservation applied to other collisions. The French physicist Mariotte demonstrated these conclusions by a series of experiments on pendulums that could strike one another. Huygens made similar experiments, and other experiments in which hard spheres in close contact were struck by a similar sphere moving in a straight-line groove. In the "Principia" Newton described these and his own careful experiments. The law of conservation of momentum was established as one of the foundations of modern physics.

14-5 Rockets

We can use the law of conservation of momentum to understand the motion of a rocket.

First, imagine a man in a space suit in interplanetary space. There is no net force on him, so his velocity is constant. His mass is m, and in his hand he holds a small object of mass Δm, which is moving through space at the same velocity that he is. Now the man throws away the small mass Δm with a velocity $\vec{v}_e$ with respect to the frame of reference in which he was at rest before throwing the mass Δm. The momentum of this small mass changes by an amount $(\Delta m)\vec{v}_e$. According to the principles of conservation of momentum, the velocity $\vec{v}$ of the man undergoes a change in magnitude Δv in the opposite direction, and his momentum will change by an amount $m\Delta v$, just equal in magnitude to the momentum change $(\Delta m)v_e$ of the small object. Therefore,

$$m\Delta v = (\Delta m)v_e ,$$

and the man's speed changes by

$$\Delta v = \frac{\Delta m}{m} v_e .$$

This man is much like a spacecraft; he changes his momentum by throwing out mass. If he wants to increase his speed toward the east, he throws out mass to the west.

In a spacecraft the mass thrown out is in the form of a stream of gas. The gas is shot out from the spacecraft at a high exhaust velocity, and the momentum of the spacecraft changes in the opposite direction. We must be careful to remember that m is the mass of the spacecraft *after* the small Δm has been thrown out. It includes the fuel, the crew, the casing—everything left on board.

14-6 The Center of Mass

Think of a Fourth of July rocket which explodes while it is moving along through the air. After the explosion, if you saw all the fragments on one side of the path you would be sure that something was wrong. Your intuition leads you to expect that the fragments into which the rocket explodes should have an average motion along the original trajectory. In our investigation of such explosions we shall find that there is a special point that moves along at a constant velocity whether there is an explosion or not.

The fragments of an exploding skyrocket are a little too complicated for a first look at this effect. Let us go back to the simple explosion we have already discussed, the two Dry Ice pucks pushed apart by a compressed spring. By measuring the photograph (Fig. 14–7), we found that the two pucks, even though of quite different mass, had equal and opposite momenta after the explosion. If we call the puck on the left m_1 and the puck on the right m_2, we can write

$$m_1\vec{v}_1 = -m_2\vec{v}_2.$$

Hence the ratio of the speeds is inversely proportional to the ratio of the masses:

$$\frac{v_1}{v_2} = \frac{m_2}{m_1} .$$

Let us assume that the explosion is over in a very short time, so that we can think of the bodies as moving with constant speeds from the start. Then, after a time Δt, they will have moved distances of $x_1 = v_1\Delta t$ and $x_2 = v_2\Delta t$ from the point of the explosion. These two distances are directly proportional to the speeds, and hence are inversely proportional to the masses of the two bodies:

$$\frac{x_1}{x_2} = \frac{v_1}{v_2} = \frac{m_2}{m_1} .$$

At all times during the motion, therefore, the starting point divides the distance between the two bodies in inverse proportion to their masses. We shall call the point which divides the distance between two bodies in the inverse ratio of their masses the *center of mass* of the two bodies. The starting point, in this case, is the center of mass.

Let us investigate the behavior of the center of mass in some more complex interactions between bodies—the collisions we have discussed in this chapter. In Fig. 14–14 we again see the collision between a golf ball and

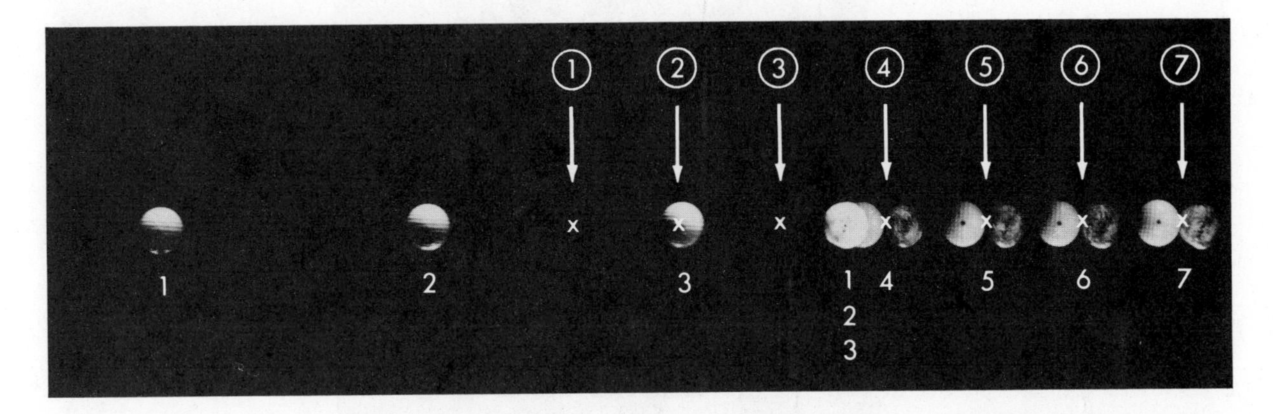

Figure 14–14
The collision of Fig. 14–13 with the positions of the center of mass marked and numbered to correspond to the positions of the two balls at the time of each flash. The center of mass is located from $m_1x_1 = m_2x_2$, using the data of the text. You can see that the center of mass moves along at the same constant velocity before and after the collision.

a putty ball shown in Fig. 14–13 and analyzed in the last section. The masses were given as $m_1 = 45.7$ gm for the golf ball and $m_2 = 69.7$ gm for the putty ball. From the definition of the position of the center of mass we find

$$\frac{x_1}{x_2} = \frac{69.7}{45.7} \simeq \frac{3}{2}.$$

That is, the center of mass divides the distance between the balls at any instant in the ratio of about 3 to 2. The positions of the center of mass at each flash are marked by arrows on the photograph and are numbered to correspond with the numbered positions of the balls. You can measure the velocity of the center of mass and see that it remains constant, just as if no collision had occurred. In this case, where the two balls stick together, the velocity of the center of mass $\vec{\mathbf{v}}_c$ is just the final velocity of the balls, $\vec{\mathbf{v}}'$. By the law of conservation of momentum, the total mass $(m_1 + m_2)$ now has the same total momentum $\vec{\mathbf{p}}'$ as the incoming ball had before:

$$\vec{\mathbf{p}}' = (m_1 + m_2)\vec{\mathbf{v}}' = m_1\vec{\mathbf{v}}_1.$$

But since $\vec{\mathbf{v}}' = \vec{\mathbf{v}}_c$, we can also write

$$(m_1 + m_2)\vec{\mathbf{v}}_c = m_1\vec{\mathbf{v}}_1.$$

Thus the total mass times the velocity of the center of mass gives the total momentum.

The center of mass of two interacting bodies behaves that way even when the bodies do not stick together. The center of mass always moves as if all the mass were concentrated there. Consider the collision shown in

Fig. 14–15. This is just the collision shown in Fig. 14–11 between two balls of masses 201.1 gm and 85.4 gm. Again we have marked the position of the center of mass for each flash. Even in this more complicated case, the center of mass continues right through the collision with constant velocity.

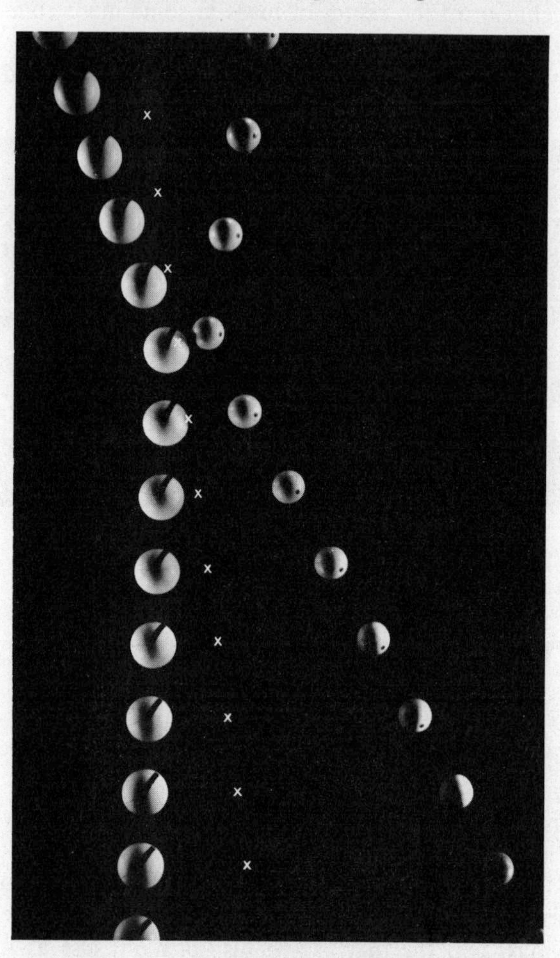

Figure 14–15
The collision of Fig. 14–11, showing the position of the center of mass at the time of each flash. The two balls had a mass ratio of about 7 to 3, so the distances between each ball and the center of mass are in the ratio of 3 to 7. Again the center of mass moves at constant velocity, quite undisturbed by the collision.

14-7 The Center-of-Mass Frame of Reference

Clearly, the center of mass is a very special point. Let us move with that point, wherever it goes, and observe the motions of the two interacting balls, m_1 and m_2, in the frame of reference in which the center of mass is at rest. To get the velocities of the balls relative to the center of mass, we must subtract the velocity of the center of mass from the velocities of each of the balls. These velocities can be measured on Fig. 14–15 and are shown in Fig. 14–16 (a). As seen from the center of mass, the velocities $\vec{\mathbf{V}}_1$ and $\vec{\mathbf{V}}_2$ of the balls before collision are exactly opposite in direction and their magnitudes are inversely proportional to the masses. The same is true of the velocities $\vec{\mathbf{V}}_1'$ and $\vec{\mathbf{V}}_2'$ after the collision. Multiplying each of the velocities by the corresponding mass, we get the momentum diagram shown in Fig. 14–16 (b). It is clear that the momenta of m_1 and m_2 are equal and op-

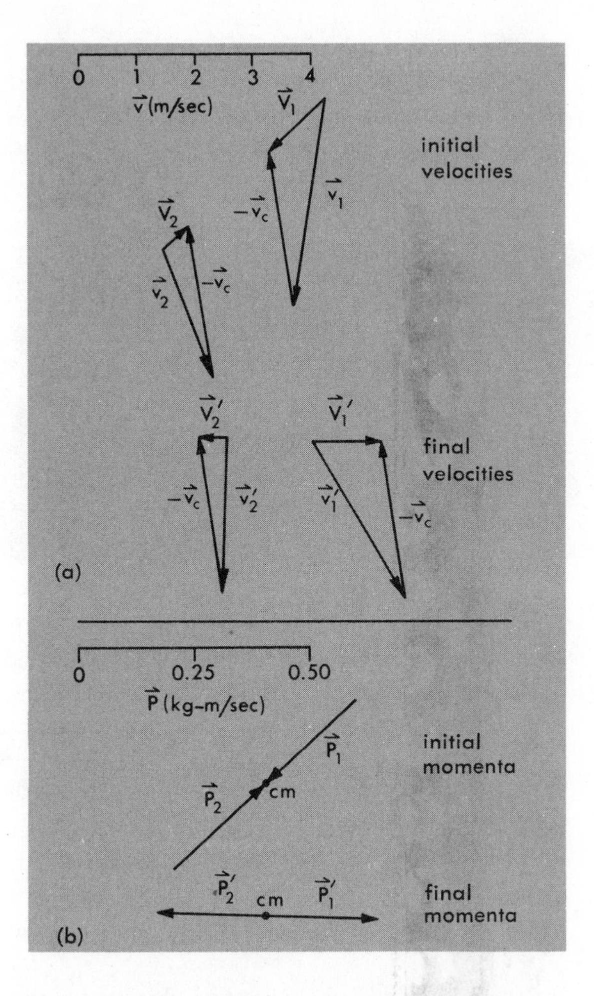

Figure 14–16
(a) The velocity vectors of the balls and of the center of mass, measured from Fig. 14–15 and plotted. The subscript "$_1$" refers to the smaller ball, and "$_2$" refers to the larger. The primes indicate velocities after the collision. The vectors marked with capital letters are the velocities of the balls from the frame of reference of the center of mass. They are found by subtracting the velocity of the center of mass from the velocities of the balls. They are opposite in direction and the ratio of their magnitudes is just the inverse of the ratio of the masses of the corresponding balls.
(b) The products mV, the momenta relative to the center of mass, are shown. They are equal in magnitude and opposite in direction, both before and after the collision. The total momentum relative to the center of mass is always zero.

posite: the total momentum of the balls, in the center-of-mass frame of reference, is always zero.

Now let us get off the center-of-mass frame of reference and return to that of the earth. The velocity $\vec{\mathbf{v}}_1$ of m_1, seen by an observer on earth (for example, the camera used to take the photograph), is made of the velocity $\vec{\mathbf{v}}_c$ of the center of mass plus the velocity $\vec{\mathbf{V}}_1$ of m_1 with respect to the center of mass:

$$\vec{\mathbf{v}}_1 = \vec{\mathbf{v}}_c + \vec{\mathbf{V}}_1.$$

Similarly,

$$\vec{\mathbf{v}}_2 = \vec{\mathbf{v}}_c + \vec{\mathbf{V}}_2.$$

Therefore the total momentum of m_1 and m_2 with respect to the earth is

$$\vec{\mathbf{p}} = m_1\vec{\mathbf{v}}_1 + m_2\vec{\mathbf{v}}_2 = m_1(\vec{\mathbf{v}}_c + \vec{\mathbf{V}}_1) + m_2(\vec{\mathbf{v}}_c + \vec{\mathbf{V}}_2)$$

or

$$\vec{\mathbf{p}} = (m_1 + m_2)\vec{\mathbf{v}}_c + m_1\vec{\mathbf{V}}_1 + m_2\vec{\mathbf{V}}_2.$$

But by riding with the center of mass, we found that $m_1\vec{\mathbf{V}}_1 + m_2\vec{\mathbf{V}}_2$ is always zero. So finally we have

$$\vec{\mathbf{p}} = (m_1 + m_2)\vec{\mathbf{v}}_c.$$

Once again, we find that the total momentum is the same as if the total mass moved with the center of mass.

The parts of a body may move with respect to each other. They may move around the center of mass or they may move in and out, toward and away from the center of mass. No matter what these internal motions, the total momentum of the body is given by its mass times the velocity of the center of mass (Fig. 14–17).

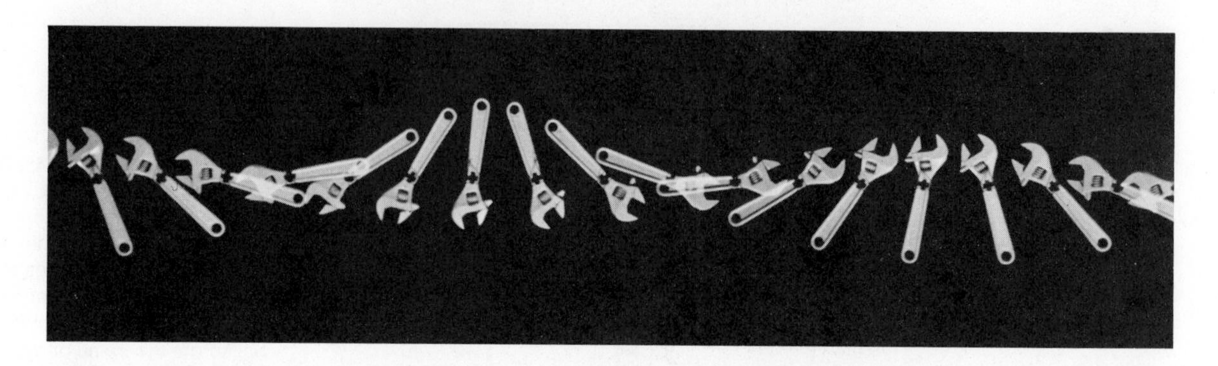

Figure 14–17
A multiple-flash photograph (1/30 second between flashes) of a moving wrench. The black cross marks its center of mass.

When the two masses interact only with each other, we know that the total momentum is constant. So we also conclude that the velocity of the center of mass never changes. The center of mass acts like a single body with no net force on it. It moves in accord with Galileo's principle of inertia. Since there is no force acting from the outside—only the interaction of m_1 with m_2—it is pleasing to find that the two interacting masses do act like a single body in this respect.

14-8 Momentum Conservation and Newton's Third Law

In every interaction between two objects that we have studied, momentum is conserved; what one mass loses the other gains: $\Delta\vec{\mathbf{p}}_1 = -\Delta\vec{\mathbf{p}}_2$. Furthermore if we indicate the force that m_1 exerts on m_2 by $\vec{\mathbf{F}}_{1,2}$ and the force that m_2 exerts on m_1 by $\vec{\mathbf{F}}_{2,1}$ then $\Delta\vec{\mathbf{p}}_1 = \vec{\mathbf{F}}_{2,1}\Delta t$ and $\Delta\vec{\mathbf{p}}_2 = \vec{\mathbf{F}}_{1,2}\Delta t$.

The Δt must be the same for both masses suffering the collision. If I push you, you automatically push me back and I cannot push you for a longer time than you push me. Thus,

$$\Delta\vec{\mathbf{p}}_1 = -\Delta\vec{\mathbf{p}}_2$$

becomes

$$\vec{\mathbf{F}}_{2,1}\Delta t = -\vec{\mathbf{F}}_{1,2}\Delta t,$$

and that now leads to

$$\vec{\mathbf{F}}_{2,1} = -\vec{\mathbf{F}}_{1,2}.$$

The idea that forces of interaction must be equal and opposite was stated by Newton (after his experiments on momentum conservation): "To every action there is always opposed an equal reaction, or the mutual reactions of two bodies upon each other are always equal and directed to contrary directions." This statement is often called his third law of motion, although you can see that it is a law about the forces in interactions and that the motions are inferred from it by applying Newton's law of motion to each body.

Even before we studied momentum, we studied one force of interaction with some care. In the last chapter we saw that the gravitational attraction between two bodies applies equal forces to each of the bodies, and the forces are directed oppositely along the line between the centers. The pull of the sun on the earth is equal and opposite to the pull of the earth on the sun. The pull of the earth on the moon is opposite to the pull of the moon on the earth, and so on. Later in this volume we shall often study what goes on when the forces are of this particularly simple kind. We shall call them Newtonian forces.

Now let us trust Newton's third law to relate the forces between two bodies even when many bodies interact. Then we can use it to give an alternative demonstration of "universal" momentum conservation. We have a number of masses moving about in any way in an isolated system, exerting various forces one on another. Since all the forces occur in pairs with $\vec{\mathbf{F}}_{2,1} = -\vec{\mathbf{F}}_{1,2}$, their effects on the momenta of the bodies also occur in equal and opposite pairs. Therefore, while one body gains some momentum, another loses an equal and opposite amount. And in the whole isolated system there can be no net change of momentum.

If Newton's third law is true, momentum conservation follows. But it is possible that momentum may be conserved for many bodies even if some of the forces are not Newtonian.

14-9 Light and the Conservation of Momentum

Every once in a while a tremendous bright light flares up somewhere out in space—a supernova. The light gets dimmer, falling off by the factor ½ every 55 nights. Compared to the millions of years light takes to reach us from a supernova, the light takes only a short period to die down. The supernova has burned out long before the light gets here.

When the light arrives here from a supernova, it gives us a slight shove. But nobody would think that we give the dead supernova a countershove at the same moment. If we think of forces of interaction between us and the supernova at the same moment, there is no reason to expect them to be equal and opposite.

The conservation of momentum is all right, however. What we must do is to include the light in our description of the universe. When the light was emitted, the star recoiled, gaining momentum, while the light carried away an equal and opposite amount of momentum. Some of the light took off toward us and when it arrived here, millions of years later, it was absorbed and its momentum was transferred to us.

This description may not seem so farfetched if we return to the people sliding around on the ice. This time we imagine them playing ball. The man throws the ball to the boy. When he throws it, he gives it momentum in one direction and he receives momentum in the opposite direction. Later on, when the boy catches the ball, the momentum is transferred from the ball to him. (Of course the mass of the ball is included with him after he catches it.) The changes in the momenta of the man and of the boy with the ball are equal and opposite. And the interaction between the man and the boy occurs with a time delay. The time delay is considerably shorter than that which occurs between the emission of light by a supernova and its absorption at the surface of the earth; nevertheless, it is an appreciable and measurable time delay. If we forget about the ball, the interaction between the man and the boy does not appear Newtonian; and momentum seems to be lost for a while, when the ball is in the air. But everything makes sense when we include the ball in the system; and in just the same way everything makes sense in the case of the earth and the supernova when we include the light.

Looking at light as a ball thrown across from man to boy helps to make the time delay in the interaction less mysterious; and it gives a sense of assurance that momentum is still conserved in the overall interaction. The ball-throwing scheme is a "model" for the transfer of light—a model in which we can use Newtonian interactions. In fact, light is not a ball—balls do not appear when thrown and disappear when stopped, but light does. So we should not take the model too literally.

By examining the emission and absorption of light (as we shall do to some extent later on), we shall find that the idea that momentum is carried by light makes a consistent picture. We can, and indeed we must, extend the law of conservation of momentum beyond those interactions in which the momentum is solely carried by material objects. We must include the momentum traveling around in radiation. The inequality in the forces between earth and supernova occurs while momentum is moving in radiation. There are many other similar examples. The same sort of thing happens when radiation is emitted by one atom and absorbed by another. When the momentum in radiation is included, the conservation of momentum spans our universe.

For Home, Desk, and Lab

1.* You push a body with a force of 3 newtons for ½ sec. What impulse do you give the body? (Section 1.)

2.* How great is the impulse that gives an 8.00-kg mass a change in velocity of 4.00 m/sec? Section 1.)

3.* Suppose you throw a ball against a wall and catch it on the rebound.
   (a) How many impulses were applied to the ball?
   (b) Which impulse was the greatest? (Section 1.)

4. A 3-kg body has been accelerated by a constant force of 12 newtons from 10 m/sec to 18 m/sec.
   (a) What impulse was given to the body?
   (b) How long was the force acting?

5. A constant force applied to a 2.0-kg object at rest moves it 4.0 m in 2.0 sec. What impulse was applied to the object?

6. A railroad freight car with a mass of 5.0×10^4 kg is rolling along a level track at 0.30 m/sec (0.7 mi/hr). A rope trails behind it.
   (a) A reasonable estimate of the largest

force you could apply to stop the car by pulling on the rope is 250 newtons. How long would it take you to bring the car to rest?

(b) Ten meters from the point where you start pulling, another car is standing. Will there be a crash?

7.* What average force is necessary to stop a hammer with 25 newton-sec momentum in 0.05 sec? (Section 2.)

8. A man is at rest in the middle of a pond on perfectly frictionless ice. How can he get himself to shore?

9. An object with a mass of 10 kg moves along a straight line at a constant velocity of 10 m/sec. A constant force then acts on the object for 4.0 sec, giving it a velocity of −2 m/sec.

(a) Calculate the impulse acting on the object.

(b) What is the magnitude and direction of the force?

(c) What is the momentum of the object before and after the force acts?

10.* An impulse is applied to a moving object at an angle of 120° with respect to its velocity vector. What is the angle between the impulse vector $\vec{\mathbf{F}}\Delta t$ and the change in momentum vector $\Delta\vec{\mathbf{p}}$? (Section 2.)

11.* In Fig. 14–12, how is diagram (b) constructed from diagram (a)? (Section 4.)

12.* Two ice pucks initially at rest were driven apart by an explosion with velocities 5 cm/sec and − 2 cm/sec. What is the ratio of their masses? (Section 4.)

13. A 20-kg cart is moving with a velocity of 2.0 m/sec. A boy whose mass is 60 kg jumps off the cart. When he hits the ground, he is

(a) moving at the same velocity as the cart

(b) not moving relative to the ground

(c) moving with twice the initial velocity of the cart.

In each case, what is the change in velocity of the cart?

14. Two heavy frictionless carts are at rest. They are held together by a loop of string. A light spring is compressed between them (Fig. 14–18). When the string is burned, the spring expands from 2.0 cm to 3.0 cm, and the carts move apart. Both hit the bumpers fixed to the table at the same instant, but cart *A* moved 0.45 meter while cart *B* moved 0.87 meter. What is the ratio of:

(a) the speed of *A* to that of *B* after the interaction?

(b) their masses?

(c) the impulses applied to the carts?

(d) the accelerations of the carts while the spring pushes them apart?

15. A proton (mass 1.67×10^{-27} kg) with a speed of 1×10^{7} m/sec collides with a motionless helium nucleus, and the proton bounces back with a speed of 6×10^{6} m/sec. The helium nucleus moves forward with a speed of 4×10^{6} m/sec after the bombardment.

(a) Can you compute the mass of the helium nucleus? If so, what is it?

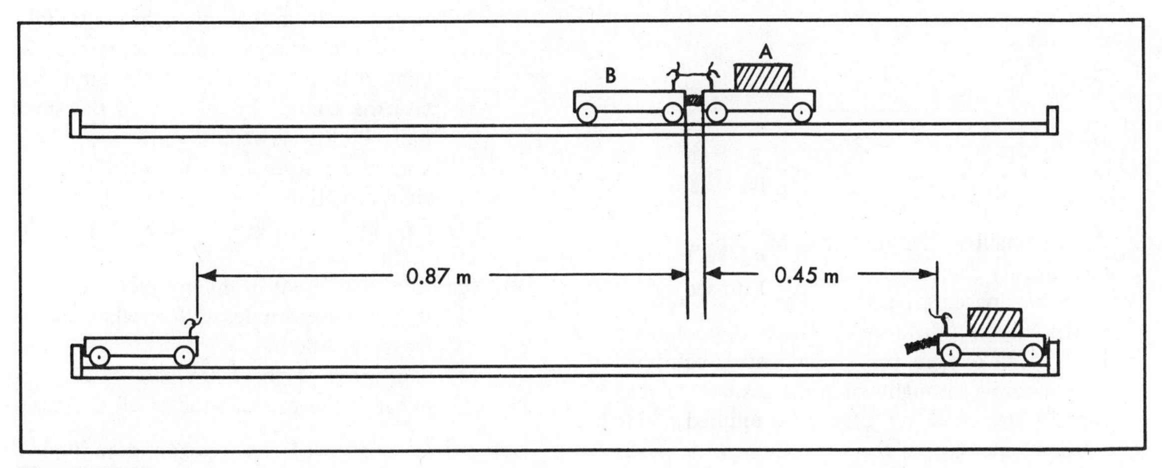

Figure 14–18
For Problem 14.

(b) Can you compute the force that acted during the collision? If so, what is it?
(c) If you answered "no" to either (a) or (b), be prepared to discuss in class why you gave this answer.

16. A stationary refrigerator car with mass 2.0×10^4 kg is rammed by a loaded gondola car with mass 3.0×10^4 kg. Before impact, the gondola car was going 1.0 m/sec. If they lock together, what is the new velocity?

17. An experimental rocket sled is slowed down by a scoop which dips into a trough of water. As shown in Fig. 14–19, the scoop is designed so that it ejects the water at right angles to the moving sled, and so that the water leaves the tubes with a speed equal to the speed of the sled at any instant.
(a) What will be the change in momentum of a small mass of water m ejected toward either side at the instant the sled is traveling with a speed v?
(b) What is the change in momentum of the sled for each mass of water m scooped up?
(c) Is any advantage gained by ejecting half of the water from each side? Why not eject all of it from one side?

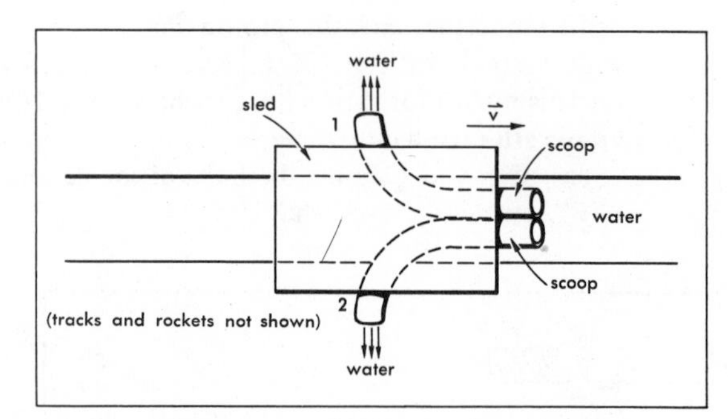

Figure 14–19
For Problem 17.

18. An astronaut with a total mass M (including his equipment) has been accidentally separated from his spacecraft and is at rest with respect to the ship at a distance d. His tank contains a mass m_o of oxygen ($m_o \ll M$) and is equipped with a nozzle through which the gas can be very rapidly ejected at an average speed of v. He must release oxygen for propulsion to get back to the spacecraft but also needs oxygen for breathing. He breathes oxygen at a rate R.
(a) If he releases an amount m of oxygen for propulsion, what speed V will he acquire? How long will it take him to reach the spacecraft?
(b) How long can he breathe on the remaining oxygen?
(c) If he is to return successfully to his craft, his breathing time, t_B, must be equal to (or greater than) his travel time t_T. What condition does this place on m?
(d) Calculate how much oxygen the astronaut may safely release for propulsion if $M = 100$ kg, $m_o = 0.5$ kg, $d = 45$ m, $v = 50$ m/sec, and $R = 2.5 \times 10^{-4}$ kg/sec.

19.* What is the direction of motion of the center of mass of the balls in Fig. 14–9 after the collision? (Section 6.)

20.† While two 1-kg carts are moving along together with a velocity of 0.5 m/sec, a spring pushes them apart. One of them continues with a velocity of 0.7 m/sec, with its direction of motion unchanged.
(a) What is the velocity of the center of mass after the spring acts?
(b) What is the velocity of the other cart after it is pushed by the spring?

21. A rocket is out in free space shooting out a stream of exhaust gases and picking up speed in the opposite direction. What happens to the center of mass of all the matter—that which is ejected and that which is left in the rocket?

22. In Fig. 14–20, the large ball came in at the top of the picture and the little ball at the bottom. As you see, a collision took place in the middle.
(a) Draw the vectors which represent the change in the velocity of the large ball and the change in velocity of the small ball. Plot these vectors to the same scale and make sure that each one is in the right direction.
(b) Are these changes of velocity opposite in direction?
(c) Are they equal in magnitude?
(d) If their magnitudes differ, what should be their ratio?
(e) The mass of the large ball is 201 gm. What is the mass of the small ball?

23. A 2.0-kg brick with no horizontal motion is dropped on a 2.0-kg cart moving across a frictionless table at 0.40 meter/sec.

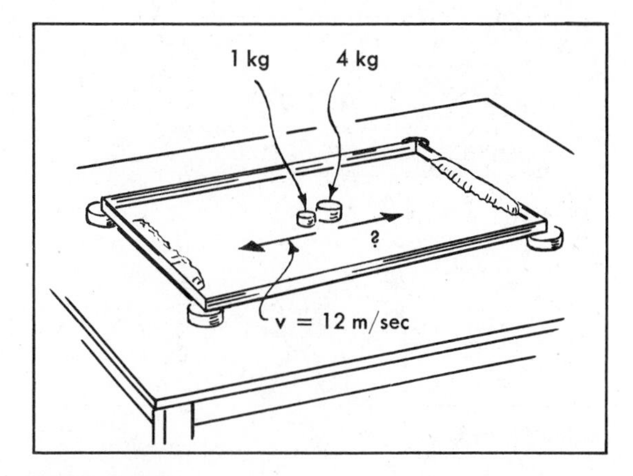

Figure 14–21
For Problem 25.

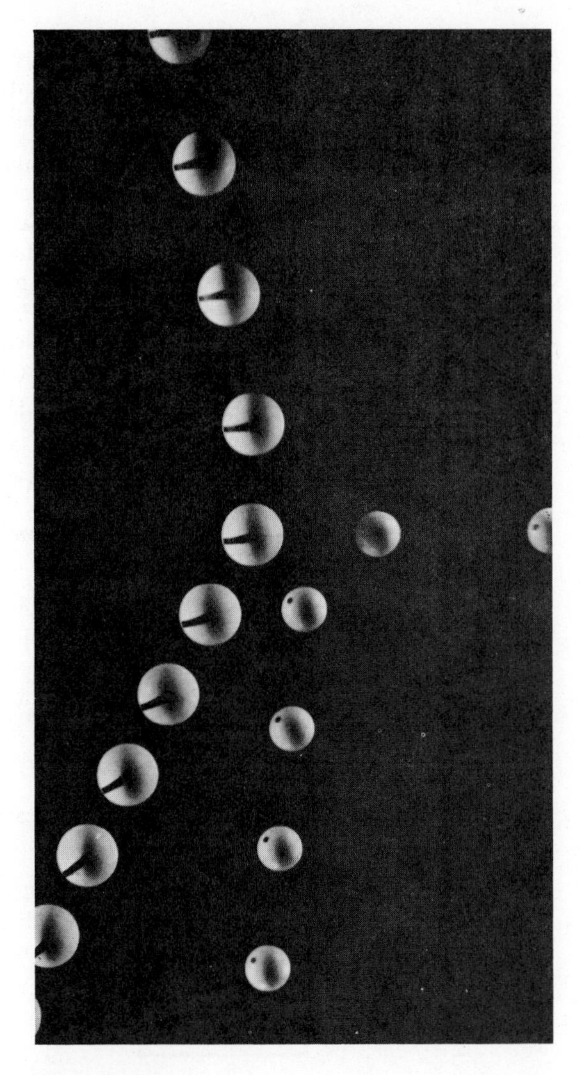

Figure 14–20
For Problem 22.

(a) What is the change in velocity of the cart?
(b) What is the velocity of the center of mass of the system composed of cart and brick before the brick is dropped?

24. An explosion blows a rock into three parts. Two pieces go off at right angles to each other, a 1.0-kg piece at 12 m/sec and a 2.0-kg piece at 8.0 m/sec. The third piece flies off at 40 m/sec.
(a) Draw a diagram to show the direction in which the third piece goes.
(b) What is its mass?

25. The system shown in Fig. 14–21 consists of a 5.0-kg frame with a 1.0-kg and a 4.0-kg mass in the middle. At either end of the frame are stops made of putty. The two masses and the frame are mounted on Dry Ice pucks. (The mass of the Dry Ice is included in the values given above.) The center of mass of the whole system is at the center of the frame. An explosion pushes the two masses apart. The 1.0-kg mass goes off with a speed of 12 m/sec. Eventually each mass is trapped by the putty.
(a) What is the speed of the 4.0-kg mass immediately after the explosion?
(b) Where is the center of mass of the whole system after 1 sec? After 2 sec?
(c) What is the velocity of the frame and the attached masses after 100 sec?
(d) Describe the motion qualitatively from the time of explosion to 100 sec.
(e) How far does the frame move?

26. A 20-kg car stands at rest on an 80-kg platform as shown in Fig. 14–22. The car can be driven along the platform by a motor and the platform, mounted on rollers, is free to move along the laboratory floor substantially without friction. A stroboscopic camera records any motions that occur.

When the motor is turned on, a study of the photograph shows that over a period of 3.0 sec the platform acquired a velocity of 0.30 m/sec.
(a) What does the same photograph show for the velocity of the car at the end of the 3.0-sec period?
(b) With what force did the wheels of the car push against the platform?
(c) What was the relative velocity of the car

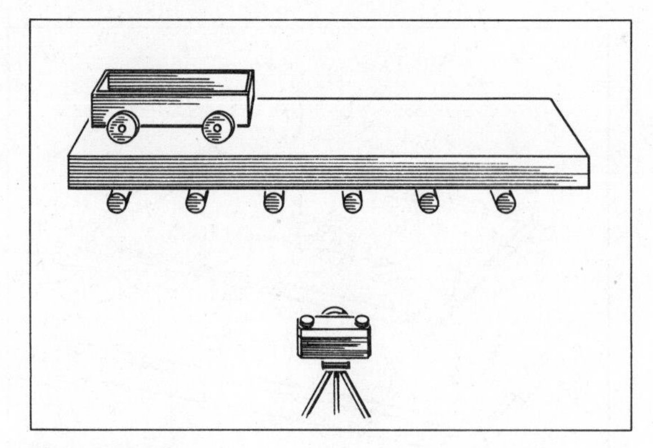

Figure 14–22
For Problem 26.

along the platform at the end of the 3.0 sec?

(d) If you know only the relative motion of the car along the platform, what force would you calculate was needed to give the car this velocity in 3.0 sec? Can you explain why the force you calculate here is different from the one you found in Part (b)?

27.* What happens to the momentum of a car when it comes to a stop? (Section 8.)

28. A double star consists of two large masses that attract each other gravitationally. By observing the motion of both masses, we can see that they rotate around each other.

(a) What do you think happens to the momentum of each of the masses in a double star as time goes on? Explain your answer.

(b) When observed carefully, the bright star Sirius seems to wobble about, instead of having a uniform motion of the center of mass. From this and other evidence astronomers believe that Sirius has a dark companion. It is really a double star. How does this explain the peculiar observed motion?

29.* Consider a collision between two particles in which light is emitted in a given direction. Since light exerts a pressure, it must have momentum. If the overall momentum (including the momentum of the light) is conserved, are the forces of interaction between the two particles equal and opposite?

30. Figure 14–23 shows the hypothetical force-time graphs between two bodies; $F_{1,2}$ and $F_{2,1}$ are the forces with which the two bodies act on each other.

(a) Is the total final momentum different from the total initial momentum?

(b) Is momentum conserved throughout the collision?

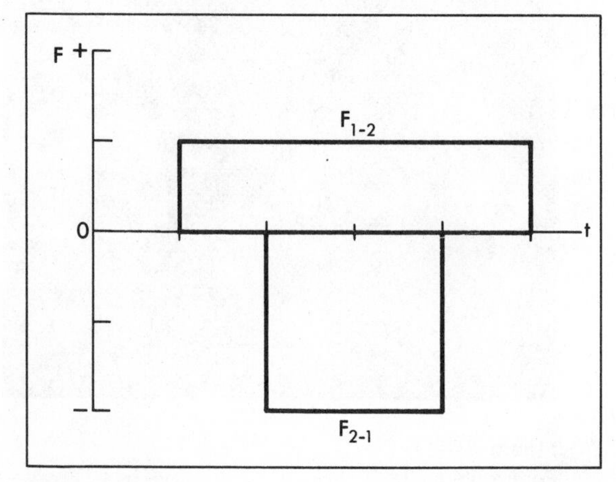

Figure 14–23
For Problem 30.

Further Reading

These suggestions are not exhaustive but are limited to works that have been found especially useful and at the same time generally available.

BONDI, HERMANN, *Relativity and Common Sense.* Doubleday Anchor, 1964: Science Study Series. (Chapters 1, 2, 3, and 4)

HOLTON, G., and ROLLER, D. H. D., *Foundations of Modern Physical Science.* Addison-Wesley, 1958. Conservation of momentum. (Chapter 17)

Kinetic Energy

CHAPTER

15

15-1 Work and Kinetic Energy

At the beginning of the preceding chapter we saw that when a constant force $\vec{\mathbf{F}}$ acts on a body for a time Δt, it causes a change in velocity given by

$$\vec{\mathbf{F}}\Delta t = m\Delta\vec{\mathbf{v}}.$$

If the body moves initially in the direction of the force, the velocity will remain in the direction of the force, making this a motion along a straight line. For such motion we do not need to bother with vector notation:

$$F\Delta t = m\Delta v.$$

Now, during the time Δt the body covered some distance Δx. Thus, in addition to saying that the force acted on the body for a time Δt, we can also say that it acted over a distance Δx. How, then, is the product $F\Delta x$ related to the mass of the body m and its velocity?

To answer this question, recall that a constant force produces a constant acceleration $a = \dfrac{F}{m}$. For constant acceleration we found in Section 9–7 that

$$v'^2 - v^2 = 2a\Delta x$$

or

$$a\Delta x = \frac{v'^2}{2} - \frac{v^2}{2}$$

where v is the initial velocity and therefore

$$ma\Delta x = \frac{mv'^2}{2} - \frac{mv^2}{2}$$

or

$$F\Delta x = \frac{mv'^2}{2} - \frac{mv^2}{2}.$$

Compare this relation with the impulse-momentum relation

$$F\Delta t = mv' - mv$$

The left-hand side depends only on the product of the force and the time interval through which it acted. Doubling the force and halving the time interval leaves the impulse unchanged. The right-hand side depends only on the mass and the instantaneous velocity of the body at the beginning and at the end of the time interval Δt, and not on what went on during the time interval.

The equation

$$F\Delta x = \frac{mv'^2}{2} - \frac{mv^2}{2}$$

has a similar character. On the left-hand side is the product of force and displacement. The product remains unchanged if the force is increased by some factor and the displacement is reduced by the same factor. On the right-hand side we have the change in the quantity $\dfrac{mv^2}{2}$ between its value at the beginning and the end of the displacement Δx. As you will soon see, the product $F\Delta x$ and the quantity $\frac{1}{2}mv^2$ are used repeatedly in our study of physics. They have been given the names *work* and *kinetic energy* respectively. We shall denote the kinetic energy by E_K. In words then, the work done by a force on a body along a given displacement equals the change in the body's kinetic energy. If the body was initially at rest, i.e., $v = 0$, then the work done equals the kinetic energy of the body.

The unit of work is a unit of force times a unit of distance, or 1 newton × 1 meter = 1 *newton-meter.* This unit is called a joule after the English physicist James P. Joule (1818–1889). For example, the work done when a force of 20 newtons pushes a body in the direction of motion for 3 meters is 60 joules.

Since the work equals the change in kinetic energy, we shall express kinetic energy also in joules. For example, a body of mass 10 kg and moving at a speed of 3 m/sec has a kinetic energy of $\frac{1}{2} \times 10 \times 3^2 = 45$ joules.

15-2 Work: A Generalization

In the last section we introduced the work done by a constant force on a body moving in the same direction as the force and the kinetic energy of the body. Now we shall extend this relation in two ways. First, what is the work done by a force when the force and the displacement are not parallel? In this case the force can be resolved into two vector components, one parallel to the displacement and one perpendicular to it (Fig. 15–1). The perpendicular vector component of the force causes a change in direction but does not change the body's speed. (See Section 12–5.) Recall that in the motion of a body in a circle, the force always acts at right angles to the displacement, the speed does not change, and so the kinetic energy remains unchanged. It is only the vector component of the force parallel to the displacement which changes the speed and hence the kinetic energy.

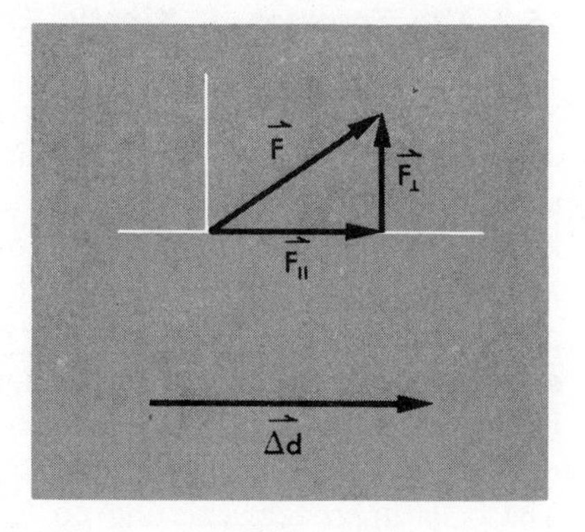

Figure 15–1
When the force is not parallel to the displacement, it can be resolved into parallel and perpendicular vector components.

To extend the work-kinetic energy relation to include forces in any direction, we have to replace the force $\vec{\mathbf{F}}$ by its vector component along the direction of displacement $\vec{\mathbf{F}}_d$. Because $\vec{\mathbf{F}}_d$ and $\vec{\mathbf{d}}$ are always along the same line, we shall dispense with the vector notation and write

$$F_d \Delta d = \Delta E_K \cdot$$

Here F_d is positive if it is in the same direction as Δd and negative if it is in the opposite direction.

If more than one force acts on a body, its acceleration is given by the net force, i.e. the vector sum of all the forces. Therefore the change in velocity and hence the change in kinetic energy are given by the work done by the net force.

Now we come to the second extension. So far we have considered only the case where F_d is constant. From your experience with the generalization of $v\Delta t$ to variable velocities (Section 9–5) and of $F\Delta t$ to variable forces (Section 14–1), you might guess that $F\Delta d$ can be generalized in a similar way. Indeed, if the component F_d changes, the work is given by the area between the F_d graph and the displacement axis (Fig. 15–2). If the area below the displacement axis is larger than the area above it, the total work will be negative; then the change in kinetic energy of the body will also be negative. This means that the body loses kinetic energy.

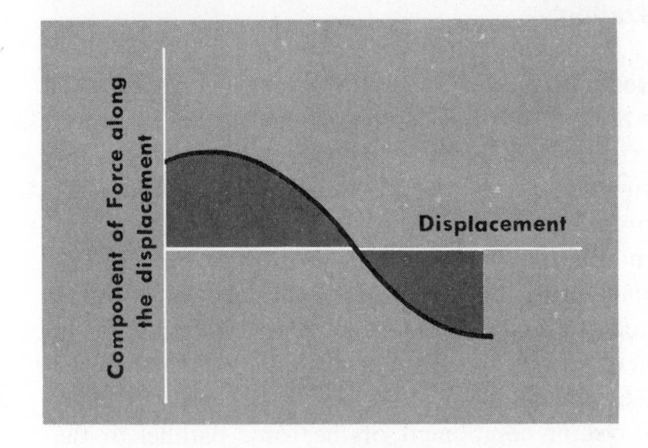

Figure 15–2
The total work done by the force equals the area between the graph and the displacement axis.

15-3 The Transfer of Kinetic Energy From One Mass to Another

After this first look at kinetic energy, let us explore what happens to it in a collision. Figure 15–3 is a flash photograph of a billiard-ball collision. By measuring the distances the balls move between flashes, we can find the kinetic energy of the ball with the dot as it came in from the left; we can find its kinetic energy after collision as it went off more slowly to the lower right, and we can also find the kinetic energy of the dotless ball after the collision set it in motion toward the upper right. We find that the kinetic energy lost by the dotted ball is almost exactly equal to that gained by the other ball.

How does the energy transfer actually occur in a billiard-ball collision? The balls approach, exerting no force on each other until they are very close, and we say they are "in contact." Then repulsion sets in: each ball pushes the other. These forces change in magnitude, increasing as the balls distort each other and decreasing as they move apart. Because the forces are changing, a billiard-ball collision is difficult to analyze in detail. We shall be able to understand the transfer better if we start with an easier example in which the force has a simpler behavior. After we understand the details in this simple case we can return to billiard balls.

To make a simple preliminary study of collisions and energy we shall study an artificial case in which two colliding bodies exert no force on each

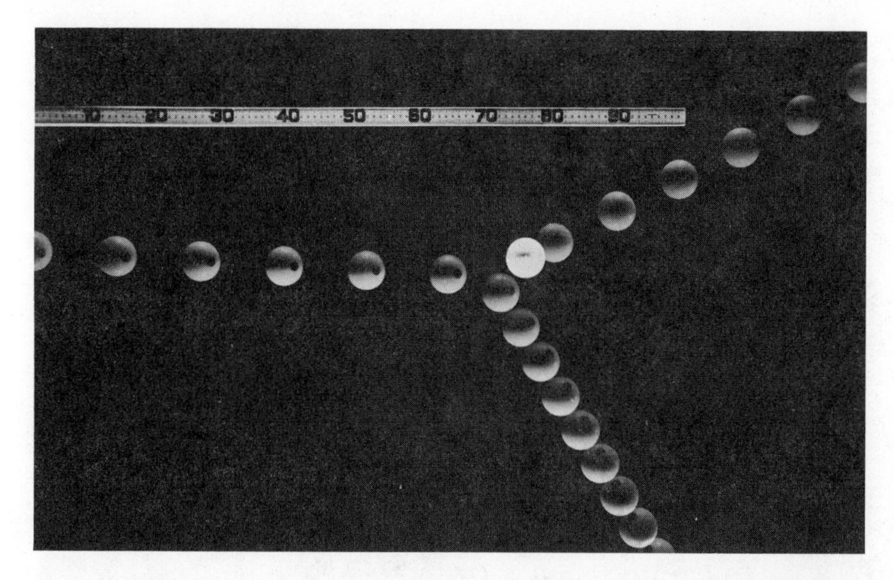

Figure 15–3
A multiple-flash photograph of a collision between two billiard balls, each of mass 173 gm. The interval between flashes was $\frac{1}{30}$ sec. The dotted ball came in from the left and struck the unmarked ball at rest. The unmarked ball appears whiter where it was at rest and was photographed during several flashes.

other until they get within a distance d and then exert a constant repulsion of magnitude F on each other as long as the separation is less than d (Fig. 15–4).

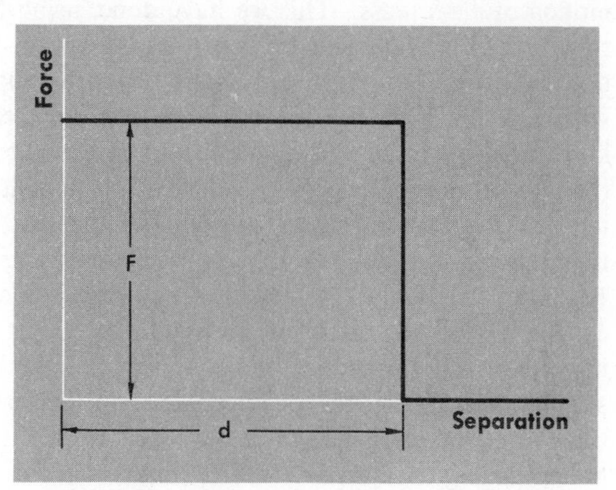

Figure 15–4
The graph of the magnitude of the interaction forces between two bodies. The force is zero when the separation s is more than d, and a constant value F when the masses are closer together than d.

Suppose the collision takes place along a line (Fig. 15–5). Moving along this line with velocity v_1, the mass m_1 approaches the mass m_2 which is at rest. Nothing happens until m_1 gets to the distance d away from m_2. From then on, however, m_2 is pushed forward with force F, and m_1 is pushed backward just as hard. Consequently m_2 speeds up while m_1 slows down. Because m_1 is moving and m_2 standing still when the repulsion sets in, the masses continue to get closer together for a while. But soon they reach a stage when they are as close as they will get and are moving along with the same velocity. Still pushed apart by F, m_2 continues to speed up and m_1 to slow down. The masses therefore separate until eventually they are d apart again, with m_2 moving faster than m_1. Then, because they are getting farther apart, the force falls to zero, and they continue their further motion without changes in velocity. The collision is over, with m_1 and m_2 now moving with definite energies that no longer change.

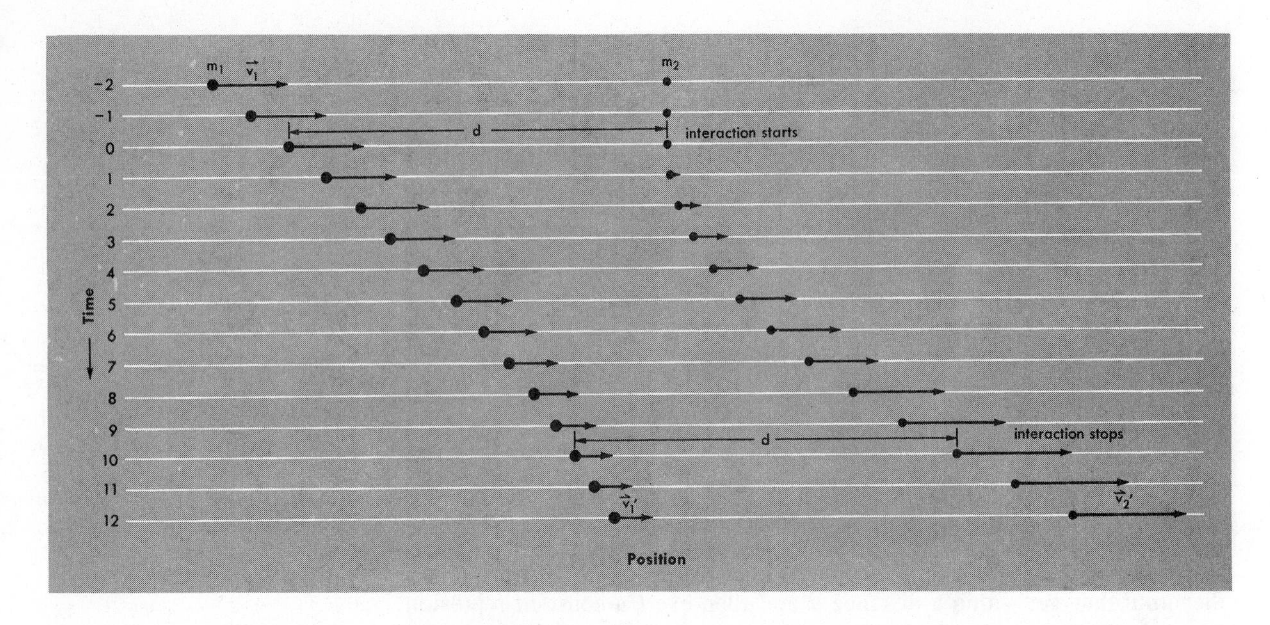

Figure 15–5
The positions and velocities of two masses m_1 and m_2 as they interact along a line, with the force shown in Fig. 15–4. The masses are shown at equal time intervals. Mass m_1 comes in from the left. When the distance between masses is less than d, m_1 slows down and m_2 speeds up from rest. After 10 time intervals, in this particular case, the separation is again greater than d; the force drops to zero; and the two masses move apart with constant velocities. The gain in kinetic energy of m_2 over the whole interaction is equal to the loss of kinetic energy of m_1.

For this collision with a steplike force, we can calculate the details of the motion of each mass. This we have done, using Newton's law, for masses such that $m_1 = 3m_2$; and we constructed Fig. 15–5 in this way. But we do not need to go through the details of calculating the motion to find out whether the transfer of kinetic energy from m_1 is the same as the transfer of kinetic energy to m_2. Because the force has the constant value F during the interaction, the energy transferred into kinetic energy of m_2 is F times the distance that m_2 moves during the collision. Also, the kinetic energy transferred from m_1 is F times the distance m_1 moves during the collision (m_1 loses kinetic energy because the force on it opposes the motion). We can show that the kinetic energy lost by m_1 is just the amount gained by m_2 if we show that m_1 and m_2 move the same distance during the collision. To see that, look at Fig. 15–6. There we show the positions of the masses when the interaction starts and when it stops. Notice that d plus the distance m_2 moves during the interaction (top line) is equal to d plus the distance m_1 moves (bottom line). Consequently the distances moved by m_1 and by m_2 are the same during the complete interaction. The force times distance moved during the collision is therefore equal and opposite for the two masses; and as we anticipated, the kinetic energy from m_1 just goes into kinetic energy of m_2.

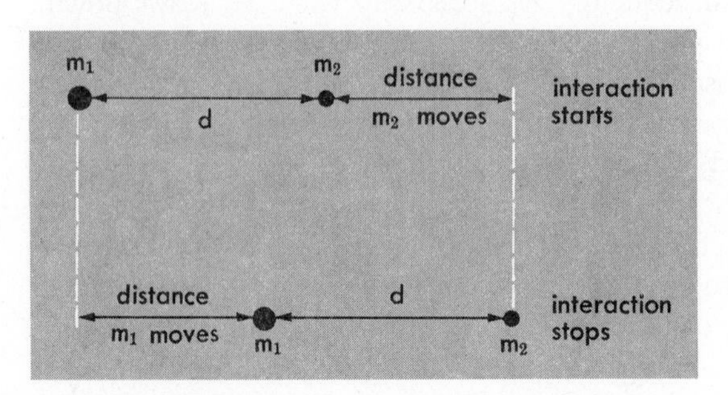

Figure 15–6
The positions of m_1 and m_2 at the beginning (top line) and the end (bottom line) of the interaction in Fig. 15–5. The scale here has been reduced. Comparing the two lines, d plus the distance m_2 moves is equal to d plus the distance m_1 moves. Thus m_1 and m_2 move the same distance during the interaction.

There is a slightly different way of looking at this special example that will make it easier to see what happens in a real collision when the pattern of force is more complicated. In order to study real collisions we might wish to find what kind of forces must be acting if all the kinetic energy lost by one mass is gained by the other. To find the nature of these forces we shall examine the sum of the two kinetic energies directly rather than each one individually. If the sum at the end is the same as at the beginning, what is lost by one mass must have been gained by the other.

However, when we look at the changes in kinetic energy during the course of a collision, we shall see that at intermediate stages the loss of kinetic energy of m_1 is not equal to the gain of kinetic energy of m_2.

To find out how the total kinetic energy changes during a collision we shall concentrate on a small time interval during a collision between two masses m_1 and m_2 (Fig. 15–7). At the beginning of this time interval, we shall call the separation between the masses s. At the end of this time interval the separation is $s + \Delta s$. Now, if m_2 moves a distance Δx_2 in the direction away from m_1, the separation is increased by Δx_2; but if m_1 moves a distance Δx_1 in the same direction—that is, toward m_2—the separation is decreased by Δx_1. Therefore, the change in separation Δs is given by the difference between the distance Δx_2, which m_2 moves, and the distance Δx_1, which m_1 moves in the same direction:

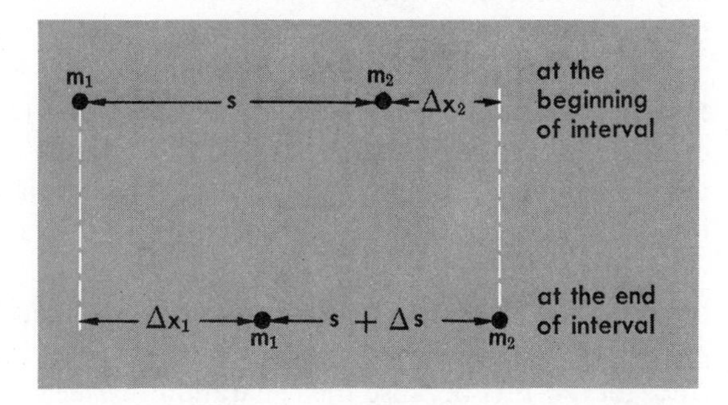

Figure 15–7
During a short time interval, while m_1 and m_2 interact, Δx_1 is the distance m_1 moves and Δx_2 is the distance m_2 moves. The original separation between m_1 and m_2 is s, and $s + \Delta s$ is the separation at the end of the interval. Therefore Δs is the change in the separation. Comparing the top line of the figure with the bottom line, we see that $s + \Delta x_2 = \Delta x_1 + (s + \Delta s)$. Then, subtracting $(\Delta x_1 + s)$ from both sides, we get the result $\Delta x_2 - \Delta x_1 = \Delta s$.

$$\Delta s = \Delta x_2 - \Delta x_1.$$

Now when the force of interaction F in this interval is repulsive, the kinetic energy transferred to m_2 is $F\Delta x_2$. This is the work done in accelerating m_2. On the other hand, the kinetic energy taken from the motion of m_1 is $F\Delta x_1$. This energy is transferred from the motion of m_1 because the force on m_1 is in the opposite direction from the motion Δx_1. Consequently, the change in the total kinetic energy in the interval when these motions take place is given by

$$\Delta E_K(\text{tot}) = F\Delta x_2 - F\Delta x_1 = F(\Delta x_2 - \Delta x_1).$$

Since $\Delta x_2 - \Delta x_1 = \Delta s$, we can write the change in the total kinetic energy as

$$\Delta E_K(\text{tot}) = F\Delta s.$$

For example, in Fig. 15–8 as the separation changes from s to $s + \Delta s$, the total kinetic energy changes by the amount given by the shaded area, F high and Δs wide.

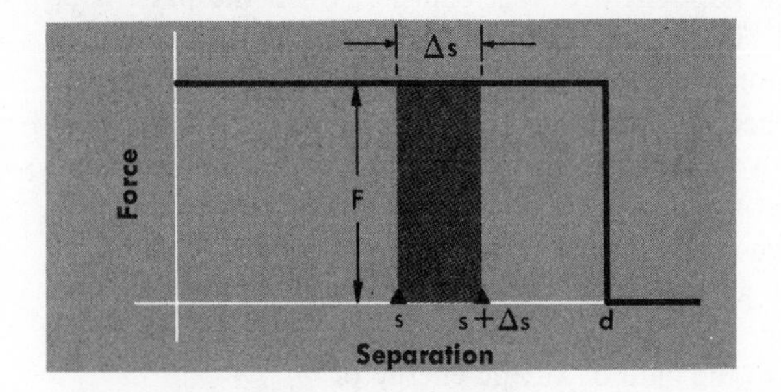

Figure 15–8
When the separation changes from s to $s + \Delta s$, the change in the total kinetic energy is given by the shaded area.

The result $\Delta E_K(\text{tot}) = F\Delta s$ is just what we need. It shows that the change in the total kinetic energy in any interval depends only on the force of interaction and the change of separation that takes place. These were the only quantities that entered our calculations. We never have to say where the masses are with respect to any other objects.

For a collision with the force shown in Fig. 15–8 (that is, F at all separations less than d, and zero at larger separations), the total kinetic energy decreases at the beginning of the collision. When m_1 first arrives at the distance d from m_2, it is moving toward m_2 while m_2 is standing still. As we see by looking back at Fig. 15–5, in the next interval of time m_1 moves considerably farther than m_2. Since the forces on m_1 and m_2 are equal and opposite, the kinetic energy taken from m_1 is greater than the kinetic energy given to m_2. There is a net loss in kinetic energy. This result follows from $\Delta E_K(\text{tot}) = F\Delta s$ because the separation between m_1 and m_2 has decreased, and so $F\Delta s$ is negative.

As long as m_1 is coming closer to m_2, kinetic energy continues to disappear. Eventually, however, the total kinetic energy is regenerated. While m_1 continues to slow down, m_2 picks up speed so that finally the separation increases again. When the separation becomes d once more, the value of Δs, measured from the beginning of the collision, is zero. Consequently the total change in kinetic energy $\Delta E_K(\text{tot}) = F\Delta s = 0$. The kinetic energy which was lost as the two bodies approached each other has been regained. After that the masses separate still farther but there is no force acting on them. Consequently E_K stays the same. The collision is over and the kinetic energy remains at its final value, the same value that it had before the collision began.

15-5 Kinetic Energy and the Center of Mass

We can look at the kinetic energy transfer in this collision in yet another way. We learned in the last chapter that the center of mass moves steadily through a collision as if nothing had happened. Also, as far as momentum was concerned, the center of mass moved as if all the mass of the system were concentrated there, and the total momentum relative to the center of mass was zero. Is there a similar behavior for the center of mass where kinetic energy is concerned?

At any instant in the collision of Fig. 15–5, the masses m_1 and m_2 are moving along a line with velocities v_1 and v_2. The center of mass moves along the same line with velocity v_c. The two masses have velocities V_1 and V_2 relative to the center of mass, also along this line. The velocity v_1 is just the velocity of m_1 relative to the center of mass plus the velocity of the center of mass: $v_1 = V_1 + v_c$. Similarly, for v_2 we have: $v_2 = V_2 + v_c$ (remember that velocities along a line add like positive and negative numbers). The total kinetic energy of the masses is then

$$\begin{aligned} E_K &= \tfrac{1}{2}m_1v_1^2 + \tfrac{1}{2}m_2v_2^2 \\ &= \tfrac{1}{2}m_1(V_1 + v_c)^2 + \tfrac{1}{2}m_2(V_2 + v_c)^2 \\ &= \tfrac{1}{2}m_1V_1^2 + \tfrac{1}{2}m_2V_2^2 \\ &\quad + (m_1V_1 + m_2V_2)v_c + \tfrac{1}{2}(m_1 + m_2)v_c^2. \end{aligned}$$

Now, we found in Section 14–6 that $m_1V_1 = -m_2V_2$: the total momentum relative to the center of mass was zero. Therefore,

$$(m_1V_1 + m_2V_2) = 0$$

and

$$E_K = (\tfrac{1}{2}m_1V_1^2 + \tfrac{1}{2}m_2V_2^2) + \tfrac{1}{2}(m_1 + m_2)v_c^2.$$

The last term is just the kinetic energy associated with the motion of the center of mass. It is the kinetic energy of a mass equal to the total mass of the system and traveling along with the center-of-mass velocity v_c. This is called the center-of-mass energy of the system. It continues along unchanged throughout the collision. The first term, on the other hand, depends on the velocities of m_1 and m_2 relative to the center of mass; it represents the kinetic energy of the masses as measured from the center of mass. It is the *internal* kinetic energy of the system. It will change throughout the collision, decreasing during the first part, reaching zero when the separation is at a minimum ($v_1 = v_2 = v_c$ at that instant so $V_1 = V_2 = 0$), and then increasing back to its original value by the end of the elastic interaction.

This separation of the kinetic energy into two parts is not limited to two bodies moving along a straight line; it can be proved for any number of bodies moving in all directions. We shall need to return to this idea of internal energy when we discuss heat and molecular motion in Chapter 17.

15-6 Conservation of Kinetic Energy in Elastic Interactions

In the collision we studied in the last sections, the transfer of kinetic energy from one mass to the other ends up without loss. This result does not depend on either the range d or the magnitude F of the interaction force between the two masses. In any interaction with any interaction range, d, and any constant force, F, we find the same total kinetic energy at the beginning and the end of any collision. The result is more general than we might have anticipated. How general is it?

We shall now show that the total kinetic energy at the end of a collision is the same as at the beginning whenever the force of interaction depends only on the distance of separation of the two masses.* The equal and opposite forces F on the two masses may be any function of the separation of the two masses as long as F is zero beyond some definite range, so that we can define a complete interaction. Such an interaction starts with the masses farther apart than this range and ends when they are again separated by more than this range (Fig. 15–9).

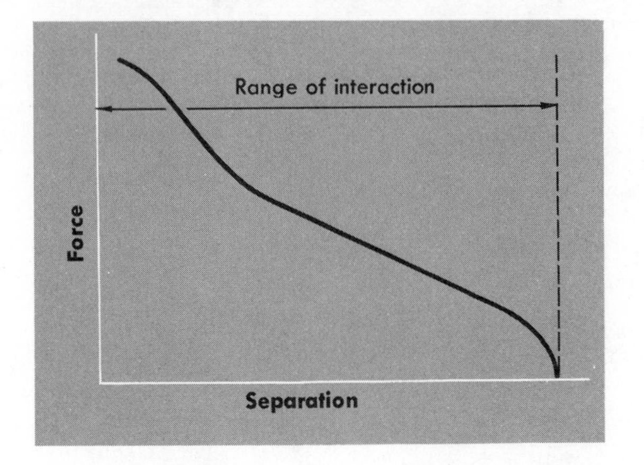

Figure 15–9
The graph of a force which is a function of the separation of two bodies. The force is zero beyond the range of interaction.

We can now apply the result $\Delta E_K(\text{total}) = F\Delta s$ to any force of interaction which depends on the separation alone—for example, the force illustrated in Fig. 15–9. Let us assume that this is a repulsive force. Consider for a moment what happens to the total kinetic energy when the separation of the two bodies decreases by the Δs shown in Fig. 15–10. The force is almost constant; and $F\Delta s$, represented by the shaded area, gives the decrease of the total kinetic energy. On the other hand, for this same separation, if the two bodies move apart by Δs, their total kinetic energy

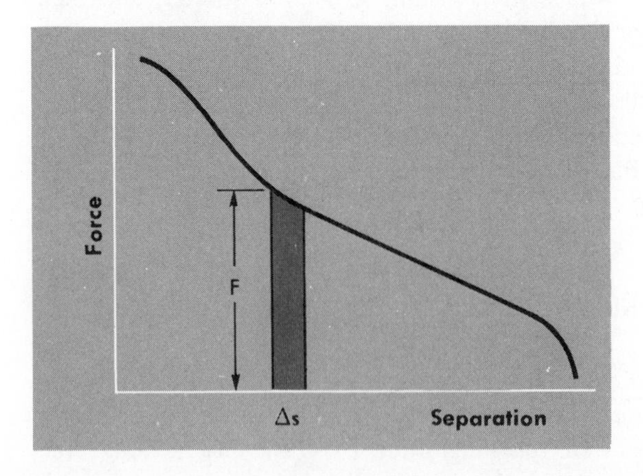

Figure 15–10
During an interaction, when the separation changes by Δs, the kinetic energy E_K changes by $F\Delta s$—the shaded area under the curve.

* This merely means that the force is the same on the way out as on the way in to the collision. This excludes forces like friction which reverse their direction on the way out.

increases by the same amount, $F\Delta s$. We see then (Fig. 15–11) that as two interacting masses move closer together from large separation to separation s they lose all the kinetic energy represented by the area under the F versus s curve. Then, in going apart again, they gain back all the kinetic energy represented by the same area. At the end of the interaction they have the same kinetic energy as at the beginning, provided only that the force as a function of separation is the same when the masses go apart as it was when they came together. We have now extended our theorem of the conservation of kinetic energy over a completed collision to apply to any interaction in which the force depends only on the separation. An interaction of this kind is called an elastic interaction or an *elastic collision*.

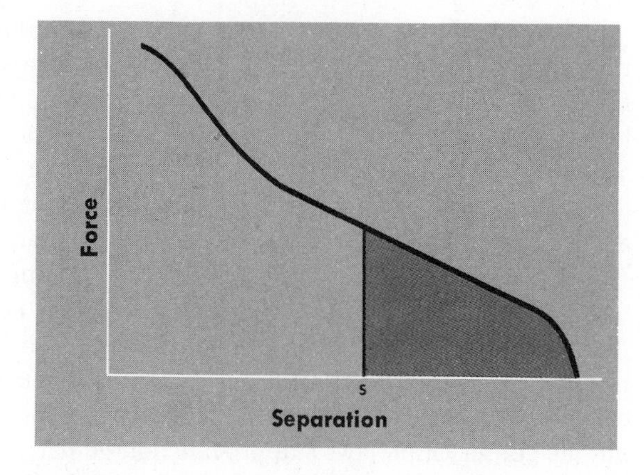

Figure 15–11
If two bodies approach each other from a long distance to a separation s, they lose a total amount of kinetic energy equal to the shaded area under the curve. When they move apart again, they regain the same amount of kinetic energy provided only that the force as a function of separation is the same on the way out as on the way in.

The interactions we see about us, such as collisions between billiard balls, or between a baseball and a bat, are never perfectly elastic, but many are very nearly so.

15-7 Conservation of Kinetic Energy and Momentum

When two bodies interact elastically, their kinetic energies E_{K_1} and E_{K_2} change by equal and opposite amounts over the whole interaction; that is:

$$-\Delta E_{K_1} = \Delta E_{K_2}.$$

This must be true because there is no change in the total kinetic energy $E_{K_1} + E_{K_2}$. Also, as we learned in Chapter 14, the changes in momenta of the two bodies are equal and opposite:

$$-\Delta \vec{\mathbf{p}}_1 = \Delta \vec{\mathbf{p}}_2.$$

This vector relation is always true for two bodies forming an isolated system; in particular, therefore, it is true for elastic collisions.

We can learn a great deal about the final motion of two interacting masses from the fact that neither momentum nor kinetic energy is lost in these two transfers. The two equations contain a large part of the information needed to determine the final velocities $\vec{v}_1'$ and $\vec{v}_2'$. In particular, if the motion takes place along a straight line, we find that

$$v_1' = \frac{m_1 - m_2}{m_1 + m_2} v_1$$

and that

$$v_2' = \frac{2m_1}{m_1 + m_2} v_1.$$

For details of finding v_1' and v_2' for head-on-collisions see the box. If the collision is not head on, we need some more information to tell us how far off the center it is and therefore in what direction the masses will go after the collision.

In making Fig. 15–5 we used a constant F within the range of interaction d; and we chose $m_1 = 3m_2$. Then from Newton's law we found that $v_1' = \frac{1}{2}v_1$ while $v_2' = \frac{3}{2}v_1$, as you can measure on the figure. If you calculate v_1' and v_2' from the equations in the last paragraph based on the conservation of momentum and of kinetic energy, you will get the same results.

The results of using either Newton's law or the conservation laws must always agree. But it is not always so easy to apply Newton's law directly. If F is a complicated function of separation, finding the motion from Newton's law is a slow numerical procedure. Often, in fact, we do not know the details of F as a function of separation, even though experience may indicate that the total kinetic energy is the same before and after every collision. In such a situation, we cannot describe the motion completely, but the conservation laws still provide a large part of the answer.

For example, billiard-ball collisions are of this kind. The detailed forces between colliding balls may not be known to us; but observation shows that the kinetic energy of the balls is almost the same after they interact as it was before. Therefore, the forces must be functions only of the separation of the centers of the balls, and the forces are zero when the centers are apart by more than the diameter. In particular, then, we can use the last two equations with $m_1 = m_2$ to describe a head-on collision between a billiard ball in motion and one at rest. From the equations we then get $v_1' = 0$ and $v_2' = v_1$ or, in other words, if one billiard ball smacks another head on, the moving ball should stop and the ball at rest take off with the velocity the other used to have. That m_1 stops and m_2 goes on with the velocity v_1 is just what we observe, as Fig. 14–8 shows.

The same kind of analysis is often of importance in modern high-energy physics. Here the colliding masses may be subatomic particles such as protons, neutrons, mesons, or hyperons. The idea of force as a function of distance may have no meaning at all on the subatomic scale. But sometimes we can pick out the elastic collisions by observing the kinetic energies before and after interaction.

Sometimes visible particles (i.e., those whose tracks we can see) collide with particles which are invisible to us. If we assume that kinetic energy and momentum are conserved, and if we know the mass and the initial and final velocities of the visible particles, we can determine the mass and the initial and final velocities of the invisible particles. In much this way James Chadwick discovered the neutron in 1932.

On the Derivation of the Formulas for v_1' and v_2'

For a head-on collision, in which the motions are along a straight line (as in Section 15–3), we can determine the final velocities v_1' and v_2' from the equations

$$-\Delta E_{K_1} = \Delta E_{K_2} \quad \text{and} \quad -\Delta p_1 = \Delta p_2.$$

(For vectors along a line, we drop the arrow and use the + or − sign to indicate direction, as in Chapter 10.) Writing ΔE_K as the difference between final and initial kinetic energies, we have

$$-(\tfrac{1}{2}m_1v_1'^2 - \tfrac{1}{2}m_1v_1^2) = (\tfrac{1}{2}m_2v_2'^2 - 0).$$

We can factor this equation to give

$$-(m_1v_1' - m_1v_1)\left(\frac{v_1' + v_1}{2}\right) = (m_2v_2' - 0)\left(\frac{v_2' + 0}{2}\right).$$

We see that each side of this energy equation is the product of the momentum change of one mass times the average of the initial and final velocities of that mass. Because we know that the momentum changes are equal and opposite, we can cancel $-\Delta p_1 = -(m_1v_1' - m_1v_1)$ on the left with $\Delta p_2 = (m_2v_2' - 0)$ on the right. We therefore obtain the additional information that

$$v_1' + v_1 = v_2'.$$

On substituting v_2' from this equation into the equation of the conservation of momentum

$$-(m_1v_1' - m_1v_1) = m_2v_2',$$

we find

$$v_1' = \frac{(m_1 - m_2)}{m_1 + m_2}v_1.$$

The energy and momentum relations thus lead to a specific prediction for the final velocity of m_1 when its initial velocity is given. Furthermore, by putting this value of v_1' into $v_2' = v_1' + v_1$, we get

$$v_2' = \frac{2m_1}{m_1 + m_2}v_1$$

which tells us the final velocity of the second mass.

Actually in getting this answer for the final velocities, we have put in quite a bit of information. Most of this information was introduced when we assumed that the collision was head on so that the masses move only along the x axis. If we had allowed one mass to approach the other off center (Fig. 15–3), the problem would be more complicated to handle, and we would have to put in the information representing the distance off center in order to get the answer. We also specified that m_2 is initially at rest. These equations do not apply if m_2 is in motion at the start of the interaction. The more general equations that apply then are derived by the same method.

Furthermore, we should note that we made on tacit assumption in finding v_1' and v_2'. We assumed that there *was* a change in momentum in the interaction. It is possible even in our head-on collision to get another answer: if $v_1' = v_1$ and $v_2' = 0$, the momentum is unchanged and no energy transfers from one body to the other over the complete interaction. We shall not worry about this possible answer too much, however, because m_1 must pass right through m_2 in this case.

15-8 The Discovery of the Neutron

A sample of beryllium was bombarded by alpha particles from radioactive polonium, and the beryllium, in turn, was found to be giving off invisible, unknown particles (Fig. 15–12).

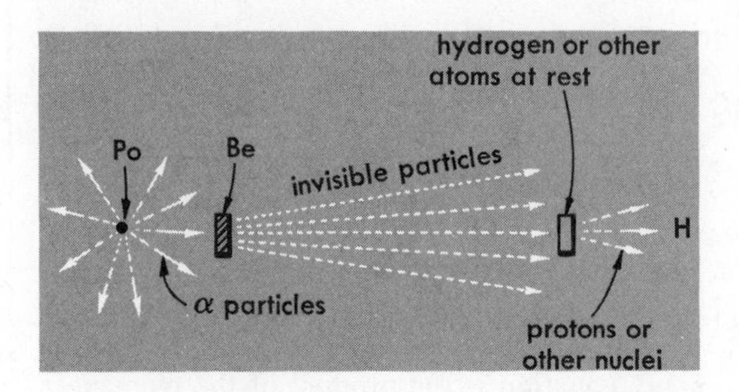

Figure 15–12
Diagram of Chadwick's apparatus.

These invisible particles then struck either hydrogen or nitrogen atoms at rest. As a result of the collisions, protons or nitrogen nuclei were knocked out, and Chadwick measured their velocities.

Suppose we select head-on collisions by looking in the region H (in Fig. 15–12), and we assume that the collisions are elastic. Then, calling m the mass of the invisible particle, and v its velocity, and calling m_p the mass of the protons and v_p' their velocity, we have:

$$v_p' = \frac{2m}{m + m_p} v.$$

In the same way, calling the mass of nitrogen and its velocity m_N and v_N,

$$v_N' = \frac{2m}{m + m_N} v.$$

In this equation we can replace m_N by $14m_p$ because, as we know, the mass of nitrogen is 14 atomic mass units and that of hydrogen is about 1 amu. After replacing m_N by $14m_p$, we divide the first equation by the second one. This eliminates the unknown velocity v of the invisible particle with the result:

$$\frac{v_p'}{v_N'} = \frac{m + 14m_p}{m + m_p}.$$

In his experiments, Chadwick measured the velocities v_p' and v_N'. He found that the ratio $\frac{v_p'}{v_N'}$ was about 7.5. Therefore

$$\frac{m + 14m_p}{m + m_p} \approx 7.5$$

or

$$m \approx 1.00m_p.$$

Chadwick repeated the experiment with other substances in place of hydrogen or nitrogen and found again that an invisible particle with about the mass of a proton fitted in with his measurements. He did a number of other experiments. All were consistent, and one of them determined the mass to within less than 1 percent. He had proved that the neutron exists.

15-9 Loss of Kinetic Energy in a Frictional Interaction

If a frictional force acts on a body—a force that does not depend just on the separation between two bodies—the kinetic energy of the overall motion of the interacting bodies decreases and finally becomes zero.

Consider the motion of a book sliding on a tabletop. The frictional force exerted on the book by the tabletop slows it down. The work done is the product of the force and the displacement of the book. Because the force is opposite to the displacement, the kinetic energy of the book decreases by the amount of the work. There is an equal and opposite force exerted by the book on the table, but the table hardly moves under the influence of this force. Consequently this force on the table does practically no work, and the kinetic energy of the table does not increase. It is, however, a fact that the surfaces of the book and of the table which have been in contact warm up a little. The energy taken out of the moving mass is associated with this increase in temperature.

Let us consider another example of a frictional interaction. When we drop a ball of putty, the interaction between the putty and the floor begins as soon as the ball and the floor come into contact. Forces then begin to act on the putty, slowing it down. Also, the shape of the ball is changed; and, on the rebound, the interaction stops when the ball is closer to the floor than when the interaction began (Fig. 15–13). The force between

Figure 15–13
An inelastic collision. The interaction begins when the ball of putty comes in contact with the floor. The ball is permanently deformed by the interaction and, on the rebound, the interaction ceases before the ball reaches the separation at which the interaction started. As a result, the speed on the rebound is less than the speed at which the ball came in.

ball and floor at this position is now zero, whereas when the ball was moving down at this position, the force was not zero. As a result, the total kinetic energy is less after collision than before and the ball of putty bounces back, moving very slowly. The putty ball seems to lose kinetic energy permanently, but we find the putty warmer after collision with the floor than it was before. This interaction is much like that between the sliding mass and the table. In this example, however, the friction may all occur inside the putty ball when one part of the ball moves with respect to another.

A collision in which the forces are smaller when the bodies separate than when the bodies come together is called an *inelastic* collision. After inelastic collisions, a temperature rise is usually observed. In Chapter 17 we shall discuss how this increase in temperature is related to energy in another form. For the present we shall keep our attention on situations in which friction and heating effects can be neglected.

1.* What is the kinetic energy of a 1-kg hammer moving at 20 m/sec? (Section 1.)

2.* How does the kinetic energy of a car change when its speed is doubled? (Section 1.)

3. A force of 10.0 newtons acts on a 2.00-kg roller skate initially at rest on a frictionless table. The skate travels 3.00 meters while the force acts.
   (a) How much work is done?
   (b) What is the kinetic energy of the skate?
   (c) What is the final speed of the skate?

4. Compare the kinetic energies of two objects, *A* and *B*, identical in every respect except one. Assume that the single difference is:
   (a) Object *A* has twice the velocity of *B*.
   (b) Object *A* moves north, *B* south.
   (c) Object *A* moves in a circle, *B* in a straight line.
   (d) Object *A* is a projectile falling vertically downward; *B* is a projectile moving vertically upward at the same speed.

5.* Two bodies of unequal mass each have the same kinetic energy and are moving in the same direction. If the same retarding force is applied to each, how will the stopping distances of the bodies compare? (Section 2.)

6.* Is it possible to exert a force and yet not cause a change in kinetic energy? (Section 2.)

7. A Dry Ice puck is acted upon by two forces $\vec{F}_1$ and $\vec{F}_2$ as shown in Fig. 15–14. Assume that in each case the body starts from rest..
   How much work is done on the body as it moves 2 meters if:
   (a) $F_1 = 10$ newtons and $F_2 = 0$?
   (b) $F_1 = 0$ and $F_2 = 10$ newtons?
   (c) $F_1 = 10$ newtons and $F_2 = 10$ newtons?

8. A Dry Ice puck has a mass of 5 kg. It is subjected to a constant force $\vec{F}$ of 50 newtons at an angle of 60° above the horizontal for 2 seconds (Fig. 15–15).
   (a) If the puck starts from rest, what is its change in momentum in the first two seconds the force is applied?
   (b) How much work was done in these two seconds?

   (*Note:* $E_K = \frac{1}{2} mv^2 = \frac{(mv)^2}{2m} = \frac{p^2}{2m}$.)
   (c) What would happen if the force were doubled?

9. A force of 30 newtons accelerates a 2.0-kg object from rest for a distance of 3.0 meters along a level, frictionless surface; the force then changes to 15 newtons and acts for an additional 2.0 meters.
   (a) What is the final kinetic energy of the object?
   (b) How fast is it moving?

10. Estimate your kinetic energy (in joules) when you ride your bicycle on the road.

11. A 2.0-kg stone whirls around on the end of a 0.50-m string with a frequency of 2.0 revolutions per second.
   (a) What is its kinetic energy?
   (b) What is the centripetal force on it?
   (c) How much work is done by the centripetal force in one revolution?

12.* In Fig. 15–5, do the two bodies repel or attract each other? (Section 3.)

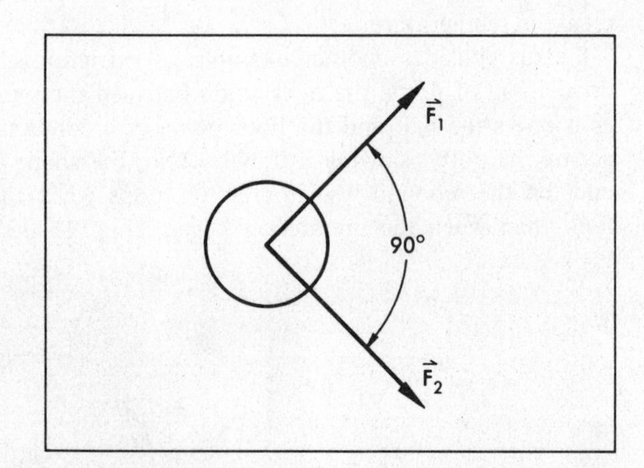

Figure 15–14 For Problem 7.

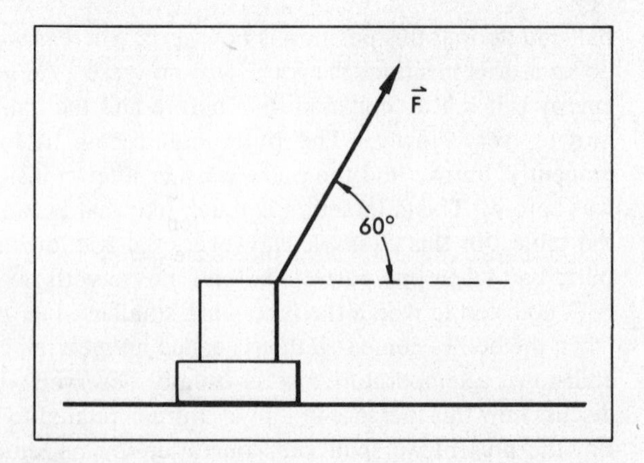

Figure 15–15 For Problem 8.

13. A 10.0-kg mass moves 2.00 meters against a retarding force that increases linearly by 4.00 newtons for every 3.00 meters the mass moves (see Fig. 15–16). If the force is zero at the beginning, how much kinetic energy is lost?

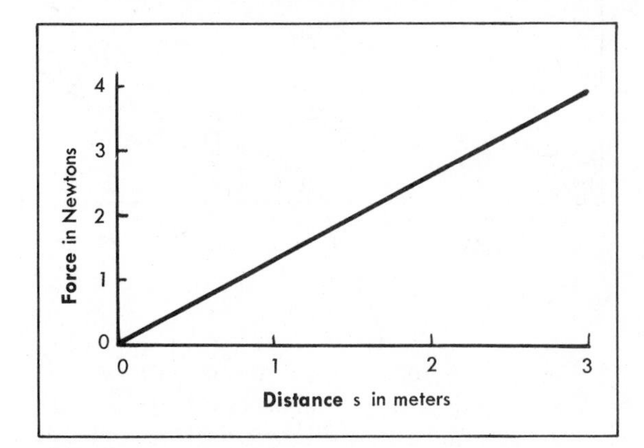

Figure 15–16 For Problem 13.

14.* When do we call a collision *elastic?* (Section 6.)

15.* When a putty ball is thrown at a wall and sticks to it, does the force between the wall and the putty ball depend on the separation alone? (Section 7.)

16.* A 3-kg ball with a speed of 6 m/sec strikes another 3-kg ball moving 6 m/sec in the opposite direction. After collision each ball has a velocity equal in magnitude but opposite in direction to its initial velocity. Is this an elastic collision? (Section 6.)

17.* A rapidly moving mass m_1 strikes a mass m_2 at rest in a head-on elastic collision. How does m_1 compare with m_2 if m_1 comes back along its path at a very low speed? (Section 7.)

18. A projectile (which does not contain an explosive charge) from an antitank gun has a mass of about 10 kg and a speed of 10^3 m/sec. A freight car being moved around in a switchyard has a mass of about 10^4 kg and a speed of 1 m/sec. What are their momenta? Their kinetic energies? Why does the projectile do much more damage than the freight car when it hits something?

19. Two 3.0-kg bodies interact. At a given moment the first body is moving to the right at 0.50 m/sec and the second body is moving to the right at 0.30 m/sec, and so at that instant the speed of approach of one body as measured from the other is 0.20 m/sec. If the force of interaction is repulsive and equal to 0.10 newton, at what rate is the total kinetic energy decreasing at that time?

20. A 1.5-kg body is at rest. It is "hit" head on by a body of mass 0.50 kg moving with a speed of 0.20 m/sec. The interaction force depends only on the separation of the two bodies.
   (a) What is the final velocity of each body?
   (b) In what direction does each move after the interaction?

21. An object of mass m_1 and kinetic energy E_K collides head on with an object of mass m_2 at rest. Assuming the force of interaction depends only on the separation, calculate the kinetic energy transferred if
   (a) $m_2 = .01\ m_1$
   (b) $m_2 = m_1$
   (c) $m_2 = 100\ m_1$
   What is the ratio of m_1 and m_2 when the transfer of kinetic energy is
   (d) a maximum?
   (e) very small?
   (*Note:* To see this relationship more clearly, you may wish to find the energy transferred for additional values of m_2, and make a graph.)

22. A 5.0-kg body is at rest. A 10-kg body approaches it with a velocity of 0.20 m/sec. The interaction force is zero when the separation is greater than 0.10 meter and is 4.0 newtons when the separation is less than this distance. (Note that some of the answers in this problem are easier to get if you use the conservation of momentum.)
   (a) What is the kinetic energy of the masses before the interaction?
   (b) What will be the kinetic energy of each mass after the interaction is complete?
   (c) What will be the kinetic energy of each mass when the separation is at a minimum? Recall that the velocities at minimum separation are equal.
   (d) What is the minimum separation? [Your answer to part (c) tells you the net loss of kinetic energy of the two masses at minimum separation.]

23. An antiaircraft shell has a kinetic energy E_K and a momentum $\vec{\mathbf{p}}$. Just then it explodes. What can you say about:
   (a) the momentum of the pieces?
   (b) the kinetic energy of the pieces?

24. In one of the experiments that led to the determination of the mass of the neutron, Chadwick measured the velocity of protons that had been hit head on by neutrons. The velocity of the protons was 3.3×10^7 m/sec.
    (a) What was the velocity of the neutrons before and after collision with the protons?
    (b) Chadwick also measured the velocity of nitrogen atoms hit head on by the neutrons. What was it?
    (c) What was the velocity of the neutrons after each kind of collision?

25. Suppose you wish to determine the force your bicycle brakes are exerting on the bicycle when you brake hard without skidding. Assume the brakes grip the rim of the wheel. At your disposal is a tape to measure length and a road of known constant slope. Also, you know the total weight of the bicycle and yourself. You do not have a watch. How would you go about it?

26. A horseshoe magnet of mass m stands on end on a frictionless table. A steel ball bearing of mass m is rolled toward the magnet from far away with velocity v and goes through the magnet and far beyond. Assume the force of attraction F changes with distance and is the same in front and behind the magnet. (See Fig. 15–17.)
    (a) What is the final velocity of the ball?
    (b) What is the final velocity of the magnet?

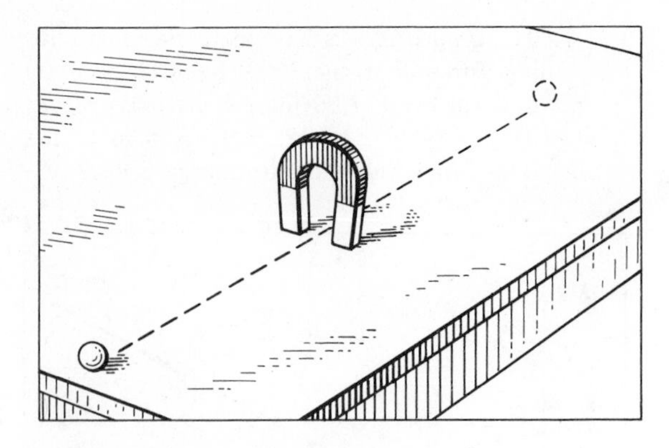

Figure 15–17 For Problem 26.

Further Reading

These suggestions are not exhaustive but are limited to works that have been found especially useful and at the same time generally available.

ANDRADE, E. N. DA C., *An Approach to Modern Physics.* Doubleday Anchor, 1956. Heat and energy. (Chapter 2)

BONNER, FRANCIS T., and PHILLIPS, MELBA, *Principles of Physical Science.* Addison-Wesley, 1957. A good overview of momentum, work, and energy. (Chapter 10)

GAMOW, GEORGE, *Gravity.* Doubleday Anchor, 1962: Science Study Series. (Chapters 4 and 6)

HOLTON, G., and ROLLER, D. H. D., *Foundations of Modern Physical Science.* Addison-Wesley, 1958. (Chapter 18)

Potential Energy

CHAPTER

16

In the last chapter we focused our attention on the transfer of kinetic energy from one moving body to another. We found that for completed collisions within an isolated system the loss of kinetic energy of one body must equal the gain of kinetic energy of the other as long as the force of interaction depends solely on their separation. Then with the help of the law of conservation of momentum we were able to calculate the final velocities of two bodies colliding head on in terms of their masses and their velocities prior to the interaction.

In this application, the law of conservation of momentum, $-\Delta\vec{\mathbf{p}}_1 = \Delta\vec{\mathbf{p}}_2$, and the law of conservation of kinetic energy, $-\Delta E_{K_1} = \Delta E_{K_2}$, seem strikingly similar; but there is an important difference. The changes in momentum are equal and opposite in any time interval, and the momentum is therefore conserved instant by instant throughout the interaction. On the other hand, even in elastic collisions the total kinetic energy is not the same at all stages of the interaction. Only at the end of the interaction does it return to its initial value. During the collision, the total kinetic energy first decreases and then increases. At intermediate stages some of the kinetic energy has disappeared.

What happens to this lost kinetic energy? Because it all comes back, we may reason that it is stored somehow in the interacting system. We call this stored energy the *potential energy* of the system.

16-1 The Spring Bumper

Here is a simple example of energy being stored. Consider a mass m sliding with constant velocity on a horizontal frictionless table (Fig. 16–1). The mass collides with a spring bumper attached to a large body so massive that it hardly moves. When the moving mass hits the spring, the spring is compressed. It exerts a force back on the moving mass, slowing it down. The kinetic energy of the moving body decreases until the speed is zero. At this point the kinetic energy of the moving body has disappeared, and the spring is compressed a maximum amount. All the energy is stored as potential energy. After that the mass picks up speed in the opposite direction. Finally it leaves the spring with its original speed and kinetic energy. All the kinetic energy lost during the compression has been regained. At intermediate compressions, the energy was partly kinetic and partly potential.

In the last chapter we saw that kinetic energy will be completely regenerated if the force depends only on separation. Here compression plays the same role as separation. When the kinetic energy is stored and can be recovered completely, we suspect that the force exerted by the spring is the same at a given separation on the way in, while the spring is being compressed, as on the way out, when it is expanding.

Measurements of the force as a function of compression confirm our suspicion. A typical curve for the restoring force exerted by a good spring looks like Fig. 16–2. The force exerted does not depend on the past history. It has the same value at a given compression if we have just pushed the

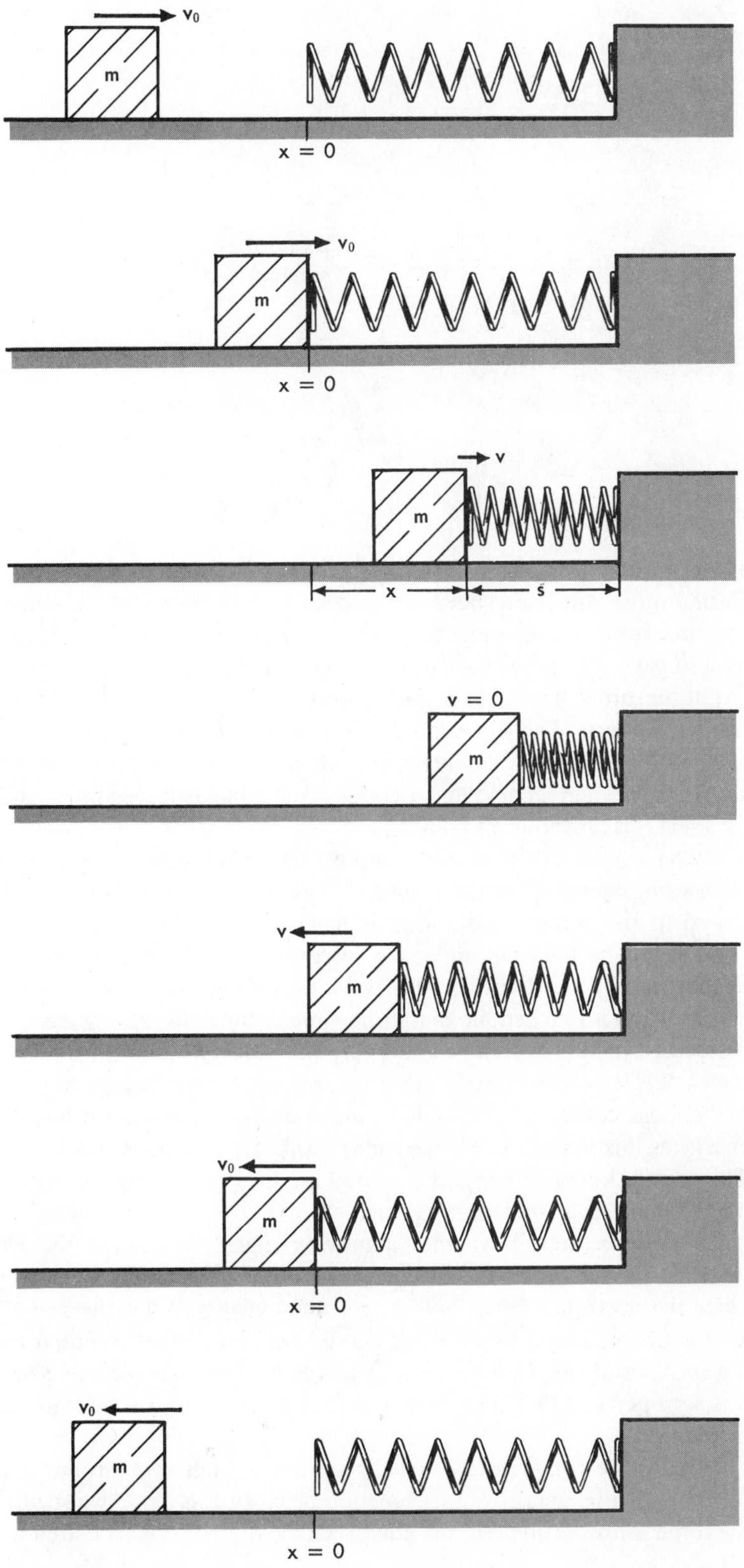

Figure 16–1

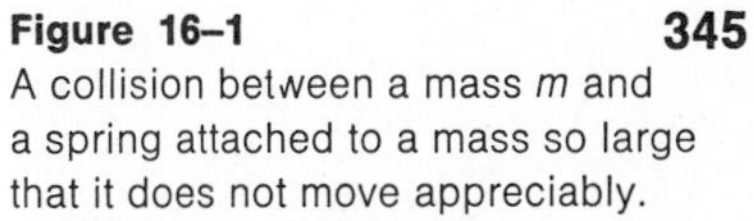

A collision between a mass *m* and a spring attached to a mass so large that it does not move appreciably.

A mass *m* is approaching a spring bumper with speed v_0.

It hits the spring and starts to compress it.

When the spring is compressed a distance *x* to separation *s*, the speed of the mass has decreased to *v*. The mass has lost kinetic energy, which has been stored as potential energy by the compressed spring.

At maximum compression, the mass has come to rest. All its kinetic energy has disappeared.

As the spring is still shoving, the mass gains speed and kinetic energy.

The mass has returned to the place where it first hit the spring. It now has its original speed v_0, and its original kinetic energy. The interaction is completed.

The mass continues to move away with speed v_0 and its original kinetic energy.

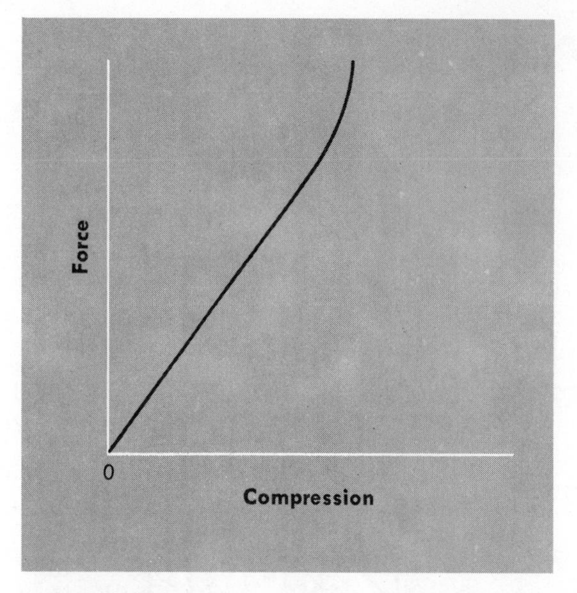

Figure 16–2
A graph of the restoring force F exerted by a good spring, as a function of its compression.

spring in that far or if we have pushed it in farther and then let it expand. Furthermore, the force does not depend on the speed of the mass. Since the force is always the same at the same compression, we can represent force by a single curve, as we have done in Fig. 16–2.

On the other hand, for a poor spring—a copper coil, for instance—the history matters. The forces on the way in and on the way out differ. When such a coil is hit by a moving mass, the mass bounces away, moving more slowly than when it came in. A copper coil is inelastic like putty and warms up when it is moved in and out.

When a mass hits a spring, compressing it and losing kinetic energy, the transfer of energy from the kinetic energy of the mass into potential energy stored in the compressed spring is measured by the work. This work is represented by the area under the force-compression curve. When we are dealing with a good spring, this work is always the same when the same compression is reached, no matter how many times the spring has been compressed. The loss of kinetic energy by the moving mass is therefore always the same. It does not matter what the original kinetic energy of the mass is. If the mass comes in with higher kinetic energy, it will have higher kinetic energy as it is passing x; but the change ΔE_K between zero and x is the same. This loss of kinetic energy, the work done in compressing the spring, is the potential energy stored in the spring.

It does not matter whether a moving mass compresses the spring or whether we compress it. If we compress the spring a distance x by hand, place the mass at its end, and let go, the spring will flip the mass off with kinetic energy equal to the work we did by hand. The potential energy of the spring is again given by the area under the force-compression curve, and it is a property of the compression of the spring without any reference to the moving mass.

If the force-compression curve is simple enough, we can find a formula for the potential energy U as a function of x. In Fig. 16–3 we see that when the force is proportional to the compression, $F = kx$, and the area under the

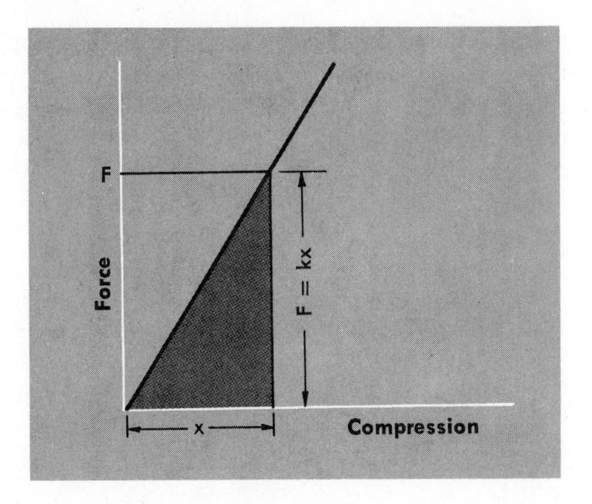

Figure 16–3
The graph of F versus x for a spring, when $F = kx$. The area under the curve is the triangle with base x and height kx. The area, and thus the potential energy, is given by $\frac{1}{2}kx^2$.

curve is a triangle with base x and altitude kx. The potential energy is therefore

$$U = \tfrac{1}{2}(\text{altitude})(\text{base})$$
$$= \tfrac{1}{2}(kx)x = \tfrac{1}{2}kx^2.$$

This formula gives the potential energy for a spring with a linear restoring force (Fig. 16–4).

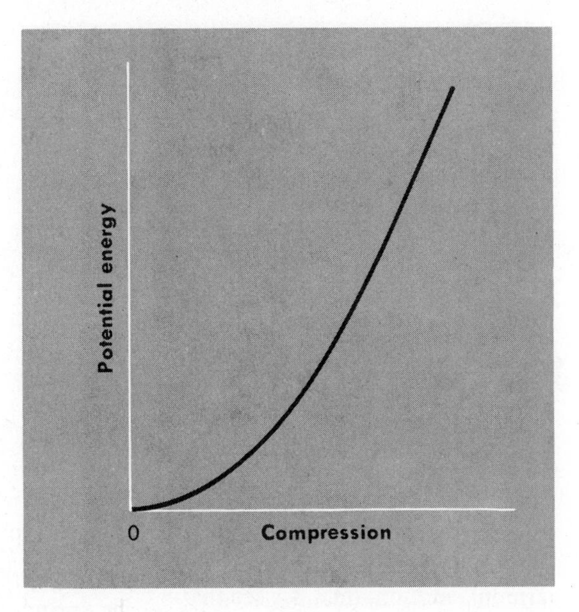

Figure 16–4
The graph of potential energy versus compression for the spring with the force-compression curve of Fig. 16–2.

The graph of U versus x can be checked experimentally (whether it was obtained from a formula or from the areas under the force-compression curve). Compress the spring by different amounts x and let it go, each time accelerating a known mass. Measure the kinetic energy which each mass acquires. Its value should be the same as U at compression x.

The kinetic energy of the mass when it leaves the spring is equal to the potential energy when the mass is at rest with the spring compressed its maximum amount. This is because all the potential energy is turned into kinetic energy on the way out. At intermediate points the gain in kinetic energy and the loss of potential energy are equal, and the sum is constant: that is,

$$\tfrac{1}{2}mv^2 + U = E.$$

16-2 Energy in Simple Harmonic Motion

We call E the total energy of the spring and the mass. It is equal to the potential energy at maximum compression when $v = 0$. It is also equal to $\frac{1}{2}mv_0^2$, the kinetic energy when $U = 0$ and the mass is just leaving the spring.

When we fasten a mass m to the end of any spring and set the system in motion, it oscillates back and forth. The expression

$$\tfrac{1}{2}mv^2 + U = E$$

enables us to calculate the speed v from the potential energy U at any stage of the motion (Fig. 16–5). For example, if the spring exerts a linear restor-

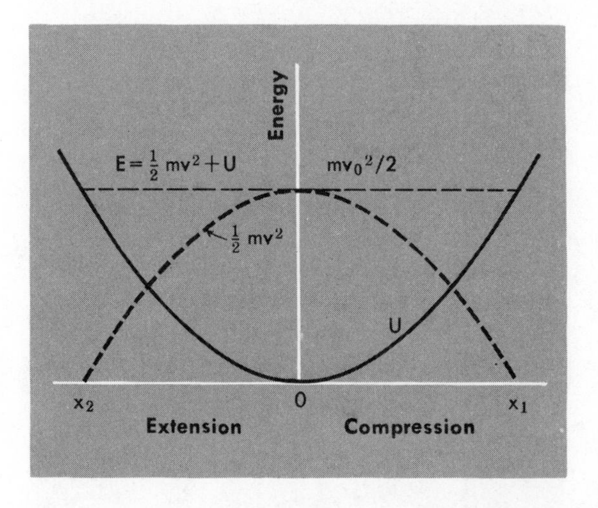

Figure 16–5
A graph of the potential energy curve U of a spring and the kinetic energy curve $\frac{1}{2}mv^2$ of a mass attached to the end of the spring as a function of the displacement x. The sum E is a horizontal line, indicating that the sum of the energies is constant. The mass has just enough energy to compress the spring to $x = x_1$ or expand it to $x = x_2$.

ing force, we know that the mass oscillates in simple harmonic motion (Section 12–8). Also we know that $U = \frac{1}{2}kx^2$ for such a force. For simple harmonic motion, then, we see that

$$\tfrac{1}{2}mv^2 + \tfrac{1}{2}kx^2 = E.$$

For example, suppose we have a spring of force constant $k = 2$ newtons/meter. We attach a mass $m = 8$ kg to it and pull it out 1 meter from its rest position. Then we let the mass go. What will be its speed as it passes through the rest position? To answer this question, we first notice that the mass is not in motion when we let it go. The kinetic energy $\frac{1}{2}mv^2$ is therefore zero, and the constant total energy E is equal to the potential energy $\frac{1}{2}kx^2$. Therefore, when we let the mass go

$$E = \tfrac{1}{2}kx^2 = \tfrac{1}{2}\left(2\,\frac{\text{newtons}}{\text{meter}}\right)(1\text{ meter})^2$$
$$= 1\text{ joule.}$$

On the other hand, when the mass passes the rest position $x = 0$ the total energy is all kinetic; that is,

$$E = \tfrac{1}{2}mv^2 = \tfrac{1}{2}(8\text{ kg})v^2 = 1\text{ joule.}$$

Consequently, $v^2 = \frac{1}{4}\left(\frac{\text{meter}}{\text{sec}}\right)^2$ and the speed is $\frac{1}{2}$ m/sec.

16-3 Potential Energy of Two Interacting Bodies

Now consider one mass colliding head on with another, as in Chapter 15. Suppose mass A is projected toward mass B, which is at rest (Fig. 16–6). As we learned in Section 15–3, when A reaches the range of interaction d, it begins to slow down and lose kinetic energy, while B starts moving faster and faster; but B does not gain as much kinetic energy as A loses. Kinetic energy is disappearing. When A and B are closest together they are both moving with the same velocity, and the total kinetic energy is at a minimum. Suppose that at just that instant a light cage (of negligible mass) is dropped over A and B to prevent them from separating. Then the whole system—A

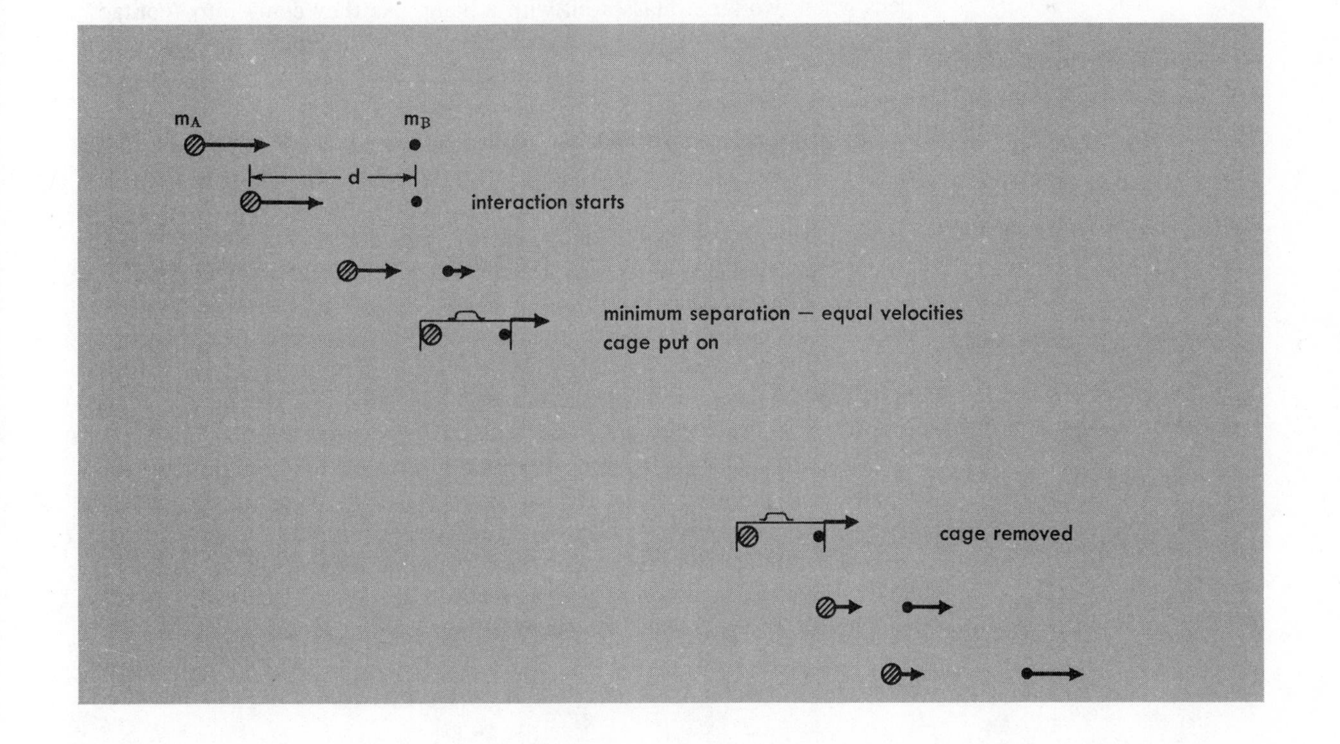

Figure 16–6
An interaction between two masses, which begins when their separation is *d*. At their closest approach, when their velocities are the same, we enclose them in a very light cage. They will continue to move along together at the same velocity, with minimum kinetic energy. We can remove the cage at any time; the masses will move apart again and regain the original kinetic energy.

and B and their cage—continues with unchanging velocity and constant kinetic energy. It will continue to move with this minimum kinetic energy until we remove the cage. If we look at kinetic energy alone, the system seems to have less energy than it had originally. However, if we remove the cage so that A and B can push apart, we get back to the original kinetic energy, provided that the interaction forces depend on the separation alone. Here we are considering only those interactions which provide the same forces at the same separation, whether A and B are approaching or moving apart. Such forces make the collision elastic.

While the cage is on, we have less kinetic energy than we had originally, and we say that the missing kinetic energy is stored as potential energy. We can keep it stored as long as we like. And whenever we allow the two interacting masses to move apart again, the kinetic energy will increase by just the amount that we stored when we trapped them in the cage.

Whenever the two masses are held at a given separation, there is a definite potential energy. This does not depend on how fast the system is moving or on how the masses were pushed together. And all this energy will come out as kinetic energy when we let them go. The whole arrangement is very much like the spring bumper—where we could have stored the potential energy by applying a latch to hold the spring compressed. However, this time we have no visible spring. All we have are the forces, and we say that this potential energy is stored in the force field of the interaction. The force field behaves like an imaginary spring.

Actually we should say things the other way around. In a spring the potential energy is really stored in the force fields of interactions between its atoms. The visible shape of the spring merely tells us where its atoms are.

To understand more fully what happens to the atoms, consider what happens when two large masses hit with a bang. As they come into "contact" atoms of one body approach the nearest atoms of the other so closely that large interaction forces arise and energy is stored in the interatomic force fields of both masses.

Obviously only some "contact" collisions (for example, those of rubber or steel balls) can store the energy in returnable form. Others leave the atomic systems in a jangle of motion; the colliding masses are left hotter and move apart with less visible kinetic energy. The collision is inelastic. When we look at one large mass colliding inelastically with another, we see that the forces are not the same on the outward trip after collision as on the inward trip. Some of the visible kinetic energy of the motion of the colliding masses disappears permanently. (We calculate it by using $\frac{1}{2}mv^2$ for each mass and do not go into atomic details.) Because this lost kinetic energy is not directly returnable, we do not regard it as potential energy.

For elastic collisions, the potential energy depends only on the separation s between the masses A and B. It is given by the work neccessary to bring them to this separation from a separation larger than the range of interaction. It is the shaded area under the force-separation curve shown in Fig. 16–7. Because the potential energy is only a function of separation, it does not depend on the method of bringing the masses to separation s. We have seen that we can let the masses collide head on with any kinetic energy or push them together by hand. The resulting potential energy is the same.

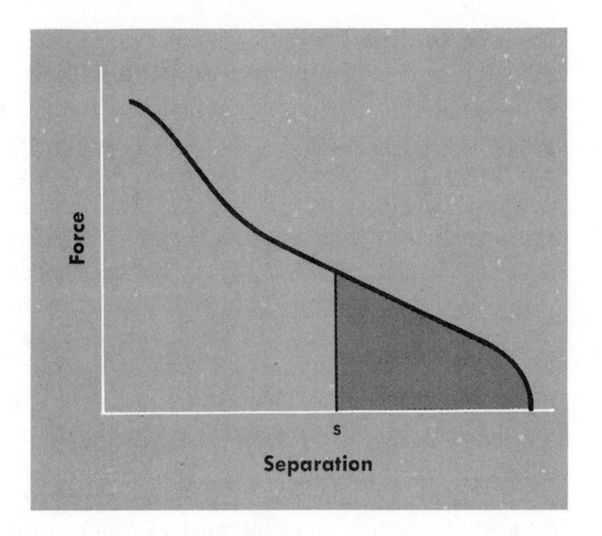

Figure 16–7
A graph of force versus separation for two interacting masses. The shaded area represents the work required to bring the masses to separation s. It gives their potential energy when they are s apart.

The potential energy is useful because it is a function of the state of the system at the moment and not of its history. Because it is independent of history, we do not need to restrict ourselves to collisions that occur along a fixed straight line. So long as the forces of interaction between the two masses depend only on their separation there is a definite potential energy at a given separation. The masses may be moving in any direction.

Let's look at such a collision in somewhat more detail. The displacement of each mass during a short time interval Δt can be split into two components, one along the line joining them and one perpendicularly across it (Fig. 16–8). One of these components is in the direction of the force acting on the mass and the other is perpendicular to it. From Chapter 15 we know

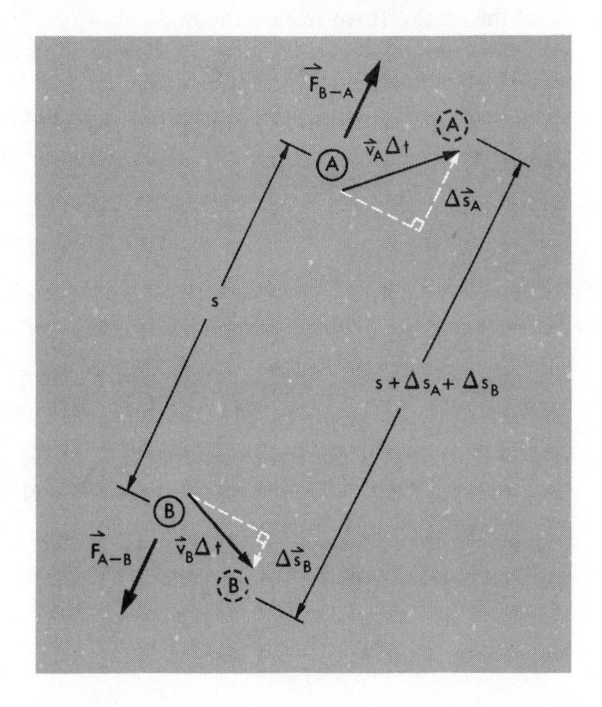

Figure 16–8
An elastic interaction between bodies A and B. When the time interval Δt is taken small enough, the force on each body can be considered essentially constant, so the work done on it is F times the component of the displacement in the direction of the force. This equals the potential energy given up by the force field and transferred to the body as kinetic energy.

that when force and displacement are perpendicular to each other, no work is done. So the components of the displacements perpendicular to the forces are not involved in any transfer of energy. On the other hand, when force and displacement are in the same direction work is done, energy is transferred. Consequently (since the forces are equal in magnitude and along the displacements $\Delta\vec{s}_A$ and $\Delta\vec{s}_B$), the magnitude of the forces of interaction times the net change in separation measures the transfer from potential to kinetic energy just as it does for a head-on collision.

We can apply the same treatment to any isolated system of masses, however complex, so long as all the forces are Newtonian forces—that is, so long as the forces of interaction between each pair of masses are equal and opposite and depend only on the separation. Under these conditions, the potential energy can be defined and the total mechanical energy (kinetic plus potential) of an isolated system remains constant.

We can now see that potential energy may be related to the energy stored in fuels: in gasoline or in the nuclei of atoms, for example. In using fuels we do not release the energy by removing a cage; but we can turn a substantial fraction of the energy stored in fuel into the kinetic energy of the motions of large masses. By using the concepts of kinetic and potential energy, we can understand many of the complex transformations of energy that go on in nature.

16-4 Gravitational Potential Energy Near the Surface of the Earth

Let us examine the potential energy associated with the gravitational attraction between the earth and other objects. Consider the system formed by the earth and by a body whose mass m is very small compared with the mass M of the earth. If we release the body above the earth's surface, it will fall. Simultaneously the earth will rise (slightly) toward the body. Each mass gains kinetic energy while the potential energy of the system decreases.

At any one instant of time the momenta of the body and of the earth are equal and opposite. Therefore the speed v of the body and the speed V of the earth satisfy the equation

$$mv = MV \quad \text{or} \quad V = \frac{m}{M}v.$$

From this relation we can compute the ratio between the kinetic energies of the earth and the body. For the kinetic energy of the earth E_{K_e} we get

$$E_{K_e} = \tfrac{1}{2}MV^2 = \tfrac{1}{2}M\left(\frac{m}{M}v\right)^2 = \frac{m}{M}(\tfrac{1}{2}mv^2).$$

Because ($\frac{1}{2}mv^2$) is the kinetic energy E_{K_m} of the mass m, we see that the ratio between the kinetic energy of the earth and the kinetic energy of the small falling mass is always

$$\frac{E_{K_e}}{E_{K_m}} = \frac{m}{M}.$$

Even for $m = 10^3$ kg, $\dfrac{m}{M} \approx 10^{-21}$.

Therefore, the kinetic energy of the earth can be neglected compared with that of the body. As m falls toward the earth, all the potential energy of the system composed of the earth and the mass m is transferred into the kinetic energy of the motion of m. We shall therefore drop the index "m" and refer to the kinetic energy of the mass as E_K.

Now assume that m moves through a distance which is small compared with the radius of the earth. This is certainly true in any laboratory experiment. Then the force of attraction between the earth and the body has a constant magnitude mg. It also has a constant direction (vertically down).

Under the influence of this force a falling body gains speed with a constant downward acceleration g. As we know from our previous work (Section 9–7), if the speed increases from v to v' as the body moves downward a distance d, then

$$v'^2 - v^2 = 2gd$$

and the change in kinetic energy is

$$\begin{aligned} E'_K - E_K &= \frac{m}{2}(v'^2 - v^2) \\ &= mgd. \end{aligned}$$

The left side of this equation is the change in kinetic energy, while the right side is the work (force times distance) which measures the transfer of potential energy *into* kinetic energy when the separation between the small mass and the center of the earth *decreases* by the distance d.

In falling from height h above the surface of the earth to height h' a body moves down the distance $d = h - h'$, and the potential energy changes by

$$\begin{aligned} U' - U &= -mgd \\ &= -mg(h - h') \\ &= mg(h' - h). \end{aligned}$$

(The minus sign in $-mgd$ shows that potential energy is lost and kinetic energy gained.) From this result we may conjecture that the potential energy at height h is

$$U = mgh.$$

which we know is the right change in the potential energy. However, we also get the right change in potential energy if we add any constant U_o to the expression for U. That is, if

$$U = U_o + mgh$$

and

$$U' = U_o + mgh',$$

then

$$U' - U = mg(h' - h).$$

The value of U_o makes no difference to our calculation of the change in potential energy.

In physical problems we deal only with changes in potential energy, and the value of U_0 can be chosen at our convenience. When we are working near the surface of the earth we often choose $U_0 = 0$. This gives the potential energy the value zero at the earth's surface. Even so, we often have to make a choice of what we shall call the earth's surface. Is it the top of a building, the surface of a street, the bottom of a hole, or sea level (Fig. 16–9)? We can make any choice which is convenient. Furthermore, when we work with spacecraft that can go far from the earth we usually choose the potential energy zero infinitely far from the earth. Basically the choice does not matter.

In the equation

$$U' - U = mg(h' - h) = -mgd,$$

mgd is the downward force mg on the mass m times the distance d it falls. This expression, therefore, gives the change in the kinetic energy as the mass falls from h to h'; that is,

$$mgd = E_K' - E_K.$$

Consequently

$$U' - U = -mgd = -(E_K' - E_K).$$

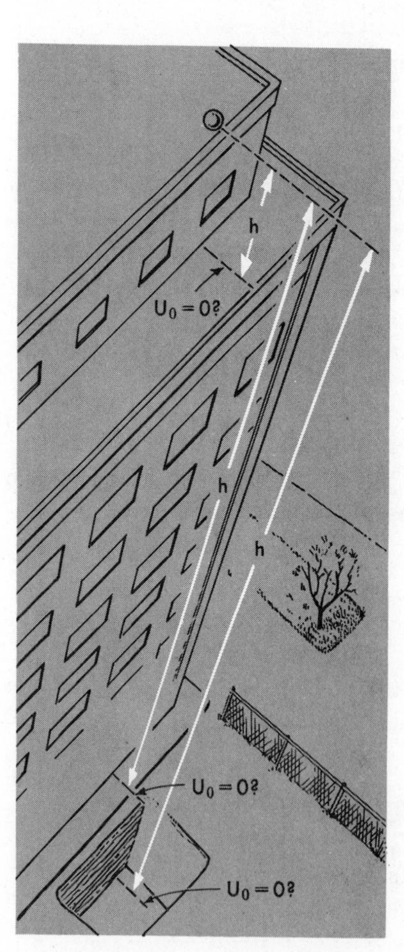

Figure 16–9
We may choose any height we want for the location of zero potential energy. The choice is arbitrary and makes no difference, since we deal only with changes in potential energy.

This equation states that the changes in potential energy and in kinetic energy are exactly opposite. As the mass falls, the potential energy decreases and the kinetic energy increases by the same amount. The work mgd measures the transfer. On the way up, the kinetic energy decreases and the potential energy increases by the same amount. Again the transfer is measured by mgd.

Since any change in potential energy is matched by an equal and opposite change in kinetic energy, the sum of potential and kinetic energies remains constant. We can see that this is so by rewriting our last equation as

$$U + E_K = U' + E_K'.$$

The left side of this equation gives the total energy E at one time, and the right side gives the total energy at any other time. The equation shows that this total energy

$$E = U + E_K$$

is the same at any two different times. It is constant, although the values of U and E_K may change. This conservation law applies to the system when it is isolated from external influences which might do work on it and thus change the total energy. It should be clearly understood that no other forces are acting. The relationship between total energy, potential energy, and kinetic energy is shown in Fig. 16–10.

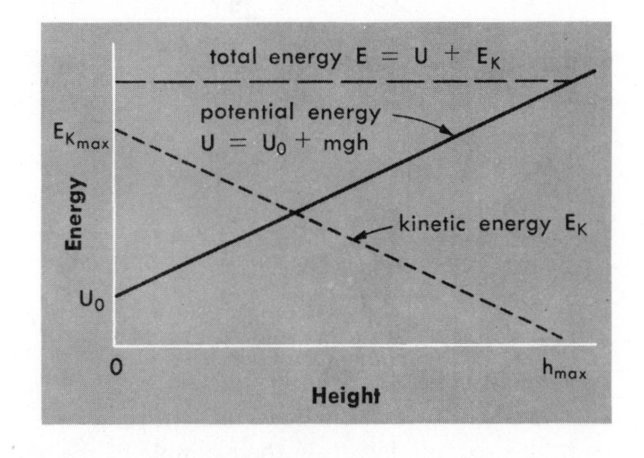

Figure 16–10
A graph of the potential, kinetic, and total energies for a mass m moving near the earth's surface. Gravity is the only force acting. The potential energy of the earth-mass system is chosen as U_0 at $h = 0$. When the total energy has the value shown, the mass can rise to a maximum height h_{max}, at which point it has no kinetic energy left.

If we choose $U_o = 0$, we can write the total energy as

$$\begin{aligned} E &= U + E_K \\ &= mgh + \tfrac{1}{2}mv^2. \end{aligned}$$

This equation is often valuable because it gives us information about the speed of m at different places without reference to the details of the motion from one place to the other. For example, when a projectile moving upward passes the height h, it has the same speed v that it will later have at the same height on its way down. This must be true because mgh is the same on the way up as on the way down, and E is always the same. Therefore, $\frac{1}{2}mv^2$ must have the same value even though the direction of motion has changed.

As a specific application of

$$E = mgh + \tfrac{1}{2}mv^2,$$

let us suppose that a mass of 1.00 kg is moving in any direction with a speed of 1.00 m/sec at a height of 3.00 meters above the earth. What will be its speed when it is 2.00 meters above the earth?

We can readily calculate its speed if we know its kinetic energy. And since the total energy E does not change, the kinetic energy of the mass at 2.00 meters above the earth will be the difference between the total energy and the potential energy at 2.00 meters. The total energy is the sum of the initial potential energy (at 3.00 meters height)

$$\begin{aligned} U = mgh &= (1.00\ \text{kg})(9.80\ \text{nt/kg})(3.00\ \text{m}) \\ &= 29.4\ \text{joules}, \end{aligned}$$

and the initial kinetic energy

$$\begin{aligned} E_K = \tfrac{1}{2}mv^2 &= \tfrac{1}{2}(1.00\ \text{kg})(1.00\ \text{m/sec})^2 \\ &= 0.500\ \text{joule}. \end{aligned}$$

That gives

$$E = U + E_K = 29.4 + 0.5 = 29.9\ \text{joules}.$$

At 2.00 meters height the potential energy is

$$\begin{aligned} U' = mgh' &= (1.00\ \text{kg})(9.80\ \text{nt/kg})(2.00\ \text{m}) \\ &= 19.6\ \text{joules}, \end{aligned}$$

and therefore at 2.00 meters height the kinetic energy is

$$\begin{aligned} E_K' = \tfrac{1}{2}mv'^2 = E - U' &= 29.9 - 19.6 \\ &= 10.3\ \text{joules}. \end{aligned}$$

When we substitute $m = 1.00$ kg and solve for v' we find

$$v' = 4.54\ \text{m/sec}.$$

In each of the above motions, the total energy E remains constant so long as the force of gravity is the only force. But we can change the total energy E by pushing the mass, thus doing some work on it. If we then release it, the mass will move with a new constant value of E.

In this section, just as in the example of the mass colliding with a spring bumper, we have been able to find a potential energy which depends only on the position of the mass. This potential energy decreases by exactly the amount that the kinetic energy increases. Just as in the case of the spring bumper, the sum of the potential and kinetic energies remains constant.

Two examples may serve to illustrate the great importance of gravitational potential energy for us. In a pile driver, for example, we increase the gravitational potential energy by raising a large mass. Then by releasing the mass and allowing it to fall freely, we let the gravitational potential energy turn into kinetic energy. This energy drives piles into the earth.

By building dams we can hold water at a greater distance from the center of the earth than that at which it would otherwise stay. By allowing the water to drop from the top of a dam to a lower level, we convert potential energy into other forms—for running mills or producing electrical energy to drive motors or light lamps. Each kilogram of water we allow to fall through 10 meters can do work equal to $1(9.8)10 = 98$ joules. By dropping a kilogram through 10 meters every second, we can keep an ordinary light bulb going.

A standard workhorse* can do 750 joules of work per second. If three of them take shifts, they can light a few light bulbs all year for only a few hundred dollars. The power company will do it for about one tenth that cost by using the potential energy from stored water or coal.

16-5 Gravitational Potential Energy in General

In the last section we determined the gravitational potential energy when a mass m is near the surface of the earth. We assumed that the strength of the gravitational field was constant. On the other hand, we know that when two masses move over larger distances, the gravitational force between them changes. What is the correct expression for the gravitational potential energy in this more general case? What, for example, is the gravitational potential energy of the system consisting of the earth and a satellite?

As we know from Chapter 13, the force of attraction between the earth M and a satellite m is GMm/r^2, where r is the distance to the satellite from the center of the earth. The force-separation curve in Fig. 16–11 is a graph of this function.

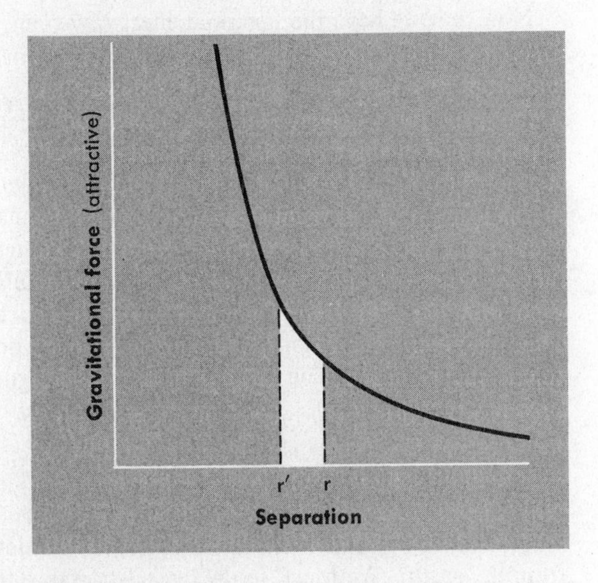

Figure 16–11
A graph of the force of gravitational attraction versus the distance to the center of the earth. The unshaded area represents the work needed to increase the separation between the earth and a body from r' to r. The dark-gray area is the work needed to bring the body from separation r to infinity.

The area under this curve between two different separations measures the work done as the separation changes. For instance, the unshaded area represents the work needed to increase the separation from r' to r. Using

* The output of workhorses was determined by James Watt in order to make a comparison with the output of his steam engines. More accurately, his standard is 746 joules/sec and is called a horsepower in honor of the horse. A joule/sec is called a watt in honor of Watt.

mathematical methods somewhat more complicated than we wish to discuss here, we can find the area under the force-separation curve from any separation r to infinity. This is the dark gray-shaded area extending infinitely to the right from r. It is

$$\frac{GMm}{r}.$$

This, then, is the work to pull the bodies apart from separation r to infinite separation. This work measures the energy transferred into potential energy of separation to force the bodies apart. The work is therefore equal to the difference $U_\infty - U_r$ between the potential energy at infinite separation and the potential energy at separation r; that is,

$$U_\infty - U_r = \frac{GMm}{r}.$$

On solving this equation for U_r, we see that the potential energy at the separation r is

$$U_r = U_\infty - \frac{GMm}{r},$$

and if we set the potential energy equal to zero at infinite separation, we get

$$U_r = -\frac{GMm}{r}.$$

Now that we have the potential energy, we can get the total energy E by adding the kinetic energy E_K to it. The result is

$$E = E_K + U_r = E_K - \frac{GMm}{r}.$$

Just as we found for masses moved by springs and for objects moving near the earth's surface, the total energy is conserved. For any motion under the influence of the gravitational attraction alone, what is gained in potential energy as the bodies separate is paid for by a reduction in kinetic energy. For instance, as a satellite goes around the earth it goes slowest when it is farthest away and fastest when it is nearest the earth. As the earth goes around the sun the same relations are true.

16-6 Escape Energy and Binding Energy

With what energy must we launch a rocket in order for it to escape entirely from the earth? We found in the last section that the gravitational potential energy for a mass m at a separation r from the center of the earth is just

$$U_r = -\frac{GMm}{r}.$$

Here M is the mass of the earth, and the zero of potential energy has been chosen at infinite separation. If we wish to launch the rocket from the surface of the earth (where $r = r_e$) with enough energy so that it will just be able to get out to very large separation with no extra energy left over, we must give it in the launching an initial kinetic energy of

$$E_K = +\frac{GMm}{r_e}.$$

Then the change in kinetic energy, from $+\frac{GMm}{r_e}$ to zero, just compensates for the change in potential energy, from $-\frac{GMm}{r_e}$ to zero, or $\Delta E_K = -\Delta U$. This amount of kinetic energy is the minimum which the rocket must have to just barely escape from the earth's gravitational pull. You will notice that this escape kinetic energy, $E_K = \frac{GMm}{r_e}$, is directly proportional to the mass of the satellite; to blast off a more massive rocket requires more energy, just as you would expect. If we put in numerical values for the gravitational constant G, the mass of the earth M, and the radius of the earth r_e from Chapter 13, we find that the required escape kinetic energy is 6.24 $\times 10^7$ joules for each kilogram of mass of the rocket. For a rocket having a mass of one metric ton (1000 kg) to escape completely from the earth requires that we give it an energy of 6.24×10^{10} joules.

We can also find the initial speed v_e which the rocket must have in order to escape from the earth. Since kinetic energy is $\frac{1}{2}mv^2$, we can write the escape kinetic energy as

$$E_K = \frac{GMm}{r_e} = \tfrac{1}{2}mv_e^2.$$

Since the mass of the rocket appears in both expressions for E_K, we can eliminate it and find

$$\tfrac{1}{2}v_e^2 = \frac{GM}{r_e} = 6.24 \times 10^7 \frac{\text{joules}}{\text{kg}}$$

which is now independent of the rocket mass. [If you work out the units, you will find that 1 joule/kg = 1(m/sec)^2.] Thus,

$$v_e = \sqrt{12.48 \times 10^7} = 1.12 \times 10^4 \text{ m/sec}$$

or 11.2 km/sec. The quantity v_e, usually called the "escape velocity," does not depend on the mass which is being launched. It is the initial speed that must be given to *any* body in order for it to escape from the earth and never return.

Some rockets are designed not to escape the earth's gravitational pull, but to place a satellite into a fairly stationary orbit around the earth just a few hundred kilometers above its surface. It is interesting to compare the energy needed to place a satellite in orbit with the energy that would be required to cause it to escape. To simplify the calculation, let us make two fairly good assumptions. First, we shall assume that the orbit is circular—this requires very delicate control of the launching conditions, but it can be done; second, we shall approximate the radius of the orbit by the radius of the earth; if the satellite is just above the earth's atmosphere, as in manned satellites, for example, these radii differ by only two or three percent.

In order to move in a circular orbit of radius r_e at constant speed v, a satellite of mass m requires a centripetal force $\frac{mv^2}{r_e}$ which will be supplied by the gravitational force of the earth $\frac{GMm}{r_e^2}$. Therefore,

$$\frac{mv^2}{r_e} = \frac{GMm}{r_e^2} \quad \text{or} \quad mv^2 = \frac{GMm}{r_e}.$$

Thus the kinetic energy of the orbiting satellite is

$$E_K = \frac{1}{2} mv^2 = \frac{1}{2} \frac{GMm}{r_e}.$$

This is just half of the kinetic energy of escape. In other words, to put a satellite into orbit just above the earth's atmosphere requires only half the energy that it takes to throw the satellite permanently away from the earth.

Now, an orbiting satellite is still in the gravitational field of the earth; so, in addition to kinetic energy, there is also a potential energy of separation $U = -\frac{GMm}{r_e}$ at the radius r_e. We can therefore write the *total energy* of the orbiting satellite-earth system as

$$E = E_K + U = \frac{1}{2} \frac{GMm}{r_e} + \left(-\frac{GMm}{r_e}\right)$$
$$= -\frac{1}{2} \frac{GMm}{r_e}.$$

What is the meaning of the negative sign on the total energy? In the previous section we chose the potential energy to be zero at infinite separation. If the kinetic energy of the satellite is also zero at infinite separation, then the total energy is zero as well. The negative total energy, then, indicates that we must do work on or supply energy to the system in order to get it to the zero total energy condition of no velocity and infinite separation. When the satellite has less than enough energy to escape from the gravitational field of the earth, we say that it is bound to the earth. The energy that we must supply to overcome the binding and just allow the satellite to escape is called the "binding energy."

For the satellite orbiting the earth just above the earth's atmosphere, we must supply the energy to bring E up to zero, so

$$\text{binding energy} = +\frac{1}{2} \frac{GMm}{r_e}.$$

For our metric-ton satellite in such an orbit, the binding energy comes out to be 3.12×10^{10} joules. For a metric-ton satellite which is stationary on the earth's surface before launching, the total energy is just the potential energy $-\frac{GMm}{r_e}$, so the binding energy needed to overcome this is $+\frac{GMm}{r_e} = 6.24 \times 10^{10}$ joules (just twice as much as for the orbiting case). Your own binding energy on the surface of the earth (the energy necessary to send you to outer space) amounts to about 4 or 5×10^9 joules.

We can even find the binding energy of the earth to the sun: it is just

$$\text{binding energy} = \frac{GM_{\text{sun}}m_{\text{earth}}}{2r_o},$$

where r_o is the radius of the orbit of the earth around the sun. The proof is identical to the one we did for the satellite orbiting the earth. Putting in values from Chapter 13, we find that the binding energy of the earth to the sun is a bit more than 2×10^{33} joules! That's a lot of energy!

For the general case, when an object in a gravitational field has a total energy

$$E = \tfrac{1}{2}mv^2 - \frac{GMm}{r}$$

which is negative, the object is "bound" by the gravitational field with a binding energy of:

$$-E = \frac{GMm}{r} - \tfrac{1}{2}mv^2.$$

There are binding energies of importance in atomic physics also, that of the electron to the proton in hydrogen, for example. However, the scale of these binding energies is quite different, and they arise from electrical rather than from gravitational forces. We shall postpone our discussion of these atomic binding energies until later in the course.

16-7 Total Mechanical Energy

In this chapter we have investigated the potential energy of systems in which the parts interacted through forces that were equal and opposite and which pointed along a line connecting their centers of mass. The parts acted like point masses; that is, each acted as if its mass were concentrated at its center of mass. Even the gravitational attraction of the earth on a mass nearby appears to come from the point at the earth's center. But it is not clear that all mechanical systems involve only such simple forces of interaction, nor that their parts can be treated as points. What happens to energy in mechanical systems where we cannot be sure that all interactions are like these?

To answer this question, suppose we have a toy merry-go-round (see Fig. 16–12) which is driven by a weight-and-pulley system. Several turns of string are wrapped around the shaft, and then the string goes through a pulley and down to the mass M. If we let the merry-go-round go, it will start to rotate, gradually speeding up as the mass M falls. No one would try to analyze this system by finding a set of interaction forces between all the atoms. But by viewing the merry-go-round as a collection of small masses and finding their speeds, we can calculate the kinetic energy of each piece. Then, adding up, we can find the total kinetic energy of the whirling merry-go-round. We can also measure the distance that the mass M has fallen and the kinetic energy of the mass. When we do such an experiment with a merry-go-round on nearly frictionless bearings, we find that the kinetic energy gained by the merry-go-round and the mass M is just equal to the de-

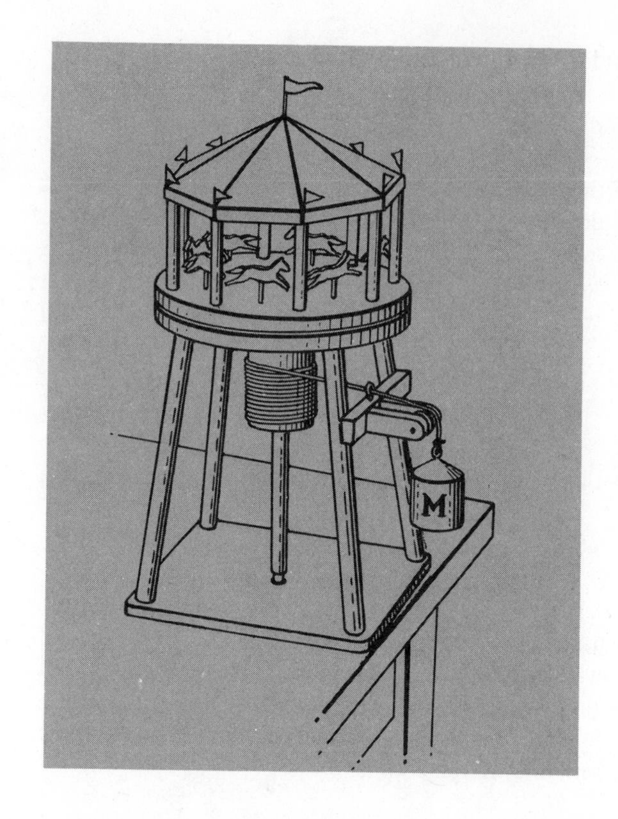

Figure 16–12
A toy merry-go-round driven by a falling weight. As the weight falls, the merry-go-round begins to rotate, turning faster and faster. The whirling merry-go-round and the falling weight gain kinetic energy which is equal to the loss in potential energy of the weight.

crease in the gravitational potential energy as the mass M falls toward the earth. Eventually the string is unwound, but the whirling merry-go-round continues to turn. The string then wraps around the center post in the other direction. The weight rises, and the merry-go-round slows down. When the merry-go-round stops turning, the mass M is almost back up to its original position. It fails to return to the original height only because of the energy transferred into other forms through frictional interactions.

We can consider more and more complex mechanical systems and, without analyzing the forces of interaction in detail, we can often evaluate the kinetic energy and the potential energy. Their sum then represents the total mechanical energy in the system. When frictional interactions are very small, the energy may change back and forth between potential and kinetic, but the total mechanical energy remains constant.

But sometimes we get into trouble; frictional interactions frequently cannot be made so small that we can neglect them. We then see the total mechanical energy disappear and the parts of the mechanical system get warmer. In the next chapter we shall learn how to take account of this energy which is transferred away as heat. Then we shall see why we believe that energy is always conserved.

1.* In Fig. 16–1, where is the acceleration of the block the greatest? (Section 1.)

2.* How could you determine the k for the spring in Fig. 16–3? (Section 1.)

3. A mass of 4.0 kg sliding with a velocity of 3.0 m/sec on a frictionless horizontal table collides with a queer kind of spring bumper. The bumper exerts a constant force of 120 newtons on the mass as it moves in (compressing the spring) and the same force on the way out until the spring is back where it was.
   (a) Is this an elastic collision? How do you know?
   (b) What is the kinetic energy at the beginning of the interaction?
   (c) How much is the spring compressed?
   (d) What is the ratio of kinetic energy to potential energy when the spring has been compressed 10 cm?

4. A 3.0-kg mass moving with a velocity of 2.0 m/sec collides with a spring bumper which exerts a force $F = 100x$, where F is the force in newtons and x is the compression in meters.
   (a) Draw a graph of F versus x from $x = 0$ to $x = 0.40$ m.
   (b) What is the potential energy stored in the spring when $x = 0.10$ m? What is the kinetic energy of the mass at this point?
   (c) What will happen if the spring is compressed 0.10 m by hand and then the 3-kg mass is placed in contact with the spring and the hand is removed?

5. The force-compression curve of a spring is shown in Fig. 16–13.
   (a) How much work is done in compressing the spring 0.3 m?
   (b) What is the potential energy of the spring when compressed this amount?
   (c) Place a 2-kg mass at rest against the spring when it is compressed 0.3 m. Let go. What is the kinetic energy of the mass as it passes the point where the spring is compressed 0.2 m?

6. Suppose you are given two springs of different sizes. How would you set them up so that the force-compression curve of the system looks like Fig. 16–13?

7. Figure 16–14 shows a roller skate with a mass M mounted on it by four hacksaw blades attached to a base. The mass M is equal to the mass of the roller skate plus the mass of the base. If you hold the roller skate and pull the mass M out to the left, and then let the whole system go, the mass moves to the right and the skate to the left. Be prepared to describe the further motion, indicating the form of the energy at various stages.

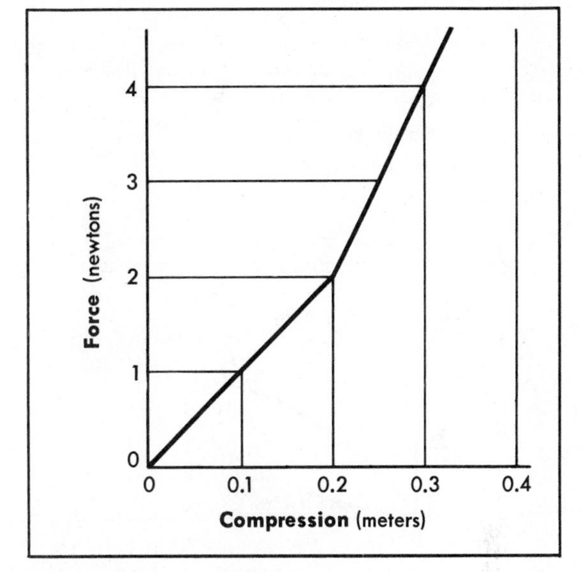

Figure 16–13
For Problem 5.

Figure 16–14
For Problem 7.

8. A linear elastic spring is compressed 0.2 m by a force of 20 newtons.
    (a) What is the force constant k (or force-compression ratio) of the spring?
    (b) What is the equation for the potential energy stored by the spring as a function of its compression?

9.* In Fig. 16–8, we see two masses which have a certain amount of stored energy at separation s and are moving to the right. If we placed the two masses at rest in the same positions and released them, which way would they go? (Section 3.)

10. A 1.0-kg mass moving with velocity of 10 m/sec

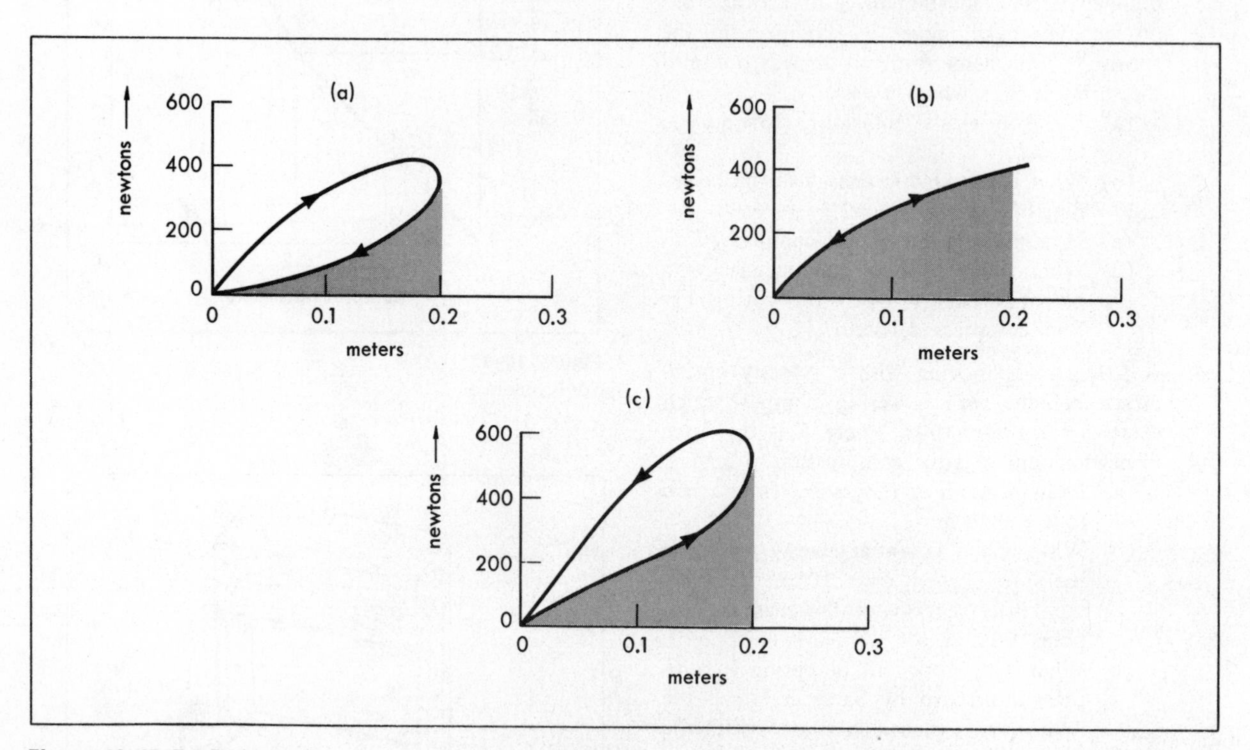

Figure 16–15 For Problem 10.

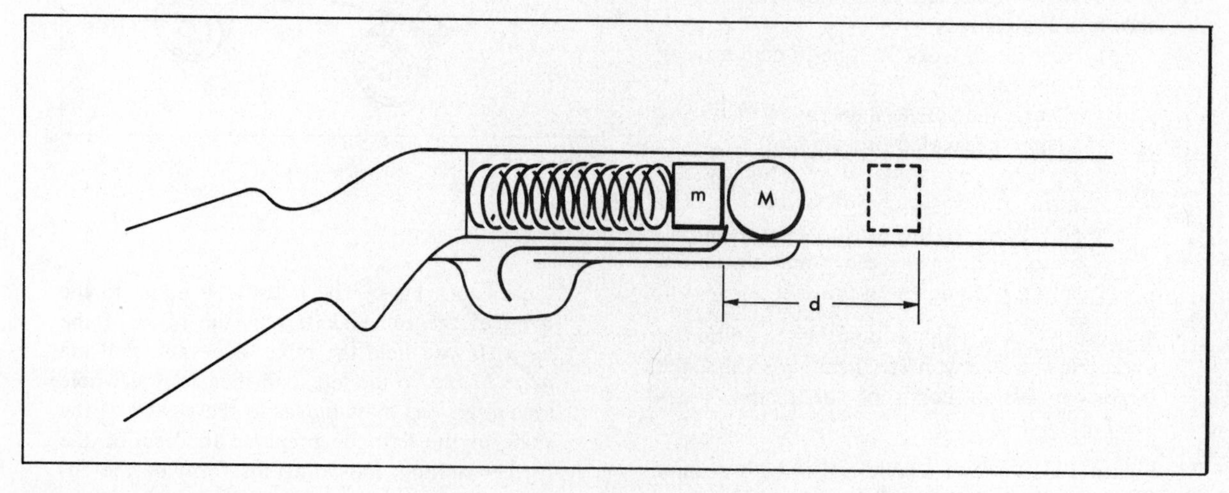

Figure 16–16 For Problem 13.

strikes a spring, compressing it a distance of 0.20 m. The mass rebounds with a speed of 8.0 m/sec.

(a) What is the loss in kinetic energy of the mass?
(b) What happens to this lost energy?
(c) Which of the force-compression curves in Fig. 16–15 do you think is most likely correct for this spring?

11. What will happen if the 3.0-kg mass in Problem 4 hooks itself onto the spring and remains fastened to it?

12. In Fig. 16–6, why does the large ball not gain kinetic energy while the force of the cage acts on it? (Section 3.)

13. Suppose we have a spring-operated popgun something like the one in Fig. 16–16. It has a piston of mass m fastened to the end of a light spring whose force constant is k, and it shoots a ball of mass M. When the gun is loaded, the piston is pushed down the barrel until the spring is compressed through a distance d, and latched to the trigger.

Leaving out friction for the sake of simplicity, work out an argument to decide whether all the potential energy of the spring will be transferred to the ball.

14. A linear spring with constant k is compressed between two Dry Ice pucks of mass m_1 and m_2. The spring is compressed a distance x and then tied with a thread. Both pucks are initially at rest.

(a) If the thread is burned, what is the total kinetic energy of both pucks?
(b) What are the momenta of the pucks, and how are they related?
(c) What is the kinetic energy of each puck in terms of its momentum and its mass?
(d) What fraction of the total energy does each puck acquire? What is their ratio?

15. A spring bumper with restoring force $F = 200x$, where F is in newtons and x in meters, is compressed 0.100 meter. A 0.500-kg mass is placed next to the end of the spring and the whole thing let go.

(a) With what momentum will the mass leave the spring?
(b) We do the same thing with masses 0.125 kg, 2.00 kg, and 8.00 kg. What is the momentum of each as it leaves the spring?
(c) What is the energy of each as it leaves the spring?

16.* A 70-kg boy climbs 5 meters up a rope. What is the increase of his potential energy? (Section 4.)

17. (a) A ball of mass 0.25 kg is thrown to the right at a speed of 2.0 meters per second. It starts along the frictionless surface at the left of Fig. 16–17. How high up the slope at the right will it go before coming momentarily to rest? What kind of motion will it perform?
(b) If the ball is released from rest at point P, what kind of motion will it perform? How high up the slope on the right will it rise?

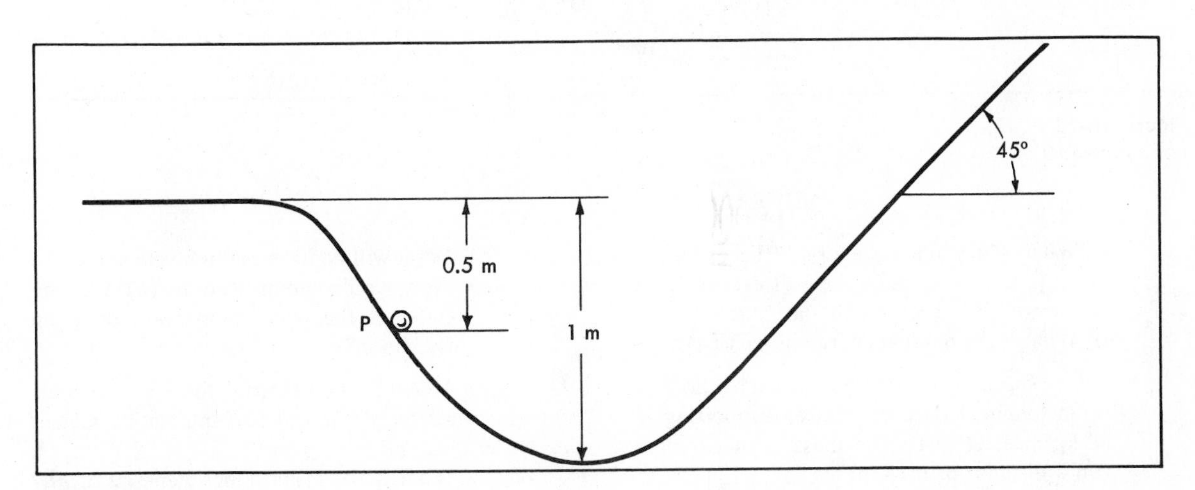

Figure 16–17 For Problem 17.

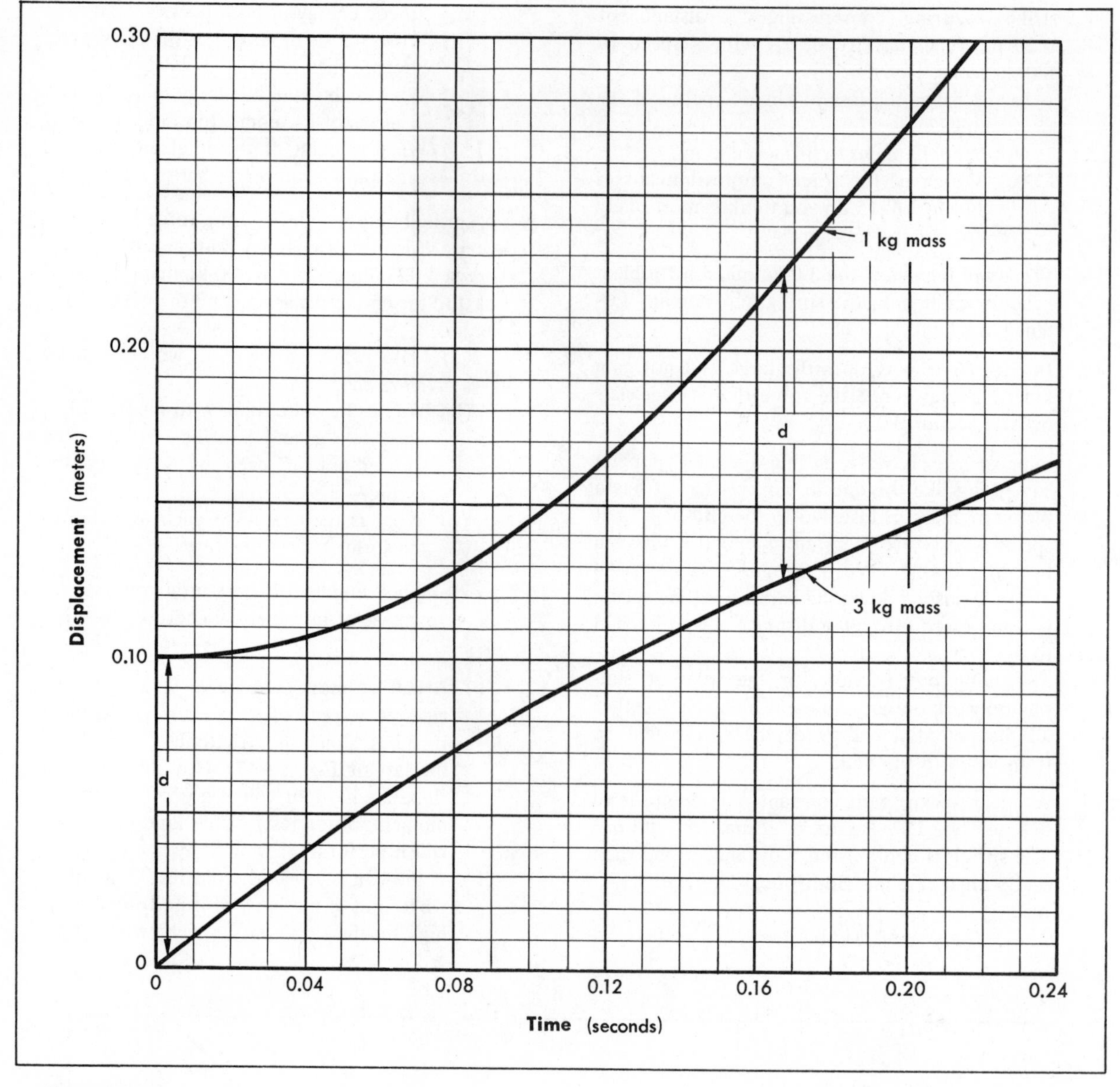

Figure 16–18
For Problem 19.

What is the binding energy (the extra energy needed to make the ball escape from the well in the center of the figure)?

(c) What is the binding energy in part (a)?

18. A linear spring whose force-extension ratio k is 40 newton/m hangs vertically, supporting a 0.80-kg mass at rest. The mass is then pulled down a distance of 0.15 m.
    (a) How high will it rise?
    (b) What will be its maximum velocity?
    (c) How would the answers to (a) and (b) differ if the experiment were done on the moon?

19. Figure 16–18 is the graph of the displacements as a function of time of two interacting masses $m_1 = 3$ kg and $m_2 = 1$ kg.
    (a) What is the initial kinetic energy of m_1?
    (b) What is the final kinetic energy of m_1?

(c) What is the final kinetic energy of m_2?
(d) What is the minimum total kinetic energy?
(e) What is the maximum potential energy?

20. A 0.200-kg stone is thrown upward from a point 20.0 meters above the earth's surface at an angle of 60 degrees with the horizontal and with a speed of 20.0 m/sec.
(a) What is its total energy?
(b) What will be its total energy when it is 15.0 m above the earth's surface?
(c) What will be its speed 15.0 m above the earth?

21. A ball of mass m drops from a height h, as shown in Fig. 16–19, and compresses the spring, of force constant k, a distance x. The mass of the spring is negligible compared with the mass of the ball.
(a) Express the maximum compression of the spring, x, in terms of m, h, and k.
(b) Evaluate x if $m = 4$ kg, $h = 3$ meters, and $k = 500$ newtons/meter.

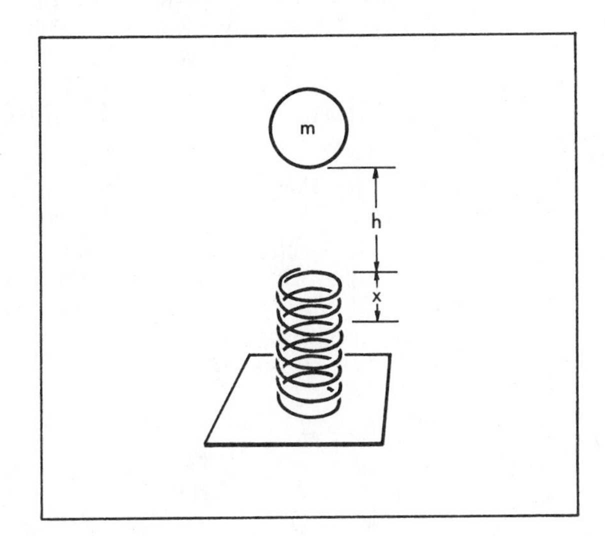

Figure 16–19
For Problem 21.

22. Describe the energy changes of the diver shown in the multiflash photograph in Fig. 16–20. Why does he first jump almost straight up?

23.† When an object of mass m moves from r to r' in the earth's gravitational field, the potential energy changes by $\Delta U = -GMm\left(\frac{1}{r'} - \frac{1}{r}\right)$, where M is the mass of the earth.

Figure 16–20
For Problem 22.

Show that, if the object moves away from the earth a distance $\Delta r = (r' - r)$ which is very small compared with its distance from the earth's center, the above expression reduces to

$$\Delta U = mg\Delta r.$$

24. The force field between a pair of protons is repulsive. Does the potential energy increase or decrease as a pair of protons are brought together?

25.* A satellite in a circular orbit about the earth has a kinetic energy E_K. How much more energy would be required to make it escape from the earth? (Section 6.)

26.* A rocket of mass m ceases to climb when it has reached a distance $10r_e$ from the center of the earth. How much more energy would it require in order to escape? (Section 6.)

27.* How do the escape speeds of a 10-mg grain of sand and a 10^3-kg rocket compare? Ignore the effects of air resistance. (Section 6.)

28. What is the binding energy of the proton and electron in a hydrogen atom if the proton and electron are 0.50×10^{-10} m apart and the force of attraction between them is given by

$$F = \frac{2.3 \times 10^{-28}}{r^2} \text{ newtons.}$$

where r is in meters?

Remember, the electron will not stand still long with a force on it. Assume it is moving in a circle around the proton.

29. Find the binding energy, to two significant figures, of:
    (a) a 70-kg man to the earth.
    (b) the moon to the earth.

30. Would the kinetic energy you would have at the equator due to the rotation of the earth make much difference in your binding energy to the earth?

31. (a) How much work would be required to launch a 100-kg satellite from the earth and cause it to move in the earth's orbit but on the opposite side of the sun, as shown in Fig. 16–21?
    (b) What work would be required for this satellite to leave the solar system from this point?
    (c) What is the ratio of the energy required to escape the solar system compared with the energy required to escape the earth, for a satellite launched from the earth?

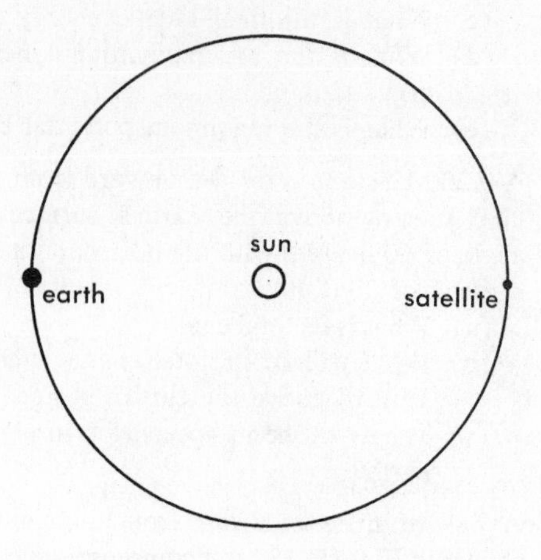

Figure 16–21
For Problem 31.

32.* You are given a very complicated mechanism with levers and wheels in it. It is known that there is no motor inside and that there is very little friction. A string comes out of each side of the apparatus. In order to lift a body 0.10 m on the right-hand string, you must pull 6.0 m on the left-hand string. With what force do you have to pull the left-hand string in order to lift a 12-newton weight on the right-hand string? (Section 7.)

33. A drainpipe pointing slightly downhill sticks out from a retaining wall at the side of a road. A ball thrown up the pipe comes back with a speed greater than that with which it was thrown. If you observed this, would you be surprised? What would you suspect?

Further Reading

These suggestions are not exhaustive but are limited to works that have been found especially useful and at the same time generally available.

ANDRADE, E. N. DA C., *An Approach to Modern Physics.* Doubleday Anchor, 1956. Heat and energy. (Chapter 2)

BONNER, FRANCIS T., and PHILLIPS, MELBA, *Principles of Physical Science.* Addison-Wesley, 1957. A good overview of momentum, work, and energy. (Chapter 10)

GAMOW, GEORGE, *Gravity.* Doubleday Anchor, 1962: Science Study Series. (Chapters 4 and 6)

HOLTON, G., and ROLLER, D. H. D., *Foundations of Modern Physical Science* Addison-Wesley, 1958. (Chapter 18)

Molecular Motion, Internal Energy, and Conservation of Energy

CHAPTER

17

A book slides along a table and comes to rest; energy seems to have disappeared, but the book and the table are slightly warmer. We fill a tire with a bicycle pump, doing a lot of work which does not seem to show up in any of the usual forms of mechanical energy, but the air and the tire and especially the pump do seem to heat up. We crush some ice in a bag by hitting it many times with a hammer; this time we do a lot of work and do not get a temperature increase, but we do find that some ice has melted as a result of our pounding. You could make a long list of such cases in which the work done on a system does not seem to balance with the changes in the mechanical energy of the system; the mechanical energy just seems to slip away. But in all such cases, if we look hard enough and measure carefully enough, we find that some other properties of the system have changed. We shall show that these changes can be related to energy in forms other than those we have yet considered, and that with these additional forms of energy included, we shall be able to retain the idea that energy is conserved. We shall begin with a simple system—a gas.

17-1 Gases, Molecules, and Boltzmann's Constant

Two of the most striking properties of gases are (1) that they are easily compressed as compared with solids and liquids, and (2) that increasing the temperature of a gas whose pressure stays constant increases its volume. But these properties, as we have just stated them, are only qualitative. By doing two experiments we can find the quantitative relationship between the pressure and the volume of a gas kept at constant temperature and between the temperature and the volume of a gas at constant pressure.

Figure 17–1 shows a test tube in a water bath that can be heated to any desired temperature between room temperature and the boiling point of water, 100°C. The test tube contains a wood piston so that as the temperature of the gas in the test tube rises, the gas can expand without change in pressure. The volume of the gas at different temperatures can be read off the scale attached to the test tube.

Using this apparatus we obtained the data shown in Table 1 for three different gases. A graph of these data is shown in Fig. 17–2. As you can

Table 1

AIR		PROPANE		CARBON DIOXIDE	
TEMP. (°C)	VOLUME (cm^3)	TEMP. (°C)	VOLUME (cm^3)	TEMP. (°C)	VOLUME (cm^3)
25.1	64.2	25.1	64.2	25.1	64.2
27.4	65.0	30.1	64.6	30.3	65.4
32.7	65.4	35.3	65.8	35.0	66.2
37.9	66.6	40.2	67.0	40.1	67.0
42.9	67.4	45.1	68.2	45.0	68.2
48.1	69.0	50.0	69.0	50.1	69.4
52.3	69.8	54.9	70.2	54.8	70.2
57.6	70.7	60.2	71.4	59.9	71.4
62.3	71.4	65.0	72.6	64.5	72.4
67.1	73.0	70.0	73.4	—	—
72.5	74.2	—	—	—	—

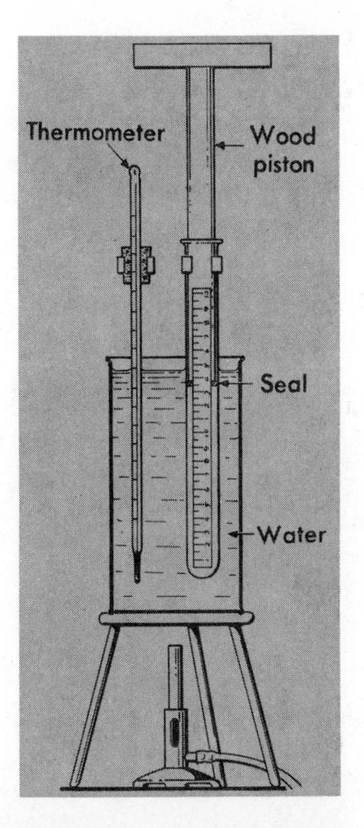

Figure 17–1
The relation between the volume and the temperature of a gas can be found by filling the test tube with different gases and measuring the rise of the piston when each of the gases in turn is heated through the same temperature range.

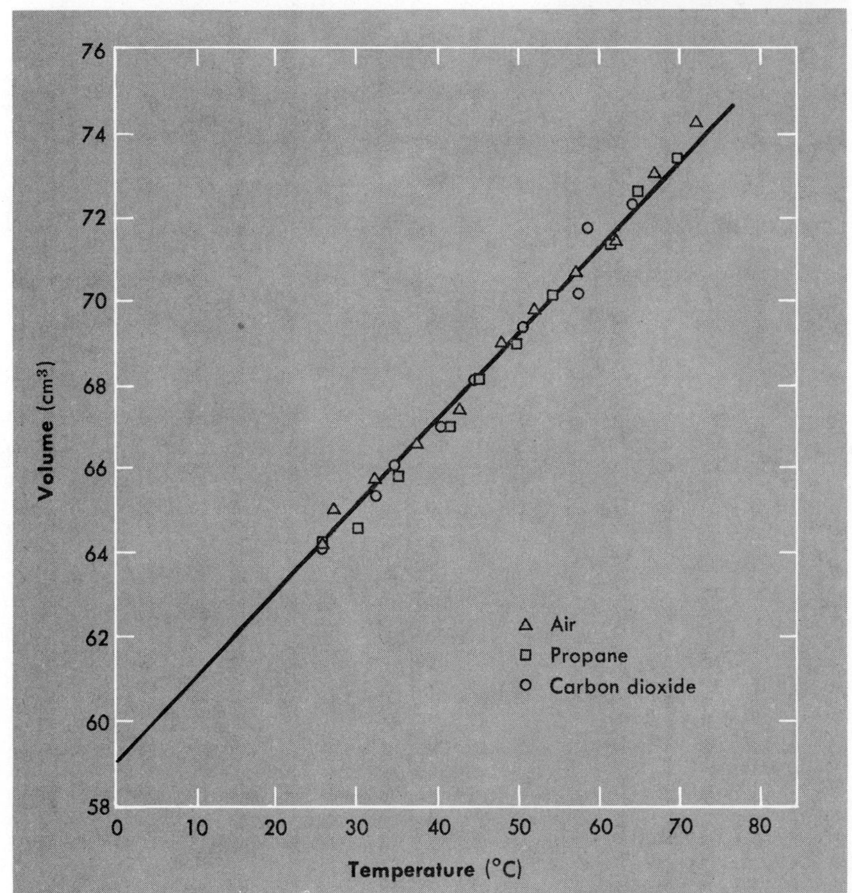

Figure 17–2
A graph of the data in Table 1.

see, the data points for all three gases lie along the same straight line. In fact, any gas that is not too high in pressure (close to its condensation point) would behave in the same way.

To find the relationship between temperature and volume we first note that at 0°C the volume is not zero. In other words, the volume is not directly proportional to the Celsius temperature, but because the graph is a straight line there is a linear relation between the volume and the temperature. However, if we define a new temperature scale we can get a simple direct proportion between the volume and the temperature. In Fig. 17–3 we

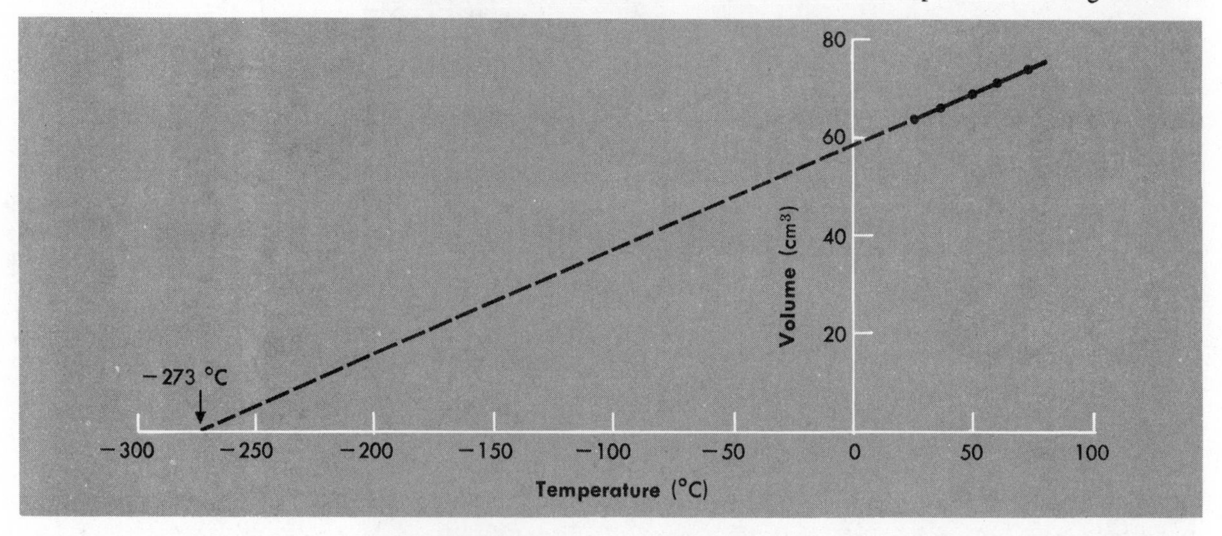

Figure 17–3
The straight line in the graph in Fig. 17–2 extrapolated to include zero on the volume scale.

have redrawn Fig. 17–2 to include zero volume. As you can see, extrapolating the straight line to zero volume gives a temperature of −273°C. So if we replace our temperature scale with one whose zero is at −273°C we shall have a direct proportion between the volume and the temperature (Fig. 17–4).*

By shifting the origin of our temperature scale to −273°C we have what is called the Kelvin temperature scale: $T(\text{in } ^\circ\text{K}) = t\ (\text{in } ^\circ\text{C}) + 273$.

To find the dependence of the volume of a gas on its pressure we can use the apparatus shown in Fig. 17–5. As more and more bricks are placed on the platform on top of the piston, the gas in the cylinder is compressed to smaller and smaller volumes and the pressure of the gas becomes greater and greater.

The pressure of the gas is defined as the force per unit area exerted on the walls of the cylinder and on the piston. In this definition the pressure P, the force F, and the area A are related by the equation

$$P = \frac{F}{A}.$$

* We extended the straight line to zero volume in Fig. 17–3 to find a temperature scale that would give us a direct proportion between volume and temperature. However, although we know that this direct proportion holds over the range of temperatures which we used in doing the experiment, we cannot expect it to hold near zero volume, because the temperature near zero volume in most cases will be far below the temperature at which the gas condenses into a liquid, where its volume changes only slightly, even with very large changes in temperature.

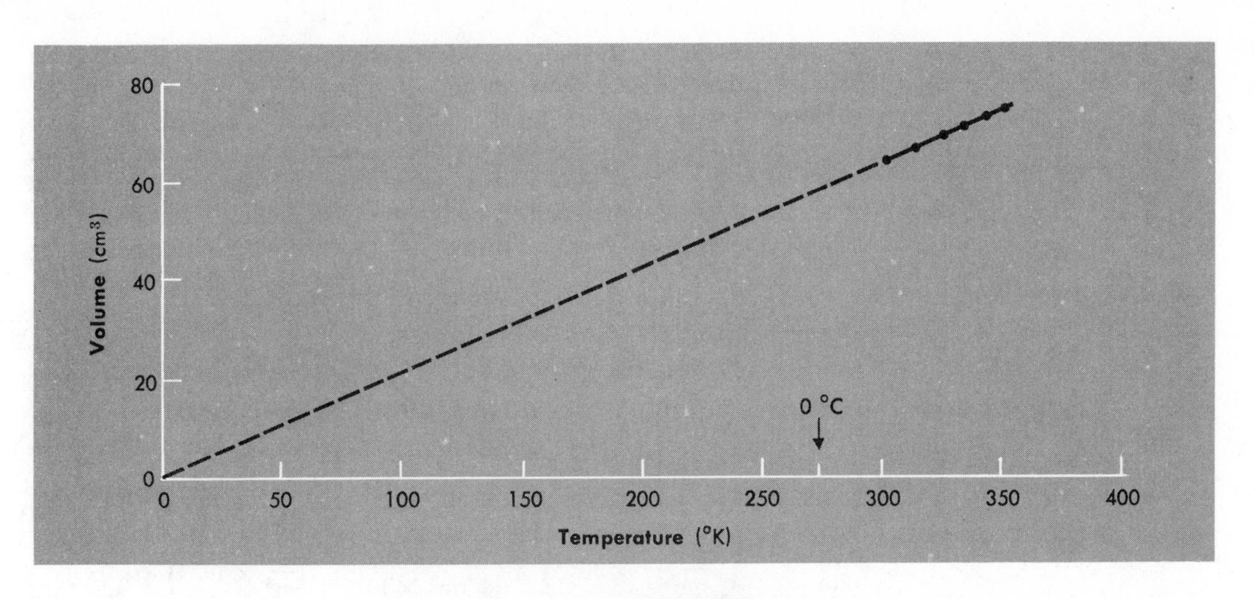

Figure 17–4
The graph of the data in Table 1 plotted using the Kelvin temperature scale.

Figure 17–5
A cylinder and piston can be used to investigate the relationship between the volume and pressure of a gas. Increasing the number of bricks on the platform increases the pressure and decreases the volume of the gas.

Since we commonly express F in newtons and A in square meters, P is given in the equation in units of newtons/m^2.

When we add a 2.00-kg brick to the platform in Fig. 17–5, the piston moves down to a new equilibrium position. Since the piston is free to move, it will be at rest only if the force exerted on it by the gas in the cylinder is equal and opposite to the force pushing on it from above. The piston has an area of 4.39×10^{-4} m^2, so adding 2.00 kg increases the gas pressure by

$$\Delta P = \frac{(9.8 \times 2.00) \text{ newtons}}{4.39 \times 10^{-4} \text{ m}^2} = 4.47 \times 10^4 \text{ newtons/m}^2.$$

The data in Table 2, obtained by using three gases separately in the apparatus of Fig. 17–5, are graphed in Fig. 17–6. Although all three gases behave the same, we do not get a simple, straight-line relationship. However, we note that in an inverse proportion, such as $y = 1/x$, one variable increases as the other decreases, as in the case in our graph. It seems reasonable, therefore, to replot the graph, plotting $1/V$, the inverse of the volume, as a function of the pressure of the gas to see if we can get a straight line.

Table 2

MASS (kg)	PRESSURE $\left(\frac{\text{newtons}}{\text{m}^2} \times 10^{-4}\right)$	VOLUME ($\text{m}^3 \times 10^5$)	1/VOLUME $\left(\frac{1}{\text{m}^3} \times 10^{-3}\right)$
	AIR		
2.20	4.9	28.5	3.51
4.22	9.4	21.5	4.64
6.26	14.0	17.4	5.75
8.36	18.6	14.8	6.76
10.40	23.2	12.8	7.81
12.47	27.8	11.1	9.01
	PROPANE		
2.86	6.4	25.5	3.92
4.88	10.7	19.7	5.08
6.82	15.2	16.4	6.10
9.02	20.1	14.0	7.14
11.06	24.7	12.1	8.26
13.13	29.3	10.8	9.26
	CARBON DIOXIDE		
3.52	7.85	23.5	4.26
5.54	12.4	18.5	5.41
7.58	16.9	15.5	6.45
9.68	21.6	13.3	7.52
11.72	26.1	11.7	8.55
13.79	30.8	10.3	9.71

The masses in the first column include the mass of the platform and piston as well as the mass of the bricks. In the case of propane we started with one brick and an additional mass of 0.66 kg. We then added bricks one at a time. The same was done in the case of carbon dioxide, except that we started with one brick and an additional 1.32-kg mass. This was done to separate the points on the graph for the different gases.

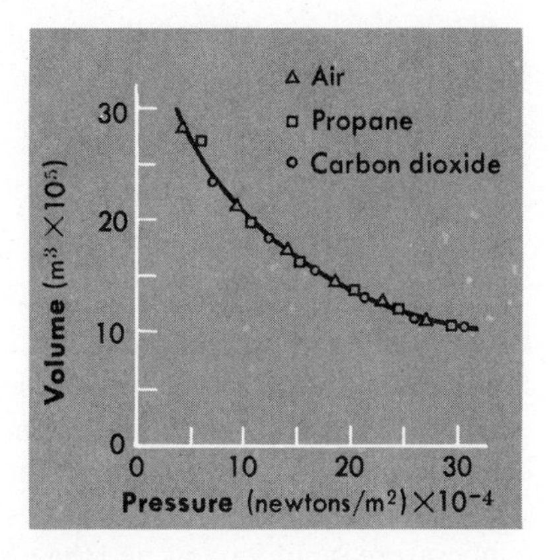

Figure 17–6
A graph of the data in Table 2.

This we have done in Fig. 17–7. As you can see, we get a linear relationship. However, as in the case of volume as a function of temperature, the line does not go through the origin, but it is easy to understand why. The air around the apparatus, the atmosphere, also exerts a force, and this adds to the force on the gas in the cylinder. As you can see, the dashed extension of the line in Fig. 17–7 goes through zero of the inverse of the volume at -10.0×10^4 newtons/m^2.* Just as in the case with the volume-temperature

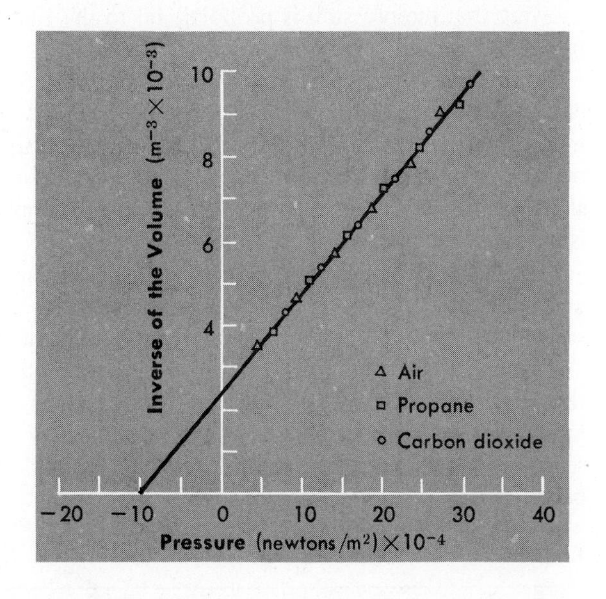

Figure 17–7
A graph of $1/v$, the inverse of the volume, as a function of the pressure.

graph if we add 10.0×10^4 to each of the values for the pressure (in other words, use what is called the *total pressure*) we will have a graph that is a direct proportion between the increase of the volume $1/V$ and the total pressure P. We can write this as

$$\frac{1}{V} \propto P$$

or

$$PV = c.$$

* The pressure of the atmosphere varies from day to day and from place to place, but the average atmospheric pressure at sea level is about 1.01×10^5 newtons/m^2.

Similarly, in the case of the direct proportion between the volume V and the Kelvin temperature T we can write

$$V = c'T.$$

We can combine these two equations in the equation

$$PV = KT,$$

where K is a proportionality constant that depends on the units we use for P and V and also on the number of molecules of gas in the cylinder. As you can see, if we keep the pressure constant in this equation, we have the volume directly proportional to the temperature on the Kelvin scale; and if we keep the temperature constant, we have the volume inversely proportional to the pressure. This general relationship applies to all gases so long as the temperature is not too low and the pressure is not too high.

You can see from the equation that for a fixed pressure and temperature the value of K is proportional to the volume of the sample of gas. In other words, its value is proportional to the quantity of gas and is, therefore, proportional to the number of molecules N in the sample. We can now rewrite the expression as

$$P = kT\left(\frac{N}{V}\right),$$

showing that the pressure is proportional to the Kelvin temperature and the number of molecules per unit volume.

If we know the density of a gas and the mass of its molecules, we can easily calculate the number of molecules in a unit volume. (A method for measuring the masses of molecules is described in Chapter 22.) The results of such calculations for several gases at 0°C and atmospheric pressure are shown in Table 3. Remarkably, the number of molecules per unit volume is the same for all gases.

Because the number of molecules per unit volume is the same for all gases at the same pressure and temperature, k must be a universal constant, the same for all gases. It is called Boltzmann's constant.

The fact that equal volumes of different gases at the same pressure and temperature contain equal numbers of molecules was first proposed as a hypothesis by Avogadro in 1811. His hypothesis was based on the chemical reactions between gases, long before it was possible to determine the number of molecules in a sample of a gas.

Table 3

	DENSITY, D (gm/cm^3)	MOLECULAR MASS, m (gm)	NUMBER OF MOLECULES IN A CUBIC CENTIMETER $\left(\frac{D}{m}\right)$
Carbon dioxide (CO_2)	1.977×10^{-3}	73.1×10^{-24}	2.68×10^{19}
Chlorine (Cl_2)	3.214×10^{-3}	117×10^{-24}	2.67×10^{19}
Helium (He)	0.1785×10^{-3}	6.65×10^{-24}	2.68×10^{19}
Hydrogen (H_2)	0.0899×10^{-3}	3.35×10^{-24}	2.68×10^{19}
Nitrogen (N_2)	1.251×10^{-3}	46.6×10^{-24}	2.69×10^{19}
Oxygen (O_2)	1.429×10^{-3}	53.2×10^{-24}	2.68×10^{19}

We can now determine the numerical value of Boltzmann's constant. We see from Table 3 that at atmospheric pressure and 0°C (273°K) there are 2.68×10^{19} molecules in a cubic centimeter of any gas, or $(2.68 \times 10^{19})(1.00 \times 10^{6}) = 2.68 \times 10^{25}$ molecules in a cubic meter. Therefore,

$$k = \frac{P}{T\left(\frac{N}{V}\right)} = \frac{1.01 \times 10^{5}\ \frac{\text{nt}}{\text{m}^2}}{273°\text{K} \times 2.68 \times 10^{25}\ \frac{\text{molecules}}{\text{m}^3}}$$

$$= 1.38 \times 10^{-23}\ \frac{\text{joule}}{\text{molecule} - °\text{K}}.$$

17-2 The Dynamics of Gases

Let us now apply what we have learned about dynamics in earlier chapters to the motion of individual molecules in a gas. For simplicity we shall begin with monatomic molecules, such as those of helium, argon, or neon. Consider first a single molecule approaching a wall, moving straight at it with velocity $\vec{\mathbf{v}}$. If the molecule has a mass m, its momentum is $m\vec{\mathbf{v}}$. We assume that the molecule bounces back elastically when it hits the wall. Now consider a small time interval Δt during which n such molecules hit the wall and bounce off. The total change in momentum is $\Delta\vec{\mathbf{p}} = nm\,\Delta\vec{\mathbf{v}}$. If the collision is elastic, $\Delta\vec{\mathbf{v}} = -2\vec{\mathbf{v}}$ and $\Delta\vec{\mathbf{p}} = -\,2nm\vec{\mathbf{v}}$. The total impulse applied by the molecules perpendicular to the wall is therefore $2nm\vec{\mathbf{v}}$. If the number n of colliding molecules is great enough, the wall responds as if to a steady force. This force multiplied by the time Δt is just another way of expressing the total impulse. Consequently,

$$\vec{\mathbf{F}}\Delta t = 2nm\vec{\mathbf{v}}.$$

From this equation we see that the magnitude of the average force exerted on the wall is

$$F = \frac{2nmv}{\Delta t}.$$

We want to relate this force to the force exerted on a unit area of wall. For this purpose we need to know how many molecules there must be in the little region of space near the wall to provide the bombardment we just considered (Fig. 17–8). We can find out by finding the volume near the wall from which the n molecules came.

We consider all the molecules that hit an area A of the wall during a short time interval Δt. All of them must have come from a little volume with the base A and altitude $v\Delta t$. No molecule that was farther away from the wall than $v\Delta t$ can have reached the wall, because $v\Delta t$ is the distance the molecules travel in time interval Δt, and if they started farther away they would fail to arrive during the time Δt. All the molecules that were approaching the wall from any distance less than $v\Delta t$ must have reached the wall, because they would travel this shorter distance in less than time Δt. We can conclude, therefore, that at the beginning of the time interval Δt, the n molecules must have occupied the volume $v\Delta tA$.

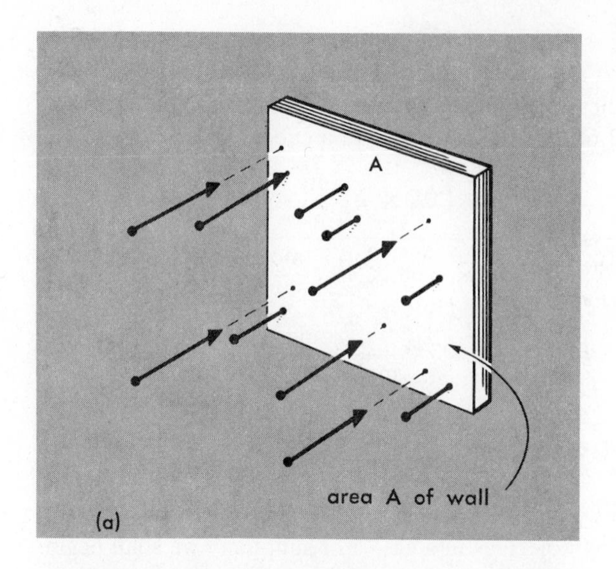

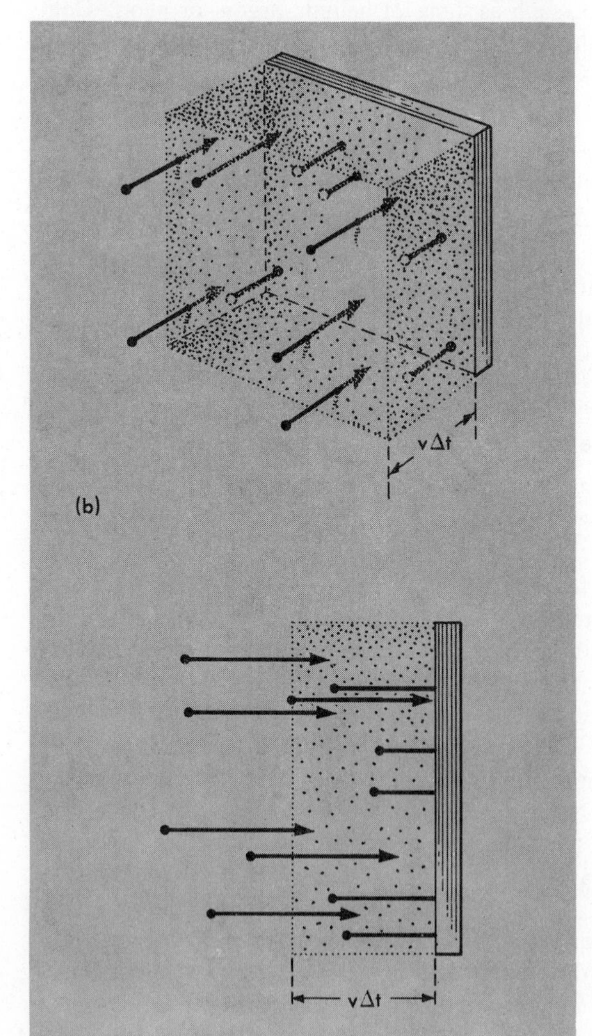

Figure 17–8
(a) The molecules move toward the wall with velocity *v*. From what volume do the molecules come that reach the area A of the wall in time *t*? (b) The large dots on the tails of the arrows show the original positions of the molecules. The arrowheads show how far they get in the time interval Δt—unless they hit the wall. Notice that all the molecules starting within the distance $v\Delta t$ of the wall hit it. All those starting farther away do not get to the wall in time interval Δt. Therefore, those molecules reaching the wall come from a volume of base A and height $v\Delta t$.

How is the number of molecules hitting an area A of the wall in the time interval Δt related to the number of molecules N in the entire volume V? The molecules of a gas in a container at rest are evenly distributed. Thus the number of molecules per unit volume is $\frac{N}{V}$, and the number of molecules in a volume $v\Delta tA$ is

$$\left(\frac{N}{V}\right) v\Delta tA.$$

Now we shall make two idealizations that will simplify the calculation. (In the next section we shall see whether these idealizations have a significant effect on the results.) First, we shall assume that all the molecules move only along three mutually perpendicular axes, say the x, y, z axes of a coordinate system. Second, we shall assume that all molecules have the same speed. Let the area A of the wall be parallel to the yz plane. Then on the basis of the assumptions only one-third of the molecules will be moving parallel to the x axis, and of these only one-half will be moving *toward* the wall in question. Thus only one-sixth of the molecules in the volume $v\Delta tA$ will hit the area A during the time interval Δt:

$$n = \tfrac{1}{6}\frac{N}{V} v\Delta tA.$$

Substituting this value of n in the expression for the magnitude of the force exerted on the wall yields

$$F = \frac{2nmv}{\Delta t} = \frac{2mv}{\Delta t}\left(\tfrac{1}{6}\frac{N}{V} v\Delta tA\right) = \tfrac{1}{3}mv^2\left(\frac{N}{V}\right) A.$$

Hence the pressure $P = \frac{F}{A}$ becomes

$$P = \tfrac{1}{3}mv^2\left(\frac{N}{V}\right).^*$$

Now since $\frac{mv^2}{2}$ is the kinetic energy E_K,

$$\tfrac{1}{3}mv^2 = \tfrac{2}{3}\left(\frac{mv^2}{2}\right) = \tfrac{2}{3}E_K.$$

We can therefore write our expression for the pressure as

$$P = \tfrac{2}{3}E_K\left(\frac{N}{V}\right).$$

This equation says that the pressure is ⅔ of the kinetic energy of a molecule times the number of molecules per unit volume. This form of the equation emphasizes the role of the molecular kinetic energy in determining the pressure.

* You may wonder how the pressure, a scalar quantity, can be a force (vector) per unit area. Here, of course, we are concerned only with magnitudes, but the relation $F = PA$ holds also as a vector equation in the following sense. An area has the property of a vector whose magnitude equals the area and whose direction is along the perpendicular to the area. Then the force PA is also perpendicular and we have the vector equation

$$\vec{\mathbf{F}} = P\vec{\mathbf{A}}.$$

17-3 The Effect of the Velocity Distribution

It is obvious that gas molecules do not in reality move only along three mutually perpendicular lines. For a gas in a container the velocity vector of a molecule is just as likely to point in one direction as in any other. Furthermore there is no reason to believe that all molecules in a gas have the same speed.

A more realistic model for the pressure of a gas, therefore, must take into account the fact that the gas molecules move in all directions at various speeds. To do this we shall review the reasoning of the last section under the more general conditions of a distribution of velocities both in magnitude and in direction.

We begin again with a single molecule. Let it have a velocity $\vec{\mathbf{v}}_1$. Upon collision with the wall only the perpendicular component of the velocity $v_{1\perp}$ will be reversed. The parallel component $v_{1\parallel}$ will remain unchanged (Fig. 17–9). Therefore, the magnitude of the change in momentum is $2mv_{1\perp}$. For a molecule with a perpendicular component of velocity $v_{1\perp}$ to collide with the area A on the wall in the time interval Δt, it has to be no farther than a distance $v_{1\perp}\Delta t$ from the wall; that is, it has to be in the volume $v_{1\perp}\Delta tA$.

Let there be N_1 molecules with $v_{1\perp}$ in the entire volume V. Then $N_1 \dfrac{v_{1\perp}\Delta tA}{V}$ molecules whose velocity is $v_{1\perp}$ will be in the volume $v_{1\perp}\Delta tA$. Therefore, the contribution of this number of molecules to the pressure on the wall will be

$$P_1 = \frac{(N_1 v_{1\perp}\Delta tA)2mv_{1\perp}}{V\Delta tA} = 2mv_{1\perp}^2\left(\frac{N_1}{V}\right).$$

If in addition to the N_1 molecules of $v_{1\perp}$ there are N_2 molecules of perpendicular velocity $v_{2\perp}$, their contribution to the pressure will be

$$P_2 = 2mv_{2\perp}^2\left(\frac{N_2}{V}\right).$$

Since the molecules do not affect one another's collisions with the wall, the pressure due to both groups of molecules will be $P_1 + P_2$. We can extend this reasoning to all possible values of $v_\perp$. The total pressure is then

$$P = \frac{2m}{V}(v_{1\perp}^2 N_1 + v_{2\perp}^2 N_2 + v_{3\perp}^2 N_3 + \cdots).$$

The sum $N_1 + N_2 + \ldots = \frac{1}{2}N$ where N is the total number of molecules in the container. The factor $\frac{1}{2}$ comes from the fact that because of the symmetry there are just as many particles with $-v_\perp$ as there are with $+v_\perp$. The sum in the parentheses is, by the definition of an average, the average of $v_\perp^2$ times $\dfrac{N}{2}$. Hence:

$$P = \frac{2m}{V} \times \overline{v_\perp^2}\,\frac{N}{2} = m\overline{v_\perp^2}\left(\frac{N}{V}\right).$$

(The bar over $v_\perp^2$ means "average.")

What is $\overline{v_\perp^2}$? If we choose the area A to be parallel to the yz plane, then $\overline{v_\perp^2} = \overline{v_x^2}$. Because all directions of motion of the molecules are equally likely

$$\overline{v_x^2} = \overline{v_y^2} = \overline{v_z^2}.$$

But:

$$\overline{v_x^2} + \overline{v_y^2} + \overline{v_z^2} = \overline{v^2}.$$

Hence:

$$3\overline{v_\perp^2} = \overline{v^2},$$

or

$$\overline{v_\perp^2} = \tfrac{1}{3}\overline{v^2}.$$

Substituting this value of $v_\perp^2$ in the equation for the pressure yields

$$P = \tfrac{1}{3}m\overline{v^2}\left(\frac{N}{V}\right),$$

or

$$P = \tfrac{2}{3}\overline{E}_K\left(\frac{N}{V}\right).$$

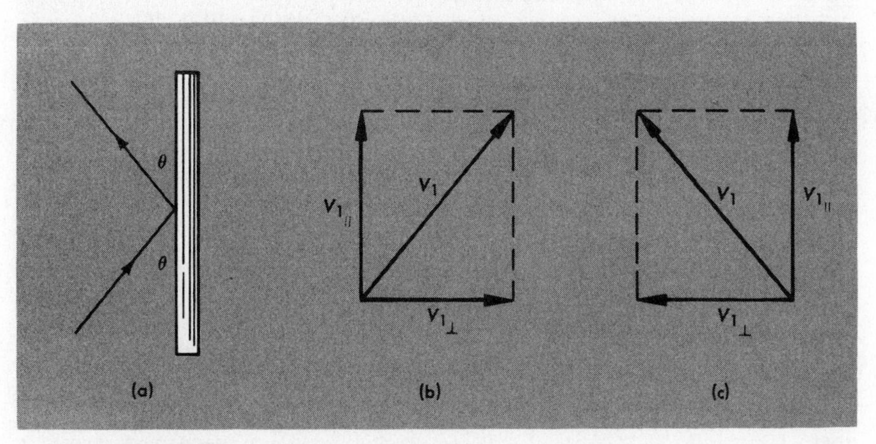

Figure 17–9
In (a) a single monatomic gas molecule moving with velocity $\vec{\mathbf{v}}_1$ upon collision with a wall leaves the wall with its speed unchanged and in a direction whose angle θ with the wall is the same as before collision. The perpendicular and parallel components of the velocity before and after collision are shown in (b) and (c) respectively. As you can see, $\vec{\mathbf{v}}_{2\parallel} = \vec{\mathbf{v}}_{1\parallel}$ and $\vec{\mathbf{v}}_{2\perp} = -\vec{\mathbf{v}}_{1\perp}$.

The only effect of replacing the idealized model of Section 17–2 with the more realistic model of this section is the replacement of the kinetic energy of the molecules (which was assumed to be the same for all molecules) with their average kinetic energy.

17-4 Temperature and Molecular Kinetic Theory; Internal Energy

We now have two expressions relating the pressure of a gas to the number of molecules per unit volume. The first is the result of experiments and involves the temperature of the gas (Section 17-1):

$$P = kT\left(\frac{N}{V}\right).$$

The second expression is the result of a theoretical prediction based on the application of the laws of mechanics to the collisions of gas molecules with the walls of the container (Section 17–2):

$$P = \tfrac{2}{3}\overline{E}_K\left(\frac{N}{V}\right).$$

Combining the two expressions yields

$$kT = \tfrac{2}{3}\overline{E}_K$$

and

$$\overline{E}_K = \tfrac{3}{2}kT.$$

This is a remarkable result. It tells us that the temperature of a gas (in °K) is proportional to the average kinetic energy of the molecules. The proportionality constant $\frac{3}{2}k$ is a universal constant, independent of the mass of the molecules. Thus, at the same temperature molecules of different monatomic gases have the same average kinetic energy.

Increasing the temperature of a monatomic gas increases the kinetic energy of its molecules. Since the gas as a whole remains at rest (provided the container remains at rest), this change in energy manifests itself by changes within the gas. We refer then to the change of energy associated with such changes as *internal energy*. For monatomic gases the internal energy is just the entire kinetic energy of the atoms.

Suppose a pendulum is hung inside a container filled with helium, a monatomic gas. The pendulum is pulled to one side and released so that it begins swinging [Fig. 17–10(a)]. After a while friction with the gas brings the pendulum to rest [Fig. 17–10(b)]. Does the loss of mechanical energy of the pendulum equal the gain in internal energy of the gas? In principle we can answer this question by calculating what the rise in temperature of the gas should be, and then compare it with the observed rise in temperature. The calculation is simple: Let the bob of the pendulum of mass m be raised initially through a height h. When the pendulum comes to rest, its loss of mechanical energy to the gas equals the number of gas molecules in the container times the average gain in kinetic energy of one molecule:

$$N\Delta\overline{E}_K = \tfrac{3}{2}Nk\Delta T.$$

If the loss in mechanical energy of the pendulum equals the gain in internal energy of the gas, then

$$mgh = \tfrac{3}{2}Nk\Delta T,$$

and we predict the rise in temperature of the gas to be

$$\Delta T = \tfrac{2}{3}\frac{mgh}{Nk}.$$

Reasonable values of the various quantities might be $m = 3.0$ kg, $h = 0.20$ meter, and $N = 4.0 \times 10^{23}$ (about 14 liters of gas at room temperature). The predicted temperature rise is then

$$\Delta T = \frac{2 \times 3.0 \times 9.8 \times 0.20}{3 \times 4.0 \times 10^{23} \times 1.38 \times 10^{-23}} = 0.71\text{°K}.$$

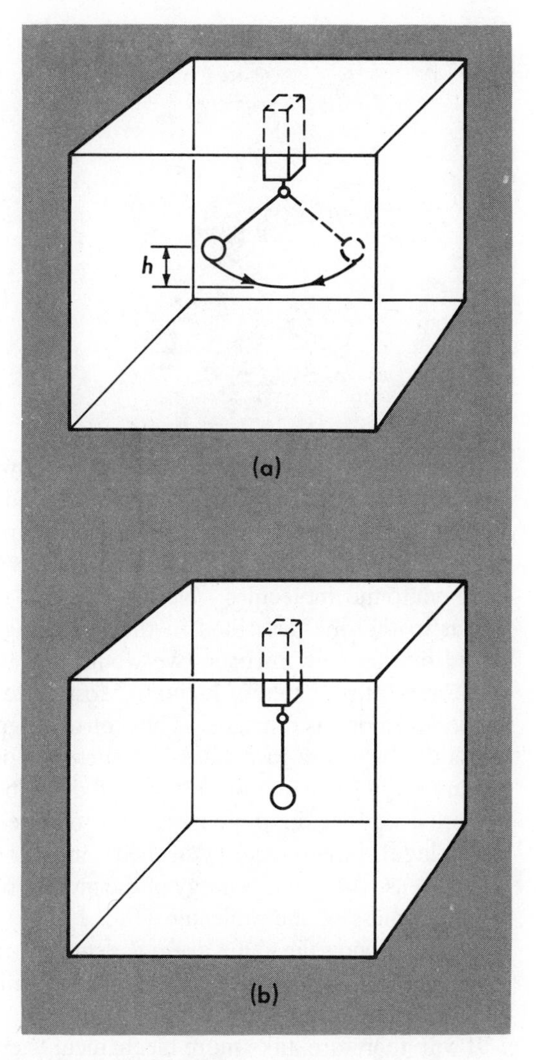

Figure 17–10
In (a) a pendulum is set swinging through a height h in a container of helium. (b) Friction between the swinging pendulum and the gas eventually stops the pendulum.

This is an easily measurable change in temperature. Nevertheless, the experiment as described has technical problems. The gas is in contact with the walls of the container. As the gas warms up, so do the container and the pendulum itself. Thus the gas will warm up less than expected. There are ways of eliminating the effects of this leakage. When this is done, the rise in temperature of the gas agrees with the prediction, i.e., that the mechanical energy of the pendulum is completely converted into internal energy of the gas.

17-5 Internal Energy of Diatomic Gases

Monatomic gases like helium and neon lend themselves readily to theoretical calculations, but you are unlikely to encounter them. Most gases have two or more atoms per molecule. In what way must we modify the reasoning of the preceding sections to relate the pressure and temperature of a diatomic gas to the average kinetic energy of its molecules?

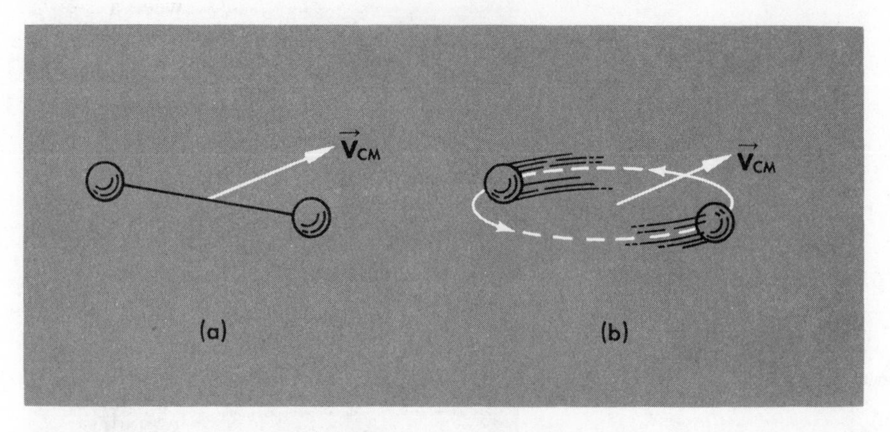

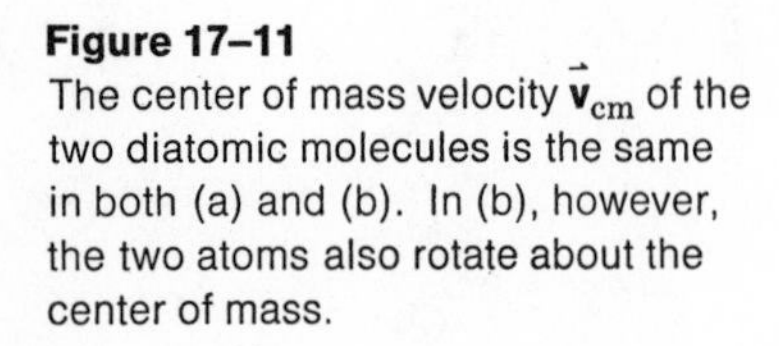

Figure 17–11
The center of mass velocity $\vec{\mathbf{v}}_{cm}$ of the two diatomic molecules is the same in both (a) and (b). In (b), however, the two atoms also rotate about the center of mass.

Reviewing this reasoning will show that nowhere have we made use of the restriction that gases be monatomic. We used the relation between impulse and momentum and the conservation of momentum in collisions. Thus all our conclusions are just as valid for diatomic (or polyatomic) molecules as for monatomic molecules. Nevertheless, if we performed an experiment similar to the one described in the preceding section, with the helium replaced by diatomic nitrogen, we would find that the rise in temperature is only three-fifths of that of helium. Most diatomic gases will show the same temperature rise as nitrogen. Thus for these gases the gain in kinetic energy is less than the loss in mechanical energy of the pendulum. Is there a net loss in energy when an object is slowed down by friction with a diatomic gas? Is there a way for diatomic molecules to increase their energy other than by increasing the kinetic energy of their center-of-mass motion? One such way is to increase the kinetic energy of the motion of the two atoms relative to the center of mass of the molecule (Fig. 17–11). A gas made up of rotating molecules having the same average center-of-mass kinetic energy as a monatomic gas will exert the same pressure but will have a higher internal energy due to their rotation.

It will, therefore, take more mechanical energy to raise the temperature of N molecules of a diatomic gas like nitrogen than the same number of molecules of a monatomic gas like helium. This is indeed observed. A quantitative model of diatomic gases is in agreement with these observations.

17-6 Internal Energy of Solids: Conservation of Energy

We were able to show that when a moving solid object is slowed down by friction with a gas, the combined energy of the solid and the gas remains unchanged. The kinetic energy of the moving object is slowed by a huge number of molecules moving in all directions so that no bulk motion of the gas is observed. The increase in the average kinetic energy of this random motion of the gas is perceived only as an increase in temperature.

When a moving solid is slowed down by friction with another solid, there is also an increase in temperature. Can a mechanical model of a solid account for the rise in temperature? The major difference between a gas and a solid is that gases fill all the space available to them, whereas solids keep their shape and are hard to stretch or compress. We must conclude that in a gas there are no forces holding the molecules together, whereas in a solid there are forces holding the atoms almost in fixed positions. That is, if somehow the distance between two atoms is larger than the "normal" distance, a force of attraction will pull the two atoms together. Similarly, if the distance is smaller than the "normal" distance, then a force of repulsion will push them apart. At least as an approximation we may consider the interatomic force to act as a restoring force under which the atoms undergo simple harmonic motion (see Section 12-8).

According to this model, therefore, the atoms in a solid have not only a kinetic energy due to their oscillation but also a potential energy that depends on the interatomic distances (see Section 16-2). To bring the model to the stage where we can make concrete predictions we must make one additional assumption.

On the basis of our discussion of gases we shall also assume that for a solid the temperature (in °K) is proportional to the average kinetic energy of the atoms,

$$\tfrac{3}{2}kT = \bar{E}_K.$$

However, for objects performing simple harmonic motion there is no way of increasing the average kinetic energy without increasing the average potential energy.*

It can be proved mathematically that for simple harmonic motion the average potential energy equals the average kinetic energy,

$$\bar{U} = \bar{E}_K.$$

But,

$$E_K + U = E,$$

the total energy of an atom in the solid. Therefore,

$$\bar{E}_K = \tfrac{1}{2}\bar{E}$$

and hence

$$\tfrac{3}{2}kT = \tfrac{1}{2}\bar{E},$$

or

$$3kT = \bar{E}.$$

Note that just as in the case of gases the mass of the individual atom does not appear. If a solid behaves like a collection of atoms performing simple harmonic motion, then the amount of energy needed to raise the temperature of a sample containing N atoms by 1°K depends on N. However, it must be independent of the total mass of the atoms. Moreover, this energy must be twice the energy needed to raise the temperature by 1°K of a sample of a monatomic gas containing the same number of atoms.

* Think of a pendulum. If you push it harder it will swing higher.

A fixed number of atoms, used primarily in chemistry, is $N = 6.02 \times 10^{23}$. This number of atoms is called a *mole.** The change in internal energy per °K of a mole of a substance is called the molar *heat capacity*. Our prediction is, then, that the molar heat capacity, C, of solids is

$$C = \frac{N\Delta\bar{E}}{\Delta T} = 3Nk.$$

Using the value of k from Section 17-1 yields

$$C = 3 \times 6.02 \times 10^{23} \times 1.38 \times 10^{-23} \frac{\text{joules}}{\text{mole-°K}} = 25 \frac{\text{joules}}{\text{mole-°K}}.$$

Table 4 shows the measured heat capacities of several solids. We see from the table that for many elements our simple model gives quite satisfactory results. In most cases the heat capacities are quite close to 25 joules/mole/°K, although the masses of the atoms vary over a wide range. Thus, by extending the laws of Newton in mechanics to the atom in a solid or a gas we are able to account for the loss of mechanical energy through friction and maintain the conservation of energy by including internal energy in the energy balance.

This is as far as Newtonian mechanics can take us. At the end of this course you will see that Newtonian mechanics runs into fundamental difficulties in accounting for the behavior of atoms. To further expand the idea of energy and its conservation requires us to go beyond Newtonian mechanics.

Table 4

OBSERVED MOLAR HEAT CAPACITIES OF SOLIDS

SOLID	MASS OF ONE ATOM (gm $\times 10^{23}$)	MOLAR HEAT CAPACITY (joules/mole/°K)
Lithium	1.2	24
Boron	1.8	12
Carbon (graphite)	2.0	9
Magnesium	4.0	24
Aluminum	4.6	24
Silicon	4.7	20
Sulfur	5.3	24
Calcium	6.7	26
Chromium	8.6	23
Iron	9.3	25
Nickel	9.8	26
Copper	10.6	24
Zinc	10.9	25
Germanium	12.1	26
Arsenic	12.5	25
Silver	17.9	26
Tin	19.7	26
Gold	32.7	25
Lead	34.4	27

* Roughly, a mole is the number of hydrogen atoms in one gram of hydrogen.

1.* If the pressure in a cylinder of gas is 3×10^5 newton/m^2 at 20°C, by how much does the pressure increase if the cylinder is placed in boiling water at 100°C? (Section 1.)

2.† A gas in a cylinder pushes a piston out, increasing its volume by ΔV. The gas exerts a pressure P on the face of the piston, which has an area A. The force exerted by the gas moves the piston a distance Δx, transferring energy $F\Delta x$ to some outside machinery. Show that the work $F\Delta x$ is equal to $P\Delta V$.

(*Note:* This is a very useful expression for work whenever we deal with a gas or a liquid pushing a piston: work equals the pressure times the change in volume.)

3.* If a molecule of mass m and velocity v hits a wall head on and rebounds with velocity $-v$, what is the impulse given to the wall? (Section 2.)

4. A machine gun fires a stream of 10-gram bullets at the rate of 400 rounds per minute. The bullets, moving at velocity 300 m/sec, hit a wall of solid rock and stop dead. Calculate:
   (a) the force on the wall.
   (b) the kinetic energy of the bullets arriving at the wall in one minute.
   (c) the kinetic energy of the bullets in a 1-meter length of the stream as they approach the wall. Compare twice this answer with your answer to (a).

5. The ratio of the mass of a neon molecule to the mass of a helium molecule is 5.0. How does the average speed of the molecules in a sample of neon gas compare with the average speed of helium molecules in a sample of helium gas if both gas samples are at the same temperature?

6. Suppose you have a cylinder of compressed helium and there is a tiny hole through which the gas leaks out. Would you expect the pressure in the cylinder to drop more rapidly, more slowly, or at the same rate if oxygen instead of helium were in the tank?

7. Consider a cylinder containing gas that is trapped by a piston near one end of the cylinder. What happens to the magnitude of the velocity of a gas molecule as a result of colliding with the piston while it is moving? (*Hint:* Think of the motion in the frame of reference of the moving piston.) What happens to the total kinetic energy of the gas while the piston is moving? What happens to the temperature of the gas while the piston is moving?

8. 6.02×10^{23} molecules of an ideal monatomic gas (in practice, helium or argon) are placed in a cylinder at temperature 273°K. The gas is at atmospheric pressure, 1.02×10^5 newtons/m^2. At this pressure and temperature the gas occupies 2.24×10^{-2} m^3.

   A piston in the cylinder is then pushed in to decrease the volume by 2.45×10^{-4} m^3.
   (a) How much mechanical work must be done to push the piston in? (Neglect the change in pressure.)
   (b) What is the final temperature of the gas if the container is completely insulated? (12.4 joules of energy raise its temperature one degree.)
   (c) By what fraction of its original value does the pressure change? Is the pressure constant?

9. (a) Estimate the speed of oxygen molecules at room temperature from the following data: 32 grams of oxygen at room temperature (20°C) at one atmospheric pressure (1.02×10^5 newtons/m^2) occupy 2.4×10^{-2} m^3.
   (b) The same volume of hydrogen at the same temperature and pressure has a mass of only 2 grams. Estimate the average speed of hydrogen molecules at room temperature.
   (c) From your answer to (a), estimate the average speed of nitrogen molecules at room temperature, within 10 percent.
   (d) What is the speed of oxygen molecules at room temperature and two atmospheres pressure?

10.* What is the total momentum of the molecules in a gram of helium gas in a container at a temperature of 300°K? (Section 4.)

11. A large bag of sand is hung from a tree by a long rope. A boy shoots a bullet into the sandbag, and the bullet stays in the bag.
   (a) Describe the energy changes.
   (b) Suppose a 10-gram bullet is moving at 300 m/sec when it hits the bag, and the bag has a total mass of 1990 grams. Calculate the amount of kinetic energy:
   (i) the bullet had originally.
   (ii) the bullet and bag have after collision.
   (iii) that disappears.
   (c) What fraction of the original kinetic energy of the bullet goes into internal energy?

12.* How does the total kinetic energy of monatomic molecules in a sample of gas depend on the number of molecules N, the volume V, and the pressure of the gas P? (Section 4.)

13.* By how much does the average kinetic energy of monatomic molecules of a gas change if the gas is warmed from 25°C to 100°C? (Section 4.)

14. In a certain gas ⅖ of the energy of the molecules is tied up in motion of the atoms around each other, and ⅗ in motion of the centers of mass.
    (a) On the average, what is the kinetic energy of the center-of-mass motion of one such molecule when the temperature is 300°K?
    (b) If the temperature is raised 1°C, what energy must be supplied to 6.02×10^{23} molecules of the gas?

15.* (a) What is the increase in the average kinetic energy of a neon molecule when the temperature of neon gas increases by 1°K?
    (b) Will your answer be different for a diatomic nitrogen molecule in nitrogen gas whose temperature is raised 1°K? (Section 5.)

16.* N molecules of helium at 100°C are mixed with N molecules of gas X at 20°C, and the final temperature of the mixture is 50°C. Can a molecule of gas X contain only one atom? (Section 5.)

17. There are 2.68×10^{22} molecules in one liter of gas at 1 atm and 0°C. How much energy must be supplied to the gas to raise its temperature by 1°K if (a) the gas is helium? (b) the gas is nitrogen?

18. Copper sulfide is a compound made of copper and sulfur. What do you predict is the heat capacity of a sample of copper sulfide containing a total of 6.02×10^{23} atoms? Do you have to know the proportions of copper to sulfur atoms in copper sulfide in order to answer this question?

19. Brass is an alloy made of copper and zinc. From Table 4 what do you predict is the heat capacity of a sample of brass containing a total of 6.02×10^{23} atoms?

20. Must gas molecules always rebound from a wall with the same kinetic energy they had before the collision? (Think of what happens when the temperature of the gas is either higher or lower than the temperature of the wall.)

Further Reading

HOLTON, G., and ROLLER, D. H. D., *Foundations of Modern Physical Science.* Addison-Wesley, 1958. Heat, molecular motion, and the conservation of energy. (Chapters 16, 17, 18, 19, and 20)

PHYSICAL SCIENCE GROUP, *Energy.* Published by Boston University, 1974. A more advanced version of the Physical Science II text.

PHYSICAL SCIENCE GROUP, *Physical Science II.* Prentice-Hall, Inc., 1972. A detailed study of energy without Newtonian mechanics. (Chapters 15–21)

ROGERS, ERIC M., *Physics for the Inquiring Mind.* Princeton University Press, 1960. A good discussion of kinetic theory. (Chapters 25–30)

Electric Charge

CHAPTER

18

You have learned that every material body attracts every other material body with a force known as the *gravitational force*. The gravitational attraction has practical consequences only when at least one of the interacting bodies has enormous mass, something like a whole planet. However, gravitational forces are not the only forces acting at a distance between material bodies. Sometimes other forces are enormously greater. A small magnet will lift a steel nail off the table against the gravitational attraction of the entire earth. A comb rubbed on your sleeve will lift bits of paper. These are examples of *magnetic* and *electric* forces, respectively.

Although the existence of such forces has been known since antiquity, it was only during the Renaissance that a systematic study of electricity and magnetism developed, and physicists did not gain a clear understanding of these subjects until the end of the last century. Hardly ever has a scientific achievement had such profound and far-reaching consequences. There have been innumerable practical applications. The harnessing of electric power and the development of electrical communications have changed our whole way of life. On the scientific side, we have learned that electric forces control the structure of atoms and molecules. Electricity is associated with many biological processes—for instance, with the action of our nerves and brain.

18-1 Electrified Objects

We shall now examine some of the basic facts of electric behavior and discuss their interpretation. Let us begin by describing a simple electrical experiment. We rub a glass rod with a piece of silk and then place it horizontally in a stirrup hung on a silk thread. Then we rub a second glass rod, and bring it near the first. The two rods *repel* each other.

We repeat this experiment with two rods of plastic rubbed with a piece of paper. These two rods repel as the two glass rods did (Fig. 18–1).

Finally we rub a glass rod with silk and a plastic rod with paper; then we place one rod in the stirrup and bring the other near it. We now find that these rods *attract* each other.

We can perform similar experiments with a number of other materials. Objects of the same material which have been "electrified" by the same procedure always repel. Different electrified objects may either attract or repel.

We find that electrified objects fall into two groups. Only two "electric states" exist, one similar to that of the glass rod and one similar to that of the plastic rod of the previous example. Following the common usage, which dates back to Benjamin Franklin, we shall say that the glass rod and all other electrified objects which behave like it are *positively charged*. Similarly we shall say that the plastic rod and all other electrified objects which behave like it are *negatively charged*. Any two positively charged objects repel each other, just as two electrified glass rods do. Similarly, any two negatively charged objects repel each other. Any positively charged object attracts any negatively charged one. The words *positive* and *negative*

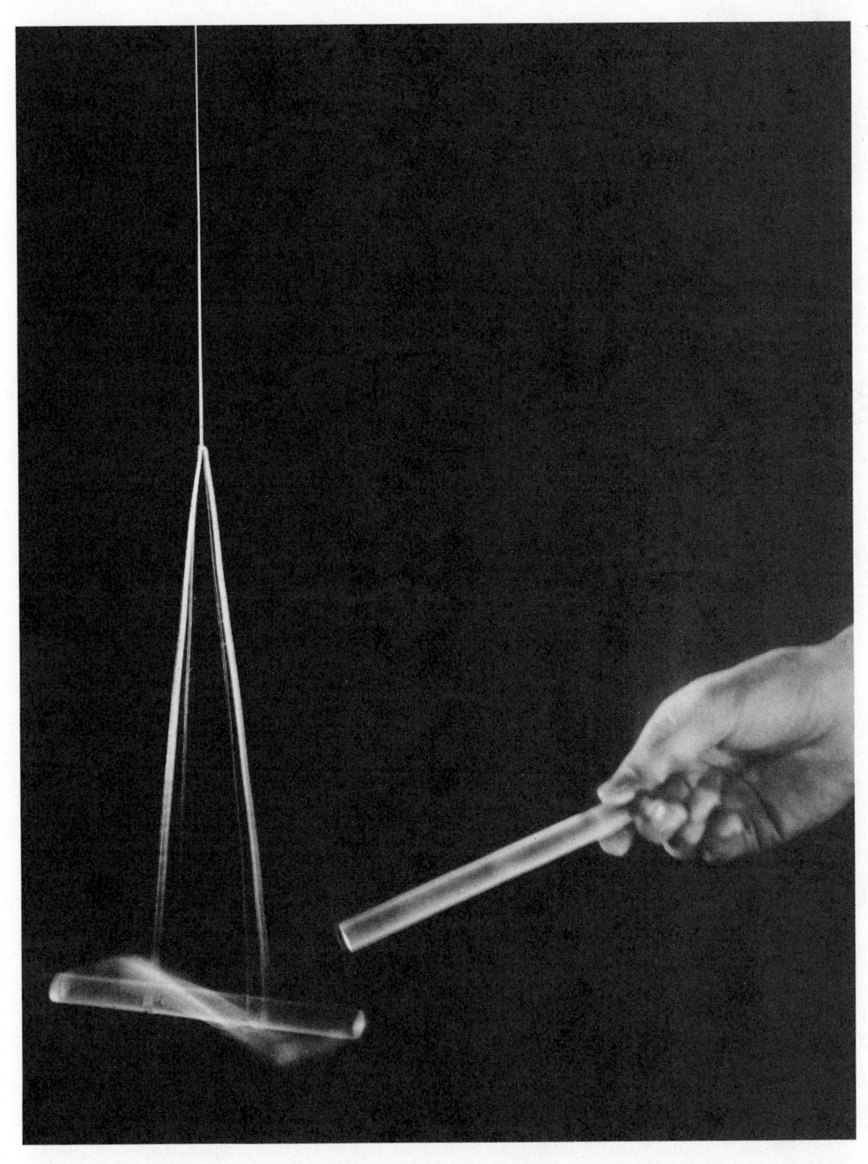

Figure 18–1
Two electrified plastic rods repel each other when they are brought close together. This photograph is a time exposure, made while one rod was brought near the other. The repulsive force pushed away the near end of the suspended rod.

are used because the forces exerted by neighboring positively and negatively charged objects on any third body tend to cancel.

18-2 Some Experiments with Charged and Uncharged Objects

The simple experiments described in the preceding section show qualitatively how two electrified objects interact. How does an electrified object interact with a nonelectrified object? Suppose we hang a light, metal-coated ball on a nylon thread and then support a metal rod horizontally on a glass stand so that one end of the rod touches the ball. Now we electrify a piece of glass and rub the glass against the other end of the metal rod. As we rub

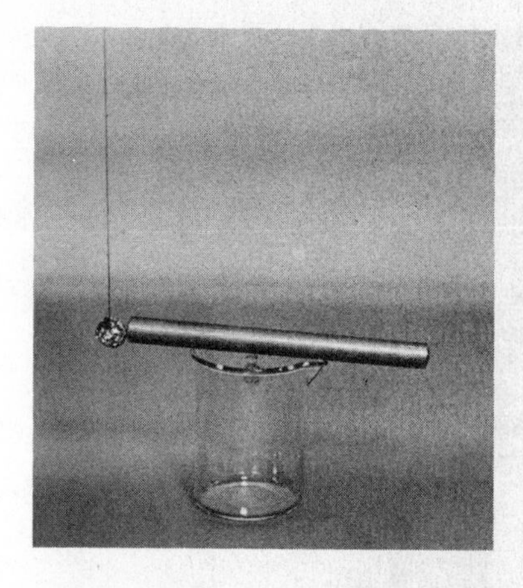

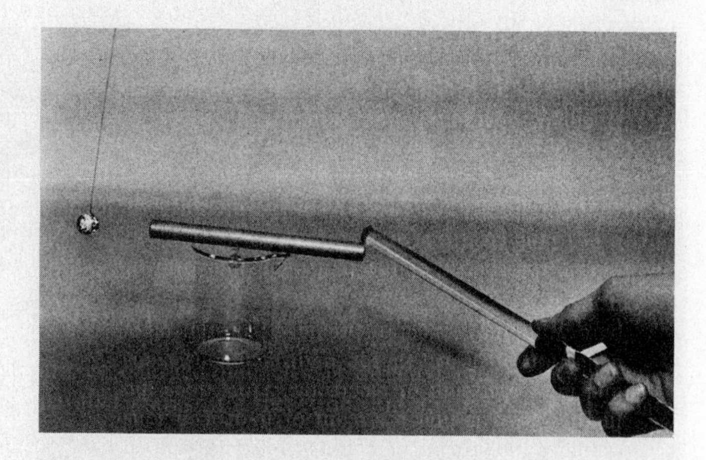

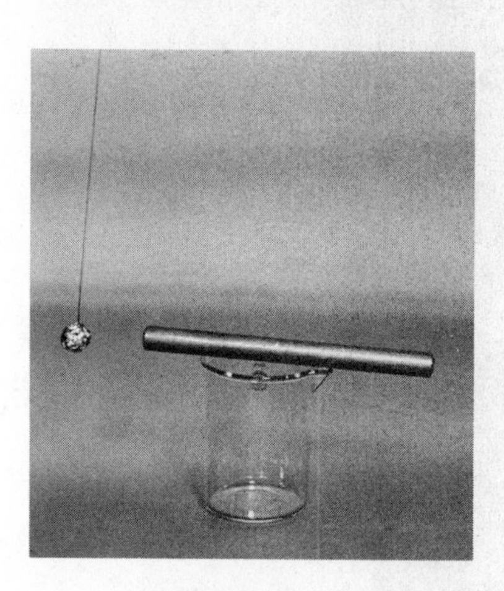

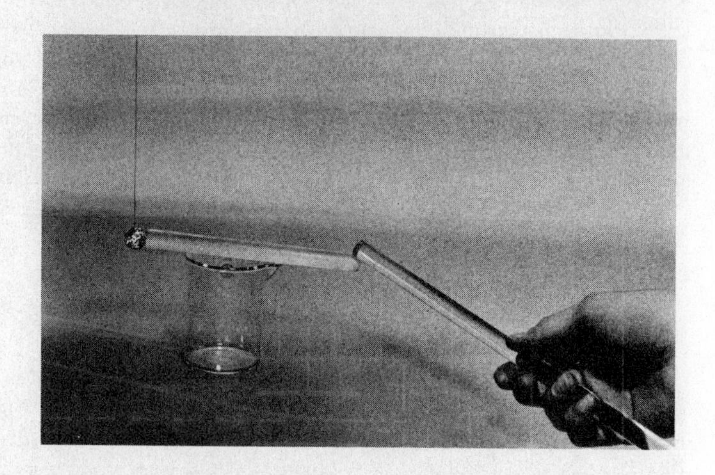

Figure 18–2
(a) A metal rod on a beaker is in contact with a light metal-coated ball.
(b) We touch the end of the metal rod with a charged rod, and the ball at the other end is repelled.
(c) After the charged rod is removed, the metal rod still repels the ball.
(d) We repeat the experiment with a plastic rod in place of the metal one. The ball stays still.

the glass against the metal, the ball swings away [Fig. 18–2(a) and (b)]. We repeat the experiment, using an uncharged glass rod in place of the metal rod. Now the ball does not move [Fig. 18–2(c) and (d)]. We, therefore, see that the metal and the glass rods act differently.

Substances that behave like the metal in this kind of experiment we call *conductors*. Substances that behave like the glass we call *insulators*.

How do we interpret what happened in the above experiments? In the case of the metal rod positive charge passed from the glass to the metal rod and spread immediately over it and the metal-coated ball. The positively charged ball and the positively charged metal rod repelled each other, and the ball swung away. Of course, if we had touched the metal rod with a negatively charged piece of plastic, both the ball and the rod would have become negatively charged, they would have repelled, and the ball would have swung away.

In the case of the glass rod, an insulator, no charge spread over the rod and the ball did not move. It is for this reason that we supported the metal rod on a glass stand and the metal-coated ball by an insulating nylon thread, so that the charge could not leave the metal rod through the stand, nor leave the ball through the thread.

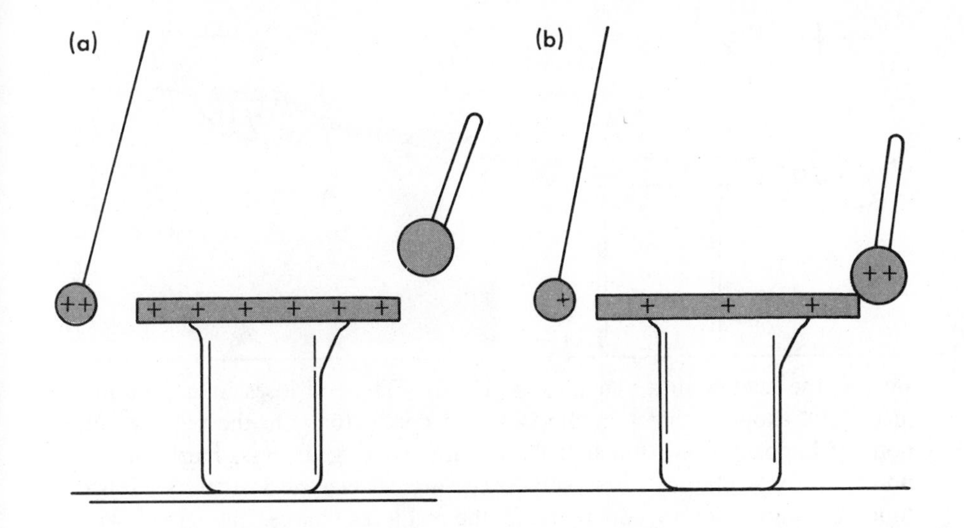

Figure 18–3
(a) A positively charged metal rod and suspended metal-covered ball repel each other. (b) We touch the rod with a large, uncharged metal ball on an insulating handle. The charge is now shared between the large metal ball and the rod. The ball swings back toward the rod.

Now suppose we have, as before, a metal rod on an insulating stand. The rod is in contact with a metal-covered ball suspended from an insulating nylon thread. We charge the rod and the ball swings away. If we touch the rod with an uncharged metal object on an insulating handle, the suspended ball swings in a bit (Fig. 18–3) because we have taken some charge away on the metal object. The charge is shared between the rod and the object. We repeat with a bigger metal object on an insulating handle. The suspended ball swings in even closer to the metal rod because the bigger object takes a bigger share of charge. The same thing happens if we connect the metal rod to a big metal object with a wire. Some of the charge leaves along the conducting wire to the big object. With a very big metal object, practically all of the charge goes to the object. The rod and suspended ball are left with so little charge that the suspended ball swings back and makes contact with the metal rod. We can use the largest available object of all, the earth. When we connect the charged metal rod to the earth, practically all the charge leaves—there is far too little left to detect easily. (This process of sharing charge with the earth is often called "grounding.")*

Although you are not as large as the earth, you too are a conducting object. If you charge the metal rod and then touch it with a finger, the suspended ball swings all the way back. If there was initially an excess of negative charge on the metal rod, some of this charge must have gone onto your body. It must have passed through your body, probably to the ground. But even if your shoes are insulators, the rod will lose practically all its charge. In this case, you are sharing the charge with the rod and suspended ball. From this experiment we conclude that the human body is an electrical conductor.

We can determine what other substances are conductors. For example, suppose we charge the metal rod again and touch it with a stick of graphite,

* The earth has some patches of nonconductor—desert sand, some rocks, etc.—but underground there is a good enough conducting material so that the charge is quickly shared all over the earth. A metal pipe, such as part of the water system, is often used to make sure of the connection to the conducting region underground.

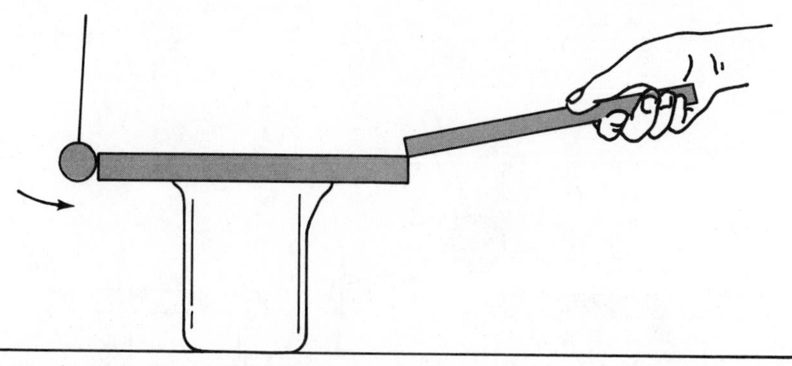

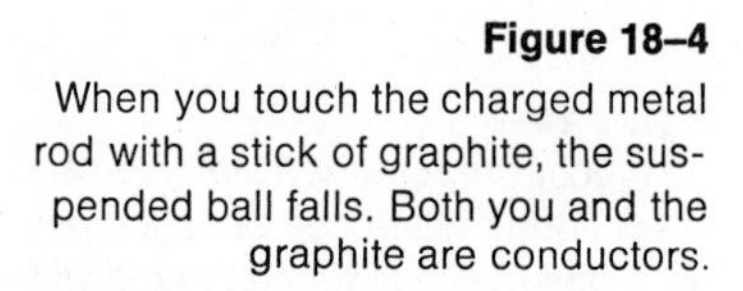

Figure 18–4
When you touch the charged metal rod with a stick of graphite, the suspended ball falls. Both you and the graphite are conductors.

such as the lead from a pencil (Fig. 18–4). The rod loses its charge immediately, showing that graphite is also a conductor. On the other hand, nothing happens if we touch the rod with uncharged glass, hard rubber, porcelain, or a plastic. These substances are all electric insulators. If we touch the knob with a wooden match, the rod loses charge, but very slowly. Apparently electric charge can move in wood, but not as freely as in metals. Wood offers a much greater "resistance" than metals to the motion of electric charge.

18-3 Electrostatic Induction

We do not need to touch a conductor in order to move charge around on it. Suppose we have two metal rods on insulating stands. We put them in contact to form one long conductor, as in Fig. 18–5 and 18–6. Then we bring a positively charged glass rod near one end of the conductor. The positive charge on the glass rod will attract negative charge on the conductor and repel positive charge. As a result, the near end of the conductor acquires an excess of negative charge and the far end is thereby positive. Now we separate the two metal rods by pulling the insulating stands apart while the positively charged glass is nearby. The near rod should have a net negative charge and the far one a net positive charge. We can verify this conclusion by taking the glass away and bringing up a light, positively charged ball on a thread. Then we see that the near rod attracts the positively charged ball and the far one repels it.

If we now move the two rods together again, we detect no charge at either end. The positive and negative charges that we separated now exactly cancel each other all over the rods and we say the rods are *neutral*, as they were before we started the experiment.

The separation of positive and negative charges on a conductor induced by the presence of a charged object nearby is called *electrostatic induction*. The local excesses of positive and negative charge which accumulate on different parts of the conductor are called *induced charges*.

Electrostatic induction allows us to detect the presence of a charge without transferring it to the metal rod and suspended ball we used in the previous section. In fact, we can even determine the sign of the charge. To do this, we first charge the rod, let us say positively [Fig. 18–7(a)]. Then we bring the unknown charge near the rod. If the "unknown" charge is positive, it induces a negative charge on the end of the rod closer to the positively

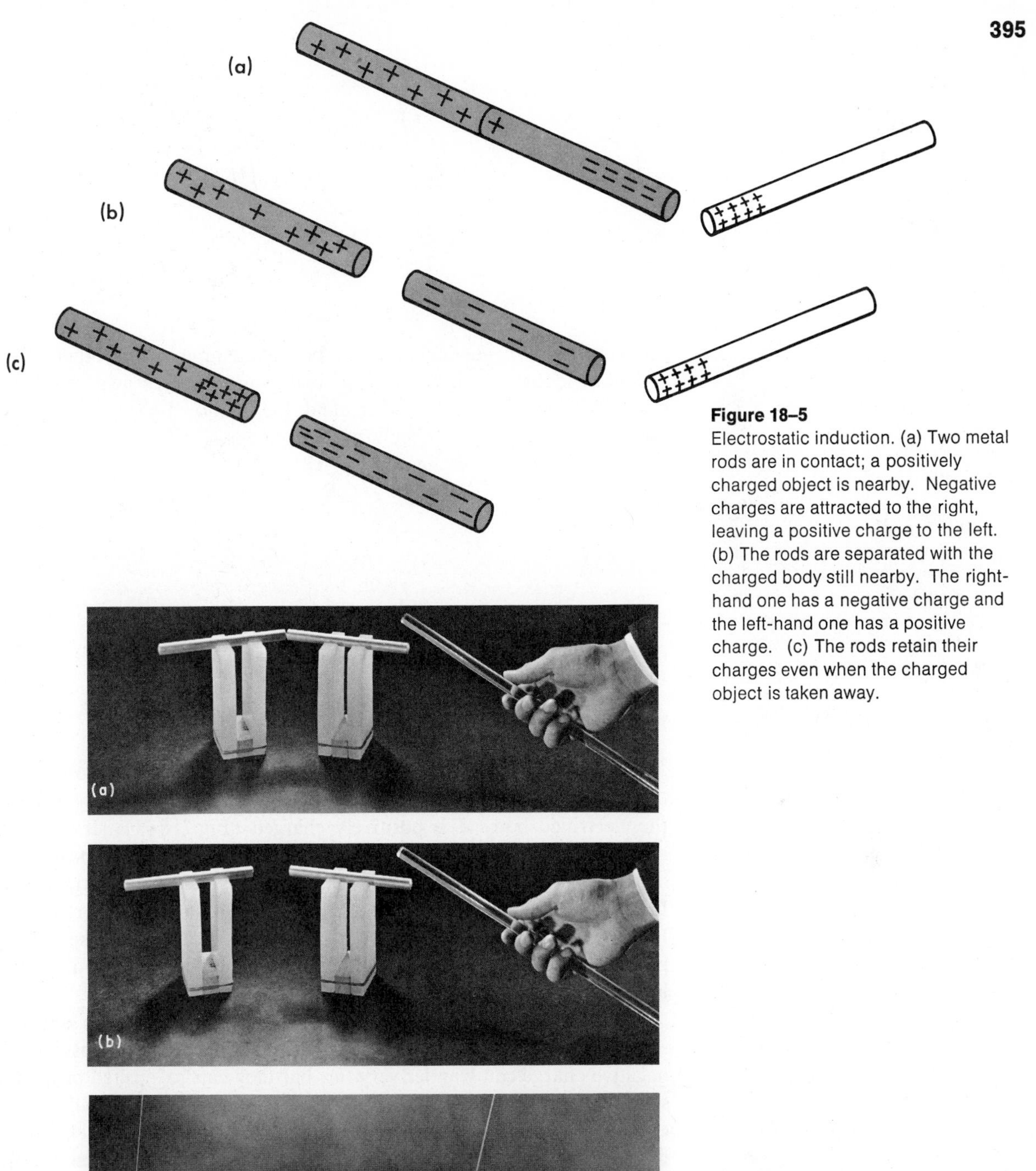

Figure 18–5
Electrostatic induction. (a) Two metal rods are in contact; a positively charged object is nearby. Negative charges are attracted to the right, leaving a positive charge to the left. (b) The rods are separated with the charged body still nearby. The right-hand one has a negative charge and the left-hand one has a positive charge. (c) The rods retain their charges even when the charged object is taken away.

Figure 18–6
Electrostatic induction. (a) Metal rods in contact, positive charge nearby. (b) The rods are separated. (c) We test the charge on each rod with a positively charged ball. The rod at left repels—it is positively charged. The one at right attracts—it is negatively charged.

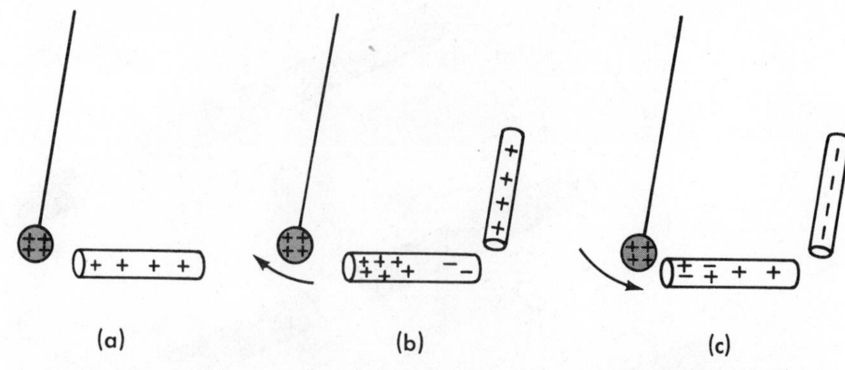

Figure 18–7
(a) A positively charged sphere hangs away from a positively charged rod. (b) The positively charged sphere swings farther away from the rod when a positively charged body is brought near the rod. (c) The positively charged sphere swings closer to the rod when a negatively charged body is brought near the rod.

charged rod and thus increases the positive charge on the other end of the rod. Consequently, the suspended ball swings farther away from the rod [Fig. 18–7(b)]. If, on the other hand, the unknown charge is negative, the induction goes just the other way. The net positive charge on the ball decreases, and the ball swings in [Fig. 18–7(c)].

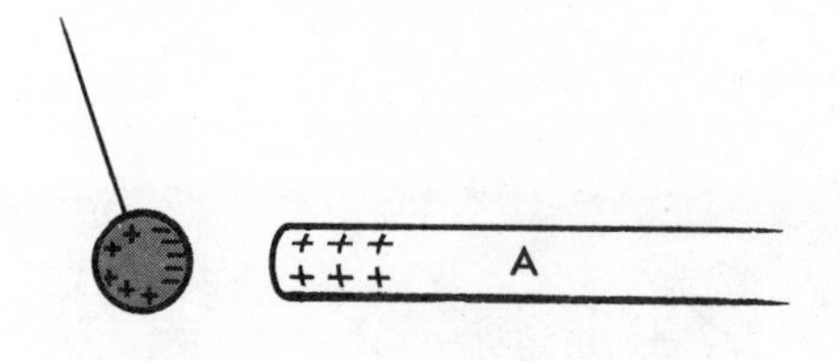

Figure 18–8
Attraction of a neutral conductor. When a neutral conductor is near a positively charged body, negative charge is attracted to the end near the positive charge. A positive charge is left on the rest of the conductor. Since the induced negative charge is nearer the positive body, there is a net force of attraction.

Electrostatic induction also enables us to understand the force of attraction exerted by an electrified body on a neutral conductor. Suppose that the electrified body A is positively charged (Fig. 18–8). It will induce a negative charge on the nearest part of the conductor and a positive charge on the part farthest away. The induced negative charge is attracted toward the positive charge on A, and the induced positive charge is repelled by it. However, the repulsion is weaker than the attraction, because the induced positive charge is farther from A than the negative one. As a result, there is a net force on the conductor pulling it toward A.

There is a similar but smaller force of attraction between a neutral insulator and a charged body. In an insulator neither the positive nor the negative charge can wander very far, but they can be pushed or pulled over short distances. For example, we may imagine that in an insulator positive charge remains fixed in place in the insulator and negative charge is held near the positive charge by springy forces. The springy forces prevent the negative charge from moving any great distance. But when we bring a positively charged body near the insulator, the negative charge is attracted toward it. Consequently, it moves a short distance away from the positive charge it is neutralizing toward the attracting body (Fig. 18–9). Because the negative charge is slightly nearer the positively charged body than its positive counterpart, the force of attraction on it is slightly greater than the repulsion exerted on the positive charge. As a result, there is a net force of attraction between the insulator and the positively charged body.

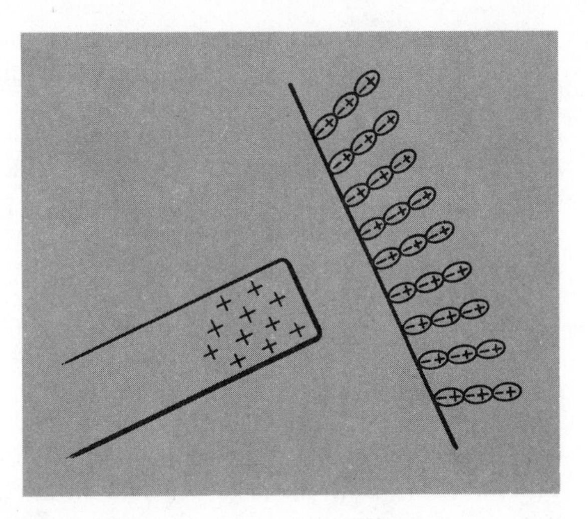

Figure 18–9
Attraction of a neutral insulator. A positively charged object is near an uncharged insulator. No electric charge is free to move through the insulators, but the positive and negative charges can be displaced slightly. The negative charge is thus slightly nearer the positively charged object, giving again a net force of attraction. Note that in fact the molecules are extremely small and so are the charges on them. This diagram is far from being a realistic picture.

18-4 A Model for Electric Charge

Now let us start to construct a model which will correlate the electrical phenomena which we have observed so far.

To begin with, it is natural to think that the force between two bodies is the vector sum of the forces acting between their individual atoms. Since two electrified objects attract or repel each other, we suppose that there are similar forces of attraction or repulsion between some of the parts of which the two objects are made. Because there are electrically positive and electrically negative objects, we assume that atoms contain two types of particles: *electrically positive* and *electrically negative.* Two particles of the same sign repel and two particles of opposite sign attract.

When ordinary "uncharged" pieces of matter are placed near each other, they exert no appreciable electric forces on each other. This does not mean we have to give up the idea that there are positive and negative electric particles in matter. Since a positive and a negative particle exert forces in opposite directions on any third electric particle, it only means that the forces exerted by the two kinds of particle can cancel. If every small volume of a body is neutral, the resultant force exerted by all the electric particles upon any electric particle outside is zero. However, a body can be neutral on the average and nevertheless contain local concentrations of charge. Then an electric particle near one of the local concentrations is subject to a force.

If we add some positive particles to an electrically neutral object, there is no longer a balance. The effect of the positive particles is greater than the effect of the negative particles; and we say the object is positively charged. We can also charge an object positively by removing some negative particles, leaving an excess of positive ones. Similarly, we can make a neutral object electrically negative either by adding some negative particles to it or by taking some positive particles away from it.

We have seen that rubbing a glass rod with silk makes the rod electrically positive. How can this happen? We may think of two possibilities. Perhaps there is a transfer of positive particles from the silk to the rod; perhaps there is a transfer of negative particles from the rod to the silk. In either case the silk must become electrically negative as the glass becomes electrically positive. We can check this conclusion by bringing the silk near the suspended glass rod. We find that the rod is attracted by the silk. Also the silk will repel a negatively charged plastic rod.

To account for the difference between conductors and insulators we need only assume that in a metal some electric particles are free to move from point to point, whereas in the plastic no electric particles can move freely. Suppose, for example, that the free particles in the metal are negative. When the positively charged glass touches the neutral metal rod as in Fig. 18–2(b), the glass pulls some of the free negative particles off the metal rod and the ball. Both the ball and the rod therefore become positively charged, and they repel each other. Even after the glass is removed, there will be a shortage of negative particles on both the rod and the ball, and they will continue to repel.

We can equally well assume that the free particles are electrically positive. Then, when we touch the metal rod with the electrified glass, positive particles will pass from the glass to the metal and spread along it to the ball. Again, the rod and the ball will both become positively charged and will repel each other.

Now let's picture what happens when we replace the metal rod with the plastic one. In the plastic neither positive nor negative particles can move freely. Thus, only the part of the rod in direct contact with the charged glass acquires an excess of positive particles. The rest of the rod remains electrically neutral, and so does the ball. Consequently, there is no force pushing the ball away from the rod.

It should be emphasized that at present in our discussion we have no way of deciding whether only positive charges, only negative charges, or both move during these experiments.

18-5 Batteries

Rubbing two things together, such as a plastic rod and a woolen cloth, is not the only method of separating electrically positive from electrically negative particles. Batteries and electric generators are really machines for separating charges.

We can coat two similar, small, light-weight spheres with a conducting coating and suspend them side by side from long, fine insulating threads so that they just touch each other. Let us connect a wire to one terminal of a battery made up of a large number of flashlight cells and touch the free end of the wire to the suspended spheres. The spheres evidently become charged, for they push each other apart and the threads stand at an angle to each other (Fig. 18–10).

We can investigate the charge on the spheres by testing with a body of known charge. For example, if the spheres have been touched by a wire connected to the positive terminal of the battery, they are repelled by a charged glass rod and hence must have a positive charge.

Now let us suspend the two spheres so that there is some space between them; then let us touch one sphere with a wire attached to the positive

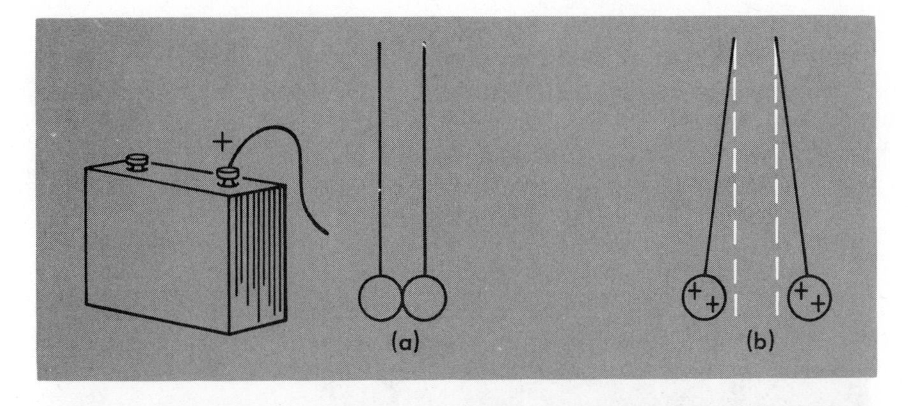

Figure 18–10
(a) Two similar light-weight spheres with a conducting coating hang so that they just touch each other. (b) After they have been touched to a wire connected to one terminal of a large battery, the spheres repel each other.

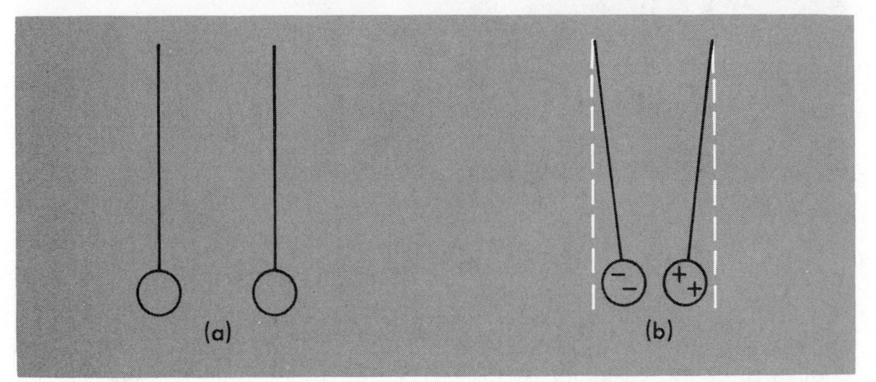

Figure 18–11
(a) Two uncharged spheres with conducting coatings hang so that they are separated. (b) They attract each other after one has been touched to the positive terminal of a large battery and the other to the negative terminal.

terminal of the battery and the other with a wire attached to the negative terminal. The spheres swing toward each other (Fig. 18–11).

If the two spheres swing close enough to each other to touch, they lose their individual charges and return to their initial vertical positions. Apparently the charges given to each of the spheres were the same size, for after touching, the spheres are neutral.

A battery, then, is a device which by chemical action maintains a positive charge on one terminal and a negative charge on the other, despite the forces of electrical attraction which tend to pull these charges together. When a battery terminal is touched to one of the spheres, charges accumulate rapidly on the sphere until after a short time the electric forces repelling the charges back toward the battery terminal compensate for the forces exerted by the chemical action driving them out of the battery.

During the time the spheres were charging, electric charges moved through the wires connected to the battery. Such a flow of electric charge is called an electric current. Electric currents have many different effects (some of which we will study later on), such as heating the wire through which they pass. Instruments can be built to detect these effects and thus electric currents can be measured.

18-6 Measuring Small Electric Forces

We constructed a model for electric charge by supposing that ordinary pieces of matter contained equal numbers of positive and negative electric particles, and with this model we have been successful in accounting for all the observations we have made so far in experiments with electricity. However, a model picturing electric charge as two different types of fluid would work equally well for these observations. We need further evidence before definitely deciding on a particle model for charge, and we should like to obtain as much additional information as possible about electric particles, provided we can demonstrate that they exist.

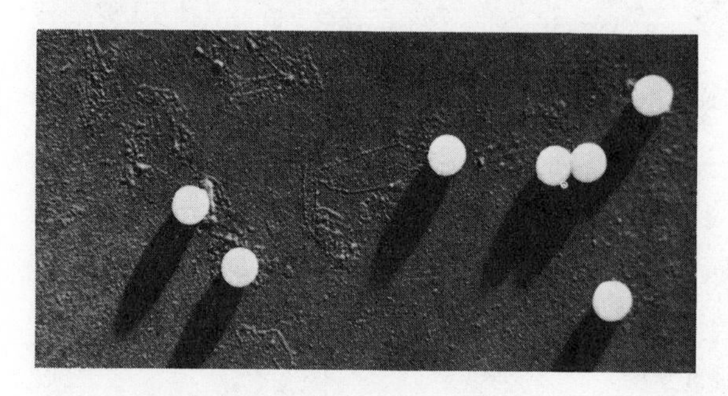

Figure 18–12
An electron-microscope photograph of a few plastic spheres magnified about 2500 times.

We begin by discussing a method of manipulating small objects and of measuring the minute forces that can be exerted on them. Then we shall be equipped to use the forces to measure electric charge.

We need objects which are large so that they will be visible, and yet small enough to be moved about by the electric forces we can exert on charged particles. Objects visible to the naked eye are too massive for our purpose. Microscopic plastic spheres about 10^{-6} meter in diameter turn out to be about right (Fig. 18–12). Although they are large enough to be seen as spots of light through an optical miscroscope, the mass of each is only about 10^{-15} kg, or a micromicrogram. Since these spheres are manufactured for calibrating the distances seen with electron microscopes, a large number of almost identical spheres are available.

In air these spheres move rather slowly. Looking at them, we see that they fall steadily (Fig. 18–13) with no apparent acceleration. The net force on them must be zero. This zero net force is the vector sum of the gravitational force, which pulls a sphere down, and the opposing force of air resistance. When a sphere starts falling, the air resistance is zero, but the

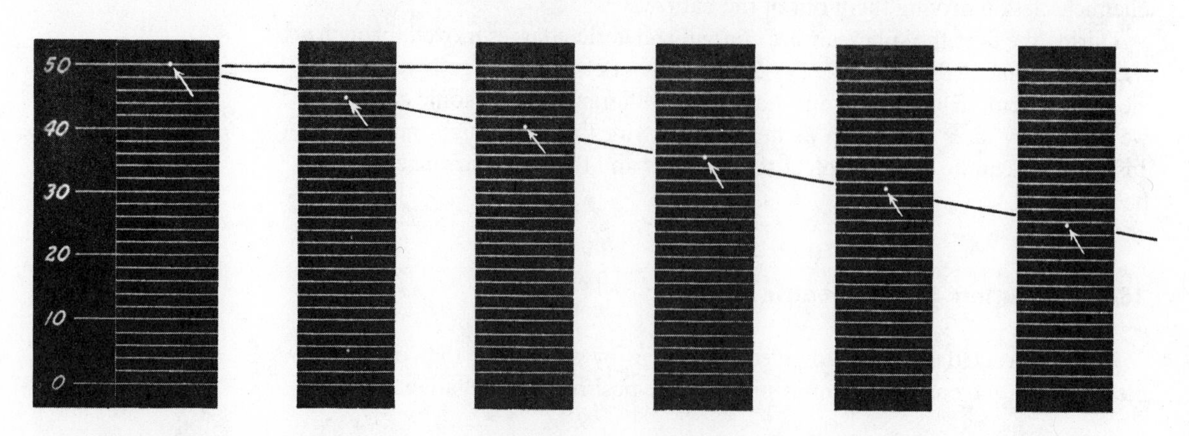

Figure 18–13
Photographs of a tiny plastic sphere driven downward by gravity, taken through a microscope at one-second intervals. The sphere falls about the same distance in each interval across a grid whose total length is slightly over 1 mm.

faster the sphere goes the greater the air resistance. So the drag of the air increases quickly until it balances the small gravitational force. From then on the sphere moves with a constant velocity, called the terminal velocity.

If we place some of these plastic spheres between two charged plates, we often find that there is an electric force acting on the spheres, in addition to the gravitational force. Some of the spheres move upward; others move downward. Apparently, such spheres carry an electric charge. As a sphere starts to move under this net driving force (the vector sum of the gravitational and electric forces), the air resistance will change until the sphere will again move with constant velocity. This terminal velocity will be different from the terminal velocity with only the gravitational force acting. However, it is the same anywhere between the plates. This shows that the force on the sphere between the two parallel charged plates is constant. It does not depend on the position of the sphere between them.*

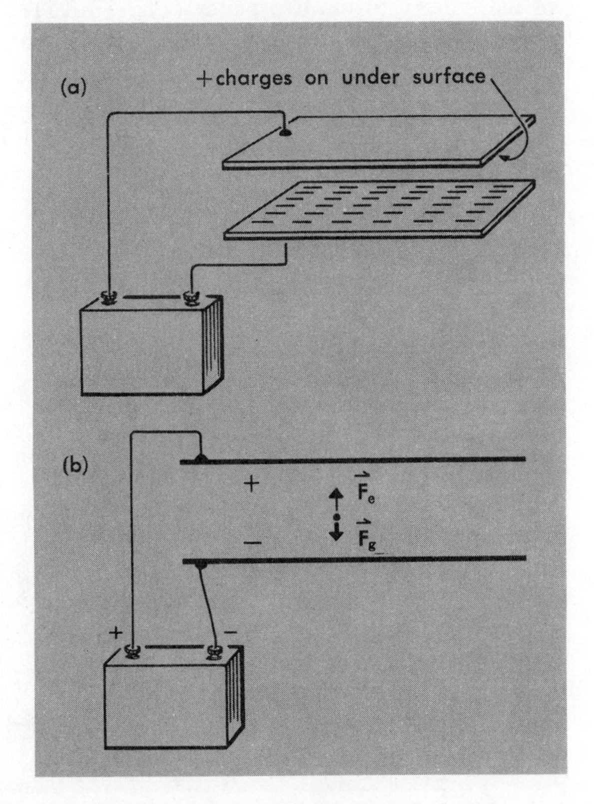

Figure 18–14
By choosing the right battery and the right plate separation we can hold a sphere motionless between the plates. The electric force and the gravitational force must be equal: (a) shows the way the plates and the battery are arranged; (b) is a sketch of the same apparatus with schematic plates.

How is the terminal velocity related to this net driving force? Let us consider a sphere placed between two charged metal plates connected to a battery, as shown in Fig. 18–14. If we vary the electric charge on the plates, we can change the electric force acting on the sphere. By choosing the right battery and plate separation, we can have an electric force acting on the sphere which will be equal in magnitude but opposite in direction to the gravitational force. Thus, the sphere will be held in balance between the plates. What will happen if we reverse the connections from the terminals

* This holds as long as the spheres are not too close to the edges of the plates and the distance between the plates is small compared to their dimensions.

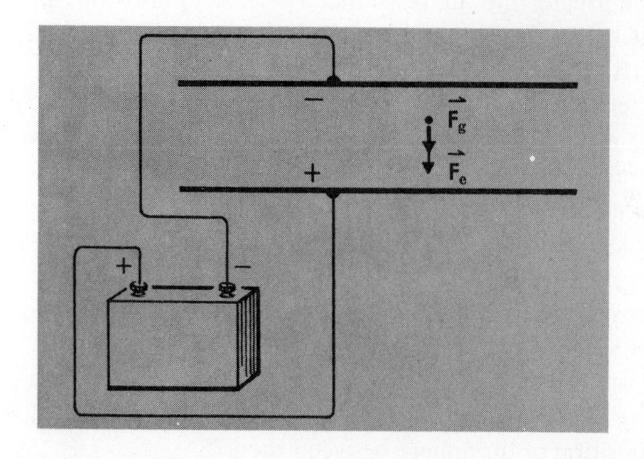

Figure 18–15
With the connections of the balancing battery reversed, the electric force now acts downward. The driving force is twice the gravitational force.

of the battery to the two plates (Fig. 18–15)? The electric force is still equal in magnitude to the gravitational force, but now it points in the same direction as the gravitational force. Thus, the net driving force on the sphere is now twice the gravitational force alone. If we measure the terminal velocity with which the sphere moves, we find it is twice the terminal velocity with which the sphere moved under gravitation alone. This result suggests that the terminal velocity of the sphere is directly proportional to the driving force. This relation has been checked and holds over a wide range of velocities.

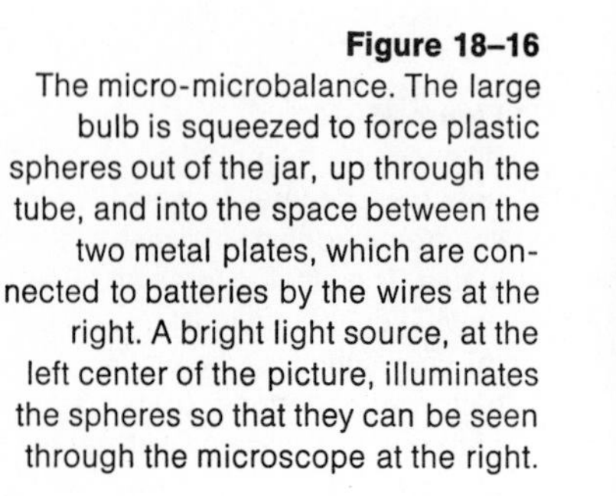

Figure 18–16
The micro-microbalance. The large bulb is squeezed to force plastic spheres out of the jar, up through the tube, and into the space between the two metal plates, which are connected to batteries by the wires at the right. A bright light source, at the left center of the picture, illuminates the spheres so that they can be seen through the microscope at the right.

Apparatus with which this experiment was actually performed is shown in Fig. 18–16 and 18–17. With this apparatus we can balance the force of gravity acting on a small sphere, whose mass is of the order of a micro-microgram, with an electric force. We shall refer to the apparatus as a micro-microbalance.

With the micro-microbalance we can show that the electric force is proportional to the number of identical batteries connected in series with the plates. Suppose we first measure the terminal velocity of a plastic sphere when no electric force acts on it [Fig. 18–18(a)]. Then we select a battery and plate separation which will produce an electric force which just balances

Figure 18–17
A close-up of the micro-micro-balance plates. The microscope has been removed to show the plates and their connections to the batteries more clearly.

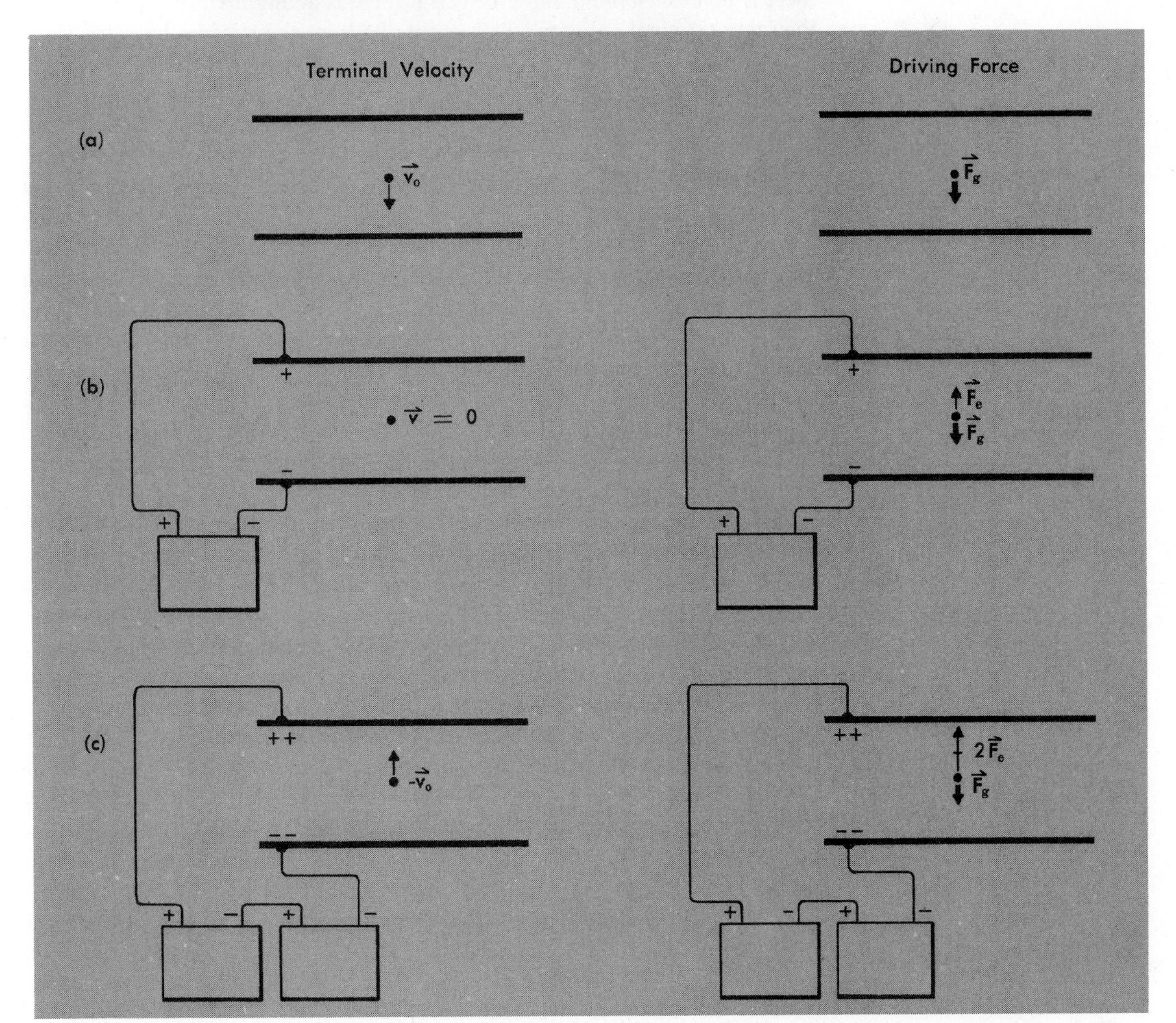

Figure 18–18
The terminal velocity (a) with no electric force, (b) with balanced gravitational and electric forces, and (c) with a second battery in series with the first, giving a terminal velocity equal and opposite to that in (a).

the gravitational force, so that the terminal velocity is zero [Fig. 18–18 (b)]. We now connect an additional battery, identical to the first, in series and find that the sphere moves upward with a terminal velocity equal and opposite to the terminal velocity with no electric force [Fig. 18–18 (c)]. This leads us to conclude that the electric force when two batteries are used is twice the electric force when only one is used. Indeed, if we add a third battery in series to the other two, the sphere moves upward with a terminal velocity of twice the magnitude of the terminal velocity with no electric force. While one battery just cancels the force of gravity, the other two produce a net force on the sphere which is double in magnitude and opposite in direction to the gravitational force.

18-7 The Elementary Charge

Suppose that we place several spheres like those shown in Fig. 18–12 between the plates of the micro-microbalance. When the plates are uncharged, there is no electric field and no electric force is acting. All the spheres have the same terminal velocity. This occurs because the spheres have the same mass, so the driving force due to gravity is the same on each sphere. On the other hand, when the plates are charged, the terminal velocities are no longer the same for each sphere. The driving force is now part electric and part gravitational, so the terminal velocity is the vector sum of the part due to the electric and the part due to the gravitational driving force. If we subtract the terminal velocity due to the gravitational driving force, we shall find the terminal velocity each sphere would have if the only driving force were electric. Since we know that the terminal velocity is proportional to the driving force, we shall be able to determine the electric force acting on each sphere.

The terminal velocities of several spheres were measured in an experiment using the equipment of Fig. 18–16 and 18–17. The terminal velocity due to the gravitational driving force alone was subtracted from each observed value, and the resulting velocities due to the electric driving force alone are plotted in Fig. 18–19. Note that the values are not distributed randomly. They cluster around certain values which are multiples of the lowest one. For this particular experiment, the velocities cluster around 6.8, 13.5, 20.4, and 27.3 (in units which were convenient to use in this experiment), as seen by rearranging them in order of magnitude (Fig. 18–20). Since the terminal velocity is proportional to the driving force, we conclude from these data that the electric force is also distributed around certain values. Now, the electric force depends on the charges on the spheres and the charge on the plates. The charge on the plates is fixed throughout the experiment; thus, the differences in the electric force on each sphere must arise from differences in the electric charge on each sphere. These differences come in multiples of a basic unit.

It is attractive to assume that these equal differences in electric force on the tiny charged spheres are the result of equal differences in electric charge on the spheres; i.e., the force on the spheres is proportional to their charge. This would be analogous to gravitational forces near the surface of the earth where the gravitational force on an object is proportional to its mass. This can be thought of as corresponding to electric charge in the case of electric forces.

If charge does come in discrete equal units, as our experiment seems to show, then there is a natural unit of charge which we can call the *elementary charge.* In fact, in 1909 R. A. Millikan perfected a balance much like the one we have described, and he performed a series of experiments which clearly showed the existence of the elementary unit of charge. Where we use plastic spheres of known mass, he used tiny oil drops—and he had to measure their masses while examining their charges. Our experiment is therefore easier than his, but otherwise it is nearly the same.

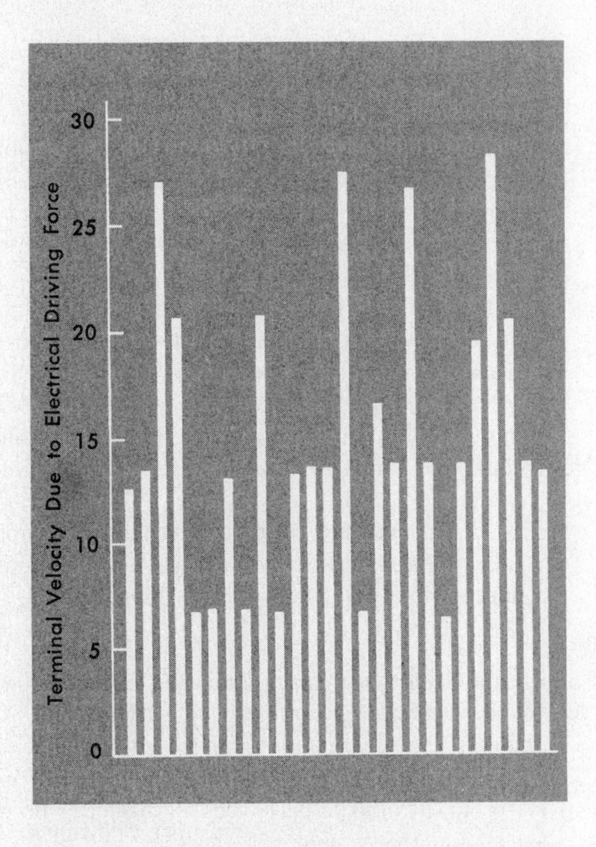

Figure 18–19
The terminal velocities (measured in units convenient for this experiment) due to the electric part of the driving force acting on several plastic spheres. The terminal velocity due to the gravitational part of the driving force has been subtracted from the observed values.

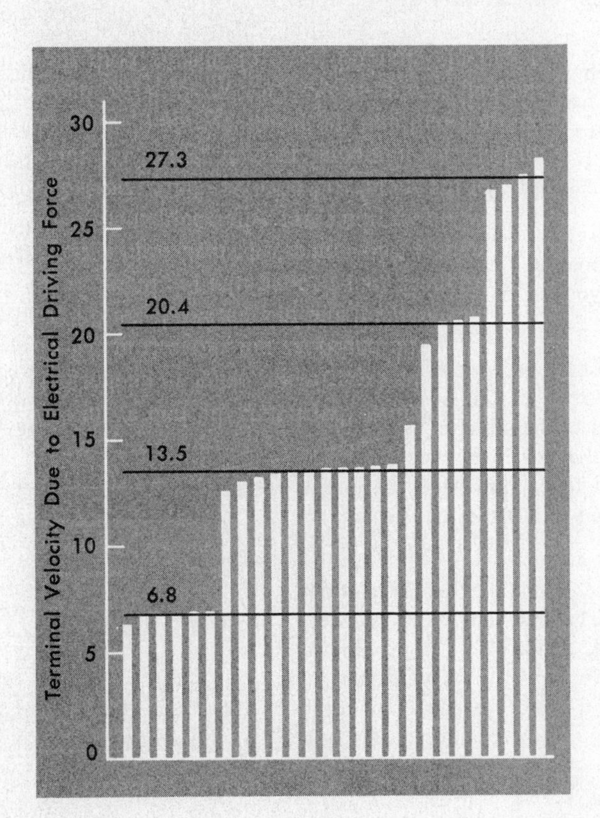

Figure 18–20
The terminal velocities of Fig. 18–19 replotted in order of increasing magnitude. Note that the magnitudes cluster around certain values which are whole-number multiples of the lowest one.

18-8 The Conservation of Charge

Our picture of particles of electric charge leads to the idea of the conservation of charge. When we charge bodies by rubbing them, we transfer charged particles from one body to another. What one body gains in charge the other one loses. When we charge the plates of an electric balance with a battery, the battery transfers charge from one plate to the other. The charges of the plates are equal and opposite.

In certain very unusual circumstances we can "create" charged particles, but we shall find that we always create them in pairs, and the charge of one particle is equal and opposite to the charge of the other. Sometimes nature "creates" charged particles automatically; for instance, a neutron turns into a proton and an electron. The proton and electron thus created have equal and opposite charges. The total charge is zero before and after the creation.

All our evidence indicates that the total amount of charge never changes. Like the conservation of energy, the conservation of charge is a law of nature which extends over everything we know.

For Home, Desk, and Lab

1.* If two charged rods repel each other, what can be said about the sign of the charge on each rod? (Section 1.)

2.* You find that object *A* repels object *B*, *A* attracts *C*, and *C* repels *D*. If you know that *D* is positively charged, what kind of charge does *B* have? (Section 1.)

3. Suppose that you have electrified a plastic rod by rubbing it with wool.
   (a) Do you expect the wool to become charged?
   (b) How could you find out if it does?

4.* (a) Why can't you electrify a metal rod by rubbing it while holding it in your hand?
   (b) What could you do to electrify a metal rod? (Section 2.)

5. When the gasoline tank of an airplane is being filled, the metal nozzle of the hose is always carefully connected to the metal of the airplane by a wire before the nozzle is inserted in the tank. Be prepared to explain why this procedure is followed and to describe how it accomplishes its purpose.

6. (a) When you touch a metal object like a door handle on a dry winter day, you are likely to see a spark and feel a definite shock. We usually explain this by saying that we have built up a static charge. How could you tell the sign of the charge you are carrying?
   (b) If you accumulate a static charge and then touch the wooden frame of the door, you often find no spark or shock—although there would have been one if you had touched the metal handle. Why?
   (c) Sometimes, if you touch the wooden frame and then touch the metal handle, there is no spark or shock in either case—although there would have been one if you had touched the handle first. Suggest an explanation.

7. Two metal rods are mounted on insulating blocks and aligned along a common axis, with a space between them. One is positively charged, the other negatively. A light sphere with a conducting coating suspended by a long insulating thread is introduced into the space between the rods (Fig. 18–21) and touched to the positive rod. What do you predict will happen? On what reasoning do you base your prediction?

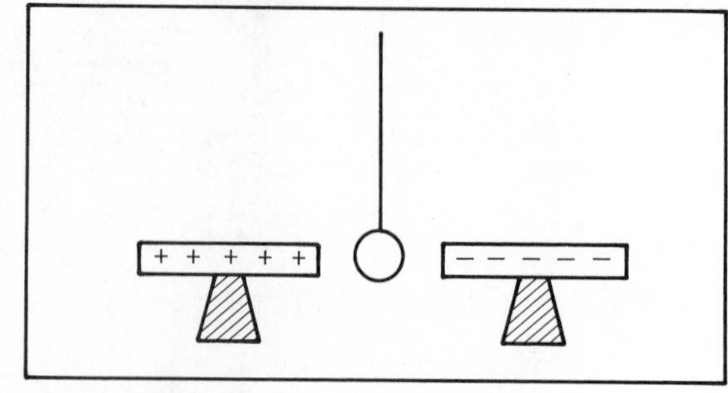

Figure 18–21
For Problem 7.

8. Three metal blocks in contact are resting on a plastic tabletop. You place two objects with strong positive charges, one at each end of the line of blocks, close to but not touching the blocks. You then poke the blocks apart with an (uncharged) insulating rod, while the objects with strong positive charges are nearby. Finally you remove the two positively charged objects.
   (a) What charge is now on each block?
   (b) Explain how the blocks acquired these charges by describing the motion of negative particles.
9. You have two metal spheres of the same size mounted on insulating stands, a strip of plastic, and a piece of paper. Describe what you would do to give the two spheres equal electrical charges (a) of the same sign, (b) of opposite signs.
10. An old-fashioned gold-leaf electroscope consists of two thin gold leaves hanging from a metal rod in a glass container [Fig. 18–22(a)]. You electrify a plastic rod, then bring it near the knob of the electroscope; the gold leaves separate [Fig. 18–22(b)]. When you take the rod away, the leaves hang straight down once again.

    Use the model for electric charge to account for the behavior of the electroscope.

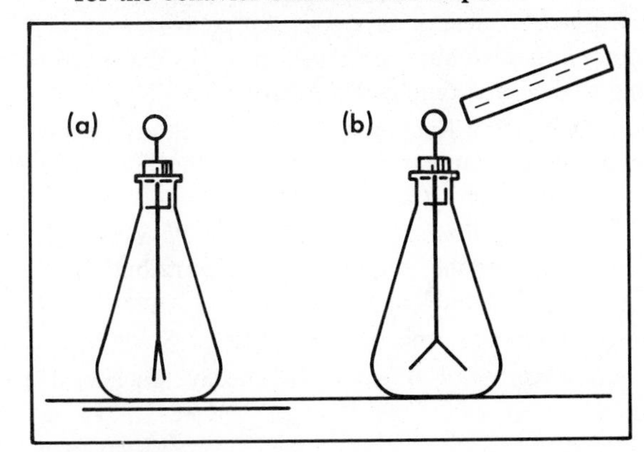

Figure 18–22
For Problem 10.

11. The leaves of an electroscope are hanging straight down. You electrify a glass rod and then touch it to the knob of the electroscope. The leaves separate and remain separated when you remove the glass rod. Use the model for electric charge to account for the behavior of the electroscope.
12. You hold a charged plastic rod near the knob of an electroscope: you touch the knob with your finger; you withdraw first your finger, then the rod. The electroscope leaves separate and remain separated. Use the model for electric charge to account for the behavior of the electroscope. What simple test could you make to verify the correctness of your explanation?
13. Must all the negative charges be removed in order for a rod to become positively charged?
14. If you continue the experiment shown in Fig. 18–18, what sphere velocities will you expect to get in the situations in Fig. 18–23?

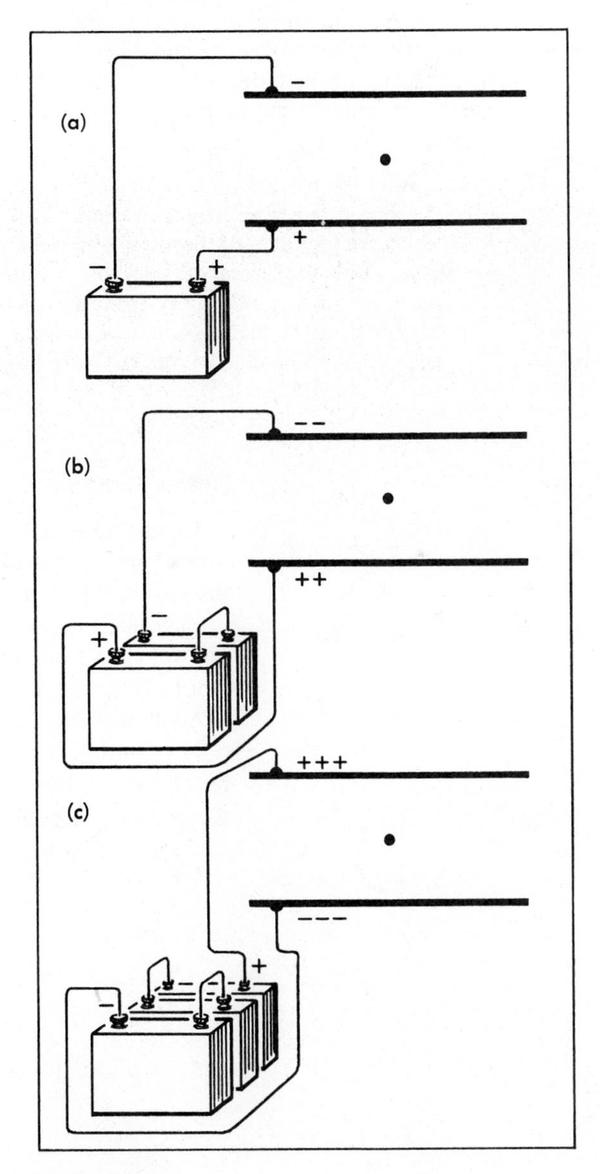

Figure 18–23
For Problem 14.

15. Suppose we measure the speed of the plastic spheres between the plates of a micro-microbalance and find it is three halves of their speed under gravity alone. The weight of a sphere is 2.8×10^{-14} newton.
    (a) What is the electrical force on them if they are moving upward?
    (b) If they are moving downward?

16. There is a relationship between the charge on two parallel plates and the force they exert on a charged object between them. There is also a relationship between the number of batteries connected and the force on a charged object between the plates. What relationship between the number of batteries and the amount of charge the batteries put on the plates would you suggest?

17. (a) Most of the plotted velocities of Fig. 18–20 are indeed clustered around certain values which are multiples of the lowest one. However, there are cases of velocities whose values lie significantly away from the cluster; for example, there is one 0.3 of the way between 13.5 and 20.4 on the arbitrary velocity scale. What is the most likely explanation of this observation?
    (b) Suppose the data shown in Fig. 18–20 had been obtained by connecting 4 batteries to the plates of the micro-microbalance. How would the appearance of the plotted data change if the experiment were repeated with 5 batteries connected to the plates?

18. Suppose that, in plotting the data shown in Fig. 18–19 and 18–20, the terminal velocity due to the gravitational driving force had *not* been subtracted. Would the figures then be different from those shown? What steps were taken during the experiments described in Sections 18–6 and 18–7 to insure that the plots of Fig. 18–19 and 18–20 have the appearances shown?

19. Suppose the experiments described in Sections 18–6 and 18–7 had been performed with spheres of double the mass of those actually used.
    (a) What observations would have been different?
    (b) How would the conclusions have compared with those actually made from the experiments described in Sections 18–6 and 18–7?

Further Reading

These suggestions are not exhaustive but are limited to works that have been found especially useful and at the same time generally available.

BOYS, C. V., *Soap Bubbles*. Doubleday Anchor, 1959; Science Study Series. Of particular interest here are those demonstrations that show the effects of electrical charge on thin streams of water. (Lecture III)

HOLTON, G., and ROLLER, D. H. D., *Foundations of Modern Physical Science*. Addison-Wesley, 1958. A very readable summary of the contributions made by Gilbert, Gray, DuFay, Franklin, and Coulomb's work. (Chapter 26)

MOORE, A. D., *Electrostatics*, Doubleday Anchor, 1969: Science Study Series.

WEISSKOPF, VICTOR F., *Knowledge and Wonder*. Doubleday Anchor, 1963: Science Study Series.

Coulomb's Law, Electric Fields, and Electrical Potential

CHAPTER

19

In the last chapter we discussed in detail the Millikan experiment in order to show that electric charge comes in discrete units that are integral multiples of a fundamental unit of charge—the elementary charge. In the experiment we made measurements of the motion of tiny charged spheres between two parallel, closely spaced, charged metal plates. We found that the force on a charged sphere (for a given battery and given charge on the sphere) was constant everywhere between the centers of the plates. It did not depend on distance. But this was a special case—two closely spaced parallel plates. In general, just as in the case of gravitational and elastic forces, the electric force between two charged bodies does depend on their separation.

19-1 Force vs. Distance

The way in which the attraction or repulsion between charged spheres depends on distance was established experimentally by the French physicist Charles Coulomb in 1785. In his experiment Coulomb used a torsion balance, like the instrument Cavendish used to study gravitational attraction (see Section 13–10). Figure 19–1 is a diagram of this kind of apparatus. When the charged sphere *A* is put in place, the electric force it exerts on the charged sphere *B* pushes the horizontal arm around. The arm then comes to rest in a new position with the suspension twisted. The more the suspension is twisted, the greater the force must be. From the angle of twist, therefore, Coulomb could measure the electric force. And by changing the distance between the charged spheres, he measured the force as a function of separation. Because the charges are distributed all over the spheres, the spheres must be far enough apart so that their diameters are small compared to the distance between them.*

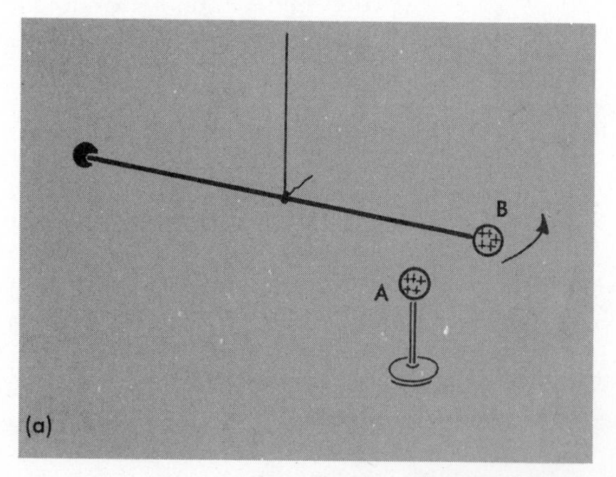

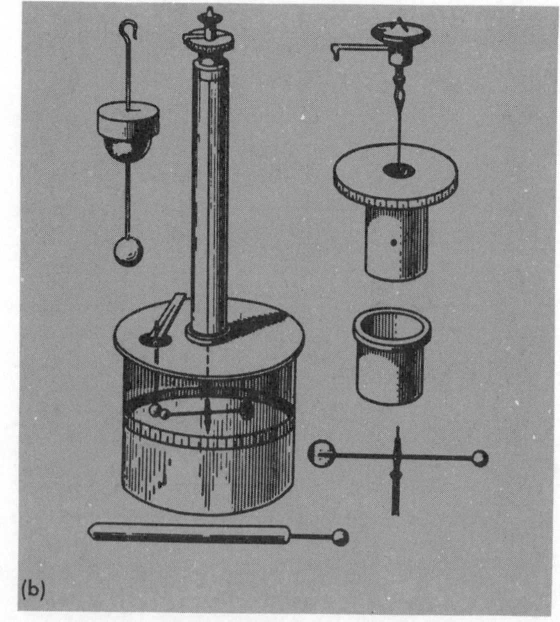

Figure 19–1
(a) Schematic diagram of Coulomb's torsion balance. (b) A sketch of Coulomb's apparatus which appeared in his original paper.

* When charged bodies are far apart compared with their dimensions, they are often called *point charges* because the detailed positions of the charges on the bodies are then unimportant.

Using both positively and negatively charged spheres, Coulomb showed that the electric force is always inversely proportional to the square of the distance between the charges. He established this result with an accuracy of about 3 percent. Later indirect tests—using the shielding effect of conductors—have shown it with much higher precision. Cavendish attained 1 percent accuracy; and in the latter part of the nineteenth century Maxwell established the exponent 2 (the *square* in the inverse square) to within one part in 40,000. It is now known to one part in 10^9. Note that electric forces and gravitational forces vary with distance in exactly the same manner. We have no explanation of this similarity, but because of it we can often understand gravitational and electrical effects in the same way.

19-2 Electric Charge and Electric Force

In an electrically neutral body the effects of positive and negative elementary charges cancel. A positively charged body contains uncanceled positive charges, and a negatively charged body contains uncanceled negative charges. Thus the charge of a body depends on the excess of positive or of negative elementary charges, measured from neutral.

The force between two charged bodies depends on their separation and increases with the excess of positive or of negative electric charges on each body. Just how does the force depend upon the excess of electric charges? To answer this question we need a scheme to divide the excess of charges in a known way—in half, in thirds, etc. Suppose we touch a charged metal sphere with an identical uncharged sphere (Fig. 19–2). Then the electric charges will move around until they are shared equally by both spheres. Each sphere will have half the original charge.

What happens to the electric forces when charges are shared? We measure the force of repulsion between two charged spheres A and C at a certain separation. Then we halve the charge on A by sharing it with an identical sphere B. The force of repulsion between A and C (still at the same separa-

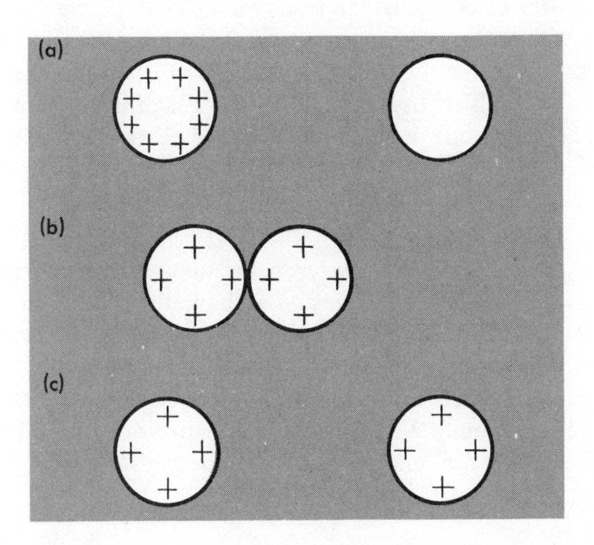

Figure 19–2
The sharing of electric charge. When a charged sphere is touched to an identical uncharged one, the excess of electric particles divides equally. The final distribution of charge must be symmetrical, as shown in (c).

tion) is also cut in half. Furthermore, we get the same force when A is replaced by B, the identical sphere with which it shared its charge. Apparently, charge and force are proportional. This confirms the assumption we made in Section 18–7 that led to our postulation of an elementary unit of charge.

Such experiments give us a way of comparing large charges quantitatively. Two charges are equal if they experience equal forces at a given distance from any third charge. One charge is twice another when it experiences twice the force. When a charge is halved by charge sharing, the force exerted on it by a third charge is also halved. In general, charges are compared by the ratio of the forces exerted on them by another charge at a given distance. This ratio does not depend on the magnitude of the "other" charge nor on the distance apart (Fig. 19–3).

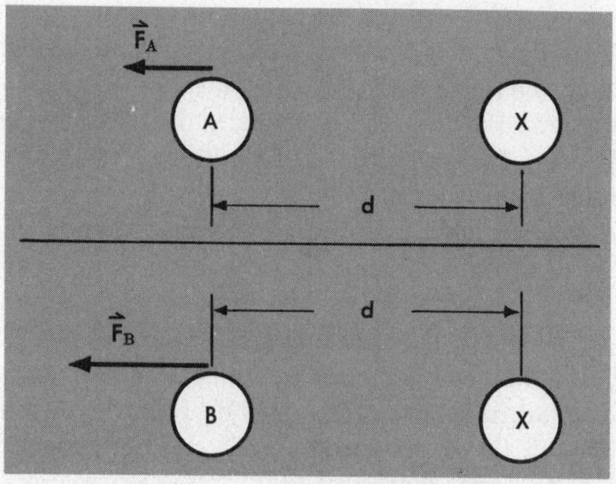

Figure 19–3
To compare two charges, A and B, we place them in turn at the same distance from any other charge X, and measure the forces. The ratio of the charges equals the ratio of the forces: $q_A/q_B = F_A/F_B$.

Now let us summarize our knowledge in algebraic language. The electric force on a charge q_1 is proportional to the charge: $F \propto q_1$. When this force is the force of interaction on the charge q_1 by another small body of charge q_2, the force is also proportional to the other charge. We can write this proportionality to both the charges as $F \propto q_1 q_2$.

Now that we know how the electric force depends on the charges, we can combine this knowledge with Coulomb's experiments. They tell us that the force is inversely proportional to the square of the separation r between the charges. So we arrive at the complete expression for the force of interaction between two charges. The magnitude of the force on either charged body is

$$F = \frac{kq_1q_2}{r^2},$$

where the proportionality factor k depends only on the units in which we measure forces, separations, and charges. If q_1 and q_2 have the same sign, the forces are repulsive and the force on each charged body points outward along the line joining the two bodies. On the other hand, if q_1 and q_2 have opposite signs, the forces are attractive and the force on each body points inward along the line joining them. We shall call the expression for the magnitude of the force *Coulomb's law*, and the forces which obey it *Coulomb forces.*

In Section 12–1 we introduced the gravitational field of the earth as the gravitational force which the mass of the earth exerts on a unit mass. Similarly, we shall define the electric field $\vec{E}$ of a charged body as the electric force which it exerts on a unit of positive charge. Thus the relation between electric field and electric force on a charge q is $\vec{F} = q\vec{E}$, just as the relation between the gravitational force on a mass m and the gravitational field is $\vec{F} = m\vec{g}$. The unit of the electric field we shall use is a newton per unit charge.

We can draw a diagram of the field vectors around a stationary charge, showing the size and the direction of the field at a large number of points. Figure 19–4 shows such a set of vectors for both a positive fixed charge and a negative fixed charge. The set of vectors is very much like the set of vectors indicating the gravitational field of the earth (Fig. 12–1), except that the gravitational field vectors of the earth all point toward the earth, whereas electric field vectors point away from the stationary charge when it is positive and the force is repulsive, and toward it when the fixed charge is negative and the force is attractive on a positive test charge.

Figure 19–4
The electric force exerted on a positive movable charge by (a) a positive fixed charge and (b) a negative fixed charge. It is the set of vectors showing the force on the movable charge at each point in space.

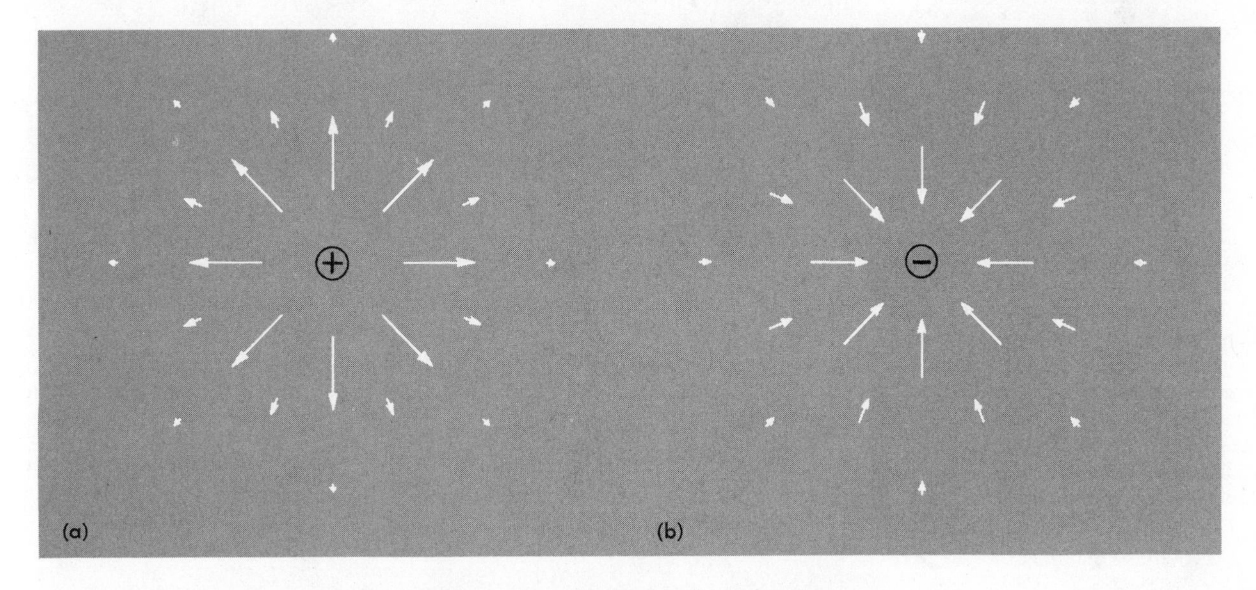

Often, instead of drawing a collection of field vectors, we represent an electric field in a slightly different way by drawing a number of electric field lines. A line is constructed by moving a point continuously in the direction of the force exerted on a unit positive charge. As the force changes direction, so does the line. The patterns of electric field lines around a single small charged object and around several separated charged bodies are illustrated in Fig. 19–5. The direction of each line is the direction of the net force on a movable positive unit charge placed on the line. Sometimes we can show the electric field lines experimentally because small bodies will line up parallel to them. For example, we can suspend grass seed in an insulating liquid. Then, if we make an electric field by putting charged objects in the liquid, the seeds line up, forming a picture of the field. Figure 19–6 shows photographs of the grass-seed patterns of several electric fields.

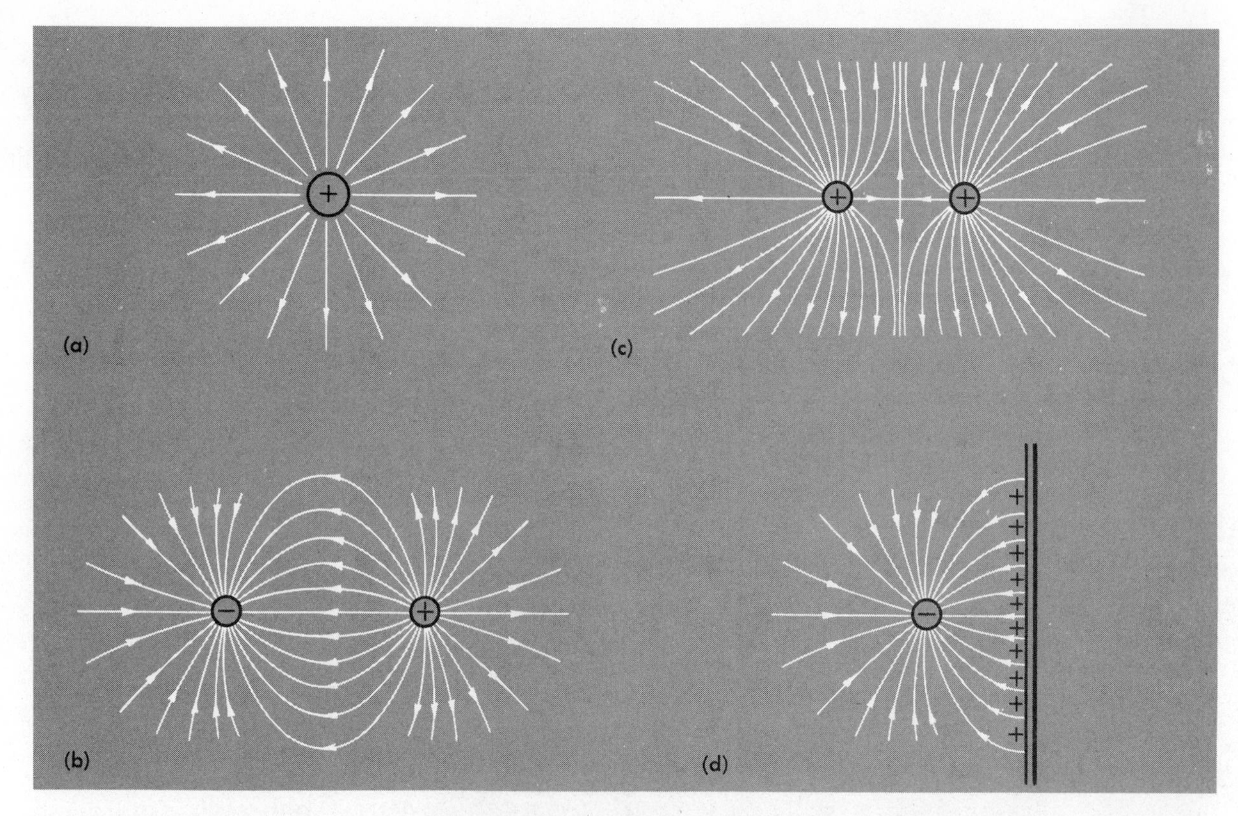

Figure 19–5
The electric field of (a) a single positively charged sphere; (b) two charged spheres of opposite sign; (c) two charged spheres of the same sign; (d) a small charged sphere near a large charged plane conductor of the opposite sign. Note that close spacing of the field lines indicates a region where the field is strong. Diverging lines indicate that the field is growing weaker.

In Fig. 19–6 the charged objects are metal conductors. Notice that the electric field lines meet their surfaces at right angles. This is easy to understand. In a metal conductor charges are free to move from point to point within the metal, but cannot easily leave its surface. If the electric forces which act on the charges in a metal surface have any component parallel to that surface, they drag the charges along the surface. This motion will continue until the charges have so arranged themselves that electric field lines become perpendicular to every metal surface. This rearrangement takes a very small fraction of a second; what we measure in these experiments is the field after the charges have stopped moving.

19-4 The Electric Field Near a Uniformly Charged Plate

In the Millikan experiment we found that the electric field between two charged plates is constant (Section 18–6). From Coulomb's experiment we know that the field around a point charge obeys an inverse square law. Are these two laws independent of one another, or can the first be deduced from the second? In this section we shall show that the constant field between parallel plates is a direct result of Coulomb's law.

Figure 19–6
Photographs of electric force patterns formed by grass seeds in an insulating liquid

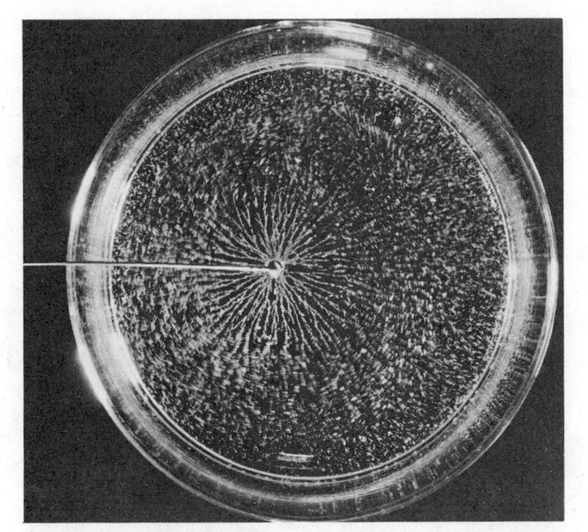

(a) A single charged rod.

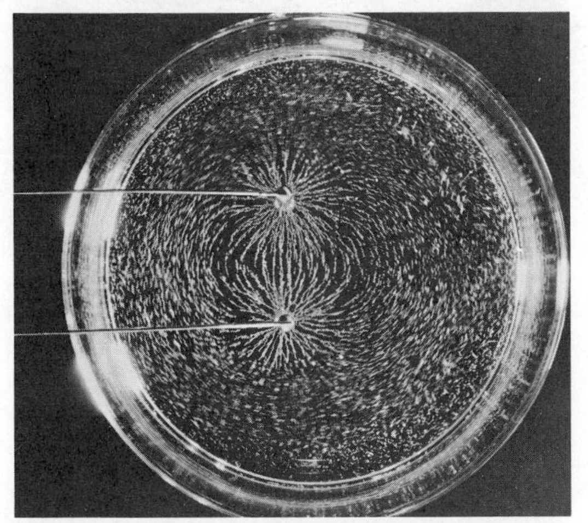

(b) Two rods with equal and opposite charges.

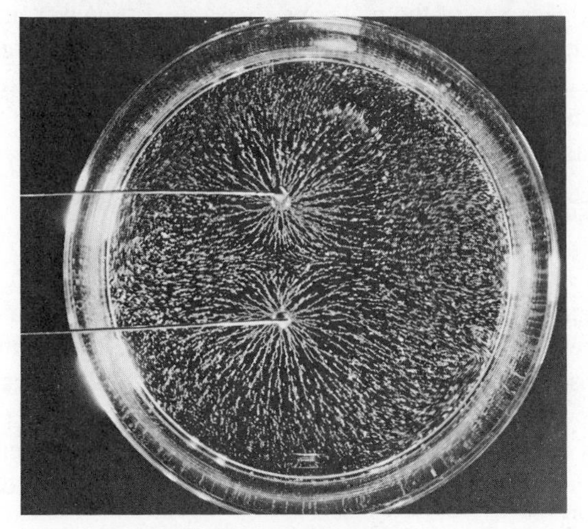

(c) Two rods with the same charge.

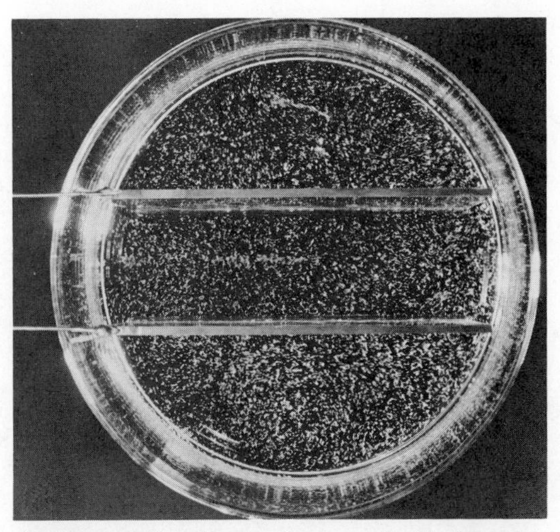

(d) Two parallel electric plates, no electric field.

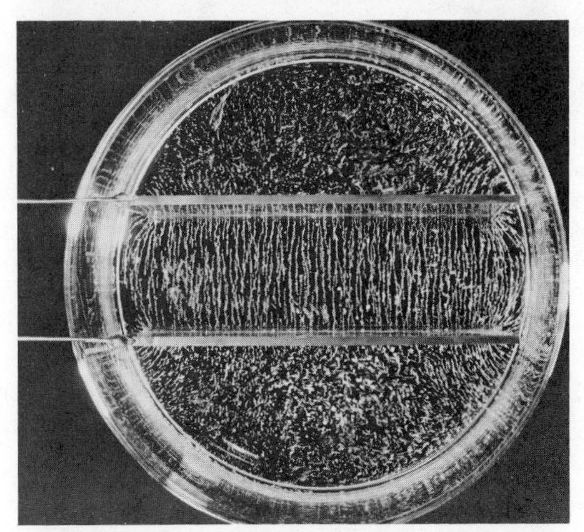

(e) Two parallel plates with opposite charges.

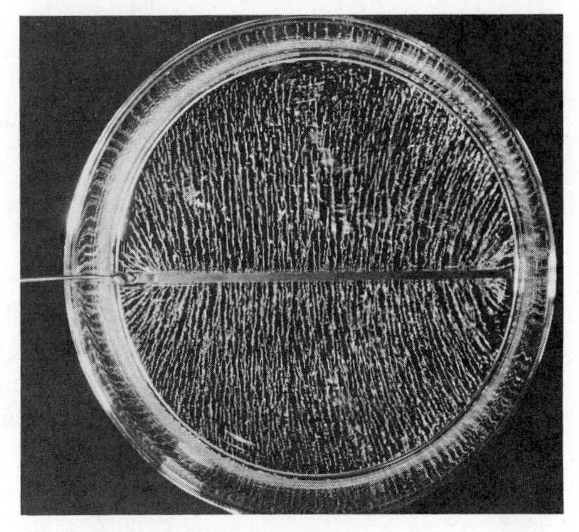

(f) A single charged metal plate.

As a first step, we shall consider the field at a point P, a distance h above the center of a single circular plate of radius R (Fig. 19–7). Suppose the plate is positively charged and the charge is spread uniformly over the surface of the plate (i.e., the charge per unit area ρ is the same everywhere). The contribution of each small charge on the plate to the field at P will depend on the distance of the charge from P.

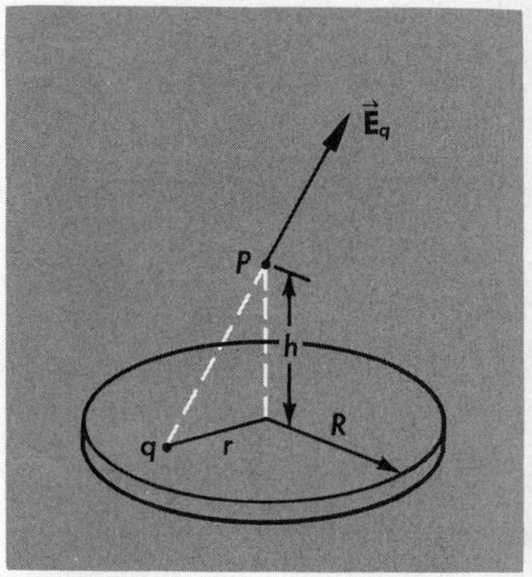

Figure 19–7
The electric field E_q at a point P directly above the center of a uniformly charged plate due to a small charge q at a point on the plate a distance r from the center.

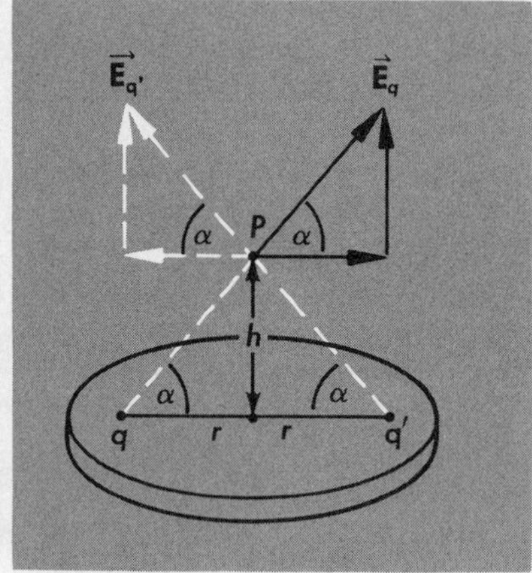

Figure 19–8
Because of symmetry, the horizontal components of $\vec{E}_q$ and $\vec{E}_q{}'$ cancel, whereas their vertical components add.

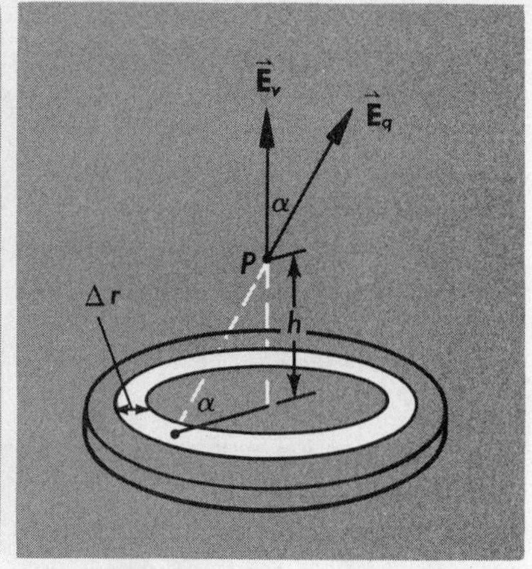

Figure 19–9
The vertical component of the field at P contributed by each charge q on a ring of radius r is, by Coulomb's law, $E_v = kqh/(r^2 + h^2)^{3/2}$.

Let us consider the electric field at P due to a small charge q at a distance r from the center of the plate in Fig. 19–7. The distance of the charge from P is $\sqrt{r^2 + h^2}$. Hence by Coulomb's law the magnitude of the field E_q at the point P is

$$E_v = \frac{kq}{r^2 + h^2}.$$

For every tiny charge q on the plate there is another equal charge q' on the opposite side of the center of the plate and the same distance from P. Because of symmetry, the horizontal component of E_q due to the charge q is canceled by the horizontal component of the equal charge q' opposite it, whereas the vertical components of the two charges add (Fig. 19–8). The magnitude of the vertical component E_v of E_q in Fig. 19–8 is

$$E_v = E_q \sin \alpha = E_q \cdot \frac{h}{\sqrt{r^2 + h^2}} = \frac{kqh}{(r^2 + h^2)^{3/2}}.$$

Now consider Fig. 19–9, which shows a narrow concentric ring on the plate whose width is Δr and whose mean radius is r. The area of the narrow ring is $2\pi r\Delta r$. (Imagine the ring cut through and straightened out into a long, thin rectangle whose dimensions are $2\pi r$ and Δr). The total charge on the ring is the charge per unit area ρ times the area $2\pi r\Delta r$. If $\Delta r \ll r$, we can consider all charges on the ring to be the same distance r from the center of the plate. The total vertical component ΔE, the resultant field due to all the charges on the ring, is

$$\Delta E = \frac{2\pi r\rho kh\Delta r}{(r^2 + h^2)^{3/2}} = 2\pi k\rho \frac{hr\Delta r}{(r^2 + h^2)^{3/2}}.$$

It now remains to find the total vertical field E. To get the total field at P means adding up the vertical contributions to the field at P of a large number of concentric rings that take up the whole area of the plate of radius R. Each of these contributions ΔE is given by the "area" of a thin rectangle $2\pi k\rho \dfrac{hr}{(r^2 + h^2)^{3/2}}$ high and Δr wide. One such rectangle is shown in Fig. 19–10, where we have plotted the function $f(r) = \dfrac{hr}{(r^2 + h^2)^{3/2}}$ as a function of r from $r = 0$ to $r = 2.5$ cm. (The value of h we used was 0.2 cm.)* This procedure is similar to getting the displacement of an object from a graph of its velocity as a function of time, by finding the "area" under the velocity-vs.-time graph (Section 9–5).

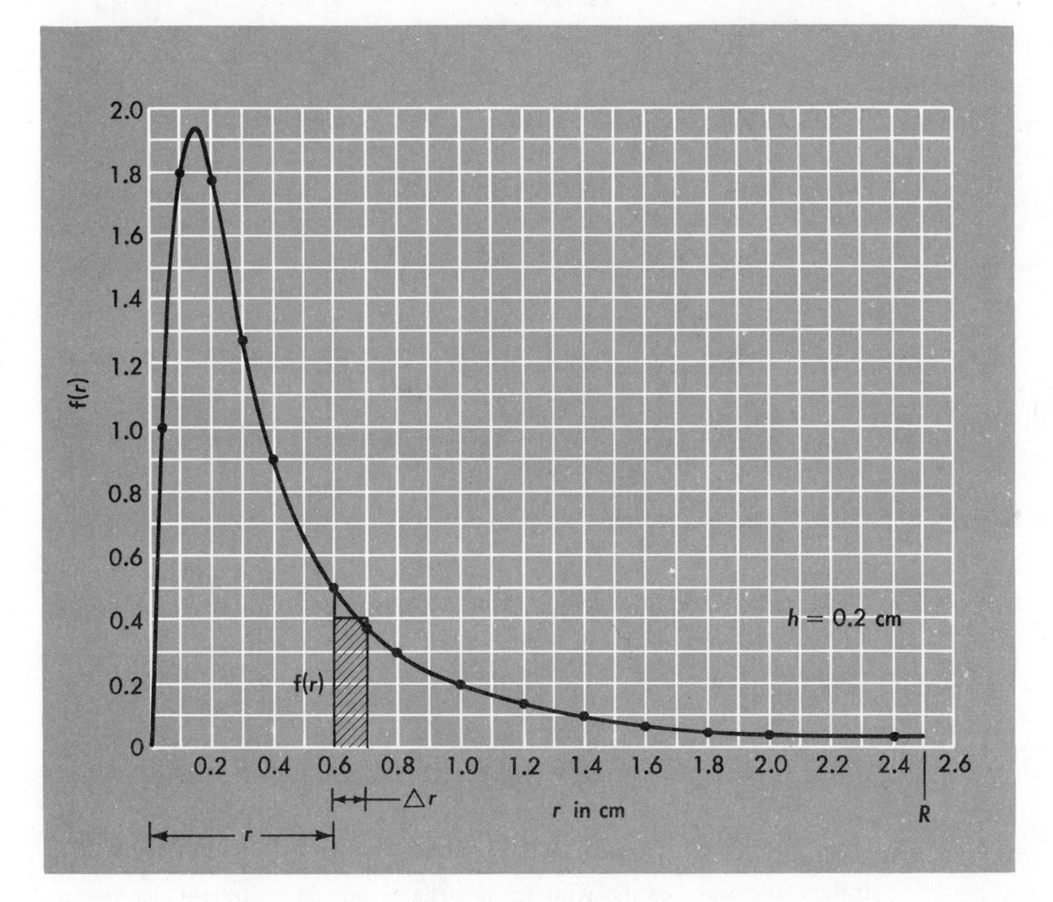

Figure 19–10
A graph of the function $f(r) = \dfrac{hr}{(r^2 + h^2)^{3/2}}$ for $h = 0.2$ cm. The shaded "area," when multiplied by $2\pi k\rho$, gives the contribution to the field at point P, in Fig. 19–9, of a uniformly charged ring of radius r and width Δr. For a plate of radius R the total field E at point P is the sum of all the narrow rectangles of width Δr, from $r = 0$ to $r = R$, times $2\pi k\rho$.

How will the total field be affected by changing the value of h? We answer this question by plotting the graph of $f(r)$ for different values of h and then, for each different curve we get, we find the area under the curve up to $r = R = 2.5$ cm, the radius of the plate. In order to get the area accurately, we chose a value for Δr that was very small.

* For simplicity, we have not used the constant $2\pi k\rho$ in calculating points for plotting the curve. Once we have found the area under the curve, we can always multiply the result by $2\pi k\rho$ to get the magnitude of the electric field.

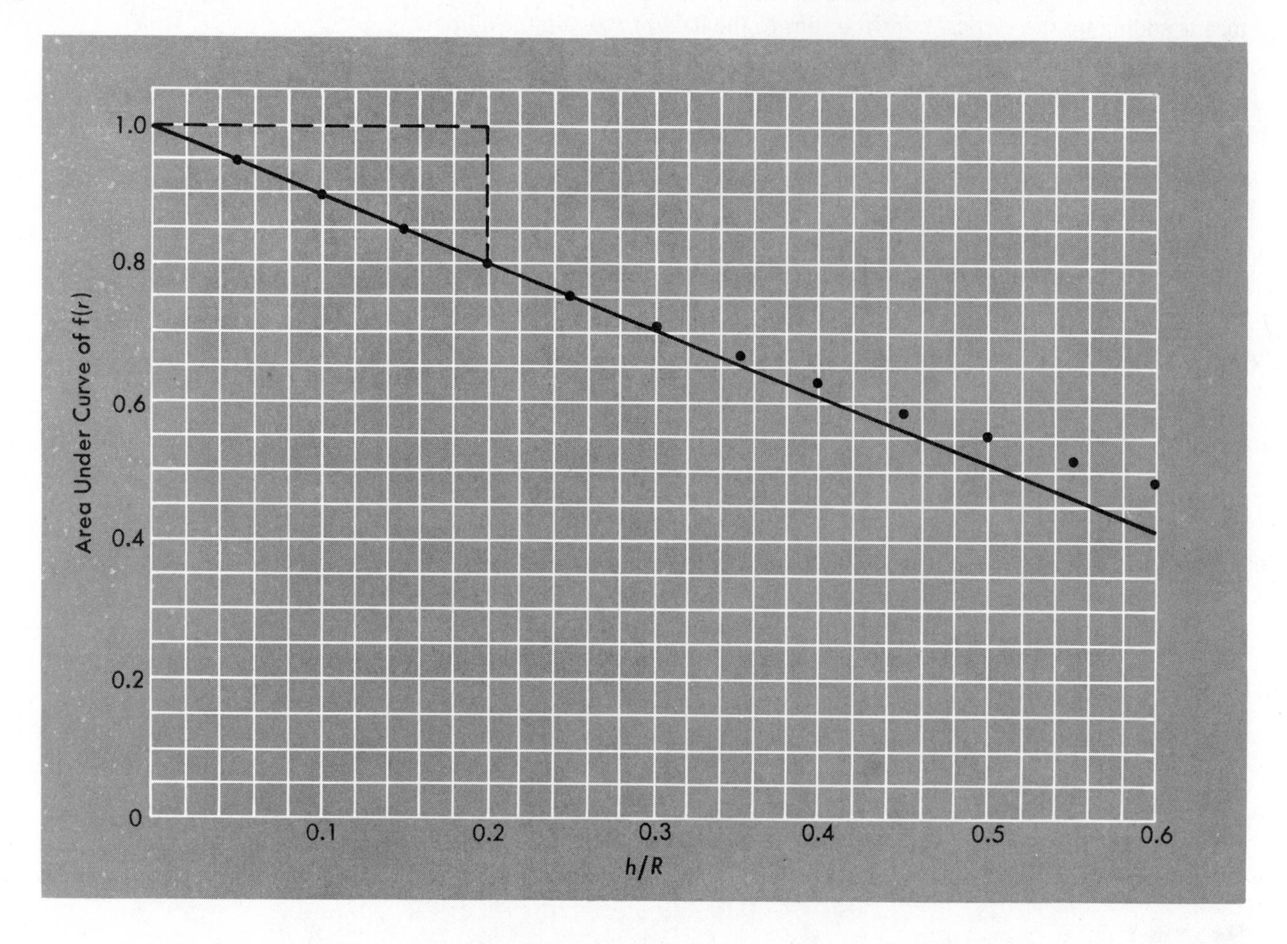

Figure 19–11
The area under the curve of $f(r)$ as a function of $\frac{h}{R}$ is very nearly a linear function for $\frac{h}{R} < 0.3$. [The curve of $f(r)$ is shown in Fig. 19–10.]

The results are shown in Fig. 19–11, where we have chosen to plot the area under the different curves as a function of h/R rather than as a function of h. The reason for plotting the area as a function of h/R is that it will enable us to get a simple, approximate equation for the field at a distance h above the plate that we can use for any value of R. Notice that for values of h/R less than about 0.3, the calculated values lie along a straight line. This means that as h is increased, the field at P must decrease linearly for any $h/R < 0.3$.

We can now express the field E as a linear function of h/R. The general equation of a linear function is $y = b + mx$, where m is the slope of the straight line and b is the y intercept. The slope of the straight line in Fig. 19–11 is the ratio of the change along the vertical coordinate to the corresponding change along the horizontal coordinate. For a positive change in the vertical coordinate of 0.20 (the vertical dashed line in Fig. 19–11), there is a *negative* change in the horizontal coordinate of -0.20 (the dashed horizontal line in the figure). So the slope is $\frac{0.20}{-0.20} = -1.0$. The y intercept in Fig. 19–11 is 1.0, so the equation for the straight line is $1 - h/R$.

Multiplying the quantity $1 - h/R$ by the constant $2\pi k\rho$ will give the electric field E at a distance h above the center of a uniformly charged circular plate of radius R:

$$E \approx 2\pi k\rho\left(1 - \frac{h}{R}\right) \quad \text{when} \quad \frac{h}{R} < 0.3.$$

19-5 The Electric Field Between Two Uniformly Charged Plates

Now we shall investigate the field at different points between the centers of a *pair* of equally but oppositely charged circular plates (Fig. 19–12). Both

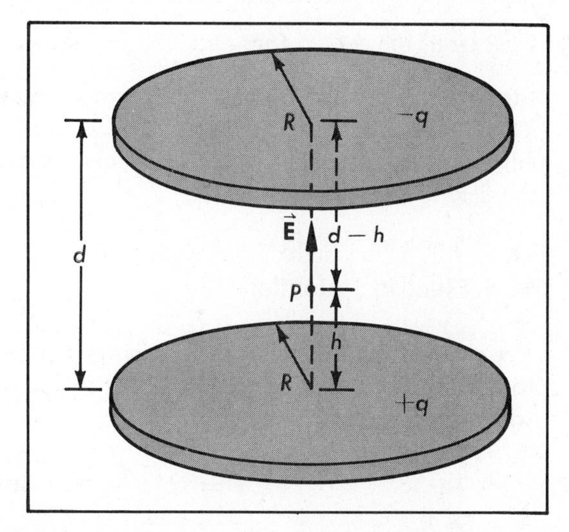

Figure 19–12
Two parallel metal plates a distance d apart that are uniformly charged. The charge on one plate is equal in magnitude but opposite in sign to the charge on the other plate.

are of the same radius R. Again we assume that the charge is uniformly distributed over the surface of each plate. The plates are a distance d apart, so the point P, which is a distance h above the lower plate, is a distance $d-h$ below the upper plate.

As we have seen, the field at P due to the charge on the lower plate is $2\pi k\rho\left(1 - \frac{h}{R}\right)$ if $\frac{h}{R} < 0.3$, and it is directed vertically upward. The field at P, due to the charge on the upper plate, is also directed vertically upward because it is charged negatively. It is $2\pi k\rho\left(1 - \frac{d-h}{R}\right)$ for $\frac{d-h}{R} < 0.3$. Adding the two contributions gives, for the magnitude of the total field at P:

$$\begin{aligned} E &= 2\pi k\rho\left(1 - \frac{h}{R}\right) + 2\pi k\rho\left(1 - \frac{d-h}{R}\right) \\ &= 2\pi k\rho\left(1 - \frac{h}{R} + 1 - \frac{d}{R} + \frac{h}{R}\right) \\ &= 4\pi k\rho\left(1 - \frac{d}{R}\right). \end{aligned}$$

We see that with two plates the term $\frac{h}{R}$ vanishes and the field at the center, for plates a fixed distance apart, does not depend on h so long as $\frac{h}{R} < 0.3$ and $\frac{d-h}{R} < 0.3$. If $\frac{d}{R} < 0.3$, then the distance of point P from either plate will always be less than $0.3R$, and the field will be uniform at the center and equal to $4\pi k\rho\left(1 - \frac{d}{R}\right)$. If $\frac{d}{R} \ll 1$, then $E = 4\pi k\rho$.

The graph in Fig. 19–10 shows that charges at distances greater than about 2 cm, or $0.8R$, contribute very little to the area under the curve and hence to the magnitude of the electric field above the center of a single plate. This means that the field is also uniform from the center to a distance of about $0.2R$ from the center, for $\frac{h}{R} < 0.3$. In the case of two parallel plates, close together, the field is very nearly uniform at greater distances from the center. This is borne out by Fig. 19–6(e), which shows that even quite close to the edges of the plates the electric field lines are very nearly parallel and hence the field is very nearly uniform.

19-6 Electric Potential

Let us continue to exploit the similarity between gravitational and electric forces. In Section 16–4 we found that two masses which attract each other with a gravitational force

$$\vec{\mathbf{F}} = G\,\frac{m_1 m_2}{r^2}, \text{ attractive}$$

have a corresponding gravitational potential energy

$$U_g = -G\,\frac{m_1 m_2}{r}.$$

The zero of potential energy is chosen to be at infinite separation, and the minus sign indicates that the potential energy is then negative at any separation r. The potential energy is negative because the masses attract each other as indicated.

Now we know that two charged bodies interact with an electric force

$$\vec{\mathbf{F}} = k\,\frac{q_1 q_2}{r^2}, \text{ repulsive (for like charges).}$$

So in exactly the same way, they have an electric potential energy

$$U_e = k\,\frac{q_1 q_2}{r}.$$

There is no minus sign in this expression because charges of the same sign repel, instead of attracting, each other like gravitating masses. The potential energy of two like charges, which has been taken as zero at infinite distance, will have a positive value for any finite value of r. The potential energy of two unlike charges is also taken to be zero at infinite separation, but is negative for finite values of r, because the charges attract.

Now, since the potential energy when the charge q_2 is a distance r from the charge q_1 is kq_1q_2/r, then the potential energy per unit positive charge due to the electric field of q_1 is

$$V = \frac{kq_1}{r}.$$

We shall call this potential energy per unit positive charge the *electric potential* V at the distance r, or the electric potential difference between r and infinity. If the charge q_1 is held in position and q_2 is moved from r to infinity under the influence of the Coulomb force alone, it will gain kinetic energy equal to q_2V. Since V is the potential energy per unit charge, it is measured in joules per unit charge. Like the electric field, the electric potential at a point depends only on the charges producing the field and on the position of the point. However, electric potential is a scalar, not a vector. It is positive when q_1 is positive, negative when q_1 is negative.

Just as we can work out the electric field $\vec{E}$ throughout space, so we can work out the electric potential V for every point in space. To illustrate this potential field around a positive charge q_1, we start by plotting a graph of V versus r. To visualize the field of electric potential at all points on a plane through the charge q_1, we swing the graph around a vertical axis passing

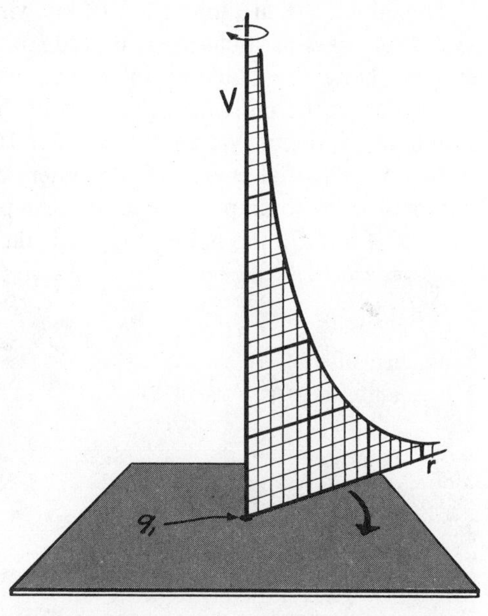

Figure 19–13
A graph of electrical potential V versus the distance r from a charge q_1 is rotated around a vertical axis passing through $r = 0$ in order to create the potential "hill" of Fig. 19–14.

through $r = 0$ (Fig. 19–13). In other words, we swing it around the position of the charge q_1. As it sweeps around, the graph generates a potential "hill" whose height above any point on the plane represents the potential at that point. The potential "hill" for a plane passing through a point charge is shown in Fig. 19–14. We shall find this potential "hill" of great value in Chapter 24 when we discuss the motion of one charged particle in the Coulomb field of another.

We can extend the idea of electric potential to include potentials which are due to more than a single point charge. At any point, the potential V of a system of several point charges is just the algebraic sum of the potentials at that point due to each point charge. The potential "hill" for such a system could be drawn, but it would have a much more complicated surface than that of Fig. 19–14. If a charge q is brought from infinity to any point P and the other charges are prevented from moving, the change in potential energy of the combined system, including q, would be qV, where V is the potential at point P.

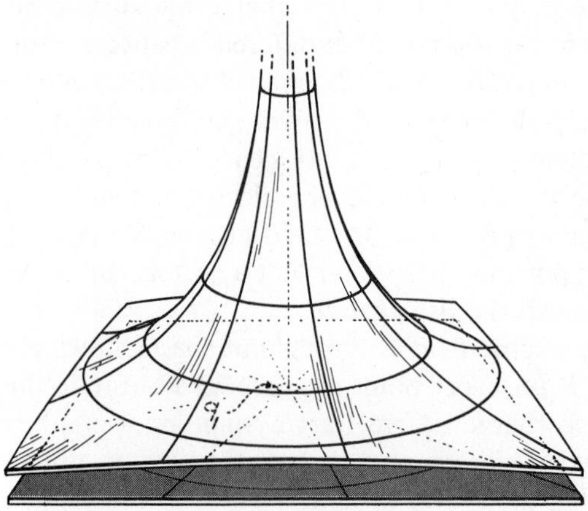

Figure 19–14
The height of the potential "hill" directly above any point on the plane represents the electric potential at that point on the plane.

Sometimes we are interested in knowing how the potential energy of a system changes as a charge is moved from one point to another within the system. Now, the Coulomb force between each pair of charged bodies depends not on the individual positions of the bodies, but only on the separation between them. We saw in Chapter 16 that, for forces of this type, the change in potential energy of the system as one of the bodies moves along any path from some point A to another point B depends only on the positions of A and B, and not on the particular path (Fig. 19–15). Suppose we

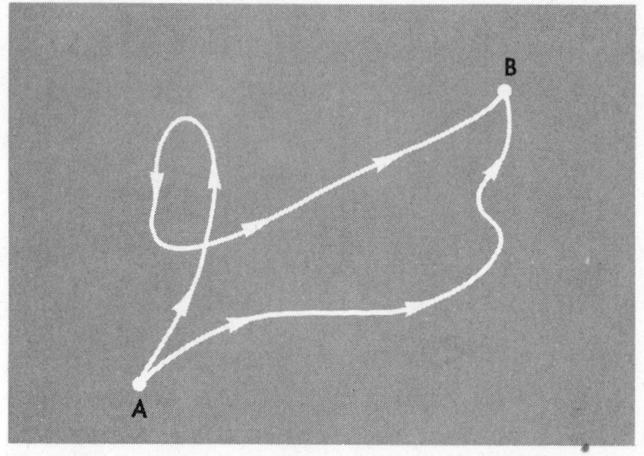

Figure 19–15
When a charge moves from a point A to a point B in an electric field produced by nailed-down charges, the change in potential energy does not depend on what path the charge takes

nail down all the charges in the system except one, of size q, and let it move from A to B. Then the change in electric potential energy, equal to the potential energy when q is at B minus the potential energy when q is at A, is just q times the difference between the electric potentials at the points B and A due to all the other charges. Thus,

$$\Delta U = q(V_B - V_A).$$

The quantity $(V_B - V_A)$ is called the *electric potential difference* between points B and A in the field of all the nailed-down charges.

With a micro-microbalance we can measure the charge of an object as a certain number of elementary charges. We know that this method will work when the number of elementary charges is small and the object is microscopic. Will it also work for large numbers of elementary charges? A really large number of elementary charges will not stay on a microscopic object. On the other hand, if the object is large enough to retain a large amount of charge, it will be too large for the micro-microbalance. Consequently, the micro-microbalance will not work for large numbers of elementary charges.

How, then, are we going to measure a large charge as a number of elementary charges? There are several ways, but the most obvious is to build a bigger balance. We want a balance large enough to accommodate a sphere from a Coulomb experiment (see Sections 19–1 and 19–2) or some similar object. If we can measure the charge on such an object in terms of the number of elementary charges, we should then be able to measure all charges as numbers of elementary charges. Indeed, we shall see in the next section that once we have measured one big charge, the Coulomb experiment will let us measure any other in terms of elementary charges.

The new balance must be large enough to accommodate a reasonably large object, and at the same time we must know the electric force on an elementary charge between the plates. Unless we know the force on an elementary charge in the balance, we can use it only to measure big charges in terms of an arbitrary unit of charge. If we could do an accurate Millikan experiment in a large balance, we could compare a big charge directly with an elementary charge. Unfortunately, to do a Millikan experiment with a large balance is impractical. However, there is a way around this difficulty: we can build a large balance in which the force on each elementary charge between the plates is exactly the same as the force on an elementary charge in the micro-microbalance. To design the large balance we must learn how the dimensions of a balance are related to the electric field between the plates. Then we can increase the size of the balance without changing the field.

In order to scale up a balance, we shall have to change both the dimensions of the plates and their separation. We can learn the effect of each of these changes by experiment, both with the micro-microbalance and with widely separated charged plates exerting forces on large charges.

When we use plates of larger dimensions at the same distance apart and connected to the same battery (Fig. 19–16), the electric field between them is the same. However, if we again keep the same battery connected but now increase the separation between the plates, the field decreases, changing in inverse proportion to the separation. We can compensate for this by using more batteries. Since the field increases in proportion to the number of batteries (Section 18–6), we must double the number of batteries whenever we double the separation of the plates. As long as the plates are big compared with their distance apart, we keep the same electric field by increasing the separation and increasing the number of batteries in the same proportion. Therefore, to scale up the electric micro-microbalance to a convenient size, we use big parallel metal plates a convenient distance apart. And we make sure that the electric force on an elementary charge (the electric field)

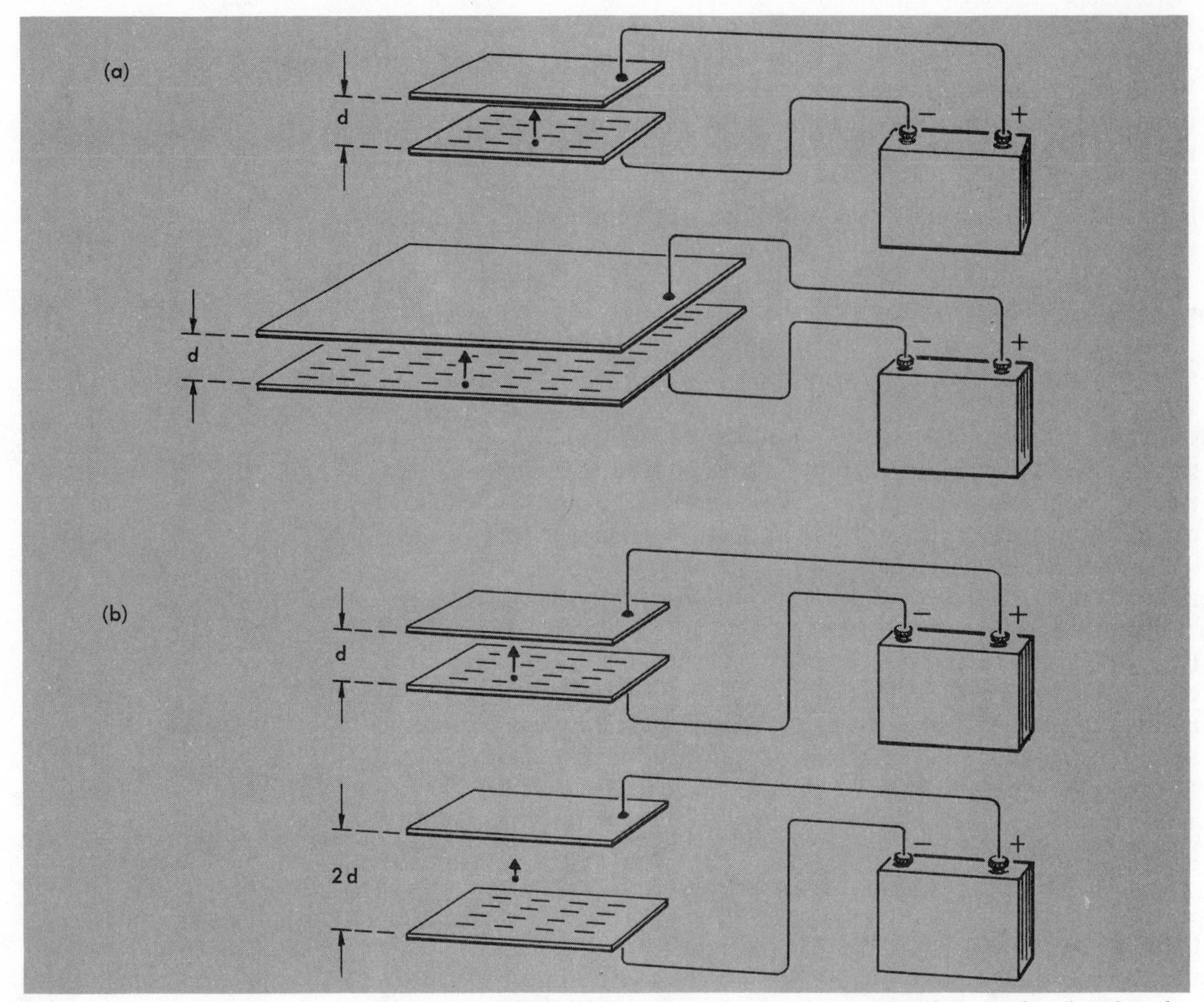

Figure 19–16
(a) If we scale up the dimensions of the plates without changing the separation or the battery, the force on an elementary charge anywhere between the plates remains the same. (b) If we double the distance between the plates with the same battery connected, the force is halved. In general, the force is proportional to $1/d$, where d is the distance between plates.

between the plates is the same as in the micro-microbalance by charging the plates with a sufficient number of batteries in series.

19-8 The Constant in Coulomb's Law

When we discussed Coulomb's law and the electric charge in Sections 19–1 and 19–2, we found that

$$F = k\,\frac{q_1 q_2}{r^2},$$

where F is the force one charge exerts on the other, q_1 and q_2 are the charges, and r is the separation between them. The proportionality factor k is a constant. Its value depends on the units in which we measure force, separation, and charge. Now, with the scaled-up electric balance, we are in a position to evaluate the constant k in Coulomb's law.

We can measure the forces that two equally charged spheres exert on each other by doing a Coulomb experiment. In one such experiment it was found that two equally charged spheres a distance 0.15 meter apart repelled each other with forces of 6.7×10^{-4} newton. A large electric balance like the one described in the last section was used to find the charge on each sphere. In this balance the force on each elementary charge was the same as that between the plates of the micro-microbalance used in a previously done Millikan experiment. The force on each elementary charge was kept the same by using plates 100 times as far apart and by using 100 times as many batteries in series.

In the Millikan experiment a charged sphere was held at rest between plates 3.1×10^{-3} meter apart connected to three 90-volt batteries in series. The mass of the sphere was known to be 2.9×10^{-15} kg, so the gravitational force on it was 2.8×10^{-14} newton. Hence the magnitude of the balancing electric force must also have been 2.8×10^{-14} newton. It was determined later in the experiment that the sphere had two elementary charges on it. Therefore, the micro-microbalance exerted a force of 1.4×10^{-14} newton on each elementary charge.

The force exerted by the plates of a large electrical balance on a charged sphere was measured by attaching the sphere to the end of an insulating balance arm [Fig. 19–17(a)]. First, with the plates uncharged, the position of the scale was adjusted so that the pointer on the balance arm was at zero. Then, with the plates connected to three hundred 90-volt batteries, the change in the pointer's position due to the electric force on the sphere was noted [Fig. 19–17(b)] and corresponded to 3.55×10^{-3} newton. Since the plates exerted a force of 1.4×10^{-14} newton on each elementary charge, the charge on the sphere must have been

$$q = \frac{3.55 \times 10^{-3} \text{ newton}}{1.4 \times 10^{-14} \text{ newton/elem. ch.}} = 2.5 \times 10^{11} \text{ elementary charges.}$$

In the Coulomb experiment described at the beginning of this section, we now have values for all the quantities we need to solve for the constant k in Coulomb's law. We have, from $F = k\dfrac{q_1q_2}{r^2}$

$$\begin{aligned} k &= \frac{Fr^2}{q_1q_2} \\ &= \frac{(6.7 \times 10^{-4} \text{ newton})(0.15 \text{ meter})^2}{(2.5 \times 10^{11} \text{ elem. ch.})(2.5 \times 10^{11} \text{ elem. ch.})} \\ &= 2.4 \times 10^{-28} \frac{\text{newton-m}^2}{(\text{elem. ch.})^2}. \end{aligned}$$

We can check on the value of k by doing other Coulomb experiments and using the large balance to measure the charges. As we expect, the value of k always comes out the same. It is one of the universal constants of nature. A great deal of work with all sorts of experiments has gone into measuring k accurately. The results of all these experiments are in close agreement, and the value given by the best experiments is almost exactly $2.306 \times 10^{-28} \dfrac{\text{newton-m}^2}{(\text{elem. ch.})^2}$. Consequently, we now write Coulomb's law as

(a)

(b)

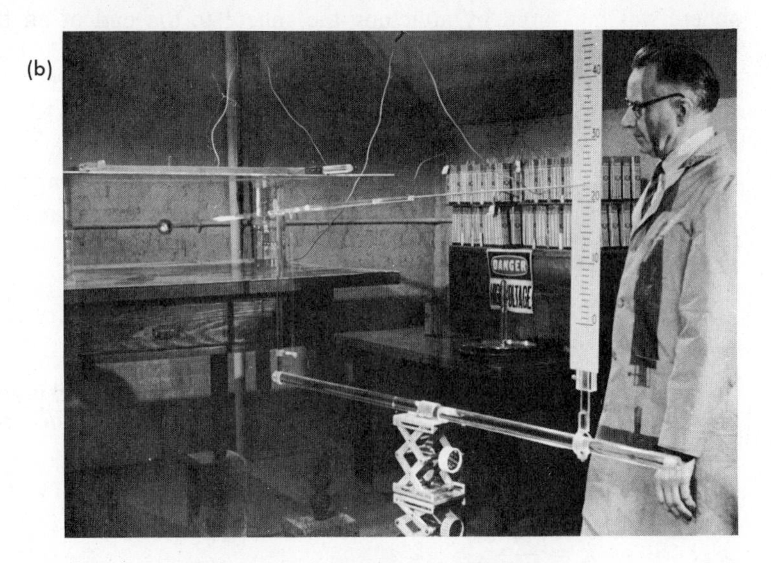

Figure 19–17
A large electric balance. (a) When no electric force is acting, the scale is set so that the pointer is at zero. (b) When the large plates are connected to the stack of batteries (to the right, behind the scale), the electric force on the sphere pushes down on it, causing the plastic balance arm to pivot on the vertical support. The entire balance arm, support, and scale are jacked up in order to keep the sphere in the center of the plates, where the force field is most nearly uniform. The scale is calibrated by placing a tiny known mass on the sphere (with the plates uncharged) and noting how many divisions correspond to the known gravitational force on the mass.

$$F = \left(2.306 \times 10^{-28}\,\frac{\text{newton-m}^2}{(\text{elem. ch.})^2}\right)\frac{q_1q_2}{r^2},$$

where q_1 and q_2 are the numbers of elementary charges and r is the separation. The expression for the electric potential energy becomes

$$U_e = \left(2.306 \times 10^{-28}\,\frac{\text{newton-m}^2}{(\text{elem. ch.})^2}\right)\frac{q_1q_2}{r}.$$

Our knowledge of the value of k is the key to measuring charges. By charge-sharing we can always get two equal charges. After measuring the force between them at a definite distance we know F, r, and k in Coulomb's law, and we can therefore work out q in elementary charges. Using one of the charged bodies as a standard, we can then measure the number of elementary charges on any other body by measuring the force between it and the standard.

In addition to giving us a prac[obscured]'s of elementary charges by measuring electric forces between ordinary-sized objects, the constant $2.306 \times 10^{-28} \frac{\text{newton-m}^2}{(\text{elem. ch.})^2}$ tells us directly the force between two elementary charges one meter apart. They exert just 2.306×10^{-28} newton on each other. This force looks absolutely negligible. But in fact it is huge. The gravitational attraction between two hydrogen atoms at this distance is about 2×10^{-64} newton. The electric force between two elementary charges is 10^{36} times as big. The force is so big that two collections of 10^{24} elementary charges (about 2 gm of hydrogen ions—hydrogen atoms with one positive elementary charge each) placed one on each side of the earth would push on each other with a force of half a million newtons—50 tons. Two similar-sized collections of the heaviest atoms placed at opposite ends of a diameter of the earth would have no measurable gravitational effect on each other.

For Home, Desk, and Lab

1.* Two small electrified objects A and B are separated by 0.03 meter and repel each other with a force of 4.0×10^{-5} newton. If we move body A an additional 0.03 meter away, what is the electric force now? (Section 1.)

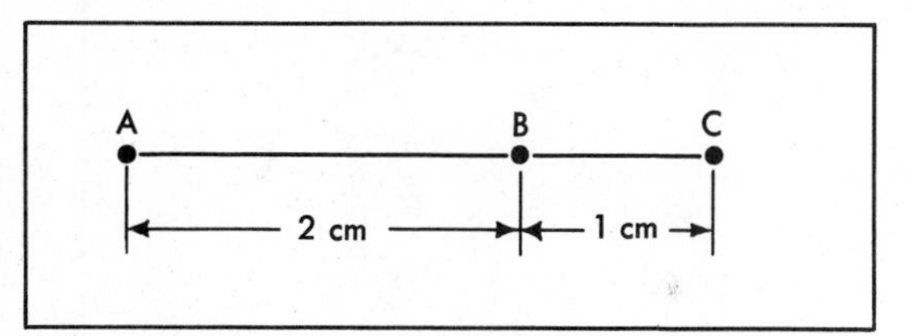

Figure 19–18
For Problem 2.

2. Three equally charged objects are located as shown in Fig. 19–18. The electric force exerted by A on B is 3.0×10^{-6} newton.
   (a) What electric force does C exert upon B?
   (b) What is the net electric force on B?

3. Suppose that we place three small charged spheres, with equal charges on them, as shown in Fig. 19–19. Sphere C exerts a force of 4×10^{-6} newton on B.
   (a) What force does A exert on B?
   (b) What is the net force on B?

4. In Fig. 19–3 what is the ratio of the forces exerted on X?

5. Four identical conducting spheres mounted on Dry Ice pucks are held on a smooth level table at the four corners of a square. Two of them carry a charge of $+q$ and the other two carry a charge of $-q$ each, as shown in Fig. 19–20.
   (a) Make a vector diagram, to scale, showing the electric forces on sphere 1.
   (b) If all the pucks are released at the same time, in what direction will each of them move?

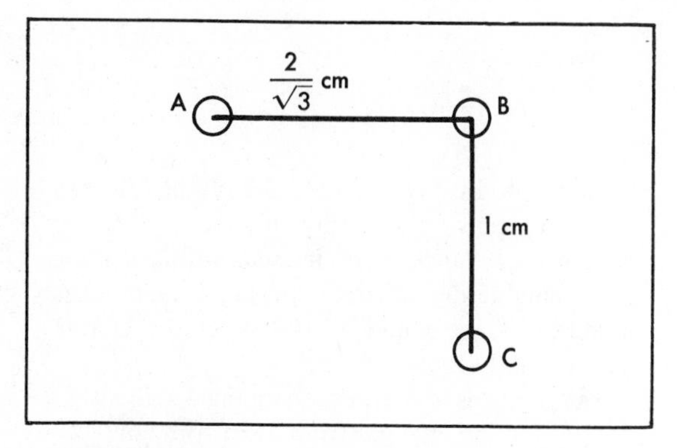

Figure 19–19
For Problem 3.

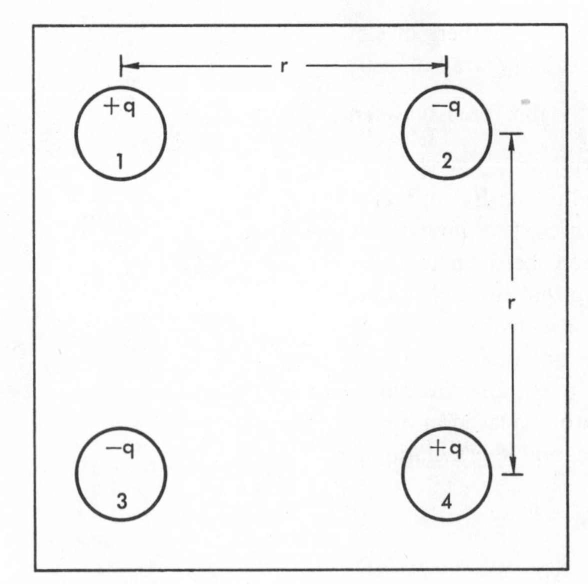

Figure 19–20
For Problem 5.

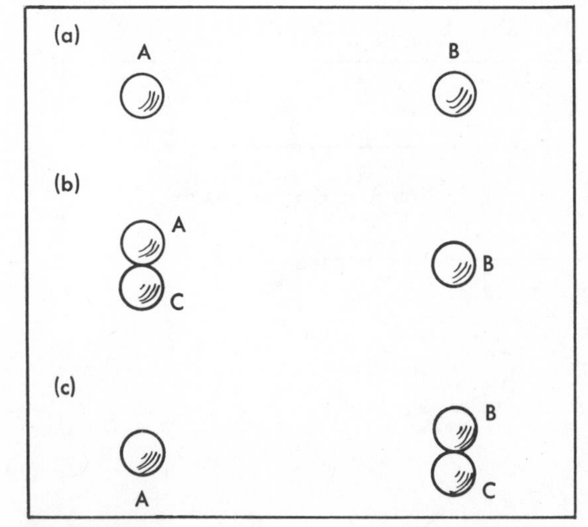

Figure 19–21
For Problem 6.

6. Two equally charged identical conducting spheres A and B repel each other with a force of 2.0×10^{-5} newton [Fig. 19–21(a)]. Another identical uncharged sphere C is touched to A [Fig. 19–21(b)] and then moved over next to B [Fig. 19–21(c)].
   (a) What is the electric force on A now?
   (b) What is the net electric force on C (after it has touched A) when it is halfway between A and B?

7.† Two charged balls A and B each with mass m are placed as shown in Fig. 19–22. The ball A is free to move; B is fixed in position.
   (a) How does the force F_e between them depend upon X?
   (b) Find the relationship between the electric force F_e and the weight mg in terms of X and l.
   (c) If the charge remains the same and the separation X is cut in half, how must the weight have changed?

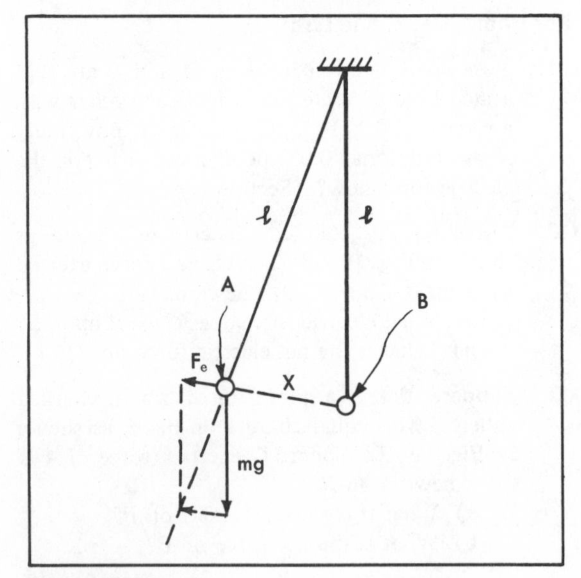

Figure 19–22
For Problem 7.

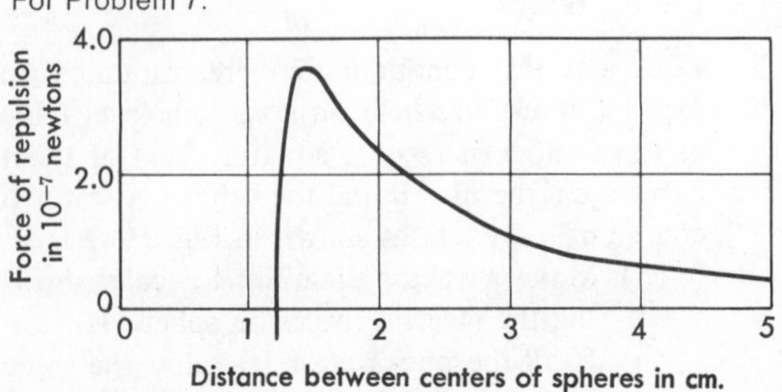

Figure 19–23
For Problem 8.

8. The graph in Fig. 19–23 shows the electric force of repulsion on a tiny charged conducting sphere as a function of its separation from a large conducting sphere. The large sphere has a radius of 1 cm and has 10 times the charge of the small sphere.
   (a) How is the force changing as the separation changes from 5 cm to 3 cm?
   (b) Explain the behavior of the force between separations 2 cm and 1 cm.

9.* Match the fields shown in Fig. 19–5 with the grass-seed patterns of Fig. 19–6. (Section 3.)

10.* Can electric field lines cross one another? (Section 3.)

11. If a charged particle is free to move, will it travel along an electric field line?

12. A square plate placed in a horizontal plane has a uniform charge per unit area (Fig. 19–24). Will the electric field at a point P above the center of the square have a horizontal component? Why?

13. Consider the electric field a distance h above the center of a square plate of side $2a$ carrying a charge p per unit area. You can find an approximate value for the field at this point by realizing that the strength of the field generated by the square is less than that generated by a circular plate of radius R that circumscribes the square, and is more than that generated by a circular plate inscribed in the square (Fig. 19–25).
   (a) Find the field strength exerted by the circular plates for $h = 0.3$ cm and $a = 3.0$ cm.
   (b) What is the largest error you can make (in percent) if you assume that the field strength due to the square plate is the average of the field strength due to the two circular plates?

14. Consider a point P halfway between two uniformly charged parallel plates carrying equal and opposite charges (Fig. 19–26). Why does the electric field at P have only a vertical component and no horizontal component?

15.* A charge q_1 is at a distance r from a fixed charge q_2. If q_1 is allowed to move freely away under the repulsion of the Coulomb field, what will be its kinetic energy at a very large separation? (Section 6.)

16. Two identical charges q are nailed in place on the y axis, equal distances a above and below the origin of coordinates.
   (a) Find an expression for the potential energy due to the interaction of a unit charge on the x axis, a distance x from the origin, with one of the nailed-down charges; with both of the nailed-down charges.

   (*Note:* Measure the potential energy with the zero chosen when the unit charge is at infinity. Express your answer in terms of the constant k in Coulomb's law, the distances a and x, and the charge q.)
   (b) What would be the potential energy due to the interaction of the unit charge if the two nailed-down charges were equal and opposite?

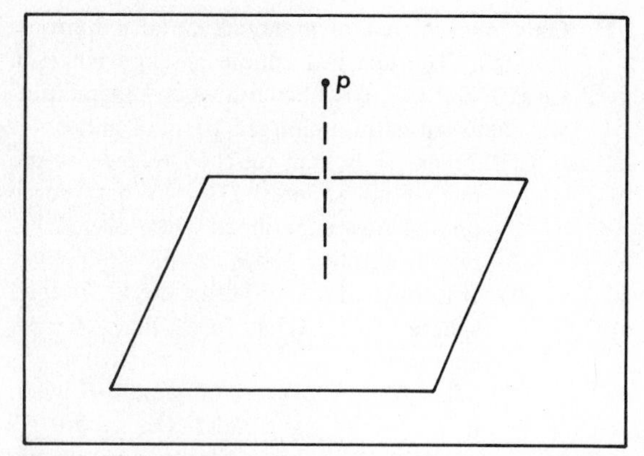

Figure 19–24
For Problem 12.

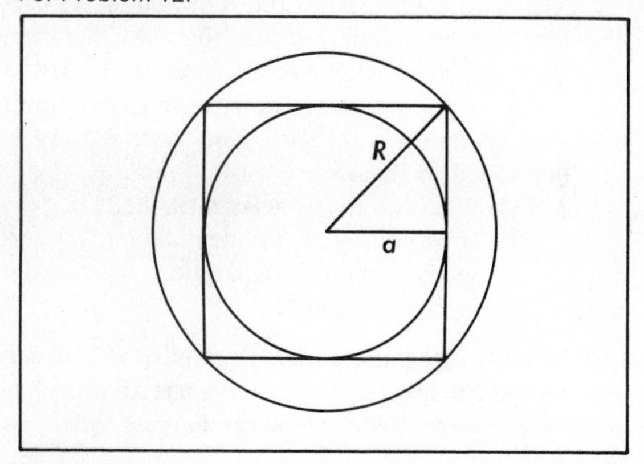

Figure 19–25
For Problem 13.

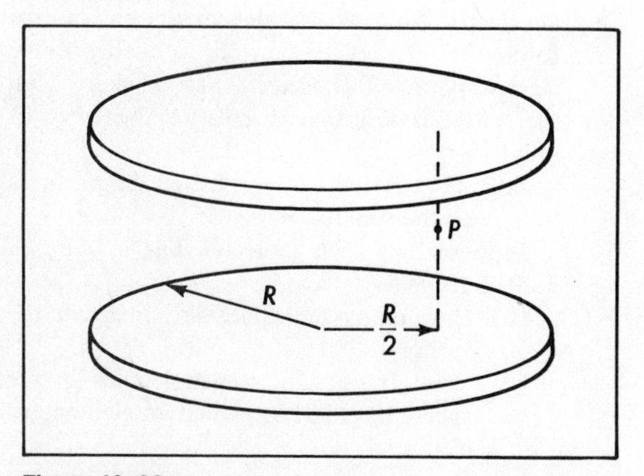

Figure 19–26
For Problem 14.

17. Only one of several identical metal spheres is charged. The charged sphere A experiences a force $F = 1.0 \times 10^{-4}$ newton when placed midway between certain charged parallel plates.
    (a) Sphere A is then touched to one of the uncharged spheres (B). What force does A now experience when placed between the same plates?
    (b) A is touched to another of the uncharged spheres (C). What force does A now experience?
    (c) What force is exerted on sphere B when it is between the plates? On sphere C?

18. Two large metal plates are separated by 0.10 m. They are connected to the terminals of a battery. A small charged ball halfway between them experiences an electric force of 3×10^{-4} newton. The plates are now moved apart until the separation is 0.15 m. The same battery is connected to them.
    (a) What force now acts on the ball?
    (b) If we add two more identical batteries in series at the new separation, what is the force on the ball?

19.* What will be the force of repulsion between two small spheres 1.0 meter apart if each has an excess of 10^6 elementary charges? (Section 8.)

20. A plastic sphere of mass 3.06×10^{-15} kg is in an electric force field which exerts an upward force of 1.00×10^{-14} newton on each positive electric particle. The electric force exerted on the sphere is sufficient to balance the force of gravity on it.
    (a) What is the excess of electric particles on the sphere?

    Suppose that each molecule has a mass of 3.00×10^{-26} kg.
    (b) How many molecules are there in the sphere?
    (c) What fraction of the molecules in the sphere have lost or gained an elementary charge?

21. A sphere of mass 4.5×10^{-3} kg is hanging on a string 2.0 meters long between two oppositely charged parallel plates as shown in Fig. 19–27.

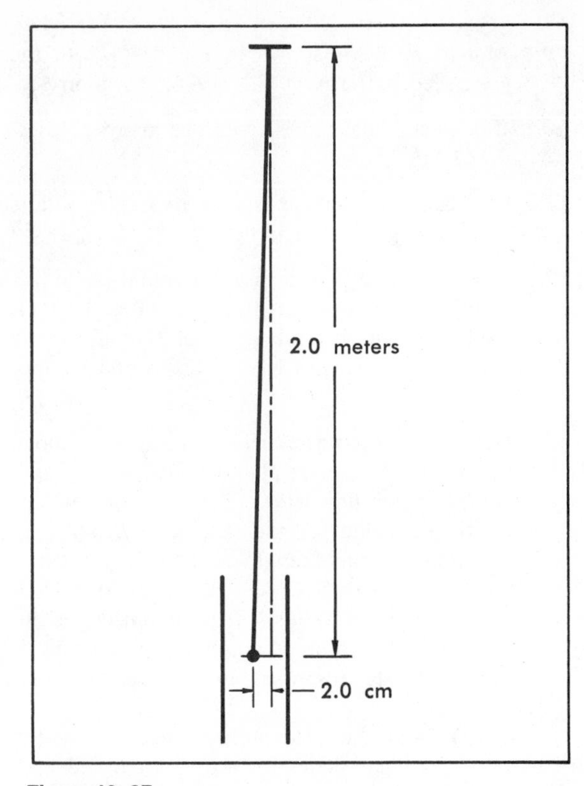

Figure 19–27
For Problem 21.

At equilibrium the ball has been pulled 2.0 cm from its original position.
    (a) What is the magnitude of the electric force on the sphere?
    (b) What is the excess of electric particles on the ball? The electric force on each elementary charge between the plates is 3.0×10^{-14} newton.

22. In an experiment two small conducting spheres (120 milligrams each) were suspended on two fine insulating fibers 82 cm long. When equal charges were placed on the two spheres, they separated and came to rest with a distance of 10 cm between centers. What was the number of elementary charges on each sphere?

23.* An excess of 1×10^{10} negative elementary charges is placed on a small object. What would be the electric field and electric potential at a point 0.5 m away? (Section 8.)

Further Reading

MAGIE, WILLIAM FRANCIS, *A Source Book in Physics*. McGraw-Hill, 1935. Extracts from Coulomb's work in 1785. (Pages 408–420)

The Motion of Charged Particles in Electric Fields

CHAPTER

20

In the last two chapters we have studied electric charge and electric forces by using objects of various masses and sizes. The smallest of these were the latex spheres in the Millikan experiment, which can be seen only under a microscope. Yet even these spheres are enormous, both in mass and in size, compared to atomic particles.

All the charged objects studied so far were either at rest (Experiment 36, The Force Between Two Charged Spheres) or moving very slowly (the latex spheres in the Millikan experiment).

The major goal of this chapter is to extend the study of the behavior of charged particles in electric fields to the atomic scale. Specifically, is the motion of such charged particles governed by Newton's law of motion, even when the speeds involved are much greater than the speed of any rocket?

In a way this question is similar to the question we raised (Section 3–12) in the study of light: Will shadows be sharp even if light passes by very narrow obstacles or through very narrow slits? The answer to this queston showed the limitations of ray optics.

The search for the answer to the question raised in this introduction will not end in this chapter but will reappear near the end of this course. It will finally give you an appreciation of the power as well as the limitations of Newtonian mechanics.

Before we can study the motion of charged atomic particles, we have to produce a beam of such particles. We shall start with electrons.

20-1 Charges in Metals: Electrons

Metals are the commonest kind of conductor. What kind of electric particles are free to move in them, making them conduct? Are these particles positive or negative or both?

Let us assume that electric particles in metals are in a state of thermal agitation like the molecules of a gas (see Chapter 17). Yet experiments show that a charged piece of metal in a vacuum holds its charge indefinitely. Consequently, we know that the electric particles cannot ordinarily escape from the metal. And we conclude that, when they are near the surface of the metal, the moving particles experience forces that prevent their escape by attracting them back into the metal.

Now suppose we heat the metal. As the temperature increases, the speeds of the particles must increase, just as the speeds of gas molecules do. At high enough temperatures an appreciable fraction of the particles should have sufficient speed to escape. Like rockets moving fast enough to escape from the gravitational attraction of the earth, these high-speed particles should not fall back into the metal. They should get out beyond the reach of the attractive forces.

With the apparatus shown in Fig. 20–1 we can do experiments to see what actually happens when we heat a metal. The essential part of this equipment is a hollow metal cylinder with a thin metal wire filament along its axis. The cylinder and the filament are sealed in a glass vessel, which is thoroughly evacuated. To heat the filament we connect it by wires to a battery as shown in the figure. We also connect the filament through an ammeter, a meter that measures electric current, to the negative terminal of another battery. The positive terminal of this battery is connected to the cylinder.

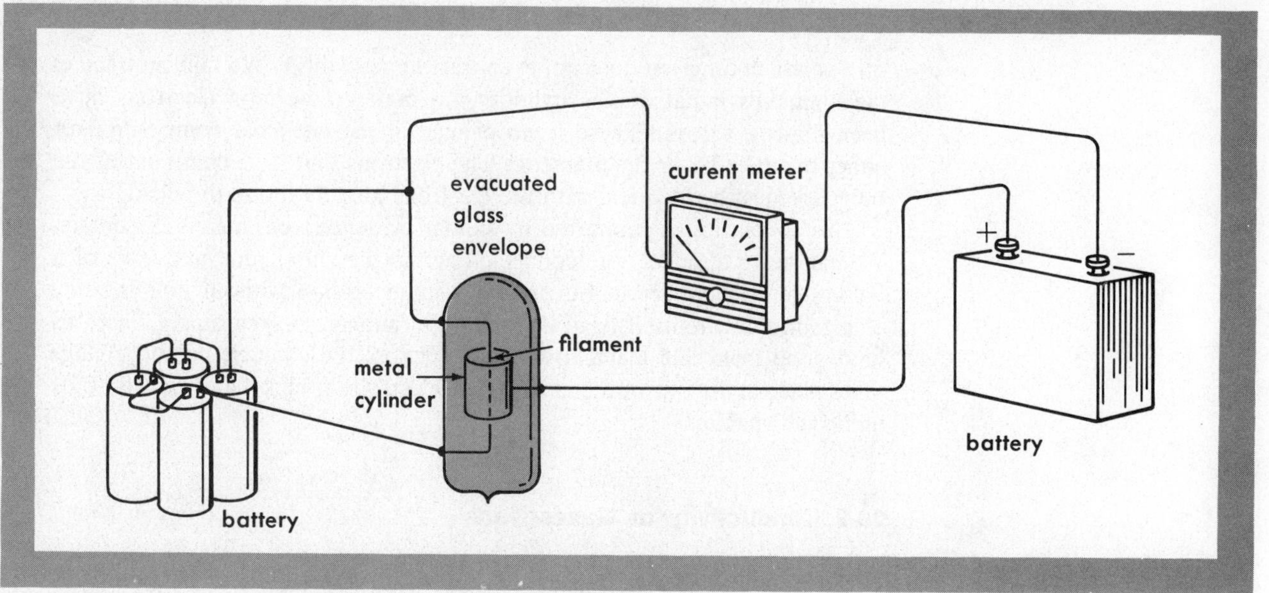

Figure 20–1
Apparatus for investigating electrical conduction in metals. The electric current from battery *A* heats the filament. The current meter connected to battery *B* shows when charge flows through the space between the filament and the cylinder.

When the filament is at room temperature, there is no current through the meter. But if we gradually raise the temperature of the filament, by increasing the number of batteries supplying current to it, the meter will eventually register a current, indicating that electric particles pass through the empty space between the filament and the cylinder. This current increases as we raise the temperature of the filament further, going from orange-hot to white-hot.

We repeat the experiment, reversing the connections to the battery connected to the cylinder. The cylinder is now *negatively* charged and the filament *positively*. In this case the meter shows no current, even when the filament is white-hot.

What do these results mean? Suppose first that the hot filament boils out negative particles. When the filament is negative and the cylinder positive, these particles are repelled by the filament and attracted by the cylinder. Therefore they travel from the filament to the cylinder; they then flow through the metal wire to the positive battery terminal and from the negative terminal through the metal wire back again to the hot filament. Thus the negative particles move across from the filament to the cylinder so that a steady current is observed.

Suppose, on the other hand, that the filament emits positive particles. They would move across the gap when the filament is positive and the cylinder negative; and we would then observe a current through the meter circuit. Consequently, the fact that there is no current with the filament positive proves that the filament emits no positive particles. And we conclude that all the electric particles boiled out of a hot metal are negatively charged.

We are naturally led to believe that we see here, boiled out into the vacuum, the same particles which move around inside a metal. These particles are called *electrons*. The emission of electrons by metals is known as *thermionic emission*. Further experiments indicate that electrons are the same in all metals. In the experiment described above, we can make the filament and the cylinder of different metals—for example, tungsten and nickel. We then run a thermionic current for a long time (but at a temperature at which tungsten does not evaporate appreciably). We find no trace of the filament's metal on the cylinder, no matter how long electrons have been flowing across. There is no change in the chemical composition of either the cylinder or the filament. The electrons that have come out of the tungsten must be identical with the electrons already inside the nickel.

This conclusion is supported by similar evidence from metals in contact. We can leave a nickel wire connected across the zinc-plated terminals of a battery and let a current run for days in succession without noticing any change in the wire itself or in the battery terminals. A great many electrons have gone from one metal to the other across the contact, but no change takes place in the zinc or the nickel. The electrons of nickel and zinc are thus indistinguishable.

20-2 Conductivity of Gases: Ions

Under ordinary circumstances an insulated, charged metal rod remains charged for a long time. The air around it is an effective insulator. This means that the molecules of air are electrically neutral. Also, if we surround the rod with another gas, such as carbon dioxide, helium, or argon, we always find the same result.

But now let us bring some radioactive material near the charged rod or direct a beam of X rays toward it or hold a burning match near it. The rod will gradually lose its charge. Furthermore, if we remove the radioactive material or turn off the X-ray beam, the rod will stop losing charge.

We can interpret these results by assuming that the X rays, or the radiations from the radioactive material, break up the gas molecules into electrically charged fragments. The charged fragments, some of which are positive and some of which are negative, are called *ions;* and a gas containing ions is said to be *ionized.*

Whereas an ordinary gas is an insulator, an ionized gas behaves like a conductor. A positively charged object immersed in an ionized gas attracts the negative ions; a negatively charged object attracts the positive ions. In either case the ions, as they come into contact with the object, gradually neutralize the original charge.

There are many experiments we can do to show that the conductivity of gas results from ionization. For example, we can make sure that the discharge of a charged rod placed in a beam of X rays is not due to a direct

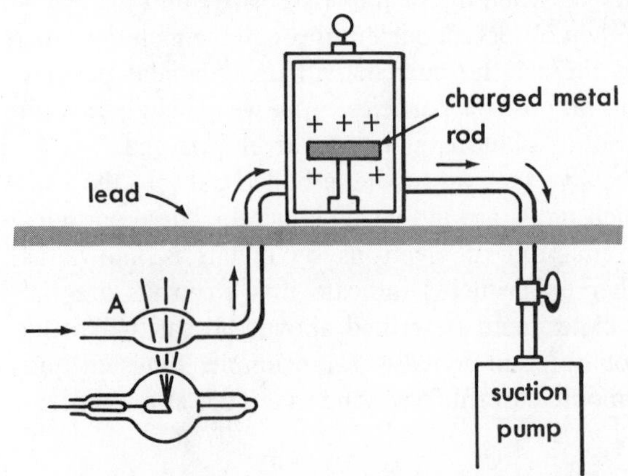

Figure 20–2
X rays ionize the air at *A*. When this ionized air with the ions reaches the charged rod, it loses its charge.

effect of the X rays on the rod. The experiment illustrated in Fig. 20–2 shows that the X rays are acting on the gas. In this experiment we pump air through a tube, part of which can be irradiated, past a charged metal rod. We charge the rod and start the pump. Then we turn on the X rays. The rod begins to lose charge, because ions formed by the X-ray beam are carried by the moving air into the box surrounding the rod. With the X-ray machine still running, we stop the pump. The flow of air stops, and the ions formed in the tube no longer reach the rod. The gradual discharge also stops.

At the beginning of this section we said that gases usually are good insulators. We might have said that in the absence of ionizing radiation they are perfect insulators. As a matter of fact, however, a small amount of radiation is present everywhere. Most materials around us contain minute traces of radioactive substances that produce radiation capable of ionizing gases. And although we can eliminate radiation from materials nearby, cosmic rays will make some ions. Even exceedingly thick shields made of materials free from radioactivity are incomplete protection against these penetrating radiations that originate outside our atmosphere. Thus, gases are always slightly ionized and therefore slightly conducting. This conductivity, however, is so small that it can only be detected with very sensitive apparatus. We see, then, that X rays, cosmic rays, and radioactive substances are able to produce both positive and negative ions in gases.

20-3 The Electric Charge of Electrons and Ions

We already know that electrons have a negative electric charge, but we have yet to find out whether their charge equals one or several elementary charges. We can determine the charge of an electron by counting electrons as we add them to a neutral body and then measuring the charge on the body in numbers of elementary charges.

With an electron gun, a device by which electrons from a hot filament can be formed into a narrow beam, we can produce a beam of high-speed electrons which will make a luminous spot on a fluorescent screen. If the number of electrons passing per second in the beam is small enough, the individual electrons will cause individual scintillations, pinpoint flashes of light. Although each scintillation gives off very little light, the scintillations can be detected by a photoelectric cell. Thus the electrons can be counted individually. The electrons can also be counted directly by using a special kind of amplifier called an electron multiplier. With such counting, we can determine experimentally the number of electrons passing per second in the electron beam [Fig. 20–3(a)].

After we have measured the number of electrons per second in the beam, we can catch the beam in a metal bottle [Fig. 20–3(b)]. The electrons which enter the bottle stop in the metal walls, and even if they eject other electrons from the inner surface of the wall, hardly any particles will find their way out of the small neck of the bottle. The bottle therefore is charged up only by the charge brought in by the electrons. After a sufficiently long time the bottle will acquire enough charge to be measured, and we can therefore determine the charge brought into the bottle in each second. By dividing the charge per second by the number of electrons per second in the beam, we then find the charge of an electron. It is exactly one negative elementary charge.

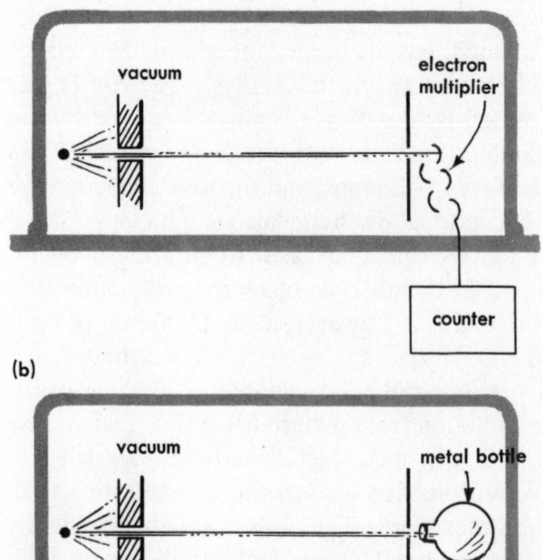

Figure 20–3
(a) We can measure the number of electrons per second passing by in an electron beam by directing the beam at a counter. (b) We can determine the rate at which the same beam carries charge by firing the beam at a "bottle" with a narrow opening for a long enough time to accumulate a measurable charge on the bottle.

We can also use the same kind of measurement to measure the charge of the particles shot out in the radioactive disintegration of polonium. Indeed, the first experiment of this kind was done by Rutherford and Geiger in 1908 in order to find the charge of these atomic fragments, which we usually call alpha particles. They found that an alpha particle carries exactly two positive elementary charges. Later experiments showed that an alpha particle can be decomposed into more basic units, into two protons and two neutrons. The neutrons are electrically neutral. Each of the protons, on the other hand, has one positive elementary charge. A number of other charged particles are now known which differ in mass from electrons or protons. But when we carry the decomposition of matter far enough, all the electric particles found always have either one positive or one negative elementary charge.

20-4 Volts and Electron Volts

In the first two sections of this chapter we discussed a number of ways of producing electrons and ions, that is, charged particles carrying one or at best a few elementary charges. When such a particle moves from point A to point B, where the electric potentials are V_A and V_B, the change in electric potential energy is given by (see Section 19–6)

$$\Delta U = q(V_B - V_A),$$

where q is the charge on the particle. If we express the charge in elementary charges and the change in energy of the particle in joules, then the unit of electric potential becomes joule/el.ch. Is this a convenient unit to use when electrons or ions move through a potential difference created by plates connected to the terminals of any reasonable number of batteries?

To get an idea of the magnitude of the quantities involved, we can go back to a specific run of the Millikan experiment reported in Section 19–8. Consider the electric force acting on a sphere carrying a charge q of 2 el. ch. The electric force F acting on the sphere was 2.8×10^{-14} newton. If the sphere had moved from one plate to the other a distance d of 3.1×10^{-3} meter, the absolute value of the change in potential energy would equal the work Fd:

$$\Delta U = Fd = (2.8 \times 10^{-14}\,\text{newton}) \times (3.1 \times 10^{-3}\,\text{m}) = 8.7 \times 10^{-17}\,\text{joule}.$$

The absolute value of the change in electrical potential would be, therefore,

$$\frac{|\Delta U|}{q} = \frac{8.7 \times 10^{-17}\,\text{joule}}{2\text{ el. ch.}} = 4.4 \times 10^{-17}\,\frac{\text{joule}}{\text{el. ch.}}.$$

Clearly, the joule/el.ch. is an awkward unit for electric potential if we wish to use it in expressing the electric potential provided by batteries.

The batteries we use are made of different numbers of individual cells connected in series; i.e., the positive terminal of one is connected to the negative terminal of the next. Most such cells, independent of the exact chemical reactions that take place, provide a potential difference of about 1×10^{-19} to 2×10^{-19} joule/el. ch. Choosing a joule/el. ch. as a unit of potential in this range would therefore be inconvenient. For technical reasons, into which we need not go at this point, 1.60×10^{-19} joule/el. ch. has been chosen as the unit of electric potential and is called a *volt*. Thus

$$1\text{ volt} = 1.60 \times 10^{-19}\,\frac{\text{joule}}{\text{el. ch.}}.$$

We have mentioned the volt in Chapter 19 simply as a way of characterizing batteries. For example, we spoke of a "90-volt battery." Now we see that this statement refers to the energy that the battery can provide per unit charge. This quantity is also referred to as the EMF of the battery.

The relation

$$\begin{aligned}\text{Energy} &= (\text{charge}) \times (\text{elect. potential}) \\ &= qV\end{aligned}$$

holds independently of the units used, provided we are consistent in our choice of units. If we express the electric potential in volts and the charge in elementary charges (i.e., the charge on the electron), then the energy will be expressed in units called *electron volts* and abbreviated ev.

From the last equation it follows that

$$1\text{ ev} = 1.60 \times 10^{-19}\,\text{joule}.$$

The electron volt is a convenient unit to express energies (potential or kinetic) of individual electrons and ions.

When we deal with large charges moving through potential differences measured in volts, we shall be interested in expressing the energy changes in joules. This will require a new unit of charge. Since energy $= qV$

$$1 \text{ joule} = (1 \text{ el. ch.}) \times \left(1 \frac{\text{joule}}{\text{el. ch.}}\right)$$
$$= (1 \text{ el. ch.}) \times \left(\frac{1}{1.60 \times 10^{-19}} \text{ volt}\right)$$
$$= \left(\frac{1 \text{ el. ch.}}{1.60 \times 10^{-19}}\right) \times (1 \text{ volt}).$$

The charge $\frac{1}{1.60 \times 10^{-19}}$ el. ch. $= 6.25 \times 10^{18}$ el. ch. is called a *coulomb*.

When the electric charge is expressed in coulombs and the electric potential in volts, the energy is expressed in joules.

20-5 Accelerating Charged Particles

To set charged particles in motion (i.e., to give them an acceleration), it is necessary to exert a force on them. The only way to do this is to establish an electric field which will exert a force on the charged particles placed in the field.

The electric field between parallel charged plates, discussed in Section 19–5, may be used for this purpose. Consider a pair of parallel plates in a vacuum, a distance d apart, having charges maintained by connecting the plates to the terminals of a battery whose EMF is V volts (Fig. 20–4).

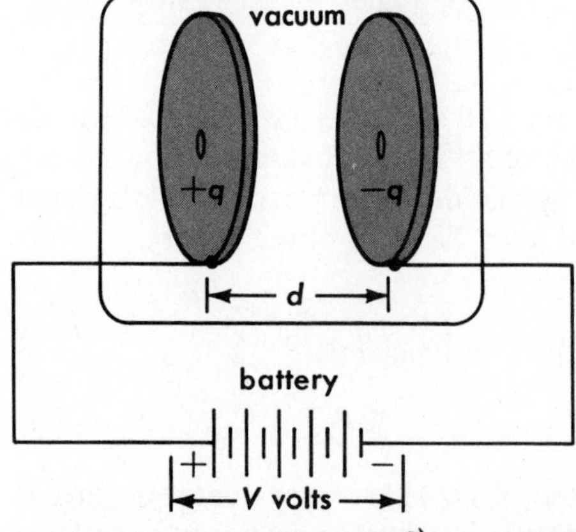

Figure 20–4
A pair of parallel charged metal plates in a vacuum having a potential difference between them of V volts and charges of $+q$ and $-q$.

Near the center, the field $\vec{\mathbf{E}}$ between the plates is uniform. Suppose a particle carrying a charge $+q$ is placed anywhere between them. It experiences a force $\vec{\mathbf{F}} = q\vec{\mathbf{E}}$. To move such a charge from the negative plate to the positive plate, work equal to Fd must be done on it. As noted in the previous section, this work is also qV.

If a positively charged particle is released from rest at the positive plate, it will be accelerated uniformly through the distance d between the plates and arrive at the negative plate with a kinetic energy $\frac{1}{2}mv^2 = Fd = qV$. Thus the particle arrives at the negative plate with a speed $v = \sqrt{\frac{2qV}{m}}$.

Now suppose that one plate has a small hole directly opposite a similar hole in the other plate and that ions just to the left of the hole in the positive plate carrying a charge $+q$ are allowed to drift with small speeds into the region between the plates. The positive ions will be accelerated between the plates, and those reaching the opening in the negative plate pass through. Consequently, a stream of particles emerges from between the plates with a speed $v = \sqrt{\frac{2qV}{m}}$.

What happens to the stream from then on? As you can see from an inspection of the photograph in Fig. 19–7(e), the field outside the plates is zero. So, if the hole in the negative plate is small, the particles of the beam simply coast from then on with their speed unchanged until they strike something or perhaps enter another region where there is another electric field.

The expression $v = \sqrt{\frac{2qV}{m}}$ was derived with the aid of the properties of the uniform electric field existing in the center region between two parallel charged plates. Does it make any difference if the electric field is not uniform? Consider once again the situation between the wire and the surrounding metal cylinder of Fig. 20–1. The electric field lines here radiate outward from the wire and terminate on the cylinder (or vice versa) in a pattern somewhat like that of Fig. 19–6(a).

An electron boiled off the hot wire filament is in a strong field because the field lines as suggested by Fig. 19–6(a) are close together at the filament. As an electron moves outward, it experiences a field decreasing in strength. Therefore, the force on the charge is large near the wire and grows smaller as the charge moves toward the cylinder. To find the work done, one needs to know how the force changes with the distance and then calculate the area under the curve on a graph of F versus d. But the work must be equal to qV, since that is the energy supplied to the electron by the battery connected between the wire and the cylinder. The particle accelerated between two charged conductors connected to the terminals of a battery of V volts acquires an energy of qV joules, regardless of the nature of the electric field.

We can check up experimentally on this fact by injecting electrons, accelerated by a nonuniform field, into a uniform retarding field between parallel plates where we know the retarding force (Fig. 20–5). The electrons will go across the space between the parallel plates as long as the retarding potential

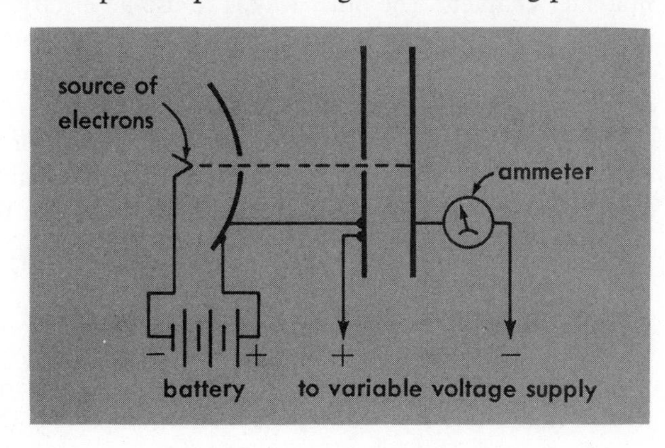

Figure 20–5
An experiment to show that the energy gained by charged particles depends only on the potential difference through which they move and not on the shape of the charged plates. The current through the ammeter stops whenever the potential difference measured across the plates becomes greater than the potential difference across the electron gun. The apparatus is enclosed in a vacuum.

difference between the parallel plates is smaller than the accelerating potential difference between the source of electrons and curved plate. They stop reaching the far end of the system, and the current through the meter stops as soon as the retarding potential difference exceeds the accelerating potential difference. This experiment shows that it is the potential differences imposed by the battery which determine the energy; the shape of the conductors is irrelevant.

The motion of charged particles in the electric field between oppositely charged parallel plates is analogous to the motion of objects in the gravitational field close to the surface of the earth; both fields are represented by uniformly distributed field lines perpendicular to the surface. Hence we expect that the dynamics of charged particles in the electric field between parallel charged plates will be similar to the dynamics of objects moving in the earth's gravitational field close to the earth's surface.

A charged particle released between charged parallel plates experiences a constant electric force $\vec{\mathbf{F}}_e = q\vec{\mathbf{E}}$. This force produces an acceleration $\vec{\mathbf{a}} = \frac{\vec{\mathbf{F}}_e}{m} = \frac{q\vec{\mathbf{E}}}{m}$ in the direction of the field. Starting at rest from one plate, a charged particle would acquire in time t a velocity $\vec{\mathbf{v}} = \frac{q\vec{\mathbf{E}}t}{m}$ and would move through a distance $d = \frac{1}{2}\frac{qEt^2}{m}$. Thus its motion is analogous to that of an object falling from rest in a uniform gravitational field (Fig. 12–2).

However, the analogy is not complete, for in the gravitational case the equivalence between inertial and gravitational masses results in the same acceleration for all objects dropped in the gravitational field. This is not true for the acceleration of charged particles in an electric field. This is because the acceleration of charged particles in a given electric field is proportional to their charge q, which can vary by integral multiples of the elementary charge on particles whose masses are the same.

20-6 Deflecting Charged Particles

Suppose we have available a stream of charged particles moving with some velocity v_0 and we wish to change its direction. There are two ways of doing this. One way makes use of a magnetic field by a method to be discussed in a later chapter. The other way is to use an electric field. Each charged particle in the field experiences, as we have seen, a force in the direction of the field; so by controlling the magnitude of the field and limiting its extent, any desired change in direction can be made.

The results arrived at (Sections 12–3 and 12–4) for the deflection of an object projected horizontally in the earth's gravitational field near the surface can be applied to the motion of charged particles. In the gravitational case the equation of the path was $y = -\frac{g}{2v_0^2}x^2$. In the electrical case each particle in a stream of charged particles is directed into a uniform field between charged parallel plates in a direction initially parallel to the plates with a velocity v_0. The particles experience a force $F = qE$ at right angles to their initial velocity. Taking the x direction parallel to the plates and the y

direction perpendicular to them, $y = \dfrac{qE}{2v_0^2m}x^2$. The path is a parabola. However, unlike objects in the gravitational field near the surface of the earth, in the electrical case the charged particles finally leave the region between the plates and pass into a region where no field exists, and their motion from then on has a velocity $\vec{v}_f$ at some angle ϕ with their initial direction (Fig. 20–6).

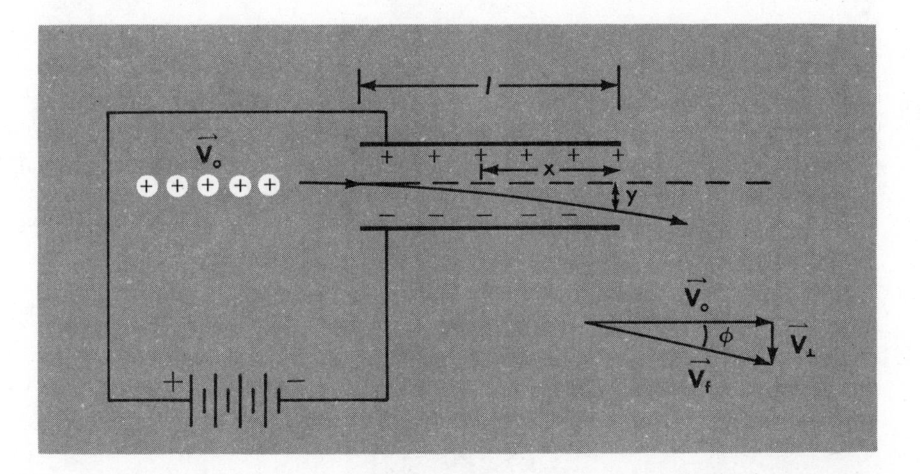

Figure 20–6
A stream of charged particles in a vacuum is deflected in a parabolic path as it passes between a pair of charged, parallel plates.

Let a stream of particles (charge q, mass m) enter the field with an initial velocity v_0 parallel to the plates and along a line halfway between them. Each particle remains between the plates a time $t = l/v_0$. During this time each particle is subject to a force $F = qE$ perpendicular to its initial velocity v_0. We have seen that a pair of parallel plates a distance d apart connected to a battery of V volts has a uniform electric field in the region between them and that $Fd = qV$, or $F = qV/d$. Hence each particle experiences a perpendicular acceleration $a = \dfrac{F}{m} = \dfrac{qV}{md}$. When it leaves the region between the plates, its velocity perpendicular to the plates is $v_\perp = at = \left(\dfrac{qV}{md}\right)\left(\dfrac{l}{v_0}\right)$. Its final velocity v_f has a magnitude $\sqrt{v_0^2 + v_\perp^2}$ and a direction such that

$$\tan\phi = \frac{v_\perp}{v_0} = \frac{qVl}{mdv_0^2}.$$

As soon as the particles emerge from between the plates, they will follow a straight line path which will appear to have originated at the midpoint of the region between the plates. You can see that this is so in the following way:

Each particle has been deflected, perpendicular to the plates, a distance $y = \frac{1}{2}at^2 = \frac{1}{2}\left(\dfrac{qV}{md}\right)\left(\dfrac{l^2}{v_0^2}\right)$ as it leaves the region between the plates. The particles will seem to have come from a point on the line along which v_0 was directed a distance x from the edge of the plates such that $\tan\phi = \dfrac{y}{x}$. Hence

$$x = \frac{y}{\tan\phi} = \frac{qVl^2}{2mdv_0^2} \times \frac{mdv_0^2}{lqV} = \frac{l}{2}.$$

20-7 Oscilloscopes

Thermionic emission, the emission of electrons from a hot filament, has an important application in an electron gun, which is an integral part of an oscilloscope. Figure 20–7 shows an electrically heated filament surrounded by a positively charged metal cylinder containing a small hole. The filament and cylinder are surrounded by a good vacuum.

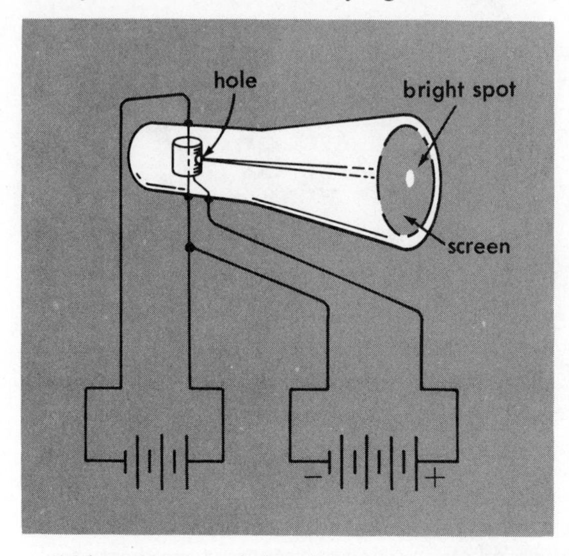

Figure 20–7
An electron gun. A cylindrical shell concentric about a hot filament is positively charged by a battery. The filament is heated electrically by the current from another battery. The right-hand end of the evacuated glass envelope, the screen, is coated on the inside with a fluorescent substance that glows where the beam of electrons strikes it.

Although most of the electrons moving across from the hot filament hit the plate, those moving straight toward the hole go through and travel in a beam until they meet some obstacle. Such an electron gun provides a beam of moving electrons in an oscilloscope. It is also used in a number of devices, such as X-ray tubes, television, radar, and computer display tubes.

Suppose we put two small horizontal metal plates in the tube so that the electron beam passes between them (Fig. 20–8). If one plate is charged positively and the other negatively, the electrons are deflected vertically to-

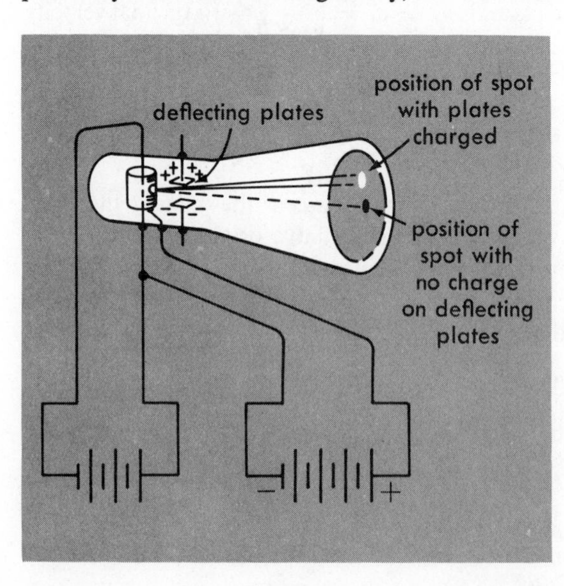

Figure 20–8
The electron gun can be made to serve as a voltmeter by putting a pair of horizontal metal plates in front of the hole. If, for example, we put a positive charge on the upper plate and a negative charge on the lower, the electron beam will be deflected upward. The movement of the spot away from its center position is a measure of the charge and voltage on the plates.

ward the positive plate and hit the fluorescent screen at a different place (Fig. 20–8). Even a single flashlight battery connected across the deflecting plates produces a visible deflection of the spot. Consequently, this tube can be used as a voltmeter to measure potential differences.

Let us investigate the factors determining the sensitivity of this "voltmeter." In Fig. 20–9 the deflecting plates are l cm long and d cm apart. The accelerating voltage in the electron gun is V_a, and the deflecting voltage applied to the horizontal deflecting plates is V_d. The fluorescent screen is at a distance R cm from the midpoint of the deflecting plates. The beam, deflected through an angle ϕ from its initial direction, strikes the screen at a distance S from the midpoint of the screen.

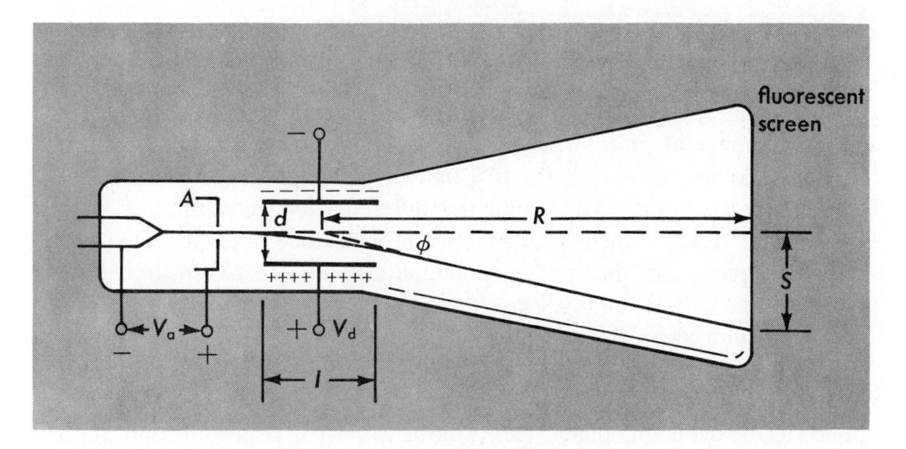

Figure 20–9
The magnitudes of V_a, V_d, d, l, and R determine the deflection S of the electron beam on the fluorescent screen.

The deflection of the beam is given by $\tan\phi = \frac{S}{R}$. From Section 20–6 $\tan\phi$ is also $\frac{qV_d l}{mdv_0^{\,2}}$; hence $S = \frac{qV_d lR}{mdv_0^{\,2}}$. From Section 20–5 $v_0 = \sqrt{\frac{2qV_a}{m}}$; hence

$$S = \frac{qV_d lR}{md} \times \frac{m}{2qV_a}$$

or

$$S = \tfrac{1}{2} \times \frac{V_d}{V_a} \times \frac{l}{d} \times R.$$

One way in which the sensitivity may be expressed is to state how many volts must be applied to the deflection plates to obtain a deflection of 1 cm on the screen. This depends on the accelerating voltage V_a. Reasonable values for the rest of the parameters could be: $l = 2.0$ cm, $d = 0.5$ cm, and $R = 20$ cm. On substituting these values and 1000 volts for V_a, we obtain a sensitivity of 25 volts/cm. For actual oscilloscope tubes, sensitivities of 10 to 100 volts/cm are common.

With a simple addition to the tube shown in Fig. 20–8 we have an oscilloscope with which we can study in detail how a changing potential difference varies with time. For this purpose we equip the tube with a second pair of metal plates arranged to deflect the beam horizontally (Fig. 20–10). A "sweep circuit" built into the oscilloscope gradually charges this pair of plates, increasing the potential difference between them, and then discharges

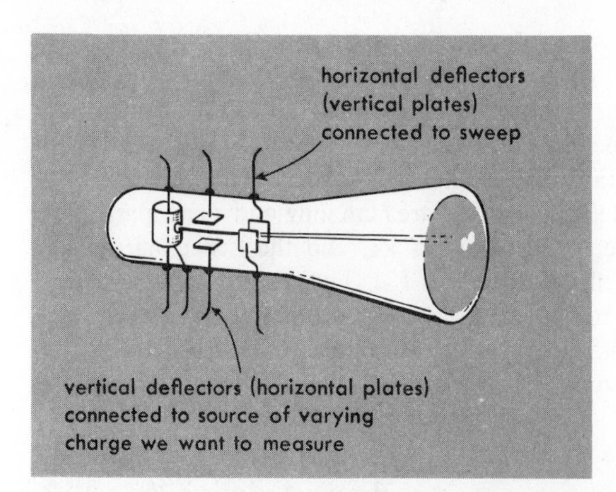

Figure 20–10
An oscilloscope tube has a second pair of vertical plates to give a horizontal deflection. Charging these plates swings the spot horizontally across the face of the tube.

them suddenly, only to start charging them again. During the gradual charging the bright spot where the electrons hit swings horizontally across the screen at a constant speed; then, on discharge, it returns quickly to the starting point. At the same time the first pair of plates produces a vertical deflection corresponding to the potential difference between them at every instant. As a result of the two deflections the spot traces a curve which gives a graphical picture of the way the potential difference between the vertical plates changes with time (Fig. 20–11).

Some oscilloscopes are arranged so that a voltage to be studied can be applied directly to the deflecting plates if it is desired to do so. However, the voltage to be studied is normally applied to an amplifier whose output is applied to the deflecting plates. Then the sensitivity is adjustable and can be

Figure 20–11
A photograph of a trace on the face of an oscilloscope tube. While the sweep circuit moved the beam from left to right, a battery rapidly charged two metal plates connected to the vertical deflection plates of the tube. A moment later they were allowed to discharge more slowly through a wire of high resistance.

made much larger than when the voltage is applied directly to the deflecting plates.

An important property of an oscilloscope is the speed of its response. Electrons have a very small mass. Consequently, the deflection of the electron beam occurs practically instantaneously. Thus the oscilloscope is capable of showing very rapid changes in potential difference.

Oscilloscopes have many other applications. Television and radar display tubes are examples. The pictures they paint are produced by deflecting the electron beam so that it traces a path over the whole face of the tube while the strength of the beam is varied to make each little region of the screen glow bright or stay dark in response to signals from a TV camera or radar antenna.

20-8 Determining the Mass of the Electron and of the Proton

In this section we shall show how the mass of an electron or proton (a hydrogen ion) can be determined experimentally by accelerating the charged particle with a constant field.

In these experiments we shall accelerate our charged particles between two charged metal plates. We shall do the experiments in a vacuum, to eliminate collisions with air molecules. From the potential difference between the plates and the charge on the particles, we can determine the kinetic energy acquired by the charged particles:

$$\tfrac{1}{2}mv^2 = qV.$$

We ionize some hydrogen near a pair of charged plates and let some of the ions drift with negligible speed through a small hole into the region between the plates (Fig. 20–12). The electric field between the plates

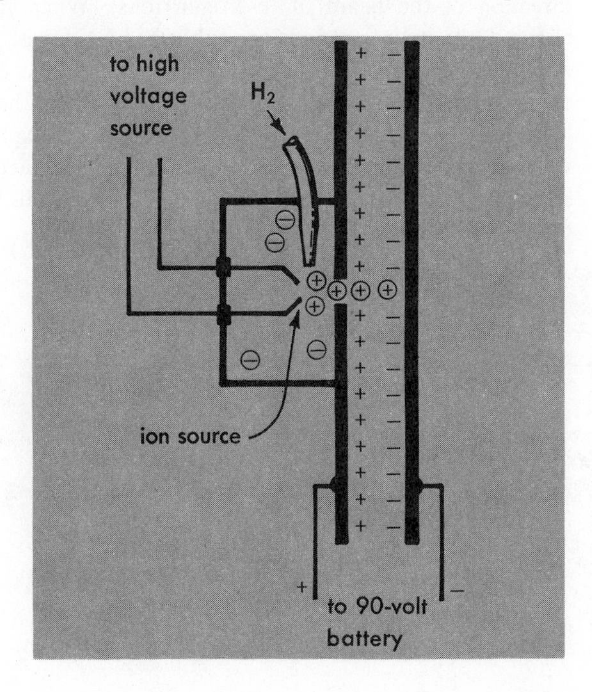

Figure 20–12
Apparatus for accelerating hydrogen ions in an electric field. The ions are formed in the box at the left, and some drift through the hole into the region between the plates. The whole setup is enclosed in an evacuated chamber.

accelerates the ions as they cross from one plate to the other, giving them a kinetic energy $\frac{1}{2}mv^2 = qV$. Now, if in the right-hand plate there is a small hole, some of the ions can pass through into a chamber 0.50 meter long (Fig. 20–13). This chamber is made of a conductor. There is no electric field inside it, so the ions move its entire length without changing their velocity. It takes the ions only a few microseconds (a few times 10^{-6} second) to go the whole distance. Although this time interval is very short, with an oscilloscope we can measure it accurately. Then we can find an accurate value for the speed v of an ion.

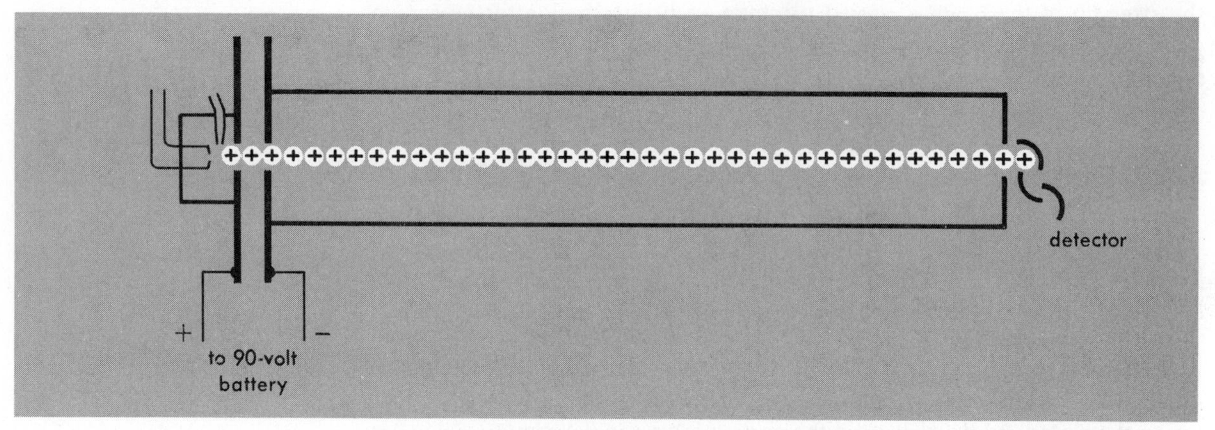

Figure 20–13
To measure the mass of hydrogen ions, we drill a hole in the right-hand plate, allowing some ions to enter a long chamber on the right, where they move at constant velocity.

In order to measure the time it takes for the ions to pass from one end of the long chamber to the other, we must record the time at which a particular ion passes a given point at the left and the time at which the same ion reaches the far end on the right. To mark the time when a given ion enters the long chamber, we place two small deflecting plates near the entrance (Fig. 20–14). With these deflecting plates we can control the direction of the beam of hydrogen ions. When the deflecting plates are charged, the hydrogen ions are subject to a sidewise electric force which deflects them out of their normal path. If the deflecting plates are then discharged, only the ions just entering the chamber have a velocity along the long axis; hence, the first ions to come out through the hole at the far end will be those which have traveled the entire 0.50 meter during the time since the plates were discharged. The arrival of these ions is signaled by a detector placed just outside the hole.

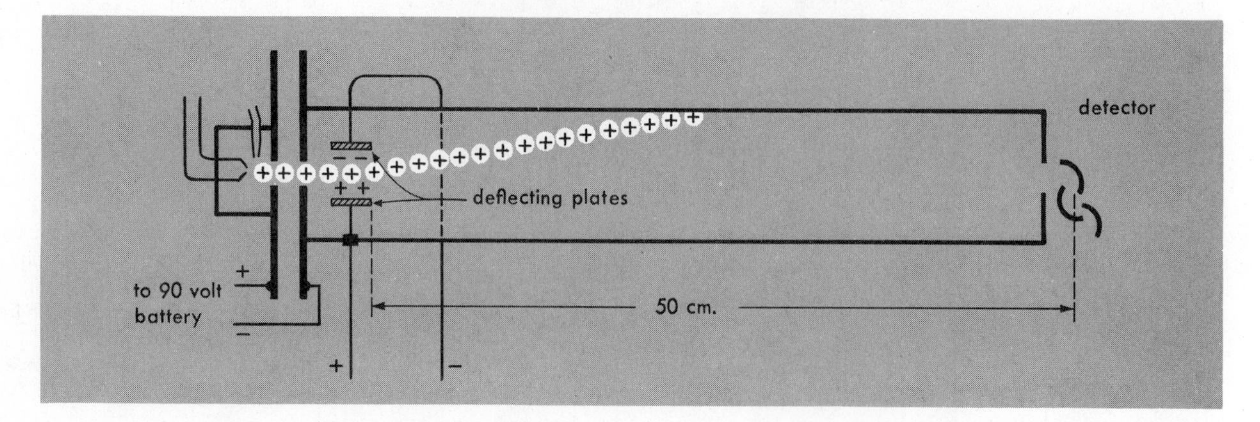

Figure 20–14
When the deflecting plates are charged, the electric force swings the beam to the side. When the plates are uncharged, ions go straight ahead and hit the detector, as in Fig. 20–13.

To measure the time interval between the discharge of the deflecting plates and the arrival of the first ions at the detector, we connect the deflecting plates in the chamber through amplifier A to the vertical deflecting plates of an oscilloscope (Fig. 20–15). When we discharge the plates in the long chamber, the time of the discharge is recorded as a pulse or spike on the line traced on the face of the oscilloscope. We also connect the detector at the far end of the long chamber through amplifier B to the vertical deflecting plates of the oscilloscope. (We make the electrical connections at both ends of the chamber identical.) When the beam of ions hits the detector, a second spike appears on the face of the oscilloscope (Fig. 20–16). The two spikes occur at different places on the tube face because they are made at different times.

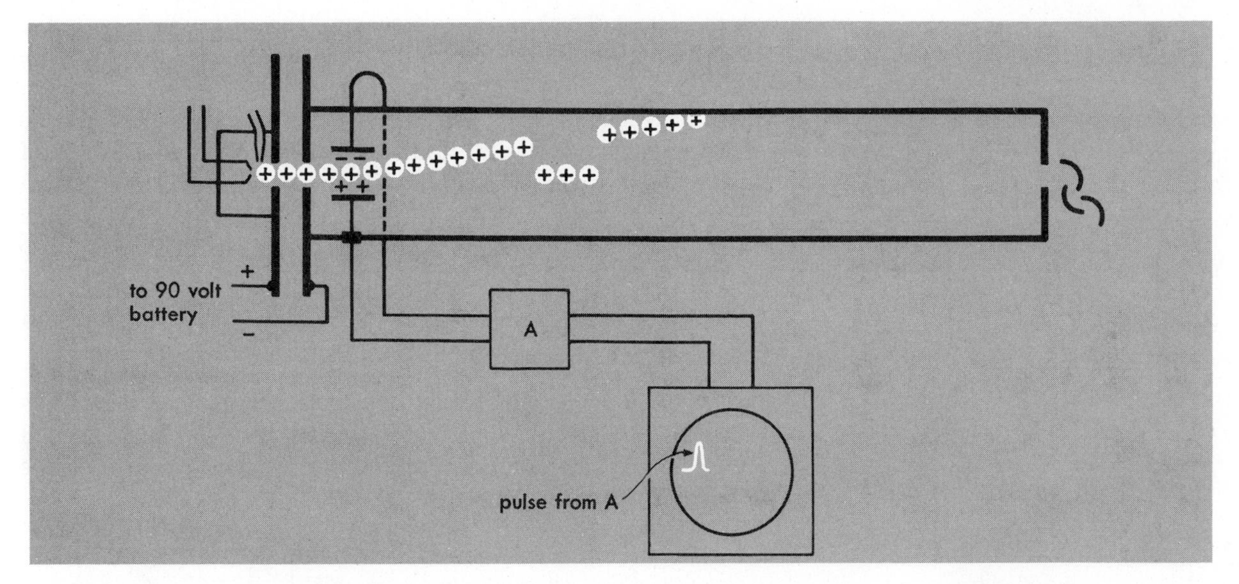

Figure 20–15
When we momentarily discharge the plates, we let through a short pulse of ions that goes straight along the chamber to the detector. The discharge is recorded as a spike on the oscilloscope, marking the time that the pulse begins its flight.

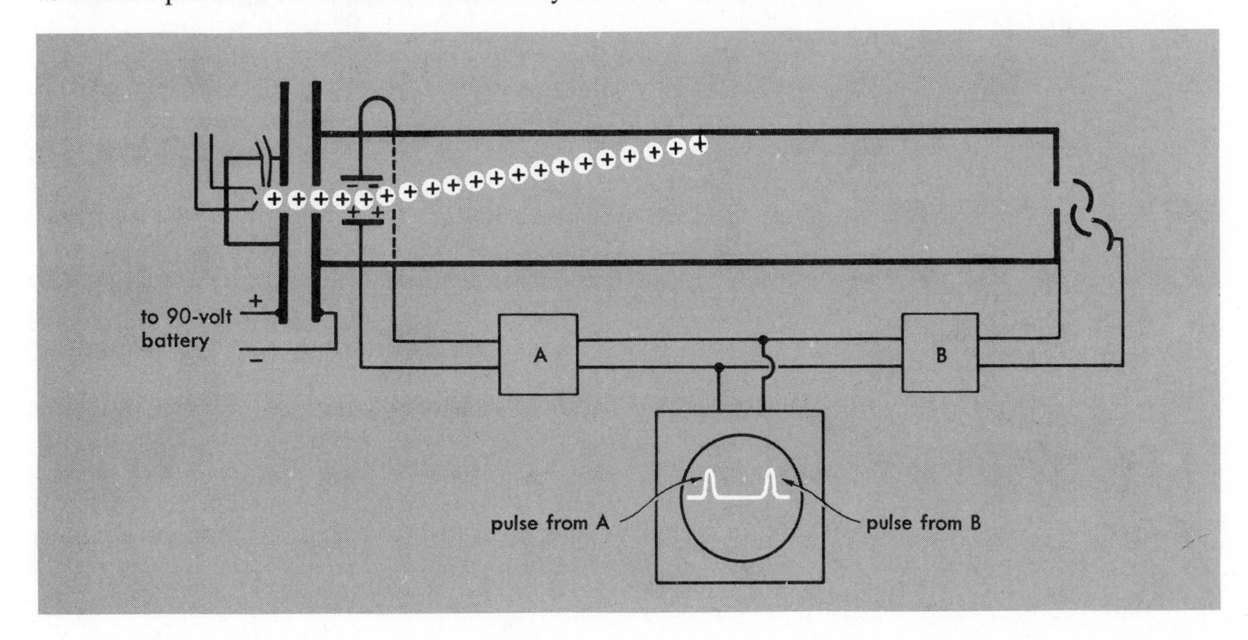

Figure 20–16
When the ion pulse strikes the detector, a second spike appears on the oscilloscope. The distance between the two spikes measures the time of flight of the ions.

During the intervening time interval the sweep circuit in the oscilloscope has moved the electron beam horizontally across the face. The electron beam in the oscilloscope was deflected the distance between the spikes in the same time that the hydrogen ions move the 0.50 m through the chamber.

With a good oscilloscope, the sweep circuit can move the electron beam horizontally across the face of the tube from one side to the other in a few hundredths of a microsecond. To measure the speed of the ions, we adjust the sweep circuit to make the whole trace on the oscilloscope represent 5 microseconds. The two spikes on the trace are then well separated on the face of the oscilloscope. By measuring the distance between the spikes, we measure the time it takes the beam to travel the length of the long chamber. We obtain the time interval from the moment when the beam is allowed to go straight ahead to the moment when it hits the detector, with an accuracy of about one-hundredth of a microsecond. With hydrogen ions and a 90-volt battery providing the accelerating electric field, the time of flight is 3.82 microseconds. From this we compute the speed v of the ions in the long chamber:

$$v = \frac{0.50 \text{ meter}}{3.82 \times 10^{-6} \text{ sec}} = 1.3 \times 10^{5} \text{ meters/sec.}$$

The measured accelerating potential difference between the plates was actually 88 volts (joules/coulomb). If we assume that the hydrogen ion carries one elementary charge (1.6×10^{-19} coul), each ion gains a kinetic energy qV when it is between the plates:

$$qV = (1.6 \times 10^{-19} \text{ coul}) \times (88 \text{ joules/coul})$$
$$= 1.4 \times 10^{-17} \text{ joule.}$$

Hence,

$$\tfrac{1}{2}mv^2 = \tfrac{1}{2}m\,(1.3 \times 10^{5} \text{ m/sec})^2 = 1.4 \times 10^{-17} \text{ joule,}$$

and solving for m, the mass of the hydrogen ion, we find

$$m = 1.7 \times 10^{-27} \text{ kg.}$$

We can use the same apparatus to measure the electron mass. If hydrogen ions are made by taking an electron out of a hydrogen atom, the source of ions at the left of our apparatus should also provide electrons. By turning the battery around, we can accelerate the electrons that drift through the hole in the left-hand plate and reject the hydrogen ions. (Alternatively we could use the electrons boiled out of a hot filament—with either source we get the same results.) Since the electrons are pushed across between the plates with the same force through the same distance, they will gain the same value of $mv^2/2$—that is,

$$mv^2/2 = 1.4 \times 10^{-17} \text{ joule.}$$

Because they are much lighter than the positive ions, they will be moving faster when they pass through the hole in the second plate. Consequently, when we measure the speed with which the electrons pass through the long chamber, we find that we are closer to the limits imposed by our apparatus. With the sweep circuit adjusted so that the distance across the tube face corresponds to a time interval of about 0.12 microsecond, we can see the two spikes recorded on the face of the oscilloscope. They are about nine one-hundredths of a microsecond apart. However, to measure the exact time interval between them to better than 10 percent accuracy requires an unusually fast oscilloscope and very sharp spikes. Taking 0.090 microsecond as a first approximation, we find that the speed is about 5.6×10^6 meters per second. Consequently, the mass of an electron (computed in the same way as the mass of positive ions) is about 0.9×10^{-30} kg. Accurate measurements of the mass of the electron give 0.911×10^{-30} kg, about a two-thousandth of the mass of a hydrogen atom.*

In measuring the masses of ions, we can change the details of the experiments so that we use different known potential differences between the plates. We then get different values of $mv^2/2$. But we always get the same masses. We can use other methods to measure the masses—and we shall describe another one in Chapter 22. The results are the same. Indeed, the masses of many ions are now extremely well established, and they lead to accurate masses of the atoms.

The experiments we have just described tell us about two of the fundamental building blocks from which atoms are made, protons and electrons. In interpreting these experiments, we naturally used the Newtonian mechanics. Because our results are sensible and consistent, we may now conclude that Newtonian mechanics can be applied over an enormous range. It gives an accurate description of the motions of large masses at large distances, including the planets in the solar system; it applies to moderate masses at the moderate dimensions of everyday life; and it also applies to the small masses of electrons moving through relatively small distances at huge speeds. Consequently, it is reasonable to try to apply the same mechanics and the same forces to the electrons within atoms. In particular, we can make a model of a hydrogen atom in which a single electron moves around the relatively massive proton under the influence of the Coulomb force. Because the Coulomb force has the same inverse-square dependence as the gravitational force between the sun and a planet, we shall call this model a planetary model of the atom. In the planetary model of hydrogen the electron moves around the proton in an elliptical orbit like a planet around the sun. The planetary model, as we shall see in Section 24–1, meets with some difficulties, but it is a good first approximation to actual atomic behavior. Since Rutherford introduced it in 1911, it has played a central role in our understanding of atoms.

* With modern fast oscilloscopes and amplifiers, we can measure time intervals to within 10^{-9} sec. Therefore it is now possible to use this technique to get an accuracy of about 1 percent. Earlier, precision measurements were made by other methods, which we shall come to in a later chapter.

1.* Why would an insulated, charged rod discharge more slowly in a vacuum chamber than in air? (Section 2.)

2. A thin layer of radioactive material emitting beta particles is painted on a small conducting sphere. The sphere is then hung on an insulating thread in an evacuated chamber (Fig. 20–17).

After 50 days we measure the charge on the sphere and find it to be 4.32×10^{10} positive elementary charges. (The sphere has been kept in an evacuated chamber, so very few charges have leaked off.) In a separate experiment with a Geiger counter we find beta particles shooting out at the average rate of 1.00×10^4 per second.

(a) What is the charge of one beta particle?
(b) What do you think beta particles may be?

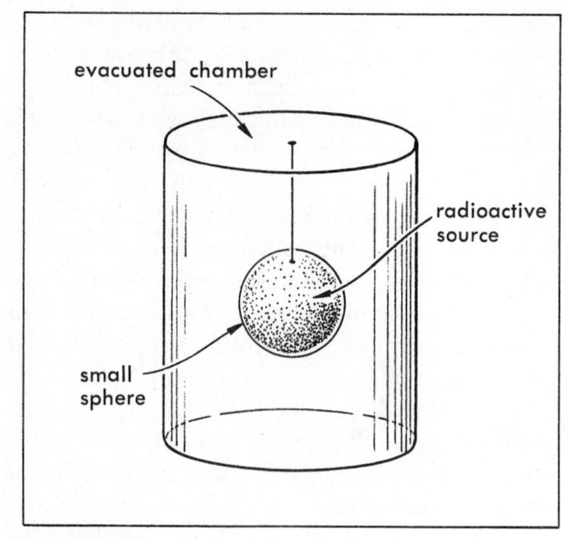

Figure 20–17
For Problem 2.

3.* In Problem 28 of Chapter 16, you found that the binding energy of the electron to the proton was 2.3×10^{-18} joule. How many electron volts is this? (Section 4.)

4.* Three 90-volt batteries are connected in series across two parallel plates 1.00 mm apart. What is the force on an elementary charge between the plates? (Section 5.)

5. Two large parallel metal plates, 6.2 cm apart in a vacuum tube, are connected to the terminals of two 90-volt batteries in series. A doubly charged oxygen ion starts from rest at the surface of one plate and is accelerated across to the other plate.

(a) With what kinetic energy $\frac{1}{2}mv^2$ does the oxygen ion crash into the other plate?
(b) If the ion starts halfway between the plates, with what kinetic energy does it hit the negative plate?

6. With the same apparatus as in Problem 5, how many doubly charged oxygen ions would have to go from the positive plate to the negative plate before that plate is heated up by one joule by the dissipation of kinetic energy of the ions?

7. Suppose the doubly charged ions in the last problem were made by ripping electrons off oxygen atoms located halfway between the plates. Then the oxygen ions are accelerated in one direction and the electrons are accelerated in the opposite direction from the midpoint between the plates. How much energy will the electrons have dissipated heating the positive plate in the time that one joule is dissipated at the negative plate?

8.* Assume that the accelerating voltage applied to the source of electrons in Fig. 20–5 is 45 volts. What will be the kinetic energy of the electrons passing through the hole in the curved plate? (Section 5.)

9.* Assume that the electrons in Fig. 20–5 are accelerated by a 45-volt battery and that the potential difference across the parallel plates is 30 volts. Describe what happens to the kinetic energy of an electron which passes through the hole in the plate at the left and strikes the plate at the right. (Section 5.)

10.* A doubly charged ion is accelerated from rest by a force of 3.10×10^{-15} newton over a distance of 9.3 millimeters. How much kinetic energy does the ion gain? (Section 5.)

11.* Other measurements on the ion of the previous problem, after it had been accelerated, show its momentum to be 1.24×10^{-21} kg-m / sec. What is the mass of the ion? (Section 5.)

12. In Fig. 20–18 electrons emitted at F are accelerated in a vacuum toward B.

(a) How much energy will the electrons have which hit plate B?
(b) What is the electric field between A and B and between B and C?
(c) Some will travel through the hole in B. Where will they go?

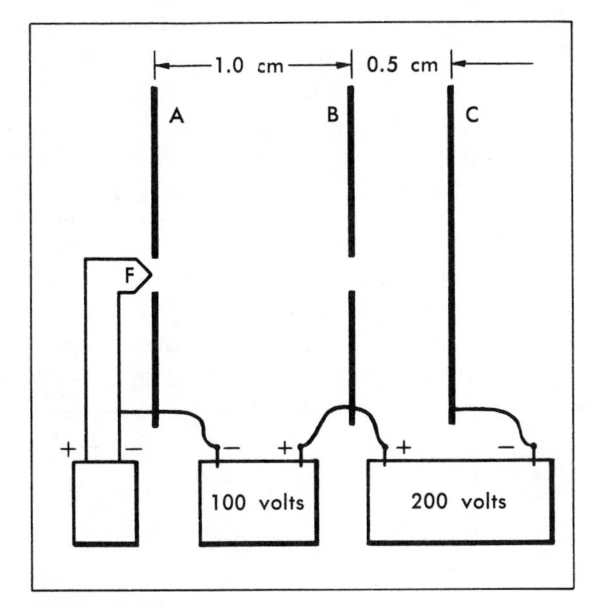

Figure 20–18
For Problem 12.

(a) If the deflecting plates are separated by 0.20 cm, what is the potential difference from one deflecting plate to the other?

(b) What is the speed of the electrons leaving the gun?

(c) How long do the electrons take to pass through the length l?

(d) As an electron passes between the deflecting plates, it picks up a sidewise velocity because of the electric force exerted on it. How big is this sidewise velocity when the electron leaves the deflecting plates?

(e) How far is the electron pushed sidewise in passing between the plates?

14.* In Fig. 20–10 what charge would you put on the horizontal deflector nearest you to make the beam move to the left-hand side of the tube face? (Section 7.)

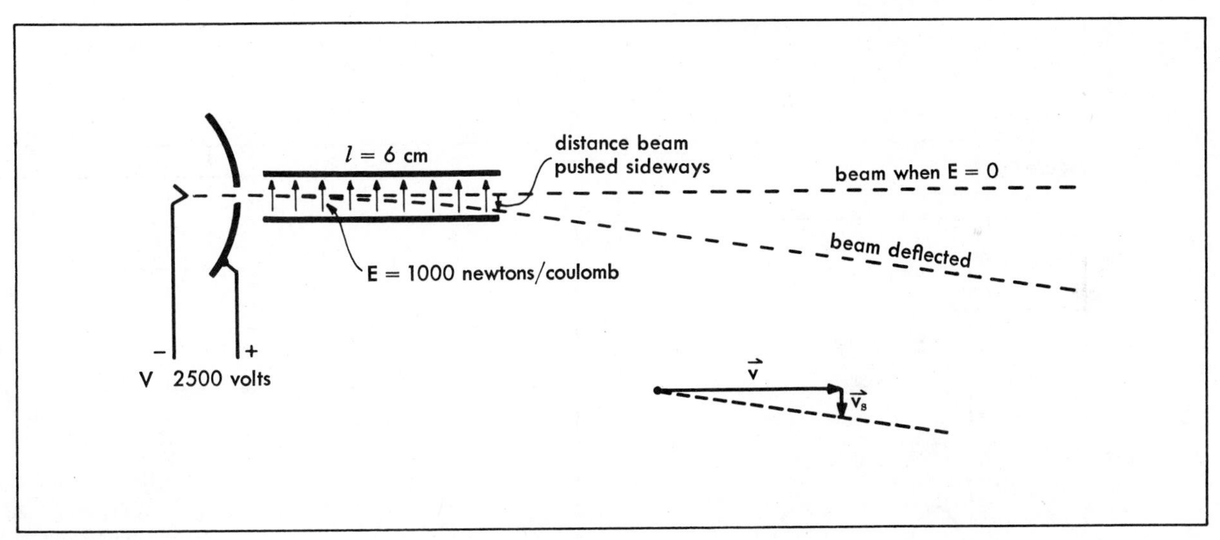

Figure 20–19
For Problem 13.

13. An electron gun shoots a stream of electrons between two deflecting plates of length $l = 6.0$ cm (Fig. 20–19). The electron mass is 0.91×10^{-30} kg. The potential difference between the filament and plate of the gun is 2500 volts ($2.5 \times 10^3 \times 1.6 \times 10^{-19}$ joule per elementary charge). The electric field E between the deflecting plates is 1000 newtons per coulomb (1000 newtons per 6.24×10^{18} elementary charges).

15.* Suppose we use two 90-volt batteries in the apparatus of Fig. 20–12 and use doubly charged ions; what energy will each ion have when it hits the negative plate? (Section 8.)

16. In a certain oscilloscope the linear sweep voltage described in Section 20–7 is applied to the horizontal deflection plates. What would you expect to observe on the face of the oscilloscope

if the time-varying voltage shown in Fig. 20–20(a) were applied to the vertical deflection plates? If the time-varying voltage shown in Fig. 20–20(b) were applied to the vertical deflection plates?

17. Graphs of three time-varying voltages are shown in Fig. 20–21; the voltage and time scales are the same in all three graphs. The voltage shown in Fig. 20–21(a) is applied to the horizontal deflecting plates of an oscilloscope. What would you expect to observe on the fluorescent screen if the voltage shown in Fig. 20–21(b) were applied to the vertical deflecting plates? If the voltage shown in Fig. 20–21(c) were applied to the vertical deflecting plates?

18. You can see from an inspection of Fig. 19–6(e) that the electric field between charged parallel plates does not stop abruptly at the edges of the plates but extends into the space beyond the plates. What do you think the effect of this extension is on the sensitivity of the oscilloscope? On what do you base your answer?

19. A beam of 1000 ev electrons enters the region between a pair of charged parallel plates as shown in Fig. 20–22 at an angle of 30° to the plates. What voltage V_d applied to the plates will just prevent the electrons from striking the upper plate?

(*Hint:* What energy would an electron released from rest at the upper plate have on reaching the lower plate?)

20. Fig. 20–23 shows an oscilloscope that has a positive ion source instead of an electron gun. If the gas admitted to the ion source is a mixture of hydrogen and helium, the positive ion

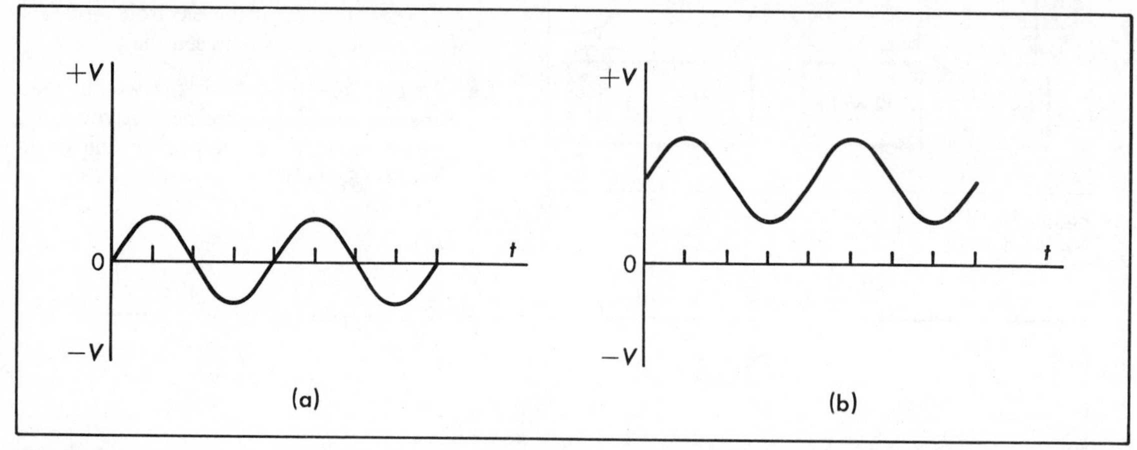

Figure 20–20
For Problem 16b.

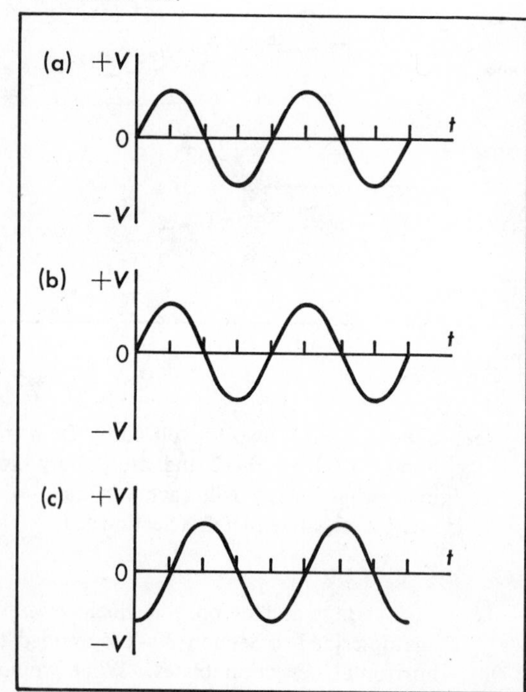

Figure 20–21
For Problem 17.

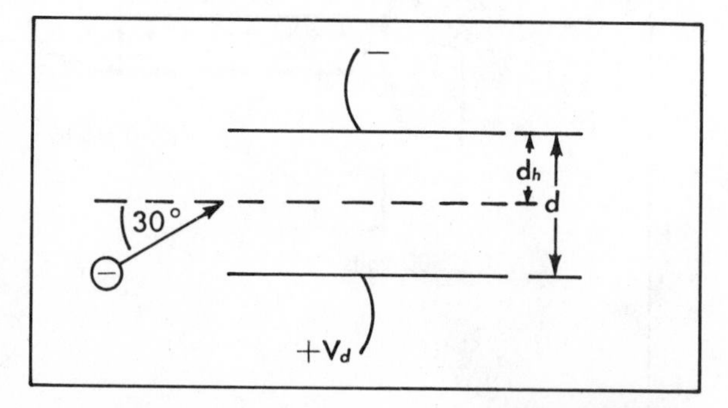

Figure 20–22
For Problem 19.

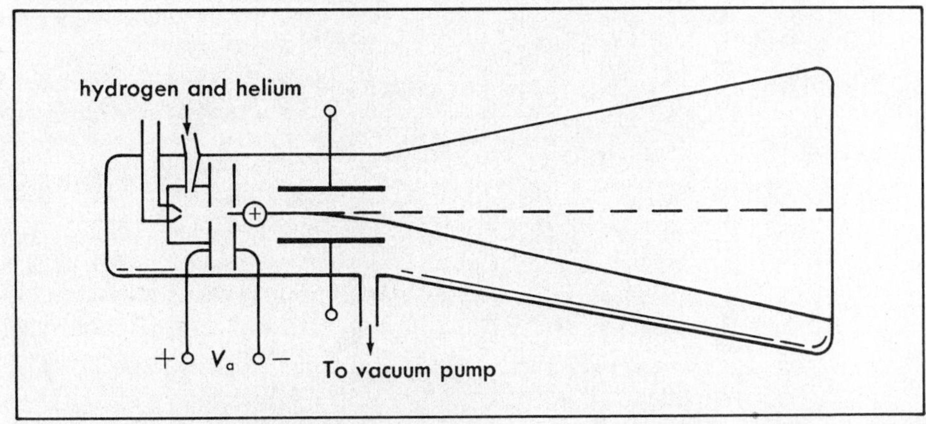

Figure 20–23
For Problem 20.

beam is a mixture of singly charged hydrogen ions, singly charged helium ions, and doubly charged helium ions. Will the deflecting plates separate the beam into its three components by deflecting different ions by different amounts? On what reasoning do you base your answer?

21. Suppose the stream of charged particles shown in Fig. 20–6 is first a stream of hydrogen ions and later a stream of electrons. How does the deflection angle ϕ_H of the hydrogen ions compare with the deflection angle ϕ_e of the electrons if
    (a) the speeds of the hydrogen ions and of the electrons are the same?
    (b) the kinetic energies of the hydrogen ions and of the electrons are the same?

22.* If singly charged helium ions (four times as massive as hydrogen ions) were put through the apparatus of Fig. 20–15, what would be the time of flight? (Section 8.)

23. In a time-of-flight experiment like that described in this chapter, you have adjusted the sweep circuit of the oscilloscope so that the beam moves across the tube in 1.0×10^{-6} sec. The total length of the trace made in that time is 10 cm. The distance between the deflecting plates and the detector is 1.2 m. If the spikes on the oscilloscope trace are 8.0 cm apart, what is the speed of the ions?

24. How does the mass of a hydrogen ion compare with the mass of a hydrogen atom?

25. In this chapter we have neglected the effect of the gravitational force in discussing the trajectories of electrons and ions in an electric field. To show that this is justified, find the ratio of the gravitational force on a singly ionized uranium atom (whose mass is 236 times that of a hydrogen ion) to the electric force on this ion in the electric field between two parallel plates 0.5 cm apart having a 1.0 volt potential difference between them.

Further Reading

These suggestions are not exhaustive but are limited to works that have been found especially useful and at the same time generally available.

BORN, MAX, *The Restless Universe*. Dover, Second Revised Edition, 1957. A complete and understandable picture of the mass, charge, and energy relationships of the electron. (Chapter II)

FINK, DONALD G., and LUTYENS, DAVID M., *The Physics of Television*. Doubleday Anchor, 1960: Science Study Series. Easily read discussion of subatomic physics showing the electron playing a basic part in television. (Chapters 2 and 3)

MILLIKAN, R. A., *The Electron*. University of Chicago Press, 1924; Phoenix Books, reissue, 1964.

THOMSON, GEORGE, *J. J. Thomson*. Doubleday Anchor, 1966: Science Study Series. A fascinating biography of the discoverer of the electron—much material on his experiments.

Electric Circuits

CHAPTER

21

So far our study of electricity can be considered to be part of dynamics; we examined the electric fields produced by some charge distributions and then studied the motion of charged particles in these fields. The approach was similar to the one we used in studying the motion of a mass tied to a spring. First we investigated how the force of the spring varies as a function of the stretch (Section 12-8) and then we determined how a mass would move under the influence of such a force. We shall use a similar approach again when we investigate the motion of charged particles in the electric field of the atomic nucleus (Chapter 24).

The motion of charged particles in a vacuum under the influence of given electric fields provides only a very limited view of the motions of charged particles. In most electrical devices electrons move in solids and not in a vacuum. The electric fields are complicated and the number of electrons per unit volume is large: therefore the effect of the motion of one electron on another electron cannot be ignored (as it can in an oscilloscope). To describe the motion of electrons in solids is, therefore, rather complicated. Fortunately, for a large variety of applications such a description is not necessary. Broadly speaking, the purpose of an electric device is to produce a desired response to a signal it receives. Often such a signal has the form of a voltage across two terminals. The response may be a voltage across two other terminals or a current through part of the device.

Most electrical apparatus, particularly electronic devices, consist of many parts or elements. How do designers of electrical apparatus know what parts to use and how to connect the parts together? By studying the characteristics of various elements separately, they can predict the properties of various combinations of elements.

In this chapter we shall get a glimpse of what is involved in such a process by studying the relation between the voltage across and the current through a number of common circuit elements and then apply the general rules for combining such elements in series and in parallel.

We shall also see how such combinations can be used in an electric circuit to perform useful functions. The circuits will include a number of different kinds of conductors, or circuit elements, such as ordinary metallic resistors, semiconductor devices, capacitors, and conductors involving ionic conduction. These circuit elements make up a large fraction of the different kinds of circuit elements that can be combined together in simple to very complex circuits that make up the many useful electronic devices, such as radios, television sets, computers, and a host of other practical applications of electric circuits. Figure 21–1, for example, shows a circuit that automatically controls two headlights, one on each end, of a model electric switch engine. When the locomotive moves in one direction, only one headlight is on. When the direction of the current through the engine is reversed, the other headlight is automatically switched on. In addition, the intensity of each headlight remains constant when it is on, until the voltage supplied to the locomotive is so low that it has come to a stop, or nearly so.

21-1 Electric Current

A flow of charge through the circuit elements in Fig. 21–1 constitutes an electric current. The beams of protons and electrons used in measuring the masses of electrons and protons are also electric currents, and the natural measure of these currents is the number of elementary charges transported per second.

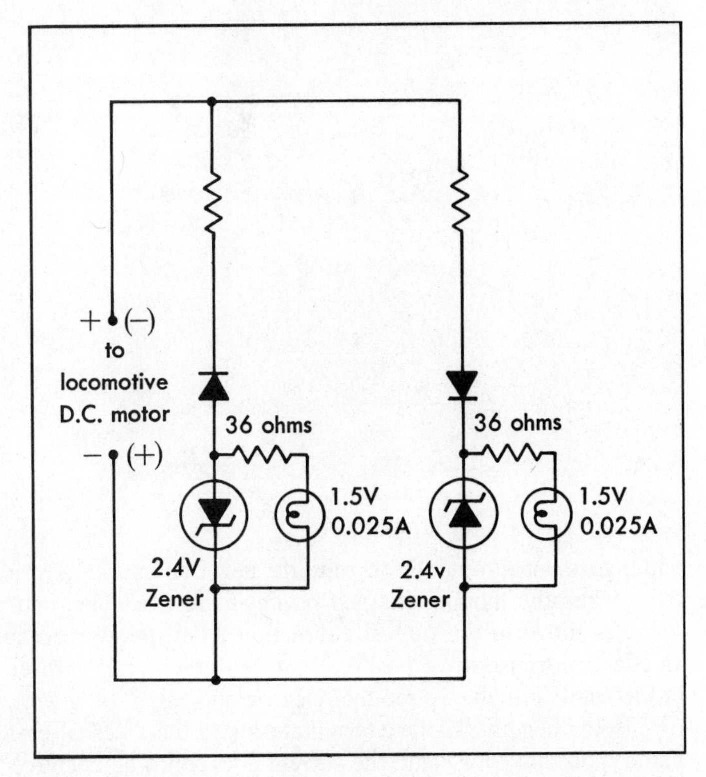

Figure 21–1
A circuit for automatically switching headlights on a model railroad locomotive.

If we catch a beam of *positively* charged particles in a metal cup [Fig. 20–3(b)], the cup becomes more and more positively charged as time goes on. For this reason we say that the direction of this electric current is into the cup because the positive charge it has received *increased* with time, i.e., has become more *positive*. The current, therefore, is in the same direction as the motion of the *positive* particles.

A beam of negative particles is also an electric current. But when we catch it in a metal cup, the charge of the cup decreases: it becomes less positive. If we catch equal beams of positive and of negative particles at the same time, they neutralize each other. When the same number of positive and of negative particles arrive in the cup per second, the charge of the cup does not change. There is no incoming or outgoing electric current. Even though the beam of negative particles moves *toward* the cup, it constitutes an electric current *away* from the cup, emptying the cup of positive charge as fast as the positive beam fills it. In other words, a beam of negative particles constitutes an electric current in the direction opposite to the motion of the negative particles. The question of the direction of an electric current can be confusing, because in conducting metal wires only negative charges (electrons) move, and the direction of the current in such conductors is always opposite to the direction of motion of the electrons.

The electric current does not begin or end at the terminals of the battery in Fig. 21–2. It comes in through the wire from the positive battery terminal

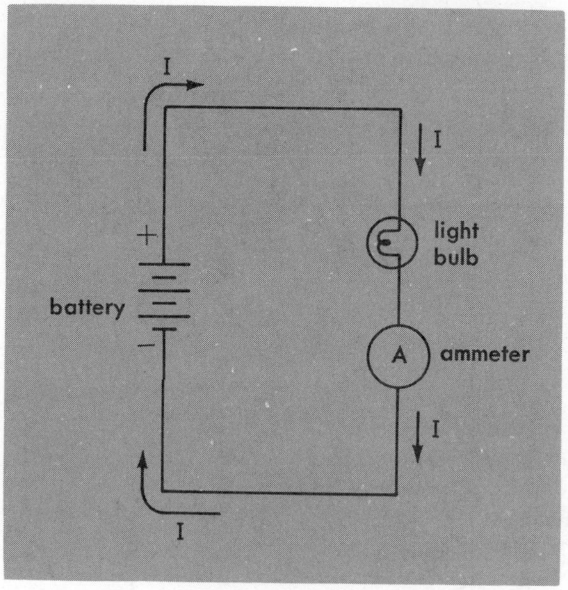

Figure 21–2
In an electric circuit, current flows all the way around a circuit—from the positive terminal of the battery, through the external circuit (light bulb and ammeter), and back through the negative battery terminal, through the battery to its positive terminal.

and it goes out through the wire to the negative battery terminal. In each of these wires the moving charged particles are electrons, and these negative particles move in the opposite direction from the electric current, causing an effective transport of positive charges in the opposite direction. Looking at the whole circuit, we see that electric current, I, is flowing from the positive to the negative battery terminal around the external circuit. Inside the battery, on the other hand, the current goes from the negative to the positive battery terminal. The battery moves more positive charges to the positively charged terminal and negative charges to the negative terminal despite the electric repulsions. In this process the battery uses chemical energy to force the charges onto the terminals.

The net rate of flow of charge which is the electric current is the same all around the circuit. Because charge is conserved (Section 18-8), the same number of elementary charges goes in one end of any part of a circuit as comes out the other.

A steady electric current cannot be measured by observing changes in charge; but we may imagine an observer who can count elementary charges passing any particular point in a circuit. If he watched the charges passing through the wire at a particular place, he would find I negative elementary charges, I electrons, passing per second.

21-2 Electrolytic Measurement of Electric Current: The Ampere

The observer who can see elementary charges passing is imaginary. But in certain circumstances we can make equivalent observations. For instance, we have already counted electrons in a weak beam from an electron gun.

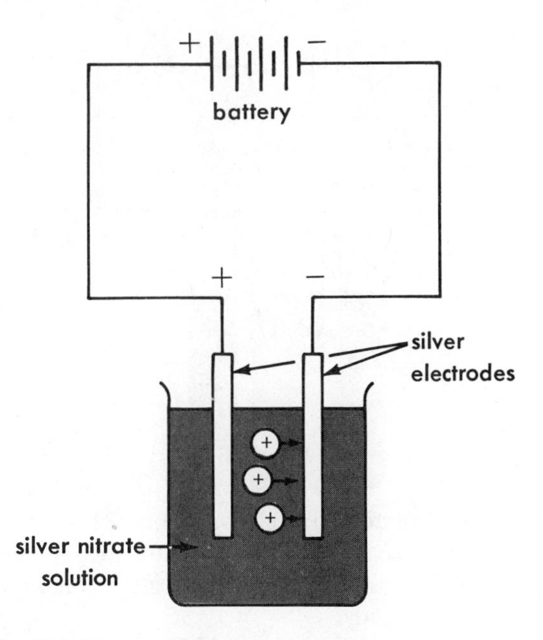

Figure 21–3
A silver plating cell. The EMF of the battery moves positive silver ions from the positive silver electrode to the negative silver electrode.

In this section we shall describe the measurement of the number of elementary charges by observing the amounts of mass transported by ions in a solution.

For this purpose we connect an electrolytic cell to a battery as shown in Fig. 21–3. The cell contains a solution of silver nitrate (the electrolyte) in which positive silver ions exist (and also negative ions), making the solution a conductor. The electrodes are made of silver.

Under the influence of the EMF of the battery, positive silver ions move through the solution—each ion picking up an electron at the negative electrode and depositing out on this electrode as a neutral silver atom. Therefore the negative electrode gains mass as silver atoms are steadily deposited. At the same time, silver atoms go into solution at the positive electrode, giving up one electron each, to become positive silver ions.

By weighing the electrodes before and after an electric current has passed through the cells, we can determine the mass of metal that has been deposited on the negative electrode. From this mass and from the mass of an atom of the metal we can compute the number of atoms deposited. (In a later chapter we will discuss how the mass of a silver ion can be measured.*)

The charge on each silver ion is one elementary charge; therefore the number of elementary charges passing through the cells is equal to the number of atoms deposited. The current, therefore, in elementary charges per second is equal to the number of silver atoms deposited per second.

An electrolytic cell of the type we have just described has been often used to calibrate current meters (ammeters). We used an ammeter to show the presence of a current in discussing thermionic emission in Section 20–1, but we did not make any quantitative measurement of electric current with it. To calibrate the ammeter, we connect it in *series* with a silver electrode and

* A silver atom has a mass very nearly equal to that of a silver ion, since the mass of an electron is only about 10^{-5} the mass of a silver atom.

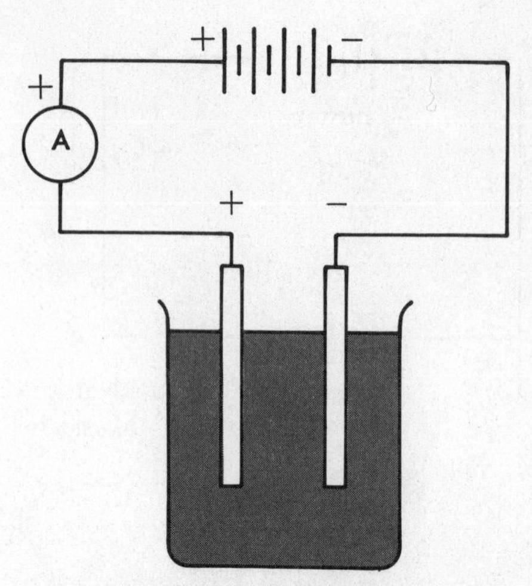

Figure 21–4
An ammeter connected in series with a silver plating cell. From the mass of silver deposited on the negative electrode and the time a steady current flowed, the current through the ammeter can be calculated.

an electrolytic cell as shown in Fig. 21–4. Thus the current through the ammeter is the same as the current through the electrolytic cell. We mark the position of the needle on the scale of the ammeter when the current in the circuit is steady.

The rate of flow of charge is measured by the number of silver atoms deposited per second in the cell, and we can put this number over the mark showing the position of the needle on the meter scale.

The common unit for measuring current is the *ampere*, which is defined as a flow of charge of one coulomb per second. To calibrate an ammeter in amperes, we simply calculate the current in elementary charges per second from the mass of silver deposited in a given time and then divide the result by the number of elementary charges in a coulomb $\left(6.25 \times 10^{18} \frac{\text{el. ch.}}{\text{coulomb}}\right)$.

When we deal with such circuit elements as light bulbs or electric irons, the ampere is a more convenient unit of current than the basic current unit of one elementary charge per second. The current that runs an ordinary light bulb is about an ampere. It would be clumsy to say that the lamp current is about 10^{19} elementary charges per second. Therefore, most ammeters are calibrated in amperes, that is, in units of 6.25×10^{18} elementary charges per second.*

21-3 Electrical Work and Power

As we have seen in Chapter 20, when a charge q is accelerated in a vacuum through a potential difference V it gains kinetic energy equal to qV. When electric charges (electrons) move in a solid metal conductor, however, they collide with metal ions in the conductor many millions of times a second. If a battery supplies a potential difference to a conductor (Fig. 21–5), the kinetic energy gained by the electrons between collisions due to the applied potential difference is quickly transferred, in successive collisions, to the metal ions, increasing the internal energy of the conductor. Thus the kinetic

* Ammeters used to measure very small currents have scales marked off in milliamperes (ma, 10^{-3} amp) or microamperes (μa, 10^{-6} amp).

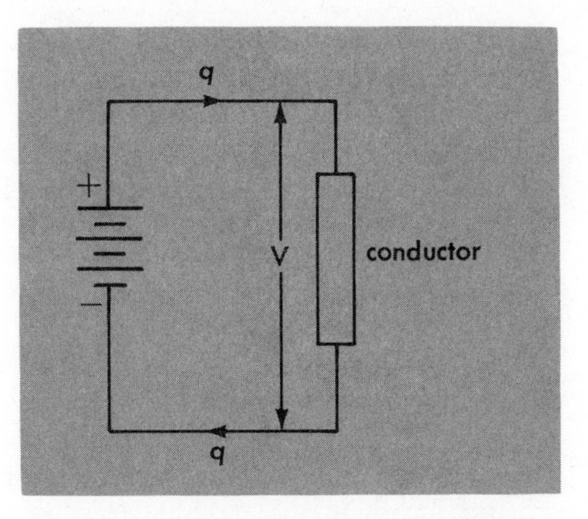

Figure 21–5
The energy supplied to a metal conductor is qV. It is transformed into an increase in the internal energy of the conductor.

energy qV gained by the electrons from the battery is converted to internal energy as fast as it is supplied by the battery, and the temperature of the conductor increases.

We can imagine an electric charge q passing through a circuit that may contain a variety of devices, such as a motor, an electric heater, a cell, and a bulb (Fig. 21–6). Nevertheless, if the voltage across this part of the circuit is V, then energy transferred to this part of the circuit is qV. This energy may take different forms, depending on the kind of device to which it is supplied. If, for example, a battery supplies energy to an electric motor that lifts a weight, then the energy qV measures both the energy transformed into internal energy in the wires of the motor and the energy that appears as gravitational potential energy $mg\Delta h$ gained by a weight of mass m lifted a vertical distance Δh by the motor.

By analogy with mechanical work, the product qV is called *electrical work*. It is electrical work that is measured by the meters electric power companies install in your house. You pay for the energy transferred to your house regardless of the way you use it.

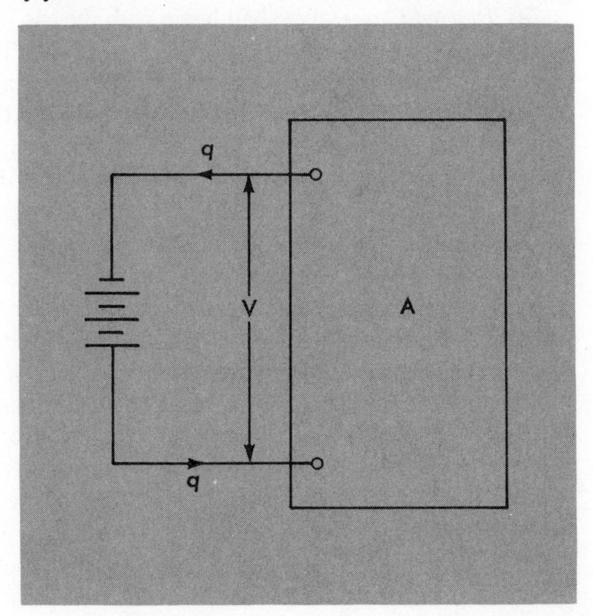

Figure 21–6
The energy supplied to a box containing a number of different electrical devices is qV regardless of what forms it may be transformed into inside the box.

In the laboratory, electrical work is not usually measured directly. Instead, it is computed, when necessary, from the readings of a voltmeter (the voltage applied to a circuit or part of a circuit), the readings of an ammeter (the current through the circuit), and the length of time the voltage is applied. The electrical work is then given by

$$\text{work} = qV = ItV.$$

If we divide both sides of this equation by the time, t, we have

$$\text{work per unit time} = IV.$$

The work per unit time, the rate at which energy is supplied to a circuit element, is called the *power* and is commonly expressed in joules/sec. A joule/sec has been given the name *watt*. Thus, electric power P is

$$P\ (\text{watts}) = I\ (\text{amp}) \times V\ (\text{volts}).$$

Motors, light bulbs, and other electrical appliances are usually designed to use energy at a particular rate. The power is usually specified, and so is the potential difference at which the appliances will use this power. From the power and the potential difference you can determine the current an electrical device will draw. For example, a 60-watt, 125-volt light bulb will draw a current of $60/125 = 0.48$ amp when an EMF of 125 volts is applied across its terminals. When power consumption is large, it is usually expressed in kilowatts or megawatts.

21-4 Current vs. Potential Difference

How do currents in ionized gases, the currents of electrons boiled out of metals, and the currents in metals depend on the potential differences maintained across these circuit elements? This is the question we shall now examine.

A. Ionized Gases. We shall begin with the case of ionized gases. The experimental arrangement is shown in Fig. 21–7. Two parallel metal plates and a sensitive ammeter are connected to a variable voltage supply. The plates are at a distance of several centimeters, and the space between is filled with the gas under investigation. When ionizing radiation (X rays or rays from a radioactive source) passes through the gas, the meter shows a current.

Figure 21–8 shows how the current depends on the potential difference between the metal plates in one typical experiment. The gas is argon at atmospheric pressure, and the radiation comes from a sample of radium. As the potential difference between the plates is increased, the current increases rapidly at first, then more slowly, and eventually reaches a certain limiting value called the saturation current.

The saturation current is reached when all the ions, as fast as they are formed, are swept out of the gas by the electric field and collect on the appropriate plates. For singly charged ions, such as those produced in argon, the ionization current (in elementary charges per second) equals the number of positive ions per second reaching the negative electrode (and this is the same as the number of negative ions—in this case electrons—reaching the positive electrode). Therefore the saturation current measures the total

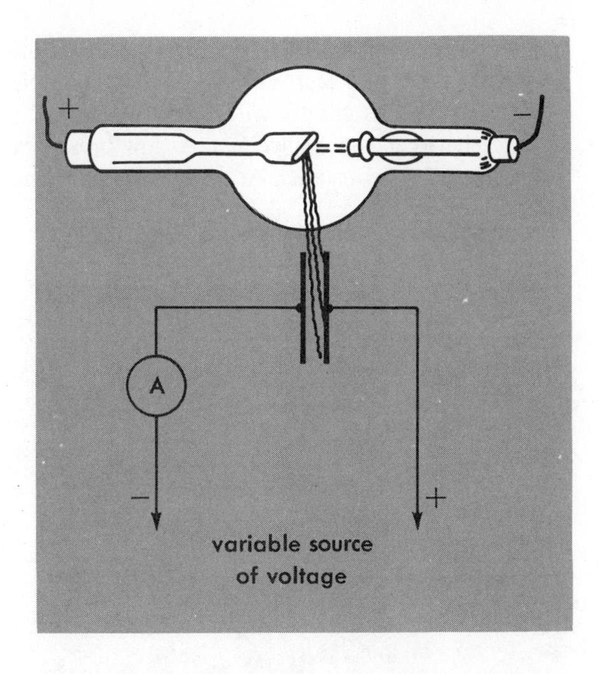

Figure 21–7
Apparatus to investigate electric currents in ionized gases.

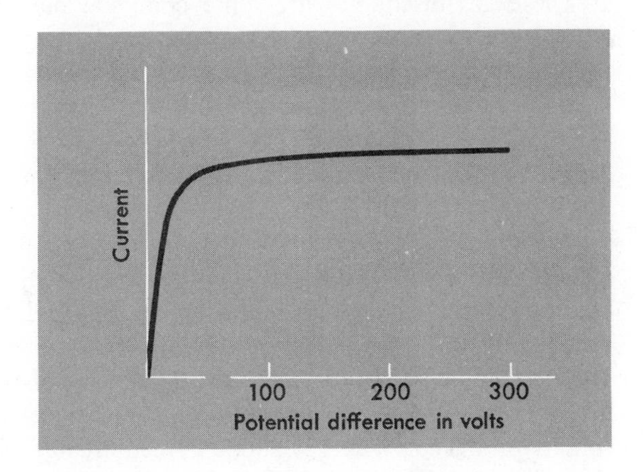

Figure 21–8
Current versus potential difference for argon gas at low voltages.

number of pairs of ions produced every second in the gas between the plates.

Why is the current less when there is a small difference in potential between the plates? The field moving the ions toward the plates is then weaker, and the ions move toward the plates more slowly. As they gradually get pushed toward the plates, they move in all directions because of their random thermal motions, and sometimes a positive and a negative ion will collide. Occasionally they will even recombine to form a neutral atom. When the systematic motion of the ions toward the plates is very slow, many positive and negative ions recombine instead of reaching the plates.

If we increase the potential difference between the plates far beyond the value necessary to establish saturation, we find that the current eventually begins to increase again. Then with a little more voltage it rises to very large values (Fig. 21–9). At these potential differences, the gas between the plates glows and there are often crackling noises. Just what we observe at this stage varies greatly with the pressure, the nature of the gas, and the distance between the electrodes. At low pressures the current and the glow are more or less constant. At atmospheric pressure, on the other hand, we have a series of sparks.

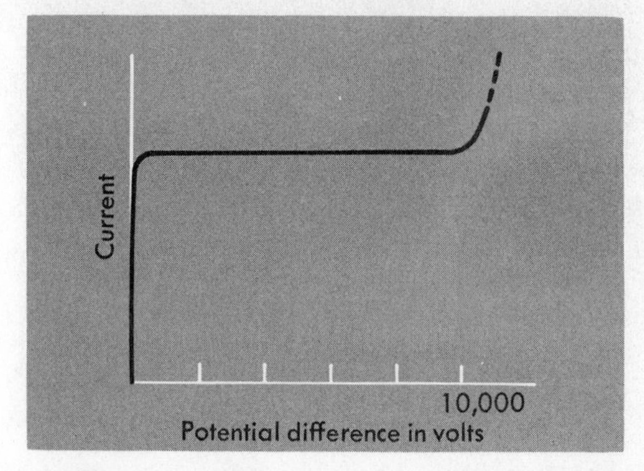

Figure 21–9
Current–potential difference curve for argon up to a high voltage.

These gas discharges are quite complex; but we can understand their fundamental cause. The electric field accelerates the positive and negative ions of an ionized gas in opposite directions. Each ion picks up energy until it collides with a gas molecule. On the average, in these collisions, the ions then lose the extra kinetic energy they have acquired since the previous collision; and this energy is transferred to the internal energy of the gas. So the gas warms up.

The situation, however, becomes different when the electric field is very strong. Then in the short time between collisions, some of the ions acquire sufficient kinetic energy to break up the gas molecules against which they collide. The new fragments are themselves ions which are accelerated by the electric field. They too become capable of disrupting the gas molecules which they encounter along their path. So an avalanche of more and more ions is produced. The gas suddenly acquires an enormous electrical conductivity, and a sudden discharge occurs. Geiger counters detect charged particles by using such an avalanche to turn the slight ionization along the track of a particle into an appreciable electric pulse.

B. Thermionic Emission. A somewhat similar relation between current and potential difference occurs for the electrons boiled out by a heated piece of metal in a vacuum (Fig. 21–10). When the potential difference between the heated filament and the surrounding collecting plate is big, we find a saturation current. Raising the potential difference does not change the number of electrons that cross from filament to plate, because all the electrons boiled out of the filament per second at a given temperature are already being pulled over to the plate. When the potential difference is low enough, however, the current drops beneath the saturation value. Apparently the electrons emitted from the filament do not move away sufficiently fast; and while they are in the neighborhood of the filament they push back electrons which would otherwise boil out.

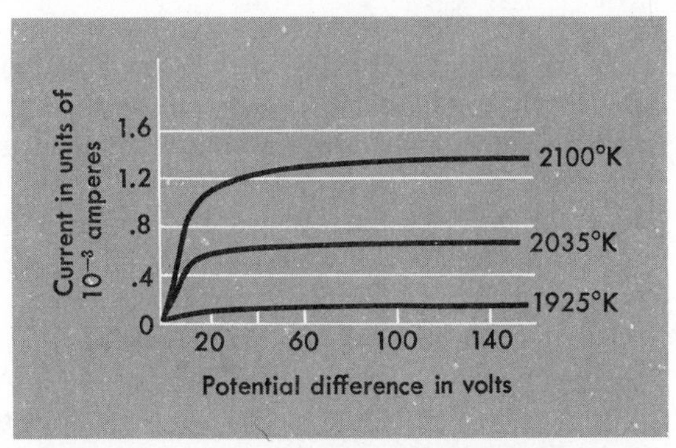

Figure 21–10
Current versus potential difference curves for electrons boiled out of a heated wire at different filament temperatures.

If the potential difference between filament and plate is reduced to zero, we observe something else of interest. A small current still exists. Even when the plate is made slightly negative with respect to the filament, a few electrons continue to get across. Apparently some electrons are boiled out of the filament with sufficient kinetic energy to go across despite a slight retarding potential difference.

C. Electrons in Metal Conductors: Ohm's Law. The relation between potential difference and current in a metal conductor is a particularly simple one: the two quantities are proportional to each other. That is,

$$V = RI.$$

The law expressed by this equation is called Ohm's law, and the proportionality constant R is called the *electric resistance* of the conductor. If V is measured in volts and I in amperes, R is measured in a unit called an *ohm*. Ohm's law says that the current in a metal wire is proportional to the potential difference between one end and the other.

Whenever one needs to introduce a particular resistance into an electric circuit to limit a current or produce a desired potential difference, one uses a circuit element called a *resistor*, a device that obeys Ohm's law.

We can understand Ohm's law by returning for a moment to the model of the conductivity of metals, in which we picture electrons running freely around inside the metal like the molecules of a gas. Because the electrons are so light, at normal temperatures they have tremendous speeds—about 10^5 meters per second. They rush around inside the metal, colliding with the positive metal ions with great frequency. The average time between collisions of a single electron with one or another of the metal ions is extremely short, only about 3×10^{-15} second. It is determined by the distance between metal ions and the average speed of the electrons. Furthermore, at each collision the electron bounces off in a new direction. So a single electron runs in all directions in its random thermal dance.

Now we add the potential difference to our picture. In the short time between collisions, it accelerates the electrons along the wire. The small extra velocity which the electrons acquire along the wire is just proportional to the potential difference and to the time between collisions. Because of the random changes in direction at each collision, or at worst after every few collisions, the extra velocity along the wire is wiped out and has to be reestablished by the action of the potential difference. Our complete picture then looks like this: the electrons rush around in all directions at huge speeds, but superposed on top of these random motions is a small systematic velocity along the wire. This systematic velocity is proportional to the potential difference and to the time between collisions. With a potential difference of a few volts across a meter of wire, the systematic velocity is about a centimeter a second. It is this systematic velocity that results in a current. The bigger the systematic velocity the bigger the current; so the current is proportional to the potential difference. Our model therefore predicts the experimental result embodied in Ohm's law.

The model also makes another prediction about the electrical resistance. It should increase with increasing temperature. As the temperature goes up, so does the speed of the electrons. Therefore the time between collisions diminishes; and the same potential difference yields less current. This conclusion agrees with the experimental observation that the resistance usually does become greater. This agreement is not perfect, however; detailed experiments show that the resistance usually increases over a wide range of temperatures approximately as a linear function of the Kelvin temperature. An example is shown in the graph of Fig. 21–11. However, the model predicts that the resistance should be proportional to only its square root. The discrepancy is explained by two facts. In the first place, because of its high density, we should not expect the electron gas to behave like an ideal gas. The electrons are so close together that the forces between them always influence their motion. Secondly, when gas of identical particles is dense, quantum effects, which we shall study later in this book, become important. When these effects are included, the improved model is in accurate correspondence with the experimental facts. Our model remains useful as one of the essential steps to a more complete description.

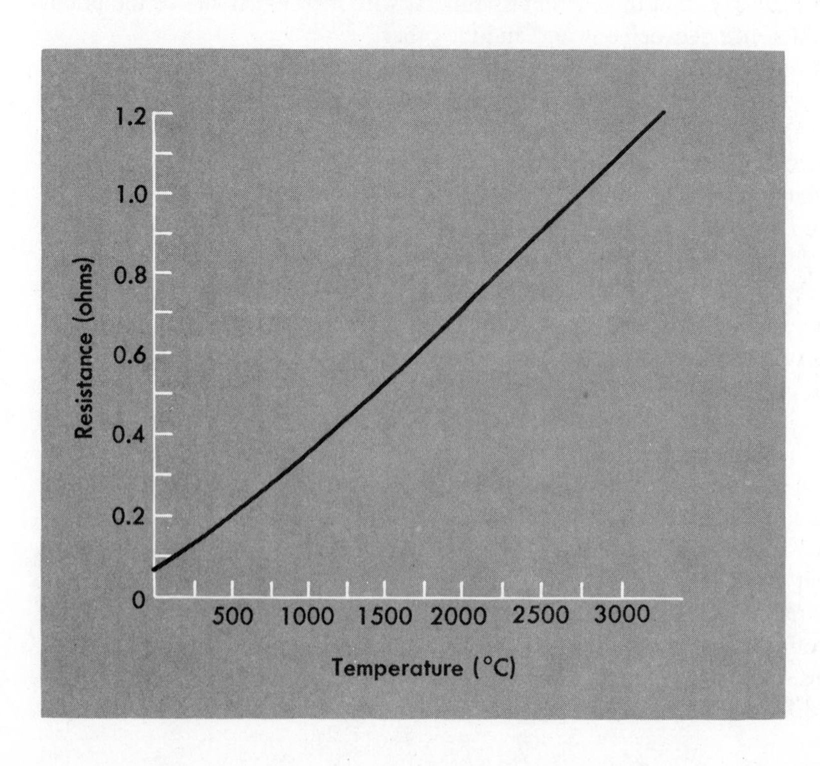

Figure 21–11
The resistance of a length of tungsten wire as a function of its temperature.

In Section 21–4 we have examined the relation between the currents that occur and the potential differences for a few circuit elements. We can now apply what we learned there to more complicated circuits containing more than one circuit element.

Suppose that we have two circuit elements for which the curves of current vs. potential difference are known, for example, the circuit elements represented by the graphs of Fig. 21–12. If these two elements are connected in parallel as in Fig. 21–13, the potential difference V across their ends from A to B is automatically the same for both elements. The current I entering the junction at A and leaving at B divides at A and part of it passes through

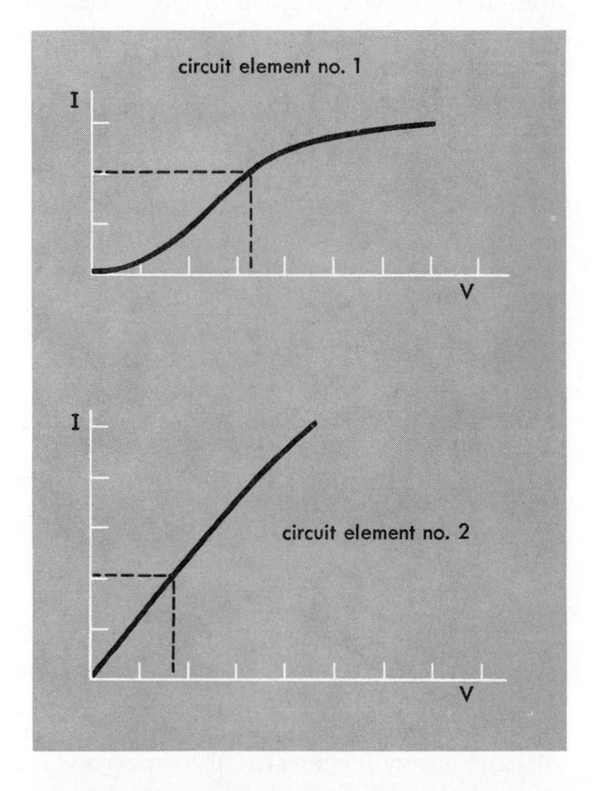

Figure 21–12
I-V curves for two circuit elements.

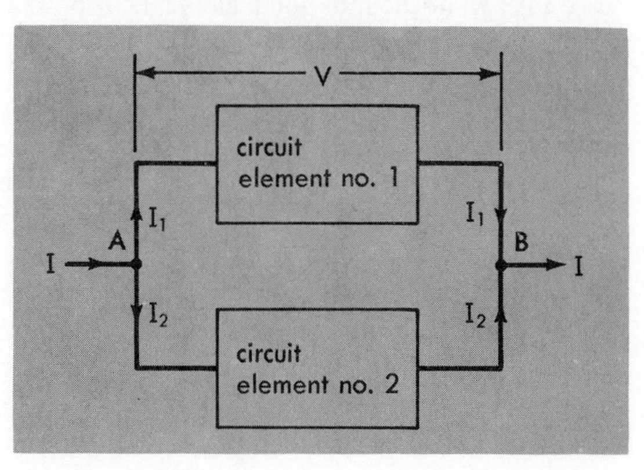

Figure 21–13
Two circuit elements connected in parallel.

each circuit element. The total current entering at A and leaving at B must be the sum of the currents through the two elements, since charge must be conserved (Section 18–8). The total current I can be found by adding the current I_1 that flows through element 1 at the particular value for the potential difference and the current I_2 that flows through element 2 at the same value of the potential difference. By adding up the two currents determined from Fig. 21–12 at each value of potential difference, we can make a graph of the total current $I = I_1 = I_2$ passing from A to B versus the potential difference V (Fig. 21–14). In this way we combine the two elements in parallel into a new single element represented by the new I versus V graph. This procedure can be extended to handle more circuit elements in parallel by adding the individual currents at each potential difference. Thus, we can always replace a group of circuit elements connected in parallel by an equivalent circuit element that acts in the same way as the whole parallel combination.

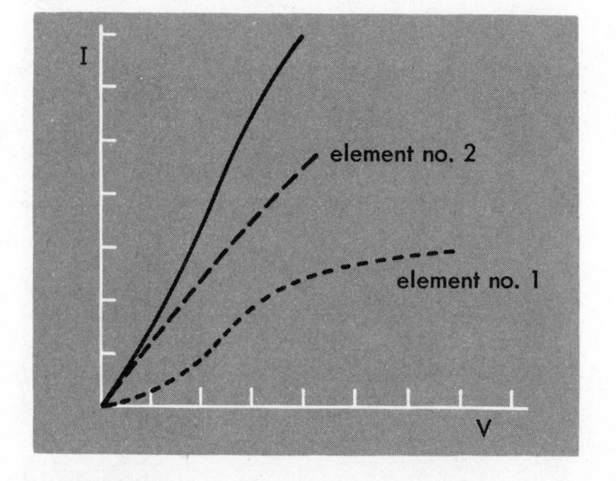

Figure 21–14
I-V curve for the equivalent circuit element made by placing the circuit elements of Fig. 21–12 in parallel. It is obtained by adding the currents of the two individual circuit elements when the potential difference has the same value for both.

For any potential difference V across the parallel combination the energy per elementary charge is still given by the potential difference. Consequently, the power used in the parallel combination is IV, where I is the total current —the current that flows through the combination viewed as a single element. Because energy is conserved, the power used must be the sum of the power used in each of the individual elements. For example, if there are two circuit elements in parallel, then:

$$IV = I_1V + I_2V.$$

If we connect individual circuit elements in series as in Fig. 21–15, the analysis is somewhat different. Since charge is conserved, the same current must flow through each of the individual elements. Let us start by assuming that this current has a particular value. By using the graphs of current vs. potential difference for each individual circuit element, we can read off the corresponding potential differences from Fig. 21–12. Then to find the potential difference from one end of the series circuit to the other, we must add the individual potential drops. By following this scheme for each possible current, we can establish the relation between the current and the poten-

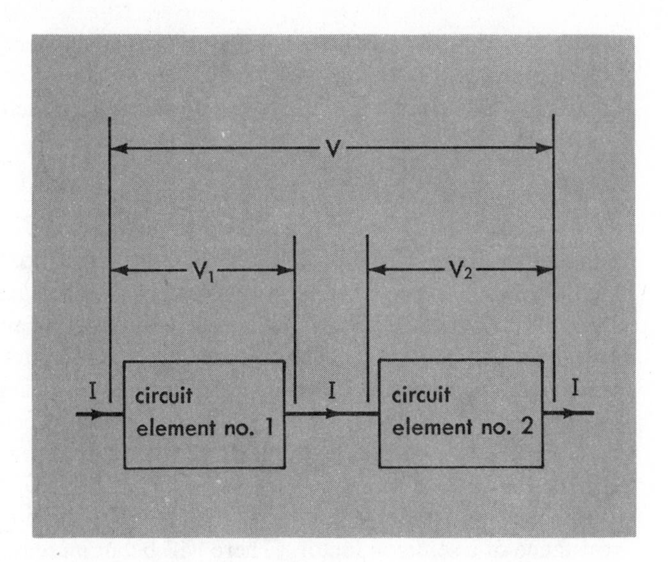

Figure 21–15
Two circuit elements connected in series.

tial difference across the whole series circuit. We can again graph it as a curve of current vs. potential difference (Fig. 21–16). In this way we can turn the series circuit into an equivalent single circuit element represented by this new curve of current vs. potential difference.*

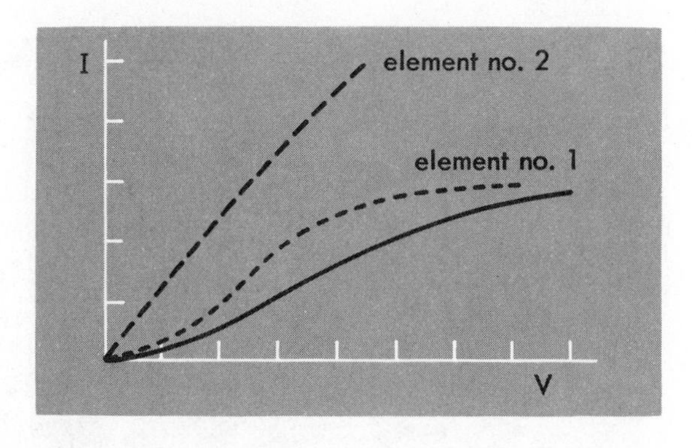

Figure 21–16
I-V curve for the equivalent circuit element when the circuit elements of Fig. 21–12 are connected in series. It is obtained by adding the potential differences across the two circuit elements when the current has the same value through both. You can visualize this addition by looking at the picture from the left. There, if you pick a given current, the potential difference needed to produce it in circuit element 1 is given by the distance along the *V* axis to the dotted curve. The potential difference needed to produce this current in circuit element 2 is given by the distance along the *V* axis to the dashed curve. The sum of these two potentials is the potential difference needed across the circuit elements in series to produce the same current.

Now that we have found the behavior of this equivalent single circuit element, we can find the current that will flow through the series circuit when any specified potential difference is applied across its ends. As always, this potential difference represents the energy used per elementary charge passing through the circuit. Again the power consumed is IV, where V is the potential difference from one end to the other of the whole series circuit. In this case, however, V is the sum of the individual potential drops across the individual elements forming the series circuit. If the circuit is made of two elements, that sum is $V = V_1 + V_2$. Consequently,

$$IV = IV_1 + IV_2.$$

* In nearly all circuit diagrams, connecting wires are assumed to have negligible resistance (and therefore negligible potential difference between their ends). Therefore, we can neglect the minute potential difference caused by the wire connecting the circuit elements in Fig. 21–13 and 21–15.

In this equation the power consumption of each element is represented by a term on the right-hand side; and because energy is conserved, the total power consumption IV is just the sum of the power used in each individual element.

21-6 Semiconductors

Suppose we have rods of copper, the element germanium (Ge), and glass of the same dimensions; and apply equal potential differences across their ends. If the potential difference results in a current of several amperes in the copper rod, there will be a very small current in the germanium rod and practically no detectable current in the glass rod. The copper is a good conductor; the glass is a good insulator. In its ability to conduct electric charge, the germanium lies between the extremes exhibited by copper and glass; it is said to be a semiconductor.

In contrast to metallic conduction, an increase in temperature reduces the resistance of a semiconductor. There will be an increase in the resistance to the motion of electrons as the temperature increases in semiconductors as there is in metals, but this effect is more than offset by an increase in the number of charge carriers brought about by the increased thermal agitation of the atoms of the semiconductor. An example is shown in the graph of Fig. 21-17.

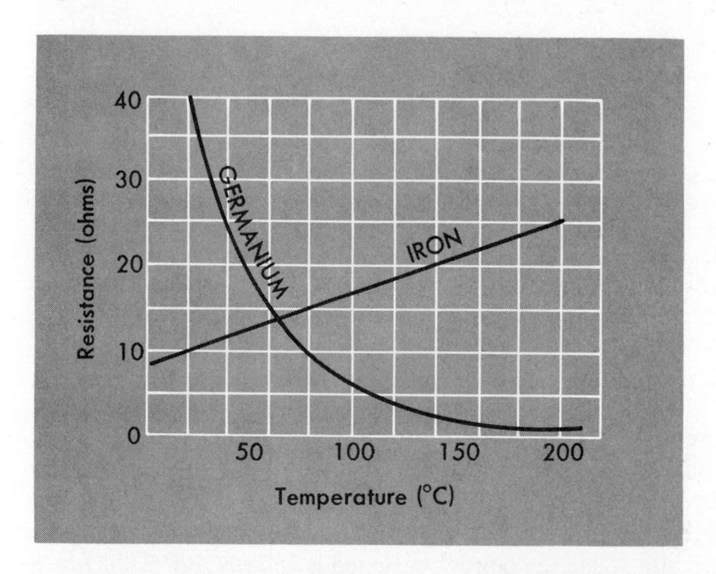

Figure 21–17
The resistance of a semiconductor (germanium) *decreases* very rapidly (exponentially) as its temperature increases. On the other hand, a conductor's resistance (iron) *increases* slowly (linearly) as the temperature rises from 0°C to 200°C.

Because of this characteristic, precautions must be taken in circuits using solid-state devices to prevent their temperatures from rising too high, especially when the current through the device is affected considerably by the resistance of the device itself. With increased temperature and reduced resistance, the current increases; the increase in current converts more electric energy into internal energy, with a further rise in temperature and drop in resistance, and so on. Without a resistor connected in series with a semiconductor, current would rise high enough to damage the device permanently.

It is common practice to give protection to some solid-state devices by attaching them with good thermal contact to a mass of metal, called a "heat sink," of a size appropriate to the power the device must dissipate. The device shares its increase in internal energy with the whole mass of metal; the latter through its relatively large surface area rids itself readily of the internal energy coming from the device, and consequently the temperature rise of the device does not increase to a harmful level.

It is also possible to change the number of electrons available for conduction by greatly purifying the germanium and adding minute quantities of some other particular substance (the process is called doping). For example, doping germanium with either the element antimony (Sb) or with the element indium (In) reduces the resistance of the substance, although it is still far larger than the resistance of true conductors.

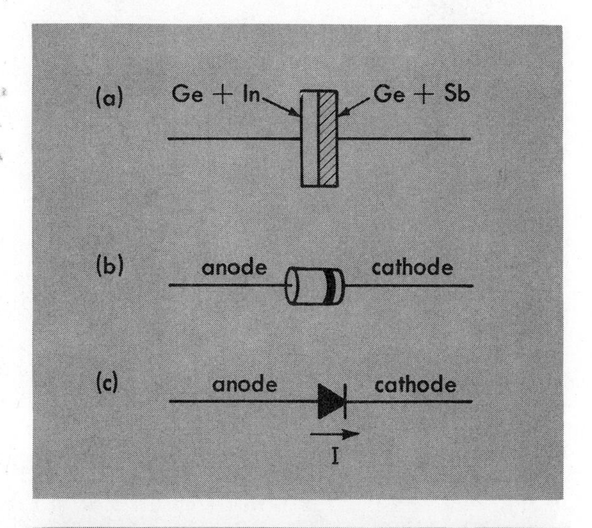

Figure 21–18
A solid-state diode (a) consisting of two doped pieces of germanium (Ge) in close contact. A common form of solid-state diode is shown in (b). The black band marks the end of the diode from which the cathode connection protrudes. (c) The schematic symbol for a diode as it appears in circuit diagrams.

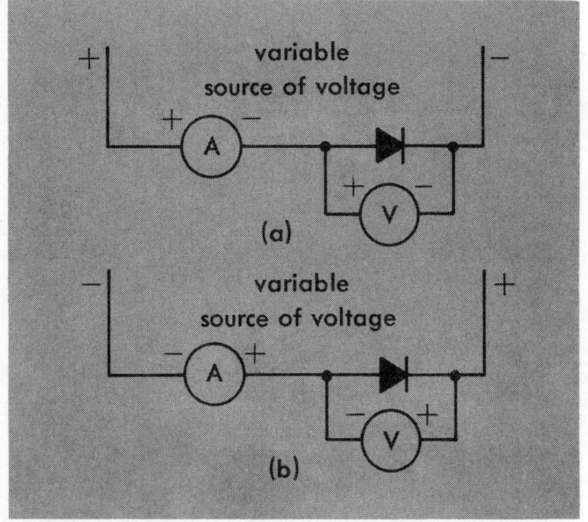

Figure 21–19
The circuit in (a) can be used to investigate the *I-V* characteristics of a diode in the "forward" direction. The circuit in (b) can be used when a variable voltage is applied in the "reverse" direction.

We can make a device called a solid-state diode in which a piece of germanium doped with antimony and a piece of germanium doped with indium are in close contact (Fig. 21–18). If we apply various potential differences to the device and measure the corresponding currents, using the circuits in Fig. 21–19, we obtain data similar to those plotted in the graph shown in Fig. 21–20.

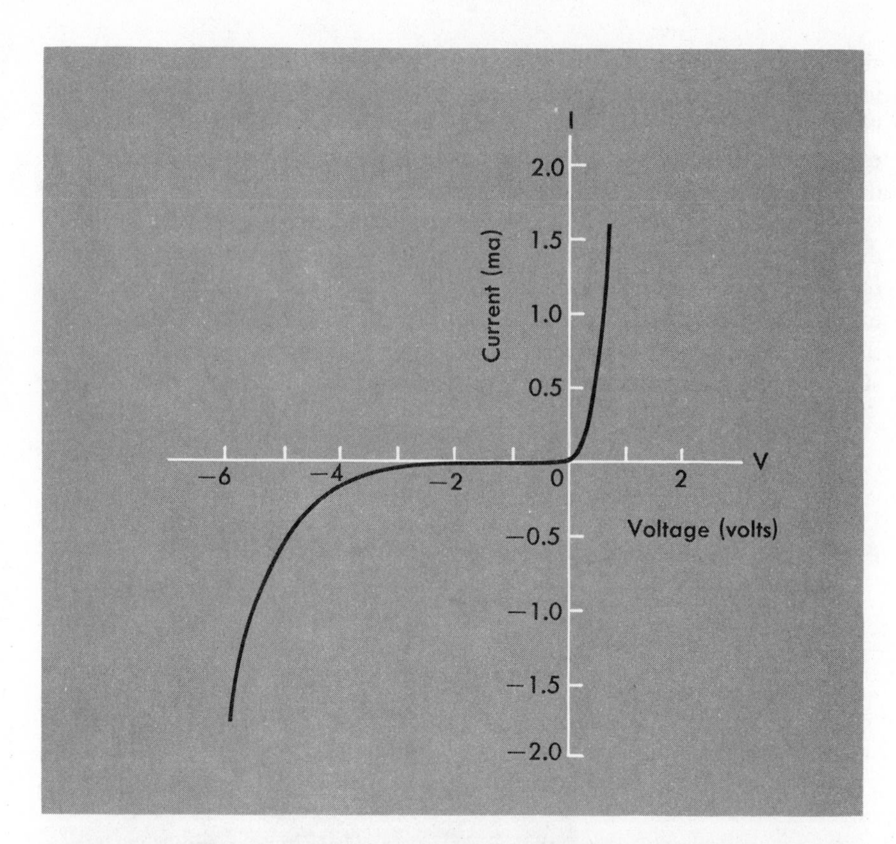

Figure 21–20
The I-V characteristic of a typical solid-state diode. Notice that it hardly conducts at all between −2 volts and 0 volts compared to its conductivity in the forward direction. At +0.5 volts the forward current is about 1.2 milliamperes.

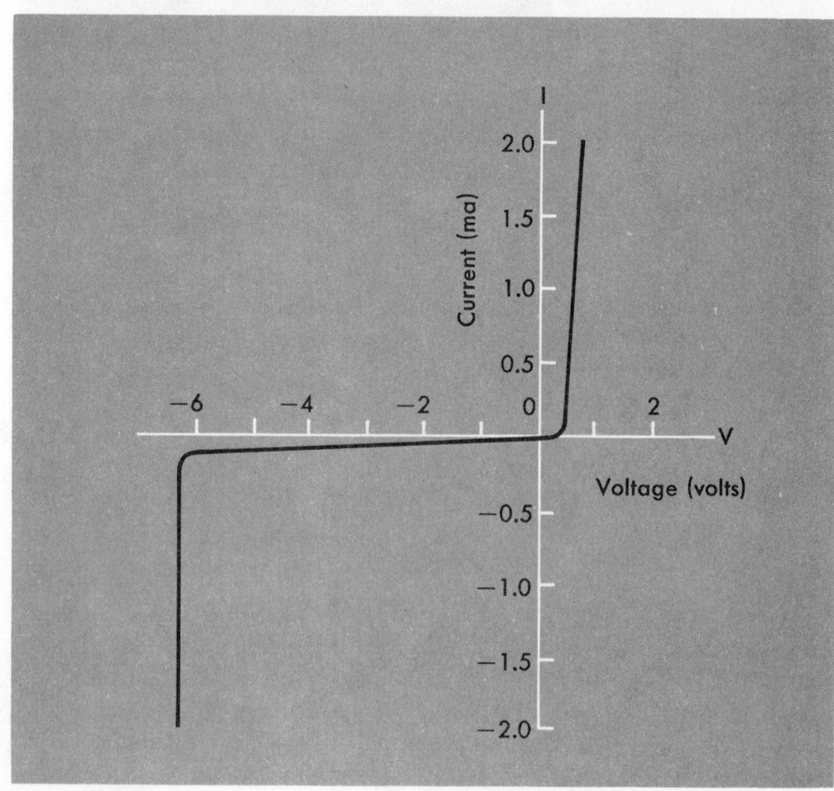

Figure 21–21
The I-V characteristics of a Zener diode. Unlike the ordinary diode in Fig. 21–20, at about −6.3 volts it begins to conduct in the reverse direction very suddenly, the voltage remaining very nearly constant over a wide range of currents.

Evidently this device passes current easily in one direction and with great difficulty in the opposite direction. It is essentially a one-way path for electric charge. When the diode is conducting readily, the potential difference is said to be a *forward* potential difference. When the diode is nonconducting or nearly so, the applied potential difference is said to be a *reverse* potential difference. (Such diodes have many uses in electronic circuits.)

The side of the diode connected to the positive terminal of the battery when the diode is conducting readily is called the anode of the diode; the other side is called the cathode. Commercial solid-state diodes usually have the cathode terminal marked with a band [Fig. 21–18(b)].

21-7 Zener Diodes

There is a special type of solid-state diode, called a Zener diode, that has a somewhat different characteristic graph of current versus voltage than that of Fig. 21–20; it is shown in Fig. 21–21.

In the graph shown there is a sharp break in the curve when the reverse potential difference is −6.3 volts; the diode conducts very little at reverse voltages between 0 and −6.3 volts, but begins to conduct abruptly at −6.3 volts; and this diode is, therefore, called a "6.3-volt Zener diode." The voltage remains near −6.3 volts over a wide range of currents. Zener diodes can be made with the break occurring at any voltage, from a few volts to several hundred, and are constructed so that they can safely dissipate wide ranges of power in the reverse direction from a fraction of a watt to 50 or more. They too find extensive use in electronic circuits, especially in maintaining constant potential differences.

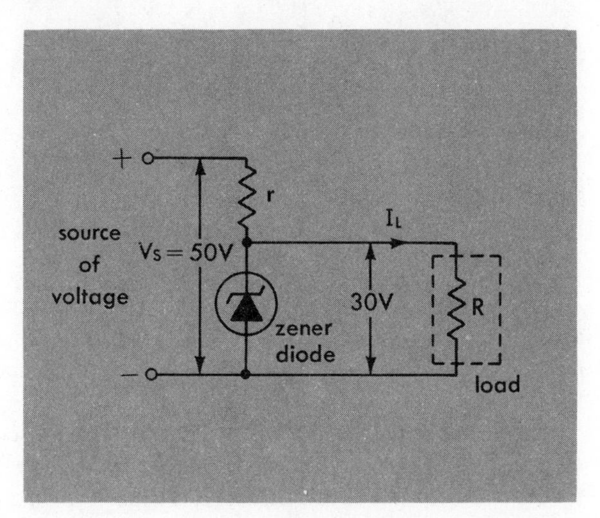

Figure 21–22
A 30-volt Zener diode used in a voltage regulator circuit. The 30-volt output to the load remains very close to 30 volts in spite of significant changes in either source voltage E_S or load current I_L. Note that the Zener diode is connected in the "reverse" direction, so that wide variations in the current through it result in only a very small change in the potential difference across it, as shown in Fig. 21–21.

An example of the latter use is shown in Fig. 21–22, where it is desired to keep the potential difference across the "load" resistor R constant at 30 volts. Suppose the potential difference of the source V_S rises slightly. That across the load would tend to rise also, but the Zener diode draws a little more current, the potential drop across the resistor r increases and the potential difference across the load resistor is maintained close to the

Zener voltage break. Or suppose that the load current I_L increases slightly. This would tend to draw more current through the resistor r and increase the potential drop across it; this would reduce the potential difference across the load. But the Zener diode would then draw less current, so that the total current through R would not change appreciably, and again the potential difference across the load remains very nearly constant at the Zener voltage.

21-8 A Simple Sweep Circuit

In Section 20–7, Oscilloscopes, you learned that the beam from an electron gun is deflected (swept) horizontally by a potential difference applied to a pair of parallel, vertical plates. If the beam is to correctly display a curve showing how a potential difference applied to the horizontal plates that deflect the beam vertically varies with time, the sweep potential difference as a function of time applied to the vertical plates must look like the graph in Fig. 21–23. The horizontal deflection potential difference must vary linearly with time, and when the beam reaches the end of its horizontal sweep it must return to its starting point in a time interval very short compared to the time required for the beam to make its linear horizontal sweep.

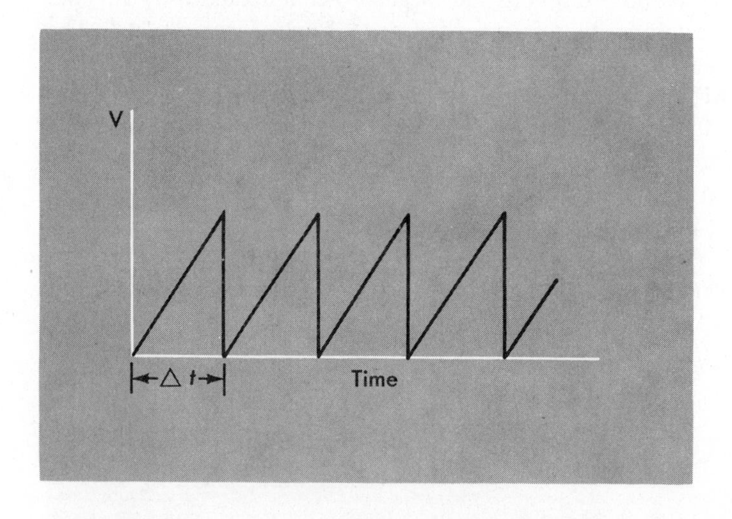

Figure 21–23
A "sawtooth" wave. The potential difference increases linearly with time over the time interval Δt and then drops to zero. The process is repeated at intervals Δt. If this voltage is applied to the vertical parallel plates in an oscilloscope that sweep the electron gun beam horizontally across the tube face, the beam spot will move at constant speed until it almost instantaneously jumps back to the start of its sweep.

The sweep circuit we shall describe in this section, using a neon bulb, is a simplified circuit. Although for a number of reasons it is not used in commercial oscilloscopes, it does have a potential difference as a function of time that is close to that in Fig. 21–23.

The sweep circuit is shown in Fig. 21–24. It consists of three circuit elements: a high-resistance resistor R, a neon bulb N containing two electrodes and neon gas at low pressure, a capacitor C, and a switch.

The capacitor is basically a pair of parallel plates of large area, very close together and separated by a thin sheet of insulating plastic. Actually the "plates" are often two long strips of aluminum foil on each side of the plastic separator. Another insulating sheet is placed on top, and this whole "sandwich" is then rolled up into a cylinder and encased in plastic to keep out moisture. Two wires, each connected to one of the "plates," project from each end.

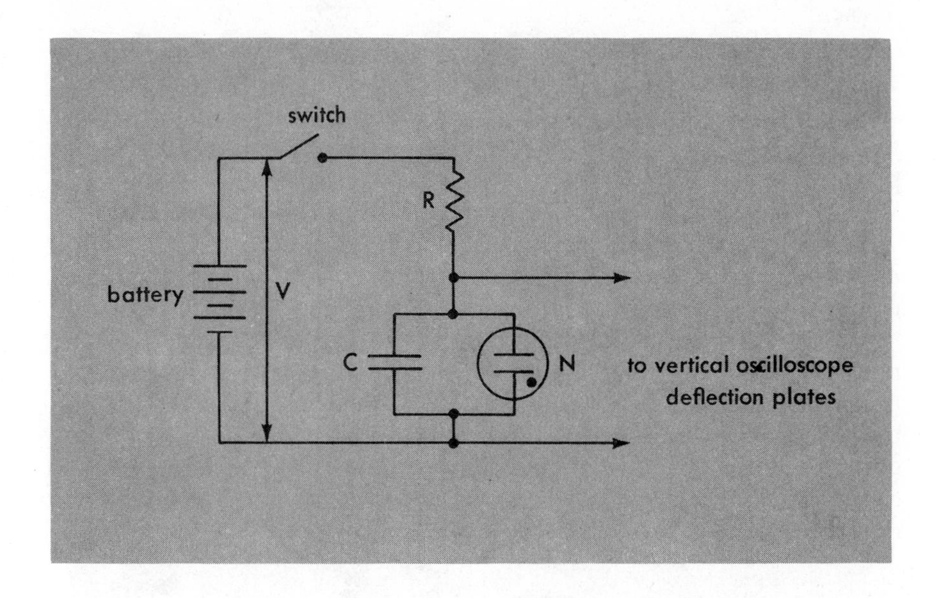

Figure 21–24
A simplified circuit using a neon bulb to generate a sawtooth wave.

Figure 21–25 shows a graph of the potential differences vs. current for a small neon bulb. As the potential difference across the bulb is increased from zero, no current flows through the neon gas until the potential difference reaches the "breakdown" voltage V_B, at which point ions are formed in the gas, which in turn start an avalanche of ions, and the current rises very rapidly in a small fraction of a second to the large value I_P. At this point the bulb lights up, giving off the bright orange-red glow characteristic of neon gas when it is conducting electric charge. The neon bulb has suddenly changed from a good insulator to a conductor.

If the voltage is now reduced, the current rapidly falls until the potential difference V_A is reached. At this point ions stop forming. The avalanche ceases, the orange glow is extinguished, and the neon bulb suddenly ceases to conduct current. It has again become an insulator. Let us now see how the current-voltage characteristics of a neon bulb affect the flow of current and the potential differences in the sweep circuit in Fig. 21–24.

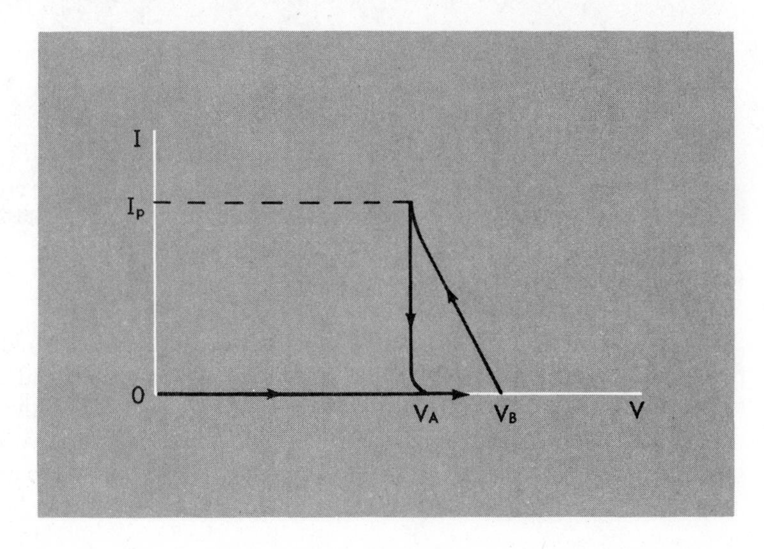

Figure 21–25
The current-voltage characteristic of a neon bulb. As the potential difference across the bulb is increased from zero, there is no current through the bulb until the voltage V_B is reached. At this point conduction begins, the bulb lights up, and the current rises very rapidly to the value I_P. During this rapid rise the potential difference across the bulb drops. If the voltage across the bulb is now decreased, the current falls until the voltage reaches the value V_A, at which point the bulb extinguishes and the conduction ceases.

When the switch is closed, the capacitor C begins to charge, the rate at which it begins to charge being controlled by the battery EMF and the resistance of R. As the charge on the capacitor "plates" rises, so does the potential difference across them and across the neon bulb which is connected in parallel with the capacitor.

When the potential difference across the capacitor and bulb reaches the value V_B shown in Fig. 21–25, the neon bulb suddenly becomes a good conductor whose resistance is much less than that of R. As a result, the capacitor rapidly discharges much faster than it was charged by the current flowing through R. The potential difference across the capacitor and bulb falls very rapidly until it reaches V_E. The neon bulb now extinguishes and stops conducting; the capacitor is slowly charged again by the current through R until the voltage across the neon bulb again reaches V_B and the bulb starts conducting and again discharges the capacitor. The resulting potential difference across the neon bulb and capacitor as it continues to rise and then fall rapidly is shown as a function of time in Fig. 21–26.

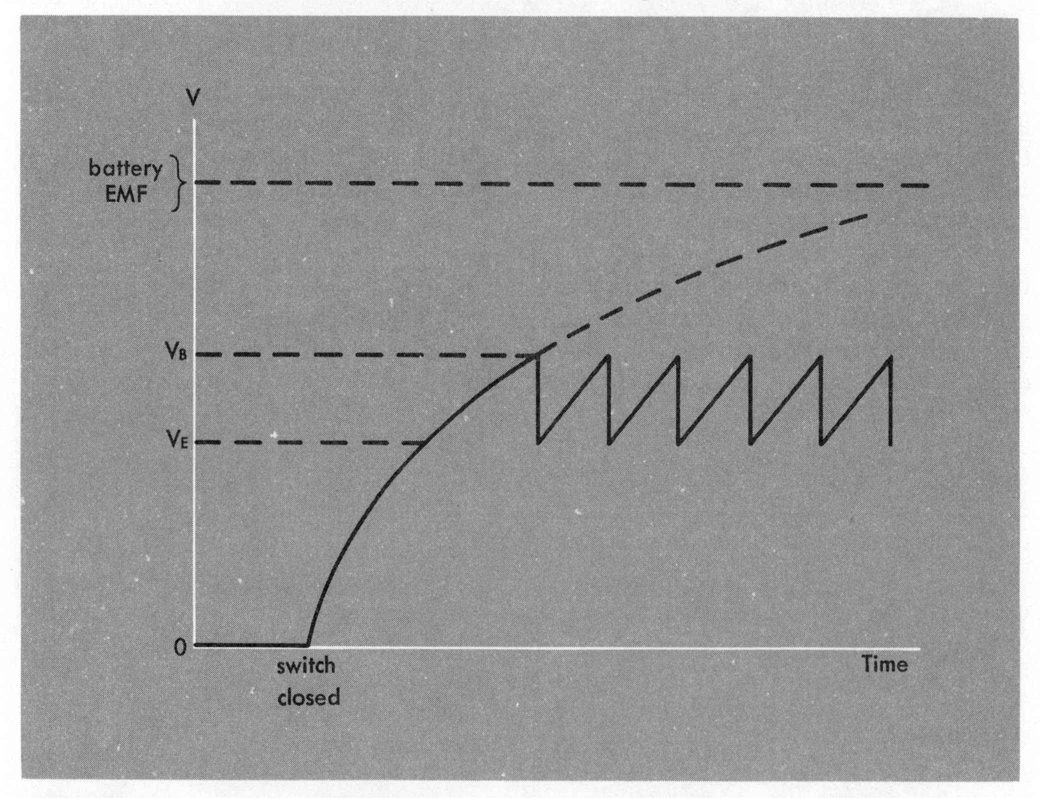

Figure 21–26
A nearly linear sawtooth voltage is generated by the circuit shown in Fig. 21–24.

If there were no neon bulb in the circuit when the switch is closed, the potential difference across the capacitor would rise as shown and continue to rise above V_B along the dashed line, approaching the EMF of the battery.

As you can see from Fig. 21–26, the voltage-time curve is very nearly linear between V_E and V_B. If this nearly linear "sawtooth" wave is applied to the vertical plates of an oscilloscope, it will sweep the electron beam horizontally across the tube face at a nearly constant speed, returning the beam almost instantly to the start of its sweep. The beam will keep repeating this linear sweep at a frequency that depends on the values of the battery EMF, R, and the construction of C.

As we said at the beginning of this section, neon bulbs are not used in modern oscilloscopes to generate sweep voltages. One reason for this is that the high sweep frequencies required to show how potential differences across the horizontal plates change in microsecond intervals (Section 20–8) can not be generated by a neon bulb. It takes too long for the neon gas in the bulb to deionize and, hence, for the neon to stop conducting. Modern sweep circuits, employing semiconductor circuit elements, can easily generate very linear microsecond sweeps.

21-9 Internal Resistance of Batteries and Power Supplies

A voltmeter connected across a new flashlight cell reads close to 1.50 volts; this is the EMF of the cell. If the meter is connected across the cell and the cell is then connected to a circuit so as to supply current to it, the voltmeter reading drops. The potential difference across the terminals of the battery is no longer the same as the EMF of the cell.

This tells us that the energy needed to transport a unit charge around the circuit external to the cell is less than the energy the cell supplies to the unit charge. What brings about this difference? The only place where we can look for the "missing" energy is within the cell itself. The most likely sign of an energy transfer within the cell would be an increase in temperature, for this would indicate an increase in internal energy of the contents of the cell. If the current in the circuit is large enough, the rise in temperature of a flashlight cell becomes evident to the touch. Careful experiments provide a way of measuring even small increases in temperature; and with a knowledge of the increase in internal energy per degree increase in temperature of the cell, one can calculate how much energy this rise in temperature represents. Taking this energy into account, it turns out that the EMF of the battery does indeed equal the sum of the energy per unit charge transferred to the external circuit and that changed to internal energy inside the cell.

By analogy with the energy needed to move charge through a resistor, we can refer to the internal resistance of the cell. Every electric cell has an internal resistance, so in every cell, part of the energy available per unit charge is always converted into internal energy within the cell itself whenever the cell produces a current in an external circuit. Cells can be designed to have a very low internal resistance, but it cannot be completely eliminated.

An internal resistance is found not only in electric cells but in every other source of EMF. For example, an electric generator has wires inside it along which charge must flow, and consequently it has an internal resistance also. Every practical source of EMF must be designed with the effect of the internal resistance in mind.

We can think of a cell that supplies EMF as being equivalent to a source

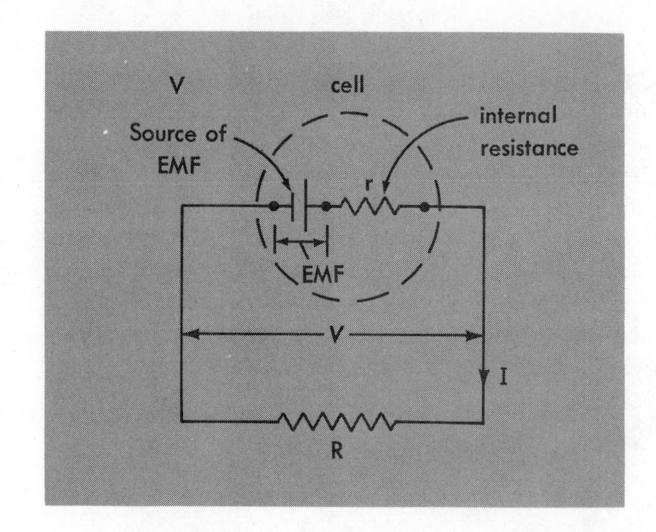

Figure 21–27
The equivalent circuit of a cell is shown inside the dashed circle. The current in the external circuit (through R) is $I = \frac{\text{EMF}}{r + R}$. The voltage supplied to the external circuit, called the terminal voltage (the voltage across the terminals of the cell), is $V = \text{EMF} - Ir$.

of EMF in series with the internal resistance (Fig. 21–27). This combination then is in series with the total resistance R of the external circuit.

The current in this circuit is given by the expression

$$I = \frac{\text{EMF}}{r + R}.$$

Evidently, if $R \gg r$, the effect of the internal resistance can be neglected; but if one attempts to supply current to a circuit in which $R \ll r$, the magnitude of the current is determined by the internal resistance of the cell rather than by that of the external circuit.

21-10 Measurement of the EMF of a Battery

If we wish to measure the EMF of a battery with high accuracy, we do not use a voltmeter. Although a good voltmeter will not disturb the voltage being measured very much, some disturbance will always take place. A voltmeter draws a small current, but any current at all drawn from the battery will reduce the potential difference across its terminals because of its internal resistance. For an accurate determination of the EMF of a battery, no current should be drawn from it.

In order to measure the true EMF of a battery, we can use the circuit shown in Fig. 21–28. The battery B produces a current in a length of uniform resistance wire and in the variable resistance R in series with the wire when the switch S_1 is closed. When the resistance R is varied, the current in the circuit varies and therefore the potential difference across the whole length of the resistance wire varies. Since the wire is uniform, equal segments of it have equal potential differences across them; or, to say it differently, different segments of the resistance wire have potential differences across them proportional to their lengths.

When the double pole switch S_2 is closed, a standard cell B_s, whose EMF is known to high precision, is connected through a sensitive ammeter (whose zero position is in the center of the scale) and to a sliding contact on the wire. If the potential difference across the resistance wire of length l_s is not equal to the EMF of B_s, there will be a current one way or the other through the

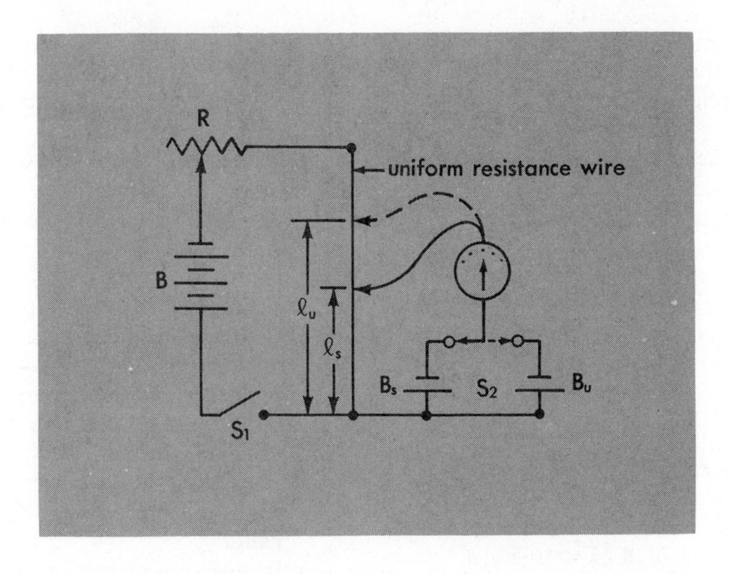

Figure 21–28
A circuit for accurately comparing the unknown EMF of the cell B_u with that of a standard cell B_s whose EMF is known very accurately.

meter; the sliding contact is moved until the meter indicates zero current, and the length l_s is recorded. The switch is then changed to connect the unknown battery B_u in the same manner; the sliding contact is again moved until the meter shows no current, and the new length l_u is recorded. From these data, the EMF of B_u can be calculated:

$$(\text{EMF})_u = (\text{EMF})_s \times \frac{l_u}{l_s}.$$

In this manner the unknown EMF can be compared with a standard EMF without drawing current from either the standard cell or the unknown battery; thus the measurement is not affected by the internal resistance of either.

For Home, Desk, and Lab

1. If a copper plating cell is added in series with the silver plating cell and an ammeter in Fig. 21–4, it is found that for a given current half as many copper atoms are deposited per second on the negative copper electrode as silver atoms are deposited per second on the negative silver electrode. What does this tell you about the charge on a copper ion compared to the charge on a silver ion?

2.* What is the current in amperes when 3.12×10^{19} elementary charges per second flow in a circuit? (Section 2.)

3*. If 3.00×10^{22} atoms of silver are deposited in 120 minutes in the apparatus of Fig. 21–4, what is the reading, in amperes, of the ammeter? (Section 2.)

4. A current meter has been built and you want to calibrate its scale. It is connected in series with an electrolytic cell containing a solution of silver nitrate and a battery. There is a steady current

which deflects the meter to the upper part of the scale for half an hour. At the end of this period, it is found by weighing that the negative electrode has 1.92 gm of silver deposited on it.

(a) How many elementary electric charges per second passed through the circuit? (A silver atom has a mass of 1.80×10^{-25} kg.)

(b) What would you do to complete the calibration of the instrument?

5. You are given some metal wire, batteries, and a calibrated current meter. How can you calibrate a second current meter?

6.† You are given a half dozen identical ammeters, each of which has a single calibration point—that is, a mark correctly indicating when a 1-ampere current (a current of 6.25×10^{18} elementary charges per second) is passing through it. You also have a variable voltage supply and a large number of identical metal wires. Answer the following questions to complete the calibration of the ammeter meter:

(a) You connect three of the meters as in Fig. 21–29. Explain how you can get a 2-ampere calibration point on one of them.

(b) Make sketches showing how to connect the current meters in order to mark one scale at 3 and 4 amperes.

(c) How would you connect the current meters so as to calibrate one of them to read ½ and ⅓ amperes? (Use sketches.)

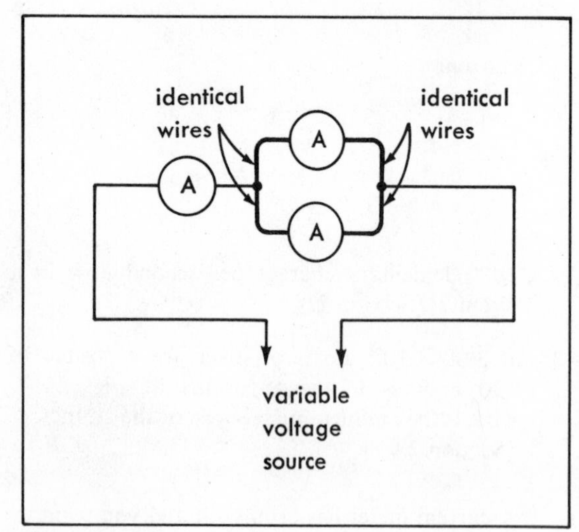

Figure 21–29
For Problem 6.

7. You ionize some gas between parallel metal plates, making equal numbers of negative and positive ions in every small volume between the plates. There is a battery and a current meter connected in series with the plates, and a steady current is flowing through the circuit. The positive plate is on the left, and the negative on the right.

(a) In what direction do you expect the electric current to go between the plates?

(b) If the current meter measures 10^{16} elementary charges per second, how many positive elementary charges are brought per second by the ions coming to the negative plate?

(c) How many negative elementary charges are coming per second to the negative plate?

(d) In one second how many positive ions cross to the right through a plane halfway between the plates?

8. A poorly made electron gun has considerable residual gas left inside. When the filament is heated and a battery connected between filament and cylinder (Fig. 20–7), there is a current consisting of both electrons and positive ions. Consider a plane close to the filament. In each second, 3.0×10^{16} electrons are emitted by the filament and move across the plane toward the cylinder. Also, 0.6×10^{16} singly charged positive ions per second pass through this surface in the opposite direction. What is the current through the plane?

9. The basic unit of electrical work is the watt-second. The practical unit, used by electric power companies, is the kilowatt-hour, and the voltage supplied to houses is 125 volts. How much current is drawn by

(a) a 2-watt electric clock?

(b) a 100-watt light bulb?

(c) a 750-watt electric toaster?

10. (a) How many 100-watt light bulbs could you keep on all day for the same cost as running a 1600-watt electric heater?

(b) How would the heat produced by the light bulbs compare with the heat produced by the electric heater?

11. Why are electric lights, toasters, television sets, etc., connected in parallel in a house circuit?

12. How much electrical work must be supplied to a 250-watt television set if it is on three hours?

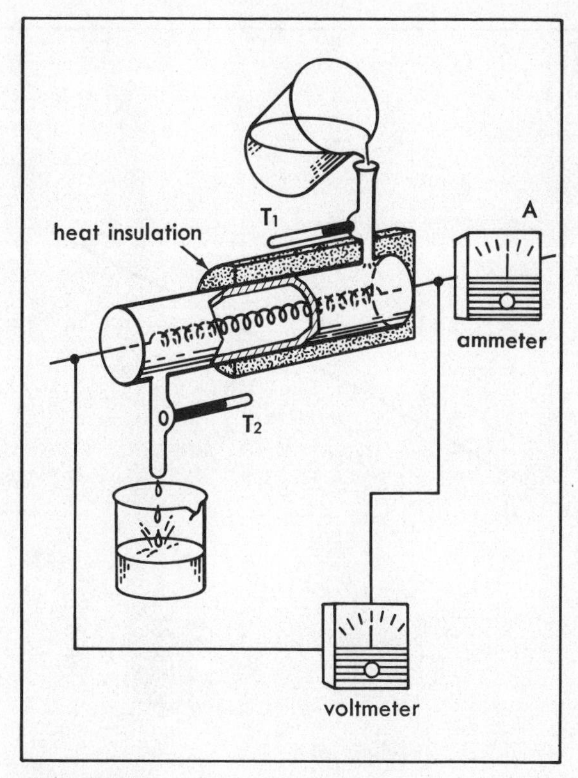

Figure 21–30
For Problem 13.

13.† Figure 21–30 shows a device for calibrating the ammeter A. It is an insulated can with a coil of wire in it. Water, whose heat capacity is 4.2 joules/gm/°C, flows in at the top of the can and passes thermometer T_1. Then, after flowing past the coil of wire, it flows out at the bottom of the apparatus past thermometer T_2. The flow of water is 1 gm/sec.; thermometer T_2 reads a temperature of 1°C higher than thermometer T_1 and the voltmeter reads a potential difference of 2.09 volts.

(a) What is the current through the ammeter?

(b) How would you calibrate the ammeter with this device?

14. You have two incandescent lamps, each marked "60 watts—125 volts."

(a) What current is drawn by one of the lamps when it is connected to a 125-volt supply?

(b) If you connect the two lamps in series and connect the combination to a 125-volt supply, do you predict that the current in the lamps will be more than, equal to, or less than half the current in your answer to part (a)? On what reasoning do you base your answer?

15. Show that the power supplied to a conductor that obeys Ohm's law is

(a) $\frac{V^2}{R}$.

(b) I^2R.

16.* A battery with an EMF of 2.9×10^{-17} joule/elementary charge is transporting charge at the rate of 6.0×10^{16} elementary charges/second. What is the power supplied by the battery? (Section 3.)

17.* A battery connected to an electric heater supplies 3.12×10^{19} elementary charges per second and produces an output of 60 joules per second (watts). What is the EMF of the battery in joules per elementary charge? In volts? (Section 3.)

18. If the conduction electrons in a metal move freely throughout it like the molecules of a monatomic gas,

(a) what is the mean thermal speed of the electrons at 27°C? (See Section 17-4.)

(b) what is their mean thermal energy in electron volts?

19.* Assume that the diode whose current vs. potential curves are shown in Fig. 21–10 is connected in series with a variable resistor to a 90-volt battery. If the filament is heated to 2100°K, what resistance in the resistor will limit the current to 1.2×10^{-3} ampere? (Section 4.)

20. (a) A copper wire has 8.5×10^{22} conduction electrons per cubic centimeter. The wire is carrying a current I produced by a systematic drift of the conduction electrons parallel to its length with an average velocity v. The area of cross section of the wire is A. How many electrons pass a marked point on the wire during the time Δt?

(b) If each electron carries a charge q, what is the current in the wire?

(c) A common size of wire has an area of cross section of 0.83 mm². A reasonable current for this wire is 5 amperes. What is the systematic drift velocity of the electrons along it?

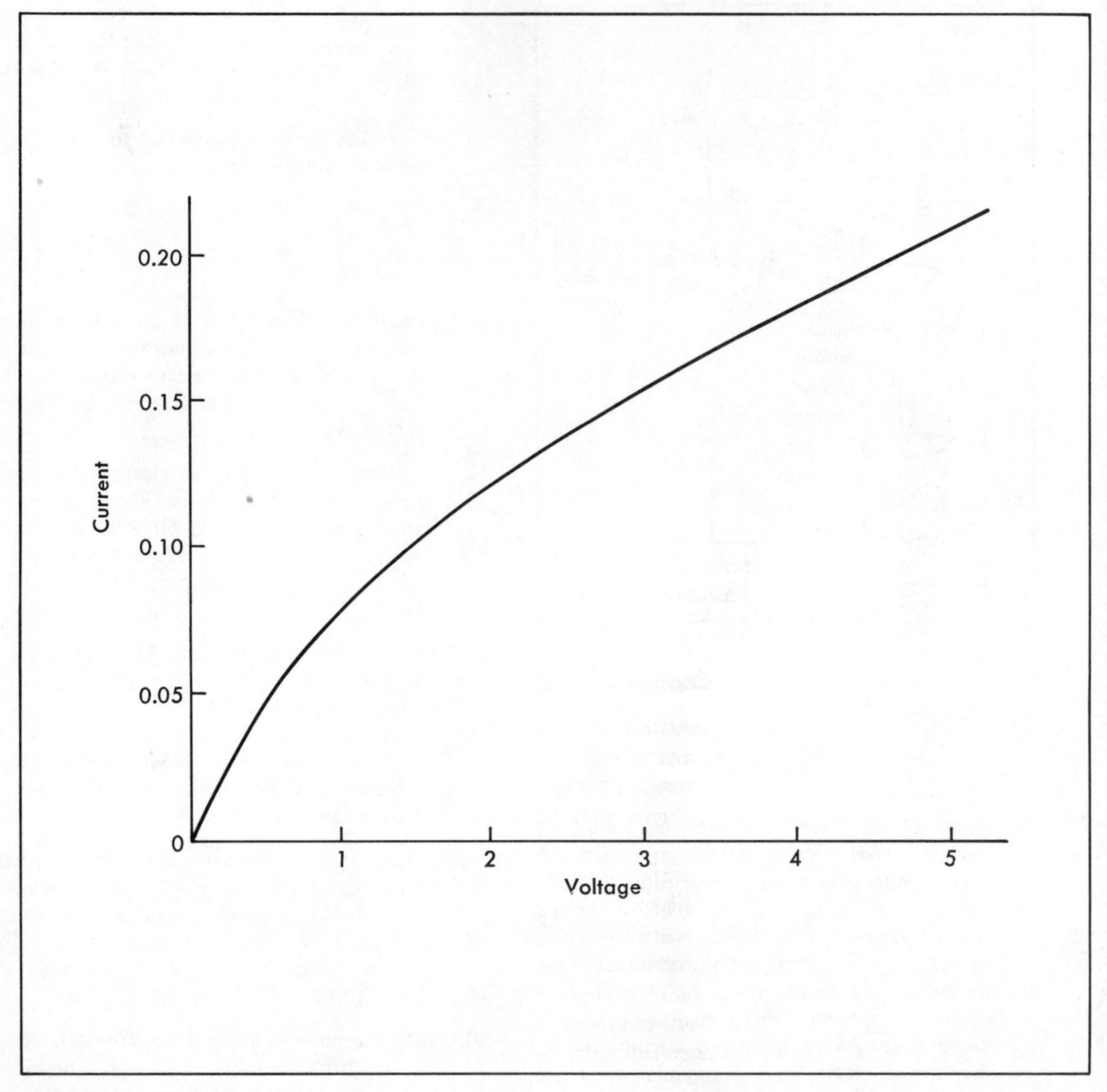

Figure 21–31
For Problem 21.

21. Figure 21–31 shows the current through a flashlight bulb as a function of voltage.
    (a) How do you account for the shape of the curve?
    (b) Sketch the graphs of the voltage as a function of current for a series circuit made up of this bulb and a 10-ohm resistor.

22.* If the scale divisions in Fig. 21–12 refer to volts and amperes and we connect the two circuit elements in series with a battery, what voltage would be needed to give a current of 2.0 amperes? (Section 5.)

23. In Fig. 21–32(a), the current through a resistor is plotted as a function of the potential difference across its ends. In Fig. 21–32(b), we plot the current through a vacuum tube as a function of the potential difference across it.
    (a) If a current of 5×10^{-3} ampere passes through both circuit elements in series,

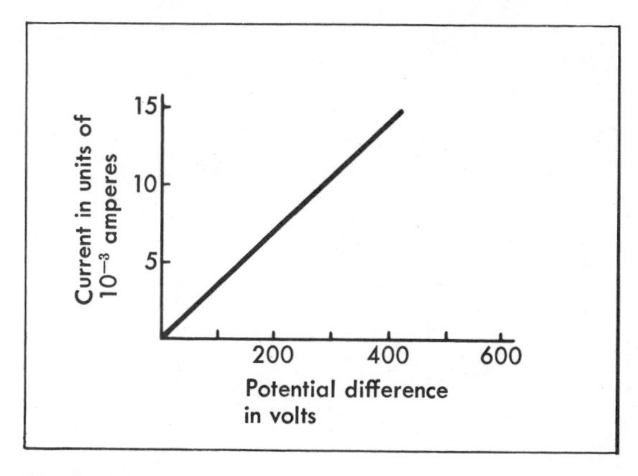

Figure 21–32a
For Problem 23.

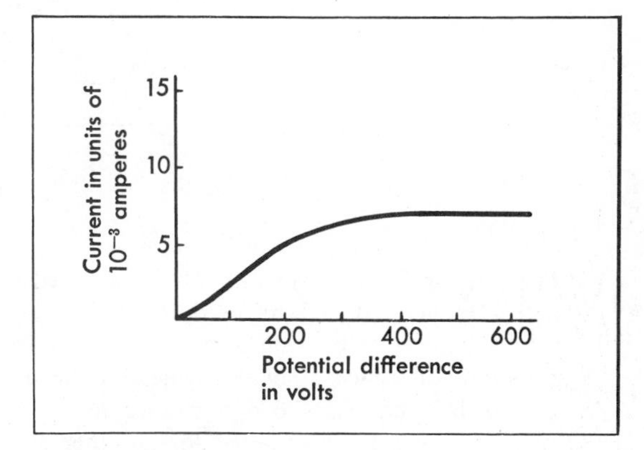

Figure 21–32b
For Problem 23.

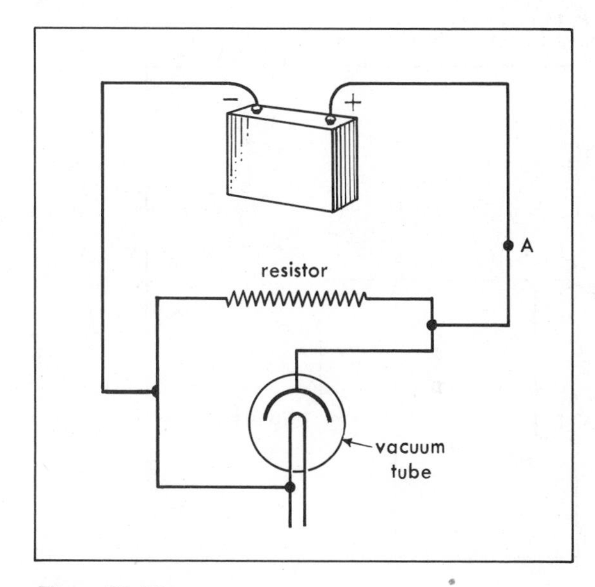

Figure 21–32c
For Problem 23.

what is the potential difference from one end to the other of the resistor? What is the potential difference across the vacuum tube?

(b) If the vacuum tube and the resistor are connected in parallel as in Fig. 21–32 (c), and the potential difference across the vacuum tube is 100 volts, what is the current through each of the circuit elements?

(c) What then is the total current passing point *A* in the figure?

(d) What is the resistance in ohms of the resistor?

24. The same vacuum tube and resistor as were used in Problem 23 are connected in series. The potential difference from one end of the circuit to the other is 1000 volts.

(a) What current flows through the circuit?

(b) What is the potential difference between one end and the other of the resistor?

25.† (a) Use the facts that:

(1) for two series-connected circuit elements the sum of the potential differences across each is equal to the potential difference applied to the combination (conservation of energy) and

(2) the current through both elements is the same (conservation of charge)

to find, if both elements are resistors that obey Ohm's law, the equivalent resistance R_e of the series-connected pair as a function of their individual resistances r_1 and r_2.

(b) What would be the equivalent resistance of three resistors connected in series?

26.† (a) For two parallel-connected circuit elements the potential difference across each is the same, and the total current supplied them is the sum of the currents through each.

Find an expression for R_e, the equivalent resistance of the pair, if both obey Ohm's law, as a function of r_1 and r_2, the individual resistances of the pair.

(b) What is the equivalent resistance of three resistors r_1, r_2, and r_3 connected in parallel?

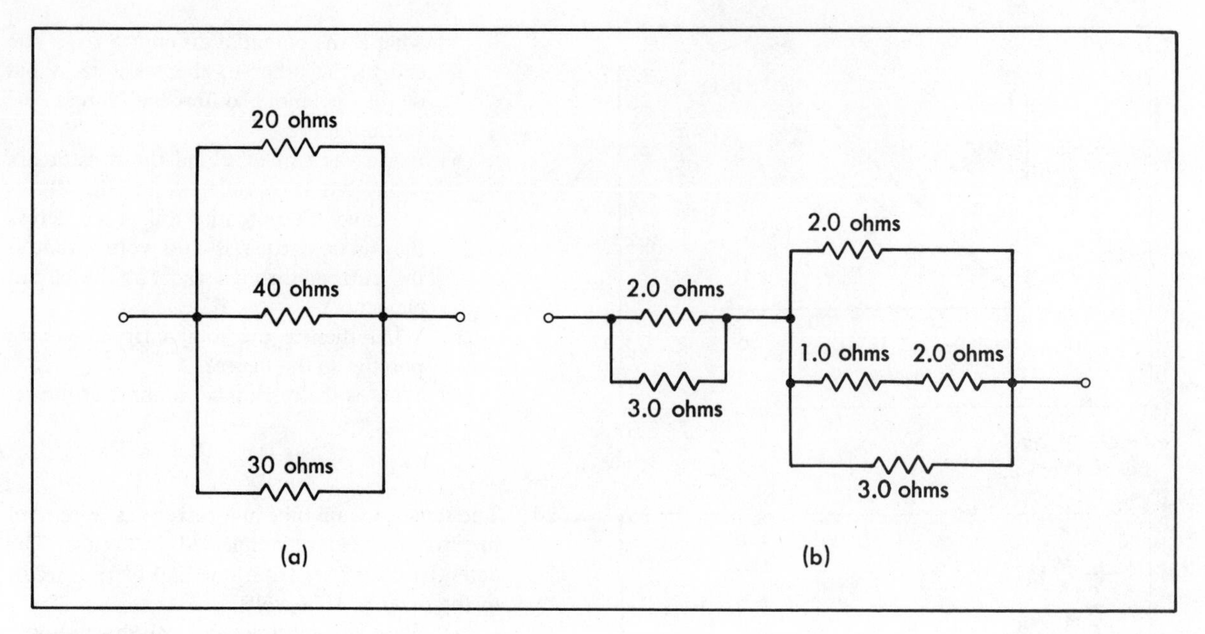

Figure 21–33
For Problems 27 and 28.

27.† What is the equivalent resistance of the combination of resistors connected in parallel in Fig. 21–33(a) and those in series-parallel in Fig. 21–33(b)?

28. Find all the different currents and potential differences in the two circuits in Fig. 21–33 if 10 volts are supplied to each circuit.

29.† The resistor R shown in Fig. 21–34 is called a potentiometer. It can be used in conjunction with a battery to provide a variable source of potential difference. The arrowhead indicates a variable connection to a resistance R which may be moved from one end to the other end of R or any place in between. Show that if the current through the light bulb I is small com-

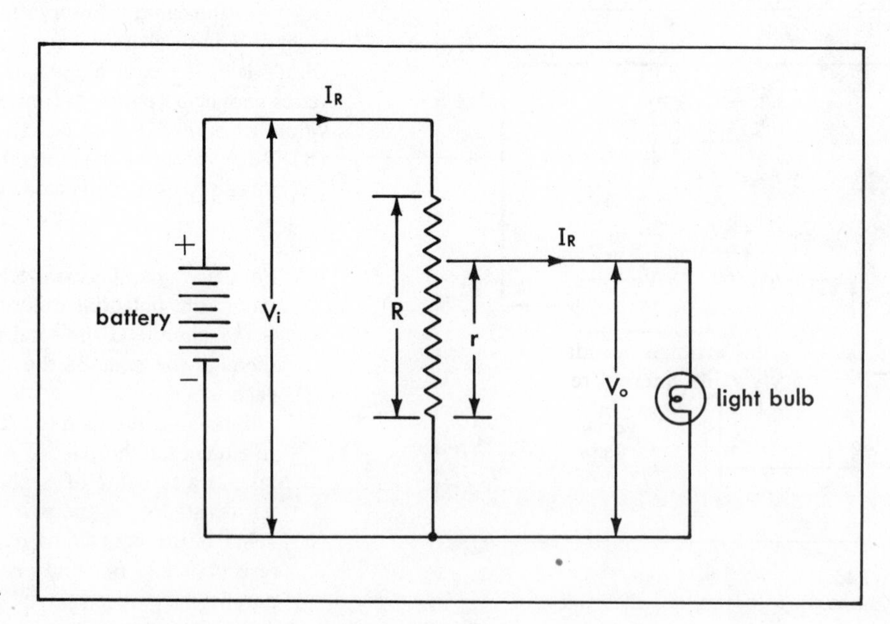

Figure 21–34
For Problem 30.

pared to I_R (the current through the potentiometer R), then V_0, the potential difference across the light bulb, equals $\frac{r}{R} V_i$.

30. Why is the I vs. V curve for neon gas in a neon bulb (Fig. 21–20) different from the I vs. V curve for the ionized gas shown in Fig. 21–5?

31.* A dry cell with an EMF of 1.5 volts and an internal resistance of 0.1 ohm is connected to a flashlight bulb whose resistance is 1.0 ohm. What will be
   (a) the current through the flashlight bulb?
   (b) the potential difference across the flashlight bulb? (Section 9.)

(b) How big is the potential difference you get?

(c) Suppose you reverse the connections to only the 90-volt battery, what potential difference do you get?

35. Suppose r in the voltage regulator circuit in Fig. 21–22 is 450 ohms and the load resistance R is 3000 ohms.
   (a) What will be the current through R, r, and the Zener diode?
   (b) Answer part (a) for the case where the supply voltage rises to 60 volts. Falls to 40 volts; to 20 volts.

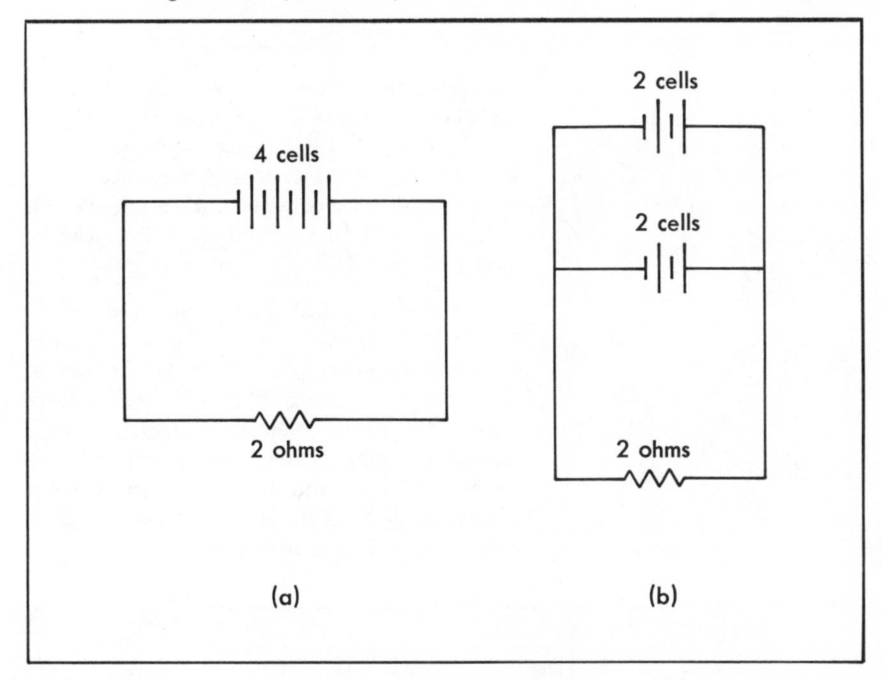

Figure 21–35
For Problem 32.

32. Find the terminal voltage and the current through the load resistor for each of the circuits shown in Fig. 21–35. Each cell has an EMF of 1.5 volts and an internal resistance of 0.1 ohm.

33.† How could you arrange standard dry cells (EMF = 1.5 volts and internal resistance = 0.1 ohm), assuming you had available a large number of them, so as to be able to turn over the starter motor of an automobile that requires a current of 100 amperes at 12 volts?

34. (a) How would you connect a 90-volt battery and a 1.5 volt flashlight cell to obtain the maximum potential difference?

36. Use Fig. 21–20 and Fig. 21–21 as a basis for sketching the important aspects of the equivalent I V graph for one of a pair of diodes (an ordinary solid-state diode and a Zener diode) connected in series as in Fig. 21–1.

37.† Explain how the circuit in Fig. 21–1 operates. Why are there four diodes instead of just two?

38.* What would be the effect on the frequency of the sawtooth voltage generated by the sweep generator in Fig. 21–24 if
   (a) the battery EMF is increased?
   (b) R is increased?

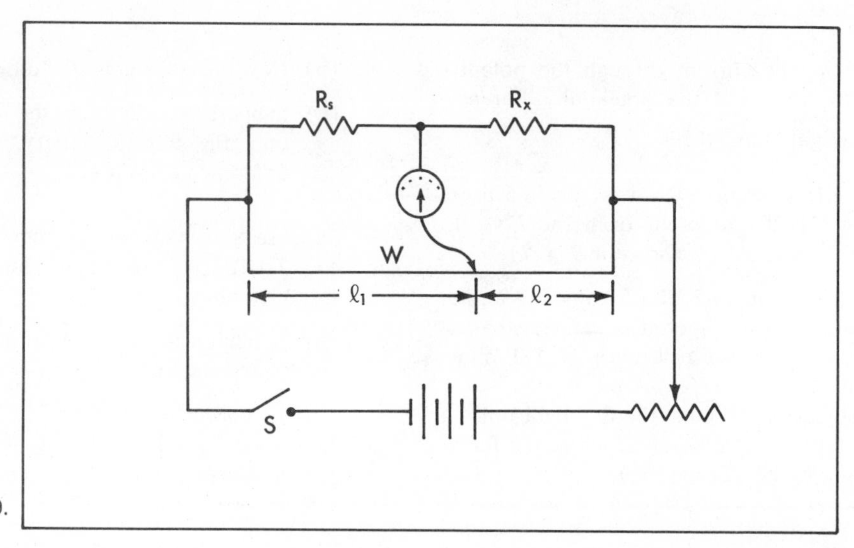

Figure 21–36
For Problem 39.

39. In the circuit shown in Fig. 21–36, R_s is a resistor whose resistance is known; R_x is a resistor whose resistance is to be determined. W is a uniform resistance wire. The switch S is closed, and the sliding contact is moved until the sensitive meter shows no current.
    (a) How do the ratios R_s/R_x and l_1/l_2 compare when the meter shows zero current?
    (b) R_s is 24 ohms; $l_1 = 60.0$ cm, $l_2 = 40.0$ cm. What is the value of R_x? (This is a simplified Wheatstone bridge, used for measuring resistance. In practice R_s would be one of a set of standard resistors that can be connected into the circuit.)

40. In the circuit of Fig. 21–28, varying R varies the current in the uniform resistance wire; this variation will cause variations in the temperature of the wire, and hence in its resistance. Would these variations in the resistance of the uniform resistance wire affect the accuracy with which the EMF of an unknown battery can be measured? Give your reasoning.

41. To the input terminals of the circuit shown in Fig. 21–37, a time-varying voltage as shown in the small graph is applied to points A and B. This voltage is large enough so that if it were applied directly to either of the lamps forming part of the circuit, the lamp would light up brightly. What do you predict you will observe if only switch S_1 is closed? If only switch S_2 is closed? If both switches are closed?

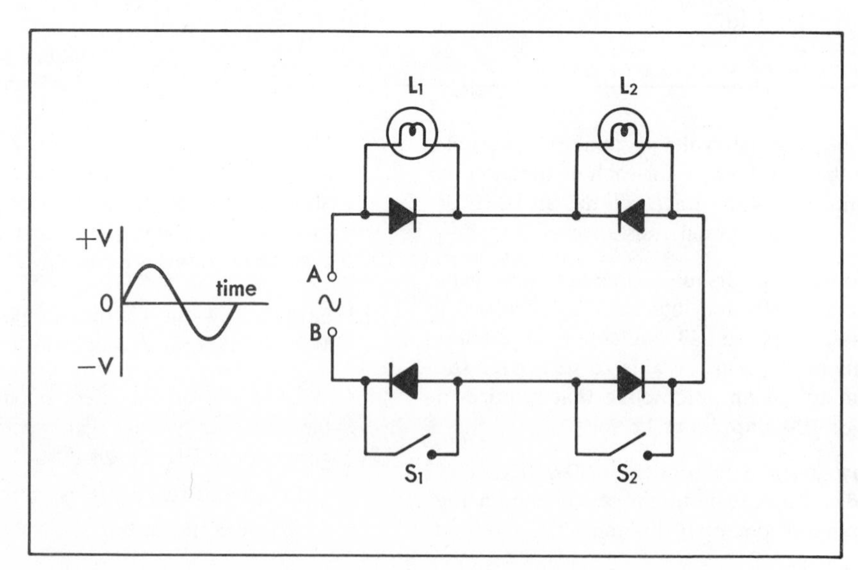

Figure 21–37
For Problem 41.

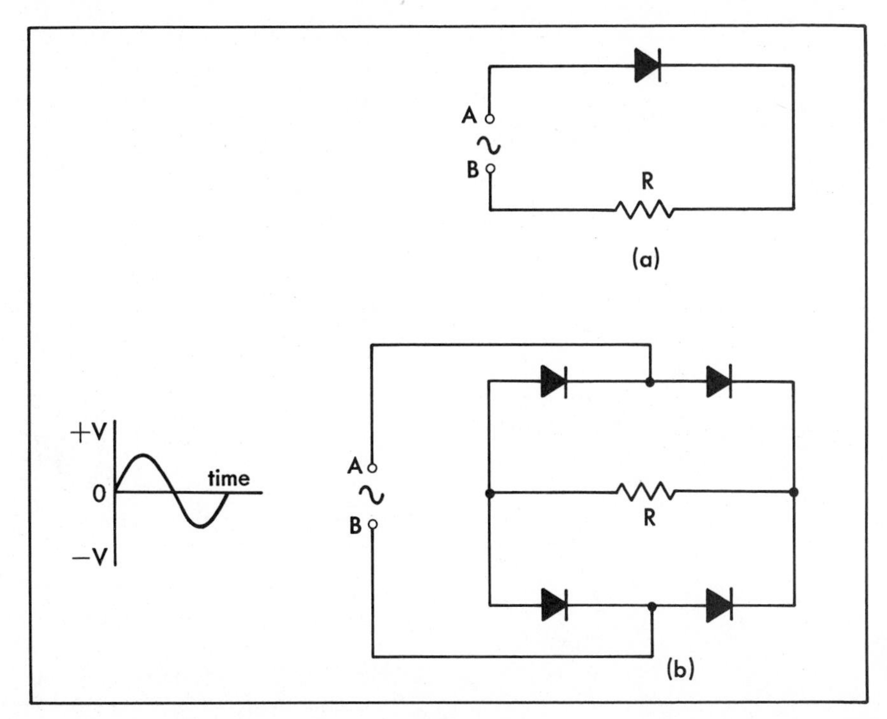

Figure 21–38
For Problem 42.

42. In each of the two circuits shown in Fig. 21–38, a voltage varying with time as shown in the small graph is applied to points A and B. In each case draw a graph (to the same time scale) of the current in the resistor R as a function of time.

Further Reading

MACDONALD, D. K. C., *Near Zero.* Doubleday Anchor, 1961: Science Study Series. (Chapter 3)

MAGIE, WILLIAM FRANCIS, *A Source Book in Physics.* McGraw-Hill, 1935. Extract from Faraday's paper of 1834 on laws of electrolysis. (Pages 492–498)

The Magnetic Field

CHAPTER

22

When Albert Einstein wrote his life story at the age of 67, he recalled the day when he was a little boy of four, happy to get a new toy from his father. That toy was a compass needle; and the wonder it aroused remained with Einstein all his life. Most of us have shared that wonder, and a good many of us as children experienced the fascination of a horseshoe magnet, which attracts iron. In this chapter we shall discuss magnetic forces and the concept of the magnetic field by which we can best interpret them.

22-1 The Magnetic Needle

A compass needle free to rotate about a vertical axis will line up in the north-south direction. We say that it aligns in the magnetic field of the earth. We call the tip of the needle that points north the northern tip.* In any region where a compass needle feels forces that tend to align it, we say that there is a *magnetic field* (Fig. 22–1). The direction of the field is the

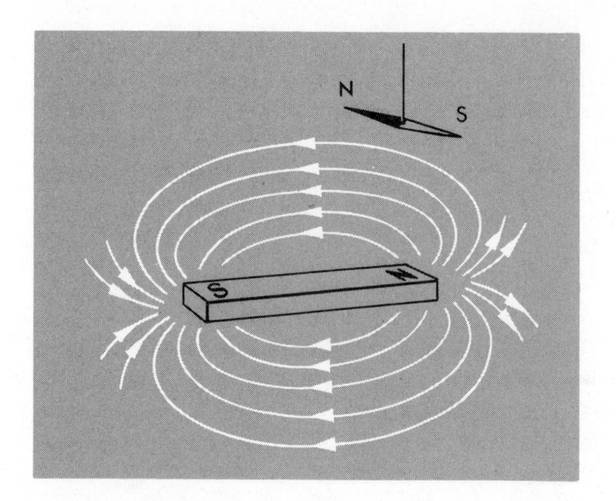

Figure 22–1
A compass needle in a magnetic field lines up in the direction of the field.

direction taken by a compass needle which is free to line up. We say that the field points in the direction that runs from the southern to the northern compass tip. Just as we used grass seeds to show the direction of an electric field, so we use a magnetic needle to explore the direction of a magnetic field. You can see that a compass needle itself has a magnetic field by bringing up another compass needle and watching the two influence each other: each pulls on the other. Each has a magnetic field, and tends to line up in the field of the other.

As early as 600 B.C. the ancient Greeks were acquainted with magnetism. Thales of Miletus, often called the father of Greek science, knew about a mineral (magnetite) which attracts ordinary iron, and found that the iron itself becomes magnetized by touching the magnetic mineral. The Chinese

* Actually, the northern tip of the needle points *approximately* to the geographic North Pole as long as we are south of the Arctic Circle (and north of the Antarctic Circle). In the Northern Hemisphere the northern tip of a compass needle pivoted on a horizontal axis also points down, measuring the "dip" of the earth's magnetic field.

discovered, probably about the eleventh century, that a magnet acts as a compass. The unifying idea that the earth itself is a magnet was provided by William Gilbert, who was working at the court of Queen Elizabeth of England in 1600, just about the time Shakespeare was writing *Hamlet.*

One of the most striking features of magnets is that if we break one in two, each piece is also a magnet (Fig. 22–2). We might ask how far we can continue subdividing before "half a magnet" is no longer a magnet. The answer is that this never happens. We can continue right down to the subatomic scale and find that electrons, protons, and neutrons themselves act like tiny magnets.

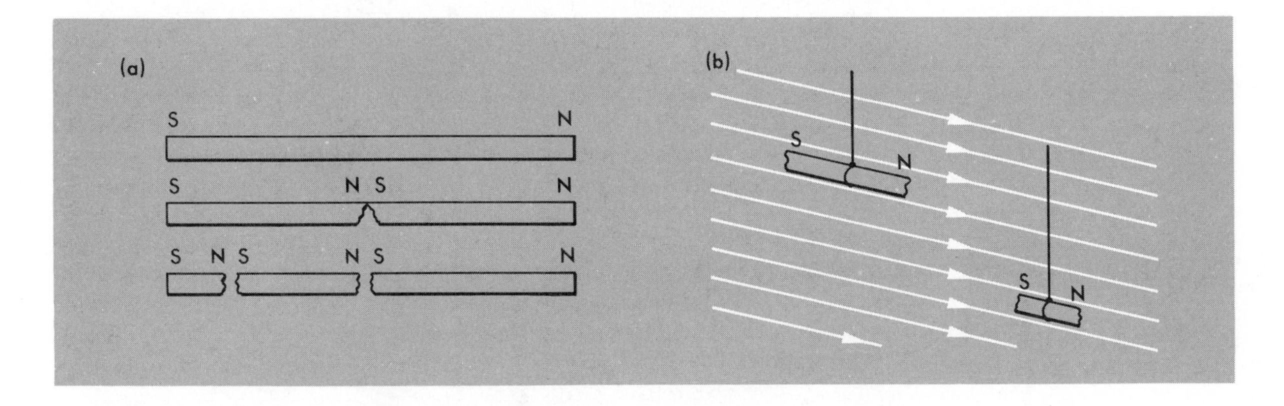

Figure 22–2
(a) When we break a magnet, each piece has opposite poles at its ends.
(b) Any of these pieces will line up in a magnetic field.

22-2 Magnetic Fields of Magnets and Currents

Magnetite and magnetized iron are not the only sources of magnetic fields. Let us perform an experiment in which we produce a magnetic field without using such materials.

We attach a long piece of wire to the terminals of a battery through a switch as shown in Fig. 22–3. With the switch open, we hold the wire above a compass needle and parallel to it. Then we close the switch. If the current in the wire is sufficiently strong, we see the compass needle suddenly deflected. It now points across the wire. We therefore conclude that electric currents produce magnetic fields in the surrounding space.

Today this fact is common knowledge, so that it is difficult to appreciate the revolutionary impact of its discovery by the Danish professor Oersted early in the nineteenth century. Until that time, electric and magnetic effects had been regarded as entirely separate. Oersted's discovery revealed an unsuspected relationship: it tied the origin of magnetic fields to the motion of electric charges.

The experiment last described is one you can do, and is probably enough to convince you that moving electric charges generate magnetic fields. But there is an even more direct experiment, though a far more difficult one. It was performed by Henry Rowland in Baltimore in 1876. He put the biggest electric charge that he could onto a hard-rubber disc about 20 cm across. Then he spun the disc at about 60 revolutions per second. In this way he could look for the magnetic effect of the moving charges directly.

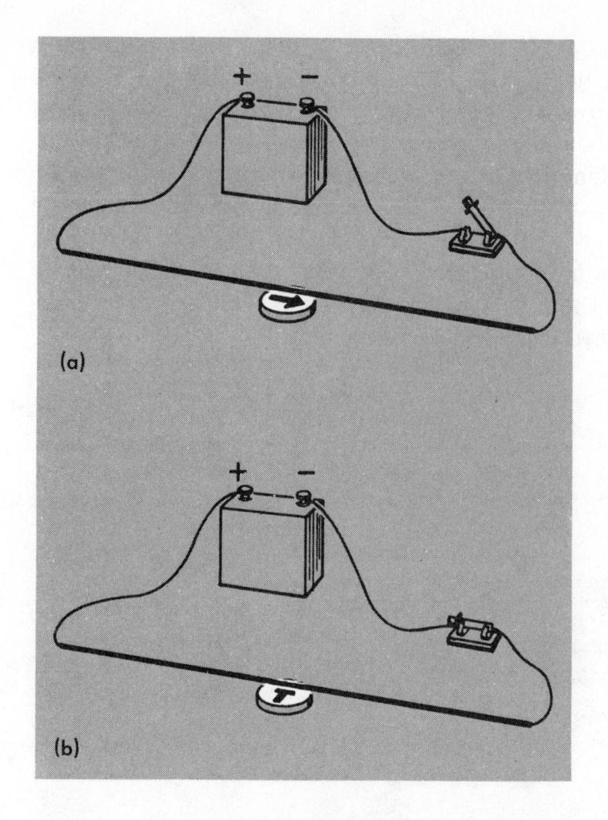

Figure 22–3
(a) A wire is placed over a compass needle and parallel to it. The switch is open, and no current flows in the wire. (b) When current flows, the needle is deflected and points across the wire.

He found that the spinning electric charge did produce a weak magnetic field.

In Chapter 21 we learned that positive charges moving in one direction have the same electrical effects as negative charges moving in the opposite direction. Experiment shows that they also have the same magnetic effects. Suppose that we replace a part of the circuit of Fig. 22–3 with an electrolytic conductor, as shown in Fig. 22–4. When the current in the electrolyte is the same in magnitude and direction as the current in the wire of Fig. 22–3, the compass needle still deflects by the same amount and in the same direction. In the first circuit the current is due entirely to the motion of negative charges, whereas in the second the current is due to the motion of both positive charges and negative charges. Thus in describing the magnetic effects of steady currents, we need not worry about the sign of the particles which move.

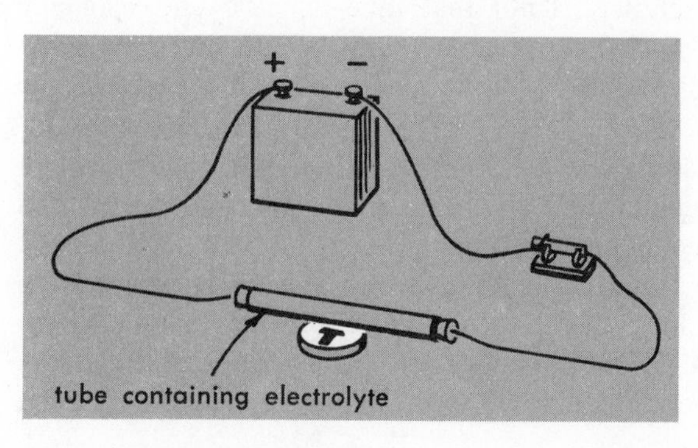

Figure 22–4
If we replace part of the circuit of Fig. 22–3 with a tube containing electrolyte, the compass needle deflects just as before if the current in the circuit is the same as before.

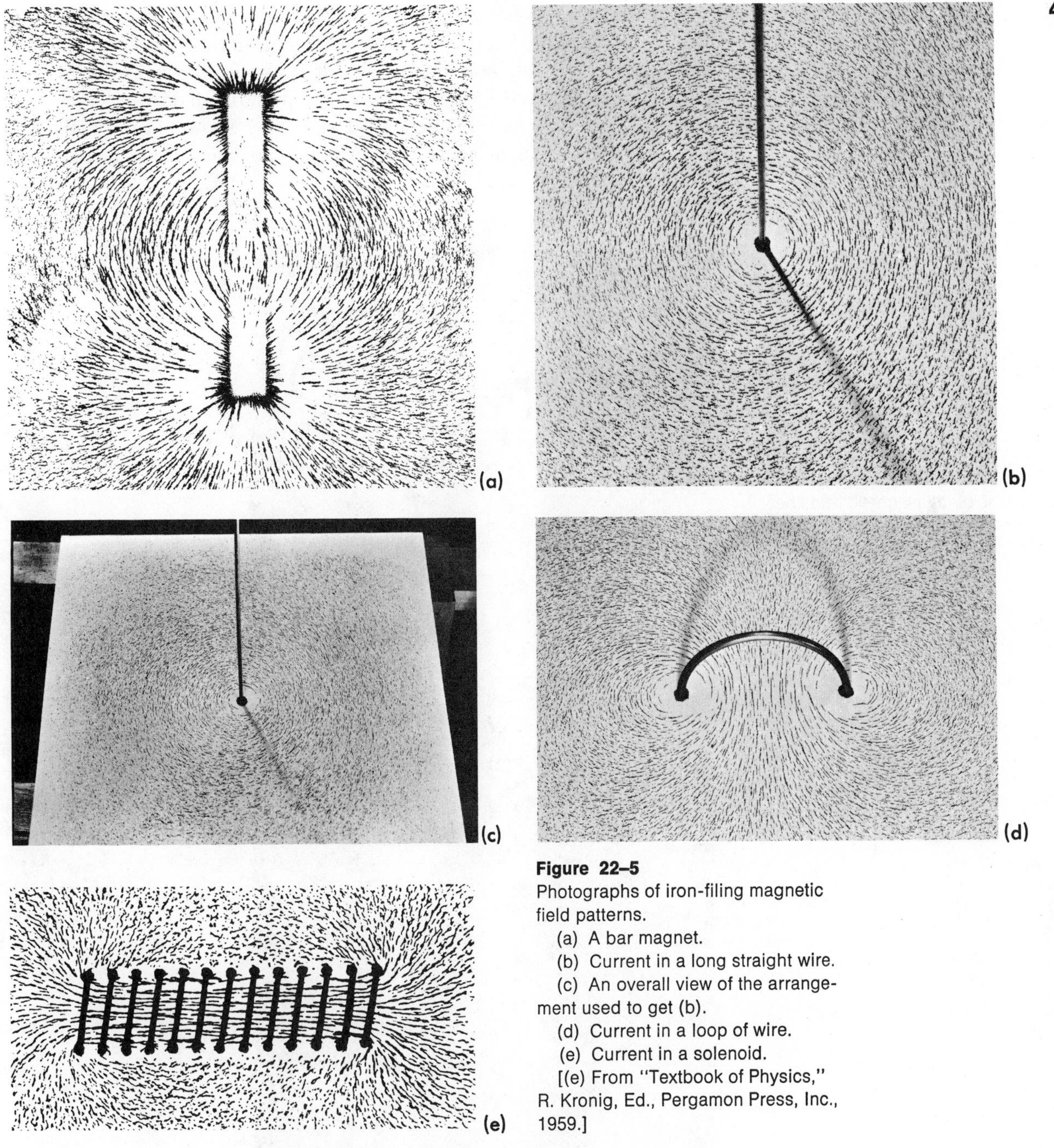

Figure 22–5
Photographs of iron-filing magnetic field patterns.
(a) A bar magnet.
(b) Current in a long straight wire.
(c) An overall view of the arrangement used to get (b).
(d) Current in a loop of wire.
(e) Current in a solenoid.
[(e) From "Textbook of Physics," R. Kronig, Ed., Pergamon Press, Inc., 1959.]

We describe magnetic fields by drawing magnetic field lines, just as we describe electric fields by drawing electric field lines. If a magnetic field is sufficiently strong, and if you are not interested in great accuracy, you can "see" the field lines by spreading some fine iron filings on a piece of paper. The grains of iron line up along field lines like grass seeds in an electric field (see Section 19–3). Tap the paper gently if the filings stick. The photographs in Fig. 22–5 are iron-filing pictures.

Of course, we do not get the direction of the field, only the pattern of the field from the iron filings. We can determine the direction by placing a small compass needle at a number of different positions in the field. Then we can draw the field lines for the various patterns, indicating their directions with arrowheads. Figures 22–6 and 22–7 show the magnetic fields produced by various magnets and currents. You may compare them with Fig. 19–5 showing various electric fields.

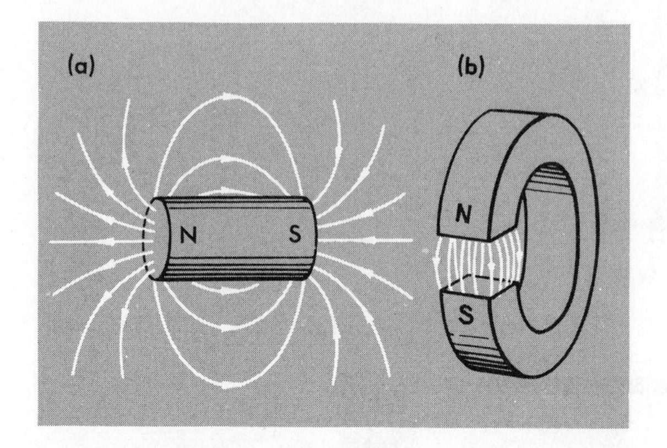

Figure 22–6
Diagrams of the magnetic field lines of permanent magnets. (a) A bar magnet. (b) A horseshoe magnet.

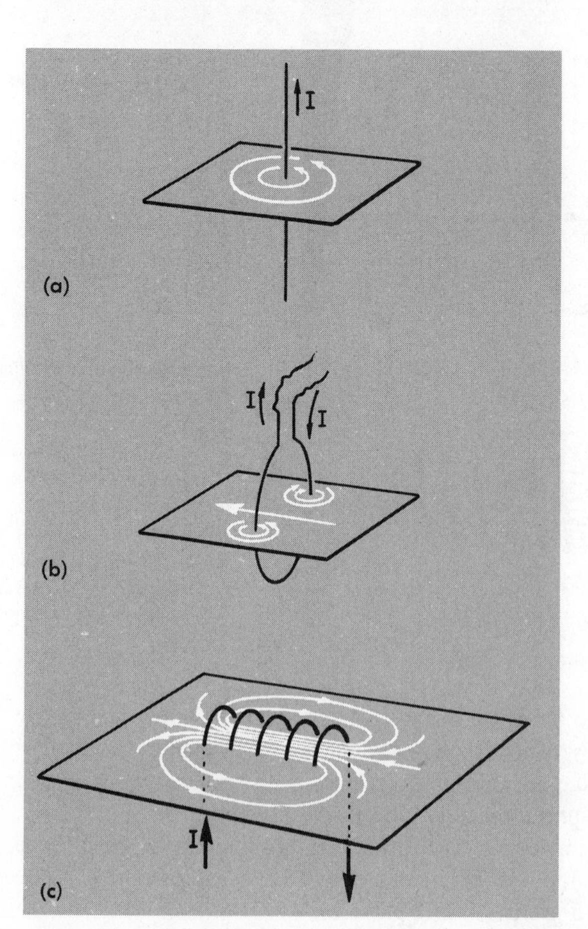

Figure 22–7
Diagrams of the magnetic field lines around current-carrying wires. (a) A long straight wire. (b) A loop of wire. (c) A solenoid.

Electric field lines from charges at rest begin and end at the charges producing the field; but the lines of the magnetic fields produced by currents have no beginning nor end. They encircle the wires carrying the current (Fig. 22–7). The lines of the magnetic fields produced by permanent magnets (Fig. 22–6) appear to begin and end at the surface of the magnets, but this is only because we have not drawn the lines inside the magnets where we cannot place an ordinary compass needle. Using neutrons as the compass needles, we can explore the inside of a permanent magnet. We then see that the lines do not stop at the surface.

You can remember how the direction of the field is related to the direction of the current by the "right-hand" rule. In imagination place your *right* hand with the thumb in the direction of the current; then your fingers curled around the wire point in the direction of the field (Fig. 22–8). Notice that this rule also gives the direction of the field through a loop of wire.

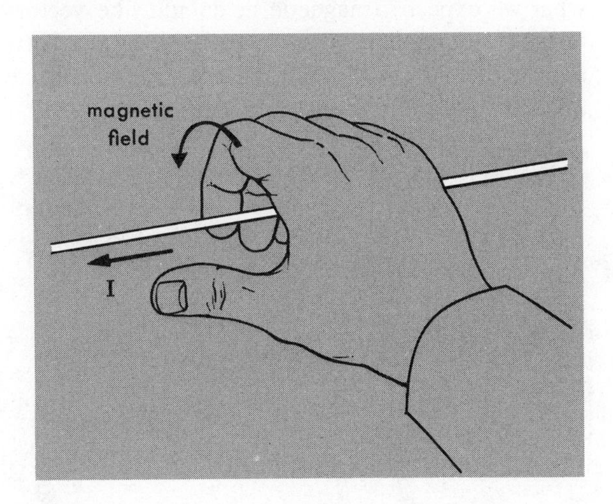

Figure 22–8
The "right-hand rule." When your thumb points in the direction of the current in a wire, your fingers curl the way the magnetic field goes around the current.

22-3 The Vector Addition of Magnetic Fields

A magnetic field can be characterized by both a magnitude and a direction. Do two magnetic fields, acting together, add like vectors? Not everything with magnitude and direction is a vector. For example, the rotations of a body in three dimensions cannot be added vectorially. Try rotating a book 90° around a horizontal axis and then rotating it 90° around a vertical axis. Now try doing the rotation in the opposite order. You do not get the same final result. The rotations do not add vectorially, since vectors can be added in any order. Let us, therefore, check up on the addition of magnetic fields experimentally.

We begin by investigating a simple case, the addition of two fields of equal magnitude. We shall use a small compass needle, pivoted on a vertical axis, to detect the resultant field. Since the earth's field is always present, let us choose this as one of our two test fields. The other test field can be produced with any convenient current arrangement for which we know the direction of the field. We shall choose a circular loop, because the field at its center is fairly uniform and has the direction of the axis of the loop [Fig. 22–5 (d)]. To make the field at the center of the circular loop equal in magnitude to the earth's field, we place the compass needle at the center of the loop and, when the current is zero, align the axis of the loop along the direction of the needle. We choose the direction of the current by the right-hand rule so that its field is opposite to the earth's field. As we increase the current in the loop, we find a value at which the compass needle swings freely. There is no force to align it in any particular direction; the needle indicates zero magnetic field. Thus the magnetic field of the earth has been just balanced by an equal and opposite magnetic field due to the current in the loop.

Now let us add these two equal fields together at right angles to each other. We rotate the loop about a vertical diameter so that its axis turns through 90°. If the current has not changed, we find that the compass needle now points at 45° to the direction of the earth's field (Fig. 22–9). This is what we expect if magnetic fields add like vectors. But one experimental check is not enough. We can try other angles between the directions of the two fields: 30°, 45°, 60°; the compass needle always exactly bisects the angle between the two equal-magnitude fields.

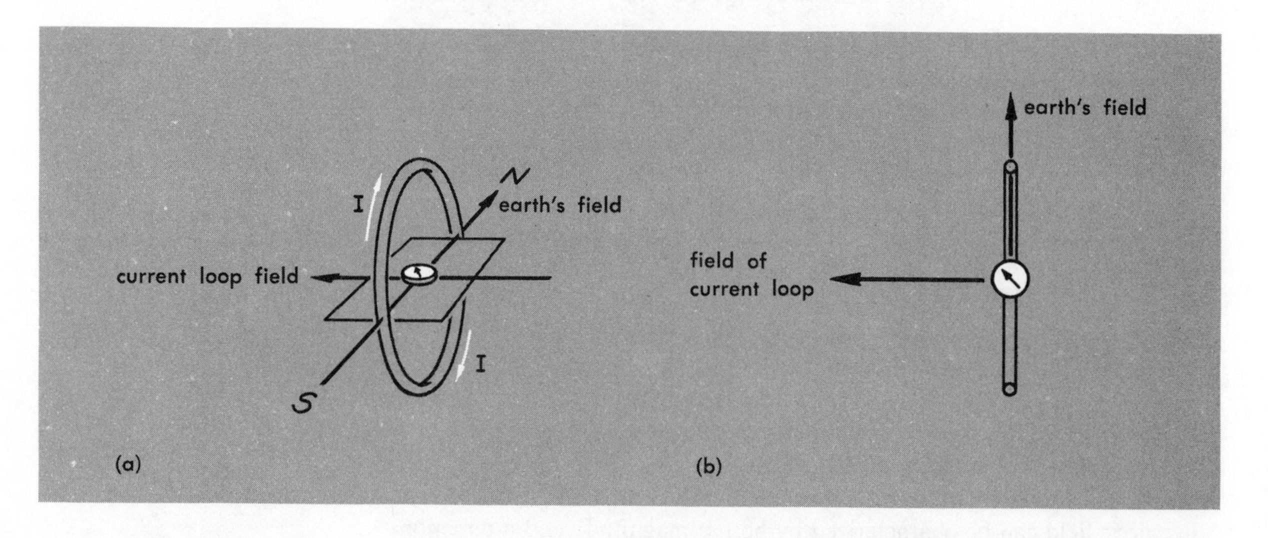

Figure 22–9
(a) A magnetic field, equal in magnitude to the earth's field (as measured by a horizontal compass needle), is added at right angles to the earth's field. (b) The horizontal needle points at 45° away from magnetic north. For any other angles between the equal-magnitude fields, the compass needle always bisects the angle. Magnetic fields appear to add like vectors.

Vector addition looks good, at least for fields of equal strength. But what about fields that have different magnitudes? Here we run into a problem, for we do not yet have a way of measuring field strength. But we might reason as follows: if one loop carrying a certain current gives a field just equal to the earth's field, then two equal loops placed right together should give twice the earth's field if magnetic fields really do add like vectors. Indeed, one loop carrying twice the current should also give twice the field strength. Let us therefore define magnetic field strength B as proportional to the current producing it, and again check on vector addition. We return to the arrangement in which the field of the loop and the field of the earth were at right angles to each other. With the original current, the compass needle stands at 45° to the direction of the earth's field; with twice this current, the angle increases to 63½°. This is just what we expect for the addition of perpendicular vectors having lengths in the ratio of two to one (Fig. 22–10). No

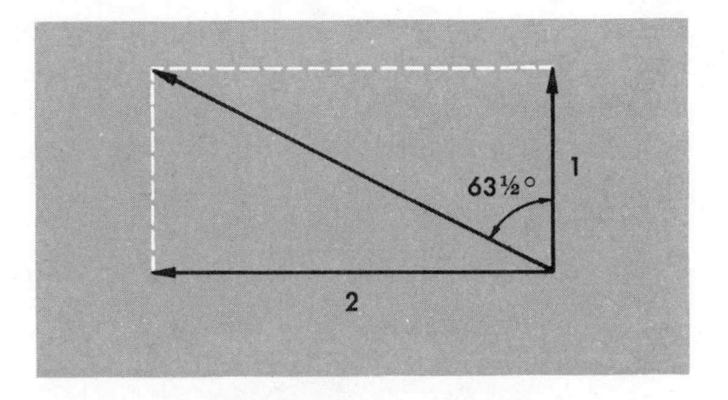

Figure 22–10
Two perpendicular vectors, with a two-to-one ratio of magnitudes, have a resultant lying at an angle of 63½° from the shorter of the two vectors. If we define magnetic field strength as proportional to the current producing it, we find that magnetic fields add in this way also. Therefore, magnetic fields are vectors.

matter how we change the current and direction of the loop, we find again and again that the direction of the total magnetic field is correctly given by the vector addition of the individual fields, if we define magnetic field strength as proportional to the current producing it. We can now represent the magnitude and direction of a magnetic field by the vector symbol $\vec{\mathbf{B}}$.

22-4 Forces on Currents in Magnetic Fields— A Unit of Magnetic Field Strength

Electric charges produce electric fields which exert forces on other electric charges. Now we find that electric currents produce magnetic fields and we might guess that, correspondingly, magnetic fields exert forces on electric currents. They do, and with ordinary fields and currents the forces are quite sizable. It is these magnetic forces which spin the world's electric motors.

Suppose, to begin with, that we hang a flexible wire in a magnetic field (Fig. 22–11). The field may be either the field of a permanent magnet or the field of a current which we control. When we turn on an electric current in the flexible wire, this wire is pushed upward in the magnetic field.

Let us try to discover the rules describing the force experienced by a current in a magnetic field. We can find this force by measuring the force

Figure 22–11
(a) A flexible wire in a magnetic field. The switch is open, no current flows, the wire hangs down. (b) When current flows, there is a magnetic force on the wire that pulls it taut.

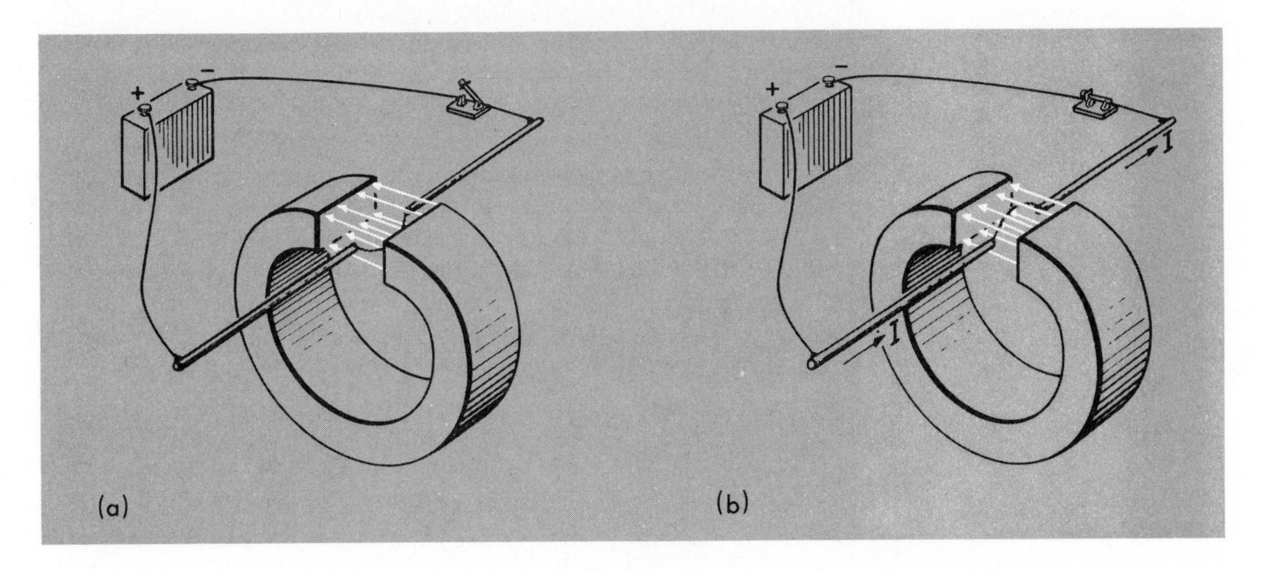

needed to hold a movable section of wire in place when a magnetic field is applied. Using experimental arrangements like that in Fig. 22–12, we find the following:

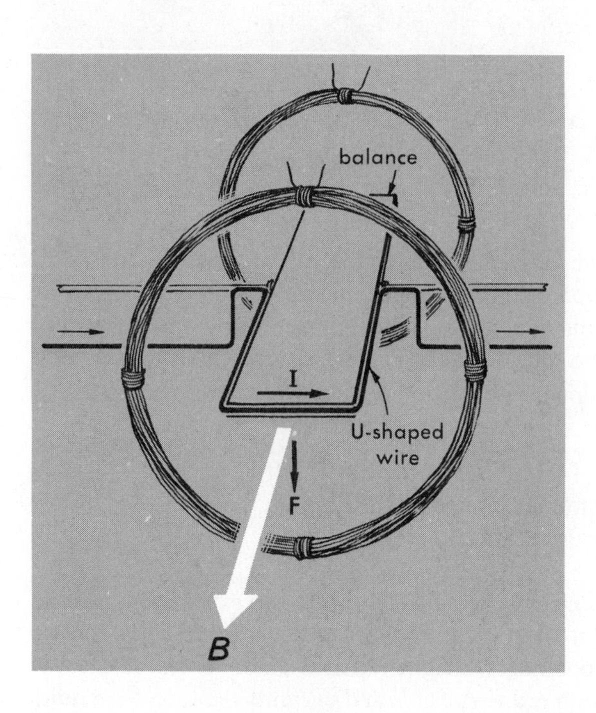

Figure 22–12
A U-shaped wire runs on the edge of a light balance in the center of the magnetic field of two large coils. When the coils are separated a distance equal to their radii, the magnetic field at the center is almost uniform. (Such coils are called Helmholtz coils.) The straight section of wire on the end of the balance is acted on by a force $\vec{\mathbf{F}}$ tending to tip the balance. By putting weights on the other end we measure $\vec{\mathbf{F}}$. By turning the balance, we find how the angle between the wire and $\vec{\mathbf{B}}$ affects $\vec{\mathbf{F}}$. Because the current runs in opposite directions in the two arms of the U, in a uniform field the forces on the arms cancel. Since the field is not perfectly uniform, these forces do not cancel exactly. However, the twist they apply to the balance is very small compared with the twist applied by $\vec{\mathbf{F}}$.

(i) *The force is perpendicular to both the magnetic field and the current.* Here is a simple rule to enable you to remember the actual arrangement of directions. Stretch out your *right* arm with your hand open and your thumb sticking out at right angles to your fingers. Turn your right hand till the *thumb points along the current* and your *fingers point along the external magnetic field.* Then the *force* on the current is in the direction pointing *straight out from the palm of your hand;* the force is in the direction in which you would push with your hand (Fig. 22–13).

(ii) By changing the strength of the magnetic field without changing its direction, we find that the force on the current is proportional to the magnitude of the field. By changing the direction of the field we find that only $\vec{\mathbf{B}}_\perp$, the component of $\vec{\mathbf{B}}$ perpendicular to the current, contributes to the force. Therefore, *the force is proportional to $B_\perp$, the magnitude of the perpendicular component of the magnetic field* (Fig. 22–14).

To find how the force depends on the current and on the length of wire, we then reason as follows. Two wires placed side by side and carrying equal currents in a magnetic field together experience twice the force experienced by each wire alone. But the two wires combined carry twice the current of one wire. Thus, doubling the current doubles the force:

The force, F, is proportional to current, I.

Next, two short straight wires of the same length placed end to end to form one longer wire experience together double the force experienced by each length alone. Thus doubling the length of the wire doubles the force:

The force, F, is proportional to length of wire, l.

To summarize our results, we now know the direction of the force and we know that the force is proportional to $B_\perp$, to I, and to l. In the form of an equation, we can write this as

$$F = (\text{const.}) I l B_\perp,$$

where the proportionality constant depends only on the choice of units.

Now we shall settle on a definite unit of magnetic field by defining it in terms of the units we have already adopted for F, I, and l. We measure F in newtons, I in amperes, and l in meters, and we choose the unit of magnetic field to make the constant equal to 1. Then

$$F = I l B_\perp$$

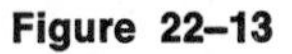

Figure 22–13
A rule to give the direction of the magnetic force on a current. Extend the fingers of your right hand in the direction of the magnetic field, and turn your hand so your thumb points along the current. The magnetic force then points out from your palm —in the direction you would push with your hand.

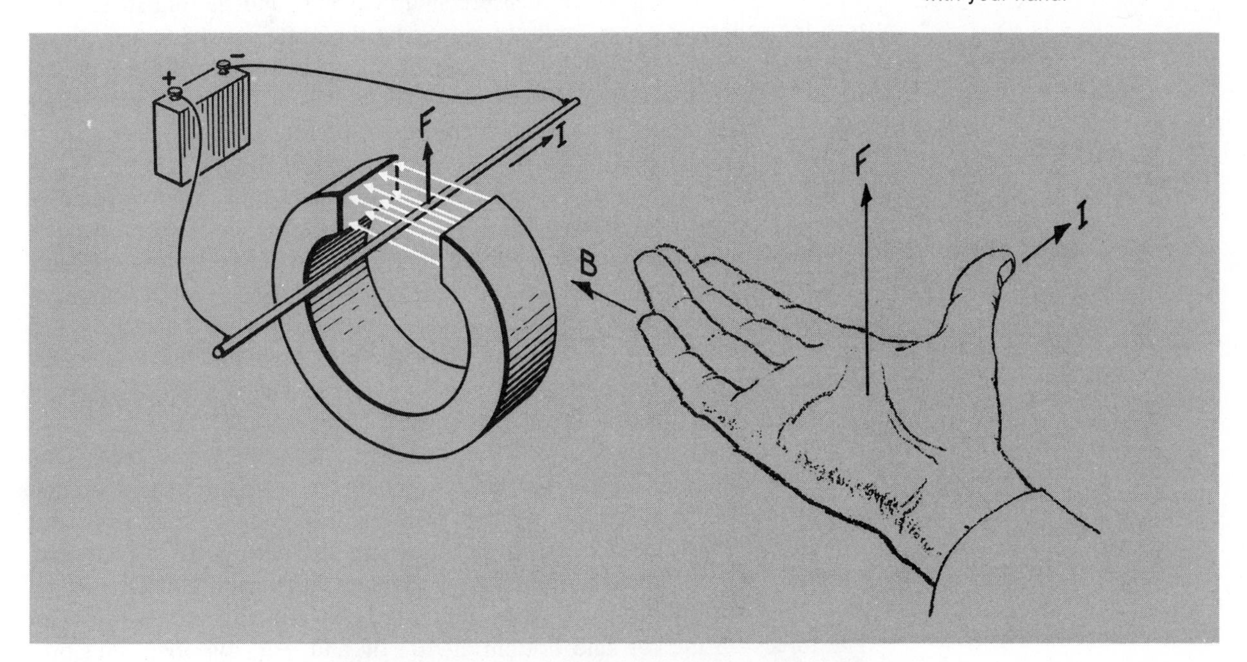

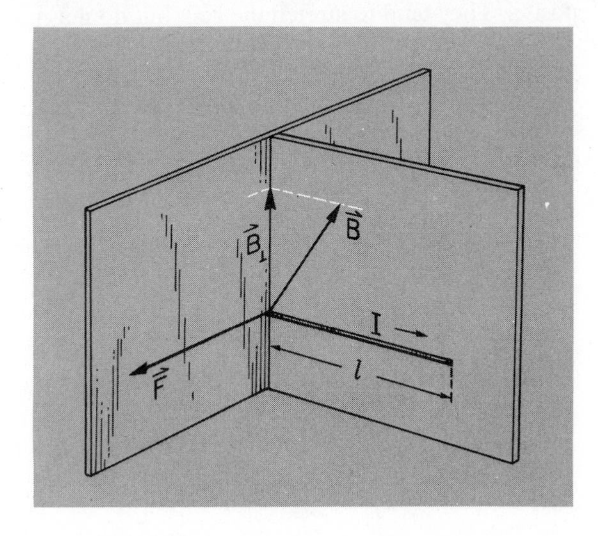

Figure 22–14
The magnetic force on a piece of wire carrying a current is proportional to $B_\perp$, the magnitude of the component of the magnetic field $\vec{\mathbf{B}}$ perpendicular to the wire. The vector representing this force is perpendicular to the wire and to $B_\perp$; its direction is given by the right-hand rule.

and the unit of B is $\frac{1 \text{ newton}}{(1 \text{ ampere})\ (1 \text{ meter})}$ A magnetic field of 1 newton/amp-meter exerts a force of 1 newton on a current of 1 ampere running perpendicularly across it for a distance of 1 meter. Another unit of magnetic field strength, frequently used in practical work, is the *gauss*. One gauss is 10^{-4} newton/amp-meter. When the magnetic field is expressed in gauss, the expression for the force on a current-carrying wire becomes $F = 1.0 \times 10^{-4}\, IlB_{\perp}$.

22-5 Meters and Motors

The delicate pointers which measure small currents and the big shafts that turn the rollers of a steel mill are moved by the same sort of forces: those exerted by a magnetic field on a loop or coil of wire carrying a current. Large currents mean large forces; small currents can exert only very small forces. Iron is found in most such devices; it provides the engineer with a way to increase magnetic fields from a given supply of current.

It is clear to anyone in our country that the use of this kind of force is very widespread, and that a great deal of understanding and skill must have been employed in inventing, designing, and building all the devices in which it is used. This is the work of electrical engineers, who have for about a century carried the applications of this force to new uses. We are not going to try to describe the ingenious and often complex ways they have found to use the force; that is another subject. But we shall give a short account of two simple devices which depend on magnetic forces.

(a) *Moving-Coil Ammeters.* Figure 22–15 shows the forces on a loop of wire in a magnetic field. The loop is rectangular with sides l and w; and its plane makes an angle with the field of strength B. When a current I runs around the loop, the forces on the sides are BIl and $-BIl$. They are parallel, but they do not cancel out because they do not pull along the same line. They form a "couple" which twists the coil around the vertical axis. The forces on the top and bottom are $B_{\perp} Iw$ and $-B_{\perp} Iw$ equal and opposite. They tend to stretch the coil, but do not move it as a whole.

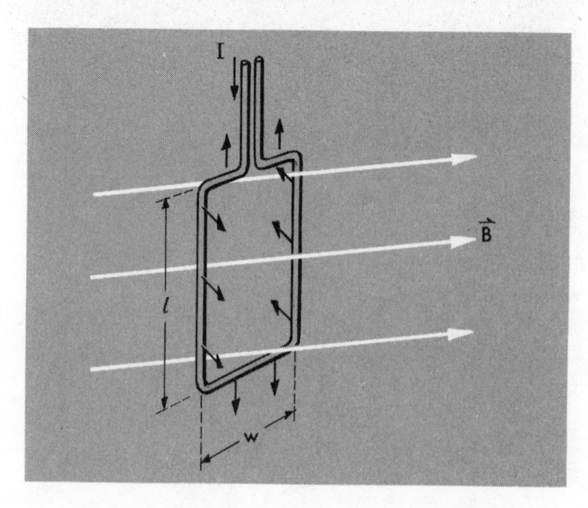

Figure 22–15
A schematic diagram of an ammeter. When current flows through the rectangular coil the magnetic force IlB and $-IlB$ tend to twist the coil. It turns until the magnetic forces are just balanced by springs. A meter is usually made so the angle of twist is proportional to the current.

In an ordinary ammeter for direct currents (D.C.) the coil is carried on an axle with pivots, and its twisting is opposed by the elastic forces of a pair of hairsprings (Fig. 22–16). The magnetic field is produced by a permanent magnet whose field gives a radial pattern. (This is produced by shaped pole pieces and a cylindrical block of iron that is held in the open space inside the coil.) Then even when the coil turns, it still finds the magnetic field perpendicular to its sides in such a direction that the magnetic forces twist it equally hard no matter how far it has turned. On the other hand, the opposing forces due to the springs increase as the coil turns. Consequently, the coil turns and reaches an equilibrium position at which the spring forces balance the magnetic forces. The greater the current, the greater the magnetic forces and the farther the coil must twist to reach equilibrium. Since the spring forces are usually proportional to the angle of twist, this angle is proportional to the current: the ammeter's scale is linear. Such meters are made in a wide range of sizes and scales.

(b) *D.C. Motors.* A motor contains a more robust moving coil, carrying current in a strong magnetic field. The D.C. motor's magnet can be either a permanent magnet or an electromagnet, an iron core inside a current-carrying coil. The coil is mounted on an axle in the magnetic field. If the current through the coil were always in the same direction, the coil would not continue to rotate; it would reach an equilibrium position. But a motor has a device to reverse the current in the coil every half turn. The coil is pulled around for half a turn; the current is then reversed and the coil is pulled around for another half turn. One arrangement of this reversing switch, or

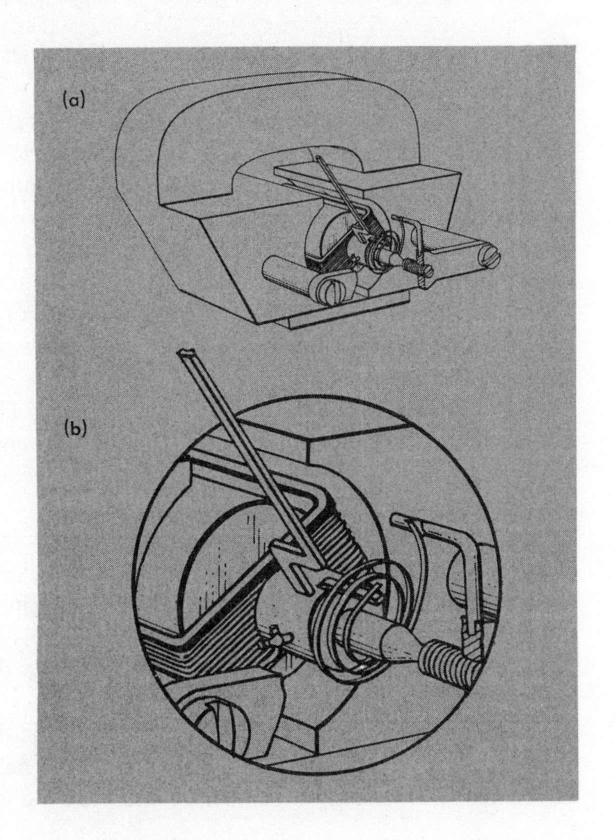

Figure 22–16
(a) An ammeter. Current enters the rectangular coil of many turns through the hairspring at the front of the shaft and leaves through a similar hairspring at the rear of the coil (not shown). When there is a current, the coil rotates around a fixed iron cylinder. The cylinder and the specially shaped poles of the large permanent magnet give a radial magnetic field of nearly uniform magnitude. (b) Close-up of the moving coil.

commutator, is shown in simple form in Fig. 22–17. Real motors do not have a single coil but rather a whole group of coils placed so as to give steadier turning forces. Also, the coils are wound on a block of iron to increase the magnetic field.

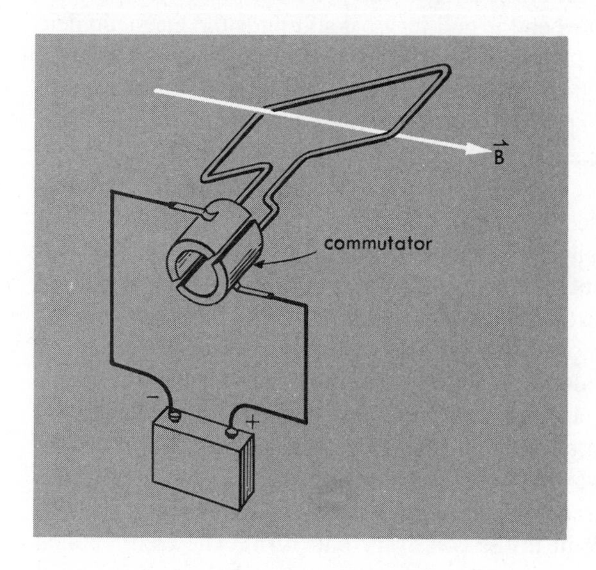

Figure 22–17
A D.C. motor is similar to a moving-coil meter. The coil is bigger, and a reversing switch or commutator is provided to reverse the direction of current flow every half turn.

22-6 Forces on Moving Charged Particles in a Magnetic Field

We have seen that a conductor carrying a current experiences a force when it is placed in a magnetic field. Since an electric current is a motion of charged particles, we naturally expect that the magnetic field acts directly on the individual charged particles—on the ions or electrons whose motion produces the current. The force exerted on the conductor as a whole is simply the resultant of the forces acting on the particles.

We can check up on this idea by firing a beam of electrons from an electron gun through mercury or hydrogen gas at low pressure. The electrons emerging from the slit make the low-pressure gas glow. Thus, the path of the beam is clearly visible in a darkened room (Fig. 22–18). Now we bring a magnet near the tube, pointed so that its magnetic field runs straight out from us across the tube. The beam is deflected at right angles to its motion and to the magnetic field.

To be specific, the electrons originally move vertically upward. Since electrons are negatively charged particles, this beam corresponds to a current directed vertically downward. Then, according to our rule for directions, the force acting on the moving particles must pull them to the right. In fact, as the beam goes up it is deflected to the right. If we reverse the magnet, the beam is deflected the opposite way.

We shall now compute the force acting on an individual charged particle moving with a given velocity in a known magnetic field. For this purpose we shall consider a stream of elementary charges evenly spaced and all moving

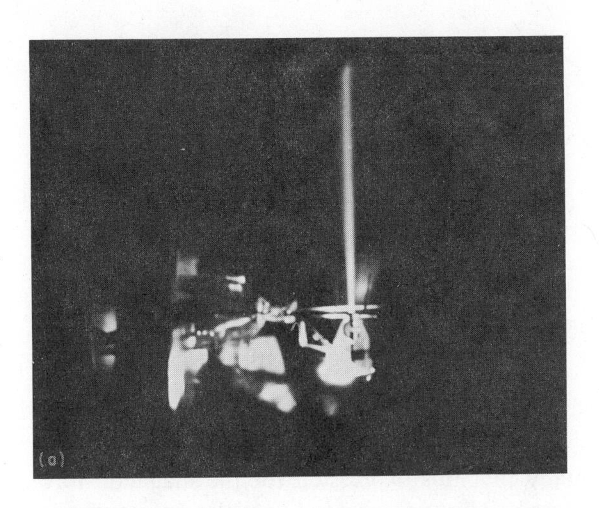

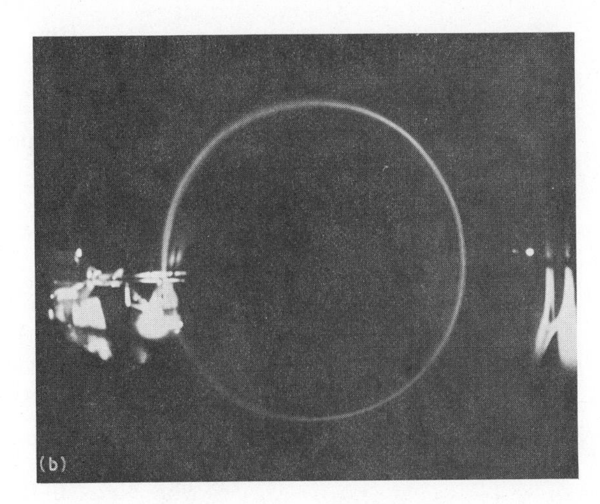

Figure 22–18
An experiment to show the deflection of moving charged particles in a magnetic field. In (a), electrons accelerated by the gun travel vertically upward, leaving a path of glowing gas. There is no magnetic field. When there is a magnetic field (b), the path of the electrons is bent into a circular arc. (We used a pair of coils rather than a bar magnet, to get a uniform field in a space into which we can look.)

with the same velocity, producing a current I. Then using $F = IlB_{\perp}$ we can find the force F_1 on one of these elementary charges. If there are N elementary charges in the length l, then

$$F_1 = \frac{IlB_{\perp}}{N}.$$

To express the force F_1 in terms of the speed v of the charges, we must evaluate the current in terms of the number of charges N in the length l. To see how I is expressed in terms of N, l, and v, we imagine an observer posted at the beginning of the length l to count the elementary charges as they go by (Fig. 22–19). Because every charge is moving at speed v, all N charges in the length l stream by him in the time $t = \frac{l}{v}$.

Now, the current I is the total charge passing per second. Since one elementary charge is 1.6×10^{-19} coulomb, the current in amperes is $\frac{1.6 \times 10^{-19} N}{t}$. By putting in l/v for the time, we can express this as

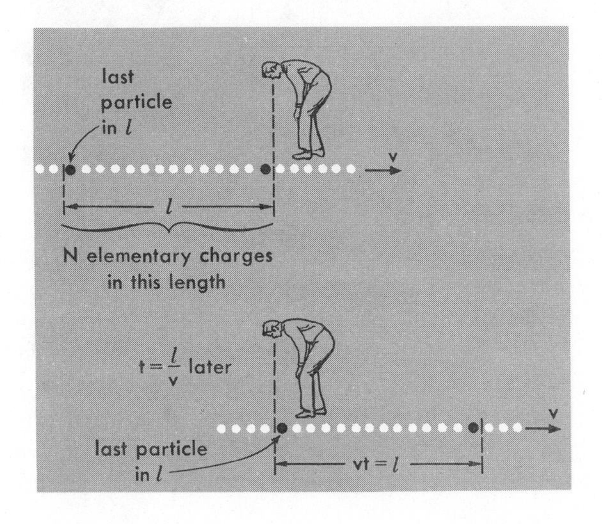

Figure 22–19
An imaginary observer is watching a stream of elementary charges moving with speed v. He counts N particles passing him when a length l of the stream goes by. This takes a time $t = l/v$.

$$I = \frac{1.6 \times 10^{-19}\,N}{l/v} = \frac{1.6 \times 10^{-19}\,Nv}{l}.$$

When we put this expression for I into our previous expression for F_1, we get

$$F_1 = \frac{IlB_\perp}{N} = \frac{1.6 \times 10^{-19} Nv}{l} \times \frac{lB_\perp}{N} = 1.6 \times 10^{-19} vB.$$

The force on an elementary charge is a function only of its speed and of the magnitude of the perpendicular component of the magnetic field. If a particle moving at speed v carries n elementary charges, the force on it will be just n times as great:

$$F = nF_1 = qvB_\perp.$$

Let us show experimentally that this expression for the magnetic force on a moving charge is correct by shooting charged particles across a magnetic field and observing their deflection by the magnetic force. Because the magnetic force is always perpendicular to the velocity, it cannot do work on charged particles shot across the field. They, therefore, move at constant speed. Consequently, $qvB_\perp$ always has the same value. In other words, as long as the field is uniform, the magnetic force is a deflecting force of constant magnitude. The particles should go around in a circle (see Fig. 22–18), and we can compute the radius of this circle by equating the magnetic deflecting force with mv^2/r, the centripetal force necessary to make a particle of mass m moving at speed v go around the arc of a circle of radius r (Fig. 22–20). From $mv^2/r = qvB_\perp$, we find that $r = mv/qB_\perp$. Therefore, we can test our expression for the magnetic deflecting force by sending particles of known mass, velocity, and charge perpendicular to a given magnetic field and measuring the radius of the circular arc on which they move.

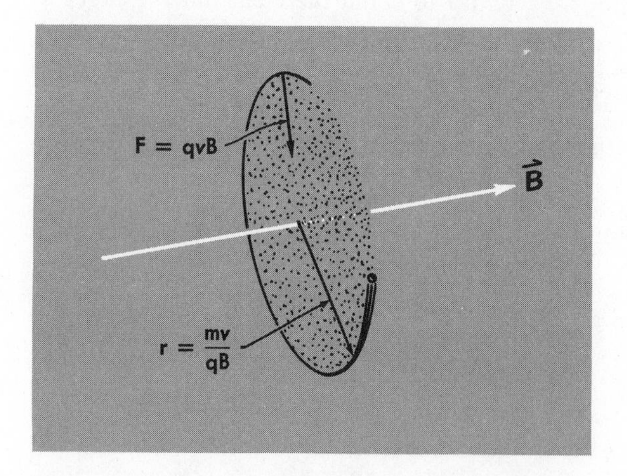

Figure 22–20
When a charged particle moves perpendicular to a magnetic field, it experiences a magnetic force $F = qvB$ at right angles to its path. If the field is uniform, the particle moves in a circle of radius $r = mv/qB$.

We already have the information for protons with which we can make this test. In Section 20–8 we measured the mass of a proton and found it was 1.7×10^{-27} kg. We also learned that when protons are accelerated from rest through a 90-volt potential difference, they emerge with a speed of 1.3×10^5 meters per second. Now we shoot test protons perpendicularly into a magnetic field. A practical magnetic field, for example, might be 1.0×10^{-2} newtons/amp-m. In this field the magnetic deflecting force should cause the protons to go around a circle of radius

$$r = \frac{mv}{qB} = \frac{1.7 \times 10^{-27} \times 1.3 \times 10^5}{1.6 \times 10^{-19} \times 1.0 \times 10^{-2}} = 0.14 \text{ m}.$$

This expected radius is actually observed when protons of 90 electron-volts energy are shot into that magnetic field. Thus, experiments show that the magnetic deflecting force is indeed given by $qvB_\perp$, in agreement with our reasoning from the magnetic force on currents.

22-7 Using Magnetic Fields to Measure the Masses of Charged Particles

To measure the masses of electrons and protons in Section 20–8, we used a known potential difference to accelerate electrons or protons to a known kinetic energy, $\frac{1}{2}mv^2$. Then we determined the speed v of these particles by doing a time-of-flight experiment.

We have another, older method of making very accurate mass measurement. Instead of measuring the speed v of the particles we can find their momentum mv by observing their deflection in a known magnetic field. If we know their kinetic energy and momentum, we can easily compute their mass. This is the method by which our most precise measurements of the masses of ions have been made in the past. It is particularly valuable for electrons because their high speeds make time-of-flight measurements difficult.

The measurement of momentum by deflecting a beam of particles in a magnetic field is basically the same measurement that we made at the end of the last section. There we used the deflection of protons as an experimental check on the expression $F = qvB_\perp$ for the magnetic force. Here we shall use the known magnetic force to find unknown masses.

To use this method to measure ionic masses, we accelerate the ions of unknown mass through a known electrical potential difference, V (Fig. 22–21). This tells us their kinetic energy:

$$\tfrac{1}{2}mv^2 = qV.$$

Then we allow the ions to go perpendicularly into the known magnetic field and measure the radius of the circular arc on which they move. The required centripetal force mv^2/r is supplied by the magnetic deflecting force $qvB_\perp = qvB$. Thus, $mv^2/r = qvB$, and so we get

$$mv = qBr.$$

Note that by measuring the radius of the circular arc, we can evaluate the momentum of the particle.

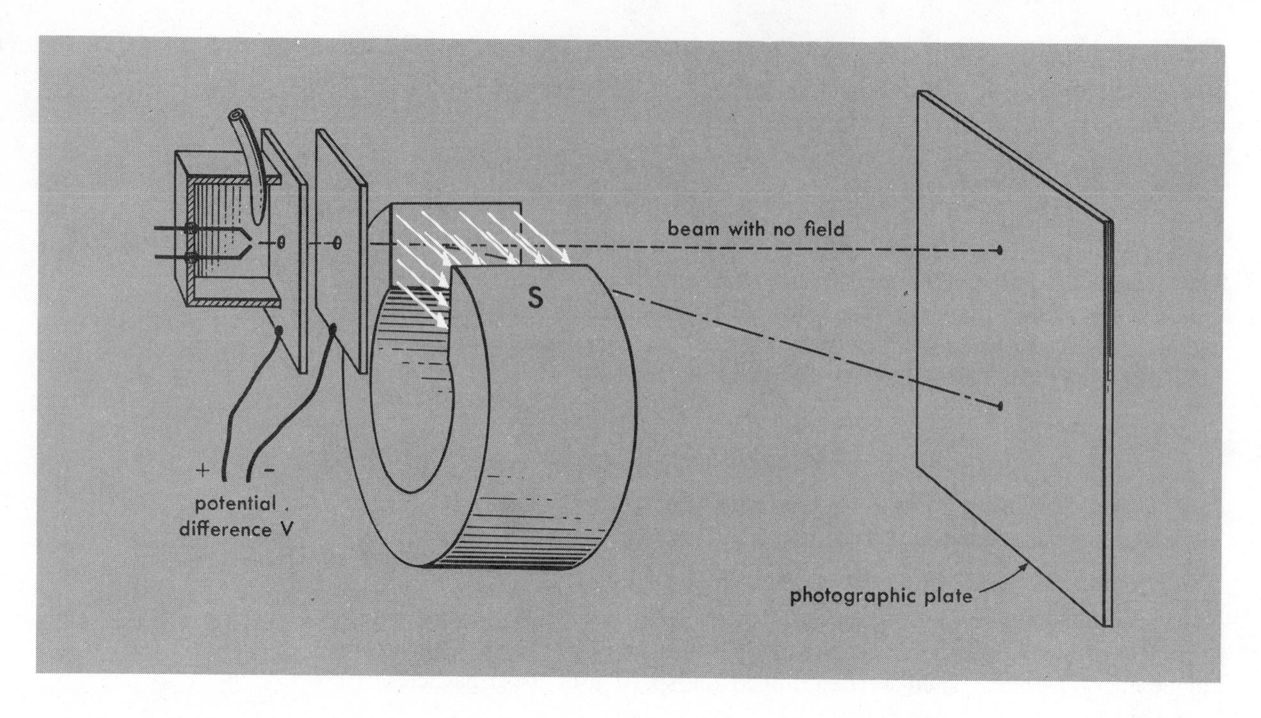

Figure 22–21
A simple mass spectrograph. Ions are accelerated to a known kinetic energy by the electric field between the plates. In the magnetic field their path is bent into a circular arc. The ions strike a screen or photographic plate, and from the deflection we can calculate the radius of the arc.

From the two equations

$$\tfrac{1}{2}mv^2 = qV$$
$$mv = qBr$$

we take $v = \dfrac{qBr}{m}$ from the second equation, square it, and substitute it into the first equation. This yields

$$m = \frac{B^2qr^2}{2V}.$$

Let us take a specific example. We accelerate singly charged sodium ions through an electric potential difference of 90 volts. The sodium ions then enter a uniform magnetic field of strength 0.100 newton/(amp-meter) at right angles to their direction of motion. Now we measure the radius of the arc through which they travel in the magnetic field, and we find that it is 0.066 meter. The ions are singly charged, each with 1.60×10^{-19} coulomb. Substituting these values into the expression for m yields

$$m = \frac{(0.100)^2 \times 1.60 \times 10^{-19} \times (0.066)^2}{2 \times 90} = 3.88 \times 10^{-26}\text{kg}.$$

This is the mass of sodium ion—good to within 2 percent. The mass of a sodium atom, which is made by adding one electron to the sodium ion, must be almost identical. So our value can be compared with the values obtained for the mass of sodium in other experiments. The accepted value for the mass of a sodium atom is 3.82×10^{-26} kg.

The apparatus that we have diagrammed in Fig. 22–21 is known as a mass spectrograph. Our mass spectrograph is very crude. One that is made for high-precision measurements usually contains auxiliary devices to "focus" the particles on a small spot and to make sure that only particles in a very narrow range of velocities can enter the magnetic field. With a good spectrograph masses of ions can be determined to many significant figures.

However, such measurements show that the masses of the elements are not exact integral multiples of the mass of the hydrogen atom. A precision mass spectrograph shows that many elements are composed or two or more kinds of atoms, differing in mass but having almost identical chemical properties; for example, 75 percent of the chlorine atoms found in nature have a mass of about 35 times the hydrogen atomic mass, while 25 percent have a mass of about 37 times the hydrogen atomic mass. Atoms of the same element which differ in mass we call different *isotopes* of the element. The two chlorine isotopes are called Cl^{35} and Cl^{37}. The superscript (called the *mass number*) labelling an isotope is an indication of the internal structure of the atoms of the isotope.

An isotope of carbon, C^{12}, has been chosen as the standard for the currently used scale of *atomic mass units.** Its atomic mass is defined to be precisely 12 atomic mass units; on this scale hydrogen has an atomic mass of 1.00797 amu and oxygen has an atomic mass of 15.9994 amu, for example (See Appendix). You will notice that the atomic mass of carbon is listed as 12.01 amu; this value, like the others in the table, is the average atomic mass of the element, when all isotopes are considered in their natural abundances. The C^{13} isotope, which has an abundance of 1.1 percent, raises the average atomic mass of carbon, as it occurs in nature, to a value slightly greater than an even 12 amu.

One advantage of the C^{12} scale over a hydrogen-based scale is that the mass of any isotope in amu is then very close to the mass number of that isotope. For example, the uranium isotope U^{238} has an atomic mass of 238.03 amu on the C^{12} scale, although it has only 236.20 times the mass of a hydrogen atom.

Mass spectrographs can be used also to measure the mass of molecules by deflecting a beam of ionized molecules in the apparatus just as we deflect a beam of ionized atoms.

Ever since J. J. Thomson introduced mass spectroscopy about fifty years ago, it has been a powerful tool of atomic research. In particular, because mass spectrographs sort the ions by their individual masses, the masses of the isotopes of a single element were largely determined by mass spectrographic measurement. Although some evidence for the existence of isotopes among the heavy radioactive elements was pointed out in the first few years of this century, it was Thomson's measurements showing the existence of the isotopes of neon that clinched the idea and started the accumulation of systematic data on isotopes. By now we have accumulated so much information that a huge table would be needed to exhibit it. Only a small sample of the known masses of isotopes is shown in Table 1.

* Incidentally, most of the tables of atomic masses printed before 1961 were based on oxygen as an even 16 amu. Due to a number of serious difficulties, the oxygen scale was officially abandoned in 1961 and replaced by the carbon-12 scale.

Table 1
A few of the isotopes of some common elements. One atomic mass unit is 1/12 of the atomic mass of the most common carbon isotope.

ELEMENT	MASSES OF THE ISOTOPES IN ATOMIC MASS UNITS	RELATIVE ABUNDANCE: PERCENT
Hydrogen	1.008	99.98
	2.014	0.02
Lithium	6.015	7.42
	7.016	92.58
Carbon	12.000 (standard)	98.89
	13.003	1.11
Nitrogen	14.003	99.63
	15.000	0.37
Oxygen	15.995	99.76
	16.999	0.04
	17.999	0.20
Neon	19.992	90.92
	20.994	0.26
	21.991	8.82
Sodium	22.999	100
Magnesium	23.985	78.70
	24.986	10.13
	25.983	11.17
Chlorine	34.969	75.53
	36.966	24.47
Iron	53.940	5.82
	55.935	91.66
	56.935	2.19
	57.933	0.33
Silver	106.905	51.82
	108.905	48.18
Lead	203.973	1.5
	205.975	23.6
	206.976	22.6
	207.977	52.3

22-8 What Alpha Particles Are

In Chapter 24 we shall take a look at some of the evidence about the structure of atoms accumulated by Rutherford and his co-workers in the first few decades of this century. Rutherford's chief experiments were done with alpha particles. We already know how Rutherford and Geiger determined that alpha particles carry two elementary charges (see Section 19–10). Now we shall describe two experiments equivalent to those done by Rutherford and Robinson to determine the mass of the alpha particle. In one of these experiments we determine the momentum of an ion by measuring its deflection in a magnetic field. Because we have just described this kind of experiment in the last section, we only mention the result. When we do this experiment using polonium as the source of alpha particles, we find that their momentum is $mv = 1.06 \times 10^{-19}$ kg-meter/sec.

Now, to obtain the mass of the alpha particle, we need another measurement which will enable us to calculate the speed. We can do this by applying an electric field across the motion of the alpha particles at the same time that we apply a magnetic one. We put on this electric field so that the electric force on the alpha particles is just opposite to the magnetic force. Then the net force on the moving particles is zero, and they move straight ahead, following the same path that they take in the absence of both fields (Fig. 22–22). The electric force on the particles is qE, and the magnetic force qBv; and when these forces balance, $qBv = qE$. The speed of the particles is therefore given by $v = \frac{E}{B}$, and since we know both the magnetic and the electric field, we know the answer. For polonium alpha particles the measured speed is $v = 1.6 \times 10^7$ meters per second. Consequently, the mass is

$$m = \frac{mv}{v} = \frac{1.06 \times 10^{-19}}{1.6 \times 10^7} = 6.6 \times 10^{-27}\ \text{kg}.$$

Within the experimental accuracy, this mass is almost exactly equal to the mass of a helium atom. It is thus natural to conclude that alpha particles are doubly charged helium ions. We can test this conclusion directly by placing a strong source of alpha particles in an evacuated glass tube. Such an experiment was done by Rutherford and Royds in 1908. After waiting for a sufficiently long time, they found a detectable amount of helium.

We shall use our knowledge of what alpha particles are to explore the structure of atoms in Chapter 24.

Figure 22–22
Measuring the speed of alpha particles. With a pair of metal plates placed horizontally in the tube we can apply an electric force equal and opposite to the magnetic force. Then the particles are not deflected, and their speed $v = E/B$. This figure is schematic. In a real experiment, we must change the dimensions to obtain known uniform electric and magnetic fields. Also, the apparatus is in a vacuum system to avoid scattering and energy loss by alpha particles hitting air molecules.

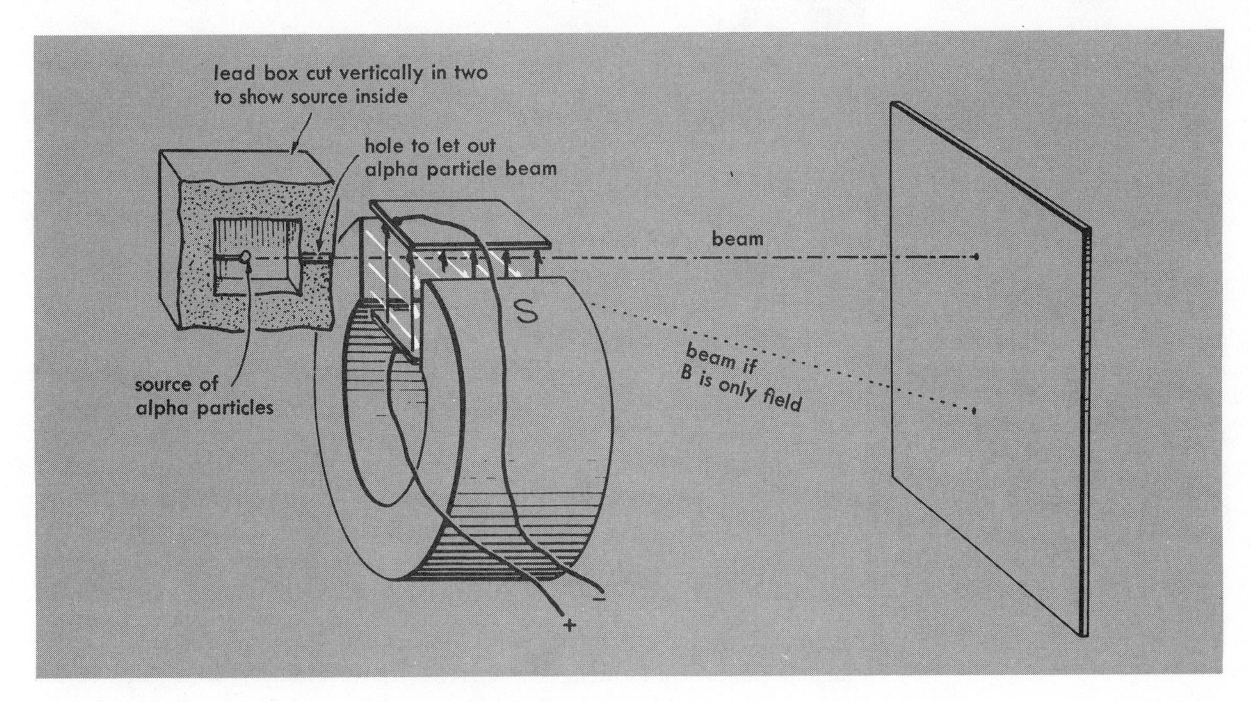

22-9 Magnetic Field Near a Long Straight Wire

When we were studying the electric field due to a stationary charge, one of the first questions we answered was how the field falls off with distance from the charge. We have not yet asked how the magnetic field of a steady current varies with distance from the current.

In 1820 the French physicists Biot and Savart announced the results of the first quantitative investigation of a magnetic field produced by an electric current. This was the field near a long straight wire (Fig. 22–23). Near the middle of such a wire, the field arises almost entirely from the current nearby. The other parts of the current do not contribute much field because the magnetic effects decrease rapidly with distance. (You can check this by moving the other parts of the circuit.) The field lines form concentric circles around the wire [Fig. 22–5 (b) and Fig. 22–7 (a)]; and the strength B of the field at a distance r from the wire was found by Biot and Savart to be inversely proportional to r. We have seen earlier in this chapter that B is directly proportional to I. Consequently,

$$B_{\text{st. wire}} = K\frac{I}{r}.$$

As usual, the proportionality constant depends on the choice of units. We shall make an experimental determination of this constant in Section 22–11.

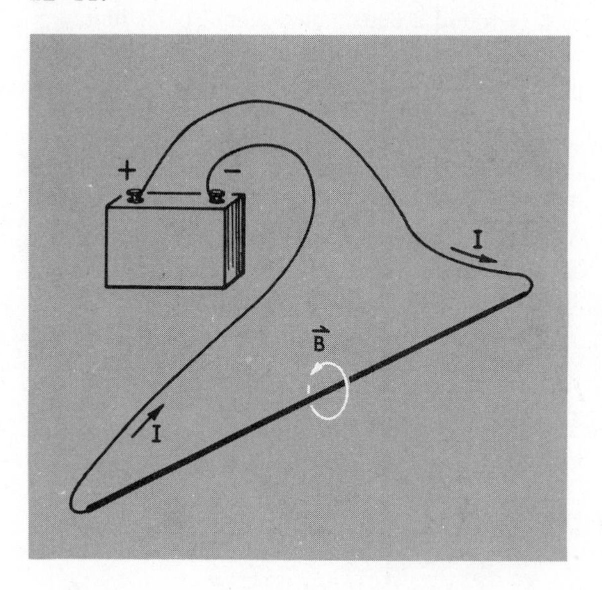

Figure 22–23
The field near the middle of the wire (where the field is shown) is uninfluenced by the parts of the circuit which are far away. See Fig. 22–5(b) and (c) and 22–7(a).

22-10 Magnetic Circulation

An important general relation can be derived from the formula for the magnetic field of a straight-line current. If we multiply both sides of the expression $B = K\frac{I}{r}$ by $2\pi r$ we obtain

$$2\pi rB = 2\pi KI.$$

The left side, $2\pi rB$, is simply the strength of the field multiplied by the length of the closed circular path to which it is tangent. And since B is inversely proportional to the radius of the circle, the product $2\pi rB$ is the same for all circles surrounding a straight-line current I. This product $2\pi rB$ is a special case of what is called the *circulation* of $\vec{\mathbf{B}}$ or the magnetic circulation. More generally, we may take any closed path, a smooth circle or an irregular loop, such as shown in Fig. 22–24, multiply the length of each small, nearly straight segment of the path by the component of $\vec{\mathbf{B}}$ along that segment and sum these products for all segments; the result is the circulation of $\vec{\mathbf{B}}$ for the path loop.

The importance of the idea of magnetic circulation is that for any loop the magnetic circulation depends only on the net current threading the loop. In Fig. 22–25, for example, the magnetic circulation of loop (1) and loop (2) are the same since the same current, I, threads both loops. The magnetic circulation of the adjacent loop (3), on the other hand, is zero even though it is in the magnetic field of I, since no current threads it.

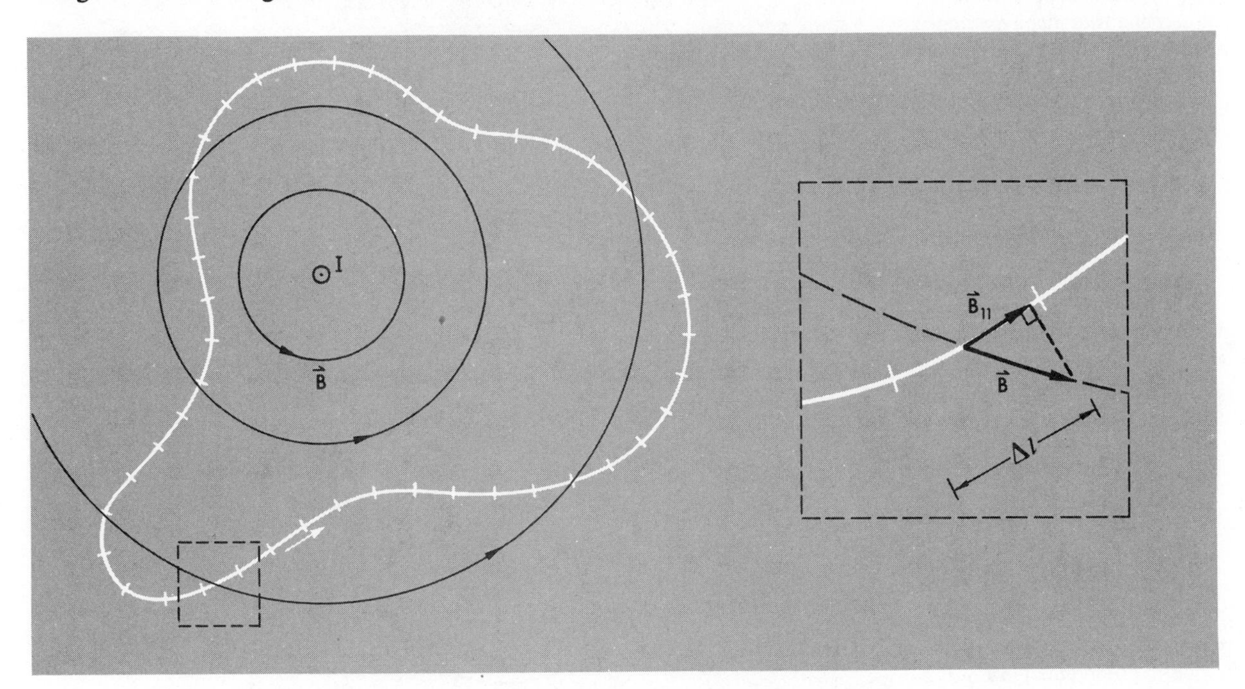

Figure 22–24
The magnetic circulation around the irregular loop is the sum around the loop of the products of each tiny segment of length Δl times $B_{||}$, the component of $\vec{\mathbf{B}}$ along the segment.

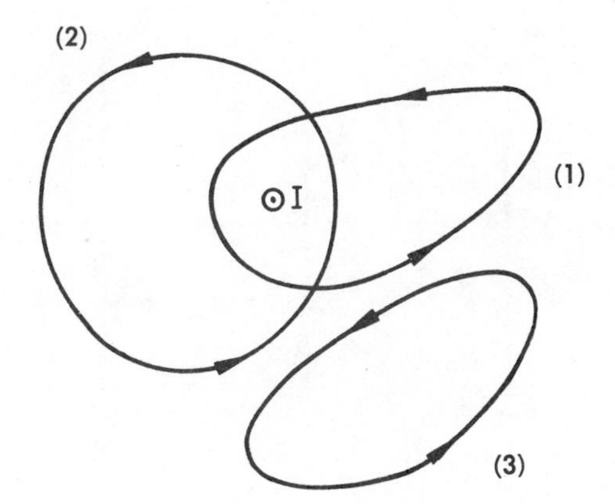

Figure 22–25
The magnetic circulation around loop (1) equals the magnetic circulation around loop (2), but the magnetic circulation around loop (3) is zero.

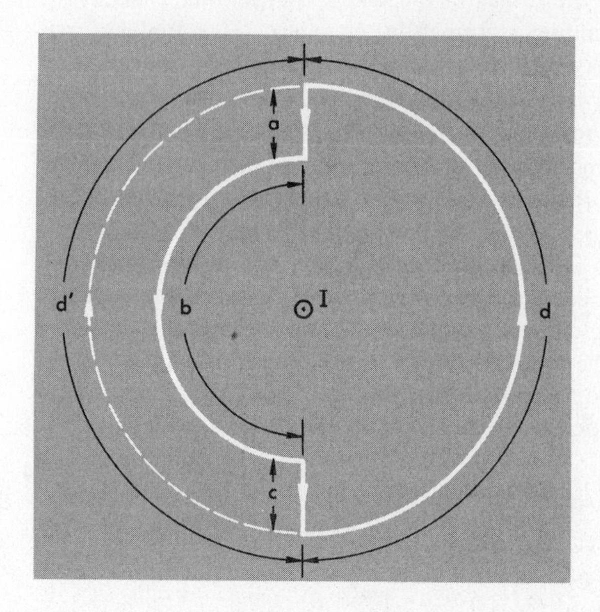

Figure 22–26
Magnetic circulation paths about a current-carrying wire that is perpendicular to the page. The direction of the current is out of the page, and the direction of the magnetic field is therefore counterclockwise.

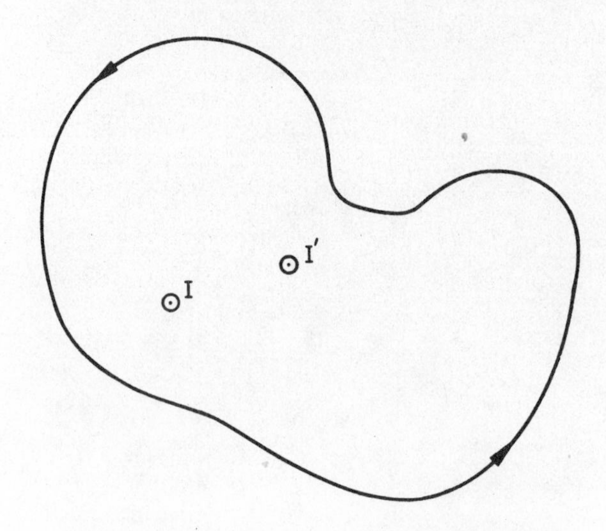

Figure 22–27
The magnetic circulation around the loop is proportional to $I + I'$.

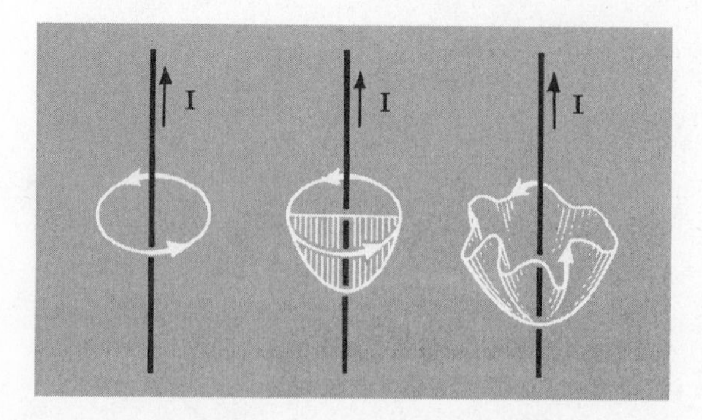

Figure 22–28
The magnetic circulation of $\vec{B}$ depends only on the total current passing through the surface bounded by the loop. This is true even when the surface is not a plane surface; indeed, even the loop itself need not lie in a plane.

To see why this is so, consider first the counterclockwise solid line of path *abcd* of Fig. 22–26. The contributions to the circulation of $\vec{\mathbf{B}}$ come only from segments b and d, since $\vec{\mathbf{B}}$ has no component along the radial segments a and c. The length of the arc b is proportional to its radius, and B is inversely proportional to the radius, so that the contribution to the circulation, bB, depends only on the angle subtended by the arc. This is true for all arcs, of any radius, so that if one goes *around* the current along the path *abcd* the total magnetic circulation is $2\pi KI$. Suppose, on the other hand, the loop is made along the path *abcd′*. The component of $\vec{\mathbf{B}}$ is opposite to the direction of the path along the dashed segment d' and the contribution from d' will cancel that from *abc*; the total magnetic circulation of the path *abcd′* is therefore zero. Thus we see that, even though a loop may lie in a region of magnetic field, it will have a zero magnetic circulation if it is not threaded by a current.

If the position of the loop with respect to the current threading it does not matter, and the magnetic circulation depends only on the current threading the loop, two currents I and I' passing through different points in the loop will simply produce a magnetic circulation proportional to $I + I'$ (Fig. 22–27). The circulation of $\vec{\mathbf{B}}$ depends only on the total current passing through the surface bounded by the loop. It need not even be a plane surface, and the loop itself need not be planar (Fig. 22–28). The currents need not even be confined to long, straight narrow filaments. The general theorem relating the property of the magnetic field to currents was the culmination of the thorough investigation by Ampère (1775–1836) on the magnetic effects of steady currents. It is called Ampère's circuital law, or the circulation theorem for magnetic fields.

22-11 Uniform Magnetic Fields

To get a magnetic field that does not change with time, all we have to do is keep the current constant that generates it. But how do we make a field that is uniform over a reasonable volume of space? From Chapter 19 you will remember that to make a uniform electric field in space we used two parallel, closely spaced metal plates carrying equal and opposite charges. With such an arrangement, the electric field was constant between the plates except near the edges.

Does the analogous prescription work here? Does a large uniform sheet of current, like a large uniform sheet of charge in the electric case, give rise to a uniform magnetic field? It does. In fact, in the case of a magnetic field we can bend the plane current sheet into a cylinder and still get a uniform field inside. In practice, the current sheet is composed of one long wire wound closely and uniformly as a cylindrical coil. The field inside is along the axis of the coil [Fig. 22–5 (e)] and uniform except near the ends.

To find the magnetic field inside such a coil, or "solenoid" as it is often called, we can use the magnetic circulation theorem discussed in the last section. For our loop we take a rectangle with one side inside the coil and parallel to its axis where the field is $\vec{\mathbf{B}}$; the rectangle is completed outside the coil, where the field is very weak (Fig. 22–29). [Notice that in Fig. 22–5 (e) the iron filings are not noticeably lined up in any direction just outside the coil, except very near the ends.] This side therefore does not contribute anything of significance to the magnetic circulation. Neither do the two short sides since the field lines are very nearly perpendicular to these segments, and only the component of field parallel to the path contributes to the magnetic circulation. So, nearly the entire magnetic circulation comes from the length l inside the coil. We conclude, therefore, that the magnetic circulation around the rectangle is just Bl.

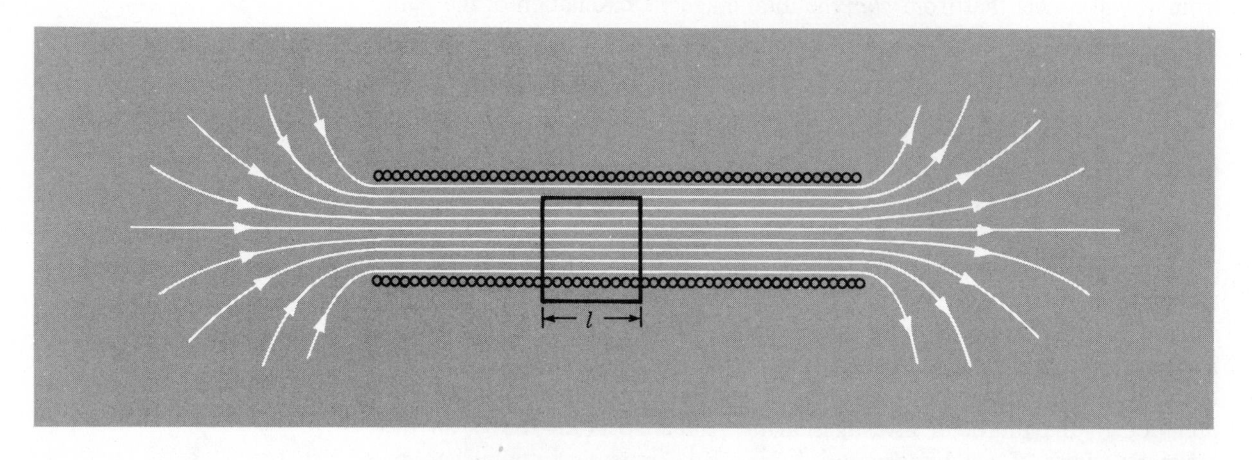

Figure 22–29
To find the magnetic field inside a long solenoid, we take the magnetic circulation around a rectangle, one side of which is inside the coil and the other outside.

Suppose the rectangle threads n turns of wire. Then, from Ampère's theorem, the circulation of $\vec{\mathbf{B}}$ around the loop is $Bl = 2\pi KnI$ and the field inside is $B = 2\pi KnI/l$. That is, it depends only on the number of windings per unit length of the coil n/l and on the current I. It does not depend on the length or diameter of the coil, as long as n/l is constant and the length of the coil is large compared with the diameter. If you displace the rectangle to the right or left (but stay away from the ends) the current it threads does not change, since the coil is uniformly wound. The same holds for a displacement of the rectangle forward or backward, up or down, as long as the segment stays inside the coil. So, according to Ampère's theorem, the magnetic circulation around the rectangle does not change. Figure 22–5 (e) shows that the field lines inside the coil are everywhere parallel to the side l. Therefore, not only is the magnetic circulation constant, but also the field B is constant inside the coil, except near the ends. If we put $N = n/l$, the number of windings per unit length in the coil, the expression for the field inside the coil is $B = 2\pi KNI$.

Now, to determine the value of K in this expression all we need to do is make a measurement of the uniform field in a long solenoid having N winding per unit length and carrying a current I. We can do this with the "current balance" shown in Fig. 22–30. The end of the balance carrying a short, straight section of wire of length l' is placed in the uniform field of the solenoid [Fig. 22–31 and 22–32]. With current I in the coil and I' in the loop, different known masses m are placed on the end of the balance outside the coil until balance is obtained. Then since $F = I'l'B_{\perp}$, the field B inside

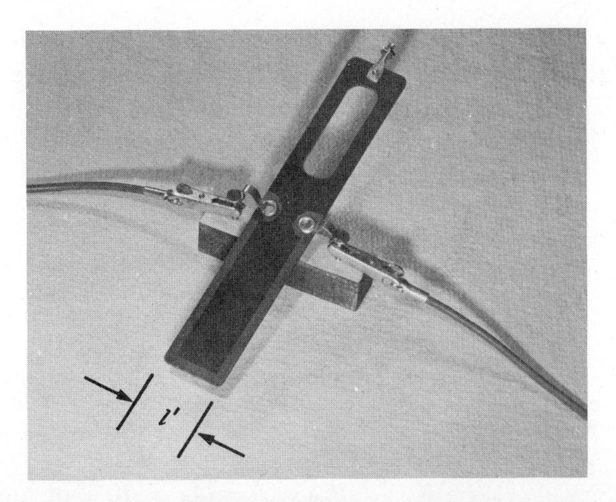

Figure 22–30
One form of the "current balance" which can be used to determine absolute units for the magnetic field, B.

Figure 22–31
The far end of the current balance, carrying the short current element $I'l'$, is in the center of the solenoid where the magnetic field is uniform. Short pieces of string of known mass are hung on the near end to balance the magnetic force on the current element.

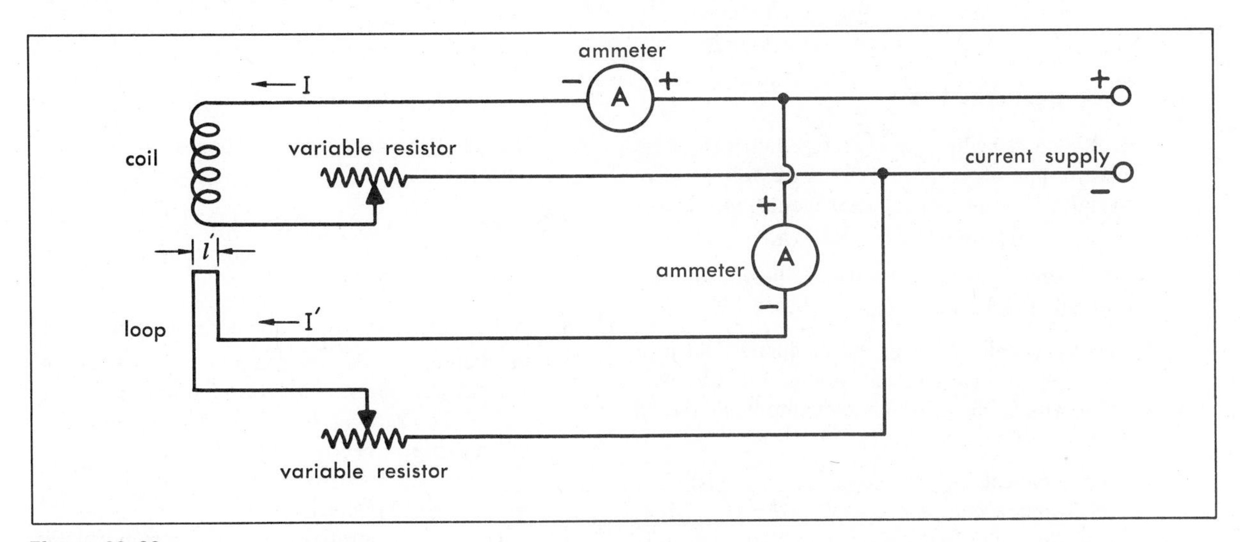

Figure 22–32
The circuit connections for the experiment.

the coil is $mg/I'l'$. Equating this value for B to the expression for B we obtained using the circulation theorem, we have $mg/I'l' = 2\pi KNI$. This gives for the value of the constant

$$K = \frac{mg}{2\pi NII'l'}.$$

When this experiment is carefully performed, the value of K is found to be 2.00×10^{-7} nt/amp². We can now include the value of the constant in the equations for the magnetic circulation, the field about a long straight wire, and the field inside a long solenoid:

$$\text{Circulation} = (4\pi \times 10^{-7}\ \text{nt/amp}^2)\, I$$

$$B_{\text{st. wire}} = (2 \times 10^{-7}\ \text{nt/amp}^2)\frac{I}{r}$$

$$B_{\text{solenoid}} = (4\pi \times 10^{-7}\ \text{nt/amp}^2)\, NI.$$

Here I must be measured in amperes and lengths in meters; then the field B comes out in $\frac{\text{Newtons}}{\text{amp-m}}$ as we expect. You may wonder why the proportionality constant, K, came out to be such a simple number as $2.00 \times 10^{-7}\,\frac{\text{nt}}{\text{amp}^2}$. The answer lies in the choice of units we made in Chapter 19 when we picked the ampere to be 6.25×10^{18} elementary charges per second. This made our unit for current agree with the "absolute ampere" which is usually determined by choosing K to be *exactly* 2×10^{-7} and doing an experiment similar to the one described here.

Looking back at the contents of this and the previous chapters, you will realize that we have dealt so far with electric fields produced by electric charges at rest, and with magnetic fields produced by steady currents. Such fields are constant in time. In the next chapter we shall again turn to experiment to learn what happens when electric and magnetic fields vary with time.

For Home, Desk, and Lab

1.* If a long straight wire carrying a current were placed flat on a paper and iron filings were sprinkled on the paper, what would you expect the iron filings to do? (Section 2.)

2.* Do magnetic field lines have a beginning and ending? (Section 2.)

3.* What is the direction of the magnetic field produced by positive charges moving directly away from you? Negative charges moving toward you? (Section 2.)

4.* Two identical circular loops carry equal currents, have a common center, and are at right angles to each other. How does the magnitude of the field they produce compare with the magnitude of the field from just one of them? (Section 3.)

5.* How does the field produced by a current-carrying wire depend on the magnitude of the current? (Section 3.)

6. A horizontal uniform magnetic field $\vec{\mathbf{B}}$ is perpendicular to the horizontal component $\vec{\mathbf{B}}_e$ of the earth's magnetic field and points east.
   (a) The ratio $B/B_e = \sqrt{3}$; in which direction will a compass needle point?
   (*Note:* the compass needle rotates in a horizontal plane.)
   (b) If the compass needle points northeast, what is the magnitude of $\vec{\mathbf{B}}$ now?

7. Two circular wire loops of equal radius are set up at right angles to each other with a common center (Fig. 22–33). If $I_1 = 3.0$ amperes and $I_2 = 4.0$ amperes, what is the direction of the magnetic field at the center, *0*?

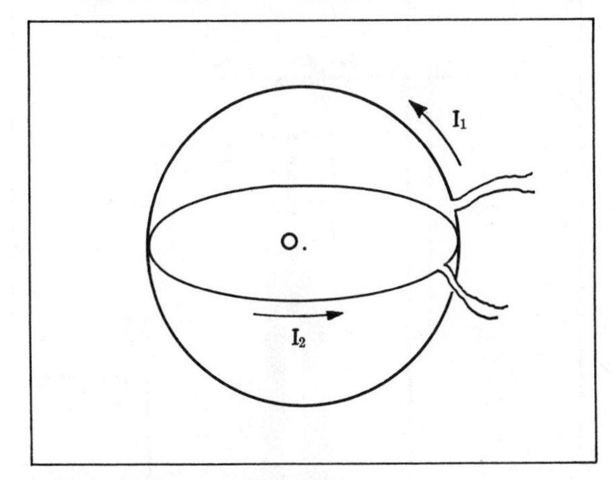

Figure 22–33 For Problem 7.

8. Suppose two loops are fixed at right angles and you can run a current in either direction through either loop. Neglecting the earth's field, how many different magnetic fields could you produce if a current of 2 amps flows whenever a loop is connected?

9.* In what direction can an electric charge move through a magnetic field without having a force exerted on it? (Section 4.)

10.* Suppose a wire carries a current from left to right through a magnetic field whose direction is toward you. In what direction is the magnetic force on the wire? (Section 4.)

11.* What is the force on a wire 0.20 meter long, carrying a current of 3.0 amps, which is placed within a uniform magnetic field of 5.0 newtons/amp-m at an angle of 45° to the field? (Section 4.)

12. The magnetic field inside a long solenoid is uniform and parallel to the axis of the solenoid.
    (a) What is the force on a current-carrying wire inside the solenoid and parallel to the axis?
    (b) What is the force on the wire shown in Fig. 22–34 which carries a current of 1.0 ampere, if the field $\vec{B}$ in the solenoid is 1.0×10^{-2} newton/amp-m and the length *CD* is 2 cm?
    (c) How much mass *m* should be placed on the other end of the balance to keep it from tilting?
    (d) Describe how you could use this setup as an ammeter.

13. (a) Show that a square loop of wire carrying a current will always tend to align itself in a magnetic field so that the plane of the loop is perpendicular to the field.
    (b) What do the magnetic forces tend to do to the loop in this position? (Hint: draw a side view.)

14.* How does the force that opposes the magnetic force in a moving-coil ammeter depend on the angle of rotation of the pointer on the ammeter's scale? (Section 5.)

15. In the moving-coil ammeter of Fig. 22–16, the magnetic force on the coil is balanced by a torsion spring. An ammeter can also be con-

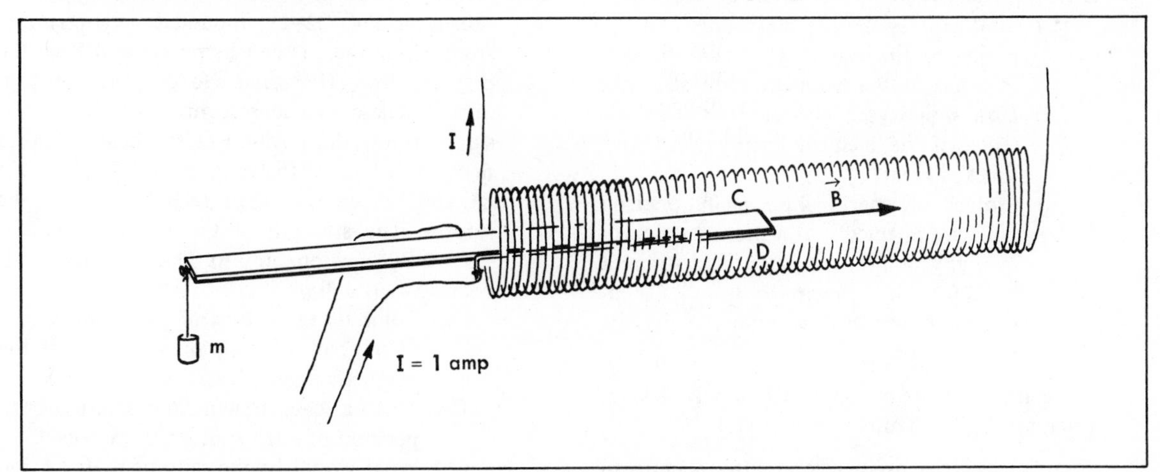

Figure 22–34 For Problem 12.

structed as shown in Fig. 22–35, using a small spring balance to measure the magnetic force on the coil. By adjustment of the position of the support, the conductors can be kept in the same position in the magnetic field for different currents.

One such coil, 4.0 cm wide by 25 cm high, was constructed of 100 turns and weighed 2.4 newtons. The spring balance had a scale of 0 to 5 newtons.

(a) When supported in a magnetic field, as shown, the spring balance changed from 2.4 to 2.7 newtons as the current was increased from 0 to 0.5 ampere. What is the maximum current that could be measured on this spring balance?

(b) What is the magnetic field in the air gap of the magnet?

16.* How could the direction of rotation of the coil in Fig. 22–17 be reversed? (Section 5.)

17.* What happens to the radius of the circle in Fig. 22–18(b) if the accelerating potential inside the electron gun is raised? (Section 6.)

18.* Upon what does the magnetic force on a moving electric charge depend? (Section 6.)

19. How do the momenta of a proton and an electron compare if both are moving perpendicular to the same magnetic field in circles of the same radius?

20.† Would it be easier for a mass spectrograph to separate the isotopes of light or heavy elements?

21. A beam of protons of kinetic energy E_K is injected into a region of uniform magnetic field $\vec{B}$ at right angles to the direction of the beam.

(a) Find expressions for the radius and the period of the resulting circular motion in terms of the magnetic field strength.

(b) How would your answers to (a) be affected if the incident energy were four times as great?

(c) Protons of energy 4.60×10^5 electron volts are injected into a proton synchrotron in which the magnetic field strength is 2.10×10^{-2} newton/amp-m. Calculate the period and the radius of the circular motion.

22.* What is the kinetic energy of alpha particles accelerated by 180 volts? (Section 7.)

23. Charged particles are accelerated by a uniform electric field and then deflected by a magnetic field. What quantities must be measured to determine the mass of the particles?

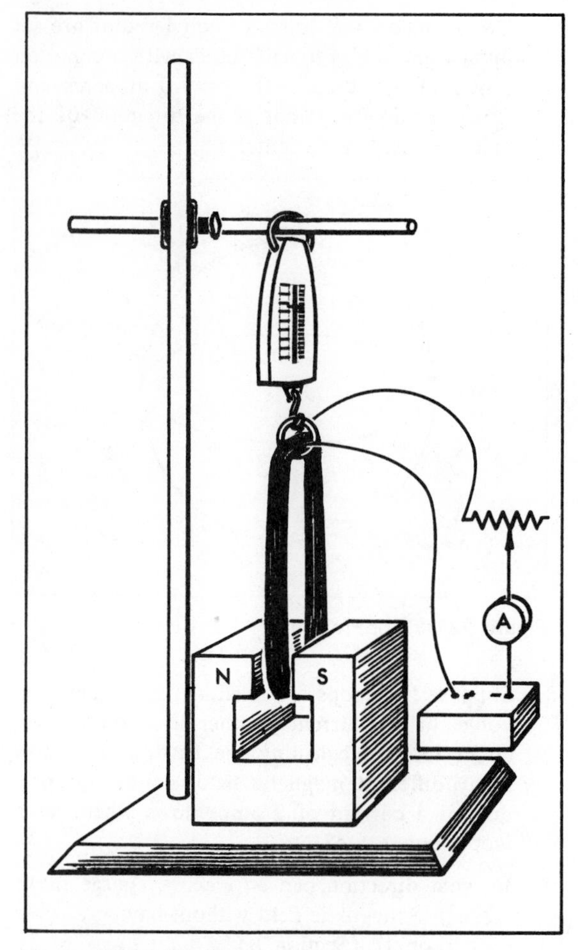

Figure 22–35 For Problem 15.

24. When helium gas is fed into the apparatus shown in Fig. 22–21, mounted in a chamber from which the gas can be pumped to keep the pressure low, He^+ and He^{++} ions are produced. These two ions form two spots on the screen below the position of the beam with no field. The face of the magnet is 3.0 cm by 1.5 cm and the center of the face is 10 cm from the screen. The strength of the magnetic field is 1.5×10^{-2} newton/amp-m. The ions are accelerated by 90 volts.

(a) Compute the radius of the circular part of the path of each ion while it is in the magnetic field.

(b) Make a scale drawing and determine the position of each spot on the screen.

(c) What would be the path of a He^{++} ion which gains one electron while in the

electric field? Describe it qualitatively.
(d) What would be the path of a He^{++} ion which gains one electron while in the magnetic field? Describe it qualitatively.

25. A nearly uniform electric field is set up parallel to the surface of a glass-topped table by charging two large parallel, vertical metal plates. The plates are 0.50 m apart and the potential difference between them is 20 kv. A small Dry Ice puck with a total mass of 60 grams carries a copper-coated Ping-Pong ball which has a charge of 2.2×10^{-8} coulomb.
 (a) What is the electric force on the ball when it is placed between the plates?
 (b) What would be the shape of its path if it were given a brief push perpendicular to the field lines?
 (c) Even in a restricted region, the largest magnetic field that can be obtained fairly easily is 0.50 newton/amp-m. If you projected the puck through such a field perpendicular to the field at a speed of one meter/second, what would be the magnetic force on the puck?

26. A beam of singly charged ions moves into a region of space where there is a uniform electric field $E = 1.0 \times 10^3$ newton per coulomb, and a uniform magnetic field $B = 2.0 \times 10^{-2}$ newton/amp-m. The electric and magnetic fields are at right angles to each other and both are perpendicular to the beam so that the electric and magnetic forces on an ion oppose each other. What is the speed of those ions which move *undeflected* through these crossed electric and magnetic fields?

27. The undeflected ions of Problem 26 are passed through a slit and move into a region where there is a uniform magnetic field $B = 0.09$ newton/amp-m at right angles to their motion. If the ions are a mixture of the neon ions of mass 20 and 22 atomic mass units, how far apart will these ions land on a photographic plate after they have moved through a semicircle?

28.* A wire carrying a current I threads a square loop 2 cm on a side. What is the circulation around the square? (Express your answer in terms of K.) (Section 10.)

29.* What is the circulation around the loop shown in Fig. 22–36? (Section 10.)

30.* What is the field in the center of a long solenoid of 10 turns/cm carrying 5 amp? (Section 11.)

31. Two very long straight parallel wires, one carrying 10 amperes and the other 20 amperes, are separated by a distance of 10 cm (Fig. 22–37). Find the magnitude and direction of $\vec{B}$ at the points P, Q, and R shown in the figure.

32. A long straight wire carrying a current $I = 100$ amperes is perpendicular to a uniform magnetic field of strength $B = 1.0 \times 10^{-3}$ newton/amp-m (Fig. 22–38). What is the strength and direction of the resultant $\vec{B}$ field at the points P, Q, R, S on a circle of radius 1 cm around the wire?

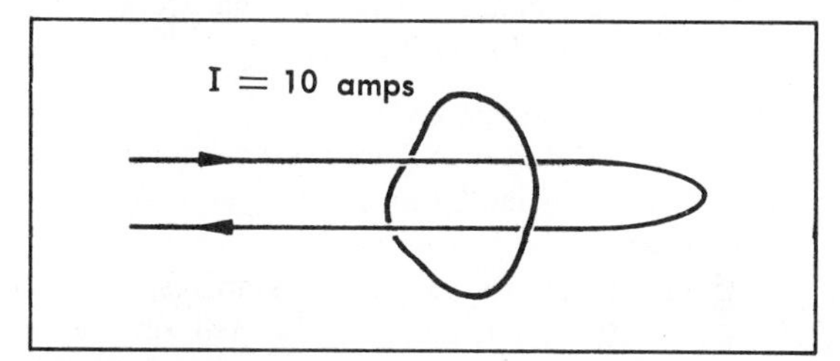

Figure 22–36 For Problem 29.

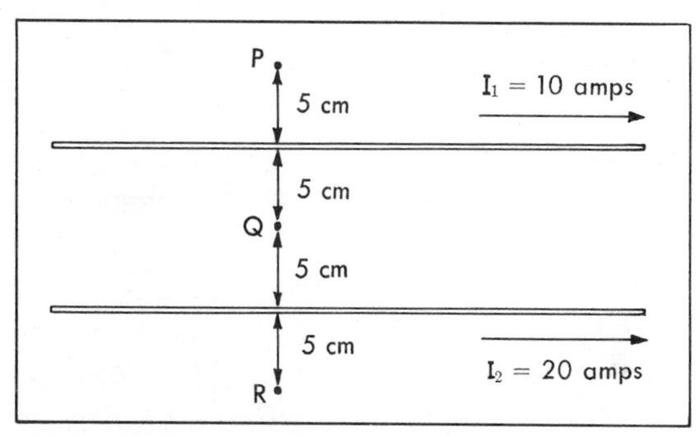

Figure 22–37 For Problem 31.

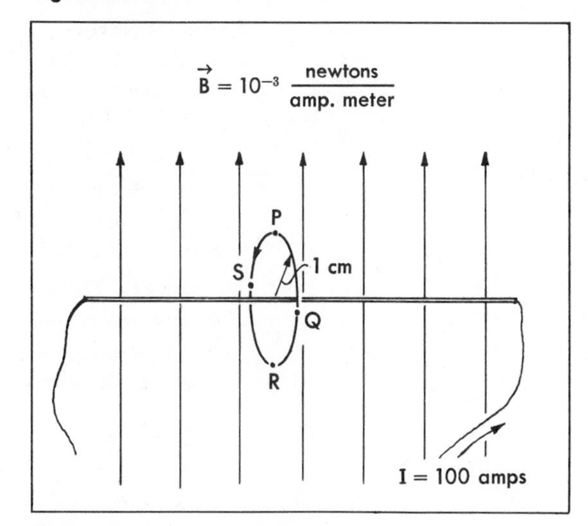

Figure 22–38 For Problem 32.

33. Suppose we have a beam of electrons moving at a speed $v = 3.00 \times 10^6$ m/sec and carrying a current of 1.00 microampere.
    (a) How many electrons per second pass a given point?
    (b) How many electrons are in 1.00 meter of the beam?
    (c) What magnetic field does the beam produce at 1.00 m distance?
    (d) What is the total force on all of the electrons in 1.00 m of the beam if it passes through a field of 0.10 newton/amp-m?
    (e) What is the force on a single electron if you assume that each electron experiences the same force?

34. Point A is 2 cm from each of 3 currents I_1, I_2, and I_3, as shown in Fig. 22–39. All three currents are carried by long wires perpendicular to the paper.
    (a) If $I_1 = I_2 = I_3 = 10$ amperes, and all three currents are directed out of the paper, what is the magnitude and direction of the magnetic field at A due to the currents in the wires?
    (b) If $I_1 = I_2 = I_3 = 10$ amperes, but I_2 is into the paper while I_1 and I_3 are out, what is the magnitude and direction of the magnetic field at A due to the currents in the wires?

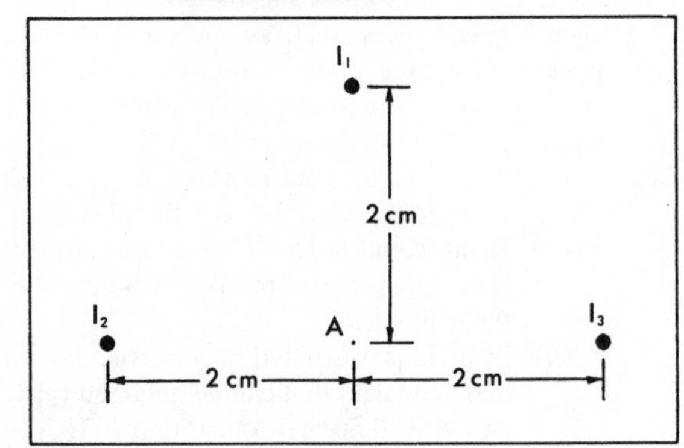

Figure 22–39 For Problem 34.

Further Reading

These suggestions are not exhaustive but are limited to works that have been found especially useful and at the same time generally available.

ANDRADE, E. N. DA C., *Rutherford and the Nature of the Atom.* Doubleday Anchor, 1964: Science Study Series.

BITTER, FRANCIS, *Magnets—The Education of a Physicist.* Doubleday Anchor, 1959: Science Study Series. (Chapters 2, 3, and 5)

BONDI, HERMANN, *The Universe at Large.* Doubleday Anchor, 1960: Science Study Series. Explains the effects of magnetic fields in space and in the earth's atmosphere and suggests the possible source of the earth's magnetism. (Chapters 8 and 12)

MACDONALD, D. K. C., *Faraday, Maxwell and Kelvin.* Doubleday Anchor, 1964: Science Study Series.

PIERCE, JOHN R., *Electrons and Waves.* Doubleday Anchor, 1964: Science Study Series. (Chapter 4)

WEISSKOPF, VICTOR F., *Knowledge and Wonder.* Doubleday Anchor, 1963: Science Study Series. (Chapter 3)

WILSON, ROBERT R., and LITTAUER, RAPHAEL, *Accelerators.* Doubleday Anchor, 1960: Science Study Series. (Chapters 5 and 6)

Electromagnetic Induction and Electromagnetic Waves

CHAPTER

23

23-1 Induced Current

After Oersted's discovery that an electric current is accompanied by a magnetic field, several people began to look for an "inverse" effect: Can a magnet produce a current?

Suppose you wanted to do some experiments to determine whether a magnet can produce a current. How would you proceed? You might wrap some wire around a magnet, connect it to a current meter to complete the circuit, and see what happens. When you found that the meter continues to read zero, you might try a bigger magnet. Disappointed, you try a more sensitive meter. Or, instead of a permanent magnet, you may want to get your magnetic field by passing current through a coil of wire. You would put a piece of soft iron inside inside the coil, making an electromagnet and thus getting a stronger magnetic field. And if you tried all of these things, you would be treading in the footsteps of that great experimenter, Michael Faraday. But, like Faraday, you wouldn't see the needle of your current meter budge.

Next you might see if what you know about the motion of charged particles in a magnetic field gives you any clues. From Chapter 22, you remember that the force on an electric charge in a magnetic field is proportional to the speed of the charge. So a stationary charge has no force on it, regardless of how strong the magnetic field. Immediately you might conclude: If I want to get a current in a wire in a magnetic field, I have to move the wire!

In Fig. 23–1, a short, straight wire is being moved to the left across the magnetic field. The right-hand rule tells us the direction of the force on negative charges in the wire. The force is directed toward Q. Positive charges would be pushed toward the near end and negative charges toward the far end of the wire. (In reality, only the electrons move in the wire.) Of course, you don't expect anything like complete separation of charges, because the electrostatic forces fight back. The positive charges near P attract the negative charges near Q and repel further positives from their

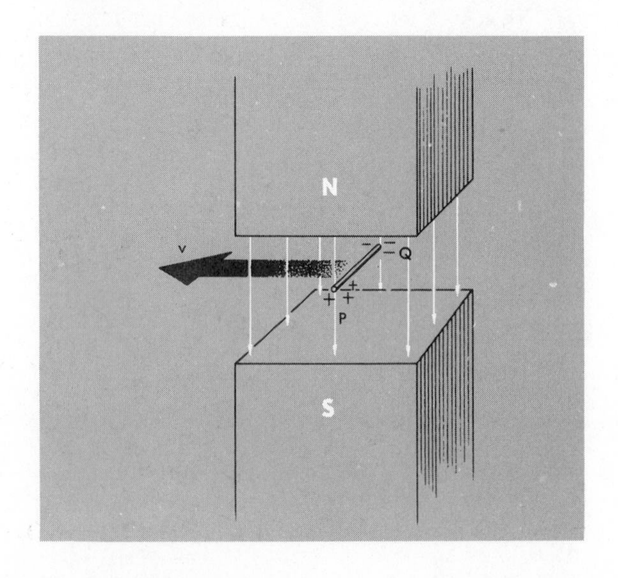

Figure 23–1
As the wire moves to the left across the field of the magnet, charges in the wire are partially separated.

end. This is just what happens in any open circuit when it is connected to a battery: charges build up on the ends of the wire.

In Fig. 23–2, the circuit has been completed outside the magnetic field, and a meter is inserted to indicate a possible current. Now charge can flow around the circuit. Will the meter register any current?

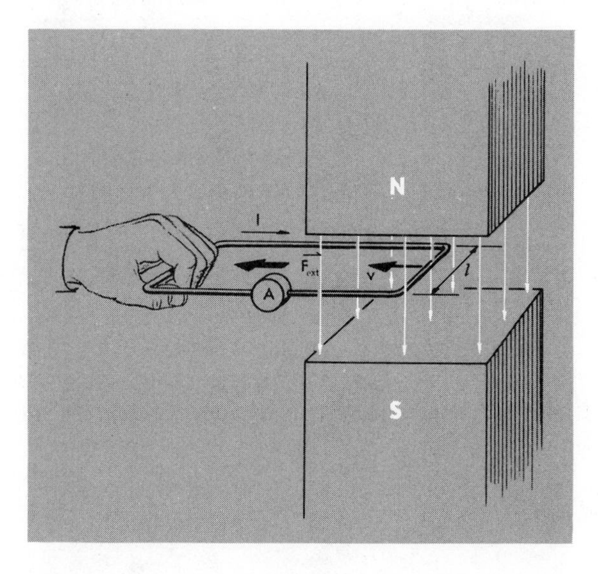

Figure 23–2
As the wire loop is pulled across the magnetic field by an external force, current is induced in the loop. Note that part of the loop is in a region of very weak field, well away from the magnet.

If you were to try this experiment in the laboratory, you would be pretty confident of seeing the meter needle deflect. This time your hopes are based on previous knowledge, not just on a hunch. If you saw nothing, you might again try a bigger magnet, or faster motion of the wire in the field, or a more sensitive meter. And eventually you would detect a current! We say that the motion of the wire loop through the magnetic field has *induced* a current in the loop.

It probably will not surprise you that doubling the velocity of the loop doubles the current; that if you move the loop at the same speed to the right instead of to the left, the current is reversed, but not changed in magnitude; or that turning the magnet upside down does the same thing, just reversing the induced current.

23-2 Relative Motion

But suppose you moved the wire loop and the magnet together, at the same velocity (Fig. 23–3). Would you expect to induce a current in the loop?

When you think about this question, you realize that nowhere in Chapter 22 was anything said about moving magnets. You might make a shrewd guess at the answer, but it would be only a guess at this stage and you would have to check it by experiment. The result is—no induced current.

What do you conclude from this? You might, of course, conclude that you have to hold the magnet stationary to get induced current. Or you

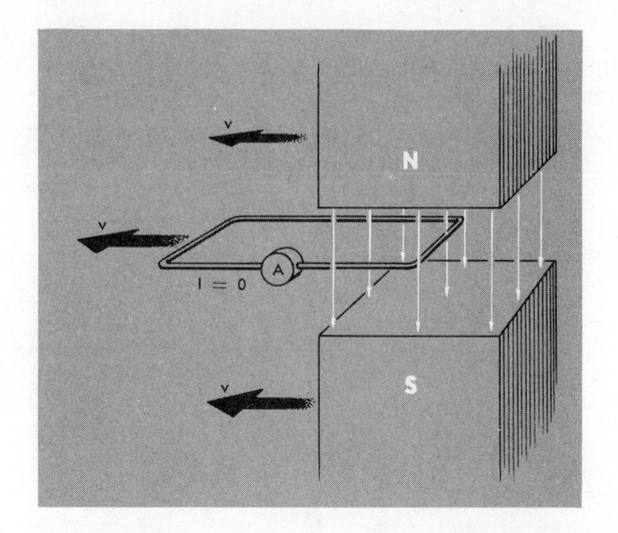

Figure 23–3
When magnet and loop move along together, no current is induced.

might be tempted to try the experiment shown in Fig. 23–4: keep the loop stationary and move the magnet. And if you did it, you would find that moving the magnet to the right and holding the loop at rest induces the same current in the loop as moving the loop to the left and keeping the magnet at rest, as in Fig. 23–2. Apparently, it is only relative motion that counts.

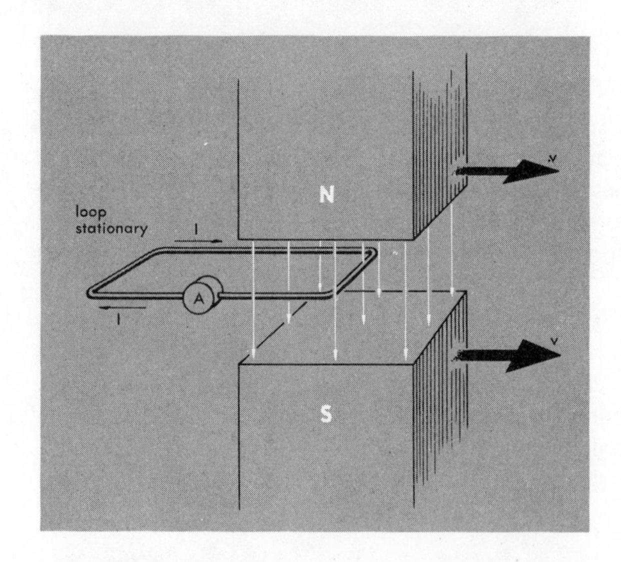

Figure 23–4
With the magnet moving to the right at speed v and the loop stationary, the induced current is the same as with the loop moving to the left at speed v and the magnet stationary. Apparently it is the relative motion that counts.

This idea may be familiar to you from your study of mechanics. You probably have had the experience of sitting in a car stopped at a traffic light and suddenly thinking you were rolling slowly backward, realizing a moment later that you were being fooled by the truck in the next lane moving slowly forward. If your car had no windows, is there any experiment you could perform inside to tell whether the car was moving forward, backward, or standing still? The laws of mechanics tell you that no dynamics experiment can detect absolute rest or uniform motion. Now it appears that the laws of electricity and magnetism show the same indifference to absolutes. They won't tell you "who is moving." They are the same in one frame of reference as in any other whose motion is uniform with respect to the first. This refusal of physical laws to give preference to one frame of reference over another became, in the hands of Albert Einstein, the cornerstone of the special theory of relativity.

23-3 Magnetic Flux Change

In reading the last section, you have frequently come across the phrase "moving the magnet." How can we say the same thing in the language of magnetic field without referring to the magnet which is the source of the field? Let us try some more experiments with induced currents and look for hints. Suppose we repeated the experiment of moving a loop in a magnetic field using a very small loop, small enough so that it stays in a region of uniform field throughout the motion (Fig. 23–5). No current is indicated.

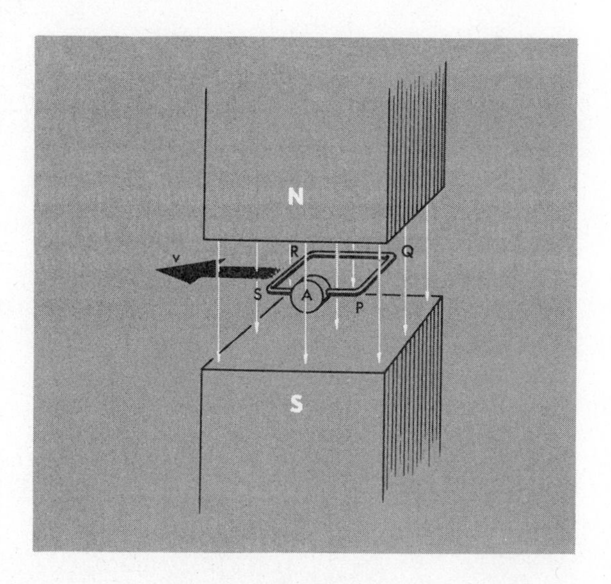

Figure 23–5
When the whole loop is in a uniform magnetic field, the motion induces no current.

It is only when part of the loop gets near the edge of the magnet, where the field gets weaker, that we see the meter needle budge. Is this a hint? Is it that the magnetic field has to be *changing* at some part of the circuit?

We can test that hypothesis without moving either the circuit or the magnet. Using an electromagnet (Fig. 23–6), we can change the magnetic field by changing the current in the coils. When we throw the switch to start the current in the coils, a current is briefly induced in the loop. And when we open the switch to stop the current in the coils, a current is induced again, but in the opposite direction.* It looks as if the hint that induced current is related to changing magnetic field has put us on the right track.

A slight modification of this last setup is shown in Fig. 23–7. Here we use a loop large enough so that no part of the wire is between the poles of

* Historically, it was through this type of experiment that induced currents were discovered in 1831, by Michael Faraday in England and, independently, by Joseph Henry, who was teaching in a boys' school in Albany, New York.

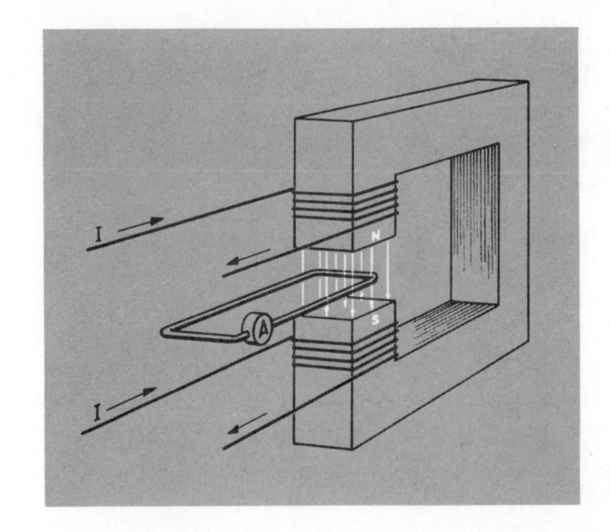

Figure 23–6
Current can be induced in a stationary loop by changing the magnetic field of the electromagnet.

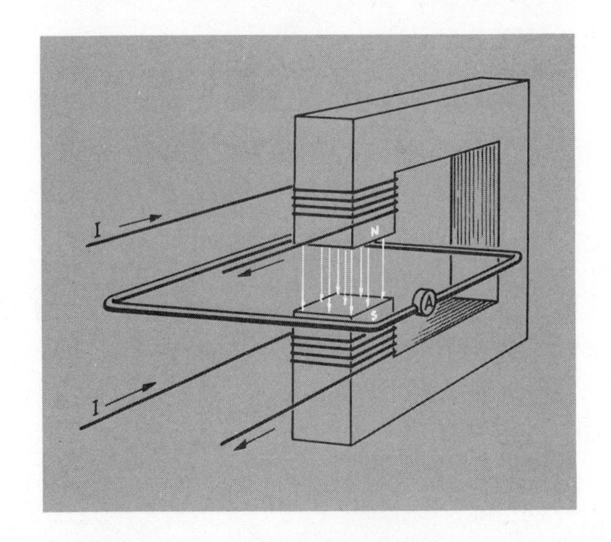

Figure 23–7
Current can be induced in a stationary loop when the magnetic field changes, even though the wire itself is in a region of very weak field.

the electromagnet. All the wire is in a region in which the magnetic field is practically zero. But when we throw the switch, the needle of the current meter jumps. So the induced current does indeed seem to depend on a changing magnetic field, but it does not require that the wire itself be in the strong part of the field. What does seem to be important is that we have a change in the "amount" of magnetic field passing through the closed loop of the circuit.

How can we describe the "amount" of magnetic field passing through the circuit loop in a way that applies to all the situations we have discussed (Fig. 23–2 through 23–7)? Certainly, the magnetic field strength, B, is one important measure of the "amount" of magnetic field, but this only describes the strength at each point. The field is spread out over an area, and we need to take account of the total magnetic field passing through the area enclosed by the loop of the circuit. We shall define a new quantity, the *magnetic flux* ϕ_B, to measure the amount of magnetic field passing through

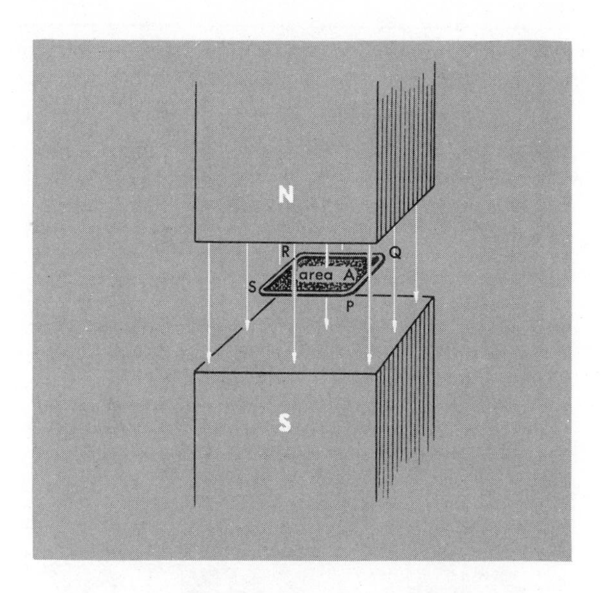

Figure 23–8
The magnetic flux ϕ_B through the circuit PQRS is the product of the field B times the area A of the circuit: $\phi_B = BA$.

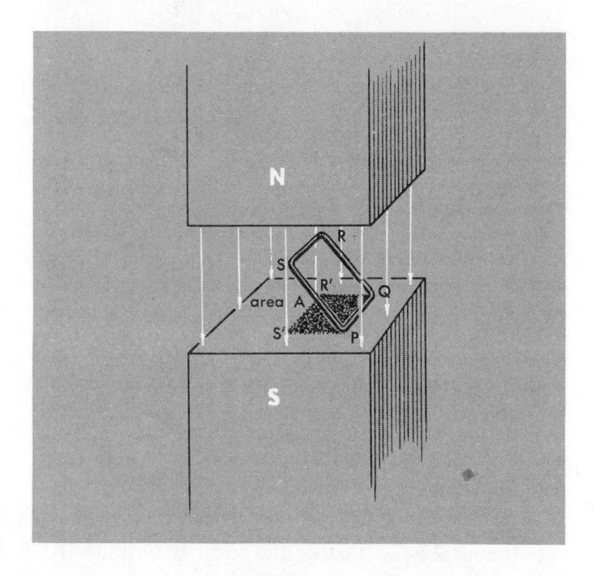

Figure 23–9
When the circuit PQRS is oblique to the magnetic field, it intercepts less magnetic flux. The effective area A is now the shaded area PQR′S′.

a given area. For uniform fields, the magnetic flux is just the field strength times the area as long as the area is perpendicular to the magnetic field (Fig. 23–8). If the area is tilted at an angle so that the magnetic field does not run straight across it, less of the magnetic field passes through it (Fig. 23–9). For example, when we look at an area which is parallel to the lines of magnetic field, no magnetic field crosses it. It is only the component of the field perpendicular to a given area that counts. Thus, we shall define the magnetic flux through the area by

$$\phi_B = B_\perp A.$$

The general case, in which B is not uniform over the area A, is treated in the special section on the next page.

The definition $\phi_B = B_\perp A$ of the magnetic flux through a circuit loop is incomplete when the $\vec{\mathbf{B}}$ field is not uniform over the area of the loop. For the nonuniform case, it would make sense to use a suitable average of $B_\perp$ over the area enclosed by the circuit loop. To find such an average, we shall cover this enclosed area by a network of tiny loops of area ΔA, chosen so small that $B_\perp$ can be considered uniform over each of them (Fig. 23–10). The total flux, then, is just the sum of the flux contributions from all of the tiny loops, or

$$\phi_B = \text{sum}\ (B_\perp \Delta A)$$

over all the ΔA's enclosed by the circuit loop. It can be shown that this sum of fluxes is independent of the way the network is drawn, as long as the ΔA's are chosen small enough.

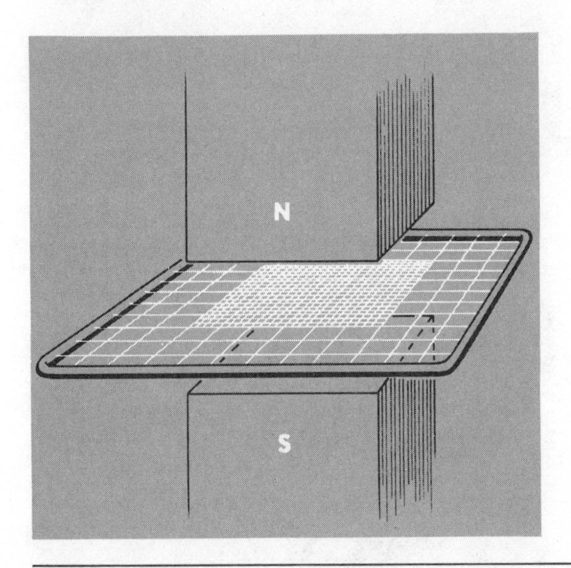

Figure 23–10

Armed with this new concept, we now attempt a general statement: To get induced current in a circuit, the magnetic flux through the circuit must be changing.

Let us test this statement to see how it applies in each situation:

In Fig. 23–2, part of the loop is outside the field. As the loop moves to the left, the effective area of loop in the magnetic field decreases, so the loop threads less and less magnetic flux, and current is induced.

The analysis for Fig. 23–4 is the same as for Fig. 23–2. It does not matter which moves; it is the change of flux through the loop that induces the current.

In Fig. 23–3, there is no relative motion, the flux through the circuit does not change; no current is induced.

In Fig. 23–6, the effective area of loop in the field stays the same (no motion). The flux threading the loop changes when the field changes; current is induced.

The same holds for Fig. 23–7, although there the flux is dense (the field is strong) only near the middle of the loop, well away from the wire.

Now let us look at some new situations:

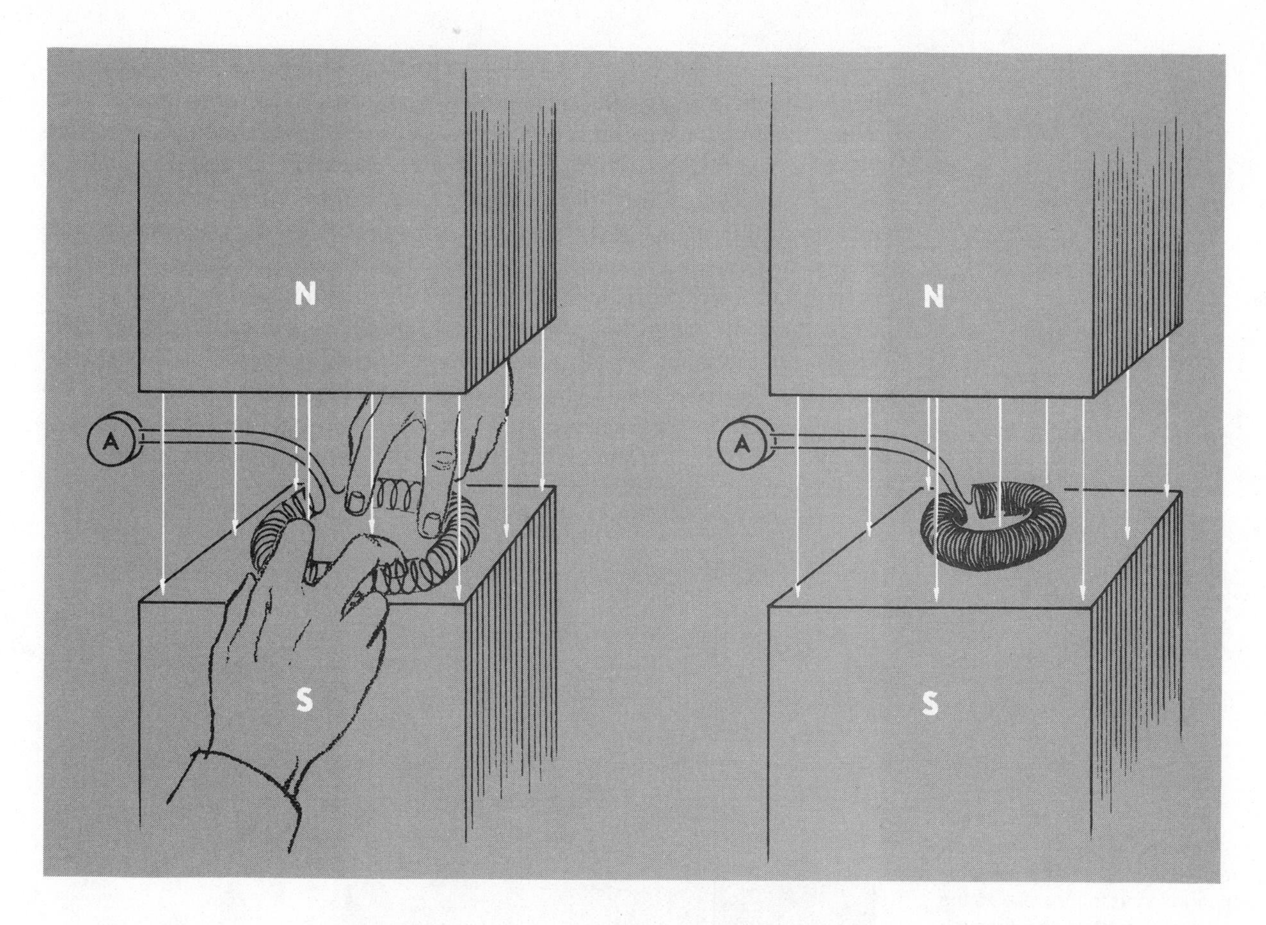

In Fig. 23–11 the field is uniform over a circuit made of a spring. As the circuit is allowed to shrink, the flux through it decreases and current is induced.

In Fig. 23–12, the loop is entirely in a uniform magnetic field, just as in Fig. 23–5, but the flux is changed by rotating the loop about a horizontal axis. The effective area A (see Fig. 23–9) is now the projection of the loop

Figure 23–11
Current is induced by changing the area of the circuit. Here an extended spring is allowed to contract to a smaller area.

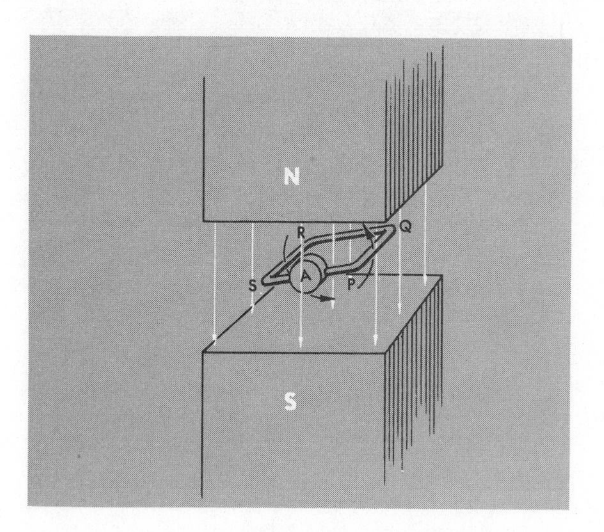

Figure 23–12
Current is induced by changing the orientation of the circuit with respect to the magnetic field. As the loop rotates, the area of its projection on the horizontal plane changes.

on a horizontal plane. As the loop is turned out of the horizontal plane, the flux through it decreases, and current is induced in the direction shown. When the loop is turned through 90°, the flux through it drops to zero (Fig. 23–13). On further rotation, the flux starts increasing. From the point of view of the magnet (and of the reader), this is flux *down* through the loop as before; but from the point of view of someone riding on the loop, this is flux in the opposite direction. So the flux change has the same sign as before, and the current induced is still in the same direction in the loop. But when the loop has turned through 180°, its plane is horizontal again, the flux through it is maximum, so the rate of change of flux is instantaneously zero. Further rotation (Fig. 23–14) decreases the flux, giving rise to induced current in the opposite direction. What we have just described is a simple alternating-current generator. The huge generators that supply electric power for entire cities operate on the same principle.

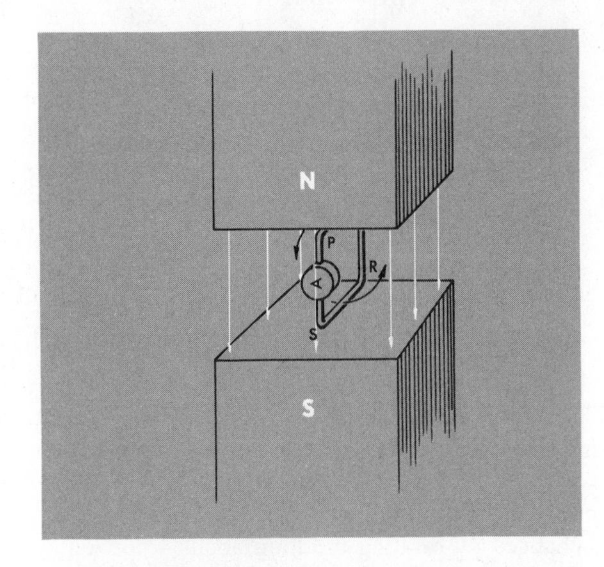

Figure 23–13
The flux through the loop is now zero. The area of the loop's horizontal projection will start to increase as the loop rotates. However, the direction of the field will now be in the opposite direction through the loop. Hence, the flux is still decreasing and the induced current is still in the same direction as before.

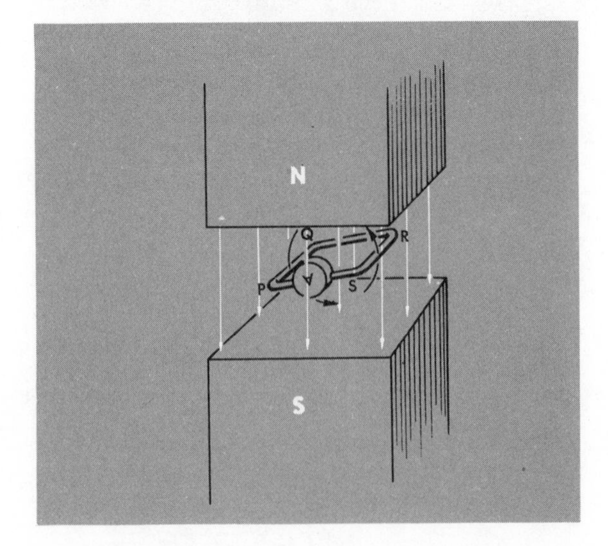

Figure 23–14
The situation is now the same as in Fig. 23–12 except that the loop now is upside down. As viewed from the outside, the current will flow in the same direction as in Fig. 23–12, but the ammeter will register a current in the opposite direction.

The credit for unifying the description of such apparently different situations by the use of the flux concept must go to Michael Faraday. He was not only a most resourceful gadgeteer and experimenter, but in addition had the creative imagination to see that the results of many very different experiments could all be explained in terms of one general principle.

23-4 Induced EMF

If we try to devise a quantitative relation between induced current and magnetic flux change, we find that there are other variables. If we choose wire of different dimensions or made of a different metal, the current also is different. Break the wire, and the current stops. The behavior of the current is just the same as if there were a battery in the circuit instead of a changing magnetic flux through the circuit. With a break in a circuit supplied by a battery, a current meter reads zero, but a voltmeter indicates a potential difference across the break equal to the EMF of the battery. This leads us to try to analyze the situation of Fig. 23–2 to see if a given motion corresponds, not to a definite current, but to a definite EMF. We shall try to relate this EMF to the rate of change of magnetic flux. The relation we obtain, put to the test of experiment in "new" situations, will be found to hold generally, regardless of how the flux change is caused.

In Fig. 23–2, suppose that uniform motion of the loop with velocity $\vec{\mathbf{v}}$ to the left across the magnetic field $\vec{\mathbf{B}}$ induces a current I in the loop as shown. Then, as you know from Section 22–4, the force on the length l of wire due to the magnetic field is

$$F_{\text{magnetic}} = IlB.$$

If you apply the right-hand rule, you will see that this force is directed to the right, tending to stop the motion. To maintain the loop in constant velocity, this magnetic force must be balanced by an equal force F to the left; that is just what the hand in the sketch is supplying. As it pulls the loop to the left, it does work. In the time interval Δt, it pulls the loop along a distance $v\,\Delta t$, doing an amount of work $W = Fv\,\Delta t$. But, since

$$F = -F_{\text{magnetic}} = -IlB,$$

then the work done, in terms of electric and magnetic quantities, is just

$$W = -Il\,Bv\,\Delta t.$$

This work provides the energy necessary to drive the current around the loop. The energy per charge needed to drive a charge around the loop is just what we call $\mathcal{E}$, the EMF. The amount of charge passing any point of the circuit in time Δt is $I\,\Delta t$, so the total energy needed to push this charge is $\mathcal{E}\,I\,\Delta t$. If we equate this to the work done in pulling, we get

$$\mathcal{E}I\,\Delta t = -Il\,Bv\,\Delta t$$

or

$$\mathcal{E}\,\Delta t = -lBv\,\Delta t.$$

The disappearance of I from the equation is important, since it tells us that the length of the wire, the field strength, and the velocity are related directly to the induced EMF, rather than to the current. Why have we not eliminated the Δt as well? Because the length l moving a distance $v\Delta t$ sweeps out an area $lv\Delta t$, the magnetic flux change is:

$$\Delta\phi_B = Blv\,\Delta t.$$

We substitute this into the right-hand side of the previous equation and solve for the induced EMF:

$$\mathcal{E} = -\frac{\Delta\phi_B}{\Delta t}.$$

We see that the induced EMF is proportional to the rate of change of the magnetic flux. That the proportionality constant turns out to be unity is due to the choice of the unit of magnetic field strength made in Section 22–4.

23-5 Direction of the Induced EMF

What does the minus sign in the last equation mean? Somewhere in the argument we must have assigned a meaning to positive or negative flux without saying so explicitly. In the identification of $+Blv\Delta t$ as the flux change, we must have introduced a sign convention for flux. But rather than analyze what we mean by positive or negative B, v, and Δt separately, let us use the conservation-of-energy principle to arrive at an understanding of the negative sign.

Induced current in the loop gives rise to a magnetic field, and thus to magnetic flux threading the loop. The flux must oppose the flux change that induced the EMF in the first place. For if it did not, if the induced current set up magnetic flux *aiding* the overall flux change, then the EMF would build up, causing the current to build up, causing the flux to build up, causing the EMF to build up, etc.—and all with no expenditure of energy. The fact that the two fluxes oppose each other was discovered by the German physicist H. F. Lenz, working in Russia. Lenz's law, in effect, allows energy conservation to limit the induced current. All the work done by the induced current has to be done by the force causing the flux change: by the hand in Fig. 23–2, by the turbine in the power-generating station that supplies your town's electricity.

23-6 Electric Fields Around Changing Magnetic Fluxes

We can get a fuller understanding of the electromagnetic induction equation of Section 23–4 if we rewrite it, this time using "field" language rather than EMF. We owe this field description to James Clerk Maxwell, who also introduced another field relation (to be discussed in Section 23–7) and was then able to predict electromagnetic waves for the first time.

We begin this train of thought with an observation: magnetic field lines around currents always close. They do not start or stop anywhere. In fact, the field lines due to current in a very long, straight wire are perfect circles around the wire. On the other hand, the electric field lines described in Chapter 20 start on positive and end on negative charges. Is it possible for electric field lines to form closed loops, too?

To answer this, let us consider a circular wire loop of radius r with a changing magnetic flux along its axis (Fig. 23–15). The law of induction tells us that an EMF is induced in the loop whenever there is a change of flux through the loop. An induced current will flow in the wire, which means that charges are being driven around the loop. There must be a net force on these charges to drive them around the circuit—to supply the energy used in heating the wire, for example. This force, measured per elementary charge, is just the electric field we are looking for. In this nice symmetrical example, there is nothing to distinguish one part of the circular loop from any other, so the electric field must have the same magnitude everywhere in the loop and must point along the wire.

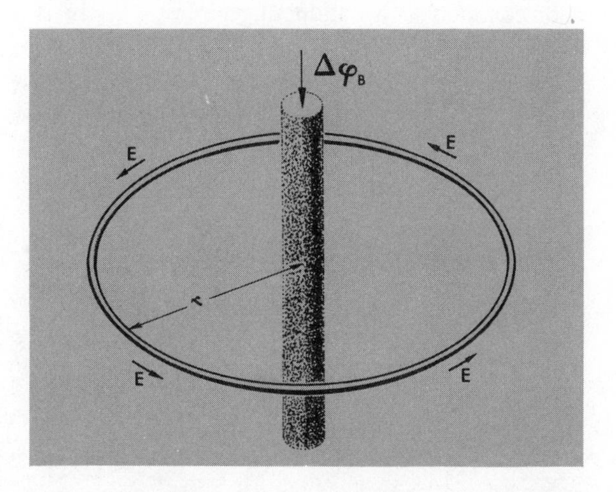

Figure 23–15
When the flux changes through the shaded area, the magnitude of the electric field that circles around it must be the same everywhere along the wire. Therefore the energy per elementary charge received by a charge pushed all the way around the wire is $2\pi rE$.

We can find the magnitude of $\vec{\mathbf{E}}$ from the law of induction. The EMF was defined in Chapter 20 as the work per elementary charge. In this case in going all the way around the circuit, a distance $2\pi r$, the work per elementary charge is then related to the force per elementary charge by the relation $\mathcal{E} = 2\pi rE$. In terms of the rate of change of flux, we get:

$$2\pi rE = -\frac{\Delta\phi_B}{\Delta t}.$$

We see that if we have a wire loop of greater radius, the E field at that greater radius must be proportionately smaller, since the product of the field times the circumference must always equal the negative of the rate of change of flux.

But in Chapter 20 the electric field was defined to describe the force per charge regardless of whether a charge is really there or not. Maxwell pointed out that the law of induction should hold whether or not there are any actual circuits, that the lines of E around a cylinder of changing mag-

Figure 23–16
The induced electric field around a changing magnetic flux runs around in circles. Its magnitude varies inversely with the distance.

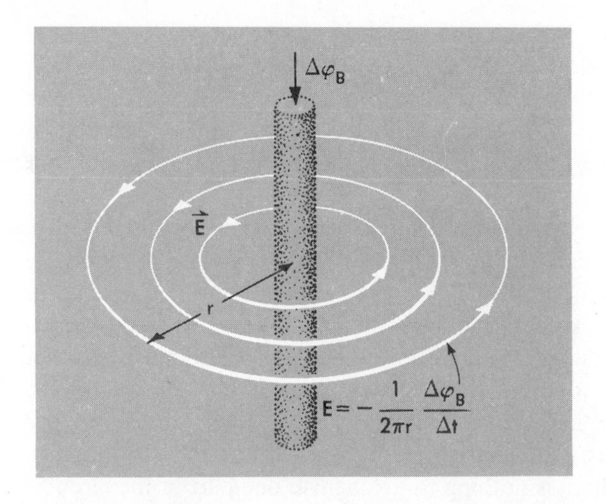

Figure 23–17
A schematic diagram of a betatron. The changing flux in the center produces an electric field which accelerates electrons. The electrons are deflected in circular paths by the outer ring of magnets. We correctly predict the paths from the deflecting magnetic field and from the induced electric field giving the acceleration along the path.

netic flux would also exist in empty space, as indicated in Fig. 23–16. That this field is really there can be tested experimentally with charged particles not confined to a conducting circuit. There is a modern high-energy ac-

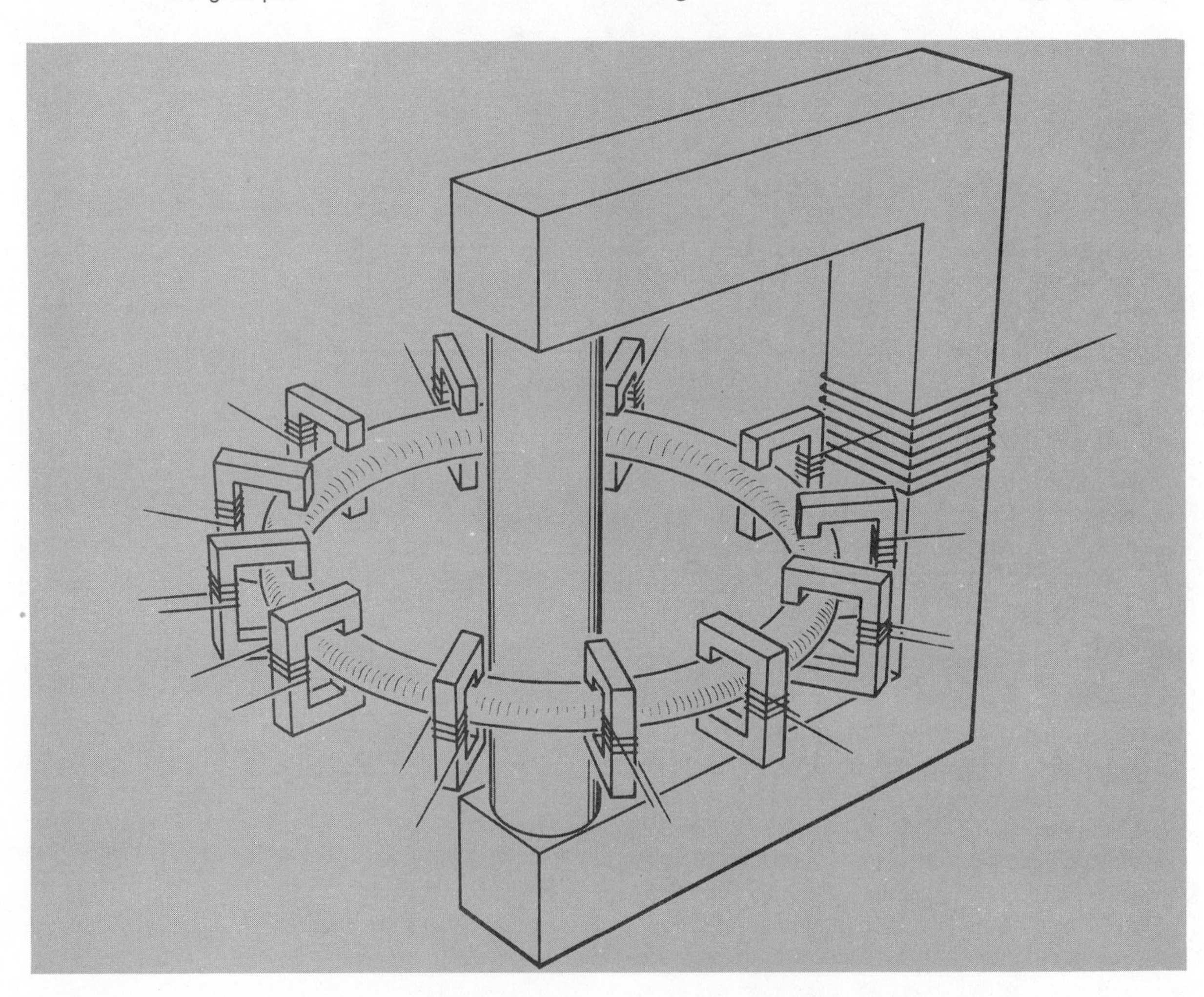

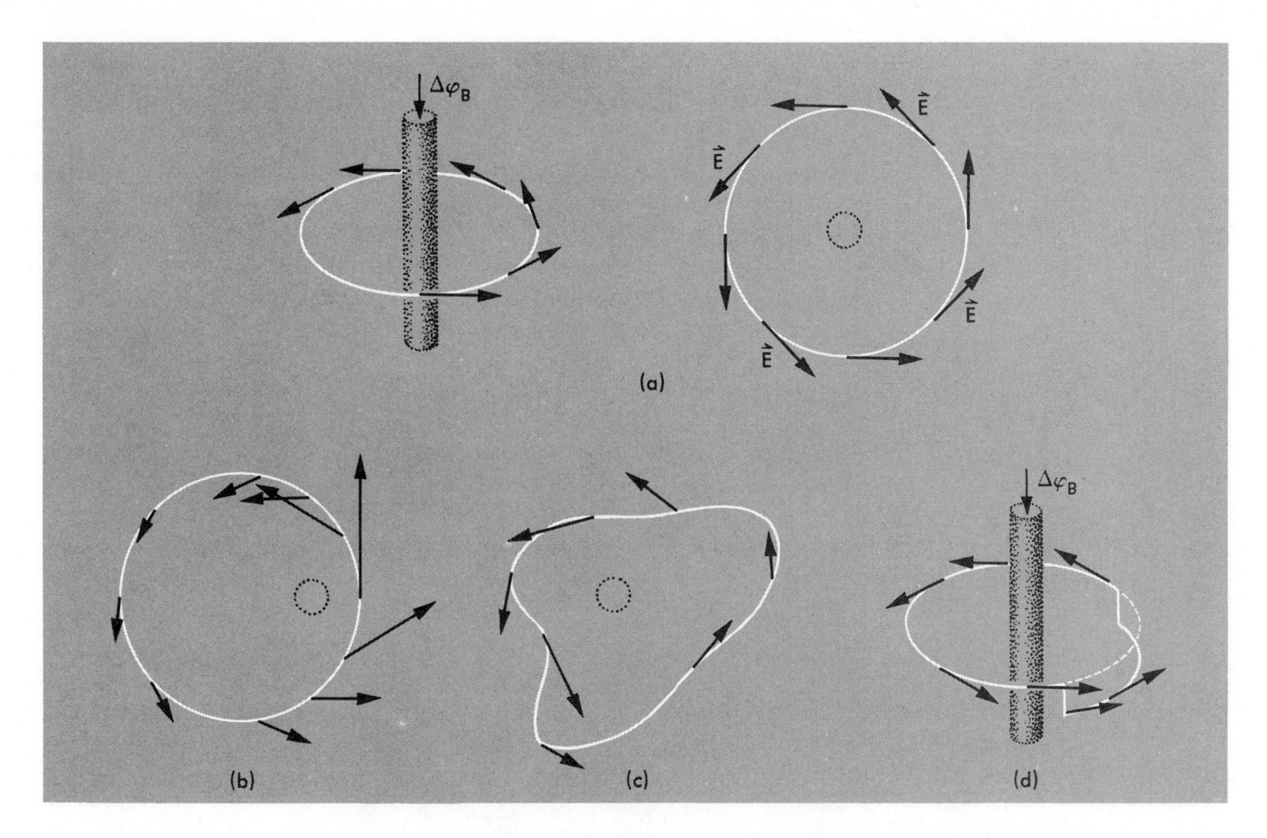

Figure 23–18
The circulation law for the induced electric field around a changing magnetic flux. Divide the path into small segments. For each one take the component of $\vec{E}$ along the segment times the length of that segment. The sum of the contributions along the entire path is the circulation of $\vec{E}$, or the induced EMF, and equals minus the rate of change of magnetic flux through the region bounded by the path, no matter what the shape of the path. In (a) the path is circular and is centered around a region of decreasing magnetic flux, as in Fig. 23–15. When viewed from directly above, the electric field vectors have the same length at every point on the path. In (b) the path is not centered around the changing flux, in (c) the path is not circular, and in (d) the path is not even in one plane. In these three cases the magnitude of $\vec{E}$ is not the same at every point on the path. Yet the circulation of $\vec{E}$ still equals minus the rate of change of the magnetic flux enclosed.

celerating machine, the betatron, which speeds up individual charged particles to high energies by using just such circular electric fields. In this machine, electrons in an evacuated ring are caused to move faster and faster around a changing magnetic flux (Fig. 23–17). They speed up in exactly the way we predict from the action of the electric field that arises from electromagnetic induction.

We have considered only a special and very simple example of the description of the law of induction in "field" language. In the general case, when we do not have the nice symmetry of Fig. 23–15, we must take the component of $\vec{E}$ along the path times the length of path for which $\vec{E}$ has that component, and add up these contributions once around the given loop. We find that the resulting EMF still equals the negative of the rate of change of flux through the region bounded by the loop. Indeed, the loop need not be circular, it need not even lie in a plane; the law of induction still holds (Fig. 23–18). It is one of the fundamental laws of physics. It is often called a "circulation law" for electric fields.

23-7 Magnetic Fields Around Changing Electric Fluxes

The electric field surrounding a cylinder of changing magnetic field (Fig. 23–16) is very similar to the magnetic field surrounding an electric current [Fig. 22–7 (a)]. It is perhaps not too surprising that the circulation laws for the two cases also came out to be very similar. We saw in Section 22–10 that the magnetic circulation

$$2\pi rB = 2\pi KI$$

is Ampère's circulation law for $\vec{\mathbf{B}}$. This is compared with the electric circulation

$$2\pi rE = -\frac{\Delta\phi_B}{\Delta t},$$

the circulation law for $\vec{\mathbf{E}}$, in Fig. 23–19 and 23–20.

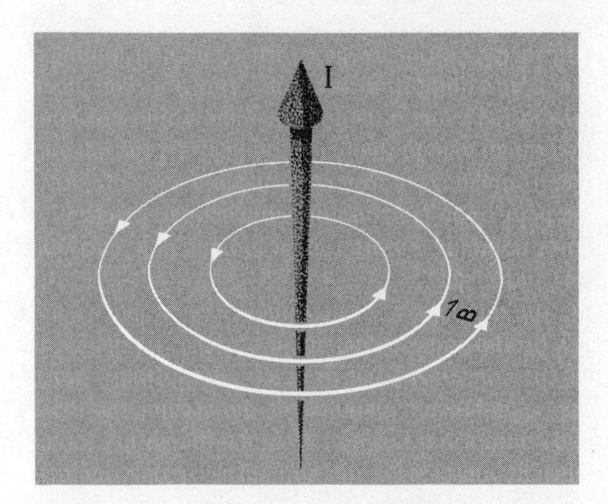

Figure 23–19
The magnetic field of an electric current circles the current, falling off in magnitude as $1/r$.

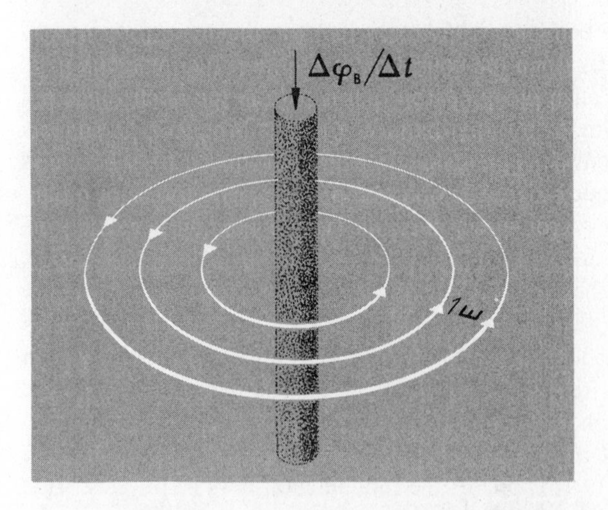

Figure 23–20
Similarly the electric field around a changing magnetic flux circles the flux and falls off as $1/r$ far from the region of strong magnetic flux.

In the first case, the magnetic field $\vec{\mathbf{B}}$ circles the current, and its magnitude falls off as $1/r$ if the current remains constant; in the second case, the electric field $\vec{\mathbf{E}}$ circles the changing magnetic flux, and its magnitude also falls off as $1/r$ if the rate of change of magnetic flux remains constant. Only the directions of the arrows along the axes of the loops are different, as expected from the minus sign of the second equation.

But what happens if the current is not steady? Let us consider the magnetic field near a long, straight wire which has been broken open so that two parallel plates, spaced a short distance apart, can be connected into the current path as shown in Fig. 23–21 (a). We can put the rest of the circuit so far away that we can assume axial symmetry, even make the plates circular so that we can be sure that any lines of the magnetic field B will be circles. Charges will be supplied to the plates by the current in the wire,

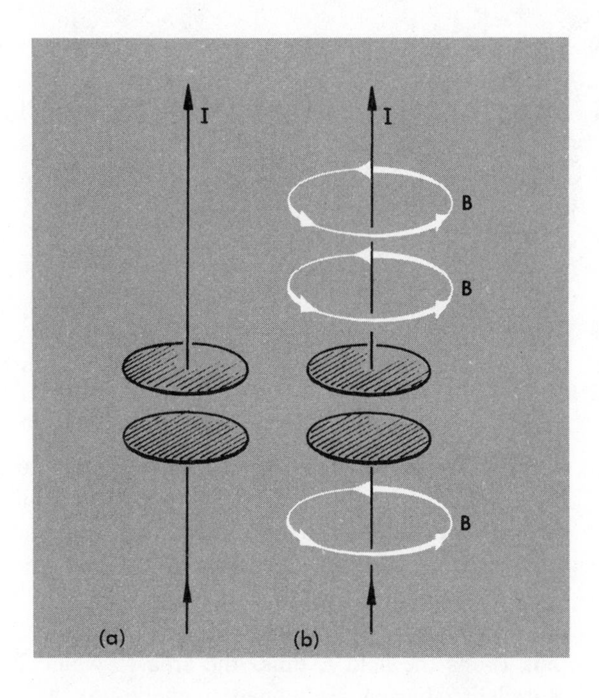

Figure 23–21
(a) A long straight wire has been broken and two parallel circular plates have been connected into the current path. (b) While the current flows, charging the plates, there will be circular magnetic field lines around the wire.

and during the process of charging there should be circular lines of B around the wire as shown in Fig. 23–21 (b). At any instant during the charging, this B should satisfy Ampère's circulation law:

$$2\pi r B = 2\pi K I$$

with I the instantaneous current through the surface bounded by the circular line of B. But here we run into a paradox! There was nothing in Ampère's circulation law that said that the surface bounded by the loop had to be plane. In fact, when we had a very long wire alone, we could readily consider a cuplike surface (or any other shape bounded by the loop) and it made no difference to the statement of the circulation law [Fig. 23–22 (a) and (b)].

Now consider the case with the plates in the circuit. Suppose that we choose the cuplike surface so that it passes through the space between the plates across which there is no flow of charge, hence no current through the surface [Fig. 23–22 (c) and (d)]. Yet, there is a $\vec{\mathbf{B}}$ field, so the circulation ($2\pi r B$) is not zero. The circulation law seems to have failed. We have a genuine paradox.

To resolve this problem we must do one of two things: either we must say that the circulation law for $\vec{\mathbf{B}}$ holds only for steady currents, or we must amend the law so that it holds also for this case where charges build up at some place in the circuit. We shall follow Maxwell's lead and do the latter.

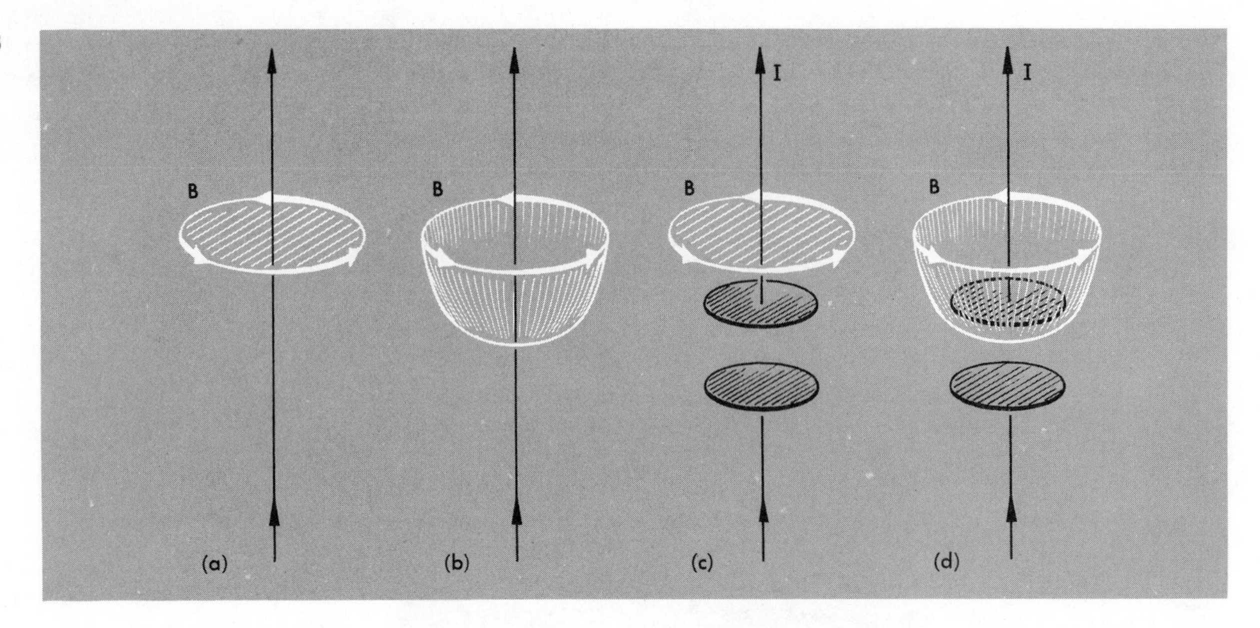

Figure 23–22
(a) and (b). Nothing in Ampere's circulation law (applied here to the field of a long, straight wire carrying a current) said that the surface bounded by the loop had to be plane. (c) and (d). Applying the same reasoning when parallel plates are in the circuit, however, leads to a genuine paradox.

First, we notice that, even though there is no current flow between the plates, there is an electric field between them at any instant after the charging begins, with lines from the positive charges on one plate to the negative charges on the other. Second, we notice that whenever there is current flow there will be a change occurring in the number of charges on the plates, and hence a change in the strength of the electric field. If we define electric flux ϕ_E as the field E times the area to which $\vec{\mathbf{E}}$ is perpendicular, exactly parallel to the way we defined magnetic flux, we can describe this change in the strength of the electric field as a change in electric flux. Now, if we say this changing electric flux is in some sense equivalent to a current, we can "save" the circulation law for $\vec{\mathbf{B}}$. What crosses the cuplike surface bounded by our loop [Fig. 23–22 (d)] is not a real current but a changing electric flux.

In the box it is shown how the change in electric flux is related to the change in the charge on the plates. We see there that the total current needed to make Ampère's circulation law correct when the currents are not steady is the sum of the true current, I, and the "equivalent current," $\frac{1}{4\pi k}\frac{\Delta\phi_E}{\Delta t}$, where k is the constant in Coulomb's law. For the cuplike surface passing between the plates, the true current is zero and the "equivalent current," due to the changing electric flux, is all-important. There,

$$2\pi rB = \frac{2\pi K}{4\pi k}\frac{\Delta\phi_E}{\Delta t}$$

$$= \frac{1}{9 \times 10^{16}\,\text{m}^2/\text{sec}^2}\frac{\Delta\phi_E}{\Delta t}$$

The importance of this term, added by Maxwell, is not so much that it resolved the paradox but that it led him to predict electromagnetic waves, and to say, many years in advance of experimental proof, that light is electromagnetic radiation.

There are many ways of seeing exactly how the electric flux is related to the charge. One of the easiest is to consider a small body with a charge Q on it, which we can surround, in imagination, with a sphere of radius r. The electric field which is directed straight outward along the radii is of strength

$$E = kQ/r^2.$$

The area of the sphere is $4\pi r^2$. Consequently, the total outward flux (field times area, which is in this case everywhere perpendicular to the field) is

$$\phi_E = 4\pi r^2 E = 4\pi k Q.$$

The flux is thus independent of the radius; it is the same for all spheres surrounding the charge Q. Hence the rate of change of electric flux is:

$$\frac{\Delta\phi_E}{\Delta t} = 4\pi k \frac{\Delta Q}{\Delta t}.$$

But $\Delta Q/\Delta t$ is just the current flowing onto the body to charge it. Since

$$\frac{\Delta Q}{\Delta t} = \frac{1}{4\pi k}\frac{\Delta\phi_E}{\Delta t},$$

the quantity $\frac{1}{4\pi k}\frac{\Delta\phi_E}{\Delta t}$ must have the magnitude of the charging current at any instant. It is a sort of "equivalent current" in a region where there is not a true current but a changing electric field instead.

We can use this "equivalent current" to amend the circulation law for $\vec{\mathbf{B}}$ in our problem with the parallel plates connected to the very long wire. To the true current we add the "equivalent current" term and obtain for the circulation of $\vec{\mathbf{B}}$

$$\begin{aligned} 2\pi r B &= 2\pi K\left(I + \frac{\Delta Q}{\Delta t}\right) \\ &= 2\pi K\left(I + \frac{1}{4\pi k}\frac{\Delta\phi_E}{\Delta t}\right). \end{aligned}$$

In Section 19–8 we found that $k = 2.3 \times 10^{-28}$ newton-meter^2/(elem. ch.)^2. Since we now know that one elementary charge is the same as 1.6×10^{-19} coulombs, we can readily determine that in practical units

$$\begin{aligned} k &= 2.3 \times 10^{-28}/(1.6 \times 10^{-19})^2 \\ &= 9 \times 10^9 \text{ newton-meters}^2/\text{coul}^2. \end{aligned}$$

From Section 22–11 we know that

$$K = 2 \times 10^{-7} \text{ newton/amp}^2.$$

Therefore the contribution to the circulation from the "equivalent current" is

$$\begin{aligned} \frac{2\pi K}{4\pi k}\frac{\Delta\phi_E}{\Delta t} &= \frac{4\pi \times 10^{-7}\text{ nt/amp}^2}{4\pi(9 \times 10^9 \text{ nt-m}^2/\text{coul}^2)}\frac{\Delta\phi_E}{\Delta t} \\ &= \frac{1}{9 \times 10^{16}\text{ m}^2/\text{sec}^2}\frac{\Delta\phi_E}{\Delta t}. \end{aligned}$$

23-8 Electromagnetic Radiation

The two circulation laws enable us to derive some important properties of electromagnetic waves in empty space. We must first rewrite the laws to include loops which are made of straight-line segments instead of symmetrical circular paths. We need the component of each field in the direction of the path at the point; we multiply by the length of path for which the field component has this particular value and then sum up the result over all segments of the loop. Thus,

$$\Sigma E_l l = -\frac{\Delta \phi_B}{\Delta t}$$

$$\Sigma B_l l = \frac{1}{9 \times 10^{16} \dfrac{\text{m}^2}{\text{sec}^2}} \frac{\Delta \phi E}{\Delta t}$$

where the capital sigma (Σ) means "sum of" around the whole loop. (We have left out the ordinary current term I because we shall apply the induction laws to a region of space containing no material at all, just electric and magnetic fields changing with time.) This is the form of the two laws which Maxwell combined to predict electromagnetic waves.

We can follow the main points of his argument if we start with the assumption of the existence of a changing electric or magnetic field pattern in space. We shall then show that if either changing field is present, it will induce the other and the two will maintain themselves in empty space only if the fields change in such a way that the pattern of both moves through space with the speed of light. Consider a changing magnetic field $\vec{\mathbf{B}}$, its pattern moving through empty space with the velocity v. In Fig. 23–23 we

Figure 23–23
The magnitude of a magnetic field $\vec{\mathbf{B}}$ at times t and $t + \Delta t$ in an electromagnetic wave traveling along the x axis at a speed v.

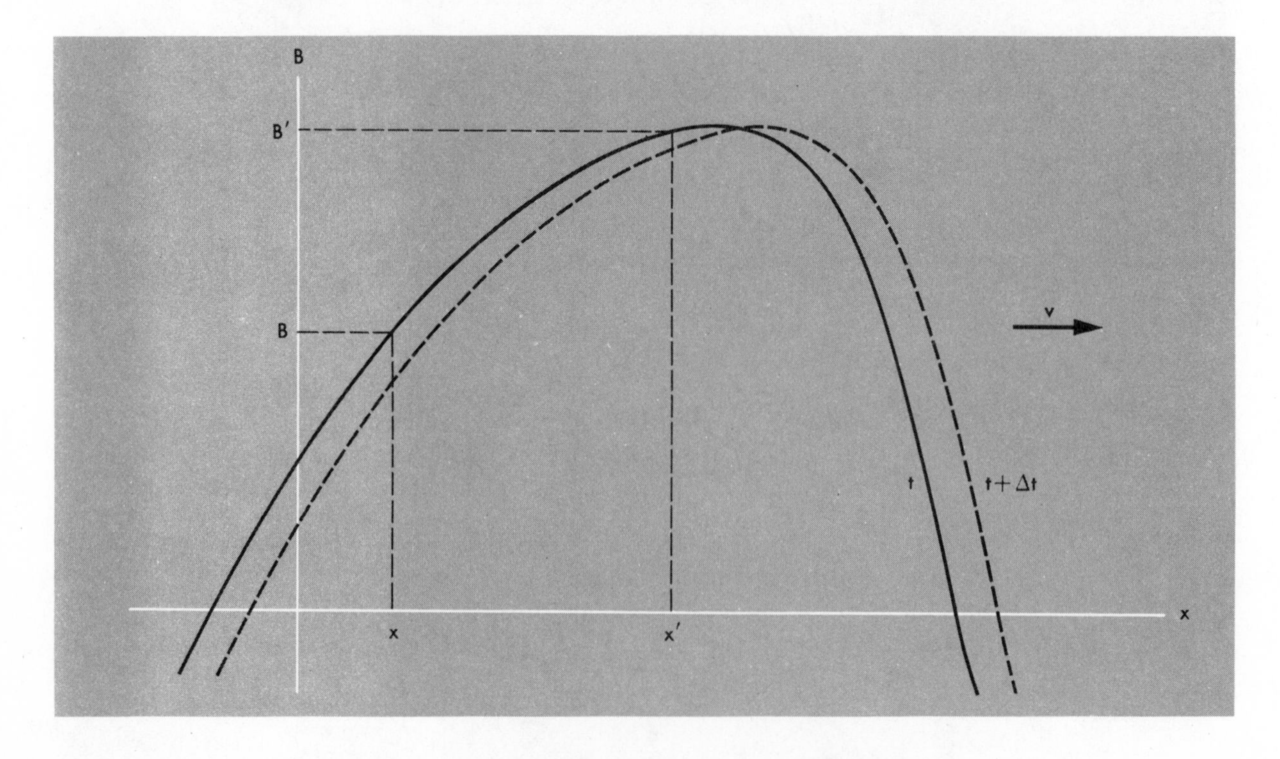

have drawn a graph of the field as a function of x, the direction in which the pattern is moving, at two different times, t and $t + \Delta t$. In Fig. 23–24 (a) we have shown the field at x and x' in three dimensions at time t. We assume for simplicity that the field is the same, for a given value of x, all over the plane parallel to the y, z plane, but varies with x. In a short time interval Δt the whole pattern will move to the right (as shown by Fig. 22–23); the magnetic field $\vec{\mathbf{B}}$ and therefore the flux ϕ_B through all parts of the rectangular loop shown in Fig. 23–24 (a) decreases during this small time interval. From the electric induction law this means that an electric field $\vec{\mathbf{E}}$ must be induced by the changing $\vec{\mathbf{B}}$ field. The direction of the electric field is shown in the three-dimensional representation in Fig. 23–24 (b). The two fields are perpendicular to each other and both are perpendicular to $\vec{\mathbf{v}}$, the direction in which their patterns are moving. We have not shown that this is true, but simple experiments (with radio waves, for example) show that it is indeed the case.

We have just shown that a magnetic field changing in time whose pattern moves with a speed v in the positive x direction gives rise to a changing electric field. We could just as well have started by considering a changing electric field whose pattern moves with a speed v. Then by examining the change in flux over a short time Δt throughout the loop in Fig. 23–24 (b) we would have found an induced magnetic field. The two changing fields whose pattern moves to the right are intimately tied together. One cannot exist without the other.

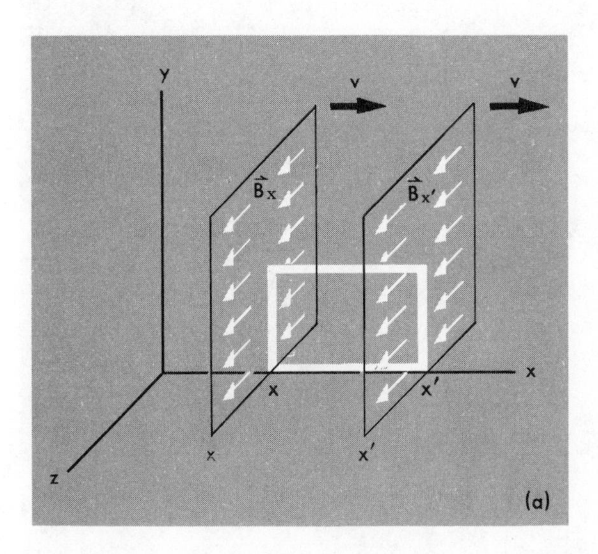

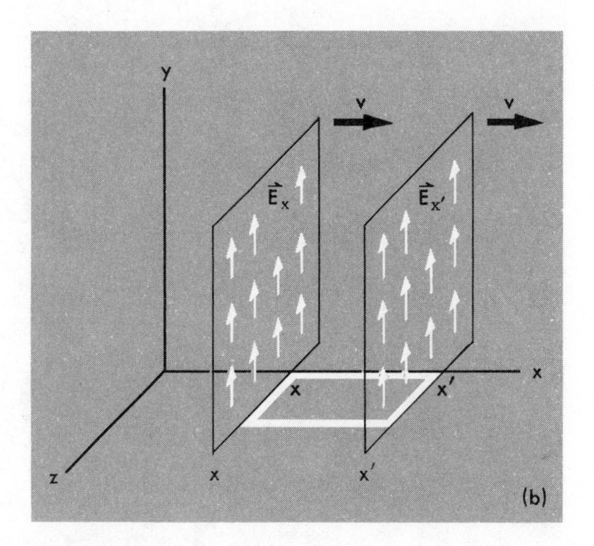

Figure 23–24
A $\vec{\mathbf{B}}$ field and an $\vec{\mathbf{E}}$ field at x and x' at time t that are constant over any y, z plane.

So far we have said nothing about the magnitude of the velocity v. To show that it must be the velocity of light we shall find the relationship between the change in the electric field $\vec{\mathbf{E}}$ and the change in the magnetic field $\vec{\mathbf{B}}$ between two points in empty space using the two induction laws and the magnetic and electric circulations around loops in the two fields.

Let us first consider the magnetic field at points x and x' at time t and $t+\Delta t$. In Fig. 23–25(b) we have drawn a rectangular loop in space in the

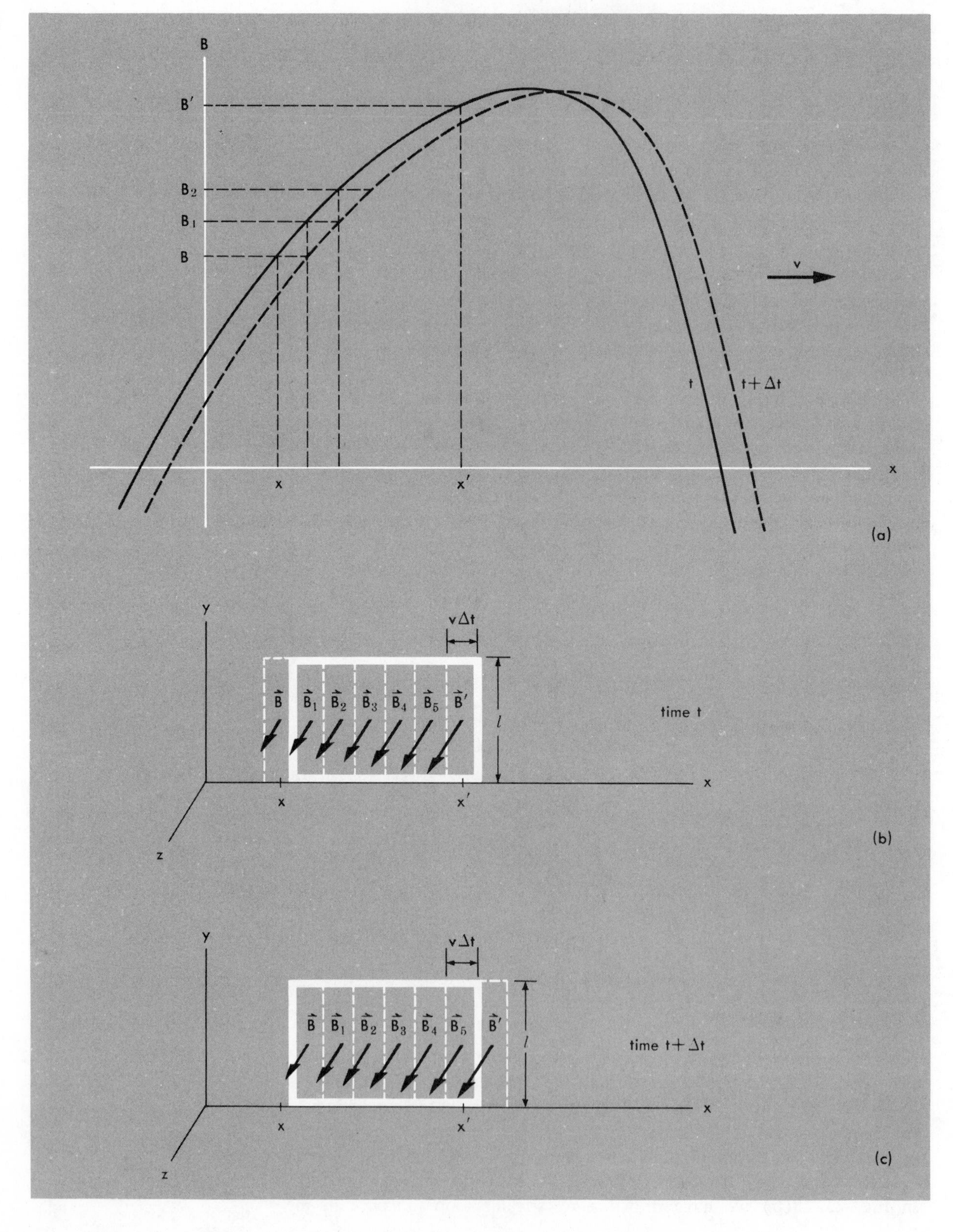
B
B′
B₂
B₁
B
v
t
t + Δt
x
x′
x
(a)
Y
v Δt
B
B₁
B₂
B₃
B₄
B₅
B′
l
time t
x
x′
x
z
(b)
Y
v Δt
B
B₁
B₂
B₃
B₄
B₅
B′
l
time t + Δt
x
x′
x
z
(c)

x, y plane perpendicular to the changing $\vec{\mathbf{B}}$ field; its length is $x' - x$. At time t the magnetic field at x is $\vec{\mathbf{B}}$ and at x' is $\vec{\mathbf{B}}'$

Now consider what happens during the small time interval Δt as the $\vec{\mathbf{B}}$ field pattern moves to the right at the speed v as shown in Fig. 23–25(a). During this short time there is a small decrease in the magnetic flux through the rectangular loop. We have chosen Δt so small that the distance $v\Delta t$ along the x axis the $\vec{\mathbf{B}}$ field pattern moves in the time Δt is small compared to the length of the loop. We can, therefore, consider the big loop to be made up of a very large number of identical small loops each of whose dimension along the x axis is $v\Delta t$ as shown in Fig. 23–25(b) and (c).

At time t the average value of $\vec{\mathbf{B}}$ through each of the loops as given by the graph in Fig. 23–25(a) is shown by the arrows in Fig. 23–25(b). At a time Δt later, the average values of $\vec{\mathbf{B}}$ through each of the small loops are shown in Fig. 23–25(c). Note that the magnetic field in any one of the little loops at time $t + \Delta t$ is the same as the magnetic field in the loop just to its left at time t. (This is because the field pattern is moving to the right at a speed v.)

To find the induced $\vec{\mathbf{E}}$ field through the large loop, we must find the change in magnetic flux $\Delta\phi_{\mathrm{B}}$ through the loop during the time interval Δt. This is the sum of the changes in magnetic flux through all the little loops which make up the big loop. It is

$$\begin{aligned}\Delta\phi_{\mathrm{B}} &= lv\,\Delta t(B + B_1 + B_2 \cdots B_5) - lv\,\Delta t(B_1 + B_2 + \cdots B_5 + B') \\ &= lv\,\Delta t(B - B') \\ &= lv\,\Delta t\,\Delta B.\end{aligned}$$

We can use the same argument to find the change in electric flux $\Delta\phi_{\mathrm{E}}$ through an identical loop in the x, z plane perpendicular to the $\vec{\mathbf{E}}$ field. As one would expect, it turns out that $\Delta\phi_{\mathrm{E}} = lv\,\Delta t(E - E') = lv\,\Delta t\,\Delta E$ and so

$$\frac{\Delta\phi_{\mathrm{E}}}{\Delta t} = lv\,\Delta E \quad \text{and} \quad \frac{\Delta\phi_{\mathrm{B}}}{\Delta t} = lv\,\Delta B$$

Figure 23–25
The $\vec{\mathbf{B}}$ field as a function of x for a moving magnetic field pattern at times t and $t + \Delta t$ is shown in (a). The average fields $\vec{\mathbf{B}}$, $\vec{\mathbf{B}}_1$, $\vec{\mathbf{B}}_2$, . . . $\vec{\mathbf{B}}'$ through each of the small loops of length $v\Delta t$ at times t and $t + \Delta t$ are shown in (b) and (c) respectively. (For simplicity we have drawn only 6 loops inside the big loop.)

Now we can find both the electric and magnetic circulation around the identical loops shown in Fig. 23–26, noting that along any side of the loop when the direction of circulation and the direction of the field are the same, the circulation is positive; when they are different, the value of the circulation is negative.

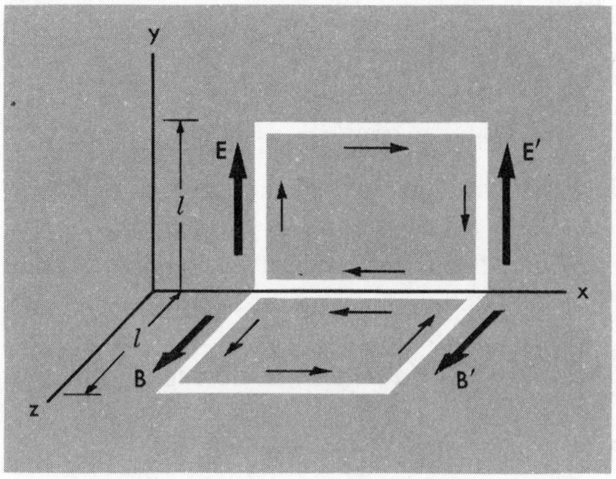

Figure 23–26
To find both the electric and magnetic circulation at time t we draw two identical circulation loops, one in the x, y plane perpendicular to $\vec{B}$ and the other in the x, z plane perpendicular to $\vec{E}$. We will use the loop in the x, y plane to find the electric circulation, and the one in the x, z plane to find the magnetic circulation.

Since the fields are parallel to the sides whose lengths are l and perpendicular to the other sides, the components E_l and B_l along the sides of the loop are E and B along the sides l and zero along the other sides. For this reason we can drop the subscript l in E_l and B_l. Furthermore, in finding the electric circulation we shall drop the negative sign in the relation $\Sigma E_l l = -\dfrac{\Delta\phi_B}{\Delta t}$, since we are concerned only with the magnitude of $\dfrac{\Delta\phi_B}{\Delta t}$ and not with its direction.

Electric Circulation	*Magnetic Circulation*
$El - E'l = \dfrac{\Delta\phi_B}{\Delta t}$	$Bl - B'l = \dfrac{1}{9 \times 10^{16}\ \text{m}^2/\text{sec}^2}\dfrac{\Delta\phi_E}{\Delta t}$
$l\,\Delta E = lv\,\Delta B$	$l\,\Delta B = \dfrac{1}{9 \times 10^{16}\ \text{m}^2/\text{sec}^2}\, lv\,\Delta E.$

Hence,

$\Delta E = v\,\Delta B$	$\Delta B = \dfrac{1}{9 \times 10^{16}\ \text{m}^2/\text{sec}^2}\, v\,\Delta E.$

Eliminating ΔE from the above two equations gives

$$\Delta B = \frac{v^2\,\Delta B}{9 \times 10^{16}\ \text{m}^2/\text{sec}^2}.$$

Thus we see that the pattern of both fields together can move through space unchanged (like a wave on a coil spring) only if $v^2 = 9 \times 10^{16}\ \text{m}^2/\text{sec}^2$; that is, $v = 3 \times 10^8$ m/sec, the speed of light c.

This conclusion from the circulation laws for electric and magnetic fields led Maxwell to conclude "that light itself (including radiant heat, and other radiations, if any) is an electromagnetic disturbance in the form of waves

. . ." In Maxwell's time there was no direct evidence of the electrical character of light, or of any other radiation. More than twenty years after this prediction Heinrich Hertz was able to generate radiation which we now call radio waves, using machinery that was obviously electrical, and to show that the waves do travel with the speed of light. Then Maxwell's theory was brilliantly confirmed. In the next section we shall look briefly at the evidence for the existence of electromagnetic radiation and its electrical nature.

23-9 Evidence for Electromagnetic Radiation; the Electromagnetic Spectrum

All electric and magnetic fields arise from charges and their motions. When should we expect these fields to radiate away from the electric particles from which they originate? An electric charge at rest has only a radial Coulomb field which falls off as the inverse square of the distance and does not propagate away from it. A charge in constant-velocity motion is really no different from a charge at rest (Fig. 23–27). If we walk alongside it, we see the charge at rest and the field around it is a Coulomb field. If we move past it, or it moves past us, because of the changes of electric flux through any region, we shall also see a magnetic field. But this magnetic field is as much attached to the moving charge as the Coulomb field. So the electric and magnetic fields stay with a charge in constant-velocity motion. They do not radiate away.

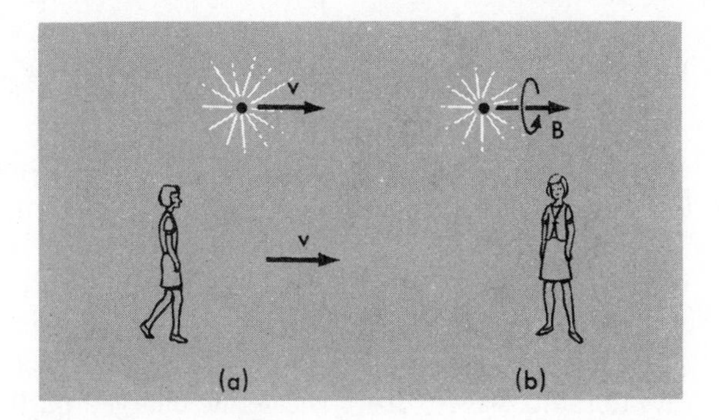

Figure 23–27
(a) If we walk along beside a moving charge, we see only its Coulomb field. (b) If the charge moves past us, we also see a magnetic field in the same region.

On the other hand, if we accelerate a charge, it cannot stay at rest in any dynamical frame of reference in which Newton's laws and the laws of electromagnetism apply. At that moment, we have our only chance to start a pulse of crossed electric and magnetic fields moving away. To find electromagnetic radiation, then, we should look for radiation from accelerated charges.

Let us check up experimentally on what happens when we accelerate charges. A radio transmitting station pumps charges along an antenna, first in one direction and then in the other direction. They do not move at constant velocity but go back and forth, accelerated first one way and then the other. The radio waves which break away from the antenna travel at 3×10^8 meters per second. Evidently they arise from the accelerated motions of the charges pushed back and forth along the antenna.

When we stop high-speed electrons on the target of an X-ray tube, we give them a rapid backward acceleration to bring them to rest. X rays radiate away from the region where the charges are accelerated. On measuring the speed of propagation of X rays, we find that it also is 3×10^8 meters per second.

In a synchrotron—a machine designed to accelerate electrons to high speeds—electrons of tremendous energies are forced around large orbits by deflecting magnetic fields. If we look around the orbit, we see light coming out. Indeed, the energy to which we can bring the electrons is limited in practice by the energy they radiate in light as they are accelerated by the deflecting magnetic forces. We feed in energy from a huge electrical oscillator as rapidly as we can, but at some speed of motion of the electrons it is radiated away as fast as we can supply it. Ordinary visible light is therefore an electromagnetic radiation with its origin in the acceleration of electric particles. The coincidence between the speed of light and the speed of electromagnetic radiation as calculated by Maxwell was no accident. All the

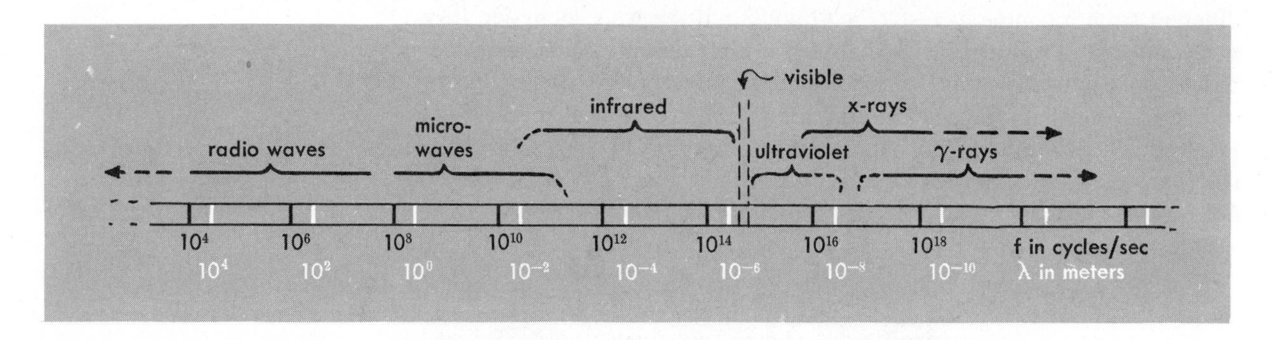

Figure 23–28
The electromagnetic spectrum: this is a continuous range of radiation spreading from gamma rays to radio waves. The descriptive names for sections of the spectrum are historical; they merely give a convenient classification according to the source of the radiation. The physical nature of radiation is the same throughout the whole range. In all sections it has the same speed, the same electromagnetic nature, and the only difference from one part of the spectrum to another is in frequency and wavelength. The regions with historical names overlap, but the names still give a hint of the common sources: radio waves and microwaves from electrons moving in conductors; infrared from hot objects; visible light from very hot objects; ultraviolet from arcs and gas discharges; X rays from electrons striking a target; gamma rays from nuclei of radioactive atoms.

radiations we have mentioned travel at 3×10^8 meters per second; all of them are electrical in origin. They are all electromagnetic radiation (Fig. 23–28).

The common speed of these radiations and their origin from accelerated charges is not the only evidence we have for their electromagnetic nature. There is other evidence connecting the whole spectrum of radiation from X rays to radio waves. If we accelerate charged particles back and forth with a given frequency, the radiant electric fields that we shake off from them should have this frequency of oscillation. Each radio-transmitting station pumps charge back and forth through its antenna at a definite frequency. It must impart this frequency to the radio waves which it sends out. The oscillating electric field in these waves should in turn drive charges back and forth in the receiving antenna at the same frequency. Are those oscillating currents really there? We detect the radio signals by tuning the radio receiver, and we pick up a particular radio station when we have tuned our receiver to exactly the same frequency that the transmitter puts out. The oscillation frequency has been carried from the transmitter to us by the electromagnetic radiation.

With various circuits and antennas we can make transmitters that will send out radio waves of frequencies from about 10^4 hertz (cycles per second) to about 3×10^{11} hertz. All of these electromagnetic waves travel with the speed of light. So the wavelengths—which we can measure by interference—run from about 30 kilometers down to the region of a millimeter. At the long-wavelength end these waves will diffract around any normal obstacle; but as the frequency rises, they behave more and more like light, traveling in straight lines and clearly reflecting and refracting as light does.

Even at the highest radio frequencies that we can generate, the wavelength of electromagnetic radiation is many times longer than the longest wavelength of visible light. In between lies the region of heat radiation. This radiation is emitted by colliding molecules. The thermal motions of the molecules involve accelerations of their charges with resulting emission of magnetic radiation of higher frequencies than we can produce with electrical circuits of man-made sizes.

At still higher temperatures—in arcs and sparks, for instance—the disturbed atoms emit light—infrared, visible, or even ultraviolet light. The visible part of the spectrum corresponds to frequencies of between 4×10^{14} hertz and 8×10^{14} hertz. Here we have long passed the region where we can directly measure the frequency, but we can identify the frequencies in light of a single spectral color by measuring the wavelengths. In this way, by using spectroscopes with the air pumped out, we can find the frequencies of the ultraviolet radiations emitted by atoms. These frequencies run up to a few times 10^{15} hertz.

Besides emitting light, atoms absorb it. An atom of a particular kind will absorb light of only certain very definite frequencies. Presumably, these are the frequencies of the natural motions of the electrons in the atom. The atom is tuned only to these frequencies and will respond only to electric fields which oscillate with them. We cannot tune atoms in the way that we can tune an oscillatory circuit that we can build ourselves. But by looking for the right kind of atom, we may be able to find an electronic motion tuned in a frequency region in which we want to absorb light.

At higher frequencies we come to the region of X rays. If the frequency of the X rays is not too high, its wavelength will be about the size of an atom. Finally, at enormously high frequencies, we come to gamma rays. They are emitted spontaneously in some radioactive decay processes.

All these electromagnetic radiations carry energy. When they are absorbed, the body absorbing them heats up. This means that electromagnetic radiation, as we expected, can only be emitted in a process in which energy is supplied. As we assumed at the beginning, the radiating electric and magnetic fields must be set up when a force accelerates a charge. The evidence that electromagnetic radiation arises from accelerated charges is now overwhelming. Electromagnetic radiation exists; it is propagated with the speed of light, and it arises from accelerated charges.

For Home, Desk, and Lab

1.* Where will the charges shown in Fig. 23–1 be located if the wire is moved in the opposite direction? (Section 1.)

2.* (a) In which direction is the force (due to motion through the magnetic field) on charges in the side arms of the loop of Fig. 23–2?
(b) Does this force help push the charges around the loop? (Section 1.)

3. (a) Analyze the situation of Fig 23–5 entirely from the point of view of the forces on charges in the wire as it moves through the magnetic field, and show why there is no induced current.
   (b) Continue this line of argument to show why there is induced current as one side of the loop gets into the weaker field at the edge of the magnet.

4.* In Fig. 23–5, would you expect induced current if the magnet were moved while the loop was held stationary? (Section 3.)

5.* In Fig. 23–12, as the loop of wire rotates, at what orientation (a) is the maximum current induced, (b) does the induced current reverse direction? (Section 3.)

6.* Why can't you induce a current in a loop of wire by moving it parallel to itself in a uniform field? (Section 3.)

7.* A loop of wire with an area of 0.5 m^2 is in a uniform field of 2×10^{-2} newton/amp-m. What is the flux through the loop if the plane of the loop is perpendicular to the field? (Section 3.)

8. What would be the magnetic flux inside a long solenoid 4.0 cm in diameter with 10 turns per cm of length when the current is 1.0 amp?

9.* If you were pulling the loop in Fig. 23–2 at a fixed speed, would you find it easier to pull it if its electrical resistance were increased? (Section 4.)

10.* Does it take more work to pull a coil out of a magnetic field at high speed or at low speed? (Section 4.)

11.* What EMF is induced in a 5-cm-long wire moving with a speed of 10 cm/sec across a field of 0.02 newton/amp-m? (Section 4.)

12.* What EMF is induced in a loop of wire if the flux through the loop changes at a rate of 0.5 newton-m/amp-sec? (Section 4.)

13. Suppose that in Fig. 23–2 the magnetic field strength is 0.5 newton/amp-m, l is 10 cm, the resistance of the loop is 0.5 ohm, and the loop is moved at a constant speed of 2 m/sec.
   (a) What is the EMF generated in the wire?
   (b) What is the current in the circuit?
   (c) What is the magnetic force acting on the wire?

14. Figure 22–17 is a diagram of an electric motor.
   (a) Explain how you could use the same arrangement of magnetic field and wire loop to make a generator that would produce current in only one direction in an external circuit.
   (b) How should the wire loop be oriented in the magnetic field when the commutator interchanges the connections between the ends of the loop and the wires of the outside circuit?

15. The rectangular wire loop shown in Fig. 23–29 is in a uniform magnetic field.
   (a) In what direction (clockwise or counterclockwise) is the induced current when the loop is in the plane of the paper and BC is moving into the page about P_1 as an axis? About P_2?
   (b) For a given speed of rotation, how would the induced currents compare when the loop is rotated about P_1 and then about P_2?
   (c) For a given speed of rotation, how does the induced EMF depend on the area of the loop?
   (d) If the magnetic field has a magnitude of 1.5 newtons/amp-m, $AB = 10$ cm, and $BC = 4$ cm, what is the maximum induced EMF (in volts) when the loop is rotating about P_1 at 19.1 revolutions per second $\left(\frac{120}{2\pi}\text{ rps}\right)$?

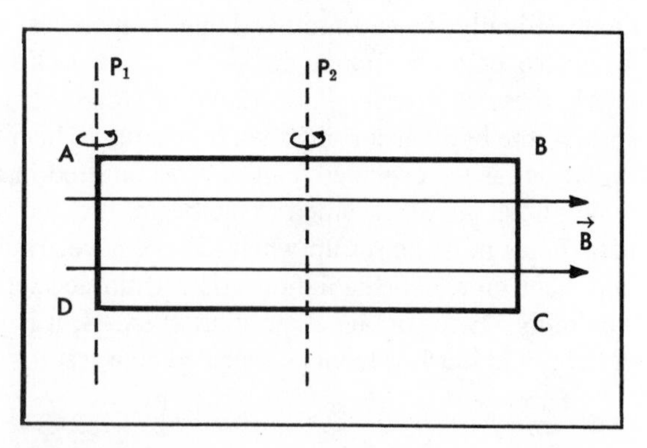

Figure 23–29
For Problem 15.

16. A U-shaped wire (Fig. 23–30) has a movable wire AB connected to it. This arrangement is in a uniform field perpendicular to, and into, the page.

(a) If the magnetic field strength is 40 newtons/amp-m, what is the induced EMF (in volts) when AB is in the position shown and moving at 20 cm/sec? Calculate this first from the rate of flux change and then from the magnetic force on the elementary charges in the wire.

(b) What is the induced EMF when AB is 5.0 cm from the left end and moving at 20 cm/sec? At 10 cm/sec?

(c) At what rate is energy fed into the loop when AB moves at 20 cm/sec and the induced current is 2.0 amperes?

(d) If the induced current is 2.0 amperes, what force is needed to move AB at 20 cm/sec?

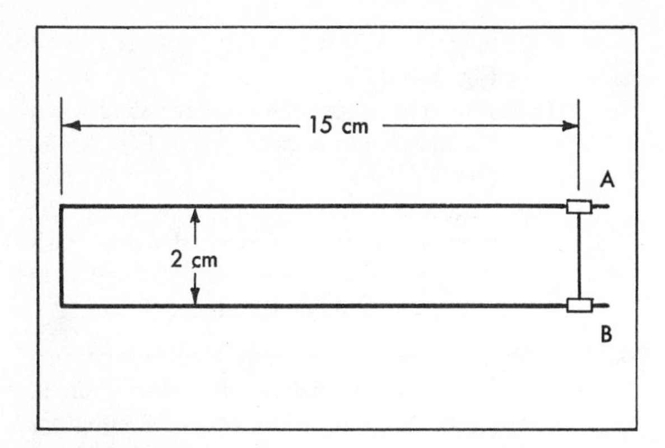

Figure 23–30
For Problem 16.

17. Suppose that the moving circuit of Problem 13 has a mass of 200 grams, and that the external force is suddenly removed when it is moving at 2 m/sec. What will its motion be like afterwards?

18.* If we applied a constant force to the loop in Fig. 23–2, would the loop accelerate uniformly as long as the segment of length l was in the magnetic field? (Section 5.)

19. A loop of wire JKL lies between the pole pieces of a magnet with the plane of the loop parallel to the pole pieces, as shown in Fig. 23–31(a). It is connected to a sensitive ammeter whose pointer stands at the center of the scale when there is no current, and which is so constructed that the pointer swings toward the terminal (P or Q) at which the current enters.

(a) Which way does the pointer swing when the loop is pulled out of the field of the

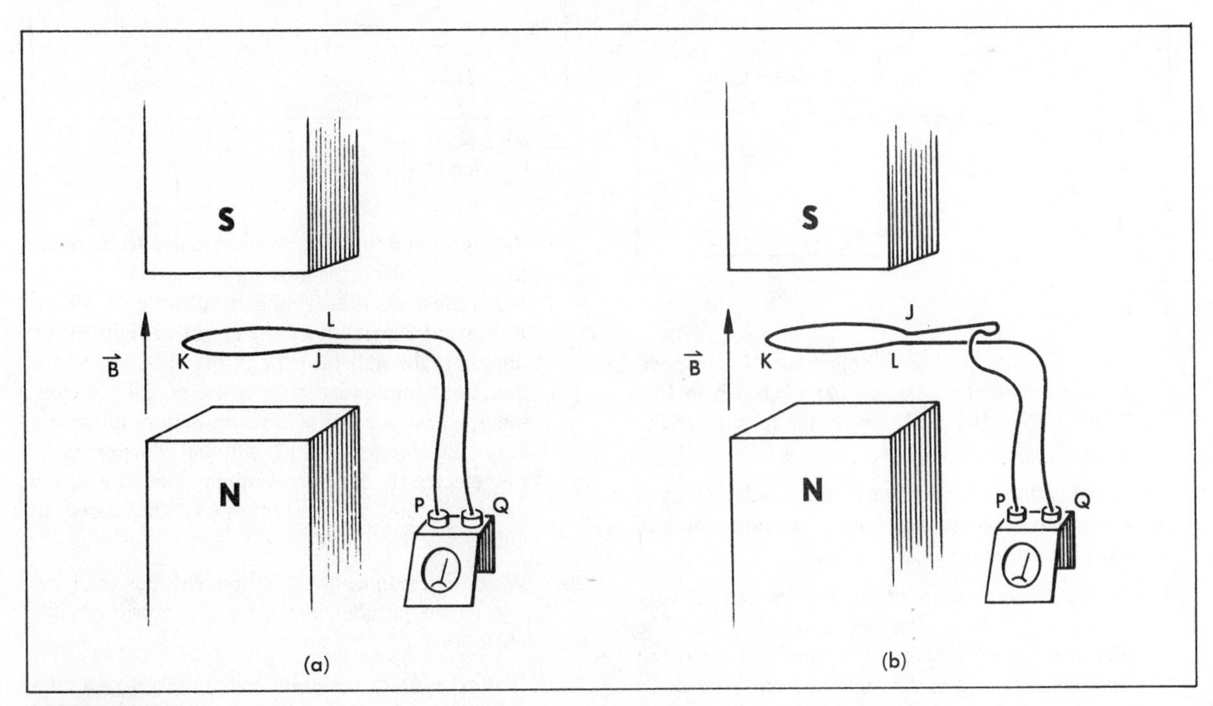

Figure 23–31
For Problem 19.

magnet? When it is put back as it is in Fig. 23–31(a)?

(b) Which way does the pointer swing when the loop is put back as it is in Fig. 23–31 (b)?

(c) How does the pointer move when the loop is turned over from the first position to the second without being taken out of the field of the magnet?

20. (a) Will the earth's magnetic field induce currents in an artificial satellite with a metal surface that is in orbit around the equator? Around the poles?

(b) If so, how would these currents affect the motion of the satellite?

21.† A bar magnet is dropped through a wire loop as shown in Fig. 23–32.

(a) Describe the changes in the direction and magnitude of the induced current in the loop as the magnet falls through the loop.

(b) Neglecting air resistance, will the acceleration of the falling magnet be constant?

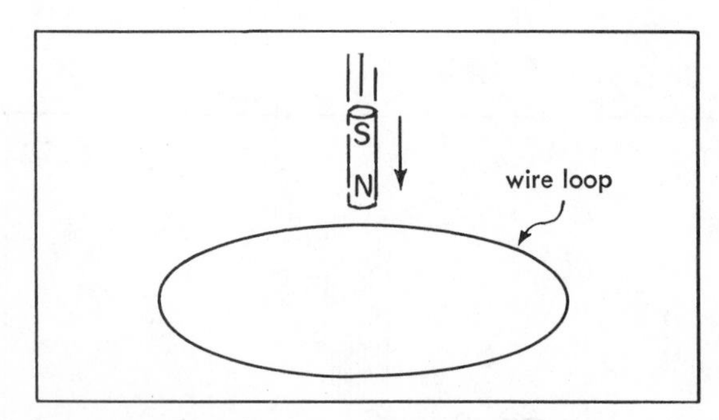

Figure 23–32
For Problem 21.

22. A rectangular loop of copper wire is dropped between the poles of a magnet as shown in Fig. 23–33. Describe qualitatively the motion of the loop. Neglect air friction.

23.* Do the lines of an electric field induced by a change in magnetic flux have a beginning and an end? (Section 6.)

24. In a betatron electrons are accelerated to very high energies by a changing magnetic flux. The electrons move in a circle around the inside of an evacuated glass "doughnut" and speed up when the flux through the hole in the doughnut increases. A separate magnetic field passes

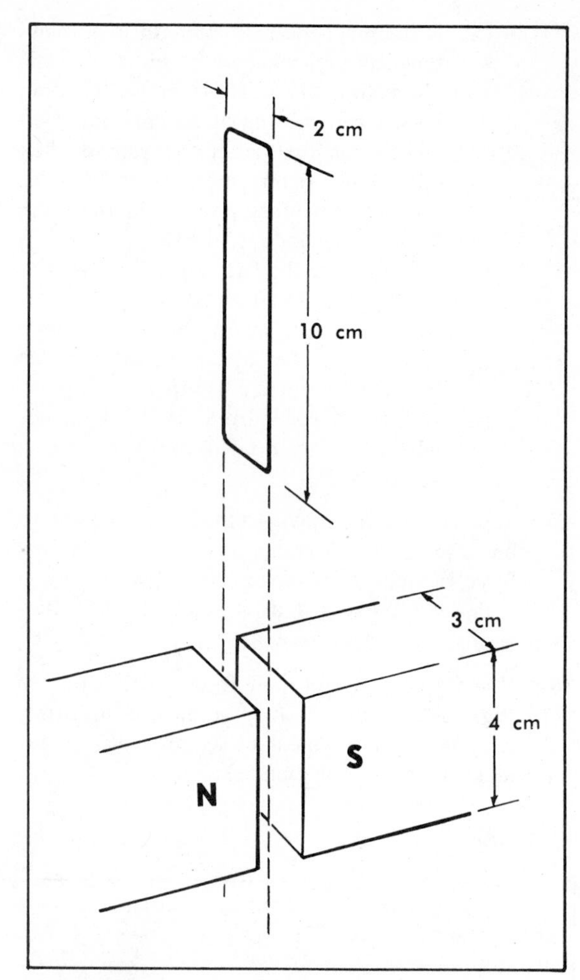

Figure 23–33
For Problem 22.

through the doughnut perpendicular to its plane to keep the electrons moving in a circle.

An electron is traveling in a circle of 50 cm radius at 98 percent of the speed of light in the doughnut of a betatron. If the flux inside the doughnut increases at a rate of 24 newton-meters/amp-sec for $\frac{1}{240}$ second, how much energy (in electron volts) will be transferred to the electron? (*Note:* Assume that the speed stays constant at approximately the speed of light.)

25.* What determines the "equivalent current" between two parallel plates in an electric circuit? (Section 7.)

26.* Is it possible to maintain indefinitely (a) a constant magnetic circulation? (b) a constant electric circulation? (Section 7.)

27.* In Fig. 23–22(c), assume that the plates started with zero charge and that the current shown has been flowing for a while. (a) What is the sign of the charges on each plate? (b) What is the direction of the electric field? (Section 7.)

28. Suppose that at the moment there is no electric field where you are, but a short time later there will be an electric field pointing vertically down. The boundary is moving straight toward you. When it has reached you, what will be the direction of the magnetic field at your position?

29.* What is the speed within a vacuum of (a) radio waves, (b) infrared waves, (c) gamma rays? (Section 9.)

30. When electrons moving at 3×10^7 m/sec hit the target in an X-ray tube, they are brought to rest in about an atomic diameter—in about 10^{-10} m.

   (a) What is the stopping time?

   (b) If you use one divided by this time to estimate the frequencies you might expect to see in the electromagnetic radiation arising from this deceleration, what frequency do you get?

   (c) What wavelength does this frequency give?

31. What evidence is there that sound is not electromagnetic radiation?

Further Reading

These suggestions are not exhaustive but are limited to works that have been found especially useful and at the same time generally available.

HOLTON, G., and ROLLER, D. H. D., *Foundations of Modern Physical Science.* Addison-Wesley, 1958. (Chapters 28 and 29)

MACDONALD, D. K. C., *Faraday, Maxwell and Kelvin.* Doubleday Anchor, 1964: Science Study Series. (Chapter 2)

PIERCE, JOHN R., *Electrons and Waves.* Doubleday Anchor, 1964: Science Study Series. (Chapters 5, 6, and 7)

WILSON, ROBERT R., and LITTAUER, RAPHAEL, *Accelerators.* Doubleday Anchor, 1960: Science Study Series. An interesting, detailed description of modern high-energy machines of physics. (Chapters 5 and 6)

The Rutherford Atom

CHAPTER

24

We already know that atoms contain electrons which are far lighter than the lightest of atoms, that of hydrogen. We can rip off electrons, leaving positive ions of almost the entire atomic mass, and sometimes we can add an electron to a neutral atom, producing a negative ion. But what does an atom look like? How is it built? About these questions we have said rather little. Now that we know some things about electric energy and charged particles, we can use the latter to probe inside the atom and learn something about its structure. As we shall see, the evidence gained by bombarding atoms with charged particles to probe their depths indicates that atoms are built on about one hundred standard patterns; and in some ways these patterns are fairly simple. For this reason, with relatively few probing experiments we can learn a good deal.

In order to bring back information about the inside of an atom, our probes must get in. Consequently, other atoms and molecules, which bounce off the atomic surface, are poor probes. Also, low-energy charged particles are not of much use, for they, too, do not penetrate very far. On the other hand, a probe that passes straight through anything will tell us nothing about the place where it has been. Highly penetrating X rays or the even more penetrating gamma rays from radioactive disintegration are hard to use because of this defect. In the early years of this century, the most suitable probes for poking in atoms were alpha particles. These come out of radioactive materials with high enough energy to penetrate well into an atom; and although they often pass on through with little change of direction, they sometimes are violently deflected. Then we assume that they have hit something relatively massive inside and bounced away.

24-1 The Deflection of Alpha Particles and the Rutherford Model of Atoms

Figure 24–1 shows the setup for an experiment to probe gold atoms with high-speed alpha particles, which are emitted in the radioactive disintegration of a piece of polonium. This is our gun. Using the method outlined in Section 20–8, we can measure the speed of the alpha particles from polonium by measuring their time of flight along an evacuated tube. We

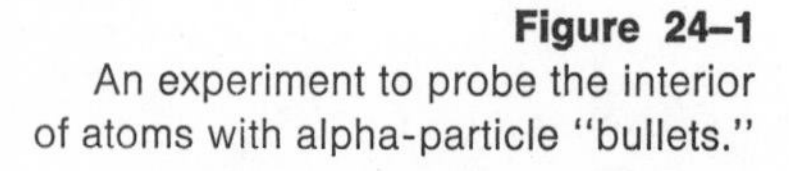

Figure 24–1
An experiment to probe the interior of atoms with alpha-particle "bullets."

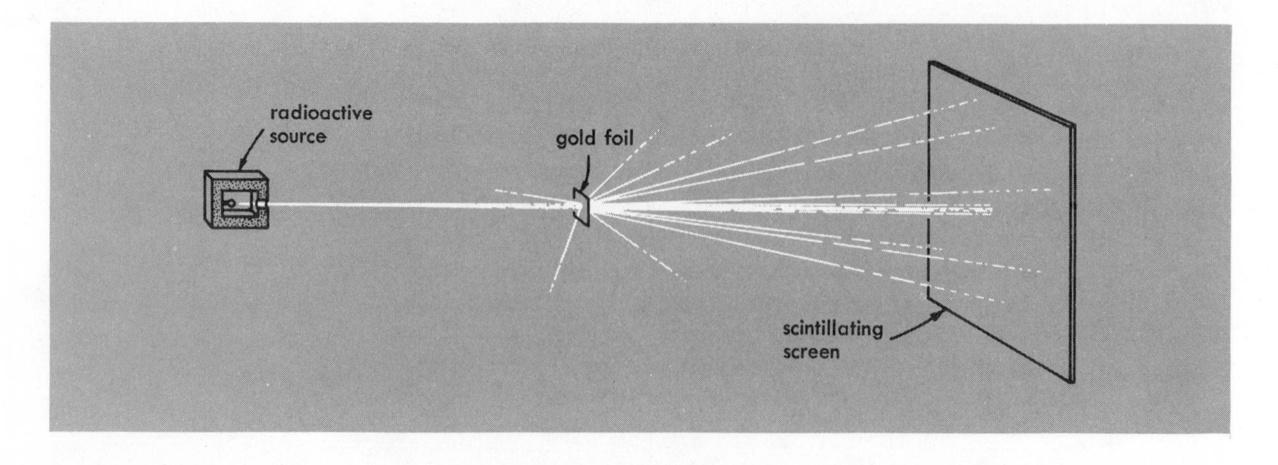

would find that the alpha particles, which are doubly charged helium ions, have a speed of 1.6×10^7 meters per second. Suppose we direct a narrow beam of these alpha particles toward a thin gold foil and on the far side of the gold foil we place a detector of alpha particles, which will give us a signal telling us where the alpha particles go after passing through the foil.

The gold foil is a big stack of gold atoms. A suitable gold foil is so thin that each square meter has a mass of only 2×10^{-3} kg. From the density of gold we can find the number of gold atoms that are in the way of an alpha particle as it passes through the foil. The density of gold is about 2×10^4 kg/m^3. The foil thickness is therefore about

$$\frac{2 \times 10^{-3}\ \text{kg/m}^2}{2 \times 10^4\ \text{kg/m}^3} = 1 \times 10^{-7}\ \text{meter.}$$

Also, we know that the mass of a gold atom is about 3.3×10^{-25} kg. In this way we learn that each atom—stacked in contact with its neighbors—has a volume of

$$\frac{3.3 \times 10^{-25}\ \text{kg}}{2 \times 10^4\ \text{kg/m}^3} = 16 \times 10^{-30}\ \text{meter}^3.$$

If we think of the atoms as boxes, the side of each of the little boxes must be about 2.5×10^{-10} meter. Therefore, the thickness of the gold foil is about

$$\frac{1 \times 10^{-7}\ \text{meter}}{2.5 \times 10{-}10\ \text{meter}} = 400\ \text{gold atoms thick.}$$

The first thing we learn with this apparatus is that most of the alpha particles pass through the 400 layers of atoms without appreciable change in their direction of motion. We can conclude that most of the inside of the atom has no hard, massive objects from which the alpha particles would bounce off at an angle. The atom does, however, have something inside, for we see that the alpha particles gradually lose their energy in passing through gold. With any ordinary thickness of gold the alpha particles would never come out of the far side. As we can easily tell in other experiments, the alpha particle energy is lost in ionizing the gold atoms. When an alpha particle moves through the atoms, losing energy as it rips off electrons, it is slowed, much as a bullet is slowed when it goes through many bales of hay.

Because we have determined the mass of atoms, however, we know that something massive must be located somewhere in the atom. We can hardly believe that this mass is made of electrons. Many thousands of electrons would be required for a light atom, and at most we can get only a few when we ionize the atom. A gold atom has a mass about 50 times that of an alpha particle, and if an alpha particle hits such a mass head-on, it should bounce straight back. Let us, therefore, look again at the beam of alpha particles passing through the gold foil. If we look carefully all around in every direction, we find that in passing through the gold foil, about 1 out of 10,000 of the incident alpha particles is actually knocked off its original course and deflected by an angle of more than 10 degrees. Every once in a while, one out of some astronomical number of alpha particles is bounced out of its course by 90 degrees or more. The same kind of event happens once in a while when alpha particles are passing through the gas in a cloud chamber. There, if we look at myriads of cloud-chamber pictures, we eventually find a few in which the alpha particles clearly ricochet from an atom, having made almost an about-face. Figure 24–2 is such a photograph.

Figure 24–2
A cloud-chamber photograph showing the tracks of alpha particles moving through nitrogen. The alpha particles are moving upward from a source in the lower left. Notice the particle which collided with a nitrogen atom near the center of the picture. It was deflected back through an angle of nearly 142 degrees, while the nitrogen atom recoiled. (From P. M. S. Blackett and D. S. Lee, in *Proceedings of the Royal Society,* 134A, 658, 1931.)

The alpha particles that are sharply deflected in the gold foil also have made just one violent collision with some part of one atom in the foil. We can be sure that it is one collision which is important, because most of the alpha particles pass straight through. Consequently, we know that the chance of even one big collision is tiny and the chance of more than one real hit on the way through is absolutely negligible.

To account for these observations any atomic model must have two features: Because almost all the alpha particles pass right through an atom, most of the atom must be devoid of anything massive. Because an occasional alpha particle bounces off backward, somewhere inside the atom there must be at least one lump of mass with which alpha particles can interact strongly. In addition, we have associated the slowing down of the alpha particles in a thick sheet of gold with the light electrons which are ripped off when alpha particles pass through atoms and ionize them. So we are tempted to associate the main mass of the atom with the positive charge which must be present in order that the atom as a whole shall normally be neutral.

In 1911 Rutherford invented an atomic model which incorporated all these features and in addition called only upon the Coulomb force to explain the interaction of alpha particles with the positive massive atomic core. His model describes an atom as a miniature solar system with a core, or *nucleus,* at the center and a number of electrons around it. The nucleus is positively charged and carries almost all of the atomic mass. The light, negatively charged electrons revolve around the nucleus, held by the Coulomb attraction, much as the planets revolve around the sun, held by gravitational attraction. Outside the atom, these negative electrons cancel the effect of the positive charge of the nucleus, so that the atom as a whole is neutral. This means that the nucleus carries a number of positive elementary charges equal to the number of electrons.

We have already concluded that the hydrogen atom is composed of a proton and an electron (Section 20–1). Thus, in the Rutherford model of the hydrogen atom (Fig. 24–3), the proton is the nucleus, and there is one electron moving around it in an orbit. Similarly, the helium atom consists of an alpha particle and two electrons. In helium the alpha particle is the nucleus, and two electron planets revolve about it.

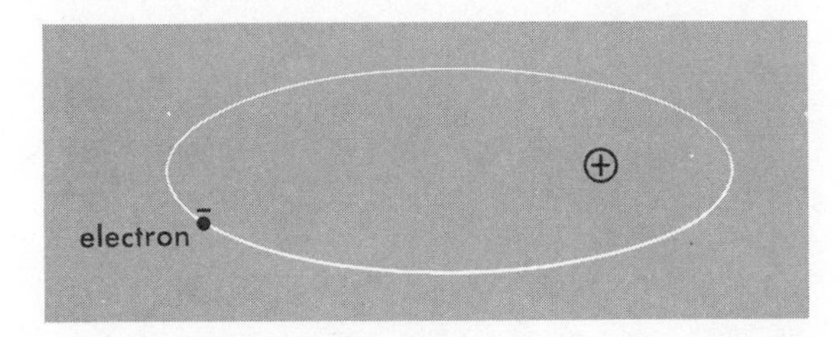

Figure 24–3
The Rutherford model of the hydrogen atom. On this scale both the electron and the nucleus are so very small that in fact you could not see them at all, even though the linear magnification is 10^9 times.

In the Rutherford model of any atom the dimensions of the nucleus and of the electrons are assumed to be very small compared with the overall size of the atom, so that most of the atom's volume is actually empty space. In this space near the nucleus, the electric field is essentially that of the nuclear charge, a Coulomb's-law field whose strength varies as the inverse square of the distance from the nucleus. The planetary electrons are found in the outer regions of the atom, and there these negative charges will also contribute to the electric field. Outside the atom these negative charges completely screen the positive charge of the nucleus, and the electric field vanishes.

In this model, an alpha particle fired through an atom some distance off center will pass through without appreciable deflection. It will just brush the electrons aside. But in a stream of alpha particles fired at gold leaf, a few will pass so near the center of an atom that they encounter the strong electric field close to its heavy nucleus—they will be "scattered" through a large angle.

We see that Rutherford's model accounts, at least qualitatively, for the experimental facts. It explains why atoms let most alpha particles go through almost unaffected, and why a few alpha particles are strongly deflected. Before we place more confidence in this model, however, we must submit it to a quantitative test. This Rutherford and his students did with great skill.

24-2 The Trajectories of Alpha Particles in the Electric Field of a Nucleus

In Rutherford's model an alpha particle near a nucleus experiences a force of repulsion, inversely proportional to the square of the distance. To test this model, we must work out the paths that an alpha particle would take in such a force field.

Examples of the computed trajectories are shown in Fig. 24–4. Although we shall not reproduce the long computations of these paths, they involve nothing more mysterious than Coulomb's law, Newton's law of motion, and the detailed geometrical arrangement. The curves are hyperbolas, all with the nucleus at one focus. (The hyperbolic orbits here are related to the elliptic orbits of planets in the inverse-square gravitational force field.

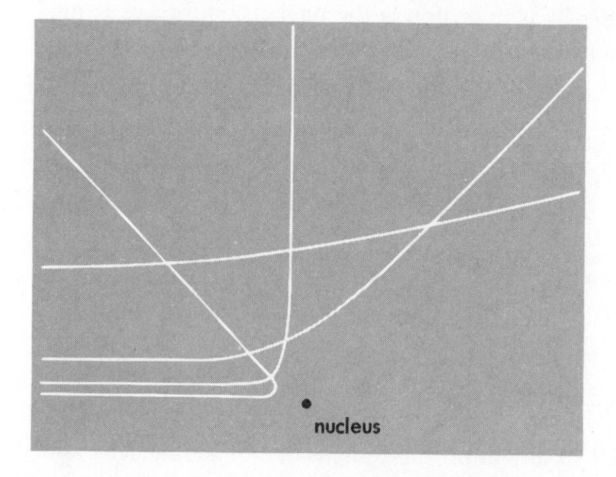

Figure 24–4
The computed paths of alpha particles approaching a nucleus. These were computed using an inverse-square force of repulsion.

You may also convince yourself that the computed results are sensible by looking at a mechanical model which reproduces the essentials of alpha-particle scattering in the Coulomb force field. You may even want to construct such a model and experiment with it. In the mechanical model the alpha particle is represented by a steel ball which rolls with little friction on a smooth curved hill rising from a level table (Fig. 24–5). We make the gravitational potential energy of the ball on the hill correspond to the electric potential energy of the alpha particle near the nucleus; so this model pictures a plane through the center of an atom with the third dimension, the height in the model, representing not space but potential energy. The electric potential energy of the alpha particle varies as $1/r$, where r is its distance from the nucleus. Therefore we build our hill so that above any point on the plane the height is inversely proportional to the distance r of that point from the center.

With this model, the motion of an alpha particle near the nucleus is modeled by the motion of a ball on the potential-energy hill in Fig. 24–6. If we look down on a ball rolling on a real hill of this shape, the path we see will closely approximate that of an alpha particle near the nucleus.

An alpha particle aimed head on at a nucleus will be scattered in exactly the backward direction. Similarly, in the mechanical model, if we start the

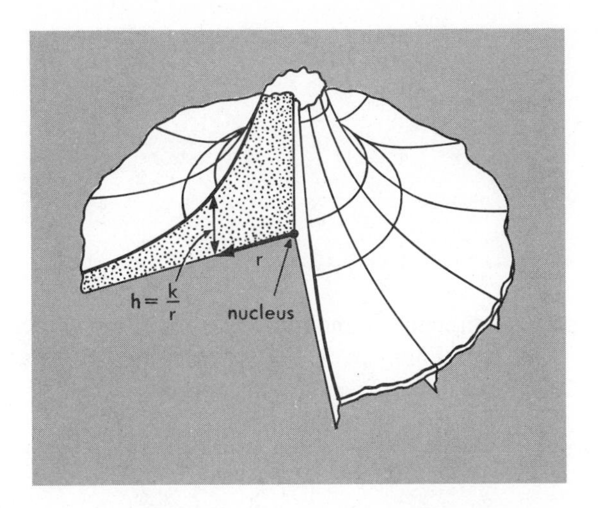

Figure 24–5
A mechanical model to illustrate the paths of alpha particles near a nucleus. Alpha particles have a potential energy proportional to $1/r$, where r is the distance from any nucleus. The hill is so constructed that the height of any point on its surface is proportional to $1/r$, where r is the distance on the plane from the center. Consequently, a ball on the hill will have gravitational potential energy proportional to $1/r$. Its motion then resembles the motion of a charge moving on the plane in the electric field of the nucleus.

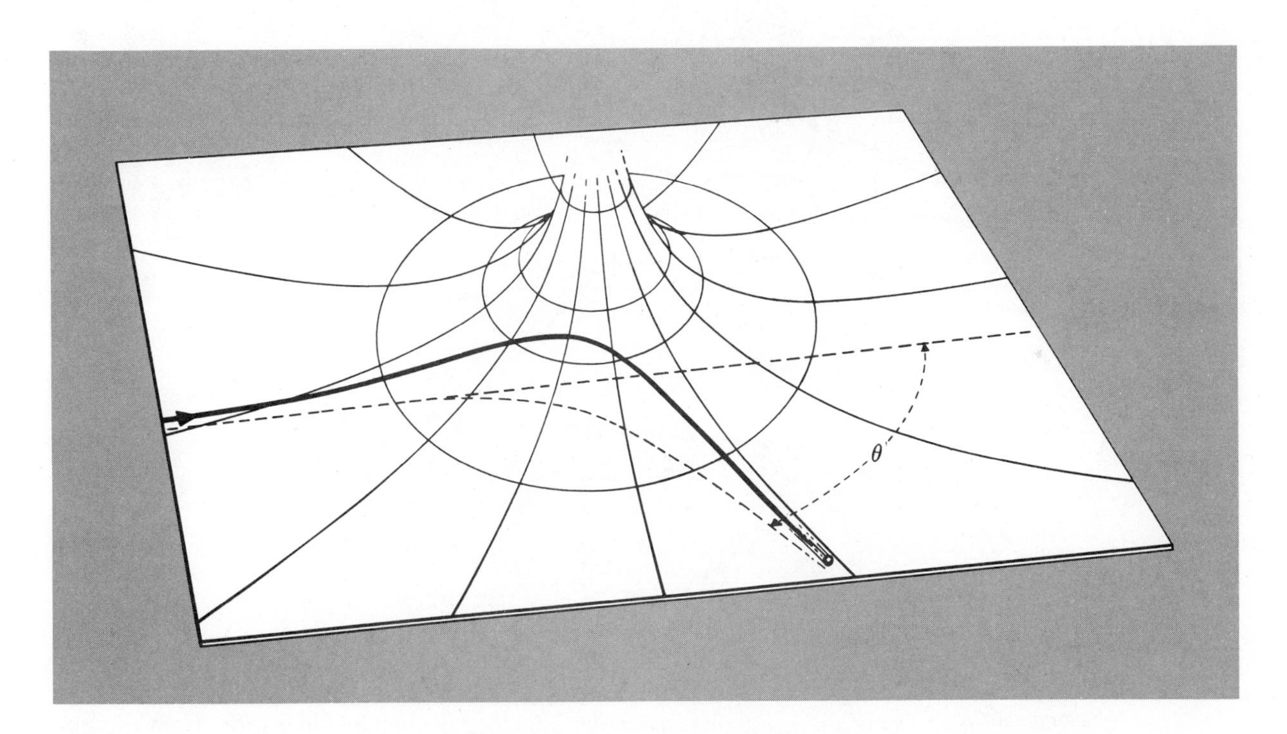

Figure 24–6
The solid line shows a possible path of a rolling ball in the "hill" model of the Coulomb field. The dashed straight line on the plane below the hill shows the direction in which one ball was originally aimed. If you look vertically down on the hill, the apparent path of a ball models the path of an alpha particle moving in the horizontal plane. This path is the dashed curved line on the plane.

ball rolling exactly toward the center of the hill, it will go straight up, to a height where its potential energy equals its original kinetic energy. There it reverses its motion and returns to the starting point. If, however, we aim the alpha particle at a point to the right of the nucleus, Coulomb repulsion pushes the alpha particle to the right, changing its direction as it passes the nucleus (Fig. 24–6). The angle θ between the final direction of motion and the original line of flight is called the scattering angle (Fig. 24–7). The closer the alpha particle comes to the nucleus, the stronger the Coulomb force it feels and the more nearly that force pushes straight backward against the

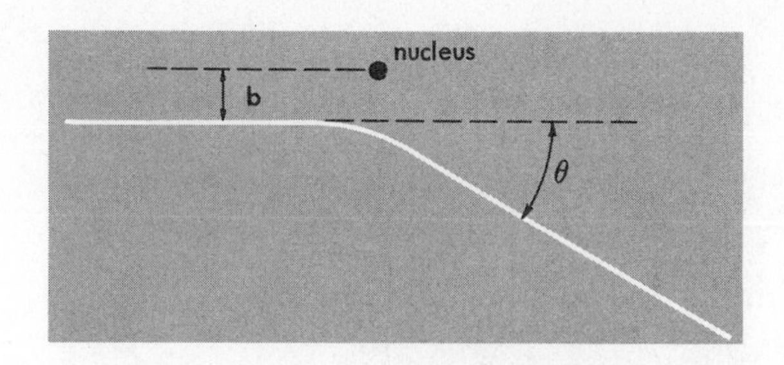

Figure 24–7
The aiming error b is the distance by which a particle would miss the nucleus if it were not deflected. The scattering angle θ is the angle between the original direction and the direction after deflection.

original direction of motion. Therefore, the scattering angle θ will increase when the "aiming error," represented by the distance b, decreases. Figure 24–8 shows the corresponding motions of the ball in our mechanical model.

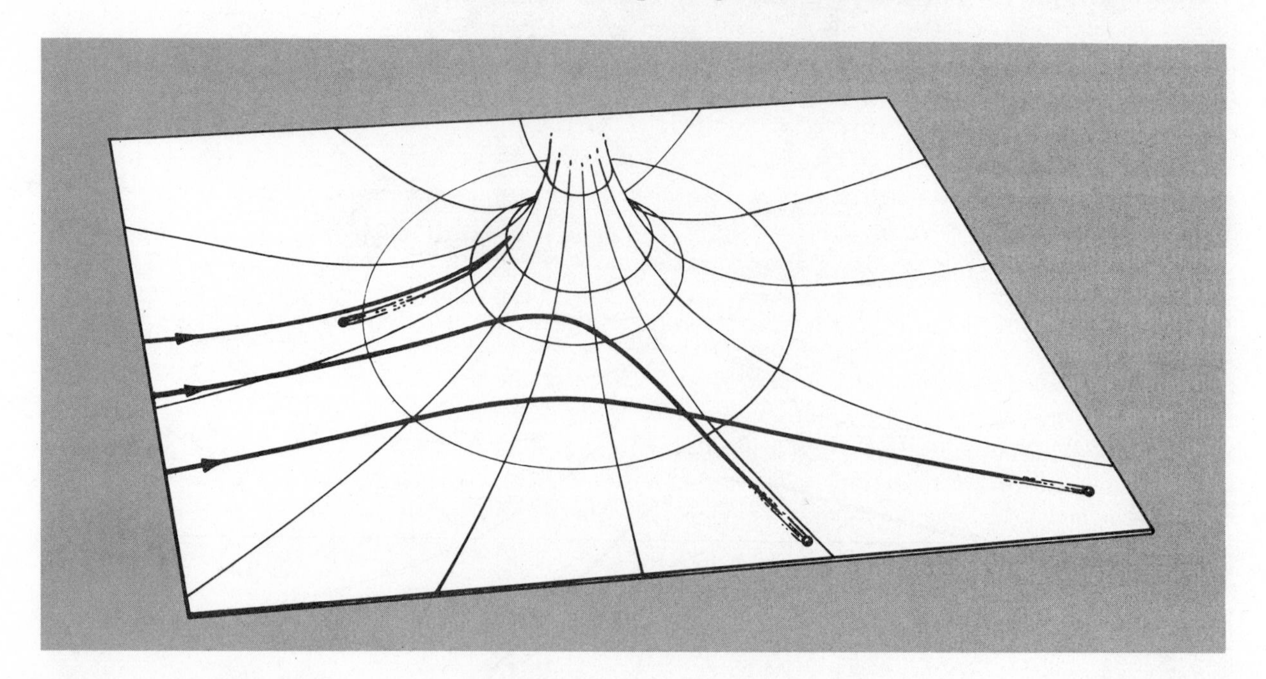

Figure 24–8
Paths of a rolling ball on the hill model. The ball is sent with the same energy each time, but with a different aiming error. Note that the smaller the aiming error, the larger is the scattering angle.

24-3 Angular Distribution of Scattering

For particles of the same energy a definite aiming error leads to a particular angle of scattering under the influence of the Coulomb force from the nucleus. By computing the paths for a number of aiming errors, b (or by more elegant methods), we can plot a graph of the aiming error b against the resulting scattering angle θ. For the Coulomb force this relation is shown in Fig. 24–9.

Now if we could see the center of an atom and aim an alpha particle with complete precision so that we picked our aiming error b, we could test the relation between b and θ by a direct experiment. However, the alpha particles in the most accurately aimed beam will hit at random all over a region many times the area presented by an atom. Consequently, we must do a slightly less direct experiment.

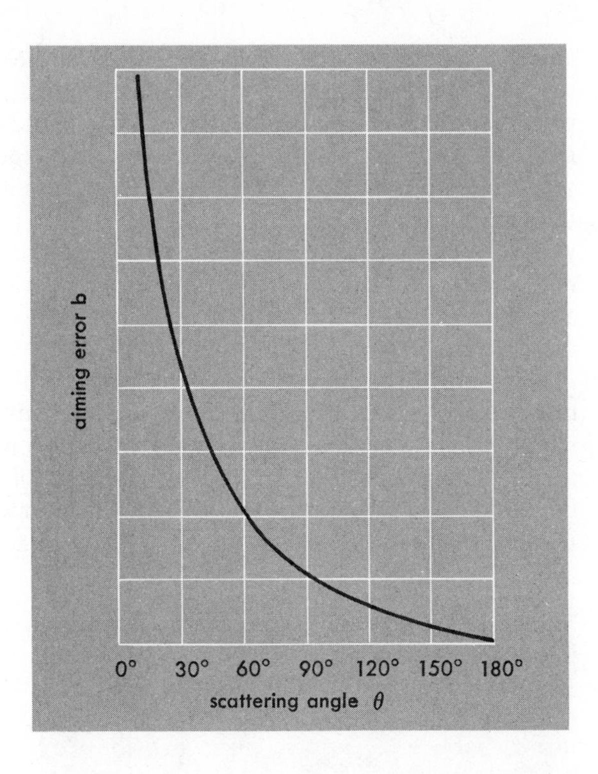

Figure 24–9
A graph of the relation between b and θ for a Coulomb force of repulsion.

Because smaller aiming errors lead to larger scattering angles (Fig. 24–10), we can take advantage of our uncontrollably poor aim by computing the number of alpha particles that will hit with less than a certain

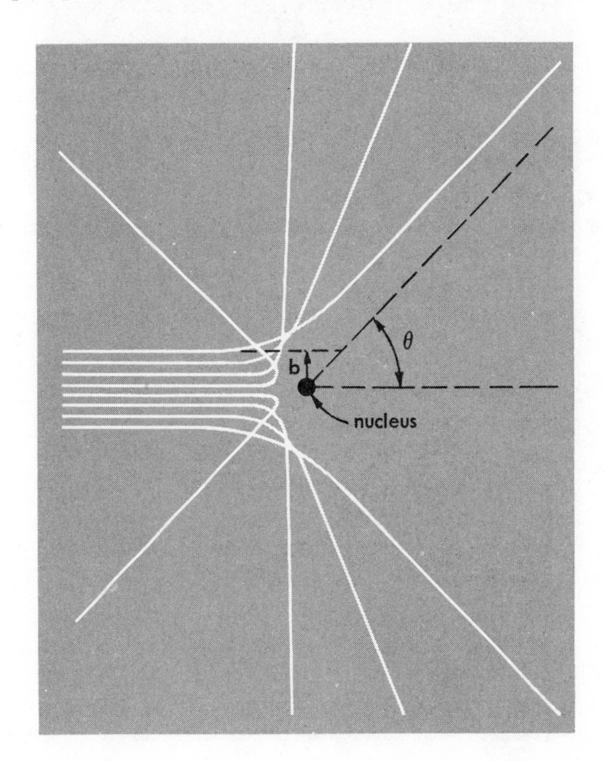

Figure 24–10
As the aiming error b decreases, the scattering angle θ increases. Therefore, all particles aimed within a distance b of the nucleus will be scattered through angles greater than θ. You see this same relation in Fig. 24–8 and 24–9.

aiming error b. All these alpha particles (and no others) should then be scattered by more than the corresponding scattering angle θ. Since we can measure experimentally the number of alpha particles N_θ scattered by more than any particular angle θ, we can then find experimentally whether the scattering is produced by a Coulomb force. Another force might produce the same number N_θ scattered by more than one particular angle θ, but as we shall see, it could not produce the same relation between N_θ and θ over many angles. In detail, therefore, our experiment will be to measure N_θ as a function of θ (Fig. 24–11) and to compare the experimental results with those we predict from the Coulomb force.

To compute the number of alpha particles that hit within a certain distance b of the center of an atom, we take advantage of the fact that the beam sprays uniformly all over the target area of an atom. Because the number of alpha particles aimed at a small area of an atom is equal to the number aimed at any other equal area in the same atom, the number of alpha particles that pass within a distance b of a given nucleus is proportional to

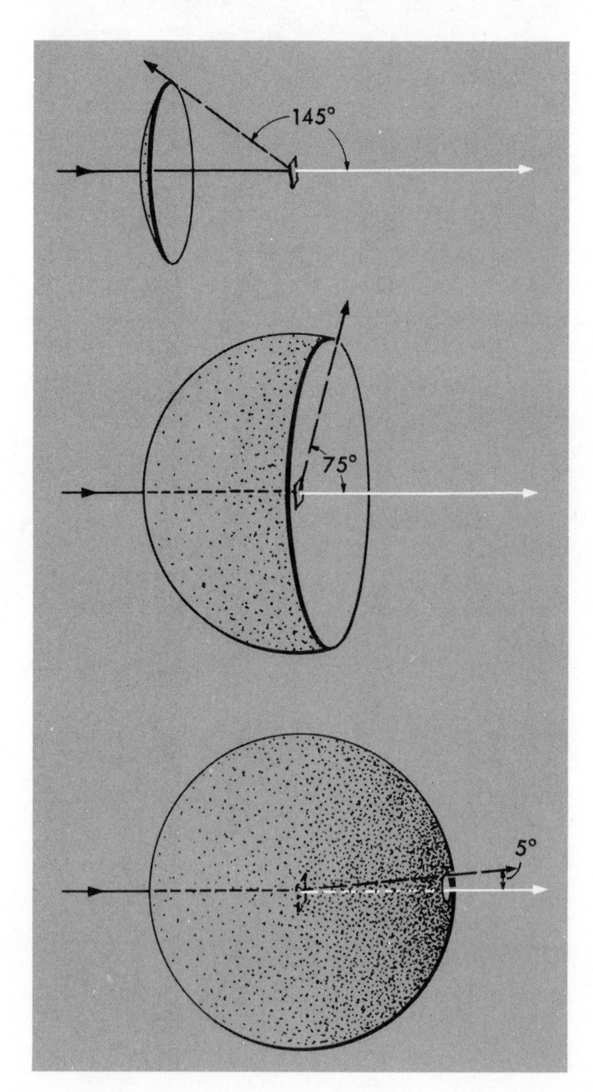

Figure 24–11
Since the particles are scattered in all directions, they spread out over the surface of a sphere. When we count all those particles scattered by more than a given angle θ, we are counting the particles that fall on a cap, on a segment of the sphere. The caps for $\theta = 145°$, $\theta = 75°$, and $\theta = 5°$ are shown. N_θ is the number of particles scattered onto the appropriate cap.

the area of a circle of radius b (Fig. 24–12). This area is πb^2. Consequently, the number of alpha particles that are scattered through more than the angle θ is proportional to b^2.

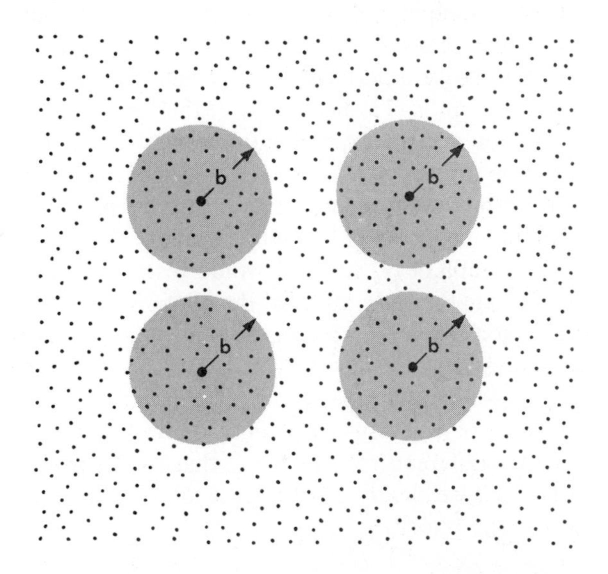

Figure 24–12
Alpha particles are sprayed uniformly over the target area of many atoms. Therefore the number of particles coming within a distance b of any nucleus is proportional to πb^2—the area of a circle of radius b.

Now we already know how b is related to θ. By using the graph of b vs. θ we can construct a graph of b^2 vs. θ. Since the number of particles N_θ scattered through more than the angle θ is proportional to b^2, a plot of b^2 vs. θ and a plot of N_θ vs. θ are the same (except for a factor to adjust the vertical scale). Figure 24–13 shows this graph.

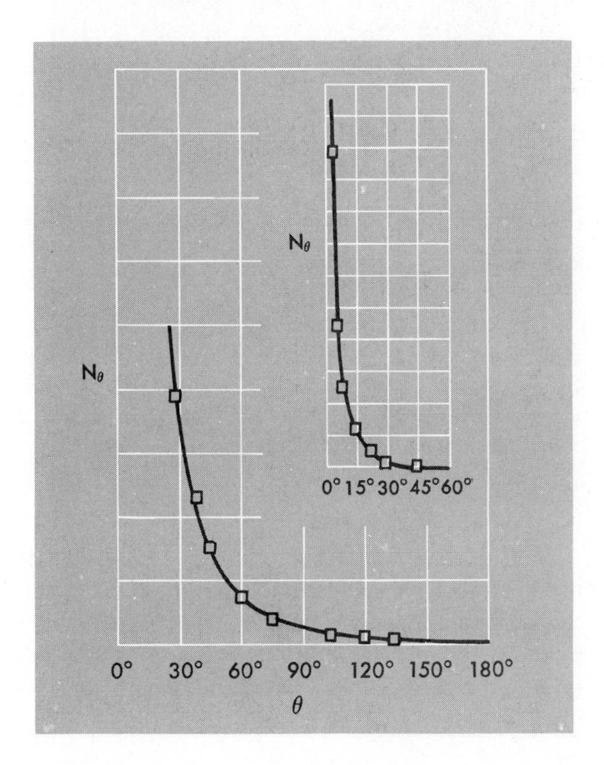

Figure 24–13
A graph of N_θ (the number of particles scattered through an angle greater than θ) vs. θ. The curved line shows what we would expect with a Coulomb force. The points in small squares represent the data collected by Geiger and Marsden in their scattering experiments. The insert shows the curve for small angles on a different scale.

We are now in a position to make an experimental test of the Rutherford model. By counting the number of alpha particles scattered out of the beam into all various directions, we can find out experimentally whether or not the number of alpha particles N_θ—the number scattered through all angles larger than θ—looks like the graph. Detailed experiments on the angular distribution were performed by Geiger and Marsden (1913). The square boxes in Fig. 24–13 show some of their results. Within the accuracy of the measurements, the experimental points fall on the computed curve for the Coulomb force.

We can now conclude that the scattering of alpha particles is the result of a Coulomb force; but to be more certain of this conclusion, we shall see what kind of graph of N_θ versus θ would result from a different force. It is easy to calculate the paths and find the relation between aiming error and scattering angle for a force which is zero at large distances from the center of the atom and then rises abruptly to large values if an alpha particle gets within a certain distance of the center. For this reason we have chosen to construct the graph of N_θ versus θ for such an abrupt force. This curve shown in Fig. 24–14 is different from the curve that results from a Coulomb force, so different that we clearly cannot get it to agree with the experimental results of Geiger and Marsden.

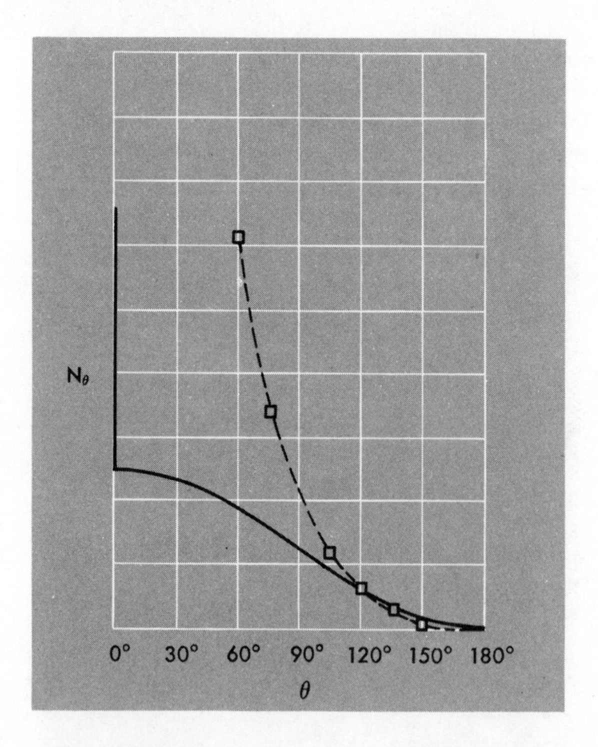

Figure 24–14
The solid line is the curve of N_θ vs. θ for a force that rises abruptly. The dashed line connects Geiger and Marsden's experimental points. The curves were adjusted so that they coincided at $\theta = 120°$. No adjustment will make them fit together.

In general we may assume that any force we fancy is acting on the alpha particles; then we can compute the graph of N_θ versus θ for that force. For each different force, we get a different graph—and we can identify the force by the shape of the graph. The shape of the graph of the experimental results is the same as that for the Coulomb force. Consequently this is the force that acts on the alpha particles.

This is indeed a most remarkable result. Using alpha particles as probes, we have explored the electric field *inside* individual atoms and found that it obeys an inverse-square law. In the first place, this means that the positive charge of the atom is actually concentrated in a much smaller volume than the complete atom. In the second place, we find evidence that Coulomb's law, which we had previously verified experimentally only down to distances of the order of centimeters, holds in fact with high accuracy down to distances much smaller than atomic dimensions. As we shall see in the next section, those alpha particles with which we probe closest to the nucleus and for which we get the largest angles of deflection penetrate far inside the atoms. They only turn around when they are about 10^{-14} meter away from the nuclear charge, about one ten-thousandth of an atomic dimension. The known domain of the Coulomb force is thus extended by a factor of 10^4 toward the very small.

24-4 More Information from Scattering

There is more information to be obtained from the scattering experiments. Suppose that we keep our source of alpha particles and our detector in fixed positions and we change to a target of a different metal (Fig. 24–15). Suppose that we use, one after the other, very thin sheets of different elements (gold, silver, platinum, etc.), choosing the thicknesses so that in each foil there are the same number of atoms in the way of the beam, that is, the same number of atoms per square centimeter of foil surface. Now, although the number of scattering centers (the number of nuclei) is the same in all cases, we find that the numbers of scattered particles observed in a given time interval are very different.

We can easily understand this result if we assume that the nuclei of different elements have different electric charges. For a larger nuclear charge, the force pushing the particle out of its original direction is greater. There-

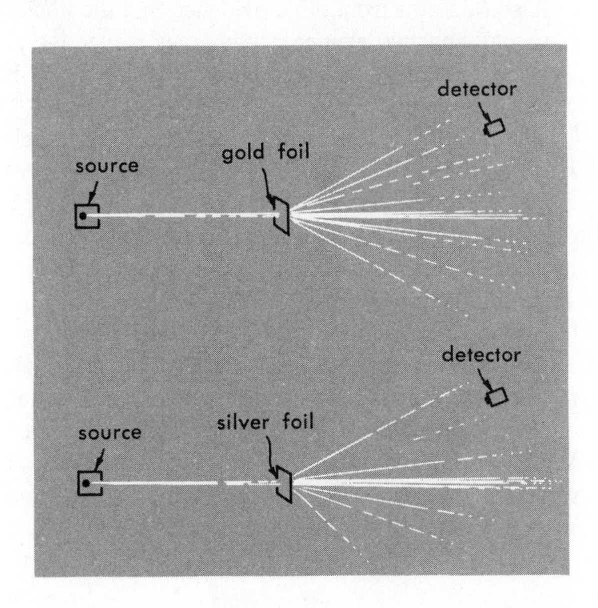

Figure 24–15
If we use foils of different metals, but each with the same number of atoms per cm^2 of surface, the observed scattering is different. Using the same beam for equal times, the number of alpha particles scattered into any given angle by the gold foil is 2.8 times the number scattered into the same angle by the silver foil.

fore, if we shoot in alpha particles with a given speed and the same aiming error, we expect them to undergo a large deflection (Fig. 24–16). Detailed computations (Newton's law, Coulomb's law, and geometry) show that the number of alpha particles scattered within a given range of angles is proportional to the square of the nuclear charge. Therefore, we can actually compare by our experiments the electric charges of the nuclei of different elements. We can go further still and compute the actual charge in elementary electric charges.

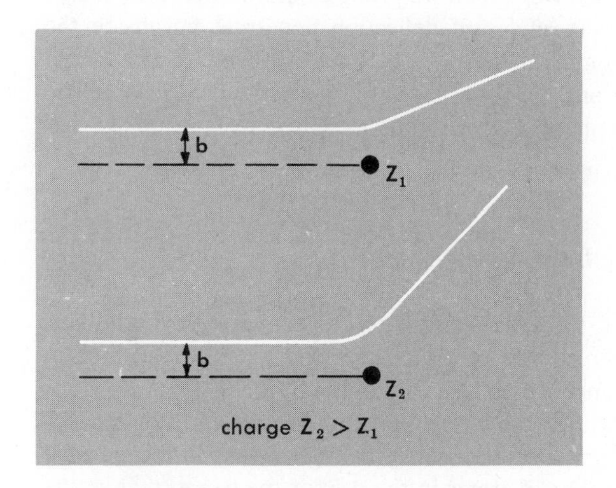

Figure 24–16
Alpha particles of a given energy and aiming error are deflected more by nuclei of larger charge.

According to other evidence, the nucleus of copper has 29 positive elementary charges, the nucleus of silver 47, the nucleus of platinum 78. From alpha-particle scattering, Chadwick found almost these results. His experimental measurements yielded 29.3, 46.3, and $77.4 \pm 1\frac{1}{2}$. This means that the nuclear charges must be 29 or 30, 46 or 47, and between 76 and 79 elementary charges respectively—a masterpiece of prying information out of atoms with simple apparatus.

Also we know from other evidence that the nuclei of hydrogen and helium have one and two elementary charges respectively. (See Chapters 20 and 22.) Scattering experiments with hydrogen and helium gases (e.g., in a cloud chamber) confirm this result.

The number of elementary charges in the nucleus is called the *atomic number* of the element. Earlier in the development of chemistry this number was used for the serial number of the element in the chemical list. It is obviously a quantity of great physical significance. As we have mentioned, it can be measured by other methods as well as by alpha-particle scattering. They all agree and they give a new meaning to the serial numbers in the table of elements. (See Appendix.) Because atoms are neutral, the atomic number tells both the number of positive elementary charges on the nucleus and the number of electrons that surround it.

We still have not exhausted the information we can get from scattering. We have seen that the radius of the nucleus must be much smaller than the radius of the atom. But how small is it? We found experimentally that in gold foils the scattering of alpha particles from a radioactive source corresponds to the predictions we made from Coulomb's law. This means that these alpha particles never "hit" a gold nucleus. For, if they did, forces other than the Coulomb repulsion would come into play and the resultant scattering would certainly differ from that computed from the inverse-square law. Thus the gold nucleus must be smaller than the distance of closest approach of the alpha particles to the center of the nucleus.

Alpha particles of a given energy come closest to the nucleus when they are aimed directly at its center. They move then on a straight line until their potential energy equals their initial kinetic energy, and then retrace their path backward. (See Section 24–2.) Let r_0 be the distance of closest approach. Since the gold nucleus has 79 elementary charges and the alpha particle 2, the potential energy at the distance r_0 is:

$$U = k\frac{(79 \text{ elem. ch.})(2 \text{ elem. ch.})}{r_0}$$

$$= (3.6 \times 10^{-26} \text{ joule-m})\frac{1}{r_0},$$

$$\text{since} \quad k = 2.3 \times 10^{-28}\,\frac{\text{newton-m}^2}{(\text{elem. ch.})^2}.$$

On the other hand, alpha particles from a polonium source have a velocity of 1.6×10^7 m/sec and a mass of 6.62×10^{-27} kg. (See Chapter 22.) Therefore, their kinetic energy is

$$\tfrac{1}{2}mv^2 = 8.5 \times 10^{-13} \text{ joule}.$$

By equating the potential energy gained to the original kinetic energy, we obtain

$$r_0 = 4.2 \times 10^{-14} \text{ meter}.$$

Thus we can assert that the radius of the gold nucleus is certainly less than 4.2×10^{-14} m. Similar results are obtained for the nuclei of the other elements. Alpha-particle experiments therefore also bear out Rutherford's hypothesis that the nucleus is a minute heavy speck.

For Home, Desk, and Lab

1. Suppose you have a large collection of small steel balls which you can fire at different targets in turn. You cannot see the targets directly, but you can observe the paths of the balls before and after striking. In each case below, tell what you know about the target from the behavior of the balls. The balls are all fired at the target in the same direction.
   (a) All the balls bounce off the target with angle of reflection equal to angle of incidence.
   (b) All the balls bounce back, but with no definite relation between angle of reflection and angle of incidence.
   (c) Practically all the balls go through the target, but a few are lost in it.
   (d) All the balls are lost in the target.
   (e) The paths of all the balls cross after reflection at a point in front of the target.

2. To help visualize the thickness of the gold foil used in the alpha-particle deflection experiment described in Section 24–1, find the number of sheets of this foil which, when stacked together, are equal to one leaf of your textbook.

3. One square meter of silver foil has a mass of 5×10^{-3} kg.
   (a) What is its thickness? (The density of silver is 1×10^4 kg/m^3.)

(b) The mass of a silver atom is 1.8×10^{-25} kg. What is the length of the side of a "box" representing a single atom?

(c) What is the thickness of the foil expressed in atomic "boxes"?

4.* What is the evidence for our belief that alpha particles which are deflected by a large angle when passing through a thin gold foil do so in one collision? (Section 1.)

5.* What are the main features an atomic model must have in order to account for the way in which alpha particles are deflected when they pass through gold foil? (Section 1.)

6.* What is a scattering angle? (Section 2.)

7. Suppose that a ball is rolled many times at the hill model of Fig. 24–6. Each time it starts along the same path, but with a different kinetic energy. How do you expect its scattering angle θ to be related to its energy?

8. In a scattering experiment using gold foil, 4 alpha particles in 10,000 are scattered through an angle of more than 5°.

(a) How many particles would be scattered through more than 5° if we doubled the thickness of the foil?

(b) If we made the foil a thousand times thicker?

9. A beam of alpha particles from polonium is shot perpendicularly into the gold foil described in Section 24–1. We observe that some of the particles are scattered through an angle of about 3° or 5×10^{-2} radian. The alpha particles have a kinetic energy of 8.5×10^{-13} joule, which corresponds to a speed of 1.6×10^{7} m/sec.

(a) What velocity $\Delta\vec{v}$ must be added at right angles to their original velocity to produce this change in direction?

(b) How long does it take an alpha particle to pass through a single gold atom in the foil?

(c) What is the average acceleration of one of these alpha particles if it gets its whole deflection in passing through a single gold atom?

10.* Why is it not possible to find directly the relation between θ and b from the Rutherford scattering experiment with alpha particles? (Section 3.)

11.* How is N_θ, the number of particles which have scattered through all angles larger than θ, related to the aiming error b? (Section 3.)

12.* What do the results of scattering experiments tell us about the electric field inside the gold atom? (Section 3.)

13.* If the alpha particles approaching a nucleus with a certain aiming error b were scattered by 90°, through what range of angles were the particles scattered which approached the nucleus with a smaller aiming error? (Section 3.)

14. Imagine a device which can drop 1000 steel balls through a circular hole 5 cm in radius and spread them evenly over the area of the circle (Fig. 24–17). In the center of the hole we fix a steel cone whose sides make an angle of 45° with its base, and whose base has a radius of 2 cm. In a group of 1000 balls, how many are scattered sideways before they reach the hole?

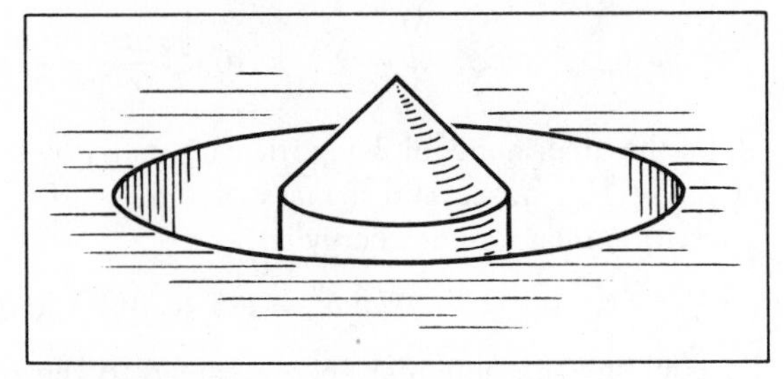

Figure 24–17
For Problem 14.

15. The point of the cone in Problem 14 is sawed off to leave a smooth plane parallel to the base and 1 cm in radius (Fig. 24–18).

(a) In each group of 1000 balls, how many are turned back and how many are scattered sideways by it?

(b) Draw a graph of the scattering angle θ versus the aiming error b, measured from the center of the hole.

(c) Draw a graph of N_θ (the number of balls scattered through an angle greater than θ) vs. θ.

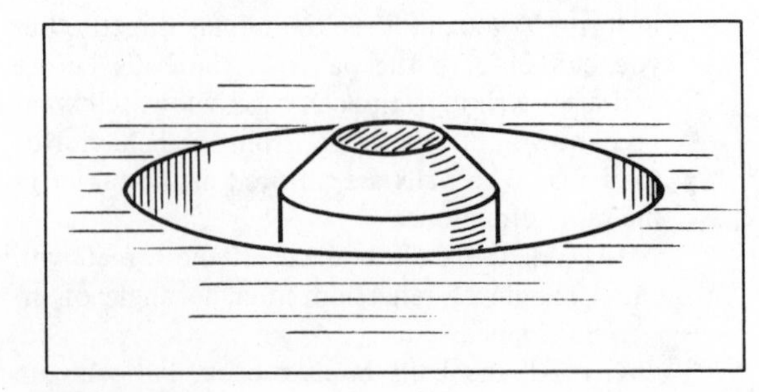

Figure 24–18
For Problem 15.

16. (a) To construct the curve of b vs. θ for an abrupt force, draw a fairly large circle to represent a plane cut through a rigid sphere and measure the scattering angles for various aiming errors. (Assume that the angle of reflection of a ball hitting the surface equals the angle of incidence. Why?)
    (b) From your curve of b vs. θ construct a graph of N_θ vs. θ and compare it with Fig. 24–14.
    (c) Explain why this relationship is independent of the energy of the particles fired at the sphere.

17. From the graph in Fig. 24–13, determine the ratio of the number of particles scattered through an angle greater than 15° to those scattered through an angle greater than 22.5°.

18.* How many times as close can an alpha particle of given energy get to the center of the nucleus of an aluminum atom as it can to the center of a gold nucleus? (Section 4.)

19.* How close can polonium alpha particles get to the center of an aluminum nucleus? (Section 4.)

20.* Which conservation law is used in calculating the distance of closest approach? (Section 4.)

21. In nuclear fission, the nucleus of an atom of uranium 238 sometimes breaks into just two fragments, owing to the Coulomb force of repulsion between its protons. Imagine that the split has just happened, and the two fragments are still touching each other (roughly 10^{-14} m apart, center-to-center) and at rest. Find their total kinetic energy when they are very far apart, if one has $Z_1 = 42$ elementary charges and the other has $Z_2 = 50$ elementary charges. Give your answer in joules and in millions of electron volts (Mev).

22. (a) Two protons are released from rest 10^{-10} m apart. What will their kinetic energy be at large separation?
    (b) Suppose one of the protons is nailed down. What will the kinetic energy be at large separation?

23. Three experiments were done with the same alpha particle beam, the same counters at the same positions, and different foil targets. Five counters were arranged with respect to each target and the beam to count particles scattered by the angles given in the table. The time interval of counting was the same for all three experiments. The foils were designed so that the beam passed through the same number of atoms per square centimeter regardless of which target was used. The results presented in the table below give the numbers of particles scattered into the same range of scattering angles. For example, the count at 30° includes all particles scattered in any direction in the range between 29° and 31°; the count at 60°, all those scattered in any direction in the range between 59° and 61°, etc.

SCATTERING ANGLE	NUMBER COUNTED		
	Target 1	Target 2	Target 3
30°	2790	35,920	102,810
60°	346	4451	12,760
90°	100	1288	3685
120°	39	496	1423
150°	14	187	532

    (a) Plot the number scattered versus the scattering angle for each target. (Plot all three graphs on the same axes.)
    (b) Target 1 was aluminum. What were the other targets?

24. (a) Explain why the gravitational force between an alpha particle and a gold nucleus has little effect on the scattering of alpha particles by gold.
    (b) What is the ratio of the electric force on an alpha particle to the gravitational force when the separation of an alpha particle and a gold nucleus is 10^{-14} m? When it is 10^{-13} m?

25. An alpha particle of kinetic energy 1.0×10^{-12} joule is scattered by a gold nucleus.
    (a) What is the minimum possible distance of closest approach?
    (b) What must be the aiming error if the alpha particle is to get this close?
    (c) How close could a proton of the same energy approach? A neutron?

26. In one scattering experiment a proton beam strikes a gold target. In a second experiment a beam of deuterons (atomic mass = 2, atomic number = 1) is used. The kinetic energy of the particles is the same in both beams. How will the minimum distance of closest approach to gold nuclei of the two kinds of beams compare?

27. (a) What fraction of its energy is lost by an alpha particle in a head-on collision with an electron? (The mass of an alpha particle is about 7200 times that of an electron.)
    (b) Approximately how many such collisions must occur before the energy of the alpha particle is reduced by 1 percent?

28.† An alpha particle with speed 1.00×10^7 m/sec is scattered by a gold nucleus at rest through an angle of 60°. Assume that so little kinetic energy is transferred to the gold nucleus that the speed of the alpha particle after the scattering is the same as it was before.
    (a) What is the momentum transferred to the gold atom?
    (b) What is the velocity given to the gold nucleus?
    (c) What is the kinetic energy given to the gold nucleus?
    (d) Was your initial assumption a good one? Explain.

29. (a) If the distance of closest approach of the alpha particle in Problem 28 was 1.65×10^{-13} m, what was its speed at this point?
    (b) At what speed was it moving toward the nucleus at that point?

30. In the hydrogen atom, the electron is on the average 5×10^{-11} m from the positively charged proton.
    (a) Estimate the average force between the electron and proton.
    (b) If the atom were placed between the plates used in the Millikan experiment of Section 19–8, how many of the 90-volt batteries would we need to put in series with the plates to tear the atom apart?

Further Reading

These suggestions are not exhaustive but are limited to works that have been found especially useful and at the same time generally available.

ANDRADE, E. N. DA C., *An Approach to Modern Physics.* Doubleday Anchor, 1956. (Chapter 8)

ANDRADE, E. N. DA C., *Rutherford and the Nature of the Atom.* Doubleday Anchor, 1964: Science Study Series.

BORN, MAX, *The Restless Universe.* Dover, 1951. A good explanation of the electronic structure of the atom. (Chapter 4)

GAMOW, GEORGE, *Mr. Tompkins Explores the Atom.* Cambridge University Press, 1945. A novel presentation of atomic structure and behavior. Includes three dreams to dramatize the properties of atoms, with accompanying lectures which provide a scientific basis for the dreams.

HUGHES, DONALD J., *The Neutron Story.* Doubleday Anchor, 1959: Science Study Series. (Chapters 1 and 4)

ROMER, ALFRED, *The Restless Atom.* Doubleday Anchor, 1960: Science Study Series. (Chapter 13)

WEISSKOPF, VICTOR F., *Knowledge and Wonder.* Doubleday Anchor, 1963: Science Study Series. (Chapters 4 and 5)

WILSON, ROBERT R., and LITTAUER, RAPHAEL, *Accelerators.* Doubleday Anchor, 1960: Science Study Series. (Chapters 1 and 3)

Photons

CHAPTER

25

25-1 The Graininess of Light

The wave theory has had convincing successes in describing all forms of electromagnetic radiation, from radio waves measured in kilometers down to the light waves which in our microscopes bring us a view of the smallest cell. We shall now give an account of experiments, some actually as they were done, some only illustrative of what might be done, in which the wave picture was subjected to a new kind of test.

That test was to see what light did to individual atoms. We cannot simply assume that light will behave in the same manner on an atomic scale as it does on a much larger scale. As a simple example, consider the experiments involving the refractive properties of light—light passing through a refractive medium is bent toward or away from the normal. If we use smaller and smaller blocks of refractive material, will Snell's law continue to hold? You know that when the thickness of the block becomes comparable to the wavelength of the light, diffraction effects become more important than the refraction. What happens if we go to a still smaller size—the size of the atom? When we ask the question, "What does light do to individual atoms?" we must experiment further.

In a kind of Millikan experiment, a good many tries were made of a related experiment in which X rays or ultraviolet light from an arc lamp falls upon large numbers of specks of material drifting between a pair of charged plates (Fig. 25–1). We ask what the wave picture of light predicts about the ejection of electrons from the tiny specks, and compare the prediction with what actually happens.

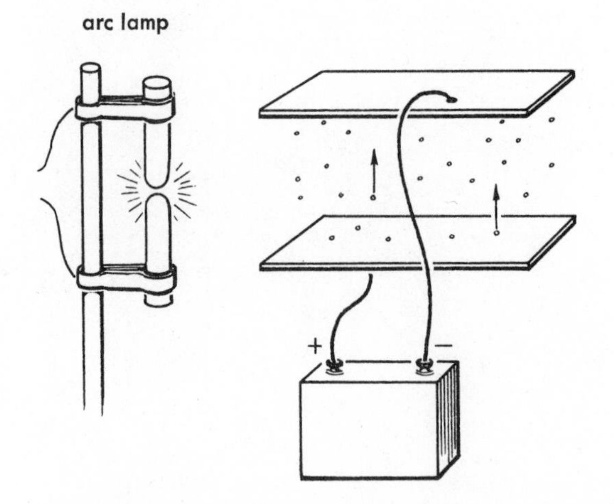

Figure 25–1
The light from the arc lamp illuminates a large number of droplets or specks of metal. These specks are floating between charged parallel plates; they are nothing but a large crowd of "Millikan oil drops." As the light strikes the specks, they may lose electrons. Each time an electron is lost, the speck will begin to drift upward toward the negative plate, for it is now positively charged by more than enough to balance its weight.

When light—that is, electromagnetic radiation—shines on a speck of matter, the electric field in the light wave pushes on the electrons in the atoms. If an electron is ejected from an atom, its ejection must be an effect of the electric field which forces it out of its normal motion within the atom. (The magnetic field exerts a force perpendicular to the velocity of the electron and therefore does no work.)

Now, the electric field is rapidly changing in time, but at a given instant it has the same strength everywhere along the front of the electromagnetic wave. When such a wave hits an atom, therefore, it starts to shake the electrons. Since the wave looks the same over a large region of space, it shakes electrons in the little specks of matter in the same way at almost exactly the same time all over the region between the plates. If the light is weak, it will take the light wave considerable time to shake any electrons loose. Then, since there are many electrons in identical atoms pushed in the same way by the electric field of the light wave, a large number of electrons should be ejected at almost the same time. We thus expect to see, for rather weak light, a time during which nothing happens. Then, at one time all over the front of the light beam, individual specks should begin to lose electrons. With stronger light, the loss should come sooner; with weak light, later.

The actual result is astonishingly different. Instead of a definite wait, after which all the specks begin to lose electrons, it is found that a speck or two may jump immediately. These jumps show that the specks have already lost electrons and are now moving like the Millikan spheres under the influence of the electric field between the plates. The first specks to jump may be found anywhere in the beam. After them others will jump—one here, one there, with no particular pattern. After a while most of the specks will have lost electrons, but in a quite random way. Some may have lost several electrons; a few will have lost none at all.

If the light is made weaker, the waiting period is sometimes no longer. If the light is stronger, the wait is sometimes no shorter. We can make the specks as similar as can be, use our best identical spheres, and yet the individual specks behave differently (Fig. 25–2). Only the average behavior

Figure 25–2
In the apparatus of Fig. 25–1, we turn the light on suddenly and watch for the first speck to start upward. This time is then recorded. The graphs show the results of many such time-lag measurements. In (a) the experiment was done with weak light; in (b) with much stronger light. Everything else remained unchanged. The total number of trials in each group of experiments was the same. On the average, the strong light ejects electrons more quickly than the weak light, but even in weak light some electrons are ejected with no appreciable lag. Experiments of a more advanced sort have pushed this minimum lag time below 10^{-8} second.

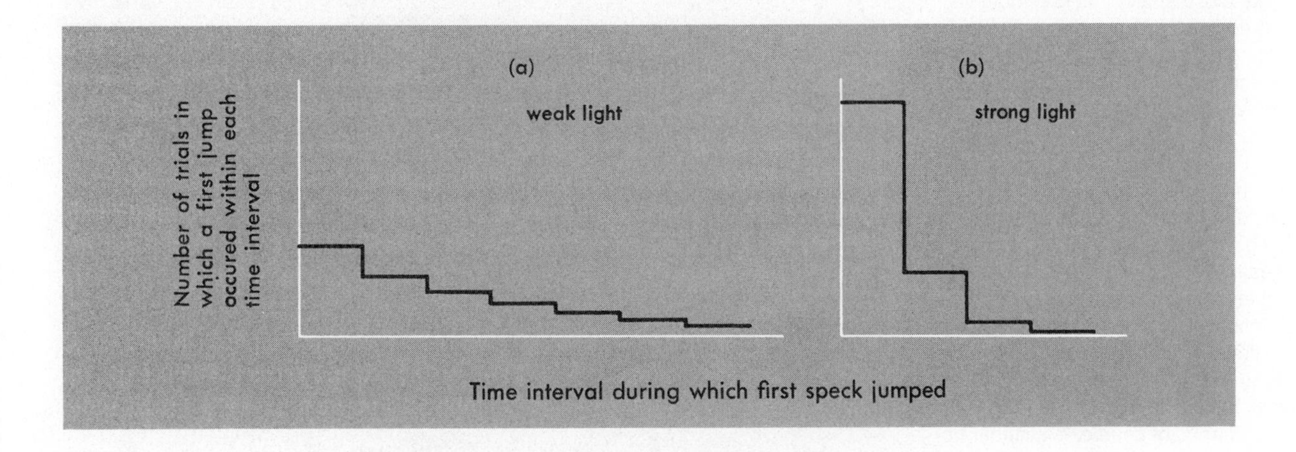

seems to be what we expect. On the average the specks will lose electrons as well in one part of a uniform light beam as in another. On the average, more electrons are driven out in one second by strong than by weak light; but this refers only to the average over many specks and many trials. In any particular case, how long it takes the first speck to respond seems to be a matter of chance. Once in a while, even the weakest light will cause the immediate loss of an electron; once in a while, even in the strongest beam, you will wait a fairly long time to see the first electron removed. The whole nature of this result seems to belie our well-founded view of the smooth oscillating wave front, everywhere uniform and moving in phase, displacing electrons in every speck by the cumulative action of the varying electric field.

The predictions of our picture of electromagnetic waves do not fit the facts, but we know of another simple model, long discarded, which we can now revive. Suppose that instead of the smooth plane wave front, the beam of light is represented by a collection of fast-moving tiny particles, like a volley of bullets or an air-driven blast of sand (Fig. 25–3). In such a beam, a speck would be hit squarely by a single particle of light only by chance.

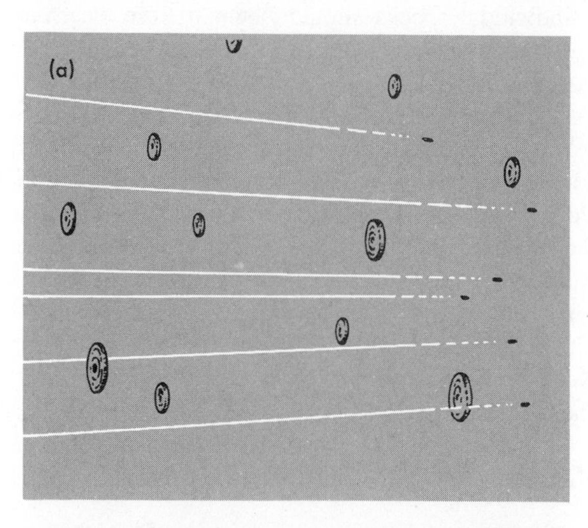

(b)

Figure 25–3
The close analogy between bullets fired in a random volley at a set of targets, and a light beam ejecting electrons. If the volley comes from only a few guns (a), the "beam" of bullets is weak. Most of the time, you would wait a long time to see a direct hit. But occasionally a lucky hit would immediately follow the start of firing. With many rifles firing (b), the "beam" is intense. Hits become frequent, and on the average you wait less time for the first one. But the first hit might come no sooner, or even later, than an early first hit in a weak beam. Just which target would be hit next, of course, is a matter of chance.

The hits would appear now here, now there, over the whole of the beam. The moment when the first speck was hit would also be a matter of chance. Even if the beam was weak—with only a few particles crossing a given area in a second—it might happen that a particle of light strikes one speck immediately after the beam is turned on. On the other hand, in a strong beam, the dense rain of particles might fail to make any direct hit for a noticeable time. Only on the average would it be true that in a strong beam the first hit would come soon, and the others follow in rapid sequence.

This picture gives an accurate and convincing description of the way light really ejects electrons from the little specks of matter.

On this picture light comes in discrete units. It arrives in little bundles which are called *photons*. The photons are the particles of light; and when they hit a speck of matter, they may eject an electron.

Our experiment directly detecting the ejection of electrons by a light beam is not the only means of displaying the grainy character of the beam of light. A photographic plate bears a layer of very small grains, each one of which can be changed by the impact of light into a form which the developer solution can blacken. The image is not formed smoothly, but by the random accumulation of blackened grains. This process is too complex to lead us to interpret light from it alone—the film grains differ in their atomic constitution, for example—but it adds support to the direct electron-ejection experiment. With other devices using ejection—the photoelectric cell, the television-camera image tube—one can watch an image being formed point by point by light. In very weak light, the successive "hits" are so slow that the eye can easily follow them. The image, whatever it may be, is painted slowly by the accumulation of random hits, which gradually pile up, many wherever much light falls, few wherever the illumination is weak. Only when very many particles of light have struck every little area—in a strong beam quickly, or in a weak beam after sufficient time—does the accumulation become so dense that all trace of its grainy structure is lost.

25-2 The Photoelectric Effect

Let us now examine the ejection of an electron by a photon, not from a speck of matter, but from a single atom. In Fig. 25–4 a beam of photons coming from the left strikes the gas in a cloud chamber. As the beam penetrates, here and there a photon ejects an electron from an atom of the gas. The little clusters of waterdrops form at the place where the electrons are ejected. After a while the beam of photons is noticeably weakened; indeed, for each electron produced, a photon has been removed from the beam. This can be verified by watching the photons penetrate so far into the gas that the original beam has all but been absorbed. The energy of the beam of light has passed into the gas. In fact, most of that energy appears as the kinetic energy of the ejected electrons.

Each electron makes a track of about equal length wherever it may start. (There are no very long tracks in Fig. 25–4; they are all about equally short and stubby.) So each of the emitted electrons must have had almost

Figure 25–4
The cloud chamber shows the absorption of photons, one by one, and the weakening of the beam as photons are removed. The photons enter from the left, and are absorbed by the argon atoms of the gas. They eject electrons, whose tracks acquire little fog droplets, which show white in the photograph. You can see that the beam is weakened if you divide the photograph in half; there are 17 tracks in the left half and only 9 in the right half. Notice the similarity of all the short electron tracks. The "light" is actually an X-ray beam of wavelength about 0.2 angstrom. (W. Gentner, H. Maier-Leibnitz and W. Bothe, "Atlas Typischer Nebelkammerbilder," Springer Verlag, Berlin, 1940.)

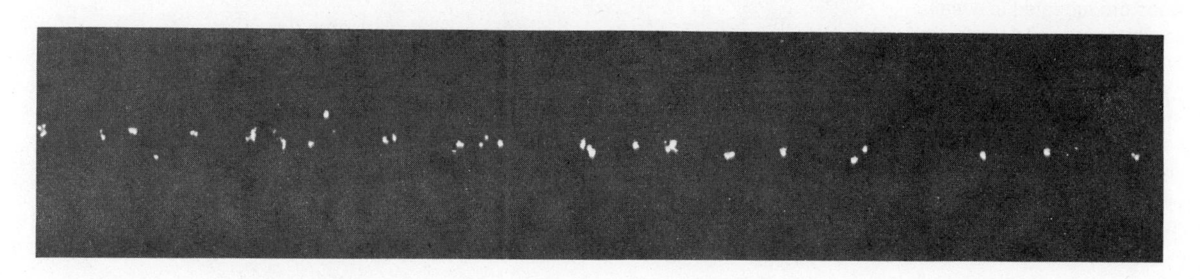

the same energy. From this point of view all the events are the same. And in particular each photon that causes an event is the same. A weaker beam means fewer photons, rather than photons of different energy. When we study the energy balance more closely we find that the energy is conserved in each individual event, and we again check that a weaker beam only means fewer photons.

Such a cloud-chamber study cannot be made with photons of visible light. In the experiment shown in Fig. 25–4, X rays were used, forming a beam of electromagnetic radiation of wavelength only a few tenths of an angstrom. At this wavelength the individual process is then made obvious. With X rays of shorter wavelength the photoelectrons are knocked out harder and the identical nature of the individual events is even easier to see.

The use of a beam of ultraviolet light allows a more detailed, if a less direct, study of the process of ejection of an electron by light. This process is called the photoelectric effect. It can occur at a metallic surface, in a liquid, or in the individual atoms of a gas.

To study the photoelectric effect, we can use a simple device (though not an easy one to build). It is a glass bulb transparent to ultraviolet light, evacuated to a first-class vacuum, with two small and really clean copper plates mounted inside (Fig. 25–5), and two leads for electrical connections to the plates. We apply a potential difference of several volts between the plates, and shine light on the inner surface of one plate. If we illuminate the positive plate, no measurable current will flow within the vacuum. But if instead we illuminate the negative plate, a current will flow. Evidently electrons are flowing across the vacuum. If the light is cut off, the current ceases. If the beam of light is turned on again, the current begins without delay—direct measurement has shown that the lag in the photoelectron emission is certainly shorter than 10^{-8} second. The current is not large, but it is within easily measurable range; a copper plate a few square centimeters in area will yield microamperes of photoelectric current in direct sunlight. The current will increase and decrease in proportion to the inten-

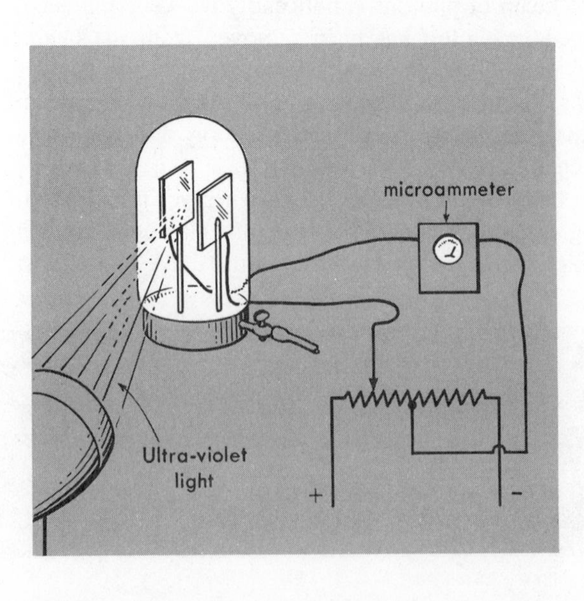

Figure 25–5
The photoelectric cell. Two copper electrodes are mounted in a well-evacuated glass bulb, and leads are brought out as shown. The circuit allows for adjustment of the potential difference between the plates; the moving contact is connected to the unilluminated plate, and can be arranged to make it either positive or negative in potential compared to the illuminated plate.

sity of light over a very wide range. Similar tubes are widely used in practical applications, from opening doors to reading the sound from movie film.

Our simple phototube confirms what we knew: when light strikes, electrons are released from the surface of a substance. The stronger the light, the greater the number of electrons ejected. The number of electrons ejected is proportional to the number of photons. All this makes sense, but we are in for a surprise when we change the wavelength of the light used. The copper photocell will give no current in even the strongest beam of red or of green light; only ultraviolet light, or light of still shorter wavelength, will work.

Perhaps this effect occurs only in copper. Let us try another substance. We continue the experiment with a tube in which the plate surfaces are coated with potassium. Now visible blue light will give a photoelectric current but red light will not. A threshold, as it is called, exists in green light. (In copper the threshold was between the visible and the ultraviolet.) Light with a wavelength slightly longer than the threshold value will give no photocurrent even if it floods the surface, while some current will flow in even the weakest beam of light with wavelength shorter than the threshold value.

All substances have photoelectric thresholds. The wavelength of the threshold depends on the nature of the substance. But there is one common point: for long waves the photoelectric effect does not take place, and for short waves—for light of high enough frequency—it does. No picture of a wave will suffice to explain these results, but they fit well with the idea of photons. An individual electron is ejected by one particular photon. The strength of the beam determines the number of photons. Therefore the current is proportional to the light intensity. But the character of each photon collision, its ability to eject an electron, does not depend at all on how many others are present. It is a property of every single photon, and must somehow depend on wavelength of the light. Light of short wavelength seems more effective than that of long. Every surface will let electrons go when hit by the photons of light which is blue enough; a few surfaces will let electrons go for the photons of red light—and those of any light of shorter wavelength. Some artful combinations, notably the alkali metal cesium dissolved with oxygen in a silver surface, have a threshold in the infrared; but metals like platinum will not respond to any of the light that passes through glass; only use of the deep ultraviolet will suffice to generate a photocurrent from them.

25-3 Einstein's Interpretation of the Photoelectric Effect

We have seen a little way into the mechanism of the photoelectric effect. We know that it involves some detail of the photoelectric surface. We also know that it always requires light of high rather than low frequency. This suggests that the photons differ in some way dependent on the frequency. That photons of red light are different from those of blue light is only reasonable. Photons must somehow carry the information which shows up as the wavelength when light waves interfere. To find out in what way the frequency is contained in the photons we must do something new. We must find the kinetic energy of the ejected electrons.

In the apparatus of Fig. 25–5 we can adjust the magnitude of the applied potential. We get the maximum current for any particular choice of surface and any given illumination by making the receiving plate slightly positive with respect to the emitting one. No increase in current follows upon increasing the potential; evidently we are picking up all the electrons ejected (Fig. 25–6). Now we can reduce the potential and reverse it, making the

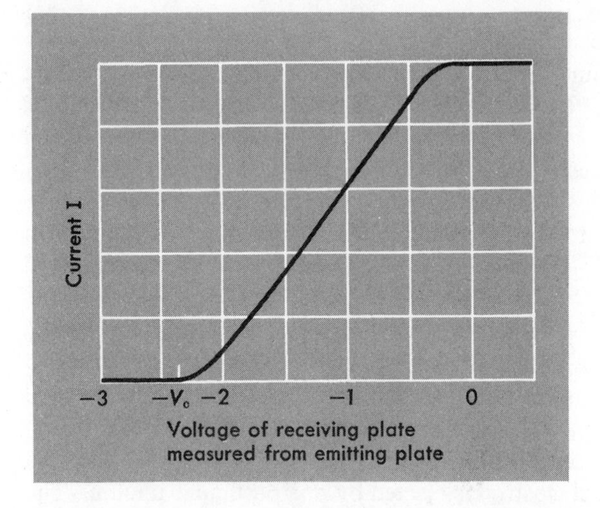

Figure 25–6
Current of ejected electrons vs. potential difference between the plates for a given kind of light. Note that the current does not stop until the receiving plate is almost 2.3 volts negative. This value is the cutoff potential V_c.

emitting plate the positive one. Then the electrons are attracted back to the plate as they seek to leave it; but the photocurrent does not vanish. Apparently the ejection of electrons by light imparts to some of the electrons enough energy to overcome the retarding electric potential difference between the plates. Finally, as the retarding potential is increased, at some particular retarding potential difference, the current stops. Even stronger light (if the wavelengths remain unchanged) will not restore any photocurrent. None of the ejected electrons have enough energy to cross between the plates against this retarding potential difference. We call it the cutoff potential. (See Fig. 25–6.)

The cutoff potential V_c times the one elementary charge of each electron tells us the energy that must be taken from the fastest electrons to stop them from going all the way across from one plate to the other. It therefore tells us the maximum kinetic energy E_{el} with which these ejected electrons are emitted:

$$E_{el} = (1 \text{ elementary charge}) \times V_c.$$

With the cutoff potential as a tool, we can measure the maximum kinetic energy of the electrons ejected in a series of experiments on light beams of different colors. When we are at the cutoff potential with one wavelength, we find that we can always restore some photocurrent by using light of shorter wavelength. This observation tells us that the shorter the wavelength of the light we use, the more energy the photons can give to the ejected electrons. We can measure just how much more by finding the cutoff potentials at shorter and shorter wavelengths.

If we plot the results of a series of experiments with light beams of different colors striking the surfaces of different metals, a simple and fundamental relation emerges. It is shown in Fig. 25–7, where we plot the maximum kinetic energy of the electrons against the frequency ν (rather than the wavelength) of the light.* For any particular substance, the curve of maximum kinetic energy versus frequency is a straight line with a slope which is called h. And this slope h is the same for all materials. Only the intercept, B, changes from one material to another. The graphs for a few typical metals are shown in Fig. 25–7. Consequently, the graphs for all substances can be described by the relation:

$$E_{\text{el}} = (1 \text{ elem. charge}) \times V_c = -B + h\nu,$$

where h is always the same, B is constant for any particular metal, and ν is the frequency of the light which changes with the spectral color.

Here we see that the experimental information clearly distinguishes between the role of the light and the role of the photoelectric surface. The surface is represented solely by the term $-B$, which is the intercept of the photoelectric curve indicated on the left in Fig. 25–7. This term is independent of the light, for we can change the frequency ν as much as we like without influencing it. Consequently, it only contains information about the

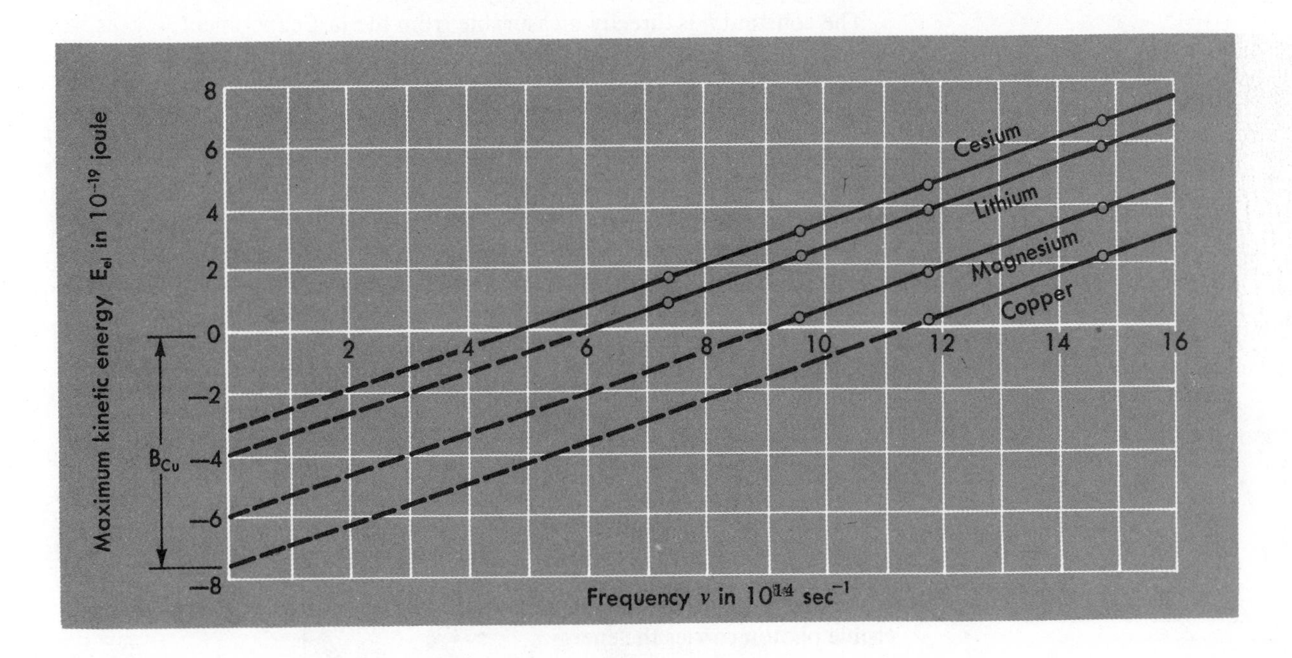

Figure 25–7
The results of experiments with the apparatus of Fig. 25–5, augmented by the use of various light sources and filters of different color. The maximum kinetic energy E_{el} is plotted against the frequency ν of the light used. The slope of the lines gives the value of h.

* In modern physics the frequency is almost always represented by the Greek letter ν (read nu) rather than by the letter f. We shall follow this standard practice throughout the rest of the book. When you read modern physics in other books, the symbols will then be familiar to you. For example, the relation between the speed of light, its frequency and its wavelength written $c = \nu\lambda$, meaning just the same things as $c = f\lambda$, which you used earlier.

nature of the surface. On the other hand, the light is described by the term $h\nu$, which is always the same for light of the same frequency. (Because the slope h *is* the same for all substances, this term has nothing to do with the nature of the photoelectric surface.) This term tells us in what way the photon carries the information about the frequency of light. Each photon is able to give more energy to an emitted electron in direct proportion to the frequency; that is, E_{el} increases with ν.

On the photon picture we can interpret the last equation completely. Each photon carries (or perhaps the photon *is*) the energy

$$E = h\nu = \frac{hc}{\lambda} \cdot$$

When the photon hits the surface it disappears, and the energy $h\nu$ is transferred. Some of this energy is needed to bring the electron out of the surface. That becomes the threshold value B. Photons with less energy than that value can eject no electrons. But if a photon has more energy than the threshold value, the surplus appears as kinetic energy of the emitted electron. Not all electrons come straight out of the metal surface; some go more deeply into the metal, and never come out at all. Others come out by indirect paths, having lost some energy in atomic collisions on the way out. A few come nearly directly out with the full surplus of kinetic energy donated by the photon; these are the electrons which still flow when the retarding potential difference is not quite so great as the cutoff value.

The constant h is directly measurable from the last experiment; for, as we saw, h is the slope of all photoelectric curves. The result is

$$h = 6.62 \times 10^{-34} \text{ joule-sec.}$$

From the value of h, we can find the energy of photons.

$$E = \frac{hc}{\lambda} = \frac{(6.62 \times 10^{-34} \text{ joule-sec})\left(3.0 \times 10^{8} \frac{\text{m}}{\text{sec}}\right)}{\lambda}$$

$$= \frac{1.99 \times 10^{-25} \text{ joule-meter}}{\lambda}$$

Also, there are 10^{10} angstroms/meter and 6.25×10^{18} electron volts/joule. Therefore,

$$E = (1.99 \times 10^{25} \text{ joule-meter})(10^{10} \text{ A/m}) \frac{6.25 \times 10^{18} \text{ ev/joule}}{\lambda}$$

$$= \frac{1.24 \times 10^{4} \text{ ev-angstroms}}{\lambda}$$

Visible light has a wavelength of about 5000 A. Consequently, an average visible photon carries the energy

$$E \approx \frac{1.24 \times 10^{4} \text{ ev-A}}{5000 \text{ A}}$$

$$= 2.5 \text{ electron volts.}$$

The whole range of visible-light photons extends from an energy of around two electron volts for the red to somewhat above three for the blue.

These energies, of a few electron volts per photon, are exactly in the lowest energy range which can cause important chemical changes, presumably because they are high enough to disturb the electrons within an atom or even to eject them. For these energies, then, but not for lower values, photographic chemicals, photoelectric surfaces, and the retina of the eye itself are sensitive. The differences among the "visible" ranges for such devices are exciting, but they are really rather small. No known photosurface will respond to infrared light which is even ten times longer in wavelength than the limiting response of the human eye.

From the value for the energy of a single photon, we can find the number of photons streaming by in any radiation beam whose total energy rate we know. Then it is interesting to compare the total number of photons striking a photoelectric surface with the number of electrons ejected. Even for a very good surface, only a few percent of the photons eject electrons. The rest send electrons deeper into the surface, or are wasted in some other way. Of course, photons with less than the threshold energy cannot eject electrons at all.

The photon interpretation of the photoelectric effect was first given by Einstein in 1905, and the equation we have found which represents the conservation of energy in the photoelectric process is known as Einstein's photoelectric equation. But the experimental evidence that we used to establish this relation was not available to Einstein. He invented the whole picture on the basis of far less evidence, and predicted the experimental results which we have used. In this work Einstein built upon a result obtained by Planck. In 1900 Planck had explained the spectrum of light inside an insulated enclosure at constant temperature.* His explanation implied that light was emitted and absorbed in energy bundles $h\nu$ in size. The proportionality constant h is called Planck's constant.

25-4 The Momentum of Photons

Energy is not all that is transferred from light to matter when a photon collides with an atom. The hail of photons impacting on a surface transfers momentum as well. From the photon point of view, we must think of the pressure of light exactly as we think of the gas pressure in the kinetic theory. The apparently smooth and continuous pressure is the result of a myriad of little collisions, each one transferring momentum to the surface. Then, by measuring the pressure of light from photons of a single "frequency" and dividing the corresponding momentum up equally among photons, we find that each photon carries momentum $h\nu/c$ in its direction of motion.

Although the electromagnetic theory of Maxwell contained no photons, it agrees with the photon momentum. According to the electromagnetic theory, any radiant energy E going in one direction should carry the momentum E/c in the same direction. Photons of energy $h\nu$ and momentum

* An insulated enclosure with a small hole in one wall is commonly called a "black body," because it absorbs radiation almost perfectly through the hole. Previous theories about the frequency distribution of radiant energy in a black body had been in violent disagreement with the experimental facts.

$h\nu/c$ "obey" this relation. Consequently, a large number of photons can lead to the same effects as a supposedly continuous wave of light.

We can verify the photon momentum directly when a single photon collides with an electron. For this experiment we need to observe the motion of the electron, and we can do this in the cloud chamber. The electrons are supplied by the atoms of the gas or by a thin sheet of matter placed in the cloud chamber. They are bound, and the photon must knock them out of the atom. So the energy of the photon must be high compared to the energy with which the electron is bound in the atom.

Even with photons which are energetic enough to remove an electron from the atom, we may not get the kind of collision we want to study. Unless we use an X-ray photon—a photon of really high energy—we shall see the ordinary photoelectric effect in which the energy of the photon is shared between the residual kinetic energy of the electron and the energy required to break the electron away from the atom, the quantity B, which fixes the photothreshold. In the photoelectric effect, the momentum also goes partly to the atom, and the momentum conservation does not take place between electron and photon alone. But if we use much more energetic photons, those of high-voltage X rays or those of nuclear gamma rays, the atomic binding energy becomes small in comparison. Then the electron is knocked out with nearly all the energy and all the momentum lost by the photon. This new process is called the Compton effect, named for the American physicist A. H. Compton.

In a Compton collision the photon is not absorbed. It appears after the collision scattered off with reduced energy and momentum, moving in some new direction. The electron recoils from the collision, also carrying off some energy and some momentum. In each collision of this kind, the energy brought in by the original photon is accounted for by the energy of the recoiling electron and that of the scattered photon. The momentum is also conserved. The momentum of the original photon becomes the total momentum of the recoil electron and of the scattered photon. Figure 25–8

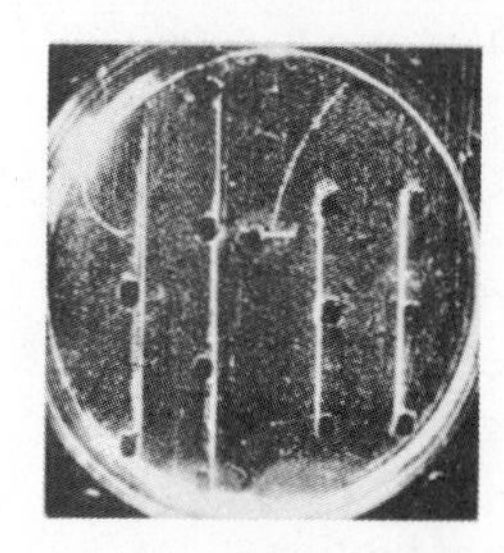

Figure 25–8
The Compton effect. To the left is a cloud-chamber photograph; next is a diagram showing the paths of gamma-ray photons and an electron; to the right is a vector diagram showing the momentum change. In the picture a gamma-ray beam, with energy of around 2 Mev, enters the cloud chamber from below. A gamma ray strikes the mica foil, and occasionally a Compton recoil electron is ejected, mostly near the forward direction. The re-emitted photon, having lost energy and momentum, goes off in another direction, passing through one of the lead strips put into the cloud chamber, and ejects an electron from a second lead strip. This reveals its path, and the measurement of the curvature of the electron track in the known applied magnetic field makes possible the check on the conservation of energy and momentum. (Photograph from: Crane, Gaerttner, and Turin, "A Cloud Chamber Study of the Compton Effect," *Physical Review*, Vol. 50, 1936.)

shows a photograph of a recoil electron which has been given momentum when struck by a photon. We see directly that the photon has brought in momentum. In addition we can check the conservation of momentum by adding up the momentum vectors as is done in the diagram of Fig. 25–8. The conservation of momentum checks out in each such collision, and in each we use the photon momentum $p = \dfrac{h\nu}{c}$. As a result we have a convincing check that

$$p = \frac{h\nu}{c} = \frac{h}{\lambda}$$

is the photon momentum.

The relations

$$p = mv \quad \text{and} \quad E_K = \frac{mv^2}{2} = \frac{pv}{2}$$

which we learned for slow-moving particles *do not apply* here. Instead $p = \frac{h\nu}{c}$ and $E = h\nu$ give $E = pc$. This is *not* something peculiar to a photon; it is a verification of the theory of relativity, which modifies mechanical laws for motions which are close to the speed of light. The photon indeed moves at the speed of light; its mechanical behavior is to be described, not by Newton's, but by Einstein's slightly modified equations of motion. Relativity is outside of our present topic; we can only pause to remark that in the atomic domain the laws of relativity mechanics have been amply verified. Of course, they must agree with Newton's mechanics whenever they are applied to bodies which move slowly compared with the speed of light, and they do.

25-5 The Orderliness of Chance

The ejection of individual electrons from a piece of metal is quite random. Some electrons are emitted immediately after the illumination, while some are delayed by various, unpredictable times. But the overall picture of electron emission is certainly one of orderliness. Recalling Section 25–1, on the average the specks will lose the same number of electrons in one part of a uniform light beam as in another. Also, on the average, more electrons are driven out in one second by strong light than by weak light.

There is no substitute for actually performing an experiment in which you watch meaning emerge from what appears to be a sequence of wholly meaningless and sporadic events. It is not practical, for example, to simulate the myriad grains of light which form a good photographic image, but we can consider a crude pattern formed by a smaller number instead.

Suppose the bombardment of some receiving plate by a light beam results in the bar pattern of Fig. 25–9 (a). More of the "grains" of light must in the end fall on the regions of the photographic plate that become dark than on those that remain lighter. We can represent this pattern by the graph of Fig. 25–9 (b), where the height of the graph is proportional to the total number of hits on that same region in the picture of Fig. 25–9 (a). Now we prepare a set of cards, marked *L*, *C*, and *R* for the left-, center, and righthand bars of the pattern. We prepare cards in a number proportional to the total number of hits expected; say, 3 marked *L*, 1 marked *C*, and 6 marked *R*. This fairly expresses the assumed probability of a hit in each of the three places we distinguish. Now draw a single card blindly from a well-shuffled mixture. Read what it is, then put it back, reshuffle, and draw again. Do this over and over again. At each draw, note the letter the card bears,

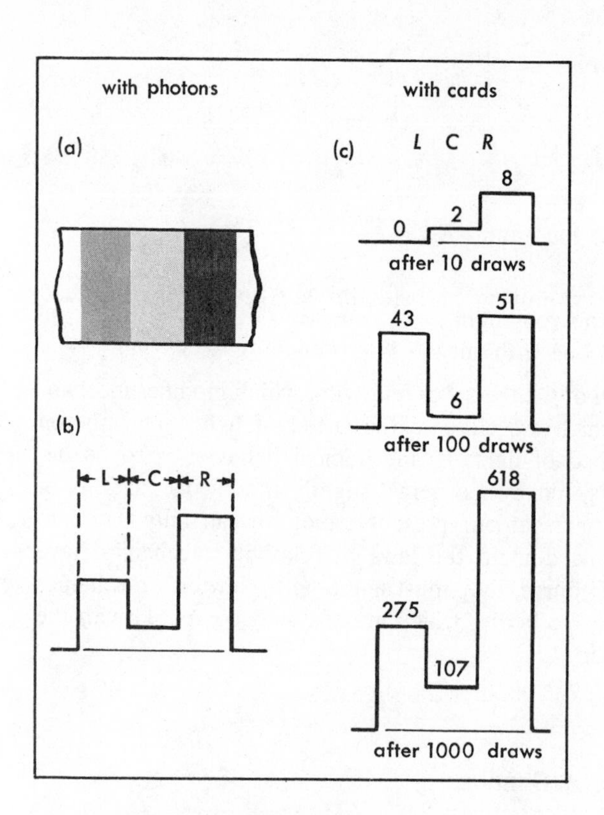

Figure 25–9
The growth of pattern out of chance. In (a) we see a bar pattern of light and dark, representing some sort of interference fringes. In (b) the same pattern has been graphed, showing how the light intensity varies with position. In (c) we see plots of an analogous experiment in which a similar pattern is built up by random chance draws from a collection of cards whose probability of draw has been correctly adjusted. The patterns made with more and more individual draws come closer and closer to the expected pattern, beginning with an apparently disorderly arrangement showing no resemblance to the final result. The ordinary light source presents us images which are painted out by some 10^{12} or even many more photons per second.

and plot the results of each draw by adding to the height of the appropriate vertical bar on a graph constructed like Fig. 25–9 (b). At first, no clear sign of the pattern will be present, except very rarely by a remarkable accident. Later on, the pattern will begin to emerge, and if you are patient enough to draw several thousand times, the pattern thus randomly built up is almost sure to be as regular as any graph you could deliberately construct. The results of one such experiment are plotted in Fig. 25–9 (c). Of course, the grains of light are not cards; we are not really experimenting with light, but with the laws of chance and probability. But many experiments have demonstrated that the formation of images by light follows in quantitative detail exactly the laws obeyed by the random draw of shuffled cards.

With this model we produced the final regular pattern and also the irregular fluctuating build-up of the pattern that can actually be observed.

It is just for this reason that the finely detailed images caught by a TV camera or a photographic camera can be formed only in sufficient light. The inviolable minimum of light needed is set by the graininess of light. No matter how sensitive the receivers of the light may be, the image cannot be made unless there are enough of the light packets to form it accurately. It is striking that the effects of the intrinsic graininess of light have been demonstrated to occur for the eye just about at the dimmest light which a man can see. Our eyes are developed about as far as any light receivers working with the same size of lens, the same time of response, and the same band of wavelengths can possibly be. They have reached very close to the natural limit.

In weak light we find the individual effects of grains. Little bundles of light come along at random, now here and now there, arriving in greater number where the intensity is higher and in smaller number where the light is weaker. This grainy picture of light is far different from our usual view of waves. It seems to have nothing to do with interference and diffraction. Can it be that the wave properties demonstrated so clearly in interference experiments apply only to strong light, where somehow the many individual "grains" of light interact one with another to produce the wavelike results? Or do the patterns built up by weak light also show interference? This question was directly tested fifty years ago by Geoffrey Taylor, a young student at the University of Cambridge, who later became Sir Geoffrey for his lifetime of successful work in physics.

In a light-tight box Taylor set up a small lamp which cast the shadow of a needle on a photographic plate (Fig. 25–10). He chose the dimensions

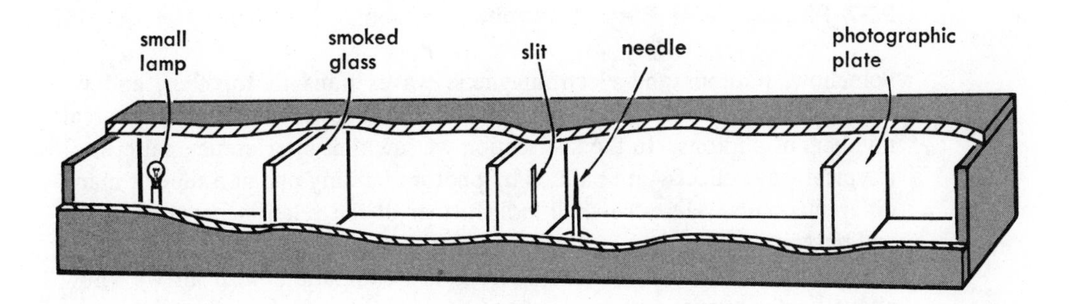

Figure 25–10
G.I. Taylor's experiment, proving that the interference pattern was correctly given even if only one photon was present in the apparatus at a time. He found good, sharp fringes in the shadow of a needle cast by light from a slit, even when that light came from a dim gas flame through such absorbers that an exposure of three months was needed.

so that the diffraction bands around the shadow of the needle were plainly visible. Then he reduced the intensity of the light. Longer and longer exposures were needed to get a well-exposed plate. Finally, he made an exposure lasting for several months (which, the story goes, he spent sailing). Even on this plate the diffraction bands were perfectly clear, though he could show for this very faint light that only one photon—one of the packets of light—was in his box at a time. Interference therefore takes place even for single particles of light. It works for every single photon.*

Taylor's experiment is not the only experiment that points to this conclusion. There are many. Nowadays a photocell working an amplifier can make the arrival of the photons audible. It is interesting to hear the clicks which signal the painting out by individual photons of a two-slit diffraction pattern. While you watch and listen, the pattern is formed by photons coming in one by one. All interference experiments work this way: the photons come one at a time. So we are forced by direct experimental evidence to believe that wave behavior applies to individual photons.

* Taylor knew that there was only one (and often no) photon in his box at one time because from the energy of one photon and the intensity of the light he could find the number of photons in any given length of light beam. He found the photons must be spaced farther apart on the average than the length of the box. So usually there was not even one photon in the box.

What do we mean by the wave behavior of an individual photon? We shall have more to say on this by no means simple point. For the moment, we shall be satisfied to say that the electromagnetic wave must somehow fix the probability of appearance of a photon. Where the field in the wave is strong, there we shall very probably find photons; where the field is weak, photons will occur with small probability; where the field vanishes, no photons occur at all. When it interacts with the photographic emulsion or the photoelectric cell, no single photon is spread out over a whole interference pattern. The photon hits at a definite place. But for each photon, the probability is highest that it will arrive where the interference pattern is most intense. The probability of arrival at a region of half that intensity is half as big, and so on. The probabilities of arrival for each photon have just the pattern of the intensity of the interference fringes (the bright and dark interference bars). So when many photons have arrived, the interference picture is painted out by their cumulative effect.

25-7 Photons and Electromagnetic Waves

Somehow, photons and electromagnetic waves must fit together, and we already noticed that this is possible for average effects involving great numbers of photons. In the last section we saw that interference patterns—a typical wave effect—are painted by photons landing one at a time at many different points. Here we shall look further at the relation between waves and photons.

First we must establish that photons are compatible with all we know about electromagnetic waves in the region of wavelengths from millimeters to miles. We make and detect these waves, not by using single electrons or atoms, but with radio tubes and antennas, through which currents flow back and forth. And we have plenty of evidence that the wave picture agrees with the facts of radio transmission.

In that range the frequencies are small, relative to atomic values. A frequency of a few kilohertz represents the low frequency limit of radio technique, while a frequency of a million megahertz is beyond the most advanced radar and microwave techniques, and lies at the margin of the deep infrared. Yet even these highest radio frequencies, with the tiny atomic constant h, give a photon energy which is really very small.

At one megahertz, for example, the photons in a radio wave have the energy

$$\begin{aligned} h\nu &= (6.62 \times 10^{-34} \text{ joule-sec})(10^{6} \text{ cycles/sec}) \\ &= 6.62 \times 10^{-28} \text{ joule.} \end{aligned}$$

Because there are 6.25×10^{18} electron volts per joule, this is

$$h\nu = 4 \times 10^{-9} \text{ electron volt.}$$

Such small photon energies mean that a radio signal cannot be detected unless it contains very many photons. The lower limit of a good communications receiver is about 10^{10} photons/second. Among so many photons the average behavior alone is detectable. The result is the smooth transfer of energy from wave to matter which we would expect on the wave theory. The large numbers of photons guarantee that the graininess will not usually show up unless we try very hard to detect it.

We might even suspect that there are no photons in electromagnetic radiation of long wavelength. But we now have experimental evidence of the absorption and emission of "radio-frequency" photons by atoms and molecules (see Section 26–4). So we know that both the photon and the wave description apply.

As we go to higher frequencies, we pass to the region of visible light. Here the number of photons is large in any light beam that appears bright to our eye. Sunlight, for example, represents a rain of some 10^{17} photons per square centimeter per second. In such a burst the individual arrivals of photons are not easily noticed. At the other extreme, however, we can still make out the glow of a faint luminous source which sends to our dark-adapted eye a hundred photons on the area of the pupil each second. So for visible light, both the wave and the photon sort of observation are possible.

The photographic plate and the photoelectric cell are, like the eye, capable of responding to the smooth flow of intense or moderate beams no less than to the fluctuating and chancy arrival of a few photons every second. It is for this reason that it was in the study of visible light, and not in the study of the longer wavelengths, that the photon idea was first conceived and found satisfactory.

As the frequency of light increases and the photon energy rises, the photon description appears more and more dominant. The photon energy in the X-ray region is large enough so that single photons are easily detectable. The arrival of a single gamma-ray photon—associated with still higher frequency and correspondingly shorter wavelength—is energetically a sizable event. It can be noted with good probability in a whole variety of detectors. Gamma-ray photons, which have energies as great as 10^{14} electron volts, have been found in the high atmosphere. There they are produced by cosmic-ray particles from outer space reacting with the atoms of the air. Such a photon produces tens of millions of recoil electrons.

As the frequency rises and the wavelength correspondingly diminishes, wave effects are less obvious. For example, interference and diffraction become more difficult to demonstrate. The wavelengths of X rays are no longer than the dimensions of atoms. For them we can hardly make slits narrow enough to exhibit interference maxima of detectable width. Although X-ray interference is widely used, the essential diffracting apparatus is rarely man-made. Instead we employ the regular atomic patterns of crystals. For gamma rays, not even crystals will work as diffraction obstacles; only the nucleus of a single atom can be employed, and the procedure has become very indirect.

We have no choice; electromagnetic radiation of every kind must be understood from two points of view. If we wish to view the overall pattern it makes in space or time, we can do so by the theory of electromagnetic waves described in Chapter 23. We can design antennas, lenses, and gratings. They all function just as the wave theory predicts. But we must admit that we cannot say exactly when or where the energy of a photon will be transferred to the electron. The wave theory gives us only the probability of finding the photon. It tells only the probability that a photon will transfer its energy to a given object within a definite time. Here we are subject to the laws of chance. The photon cannot be pinned down by any wave prediction. In an interference experiment, a particular photon may occur in any bright bar, and we cannot trace its path through the slits of the apparatus. But wherever and whenever it interacts, there and then we can be certain that energy is precisely conserved.

In the older mechanics and classical physics, we thought that we could predict with certainty when and where it was going to hit. That was naïve, and the study of the photon was the hard school in which physicists became more sophisticated in the ways of the world. For photons, at least, we cannot do any more than assign probabilities to the time and place of arrival. These probabilities follow from the associated wave properties of light. We can therefore use only part of the ideas that we formerly associated with particles and their motions, and at the same time we must add new ideas from the wave picture and the probability interpretation to fill out our description of what actually happens.

Sometimes only one of the older pictures is needed. When many photons take part in any detectable change, it is the wave picture alone. On the other extreme, when a single photon is so energetic that it can easily be observed, and when it has as well so short a wavelength that interference studies are nearly impossible, then the photon picture is sufficient. But the real nature of light is more subtle. It shows us both wave and photon behavior. We have therefore been forced to make a new picture of light, one which combines some aspects of wave behavior as we previously understood it with some aspects of photon or particle behavior. This new picture is forced upon us, for this is the way the world is. Newton's corpuscular picture of light is by no means restored by the photon. It is no more true than the pure electromagnetic picture, with its waves of continuously distributed energy. Light is to be understood fully only by a novel scheme of the kind we have tried to outline.

For Home, Desk, and Lab

1.* How would the two graphs in Fig. 25–2 look if the electrons were ejected according to the wave model of light? (Section 1.)

2. In a certain camera, at dusk, the light that gets through the lens is so dim that in $1/100$ second one photon on the average lands in 10 mm^2 of the film. If you want to take a finely detailed, sharp picture, what should you do?

3. With a motion-picture camera you could determine the time at which the charge on each speck was changed for a large number of specks in the apparatus of Fig. 25–1. What relationship would you expect between the number of specks on which the charge changed and the time after the light was turned on? Does the data given in Fig. 25–2 support your prediction?

4.* A surface ejects electrons when hit by green light, but none when hit by yellow light. Do you expect that electrons will be ejected when the surface is hit (a) by red light? (b) by blue light? (Section 2.)

5. Some photographic materials can be handled safely in red light but are spoiled instantly when white light is turned on. How do you account for this?

6. Suppose that a photograph similar to Fig. 25–4 had been made with X rays of wavelength about

0.1 angstrom. In what respects would you expect the new photograph to resemble Fig. 25–4? In what respects would it be different?

7. Assume that the entire negative plate in the cell illustrated in Fig. 25–5 is always illuminated by a constant source of light capable of producing a photoelectric current. What would be the effect on this current of
   (a) doubling the area of the plates?
   (b) halving the distance between the plates?

8.* In Fig. 25–7, which points on the curves were obtained by using violet light? (Section 3.)

9.* What is the minimum energy, in electron volts, needed for photons to eject electrons from the cesium photoelectric surface used in Fig. 25–7? (Section 3.)

10. A good mirror reflects about 80 percent of the light which hits it. How could you determine whether 20 percent of the photons were not reflected, or whether all of the photons were reflected but each with 20 percent less energy?

11. Red light has a wavelength of about 6500 angstroms.
   (a) Estimate its frequency.
   (b) Estimate the energy in joules of one photon of red light.
   (c) Estimate the energy in electron volts of one such photon.
   (d) Find out approximately the longest and shortest wavelengths you can see, and compute the energy of the corresponding photons.

12. Light of 5000-angstrom wavelength illuminates a surface. What voltage is needed to stop all the electrons emitted from the surface if their binding energy to the surface is 2.0 electron volts?

13. Sketch the graph of Fig. 25–6, then upon it sketch graphs that will show the effect of
   (a) increasing the intensity of the light source.
   (b) decreasing the wavelength of the light.

14. (a) What is the ratio of the energy of a photon of wavelength 4500 A (blue light) to the energy of a photon of wavelength 6300 A (red light)?
   (b) What is the ratio of the maximum energy of electrons emitted from a surface with a binding energy of 1.8 electron volts using the same red and blue lights?

15.† (a) How much energy is carried by an "average" photon of visible light with wavelength of about 5000 angstroms? How much momentum?
   (b) Estimate the number of photons of visible light emitted per second from a 100-watt light bulb emitting 1 percent of its power in the visible region.
   (c) What is the light pressure exerted by the photons from the bulb when they strike a black body 2 meters away head on?

16. (a) Take a pack of ordinary playing cards (from which the jokers are removed), thoroughly shuffle them, and draw a card. After noting which card it is, return it to the pack, shuffle, and draw again. After ten draws, note how many times each of the following were drawn:
   (i) A face card.
   (ii) A card on which a 3 is printed.
   (iii) A card on which an even number is printed.
   (b) Make a bar graph, after the fashion of Fig. 25–9, to display the results of ten draws.
   (c) Make a bar graph to display the results you would expect to obtain as the result of about 5200 draws.
   (d) In what respects are your graphs like those of Fig. 25–9?

17. In the three-month experiment described in Section 25–3, G. I. Taylor estimated the energy of the light reaching his photographic plate. He found it was 5×10^{-13} joule per second. (He obtained that value by comparing the average blackening of the plate with that produced in 10 seconds by a candle 2 yards away, without any absorbing screens of smoked glass.) From that estimate, calculate the average distance between photons, as follows:
   (a) Assume the wavelength of the useful light was about 5000 angstrom units. What energy did each photon carry?
   (b) Calculate from the given flow of energy the average time that elapsed between the arrival of one photon and the next.
   (c) From the average time, calculate the average distance between one photon and the next in the beam of light.
   (d) If you were asked, "How many photons are in the box at any chosen instant?" you would have to answer "None" most of the time, but on looking again and

again, you should expect to "see" one photon in the box about once in so many trials. How many? Assume the box was 1.2 meters long.

18. Use the language of photons to describe how the pattern of Fig. 8.5 is formed with very dim light.

19. An ordinary flashlight has a bulb which draws 0.25 amp at 3.0 volts, and concentrates the emitted light into a beam about 10 cm^2 in area. Assume that 1.0 percent of the energy consumed by the bulb is turned into visible light of wavelength 5.8×10^3 angstrom units.
    (a) How many photons are there in each meter along the axis of the beam?
    (b) How many photons per second arrive on each square centimeter of a surface illuminated by this flashlight?

20. A typical FM radio station broadcasts with a wavelength of 3 meters. Estimate:
    (a) the energy in one photon of this radiation.
    (b) the number of photons broadcast per second if the radiated power is 10 kilowatts.
    (c) roughly the number of photons received by a receiving set during one vibration of the letter "s" in the audio pattern, if the receiver is 100 km from the station. (Make the following simplifying assumptions:
        (i) the radiation spreads uniformly in all directions along radii from the source so that an inverse-square law holds for energy flow. (This is not true for any ordinary antenna.)
        (ii) the receiving antenna collects radiation from an area of 1 $meter^2$ perpendicular to the line of travel of waves.
        (iii) the sound of the letter "s" involves air oscillations of frequency about 4000 cycles per second.)
    (d) Will you ordinarily notice the individual photons in receiving the letter "s"?

21. In a particular X-ray tube electrons hit a target after they have been accelerated by a potential difference of 20,000 volts. As they decelerate to a stop in the target, a few of them emit X-ray photons.
    (a) Why is there a definite minimum wavelength of photons observed from this tube? (This limit is known as the Duane-Hunt limit.)
    (b) Calculate that minimum wavelength.

Further Reading

These suggestions are not exhaustive but are limited to works that have been found especially useful and at the same time generally available.

FINK, DONALD G., and LUTYENS, DAVID M., *The Physics of Television*. Doubleday Anchor, 1960: Science Study Series. (Chapter 2)

HOFFMANN, BANESH, *Strange Story of the Quantum*. Dover, 1959.

HOLTON, G., and ROLLER, D. H. D., *Foundations of Modern Physical Science*. Addison-Wesley, 1958. (Chapters 31 and 32)

HUGHES, DONALD J., *The Neutron Story*. Doubleday Anchor, 1959: Science Study Series. (Chapters 3 and 5)

WEISSKOPF, VICTOR F., *Knowledge and Wonder*. Doubleday Anchor, 1963: Science Study Series.

WILSON, ROBERT R., and LITTAUER, RAPHAEL, *Accelerators*. Doubleday Anchor, 1960: Science Study Series. (Chapter 3)

Atoms and Spectra

CHAPTER

26

If we look back over the wide range of phenomena which can be summarized in terms of Newtonian mechanics, we find the record truly impressive. The conservation laws of momentum and energy span the field from the motion of the planets down to the motion of molecules in a gas. The inverse-square law for the interaction between masses or electric charges governs the motion of a spacecraft to the moon as well as the motion of an alpha particle passing near a gold nucleus.

Let us now try to pull together all our knowledge of mechanics and electromagnetic waves and apply it to individual atoms. That is, let us see what happens when we try to apply the laws of mechanics and electromagnetism to objects whose dimensions are 10^{-10} m. We shall do this using the Rutherford model described in Chapter 24.

26-1 The Stability of Atoms

The Rutherford model, describing the atom as a tiny but massive positively charged nucleus surrounded by a number of electrons, seems essentially correct. Moreover, we have learned how to measure the charge on the nucleus and how to estimate at least an upper limit for its "radius." Knowing the charge of the nucleus, we also know the number of electrons which complete the neutral atom.

And yet, as we continue our study, our complacency begins to be shaken. A serious flaw in the Rutherford model becomes evident as soon as we start considering the behavior of the electrons. In the atom, as imagined by Rutherford, the electrons move around the nucleus like planets around the sun. In fact, under the urging of the Coulomb force, they can avoid falling immediately into the nucleus only by continuously revolving around it. This motion is an accelerated motion, and therefore should be accompanied by the emission of electromagnetic waves. But the electromagnetic waves drain away energy. So, like an earth satellite which loses energy in the atmosphere, the electrons must spiral into the nucleus. By imagining that they move like planets, we have only stretched the time for electrons to fall in. Instead of falling into the nucleus in about 10^{-17} sec, as an electron would if it started from rest at the outside of an atom, an electron spiraling in while radiating away its energy should take about 2×10^{-11} sec to get to the nucleus. This time is still far too short to make the atom look stable at the dimensions we know. Consequently, Rutherford's model is not consistent with our knowledge of the process of electromagnetic radiation.

Atoms do emit light, and that light must come from the accelerated motion of the charge in the atom. Therefore our whole picture of the atom is clouded by the conflicting demands. The emission of light is consistent with the Rutherford model—but it also destroys that model, for the atoms should all shrink to nuclear dimensions long before we can blink at their light. The normal atom should have the electrons stuck in the nucleus into which they have fallen. But this is in direct conflict with the evidence of alpha-particle scattering and with the size of atoms as seen in collisions in gases and in their packing in solids and liquids.

The more we probe into these questions, the more we become aware of serious troubles. Let us consider for simplicity the hydrogen atom, consisting of a single proton and a single electron. The frequency of the light emitted by the atom should be related to the number of revolutions per second of the electron around the nucleus. But the period of revolution depends on the diameter of the orbit, just as in the case of the planets in the solar system (Chapter 13). Electrons rotating in a smaller orbit have a shorter period and therefore should emit light of a higher frequency than electrons rotating in larger orbits.

As an electron emits light, its energy must decrease. Therefore, the diameter of its orbit must decrease and the frequency of the emitted wave must increase. A light source contains a huge number of atoms, and many are emitting light at one time. Some of these atoms are at one stage in the process of light emission, and some at another. Consequently, the source should be emitting light of practically all frequencies. Thus, hydrogen gas made to glow, for example, by an electric discharge should emit a continuous spectrum of light. In contrast to this prediction, spectral analysis of the hydrogen light reveals a number of sharp "lines"—that is, a number of separate wavelengths or frequencies (Fig. 26–1).

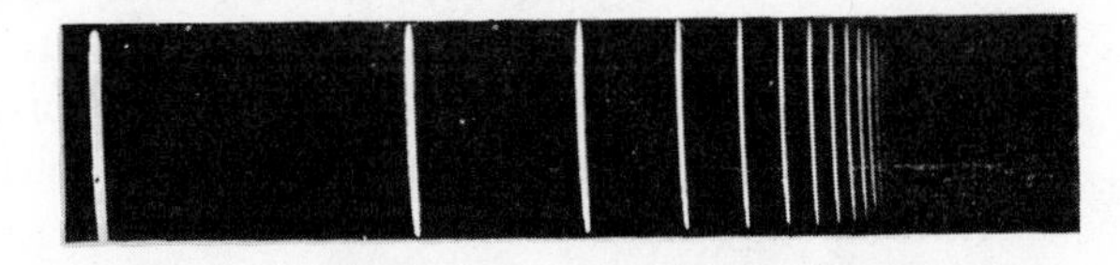

Figure 26–1
A photograph of the radiation from glowing hydrogen gas, taken through a spectroscope. Instead of a continuous spectrum of wavelengths, only certain discrete wavelengths are produced. (Photo from: G. Herzberg, "Über die Spektren des Wasserstoffs," *Annalen der Physik,* Vol. 84, 1927.)

Apart from this basic trouble—energy loss by radiation—there is another reason to fear that the Newtonian laws of mechanics can never explain the Rutherford atom. We can picture some agent (such as a visiting star) changing our own solar system by any amount, small, medium, or large—short of a major breakup—and when the disturbing agent was removed the new arrangement would just continue. We can imagine a solar system of planets built to any size over a wide range—with orbits, masses, and numbers of planets different from ours, and yet equally stable. A collection of similar solar systems, such as we might find around other stars, would be like the other things we see in nature, like a set of roses or of raindrops; they would bear a strong family resemblance, but with different-sized suns, and planetary orbits of different sizes, no two would be precisely alike.

Atoms are not like that. Their most remarkable property is the likeness they bear one to another. All hydrogen atoms act the same, all helium atoms are alike, and so on. Atoms of one kind are far more alike than any Grade A peas. Yet they have different histories; different stories of collisions and ionizations and chemical combination lie behind each one. Through it all they remain one and the same, showing the same spectral lines and making the same chemical reactions with the same amounts of energy. The atom may be a planetary system, but it is a planetary system that comes in myriad identical copies.

26-2 The Experiments of Franck and Hertz; Atomic Energy Levels

For a planetary system, according to Newtonian mechanics, motions at all total energies are possible. For an atom, on the other hand, apparently only a limited set of motions can take place. This limitation is suggested by the repetitious behavior and by the stability of atoms. The identity of the members of an atomic species suggests the same thing.

If only certain distinct patterns of motion are possible, we would expect that an atom can take on only the energies corresponding to those motions. Then there should be gaps between the possible energies; and in that case the internal energy of an atom can be changed only by certain definite amounts of energy transferred from the outside.

Are there really gaps between the possible energies that an atom may have? A direct test of this idea is important. It can be made by trying to change the energy of atoms directly by bombarding them with electrons. Can an atom that is hit by an electron take in some energy and keep it as a gain of internal energy? And if so, can it retain *any* gain from small to large, or is it limited to a selection of sharply defined amounts? In Germany in 1914 such a test was made. James Franck and Gustav Hertz performed an electron-bombardment experiment for which they later received the Nobel Prize. This experiment—and the many similar ones that stem from it—are a key to the problems of atomic stability and structure. Here we shall describe an experiment similar to the original Franck and Hertz experiments. We choose this particular experiment because it is easier to interpret though harder to carry out.

Like Franck and Hertz we shall use electrons to bombard the atoms. We can do this by using an electron gun like the ones we described in Chapter 20. The electron gun gives kinetic energy to the electrons boiled out of the filament. If the accelerating voltage in the gun is V_A, the electrons leave the gun with kinetic energy of one electronic charge times V_A.

When the electrons leave the gun they pass into a chamber containing a gas. They enter this chamber through a small hole, and the electron gun operates in a good vacuum because we continually and rapidly pump away any gas that escapes from the chamber into the gun (Fig. 26–2). The chamber containing the gas is made of a conducting material so it is all at one potential and there is no electric field inside it which could change the energy of the electrons. Consequently, any change in the electrons' energy arises because of their interaction with the gas atoms.

Now to investigate what happens to the electrons when they collide with the gas atoms, we allow some of the electrons to come out of the gas chamber through another small hole leading into a second evacuated space. There we measure the energy of the electrons which have passed through the gas. We can make this measurement with any one of a number of devices. For example, we can use a magnetic mass spectrograph like that described in Chapter 22. Then, since we know the mass of the electron, the curvature of the electron path in the magnetic field allows us to determine the energy. Alternatively, we can determine the electron's energy by making the electron move against a retarding potential difference and mea-

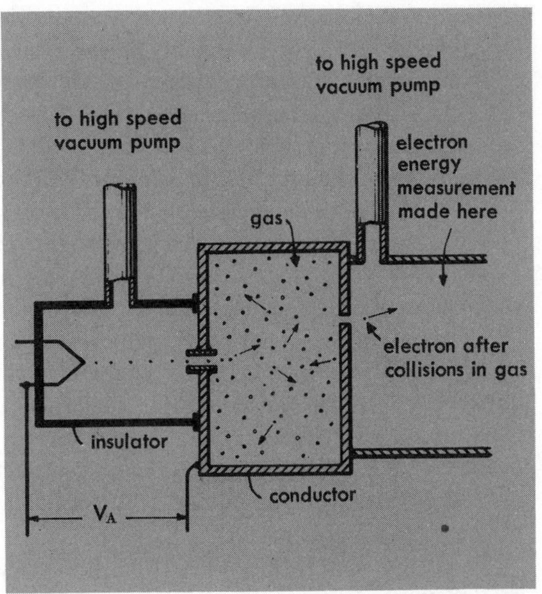

Figure 26–2
The general idea of an experiment to measure the changes in the energy of electrons when they collide with gas atoms. The electrons pass through a sample of gas (mercury vapor) in the middle chamber. Electrons leave the gun with energy given by the accelerating voltage. Their energy remaining after collisions is measured in the right-hand chamber. Note that the opening into the energy-measuring region is not in the direct line of the original beam of electrons. This is to insure catching only those electrons that have been deflected by collisions.

suring what potential difference just prevents the electrons from reaching the collector. We could even use time-of-flight measurements in the detecting chamber to find electron speed and therefore kinetic energy. The practical details of these measurements turn out to be difficult and complicated, but despite the complications we emerge with the knowledge we want of the energy of electrons after interaction with the gas atoms.

What kind of information do we get from this apparatus? Let us take a specific example. What happens when we have mercury vapor in the gas chamber? As long as the accelerating potential in the electron gun is only a few volts, we find that the electrons entering our detector after colliding with the gas atoms have almost exactly the same energy that is imparted to them by the gun. The collisions with gas atoms at this low energy are elastic. In such collisions the massive atoms take a negligible share of kinetic energy. However, when the accelerating potential in the gun passes 5 volts, so that the electrons bombard the atoms with 5 electron volts or more kinetic energy, a dramatic change occurs. The electrons entering the detector no longer have the energy with which they left the gun. Instead the electrons entering the detector have practically no kinetic energy.

Now we increase the gun potential further so that the kinetic energy of the bombarding electrons is greater. We find that the energy of electrons reaching the detector is also greater. When we raise V_A from 5 volts to 6 volts, the kinetic energy measured in the detector rises from a small fraction of an electron volt to one electron volt more than that. Careful measurements in which we determine the gun voltage and the kinetic energy in the detector accurately show that in their collisions with the atoms the electrons are losing almost exactly 4.9 electron volts. This loss only sets in when the bombarding electrons have 4.9 electron volts of kinetic energy, and the amount of the loss stays the same while we increase the bombarding energy through the next couple of electron volts. The mercury atom cannot accept anything less than 4.9 electron volts of energy, and even when it is offered somewhat more energy it still accepts exactly this amount.

However, 4.9 electron volts is not the only amount of energy the atom can take in. If the bombarding energy is 6.7 electron volts or more, the electrons can lose either 4.9 electron volts or 6.7 electron volts. On increasing the bombarding energy still further, we find other thresholds at which greater losses can occur and beyond which any of several different amounts of energy can be robbed from the electrons. This is the essential property Franck and Hertz looked for: atoms can change their internal energy, but the changes are restricted to sharply defined steps.

The smallest amount of energy which can be accepted by an atom is called its first *excitation energy*. For mercury the excitation energy is 4.9 electron volts. For helium, which has the highest excitation energy, the value is 19.8 electron volts. The lowest excitation energy of all the atoms goes to cesium, which will accept a bundle of 1.4 electron volts. Other atoms will also accept certain definite bundles of energy, and these bundles are characteristic of the kind of atom.

The experiment we have described is an idealization; the results we have quoted were collected in a number of similar experiments, all of which agree. Thus the Franck-Hertz experiment and its later modifications have shown us that atoms can accept only parcels of energy. The internal energy content of an atom cannot be continuously changed. It changes only in steps. The successive internal energies which an atom can take on are called its *energy levels.* Often they are diagrammed as we show them in Fig. 26–3.

The ground state is the state in which we normally find the atom before any excitation energy has been fed into it. Above it lie the various excited

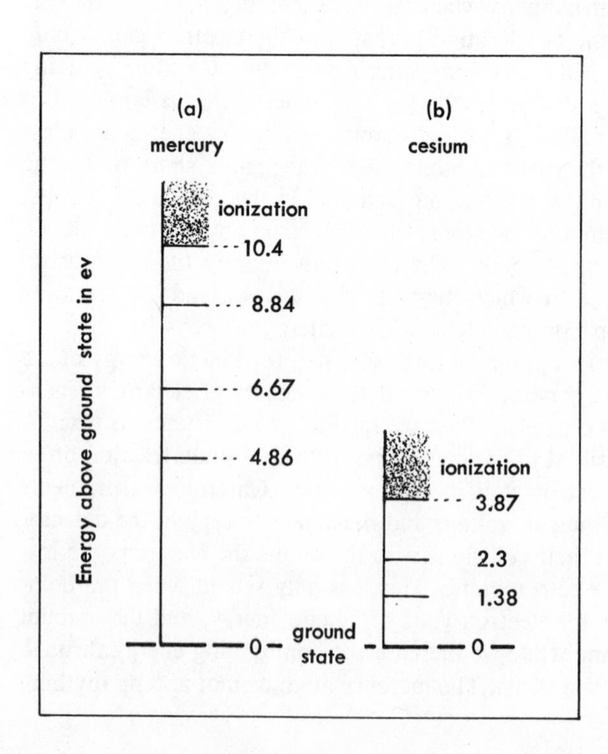

Figure 26–3
(a) Some of the energy levels of the mercury atom. In this diagram the normal "ground state" is taken as the zero of the energy scale; and we have plotted a few of the bundles of energy that the atom will accept. (b) A similar diagram for the cesium atom.

states, separated by gaps. These excited states are formed when enough energy is supplied by the impinging electrons. Finally, when an atom is hit by an electron of sufficient energy, the atom is in fact disrupted. An electron is ejected from the atom and a positive ion left behind. Since the energy of the ejected electron can take on any value, the atom can accept any parcel of energy larger than the ionization energy. For mercury, as we have indicated in Fig. 26–3, the ionization energy turns out to be 10.4 electron volts.

The atoms of many elements have been studied by electron collision. From these experiments, we find that every kind of atom has its own set of energy levels and its own ionization energy. They identify the atom much as its chemical reactions or its optical spectra identify it.

26-3 Dissecting Atomic Spectra: Excitation and Emission

A completely different means of measurement fully confirms the existence of energy levels. While we are bombarding a gas with electrons, we observe the light emitted by the gas. At low bombarding energies, we see no appreciable light emitted by the gas. But when we pass the first excitation energy, light suddenly appears. Examining the light with a spectroscope, we see a sharp line indicating light of a definite wavelength, photons of a single energy.

With mercury, for example, when the energy of the bombarding electrons passes 4.9 electron volts, a spectroscope with which we are observing the gas will suddenly show the presence of ultraviolet light having a wavelength of 2537 angstroms. This strong spectral line is one of the lines by which the mercury spectrum is recognized. But this single-line mercury spectrum is most unusual. From a discharge tube filled with mercury vapor we usually see many characteristic mercury spectral lines, many lines of definite wavelength which identify the element mercury. We have produced only one part of the normal mercury spectrum.

The experiments we described in the last section tell us that we have succeeded in feeding energy bundles of 4.9 electron volts to some of the mercury atoms in the gas. Only when such energy bundles have been transferred do we see any light from the mercury. Apparently the excitation energy which we feed in from the kinetic energy of the electrons is radiated as photons by the individual atoms as they return to the ground state. (Remember that the ground state is the lowest energy state, the state in which the mercury atom is normally found before the 4.9 electron volts of energy are transferred to it by a colliding electron.) If all the energy taken in by an individual atom is radiated in one lump, the photons should have 4.9 electron volts energy, and the wavelength of the corresponding light should be

$$\lambda = \frac{12{,}400 \text{ ev-A}}{E} = \frac{12{,}400 \text{ ev-A}}{4.9 \text{ ev}} = 2530 \text{ A}.$$

If a mercury atom could radiate its extra energy in more than one photon, light of longer wavelengths would be visible. The experiment with the spectroscope therefore shows that the atoms not only take a definite amount of energy from the electrons but that they can only radiate exactly this amount of energy in returning to the ground state. Both in accepting energy from electron bombardment and in emitting energy as light, the atoms change in one jump between two energy states a little less than 4.9 electron volts apart (Fig. 26–4). This is a magnificent confirmation of the idea that atoms can make only definite shifts of internal energy.

Let us continue to experiment with the spectroscope, increasing the energy with which the electrons bombard the mercury atoms and observing the light they emit. As we increase the bombarding energy, the rest of the familiar lines of the mercury spectrum appear. They enter in groups as the bombarding energy rises successively above the thresholds for excitation of higher and higher energy states. This procedure therefore allows us to dissect the mercury spectrum and to see how it arises.

Before an atom is hit by an electron, the atom is in its lowest internal energy state, the ground state. When an atom is hit by an electron of less than the energy difference between the ground state and its first excitation

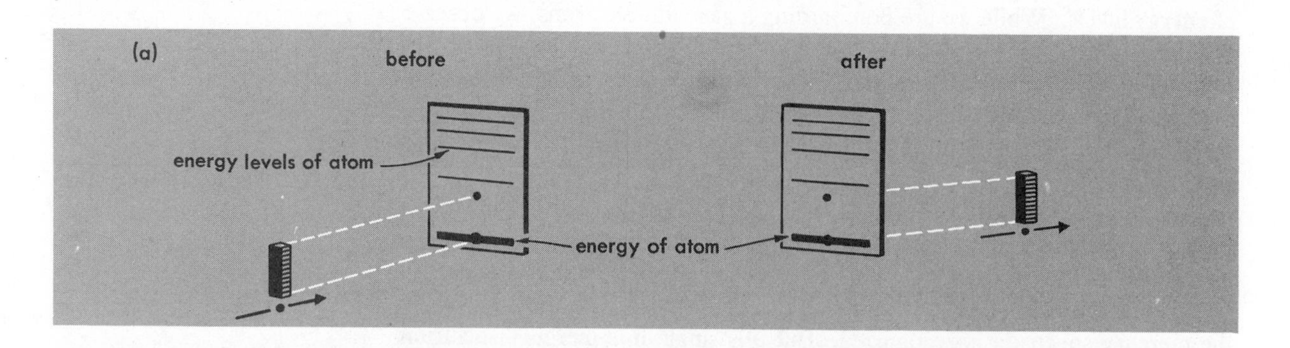

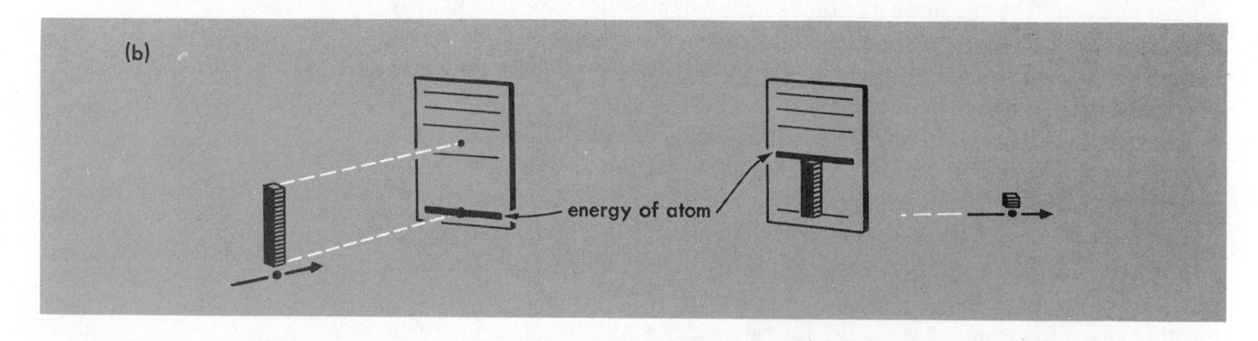

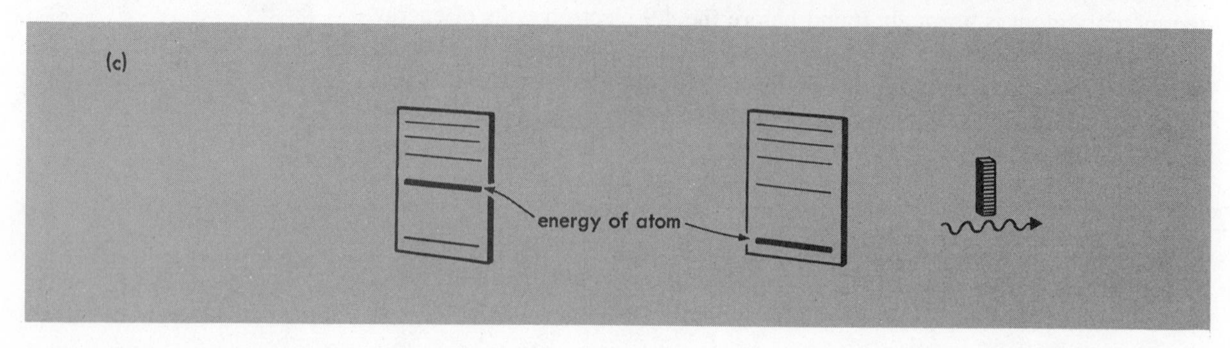

Figure 26–4
The "load" above the electron represents its kinetic energy. The atom and its energy levels are represented by the rectangular "billboards." The electron does not have enough energy to raise the atom to the next higher energy level so it scatters off elastically. (b) The electron has more energy. The atom absorbs just the right amount of energy to reach its second level, and the electron keeps only the surplus kinetic energy. The collision is inelastic. (c) The atom is at its second energy level. In returning to the ground state, it emits a photon with energy equal to the difference in energy between the two states.

state, nothing happens. A higher-energy electron, however, may transfer energy into the atom. If there is only sufficient energy to reach the first excited atomic state, just the first excitation energy will be taken from an electron, and the atom will then radiate a photon which helps to make the 2537-angstrom line which we first saw. Many atoms contribute photons, and each photon has the same energy because each of the individual atoms falls from the first excitation state back to the ground state. When the electrons can supply still more energy, individual atoms can be excited to still higher internal energy states. Then as an atom returns to the ground state, it can radiate a photon of greater energy or possibly several photons with energies equal to the energy differences between the several internal energy states at which the atom can exist.

In radiating photons—as in every other process that we know—energy is conserved. The energy of the emitted photon hc/λ is equal to the difference between the initial internal energy and the final internal energy of the atom as it jumps from one state to another:

$$E_{\text{photon}} = E_{\text{initial}} - E_{\text{final}}.$$

Each spectral line consists of the radiation from many atoms, all jumping between the same internal energy states independently of each other. Figure 26–5 on page 600 shows the analysis of the mercury spectrum and its connection with the energy levels of the mercury atom. The vertical lines on the energy-level diagrams show the energies of the photons that can be emitted. For each photon emitted, some single atom must jump from a higher state down to a lower state, and the photon carries off the energy given up.

Notice that the longer wavelength lines of mercury correspond to jumps between states both of which are above the ground state. Their energy differences are smaller, so the photons emitted have lower frequency and longer wavelength. To make such long wavelength lines, however, some of the mercury atoms must first be sent into the higher energy states. To get them there takes more than 5 electron volts. Thus we can explain what otherwise would seem strange: the longer wavelength lines of mercury consist of photons of lower energy than 5 electron volts. Yet more electron energy is needed to produce them than to produce the strong line at 2537 angstroms. The reason is clear: spectral lines are formed by the transitions of atoms between particular internal energy states. And none of the initial states required for these long wavelength transitions can be reached from the mercury ground state without energy transfers well above 5 electron volts.

So far in our discussion we have assumed that the mercury atoms are normally found in the lowest energy level, or ground state. This is true at room temperature. As temperature increases, however, the collisions among mercury atoms become more violent. At sufficiently high temperature, some atomic collisions can be inelastic just as electron collisions can. They leave one or both of the colliding atoms in an excited state. So as the temperature rises, more and more atoms are moved into excited states. When enough atoms are in excited states, the gas glows. The light is emitted as the atoms rapidly return to the ground state, sometimes stepping from

Figure 26–5
The photographs show the spectral lines that appear when mercury vapor is bombarded with electrons. In each case the bombarding electrons have a definite energy which is recorded at the side of the photograph. (Photos from: John A. Eldridge, "The Spectrum of Mercury Below Ionization," *Physical Review*, Vol. 23, Series 2, 1924.)

(a)
7.0 ev

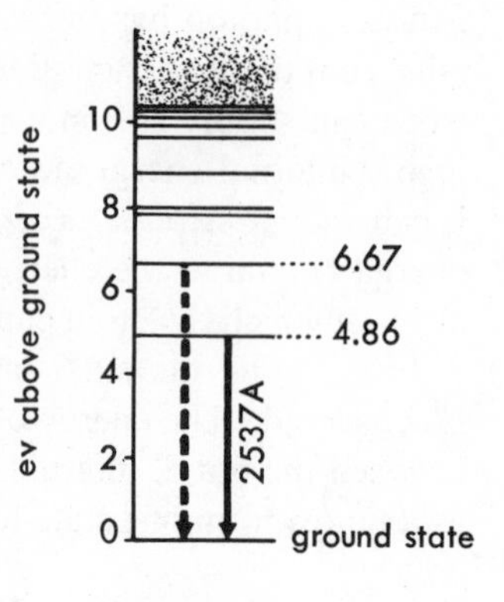

(a) With electrons of any energy above 4.9 ev, the strong ultraviolet line of wavelength 2537 angstroms is emitted. This corresponds, as shown at the right, to an energy change in the atom from the first excited state to the ground state. By the time the energy is 7 ev, there is also an energy change from 6.67 ev to the ground state, giving a line of wavelength 1849 angstroms. The wavelength of this light, however, is too short to affect the photographic plate. Here this transition is shown by a dotted arrow. In the subsequent figures that arrow is omitted.

(b)
8.4 ev

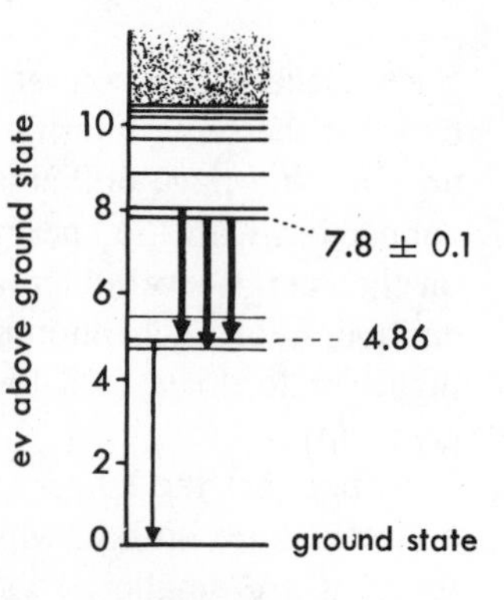

(b) At 8.4 ev three new lines of longer wavelength show up in the region photographed. The photons of this light have energy given by the transitions from the levels around 7.8 ev to the levels around 4.9 ev.

(c)
8.9 ev

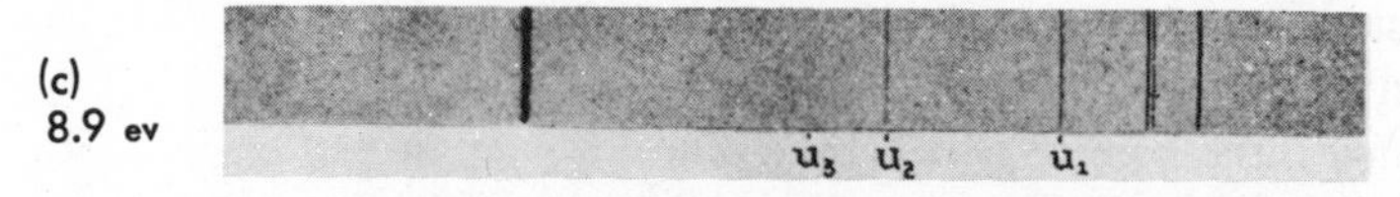

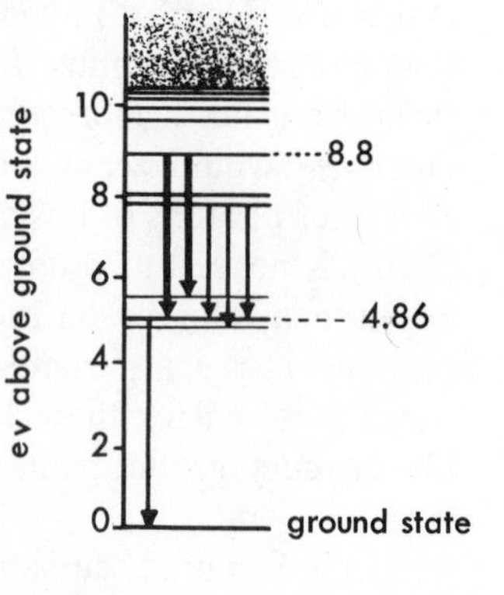

(c) With bombarding energy raised to 8.9 ev, two more lines of intermediate wavelength appear. The photons have energies given by the transitions from the levels around 8.8 ev to the levels around 4.9 ev.

(d) As the electron-accelerating voltage is further increased, more and more lines appear. At 10.4 ev the atom is ionized and the complete spectrum is produced. In this case the classifications above and below the photo help to sort out the various sets of lines.

(d)
9.9 ev
10.4 ev

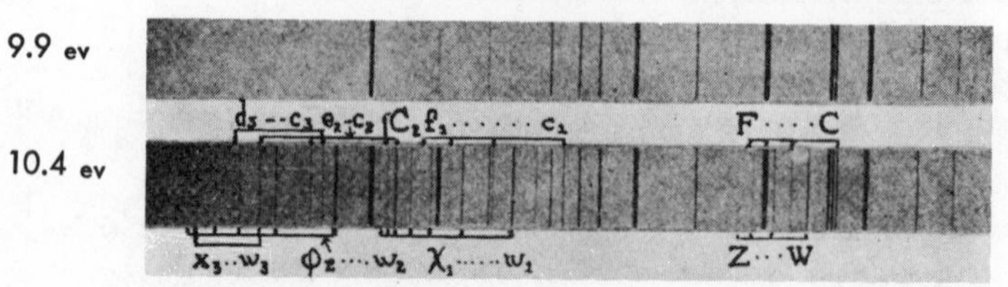

energy level to energy level, emitting photons as they go, sometimes jumping from a high level to a low one emitting a high-frequency photon.

High temperatures, electron collisions, and every other way of producing emission of light from gas atoms are all means of promoting atoms to excited states. The atoms then emit the photons whose energies correspond to the energy differences between the atomic energy levels.

$$E_{\text{photon}} = E_{\text{initial}} - E_{\text{final}}$$

These emitted photons make up the atom's spectrum. A given kind of atom can emit only the lines which have the characteristic wavelengths:

$$\lambda = \frac{hc}{E_{\text{photon}}}$$

determined by the differences between its energy levels.

Of course, different ways of putting energy into the gas can result in different proportions of atoms in the various excited states. As we shall see in the next section, photons from light of a single wavelength can excite atoms to a single high energy state; and only a few other states may be reached as the atoms radiate their way back to the ground state. The carefully adjusted electron beam of 5 electron volts excites only the first excitation state of mercury. High temperatures act differently. The collisions of the randomly moving atoms raise the internal energies of a few more atoms to high energy levels, but there are more atoms in lower states. Thus the brightness of lines can vary, but an individual line of an atom always has the same wavelength whenever it is present.

26-4 Absorption Spectra

Emission spectra—the bright lines we see in the spectra of excited atoms—are not the only spectra which reveal the internal energy states of the atoms. We get similar information about those states when we send white light through as gas and then analyze it with a spectroscope. The white light is a mixture of all frequencies; it contains photons of all energies. Most of these photons pass through the gas and show up in the spectrum on the spectrographic plate. They are not absorbed, because the energy of the photon cannot be taken up by the atom as it makes a transition from one internal energy level to another. But some of the photons are taken up. Their absence is shown by narrow gaps in the spectrum of the light which passes through the gas. Each missing photon has been absorbed by an atom, causing a transition from one internal energy level to another. These gaps in the spectrum then tell us about the internal energy states (Fig. 26–6).

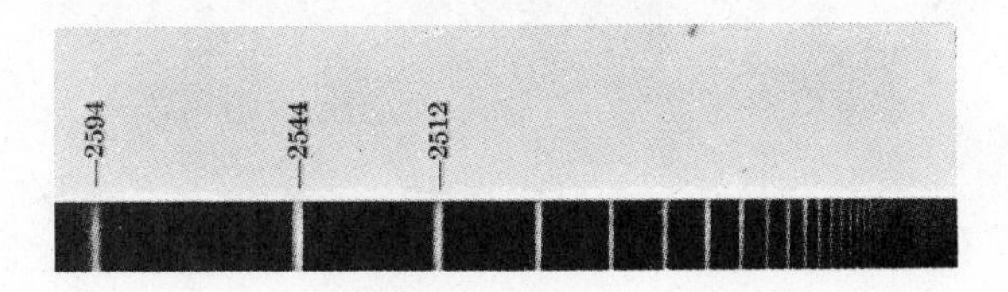

Figure 26–6
The short-wavelength region of the absorption spectrum of sodium. Since this is a reproduction of a photographic negative, the wavelengths absorbed appear as light lines on a continuous dark background. (From: H. Kuhn, "Über-Spektren von unecht gebundenen Molekülen (Polarisationsmolekülen) K_2, Na_2, Cs_2, und Verbreiterung von Absorptionslinien," *Zeitschrift für Physik*, Vol. 76, 1932.)

Here the same interpretation and the same rule of energy conservation apply; only the order of the process is different (Fig. 26–7). In this case the initial state of the atom is the lower internal energy state; the final state is the higher, because it must have the full energy of the original atom plus that of the absorbed light photon:

$$E_{\text{final}} = E_{\text{initial}} + E_{\text{photon}}$$

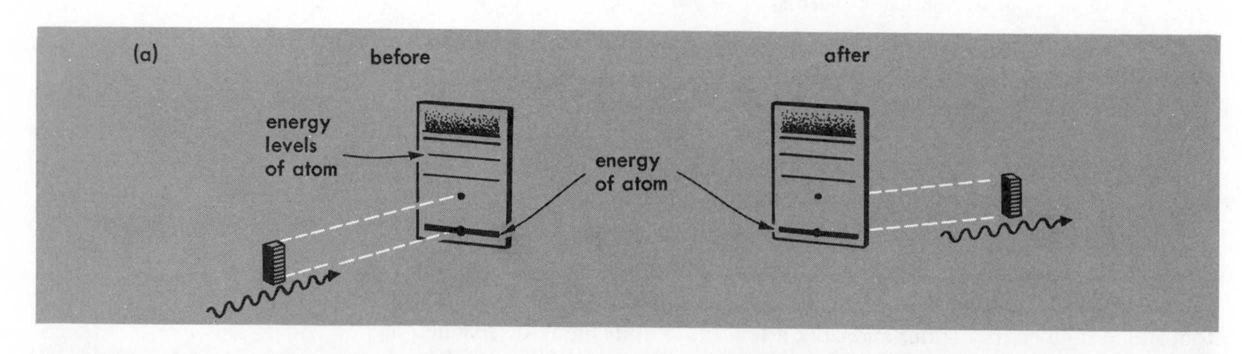

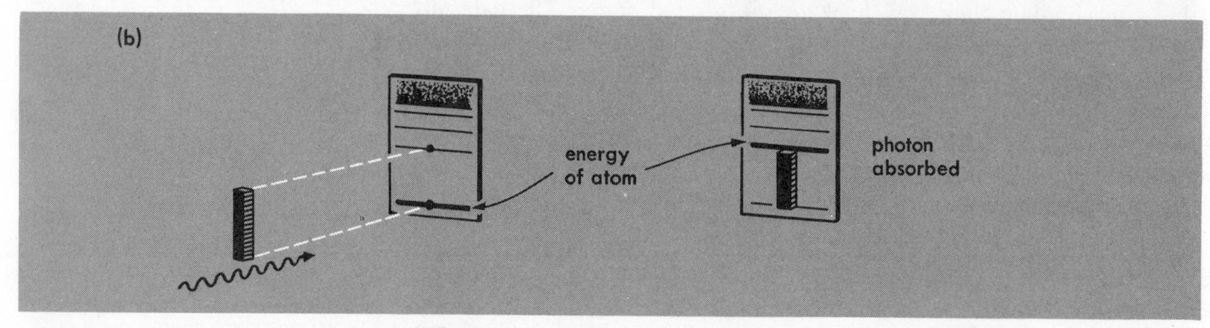

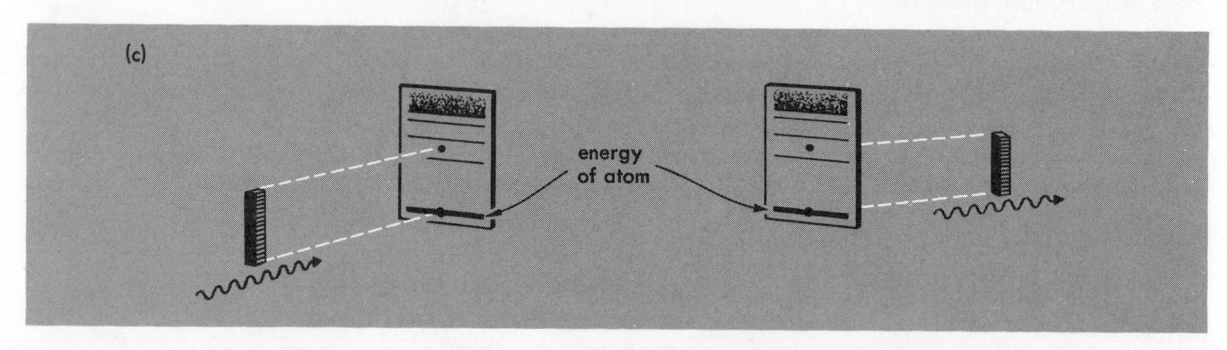

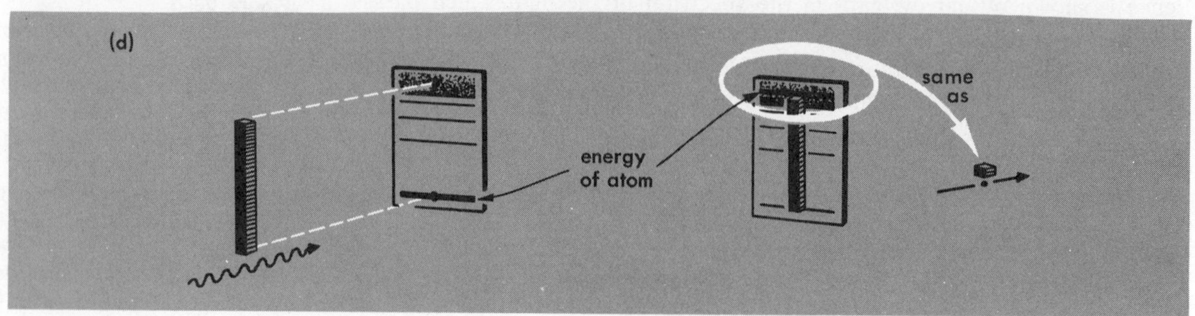

In this way one can see, for example, why mercury vapor at room temperature is transparent to visible light. No photon in visible light has enough energy to promote the mercury atom from the ground state to its first excited state. The right photons come only in ultraviolet light of 2537 angstroms wavelength. As we should expect, that particular light is very strongly absorbed. The same argument applies to the other common colorless gases—oxygen, nitrogen, helium, and the rest. They can absorb light only in the ultraviolet. And there we find a series of absorption lines.

Once the frequency of the light is sufficiently high, in the deep ultraviolet or beyond, a photon can take an electron completely out of an atom. The atom is ionized, and any surplus energy is carried away by the electron as kinetic energy. This is the photoelectric effect. There is no limitation on the amount of kinetic energy the electron can be given. So there are no restrictions on the frequency of the light that ejects electrons provided it exceeds a certain minimum "threshold" value. The observed threshold of the photoelectric effect in atoms is just the minimum frequency required to break an electron loose from the atom. The evidence of spectra in this matter agrees perfectly with the evidence obtained by observing the photoelectrons.

All light, all electromagnetic radiation can excite atoms, or be emitted by them in jumps or transitions between energy levels. Consider one example far outside the range of visible light. Although we are apt to think of radio waves interacting with antennas rather than with atoms, the study of atomic spectra at radio and microwave frequencies is a modern and powerful tool for physicist and chemist. Because radio waves are long, their photons carry little energy, so they arise from levels which lie very close together in energy. All the alkali metals, lithium, potassium, rubidium, and cesium, have pairs of close-lying energy levels. The photons from transitions between these pairs of energy states are in the radio-frequency part of the spectrum. In particular, the ground state of the element cesium is really a pair of energy levels separated by about 4.14×10^{-5} electron volt. Precisely, the radiation from this transition between energy levels has a frequency of

$$\nu = 9.19263177 \times 10^{9} \text{ cycles per second.}$$

This frequency is now used as a standard of frequency and time. Clocks working directly with it are commercial available.

Figure 26–7
The change in energy when an atom is bombarded by light photons of various energies. The "load" sketched above the photon represents its energy.

(a) The atom is in the ground state, and the photon has too little energy to change the atom to the next excited state. The photon simply passes through or scatters off elastically.

(b) The photon is one that has exactly the right energy to raise the atom to its first excited state. The atom absorbs the photon, taking on all the energy of the photon.

(c) The photon is one that has more than enough energy to excite the atom but not enough to reach another excited state. This photon, too, must scatter elastically, for if it were absorbed the atom could not both conserve energy and be in an allowed energy level. The only alternative is the reemission of the photon.

(d) In this case the photon is one with enough energy to ionize the atom. It raises the atom to a region filled with positive energy levels. This is another way of saying that the atom is broken up and the electron carries away the energy above the ionization threshold as kinetic energy. (Note that an atom which is already excited can be ionized by a photon of lower energy which would simply have been scattered elastically from an atom in its ground state.)

The spectrum has become the best clue to the energy levels of any radiating atom. But realizing this and establishing the energy levels from these clues required great ingenuity. Once the spectral wavelengths are known, they must be converted to frequencies, and the frequencies examined and fitted together to find the energy levels themselves. Each spectral line gives a difference of two energy levels, and a few levels can give quite a complicated collection of lines. So the job is not easy. To disentangle the levels, absorption spectra help. So does excitation by photon absorption, and other special forms of excitation. Following up all these clues, the spectroscopist finally presents his results in the form of a table or diagram of energy levels belonging to the atom he is studying. As the result of his labor, many internal energy states have been precisely established for practically all kinds of atoms.

26-5 The Energy Levels of Hydrogen

The energy levels of atomic hydrogen and some of the spectral lines with which each one is associated are displayed in Fig. 26–8. It is the fruit of the work of a generation or two of spectroscopists; in it we see displayed the level structure of the simplest of atoms.

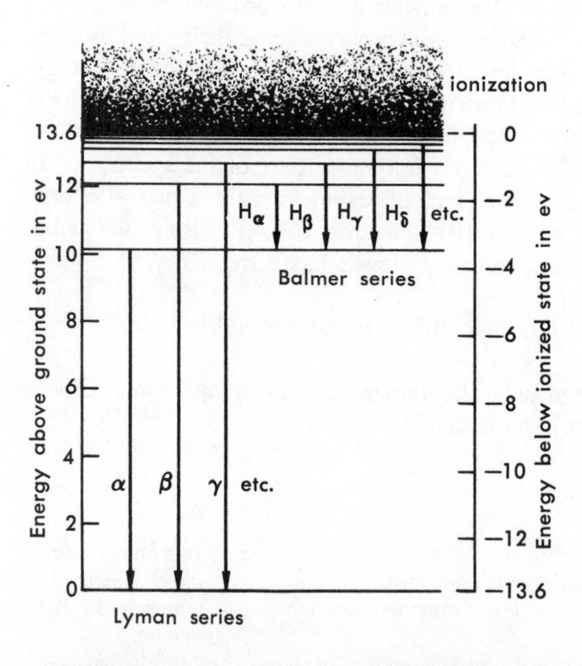

Figure 26–8
(Top) The energy level diagram of the simplest atom, hydrogen. The arrows show the energy changes for two of the sets of spectral lines characteristic of hydrogen. The jumps marked H_α, H_β, etc., are called the Balmer series. The lines of the Lyman series in the ultraviolet are also shown. (Bottom) A photograph of the Balmer series, from n = 6 to as high as n = 20 or more. Their positions fit the predicted values to within one part in ten thousand or even better. (Photo from: G. Herzberg, "Über die Spektren des Wasserstoffs," *Annalen der Physik*, Vol. 84, 1927.)

What is remarkable is that these many energy levels, which gave rise to still more numerous spectral lines, all can be represented by a very simple formula

$$E_n' = E_I - \frac{E_I}{n^2},$$

where E_I is the ionization energy of 13.6 electron volts, and n is any positive whole number. When $n = 1$, this formula gives $E_1' = 0$, setting our zero of energy at the ground state. When $n = 2$, we get

$$E_2' = 13.6 \text{ ev} - \frac{13.6 \text{ ev}}{(2)^2} = 10.2 \text{ ev},$$

which is the energy of the first excited state. For other values of n we get the other observed energy states. Scores of levels have been identified at their predicted positions. The formula works wonderfully well, fitting experiments to one part in ten thousand or better.

In all these energy states, the electron and proton are bound together. They cannot get infinitely far apart. For higher values of n, however, the energy states approach the ionization energy, and as n becomes infinitely large we reach the energy at which the electron can just escape from the proton.

At still higher energies the electron and proton are no longer bound together. The electron is just moving past the proton. At these energies, even when the electron gets infinitely far from the proton, there will be kinetic energy left over. On the energy-level diagram you will notice the essential difference between these energy states in which the electron moves freely away with any energy and the bound states (beneath the ionization energy). There are free states at all energies above ionization, while there are gaps between the individual bound energy states.

In Chapter 24, we saw that the atom resembled a planetary system. In particular, on the Rutherford model the hydrogen atom consists of a proton sun with a single electron planet. The Coulomb force between them goes with the inverse square of their separation, just like the force between the sun and a planet, and the force between the earth and a satellite. In Section 16–6 we studied the binding and the escape of satellites in just these systems. There we found that it was natural to choose the zero of energy when the sun and planet, the earth and satellite, or the proton and electron are standing still infinitely far apart. That energy marks the border between bound states and free states. For the atom it is the ionization energy.

When we raise the zero of the energy scale by 13.6 ev, the formula for the energy states of hydrogen becomes even simpler. It is

$$E_n = E_n' - E_I = -\frac{E_I}{n^2}.$$

Now $E_1 = -13.6$ ev, which tells us directly that the ground state is 13.6 ev lower than the ionized state. The atom is bound together by 13.6 electron volts. The equality

$$E_2 = \frac{-13.6 \text{ ev}}{(2)^2} = -3.4 \text{ ev}$$

says the first excited state is only 3.4 electron volts below ionization level; and so on. On the right-hand edge of Fig. 26–8 we have added an energy scale with its zero at the ionized state. You can compare these answers with the energy levels measured against that scale.

For Home, Desk, and Lab

1.* What are the discrepancies between the predictions of the Rutherford model about the emission of light by atoms and the observed emission of light? (Section 2.)

2.* An electron whose kinetic energy is 4.2 ev collides with a mercury atom. What will be the kinetic energy of the electron after the collision? (Section 2.)

3. Figure 26–9 is a schematic sketch of a vacuum tube in which electrons emitted by S can be accelerated by either of the potential diffrences V_1 or V_2. Suppose the tube contains some mercury vapor and is used to perform the Franck-Hertz experiment.
   (a) If $V_1 = 0$ volts, $V_2 = 5.0$ volts, and the field between G_1 and G_2 is uniform, in what region of the space between G_1 and G_2 can electrons give up 4.9 ev in inelastic collisions?
   (b) If $V_1 = 0$ volts and $V_2 = 10.0$ volts, in what region of the space between G_1 and G_2 can electrons give up 4.9 ev in inelastic collisions? In what region 6.7 ev?
   (c) If $V_1 = 10$ volts and $V_2 = 0$ volts, how far do the electrons have to go through the gas before they can lose 4.9 electron volts in inelastic collisions? How far before they can lose 6.7 ev? What electron energies should it be possible to measure to the right of G_2?
   (d) The 6.7-ev transfer of energy is observed when the apparatus is operated as in (c), but it is not observed when the apparatus is operated as in (b). Explain this result in terms of your answers to (b) and (c).

4. (a) What is the significance of the constant B that appears in the equation for the photoelectric effect in Section 25–3?
   (b) How do the values of B for the four metals exhibited in Fig. 25–7 compare with their

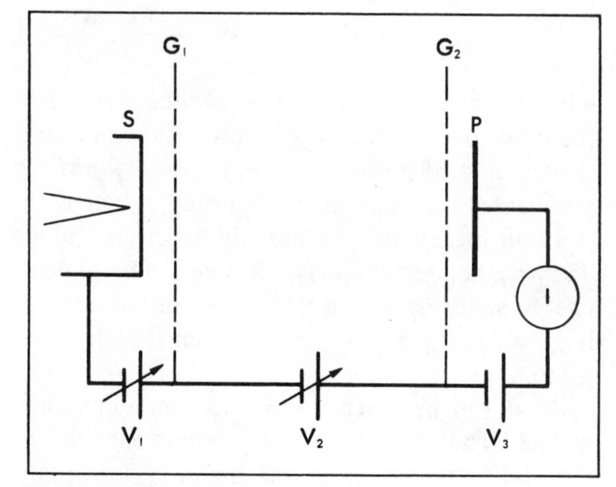

Figure 26–9
For Problem 3.

ionization energies? (Ionization energies of Li, Mg, and Cu are 5.36 ev, 7.61 ev, and 7.68 ev, respectively.)
   (c) What conditions seem to influence the amount of work required to remove an electron from a metal?

5. In the experiment of Franck and Hertz, schematically represented in Fig. 26–2, electrons are used to bombard the mercury atoms. What advantages do the electrons possess for this purpose over protons?

6.* Mercury vapor is bombarded by electrons whose kinetic energy is 6.67 ev. What are the possible energies of the emitted photons? (Section 3.)

7.* Suppose a certain atom emits a photon with a wavelength of 6838 A. How much energy does the atom lose? (Section 3.)

8. Figure 26–10 shows a mechanical model of energy levels with platforms at four different levels above the ground. A number of uniform putty balls are located on top of A. When balls are pushed over the side from A and land some-

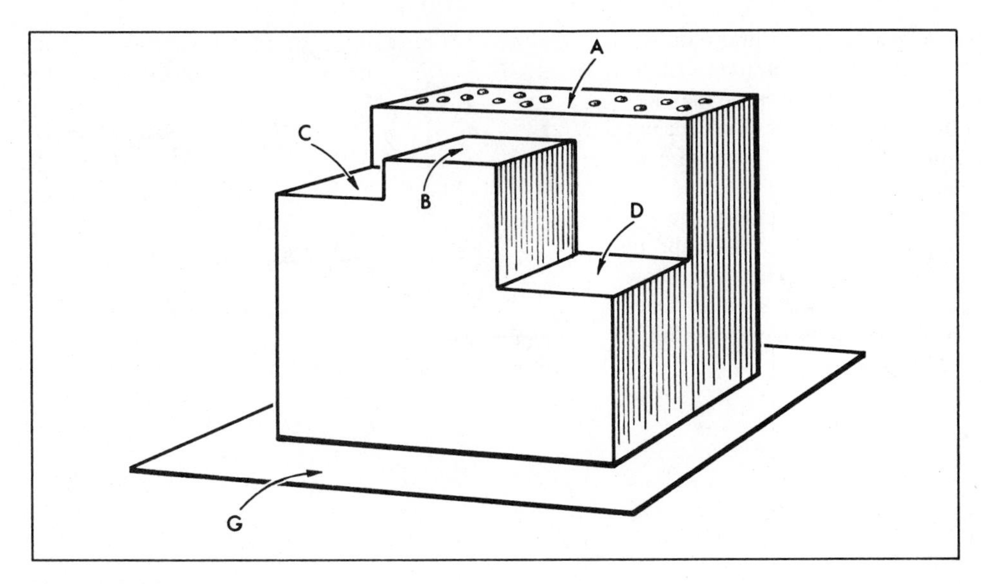

Figure 26–10
For Problem 8.

where below, temperature will rise in direct proportion to their loss in potential energy. Balls which land on *B*, *C*, and *D* are again pushed until they reach the ground. How many different changes in temperature are possible? Draw the "energy levels" and mark the possible jumps.

9. Suppose you are going to bombard a vapor of cesium atoms with 4.00-electron-volt electrons. Use Fig. 26–2 to predict some of the things that you would expect to happen.

10. If ordinary table salt is introduced into a gas flame or a soft glass rod is heated in it, the flame is colored with the characteristic light from sodium atoms, whose wavelength is 5890 angstroms. This light is emitted by electrons returning to the ground state from the first excited energy level.
    (a) What is the energy of this level above the ground state, in electron volts?
    (b) What would be the temperature of a gas whose molecules had an average kinetic energy of this amount?
    (c) The average temperature of the molecules in a gas flame is about 2100°K. How is it possible for a gas flame to raise sodium atoms to the first excited energy level?

11. If we heat helium gas to a high enough temperature, the average kinetic energy of an atom in its thermal motion will be so high that an inelastic collision between two helium atoms can easily excite one of them to its first excited state at 19.8 electron volts above the ground state. Estimate this temperature. (See Section 17-4.)

12.* Compare the amount of energy absorbed when mercury vapor is bombarded with a 6-ev electron and when it is bombarded by a 6-ev light photon. (Section 4.)

13. White light, having passed through sodium vapor, is examined with a spectroscope, and the dark lines characteristic of the absorption spectrum of sodium are observed.
    (a) What happens eventually to the energy absorbed by the sodium atoms?
    (b) Why, then, are the absorption lines dark?

14. If you point a good spectroscope at an arc lamp, you will see a continuous spectrum. If you point it out the window on a bright day, you will again see a continuous spectrum; but this time, with a narrow enough slit, you will see a number of dark lines crossing the spectrum parallel to the slit. (If you have access to a good spectroscope and have never done this, it is worth doing.) What is the origin of these lines? What can be deduced from a study of them?

15. Figure 26–11 is a schematic diagram of another vacuum tube in which electrons can be accelerated by V_1 and the energies of the electrons measured by adjusting V_2 until no current is shown on the meter. Suppose this tube contains mercury vapor.

As in the Franck-Hertz experiment, the energy of electrons arriving at P shows an abrupt decrease when V_1 is 4.9 volts. They also show an abrupt decrease when V_1 is 9.8 volts. Does this mean that mercury atoms have an energy level at 9.8 volts? Explain.

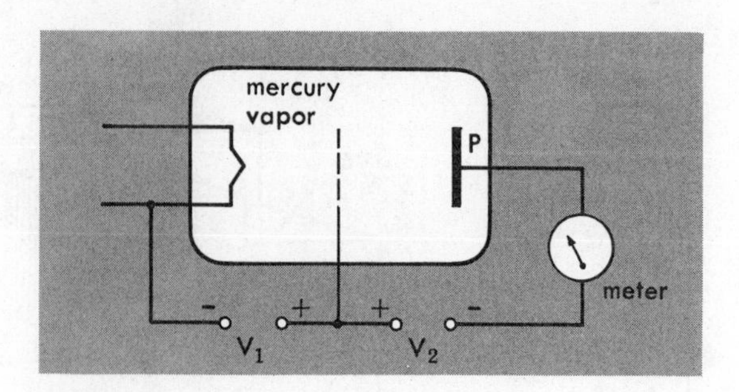

Figure 26–11
For Problem 15.

Further Reading

These suggestions are not exhaustive but are limited to works that have been found especially useful and at the same time generally available.

BITTER, FRANCIS, *Magnets—The Education of a Physicist*. Doubleday Anchor, 1959: Science Study Series. (Chapters 3 and 7)

GAMOW, GEORGE, *Thirty Years That Shook Physics*. Doubleday Anchor, 1966: Science Study Series. The story of the quantum theory—very interesting.

HOFFMANN, BANESH, *Strange Story of the Quantum*. Dover, 1959. Understandable and helpful. (Chapters 4, 5, and 8)

HUGHES, DONALD J., *The Neutron Story*. Doubleday Anchor, 1959: Science Series. (Chapters 6 and 7)

ROMER, ALFRED, *The Restless Atom*. Doubleday Anchor, 1960: Science Study Series. (Chapter 16)

WEISSKOPF, VICTOR F., *Knowledge and Wonder*. Doubleday Anchor, 1963: Science Study Series. (Chapters 5, 7, and 9)

Matter Waves

CHAPTER

27

The identity of the atoms, the stability of their internal motion, and the discrete spectrum of the light which they emit and absorb present severe difficulties for the atomic model, which cannot be resolved by a simple modification. The understanding of atoms requires a fundamental change in our approach to the subject.

To see in which direction this change has to be made, we follow the work of Louis de Broglie, who was intrigued by the fact that although a photon behaves like a particle when it collides with other particles, *where* it is likely to go is described by a wave. The energy of a photon is related to the frequency of the wave by $E = h\nu$ (Section 25–3).

Hence, the wave length $\lambda = \frac{c}{\nu} = \frac{hc}{E}$.

In Section 25–4 we have seen that a photon of energy E has a momentum $p = \frac{E}{c}$. Substituting this expression in the relation for λ we find a relation between the wavelength and momentum of a photon,

$$\lambda = \frac{h}{p}.$$

De Broglie asked the following question: Is it possible that this relation holds for all particles and not just for photons? The best way to answer this question is to perform an experiment with beams of various particles to see if they give rise to an interference pattern, since interference is the most characteristic property of waves. We might try, for example, to pass a stream of particles through a double slit or reflect them from some sort of grating. To see a sizable effect, the distance between the two slits (or the distance between the lines of the grating) must be of the order of magnitude of the wavelength of waves associated with the particles. Table 1

Table 1

PARTICLE	MASS (KG)	SPEED (M/SEC)	WAVELENGTH, $\lambda = \frac{h}{mv}$ (A)
Marble	2.0×10^{-2}	1.0×10^{-2}	3.3×10^{-20}
Latex sphere	1.0×10^{-15}	3.0×10^{-4}	2.2×10^{-5}
1 Mev alpha particle	6.7×10^{-27}	6.9×10^{6}	1.4×10^{-4}
1 Mev proton	1.7×10^{-27}	1.4×10^{7}	2.8×10^{-4}
100 ev electron	9.1×10^{-31}	5.9×10^{6}	1.2
0.1 ev neutron	1.7×10^{-27}	4.3×10^{3}	0.90

shows the magnitudes of the wavelengths associated with various moving particles, assuming the wavelength is given by the relation $\lambda = \frac{h}{p}$. As you can see, even if there is a wave associated with the motion of the marble or the small latex sphere, we cannot make a grating or a pair of slits fine enough to allow us to detect any interference effects. The best way to answer de Broglie's question is to reflect beams of electrons or slow neutrons from a crystal, which is a natural grating with a distance between lines of atoms of the order of 1A.

If a neutron beam is directed at a salt crystal mounted so that it can be rotated about an axis, the beam will go right through it at most angles of incidence. However, if we rotate the crystal, keeping the beam direction constant, there will be an angle of incidence at which a reflected beam will suddenly appear (Fig. 27–1). If we rotate the crystal further, reflection will cease and the neutrons will again go through. The reflection appears when the path difference between the regularly spaced layers of the crystal give constructive interference for the waves connected with the motion of the neutron (Fig. 27–2). This is analogous to the thin-film interference described in Chapter 8. Figure 27–3 shows an actual experimental arrangement used in an attempt to produce an interference pattern with neutrons.

Figure 27–1
An interference experiment to show the wave properties of neutrons. Neutrons are reflected from a salt crystal in which the regular lines of atoms act as a diffraction grating.

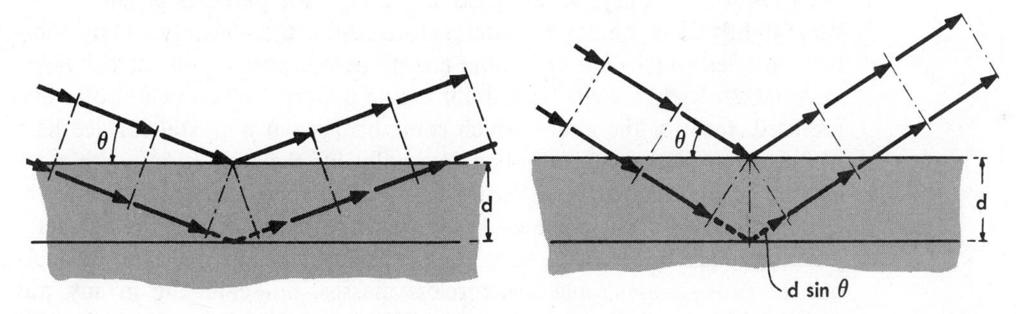

Figure 27–2
The dashed line shown is the extra distance traveled by a neutron wave that reflects from the second layer of the crystal instead of the first. If the beams from each layer are to interfere to give a maximum reflection, this dashed distance must be a whole number of wavelengths. The dashed distance is also $2d \sin \theta$, where d is the spacing between the reflecting layers of atoms. Therefore the first flash-out should occur when $2d \sin \theta = \lambda$.

A beam of neutrons having a speed of 4.04×10^3 m/sec falls on the crystal from the left. When the beam is reflected, the detector responds by sending a signal to the counter for each incident neutron. The graph in Fig. 27–3 shows the neutron count as a function of the crystal angle as measured by the pointer on the crystal mounting. As you can see from the graph, the flash-out angle is $\theta = 10.01$ degrees. The spacing D in Fig. 27–2 for a salt crystal is 2.815A. Thus $\lambda = 2d \sin \theta = 2 \times 2.815 \times 0.1738 = 0.9785$A, which checks very closely with the predicted de Broglie wavelength of

$$\lambda = \frac{h}{p} = \frac{6.63 \times 10^{-34}\text{ joule-sec}}{1.67 \times 10^{-27}\text{ kg} \times 4.04 \times 10^3\text{ m/sec}} = 0.981\text{ A}.$$

The results of such experiments with neutrons are so good that we now use neutron interference patterns to find out more about the positions of atoms in crystal gratings. Hydrogen atoms in crystalline compounds, for instance, whose regular spacing will not show up well in X-ray interference,

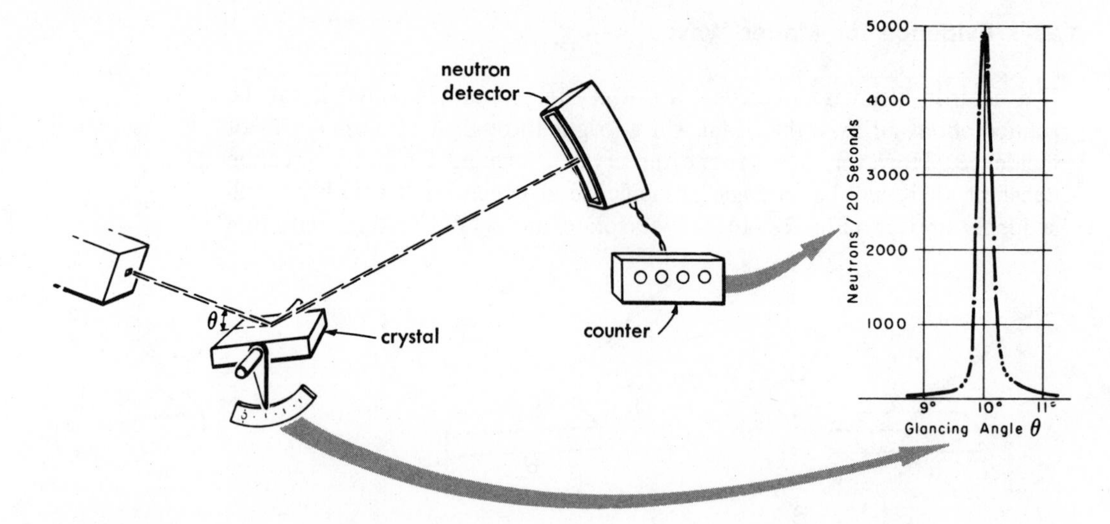

Figure 27–3
The experimental arrangement for detecting the interference patterns produced by neutrons reflected from a salt crystal. (Experiment by Prof. Clifford G. Shull.)

form a far more effective part of the grating as seen by neutrons. Locating them in their regular repetitive array in a crystal is only one of the jobs neutrons can do because of their wave nature and their own special way of interacting with other particles of matter.

Neutrons will travel a long way in air, but experiments with atoms and electrons must be done in vacuum. Otherwise the air molecules would scatter and diffuse the beam beyond any possible use for sharp measurement. Also, when crystals are used as gratings, the particles penetrate only very slightly into the solid matter before losing their energy. Only those few particles which lose very little energy before passing out of the target material can yield a well-marked diffraction pattern. When beams of atoms are used, most of the atoms which come back from a crystal surface have in fact been somewhat disturbed internally by their collisions within the crystal, or have imparted differing fractions of their kinetic energy to the crystal. These do not make clear the wavelike behavior. In the diffracted wave there are to be found only those atoms which have behaved throughout the process as simple structureless masses, not changing in any way their internal condition. All the particles diffracted into a definite direction have the same velocity; this was directly checked to verify the point. Every moving particle, as long as it behaves as a whole and has no change internally, or in its kinetic energy, diffracts as though it has a single wavelength given by the de Broglie relation:

$$\lambda = \frac{h}{p}.$$

The de Broglie wavelength applies to the motion of the center of mass of the whole "particle."

Many different kinds of interference experiments involving particles have been performed. In an early investigation of matter-wave interference helium atoms were used. Figure 27–4 shows the pattern they make when they reflect from the atomic grating of a lithium fluoride crystal. Here the incident beam did not have a single well-defined momentum. It was made of atoms coming out of a warm box of helium gas. Nevertheless, the inter-

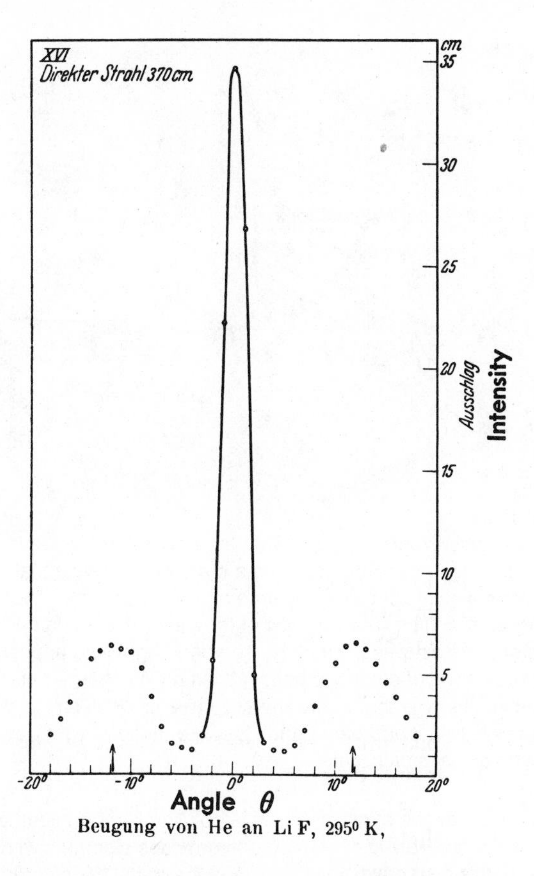

Figure 27–4
The graph shows the result of an experiment with a beam of helium aimed directly at a freshly cut surface of a lithium fluoride crystal. The existence of interference shows the wave nature and the angle measures the wavelength. (From "Beugung von Molekularstrahlen," I. Estermann and O. Stern, Zeitschrift für Physik, Volume 61, p. 107, Springer Verlag, Berlin, 1930.)

ference maxima are clearly visible, humped up around the wavelength expected from the thermal energy and the corresponding momentum of the atoms. From the crystal lattice spacing of 2.85 angstrom units, and from the measured angle, about 11.8°, the de Broglie wavelength is calculated to be about 1.2A. The de Broglie wavelength calculated from $\lambda = h/mv$ for helium at speeds corresponding to room temperature is also 1.2A.

Figure 27–5, on the next page, describes another matter-wave interference experiment involving the diffraction of electrons by a "biprism."

27-2 When Is the Wave Nature of Matter Important?

To observe effects which show us the wave nature of matter, we must have de Broglie wavelengths of the same order of size as the object we can use to diffract them. If we use moving particles whose speed or mass is large, $\lambda = h/mv$ will give us a wavelength too short to be diffracted by atomic particles. How can we lengthen the de Broglie wavelength? We can try to reduce the speed; but this is not practical. Thermal agitation itself limits the slowness of motion which can in fact be realized, but we can use smaller masses. The smallest masses are the particles of which matter is composed: the molecules, the atoms, the electrons, and the particles within the nucleus. As we have seen, their wavelengths can be long enough that we can see their diffraction patterns. Because an electron has a mass of only 10^{-30} kg, at a

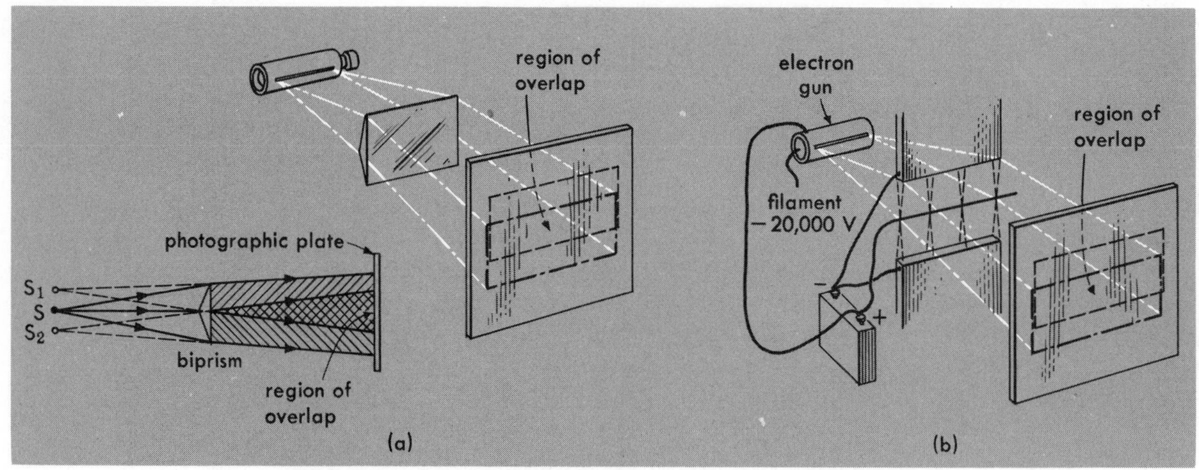

Figure 27–5
Two analogous experiments showing the interference of waves. Both show interference fringes from a biprism—a device that takes waves from a single slit as the source and produces the effect of waves from two slits. The likeness of the patterns obtained with light and with electrons is striking. The bending would happen in just the same way if electrons behaved simply as charged particles; but the peculiarity is that where the two streams of electrons overlap there are bright and dark bands just like those of light waves, though on a very small scale.

(a) With visible light, a glass biprism (two wedges of glass pointing in opposite directions) bends the two halves of the beam from the slit S_1. The upper half is bent down so that it appears to come from S, and the lower half is bent up so that it appears to come from S_2. The two halves of the original beam overlap on a photographic plate. Interference is observed in this region of overlap. At some places on the photographic plate, crests that seem to come from S_1 arrive along with troughs that seem to come from S_2, making a dark fringe. At other places crests arrive together, making a bright fringe.

(Left) The interference pattern painted by light. (Valasek, "Introduction to Theoretical and Experimental Optics," John Wiley & Sons, New York, 1949)

(b) The electron source is a 20,000-volt gun which provides electrons with a de Broglie wavelength of 0.086 angstrom. The electric field "biprism" is made by a fine silvered quartz thread, a few microns in diameter, suspended in a wide slit in a metal screen. The thread is kept about 6 volts positive with respect to the slit. The diagrams show the electrostatic biprism and its electric field lines. The electric field bends the streams of electrons that pass each side of the wire. They overlap on the photographic film, arriving there as if they came from two sources.

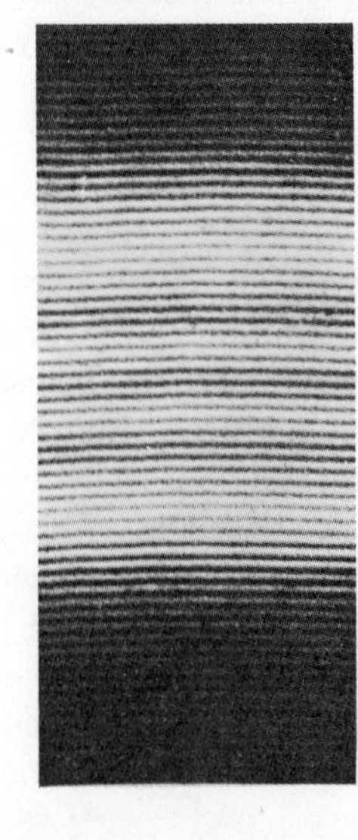

(Left) The interference pattern painted by electrons. It has been greatly magnified by an electron microscope. (From: H. Düker, "Lichtstarke Interferenzen mit einem Biprisma für Elektronenwellen," *Zeitschrift für Naturforschung,* Volume 10A, 1955.)

reasonable speed it will have a wavelength of the size of an atom. An electron moving with 100 electron volts of kinetic energy moves at a speed of 5.9×10^6 meters per second and its de Broglie wavelength is therefore about 1.3×10^{-10} meter (Table 1), which is in the range of atomic dimensions. A reasonably slow-moving electron with one electron volt of kinetic energy moves at about a tenth of the speed, and has a wavelength of about 10 atomic dimensions. Atoms are small obstacles for such an electron. The associated waves can easily diffract around them (Fig. 27–6). These examples show that wave behavior is essential whenever an electron of small kinetic energy having a wavelength close to atomic size moves in or near an atom. For these motions our ordinary experience with Newton's particle mechanics is not an adequate guide.

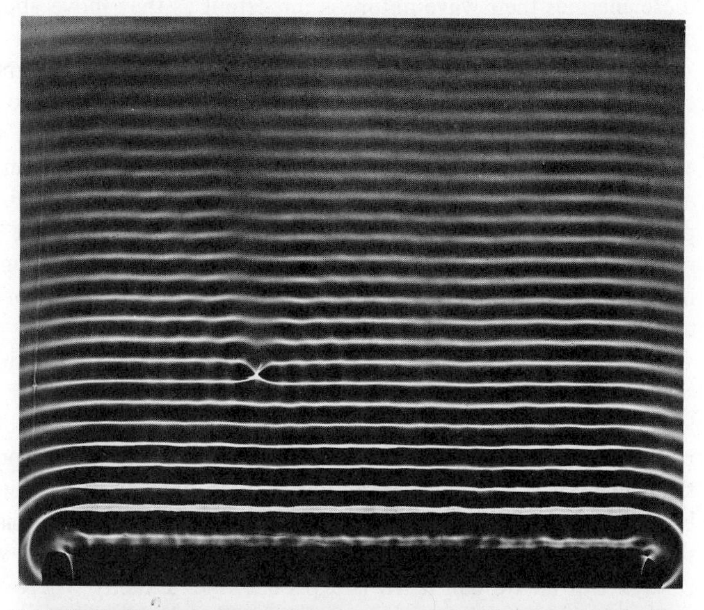

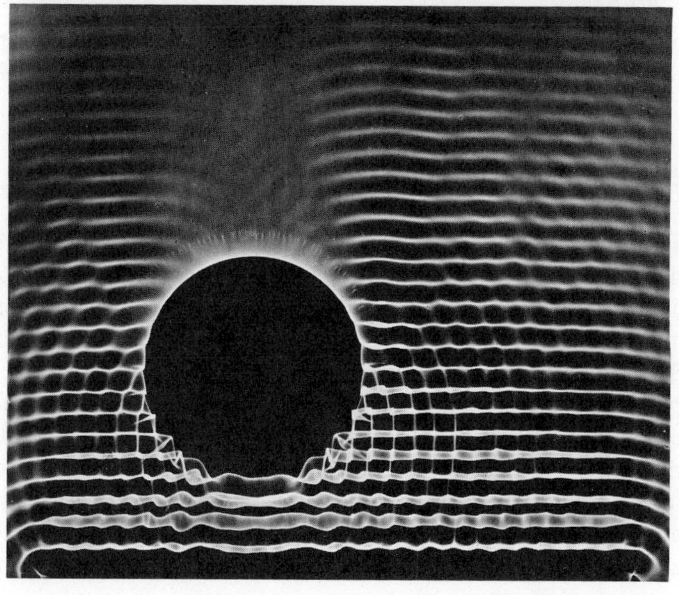

Figure 27–6
Ripple-tank photographs of waves meeting obstacles. In the upper picture a small object is located where the waves seem to bend together. Notice that this small barrier disturbs the waves very little. In the lower picture there is a large obstacle in the ripple tank. The waves are strongly reflected and there is a definite "shadow" region.

The situation is the same for protons and neutrons with respect to the size of the atomic nucleus. The wavelength of a proton or neutron with 100 electron volts of kinetic energy is about 3×10^{-12} meter. This wavelength is large compared to nuclear dimensions, which are 10^{-14} meter or less. Consequently, such a neutron approaching the nucleus from the outside will find it a small obstacle around which it diffracts. Neutrons with ten million electron volts have a wavelength of 10^{-14} meter, comparable to the nuclear size. Consequently, neutrons of this energy just get inside the nucleus. A neutron that stays inside a nucleus, forming a part of it with practically no chance of being found outside, must have a wavelength of this size or smaller. Consequently, we expect that the neutrons and protons in nuclei have kinetic energies of about ten million electron volts or more. Even at these energies their wave nature is important as they move about inside the nucleus.

The alpha particles emitted from nuclei also show kinetic energies of the order of magnitude of 10 Mev, and their wavelengths are also of the order of 10^{-14} meter. When we use them in scattering experiments, a wave effect explains why the whole pattern of alpha-particle scattering can be modified at energies at which the alpha particles get near certain nuclei. The alpha-particle waves may then diffract around the nucleus so that scattering at many angles differs from the Coulomb scattering, although only scattering in the backward direction would be modified according to ordinary Newtonian mechanics.

Once again we have found that the atomic world cannot be fully described by the simple models of the older physics. The little particles cannot be followed like so many bullets as they fly about within the atom; they obey rather the laws followed by waves. The waves determine the probable location of the particles, and the motion of wave packets determines the probable motion of the particles. Where the amplitude of the wave is large, there the particle is probably to be found; where it is small, there it will be found rarely. But all we can learn is a probability. Exactly where the electrons will be found in an atom we cannot predict. Inside the atom we must be prepared to use a new version of mechanics which combines parts of Newton's particle mechanics with the effects of the wave nature of matter.

27-3 Light and Matter

The circle has closed. Waves of light proved to describe the probability of the appearance of certain particlelike photons; the particles of matter turn out to be governed by a wavelike quantity. Planck's constant controls the magnitudes involved in both cases. We cannot follow the path of a particle through the slits of a diffraction grating any better than we can follow the path of a photon. The associated waves diffract and from these waves we can calculate probable locations of a particle and probable numbers of arrivals. But, just as in the photon picture, the particles transfer energy and momentum in every single collision in the same way as they are expected to do from the mechanics of Newton. The wave nature does not modify that at all. Therefore, whenever the wavelengths are small compared to the dimensions of slits or obstacles, small compared to the dimensions of atoms or of measuring instruments, we get results identical with the predictions of ordinary mechanics. Indeed, it is for this reason that we can use ordinary measuring instruments, and it is for this reason that ordinary measurements will only allow us to measure the same quantities which are dealt with in Newtonian mechanics.

For all objects of everyday size, or for objects like the planets, the mechanics of Newton is without any important error, but if we look with great refinement, or if we ask questions on the scale of the atom, the classical physics (as physics before 1900 is often called) is found wanting. It remains the foundation of physics, because the very instruments and apparatus with which we measure are well understood in classical terms, but it is powerless to explain the atom. Only the wave theory of matter (generally called quantum theory) can do that, with its orders of magnitude proclaimed by Planck's constant, and its more subtle picture of the atomic world. The photon governed in space and time by its electromagnetic wave of probability, the particle by its matter wave of probability—all with the de Broglie wavelength—are the keys with which we propose to enter the interior of the atom.

In all that we have said, the analogies between the behavior of light and matter have been striking. Light behaves both as waves and as particles—photons. Matter behaves both as particles and as waves. But the similarities have their limitations. There is one sharp distinction between a photon and a particle of the sort we usually think of, between a photon and a little bit of matter. Photons are not conserved in number; they are easily created and easily destroyed. To turn on the light creates photons by the swarm; they rush to the walls. Some are reflected, but in the end all die, giving their energy to the atoms they strike. Energy is conserved, for heat or chemical changes will reveal the photon energy content, but the number of "particles" is not. This behavior is sharply different from the behavior of electrons and atoms. Those particles cannot be created or destroyed (not in any commonplace way, at least). They are permanent as a photon is not. In this sense photons differ very much from little grains or bullets of matter.

In physics today we regard the wave-particle relationship as fundamental for every one of the objects we deal with on the atomic scale; the main distinction between light and matter is between the permanence of particles and easy creation or absorption of photons.

27-4 What It Is That Waves

In the ripple tank it is quite plain that the "waves" are waves of water. The water moves, the shape of its surface changes, the orderly nature of the patterns is in the end the orderly behavior of little masses of water. But in the electromagnetic wave, there is no material changing its shape or its motion or its density. The emptiest of vacuums will transmit light or radio waves just as well as—indeed, better than—a region filled with any material we know. The waves are patterns of electric and magnetic fields formed in space. In the last century there was for a while a fashion of thinking that the empty space was filled with some strange substance—the "luminiferous" or light-bearing aether—and that waves of light were motions of that curious substance. Gradually physicists came to feel that it added very little to say that space was filled with aether. We now regard it just as a property of space itself that electromagnetic disturbances can travel within it. "Empty" space is devoid of atoms, of any material things at all, but that no longer means that we try to think of it as empty of properties. One of the properties of space is its ability to carry energy in electromagnetic form, and to support the waves of electric and magnetic field which we find to make up radio or light waves.

The same view holds for the de Broglie waves. The de Broglie waves are not changes of shape or of motion of the electrons whose positional probabilities they describe. They are simply patterns of probability which behave in most respects as analogies to the waves we studied in the ripple tank. There is no reason to say that any substance is waving to and fro. We have been able to show that the fundamental properties of the world resemble some simple analogues which we can build up in the laboratory. But we do not need to carry that to extremes: electromagnetic waves and matter waves are not, as far as we now know, waves of any substance. There is no point in insisting that every feature of the ripple tank be found in an atom or in a photon. The wonderful thing is that these things are so much like the waves we can see in the ripple tank: they obey the principle of superposition and therefore form interference patterns.

The physicists of the early twentieth century who lived through the development of modern physics found it a magical but a terrible time. It seemed full of paradox: things were both waves and particles, waves of no substance at all, particles without definite paths through space. The novelty of combining these ideas in a new, consistent picture was often mistaken for impossibility. But we have learned that the old models based on experience with matter more or less on our own scale do not completely suit the world on the scale of the atom. The new picture is much better. The astonishing thing is that we can use a large part of our everyday experience as a guide to the atom. The quantum theory—the combination of wave and particle—was the greatest of triumphs. For a physics willing to face the facts squarely and to be neither frightened nor dazzled by novelty, it opened up a whole new domain.

27-5 Standing Waves

How can we understand the presence of definite internal energy levels of atoms and the simple numerical relations they sometimes have? We hope to do this by using the idea of matter waves, waves whose intensity at any place indicates the probability of finding a particle there. How does that idea explain separate, well-defined energy levels?

The matter wave for a particular energy level must stay the same for a very long time. We know from Franck-Hertz experiments and from the definite line spectra of elements that there are definite separate energy states in atoms. We should not find that behavior if the matter wave for a particular energy level wobbled around and did not remain the same for a long time. Also, there must be a different matter wave corresponding to each separate energy level, because the energy levels are different. We must therefore look for several distinct wave patterns, each one different from the others and each unchanging. The only waves which show the same patterns all the time and do not travel away are standing waves. These are the stationary waves that you find on vibrating strings or in tubs of water. They do not move about. Instead, they stay in place and keep their shape while the displacements periodically grow and collapse without moving along. Look at Fig. 27–7, for example, at the second picture from the left. At the two ends of the long rubber tube and in the middle there are nodes—points which do not move. In between each pair of nodes are regions of maximum displacement. These regions always stay at the same place along the tube, but as time goes on the displacement ranges from maximum in one direction through zero to maximum in the other direction. Everywhere, the displace-

Figure 27–7
Standing waves. As one end of the long rubber tube is moved from side to side with increasing frequency, patterns with more and more loops are formed. There are, however, only certain definite frequencies that will produce fixed patterns.

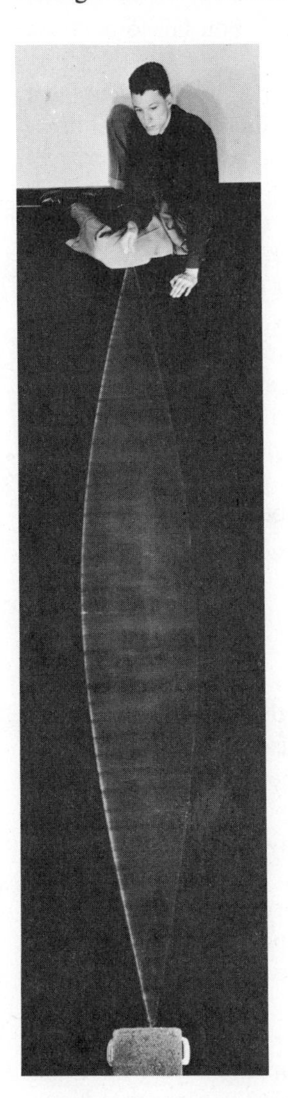
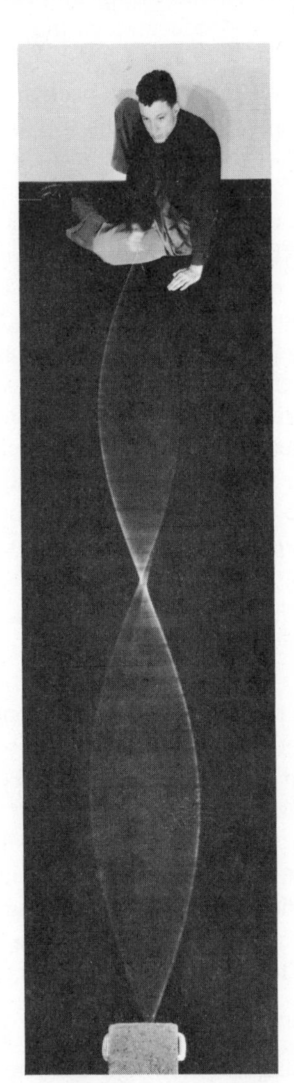
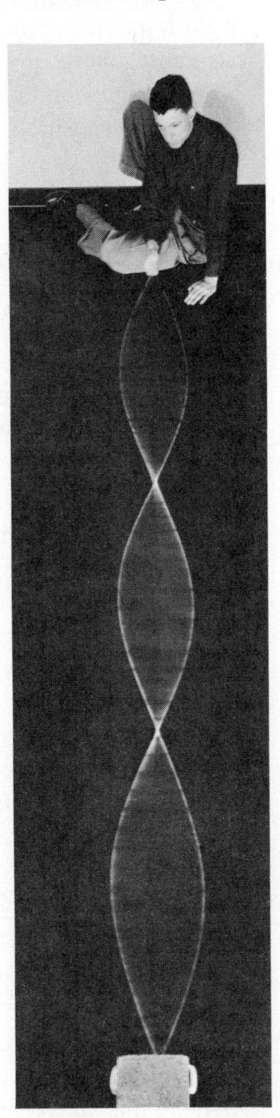
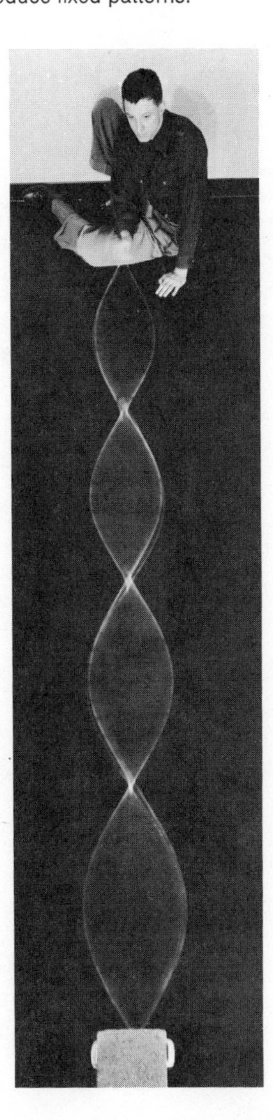

ment oscillates back and forth in the same way, but nothing appears to move along the tube.

There is no substitute for making standing waves yourself. Tie a rubber tube between two firm supports. Wiggle the tube up and down with your hand near one support. By choosing the right frequency for the motion of your hand, you can set up a standing wave pattern in which the tube oscillates up and down but the pattern does not travel along. It may be a pattern with one loop between the end supports, or two loops, three loops, four loops, and so on (Fig. 27–7). This is the kind of wave motion we are going to picture for a stable state of an atom.

With the ends of a rubber tube held fixed, only the patterns with a whole number of loops (and corresponding definite frequencies) will stand on the tube. If you try to drive the tube with any other frequency, you will find that the motion of your hand undoes some of the tube's existing motion rather than being in step and increasing the motion. Unless you use the right frequency, your efforts produce very little effect. You fail to build up a standing wave.

Although standing waves just oscillate back and forth, we can view them as the superposition of two equal traveling waves moving in opposite directions. If you draw two wave patterns, move them across each other, and add their displacements to find the resultant, you will find that they make a standing wave. Also, you will find that the loops of the standing wave are just $\frac{1}{2}\lambda$ long, where λ is the wavelength of either wave.

As you can see from Fig. 27–7, if we tie one end of a rubber tube to a rigid support and wiggle the other end very slightly with the "right" frequency, a standing wave is established. We may imagine that two rigid reflecting walls are suddenly installed at nodes in the standing wave pattern. Because we install the walls at places of no motion, they would trap a section of standing wave between them. In any such section we would see a pattern of oscillations, a standing wave, instead of a moving wave pattern. To trap a standing wave pattern the supports must be one half-wavelength apart, or two half-wavelengths, or three, and so on.

27-6 A Particle in a "Box"

We can also imagine a standing matter wave along the x axis confined between two points beyond which its amplitude is also zero. Suppose this matter wave is associated with an electron; then the behavior of this electron would be characterized as follows:

(a) The x coordinate of the electron is always between the points A and B, since we never find the electron outside that range.

(b) We can look at the standing wave as the superposition of two traveling waves of the same wavelength and moving in opposite directions. Both of these traveling waves together describe an electron with an x component of the momentum having the magnitude $p_x = \dfrac{h}{\lambda_x}$. The electron which is described by the standing wave must have a component of the momentum which is both $+p_x$ and $-p_x$ at the same time. Within the context of Newtonian physics such a statement makes no sense, but then, the wave nature of the motion of a particle is important precisely when the Newtonian picture fails. Thus, in the quantum world a particle can have two equal and opposite momenta at the same time; but, since its energy is

$$E = \tfrac{1}{2}mv^2 = \frac{\tfrac{1}{2}(mv)^2}{m} = \frac{p^2}{2m},$$

its energy is not zero.

The energy of a particle described by a standing wave equals the energy of the same particle described by a traveling wave of the same wavelength. But, unlike the traveling wave, the standing wave which vanishes outside a given range can have only certain discrete wavelengths and hence only discrete energies.

Consider now a particle of mass m confined to a limited volume. To get a first idea of the effects of this confinement on the energy of the particle, we can ignore the exact nature of the force field which binds the particle between two points A and B, and assume that there is no force at all acting on the particle inside the volume but an extremely strong force acting on it at the boundaries A and B of the volume and keeping it inside. Furthermore, we shall take the volume to be a cube of side length d. This highly idealized picture is usually called a particle in a "box."

Taking the potential energy reference level as zero inside the box, the total energy of the electron is all kinetic energy:

$$E = \frac{p_x^2 + p_y^2 + p_z^2}{2m}.$$

The electron cannot escape from the box, so its matter wave must be a three-dimensional standing wave. Such a wave can have different wavelengths along the three axes of the box. For a cubical box of edge length d, they must fulfill the conditions

$$\frac{n_x\lambda_x}{2} = d, \quad \frac{n_y\lambda_y}{2} = d, \quad \text{and} \quad \frac{n_z\lambda_z}{2} = d.$$

The three wavelengths λ_x, λ_y, and λ_z then determine the three components of the momentum, $p_x = h/\lambda_x$; $p_y = h/\lambda_y$; $p_z = h/\lambda_z$, and the kinetic energy then becomes

$$E = \frac{p_x^2 + p_y^2 + p_z^2}{2m} = \frac{(n_x^2 + n_y^2 + n_z^2)h^2}{8md^2}.$$

The numbers n_x, n_y, and n_z, called quantum numbers, are completely independent of one another and can each take the values 1, 2, 3, etc.

The energy levels which arise from such an analysis clearly depend on the mass of the particle and the size of the box. The quantity $(h^2/8md^2)$ is a measure of the order of magnitude of the energy scale of the system and the approximate energy difference between adjacent energy levels. For an electron ($m = 10^{-30}$ kg) in a box the size of an atom ($d = 10^{-10}$m), this factor is numerically equal to about 30 electron volts, about what you would expect for an atomic sized system.

For the ground state, the quantum numbers have their smallest values ($n_x = n_y = n_z = 1$)

$$E_0 = 3h^2/8md^2.$$

Now when we think back to the hydrogen atom, we can see that we are following a promising road. For just such reasons, the hydrogen atom must have a ground state—although Newtonian mechanics would allow it indefinitely lower energies. Thus an explanation of the stability of atoms is already in sight. Furthermore, there are many matter waves of higher energy that can stand in the box. There are the standing waves with 2 loops, 3 loops . . . n loops, and so on. These stationary wave states, with their definite values of energy, provide a model for the excited states of the atom. In the next section we shall apply this model to the hydrogen atom.

27-7 The Standing Wave Model of the Hydrogen Atom

Our first step is to look at the motion of an electron moving in the electric field of a nucleus containing one elementary charge (a proton). From the point of view of Newtonian mechanics the hydrogen atom resembles the earth-moon system: at a distance r from the nucleus the force acting on the electron has a magnitude k/r^2. If the electron is to move in a circle at that distance with speed v, the electric force must equal the centripetal force necessary to keep the electron in the circular orbit:

$$\frac{k}{r^2} = \frac{mv^2}{r}.$$

Multiplying both sides of this equation by ½ r yields

$$\frac{1}{2}\frac{k}{r} = \tfrac{1}{2}mv^2 = E_K.$$

You may recall that the potential energy of an electron in the field of the positively charged nucleus is

$$U_r = -\frac{k}{r}.$$

Thus, for an electron in a circular orbit we find that

$$-\tfrac{1}{2}\,U_r = E_K.$$

Using this relation, we can express the total energy of the electron in two forms

$$E = U_r + E_K = -\frac{k}{r} + \frac{k}{2r} = -\frac{k}{2r}$$

and

$$E = U_r + E_K = -2E_K + E_K = -E_K.$$

This is as far as Newtonian mechanics will take us. Neither of these expressions suggests any restriction on the values of the total energy, and therefore they cannot account for the discrete lines in the hydrogen spectrum. Moreover, the hydrogen atoms and many other atoms in their ground state behave like a positive nucleus surrounded by a spherical cloud of negative charge rather than an electron revolving in a circular orbit around a positive nucleus. If, for example, the latter picture were correct, we should be able by some experiment to determine in which plane the electron moves, yet experiments do not show a preferred plane. We must conclude that a spherically symmetric wave pattern is a more appropriate description of the electron in the ground state of the hydrogen atom than a circular orbit. How can we, then, bridge the gap between this description and the orbital description of motion which holds on a larger scale?

We have seen in Section 27–2 that the wave nature of particles does not present itself as long as all distances which appear in the experiment are large as compared to the de Broglie wavelength. In our case we may expect that Newtonian mechanics will be adequate as long as the circumference of the orbit, $2\pi r$, is much greater than the de Broglie wavelength, λ.

Now, from our first expression for the total energy of the electron we have

$$2\pi r = -2\pi \frac{k}{2E} = -\frac{\pi k}{E}$$

and from the second expression we find

$$\lambda = \frac{h}{p} = \frac{h}{\sqrt{2mE_K}} = \frac{h}{\sqrt{-2mE}}.$$

Thus, Newtonian mechanics will be adequate if

$$\frac{2\pi r}{\lambda} = -\frac{\pi k}{E} \cdot \frac{\sqrt{-2mE}}{h} = \frac{\pi k}{h} \sqrt{\frac{2m}{-E}} \gg 1.$$

We can now try to extend the validity of the relations based on Newtonian mechanics by admitting that even when $2\pi r/\lambda \gg 1$ the standing wave in a somewhat fuzzy circular channel will be a better description than the classical orbit, just as taking into account the very small diffraction of a light beam passing through a wide slit is an improvement on geometrical optics.

For the wave pattern to form a standing wave in this circular channel, the circumference of the channel will have to be an integral multiple of λ;

$$\frac{2\pi r}{\lambda} = n$$

where n is a large integer (Fig. 27–8). If n can have only integral values, then E becomes discrete too; we denote the energy of the nth level by E_n:

$$\frac{\pi k}{h} \sqrt{\frac{2m}{-E_n}} = n.$$

By squaring and solving for E_n we find

$$E_n = -\frac{2\pi^2 k^2 m}{h^2} \cdot \frac{1}{n^2} = -\frac{R}{n^2}.$$

The constant

$$R = \frac{2\pi^2 k^2 m}{h^2} = 13.6 \text{ ev}$$

is called the Rydberg constant.

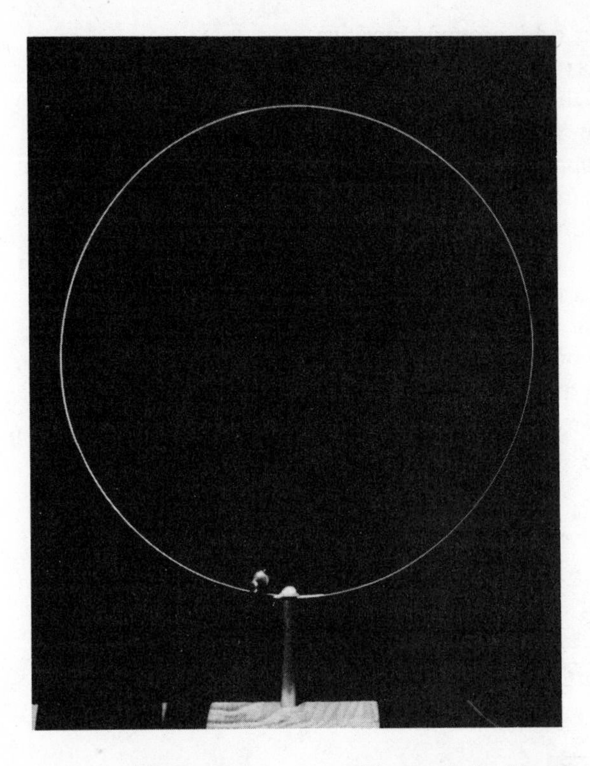

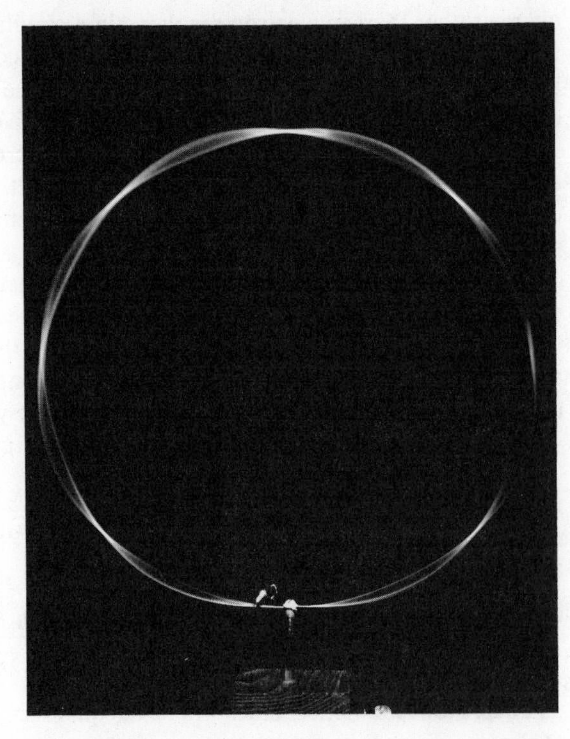

Figure 27–8
An illustration of a circular standing wave. (Left) A wire loop stationary with no waves on it. Just to the left of the support you can see the shaft of a small electric motor protruding through the black cardboard background. (The motor is behind the cardboard.) The shaft is not circular, so that when it rotates it pushes on the wire, sending waves around the loop. (Right) A standing wave with $2\pi r = 4\lambda$.

This result is identical to the expression for the energy levels of the hydrogen atoms which was found in Section 26–5. There it was found empirically by looking for a set of energy levels which would account for the observed spectrum. Here we arrived at the same result by modifying Newtonian mechanics as applied to atoms to include the requirement of a standing matter wave.

For Home, Desk, and Lab

1.* What is the wavelength of an electron whose speed is 1.0×10^7 m/sec? (Section 1.)

2.* As the angle θ in Fig. 27–2 is increased, the beam will again reflect very strongly out of the crystal. What must $2d \sin \theta$ be equal to at this angle? (Section 1.)

3. The layers of atoms in crystals are known to be a few angstroms apart, say 3×10^{-10} m. X rays from ordinary commercial X-ray tubes are found to be diffracted at quite large angles, such as 10°, by such crystals acting as interference gratings. Make a rough estimate of the energy of photons in such X rays.

4. Electrons passing through a double slit (fantastic but possible, as discussed in the next question) make an interference pattern on a remote screen. If the gun voltage used to provide the original stream of electrons with kinetic energy is changed from 50 volts to 5000, what does this do to the spacing of the fringes?

5. A narrow stream of 100-ev electrons is fired at two parallel slits very close together. The distance between the slits is estimated to be 10 angstrom units. The electrons passing through the slits reach a screen 3 meters away, and form a pattern of interference fringes.

[It seems almost impossible to make such a pair of slits in any real sheet of matter, made of atoms which are themselves 1 angstrom or more in diameter, and even more impossible to measure their separation, since light has a wavelength of thousands of angstroms. Yet the equivalent of this pair of slits can be made—has been made and used. It is the "biprism" of Fig. 27–5 (b).]

(a) Estimate the distance between one bright fringe and the next.

(b) The bright fringes would appear as bright marks on a fluorescent screen sufficiently sensitive to glow when bombarded with 100-ev electrons (or as dark marks on a photographic film, or as large pointer readings on some meter attached to an electron collector). What is the essential difference between the bright-fringe electrons and those that reach the screen in a dark fringe? Are they bigger? More massive? Of better quality? Of greater charge, or what?

6. In a certain type of vacuum tube, electrons escaping from the hot cathode pass through a grid maintained at a potential of −10 volts at a distance of 0.5 mm from the cathode, then hit a plate maintained at a potential of +110 volts at a distance of 2.0 mm from the cathode.

   Must the wave behavior of the electrons be taken into account in describing their passage from the cathode to the plate? Explain.

7. (a) What is the wavelength of X rays whose photons each carry 40,000 electron volts of energy?

   (b) At about what energy do electrons have a de Broglie wavelength equal to that of 40,000-volt X rays? (Give your answer in electron volts.)

   (c) At what energy do baseballs have that wavelength?

   (d) What is the wavelength of a baseball moving at 10 meters per second?

8. The sizes of atomic nuclei are hard to measure with alpha particles, because we have to look for small differences from Coulomb scattering. Neutrons are often used.

   (a) What are the de Broglie wavelengths of neutrons with kinetic energy 10^4 electron volts? 10^6 electron volts? 10^8 electron volts?

   (b) Compare these wavelengths with the diameter of a gold nucleus (15×10^{-15} m).

   (c) Look back at Fig. 27–6 to see how waves behave when they meet an obstacle. Suppose we fire a stream of neutrons at a small target of gold. For which of the neutron waves in (a) do you expect to find

      (i) most of the neutrons pass by the nuclei undisturbed while some are absorbed and some bounce off, all like balls in Newtonian mechanics?

      (ii) most of the neutrons pass the nucleus apparently undisturbed and the rest come out of the gold target equally in all directions, regardless of the shape of the nucleus?

   (d) With neutrons of which energy would you prefer to try measuring the nuclear size?

9. The relation between photon energy in electron volts and the wavelength of light in angstroms is shown in Section 25–3.

   (a) Work out a similar relation between the kinetic energy of a particle and its de Broglie wavelength.

   (b) If the particle is an electron and its kinetic energy is generated by accelerating it through a potential difference, relate the potential difference in volts to the de Broglie wavelength of the electron in angstroms.

   (c) What is the energy of electrons whose wavelength is the same as that of yellow light?

10. Borrow a cymbal from the school band (or use your own). Hold the cymbal with the handle down and strike the rim with your finger, or a small rubber hammer. With a little practice, you should be able to form standing waves along the rim. Describe them.

11. Waves travel at 2 meters/sec in a spring fastened firmly at both ends to rigid walls 8 m apart.

   (a) What frequencies of vibration could be found in standing waves formed on this spring?

   (b) Could you produce standing waves corresponding to all these frequencies by shaking the spring at its midpoint?

12.* When a length d is chosen for the "box" discussed in Section 27–5, does this restrict the particle in the "box" to only one fixed energy? (Section 5.)

13.* What is the order of magnitude of the kinetic energy of an electron (mass 10^{-30} kg) bouncing back and forth in a "box" of length 1 A. (Section 5.)

14. (a) Estimate the kinetic energy of a neutron in a nucleus of radius

$$R = 6 \times 10^{-15} \text{ m.}$$

(This is a big nucleus.) Assuming the nucleus is a box of length R will be good enough.
(b) If the nucleus is smaller, should the kinetic energy be smaller?

15. Suppose that an electron is confined within a small cubic crystal about 1 micron (10^{-6} m) on a side. Would quantum effects be detectable if you observed changes in the energy of the electron? (Hint: Calculate the lowest possible energy, considering the electron as a standing wave within the crystal.)

16. Suppose the particle between fixed walls (Section 27–5) is charged and can, therefore, radiate. What is the frequency of its radiation as it goes from the first excited state to the ground state? Express your answer in terms of h, m, and d.

17. Suppose an electron moves so as to remain always in one plane within a square of side length d.
(a) How would its minimum energy compare with that of an electron confined to a cubical box of edge length d?
(b) Compare the wavelengths of the radiation emitted when each of these electrons makes a transition from its first excited state to its ground state.

18. Suppose the particle described in Section 27–6 were confined in a rectangular box whose dimensions were d, d, and $2d$. How would the minimum energy of the system compare with that when it is confined to a cube of side d?

Further Reading

These suggestions are not exhaustive but are limited to works that have been found especially useful and at the same time generally available.

HOFFMANN, BANESH, *Strange Story of the Quantum.* Dover, 1959.

HOLTON, G., and ROLLER, D. H. D., *Foundations of Modern Physical Science.* Addison-Wesley, 1958. (Chapters 31 and 32)

HUGHES, DONALD J., *The Neutron Story.* Doubleday Anchor, 1959: Science Study Series. (Chapters 3 and 5)

WEISSKOPF, VICTOR F., *Knowledge and Wonder.* Doubleday Anchor, 1963: Science Study Series.

WILSON, ROBERT R., and LITTAUER, RAPHAEL, *Accelerators.* Doubleday Anchor, 1960: Science Study Series. (Chapter 3)

Appendix 1

HISTORY and ACKNOWLEDGMENTS

The Development of This Book

This book and the coordinated development of laboratory materials, teachers' resources, and films are the cooperative work of many people. Any brief summary of its development is bound to be unsatisfactory. Even the basic outline and aims were formulated by a large number of people at several centers, and teachers all over the country shared the desire to make a fundamental change in the presentation of physics for beginning students.

The present book is the result of innumerable individual contributions, of extensive trials in many schools, and of a process of revision over three years. To assign detailed credit for the creation and formulation among the hundreds of contributions is impossible. Nevertheless, as the person who took the responsibility for the selection that now appears, I should like to sketch some of the stages. Above all, both on behalf of the PSSC and also personally, I wish to thank my many co-workers for their part in the difficult but pleasurable experience of working together to bring forth a new physics course.

During the fall of 1956 and the winter of 1957, under the leadership of the PSSC Steering Committee, research physicists and physics teachers—often they are the same people—outlined, drafted, and discussed many of the ideas that now appear in this book. Then at Massachusetts Institute of Technology during the summer of 1957 some 60 physicists, teachers, apparatus designers, writers, artists, and other specialists pooled their knowledge and experience to produce a pilot model of the PSSC physics course.

In common with every section of the course, Part I of this book benefited from the work and the discussions of the group as a whole. In particular, it arose largely from the initial discussions of a group at Cornell University, especially Professors K. I. Greisen, Philip Morrison, and Hans A. Bethe. The job of making a complete first draft was carried out during the summer by Professor Morrison, with the help of George L. Carr now of Milford Mill High School, Baltimore, Maryland, and John Marean of Reno High School, Reno, Nevada. With slight rewriting, this draft, edited by Judson B. Cross of Phillips Exeter Academy, Curtis Hinckley of Woodstock Country School and me, formed the basis of the first school tryouts.

From the beginning, work in school has been a major part of the program. The teachers have been among the authors. To keep in touch, there has been an extensive program of school visiting and meetings between teachers and editors. Every part of the book has been improved by this process of testing and revision. For example, several sections of Part I have been revised three times. On the whole, the revisions, again tested in school, have proved their worth, and at the same time an astonishing amount of the original conception remains basically unchanged. We are particularly grateful for the efforts of the teachers who tried the earliest versions and who spent large amounts of time analyzing their experience. By now we have benefited by the experience of over 600 teachers and innumerable students. Their impressions and suggestions were collected and digested by a feedback team. Professor Gilbert Finlay of the College of Education at the University of Illinois has been in charge of this part of the job, and has contributed to the whole project in many ways.

In revising Part I at various times the special work of the group at the University of Illinois, that of Walter Michels at Bryn Mawr, and Sherman Frankel at the University of Pennsylvania was combined with more contributions from Professor Morrison and with the work of the staff at the central office of PSSC. For the present edition, further revision based on more experience in school has been provided by Malcolm K. Smith, Thomas Dillon of Concord High School, Concord, Massachusetts, Professors Eric M. Rogers of Princeton University, Nathaniel H. Frank of MIT, and me.

Most of the discussions and the preliminary development of apparatus that led to Part II of this book took place at the Massachusetts Institute of Technology. There, with the aid of Professor Walter Michels and Elbert P. Little, later assistant to the president of Educational Services Incorporated, I prepared an extensive outline. Following this, the first half of the volume was drafted by Professor Michels and Charles Smith of the Radnor High School, Radnor, Pennsylvania; the second half by Professors Uri Haber-Schaim and Arthur Kerman of MIT with Richard Jones of Indian Springs School, Helena, Alabama, and Darrel Tomer of Hanford High School, Hanford, California. Judson B. Cross and I edited the preliminary edition. Since that time it has had a history of tryout and revision similar to that of all the other parts. In particular, revisions for the present edition were carried out by Professor Haber-Schaim, Malcolm K. Smith, Professor Kerman, and me.

Most of the preliminary discussions leading to Part III took place in a group working at the University of Illinois. Then at MIT in the summer of 1957, Professors E. L. Goldwasser, Peter Axel, David Lazarus, Leon Cooper, and Allen C. Odian of the Illinois group worked with Thomas J. Dillon, Richard G. Marden of Worcester Classical High School, and John H. Walters of Browne and Nichols School to produce the first complete draft. Later Part III was considerably redrafted at PSSC to take advantage of the first draft, of suggestions from many physicists and teachers, and of new laboratory developments carried out by the PSSC staff in Cambridge. In this process Professor Bruno B. Rossi, Professor Frank, and I (all of MIT), were joined by Professor Eric M. Rogers and Malcolm K. Smith. So many members of the PSSC staff, both at MIT and the University of Illinois, contributed that it is even less possible here than elsewhere to give a fair impression of their joint efforts.

The first half of Part IV, dealing with electricity and magnetism, was originally drafted by Professor Rossi with the help of Alexander Joseph of the Bronx Community College, Bronx, New York, Thaddeus P. Sadowski of North Quincy High School, Quincy, Massachusetts, and Edwin Smith, Withrow High School, Cincinnati, Ohio. Also, starting in the summer of 1957, Professors Herman Feshbach and Roy Weinstein of MIT worked on the job of bringing as much modern atomic physics as possible from the last years of college physics to within reach of beginning students. Following after this and other preliminary work, Professor Morrison, Professor Rossi, and I laid out most of the present structure of Part IV, and drafts of the text were produced mainly by Rossi and Morrison. Malcolm K. Smith, Professor Rogers, Professor Frank, and I have been responsible for bringing the comments of teachers to bear on these drafts and for such rewriting and editing as has resulted in the present edition. Here we must particularly thank Professor James H. Smith of the University of Illinois, who tried some of these materials in preliminary form and pointed out the advantage of one radical change in order. We are also especially indebted to Richard Brinckerhoff of Phillips Exeter Academy for detailed commentary not only in this part but throughout the text, and to David A. Page of the University of Illinois, whose careful classroom observations were of great value especially in improving the earlier parts. To their names a great number should be added if space permitted.

In this kind of sketch, it is clear that many people are left out, especially if there is no one thing to which their names are obviously attached. Such omissions are painful: often enough, wise advice has been as valuable as an identifiable draft of text. Along these lines, then, I would like to mention the overall contributions of many others. Professor I. Bernard Cohen of Harvard University has read successive drafts and supplied historical materials. Stephen White, whose main concern has been with the PSSC films, has helped with other jobs from time to time since the beginning of the Committee's work. Paul Brandwein of Harcourt, Brace & Company and the Conservation Foundation has supplied detailed criticism, encouragement, and aid at many times. For instance, along with George H. Waltz, Jr., he helped to set up the effective system which turned authors' scrawls into respectable preliminary editions. In this process we benefited from the editorial experience of Judy Meyer and Lee Wertheim. Judson B. Cross and Malcolm Smith acted as executive editors of the preliminary volumes. For three years their work and mine has been lightened and made effective by the efforts of Benjamin T. Richards, who supervised the production of the texts for schools. Over the last year we have been joined by Richard T. Wareham, of D. C. Heath and Company, who smoothed the transition into this edition.

In this book the illustrations are essential. They were created by cooperative work between the illustrators, photographers, and physicists. Peter Robinson and Percy Lund have worked hard to make their illustrations meet our needs. James Strickland and Berenice Abbott have worked together to produce many of the excellent photographs; others were taken by Charles Smith, Ben Diver, Phokion Karas, Robin Hartshorne, Paul Larkin, and (in Part III) by Professor Chalmers Sherwin and Louis Koester of the University of Illinois. Special film clips were supplied by the Educational Services Incorporated film studio, where work on the related films has gone on in close collaboration.

The preliminary editions and the classroom work of teachers were essential to the process of developing this course. In utilizing them to bring about further improvements, the work of Gilbert Finlay and his co-workers, who stayed in close touch with teachers, was supplemented by information gathered directly from testing of students. Here Walter Michels made another of his many contributions. With Frederick L. Ferris, Jr., of Educational Testing Services and a group of physicists and teachers from the region around Philadelphia, he is responsible for a large set of standardized tests from which we learned much. Professor Hulsizer of the University of Illinois provided a teacher's guide to the test answers; and he and several others at Illinois and MIT have helped with test revisions. The Illinois group has taken the major part in development of detailed resource books for teachers, and in this they have been joined by the PSSC staff in Cambridge, who (with the usual aid from all sources) worked out both the student laboratory and the laboratory parts of the teacher's materials.

It is impossible for me to express my thanks adequately to those people whose continuing wisdom has been my chief reliance. The trouble—and pleasure—is that there are too many. Professor J. R. Zacharias has restrained himself with admirable control when, as often happened, the text or laboratory was not sufficiently settled to ease his job of making an accompanying film. He has encouraged us with his confidence despite the inevitable problems that arose in some of the new developments. Professors Morrison, Rogers, Frank, and Rossi have been among the major creators and selectors of material at every point.

The whole process has been a continuing cycle and I can only stop where I began, by acknowledging our indebtedness to more people than I could make clear with ten times this space.

Francis Lee Friedman

September 1960

The second edition, like its predecessor, is the fruit of the thought and effort of many people. It is impossible even to name the many persons—college and university professors, high school teachers, and high school students—who have contributed helpful criticisms and suggestions for revising the course. The members of the PSSC Steering Committee have taken an active role in evaluating these suggestions, generating new ideas, and approving the changes made for this edition. They are Judson B. Cross, Educational Services Incorporated; Frederick L. Ferris, Jr., Princeton University; Dr. Uri Haber-Schaim, Educational Services Incorporated; Ervin H. Hoffart, Educational Services Incorporated; Professor Robert I. Hulsizer, Jr., Massachusetts Institute of Technology; Professor Aaron Lemonick, Princeton University; Professor Philip Morrison, Massachusetts Institute of Technology; Professor Melba Phillips, University of Chicago; Professor Byron L. Youtz, Reed College; and Professor Jerrold R. Zacharias, Massachusetts Institute of Technology.

The complete rewriting of Chapter 31 and considerable revision of Chapter 30 were done by Professor Phillips and Professor Stefan Machlup of Western Reserve University. Helpful criticisms of their draft chapters and additional suggestions were contributed by Professors Anthony French of Massachusetts Institute of Technology and Arnold Arons of Amherst College.

Professor Philip Morrison was most helpful in reviewing the proposed changes for Part I of the text; he also contributed the essay on molecules and chemical structures at the end of Chapter 8. Professor Stephen Kline of Stanford University made many useful criticisms and suggestions on the treatment of the subject of heat and the conservation of energy.

The development by PSSC staff members of an effective, low-cost Millikan apparatus now allows the student to find his own evidence for the existence of the elementary charge instead relying entirely on the film. Relevant portions of the text have been rewritten by Dr. Uri Haber-Schaim to take account of this new laboratory development.

The final choice of material for this edition was, of course, my responsibility.

In addition to Dr. Haber-Schaim, who has been a constant source of advice in all phases of this revision, I am indebted to many other members of the PSSC staff. Ervin H. Hoffart has collected and compiled feedback for the past six years and has distilled it into a form useful for the revision process. Judson B. Cross has again offered useful editorial advice on several chapters. Over the past several years a group of staff members have generated and tested new Home, Desk, and Lab problems. James Walter was particularly helpful in selecting and organizing the material for that all-important part of the text.

Paul Larkin has guided the many revisions of art work necessitated by pedagogic or artistic considerations. In this he was assisted by George Cope and his staff in the photographic studio of Educational Services Incorporated and by Dr. James Strickland of the move studio, which generously supplied frames from existing films. John DeRoy of the PSSC shop lent his skill to setting up experiments and demonstrations for photographing. The lists of "Further Reading" at the ends of chapters have been expanded and brought up to date by a team of experienced high school teachers under the direction of Bruce Kingsbury.

As with all previous versions of this work, Benjamin T. Richards has assumed responsibility as production editor. For cooperation and assistance we are indebted to Louis Vogel and other members of the editorial and art departments of D. C. Heath and Company. Finally, I wish to acknowledge my indebtedness to my editorial assistant, Andrea Julian, whose sharp eye and understanding of physics have detected many lapses in consistency and clarity.

The fact that this course has been so widely and successfully used and has required so little real change in the revision is the highest possible tribute to the judgment and taste and insight of the man who, above all others, was responsible for the first edition: the late Francis Lee Friedman. It is my earnest hope that this edition has retained the spirit and the style which he established, while at the same time reflecting the light of experience and showing a sensitivity to the valid criticisms which we have received.

Byron L. Youtz

August 1965

In planning the revisions for the third edition we benefited from general comments of many teachers. In the preparation of the manuscript we wish to acknowledge in particular the help of Mrs. Bette LeFeber and Edward A. Shore in drafting answers to "daggered" problems, and to Dr. Poul Thomsen for valuable suggestions. Stephen McKaughan prepared the apparatus for photographing. The new photographs were taken by George Cope and Victor Stokes.

As with all previous editions of this work, Benjamin T. Richards has given most valuable support as production editor. We enjoyed the full cooperation of Bruce F. Kingsbury and other members of the editorial and art departments of D.C. Heath and Company.

Uri Haber-Schaim

Judson B. Cross

John H. Dodge

James A. Walter

August 1970

Appendix 2

Answers to problems marked with a dagger (†)

Chapter 1, Problem 9

Begin with a drawing. Imagine yourself looking down on the North Pole of the earth; you would see the earth moving counterclockwise underneath you. Show the direction of the sunlight, and locate the satellite above the dark side of the earth, where it can just be illuminated by sunlight grazing the earth's surface. Draw lines from the center of the earth to the satellite and to the point of contact of the grazing sunlight (Fig. 1).

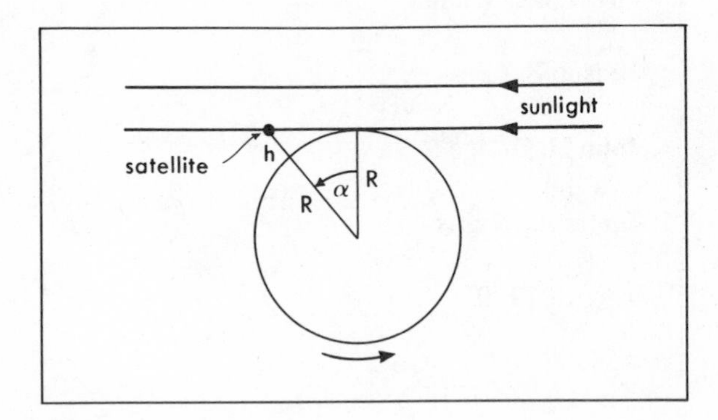

Figure 1

At a particular time after sunset, the earth will have turned through an angle α; from the diagram you see that $\cos\alpha = \dfrac{R}{R+h}$.

Since you are trying to find h, it will be necessary to rearrange this equation. It is best to solve an equation for the unknown in terms of letters before substituting the numerical values. In the above equation multiply both sides of the equation by $(R+h)$. You then have

$$(R + h)\cos\alpha = R.$$

Multiplying out and subtracting $(R\cos\alpha)$ from each side gives

$$h\cos\alpha = R - R\cos\alpha.$$

Dividing by $(\cos\alpha)$ and factoring gives

$$h = R\left(\frac{1}{\cos\alpha} - 1\right).$$

You are now ready to substitute the values for R and α to find h. In two hours,

$$\alpha = \frac{2}{24} \times 360 = 30°,$$

R is given in the problem as 6.38×10^6 meters.

$$\text{Hence } h = 6.38 \times 10^6\left(\frac{1}{\cos 30°} - 1\right)$$

$$= 6.38 \times 10^6\left(\frac{1}{0.866} - 1\right)$$

$$= 9.9 \times 10^5 \text{ m}.$$

Chapter 2, Problem 12

Start with a sketch so that you can visualize the situation. It does not have to be exact, but try to make it roughly to scale. Show the mirror with its axis and principal focus F at 20 cm. Next locate the position of the nail in front of the mirror 15 cm from the principal focus. Actually there are two points, A and B, where the nail can be 15 cm from the focal point and still be in front of the mirror. Mark these on the diagram (Fig. 2).

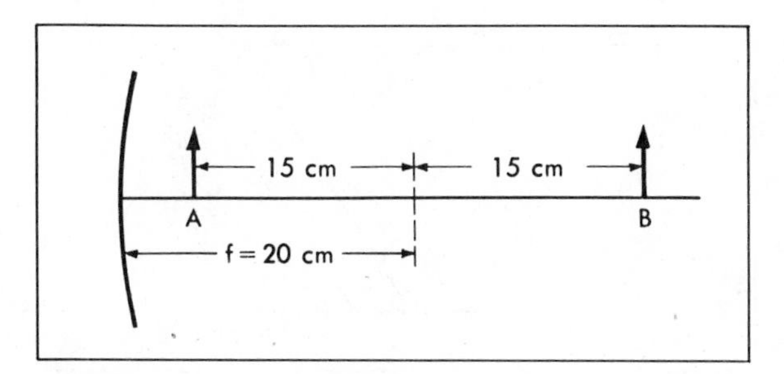

Figure 2

You can now draw ray diagrams to find the location and height of the image H_i for each of the two positions of the nail, or refer to Sections 2–5 and 2-6, where this is done, and apply the result:

$$\frac{H_i}{H_o} = \frac{f}{S_o}$$

or

$$H_i = \frac{f}{S_o} H_o = \frac{20 \text{ cm} \times 4.0 \text{ cm}}{15 \text{ cm}} = 5.3 \text{ cm}.$$

Chapter 3, Problem 24

Begin by drawing a diagram showing a ray of light passing up through the liquid, through the surface, and out into the air (Fig. 3). Since the index of refraction of carbon disulfide is larger than that of air, the light must bend away from the normal N as it leaves the liquid. At the same time some light is reflected internally back down into the liquid. What happens at larger angles? At angles of incidence i larger than the critical angle θ, you know that the light will cease to get out of the liquid; it will all be reflected (Fig. 4). Therefore, only the light within the cone ASC in Fig. 5 will escape.

This cone will form a circle on the liquid surface

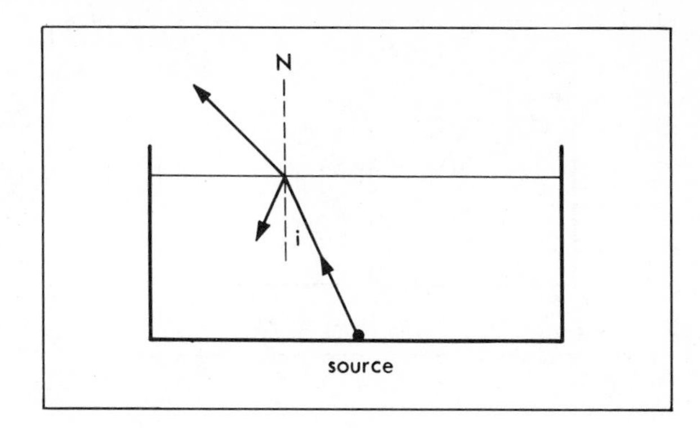

Figure 3

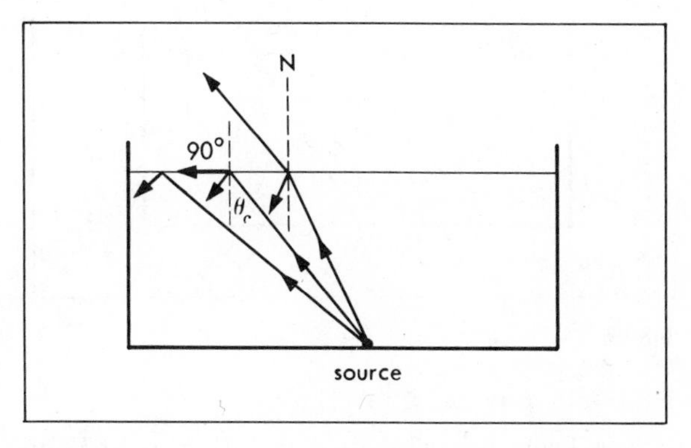

Figure 4

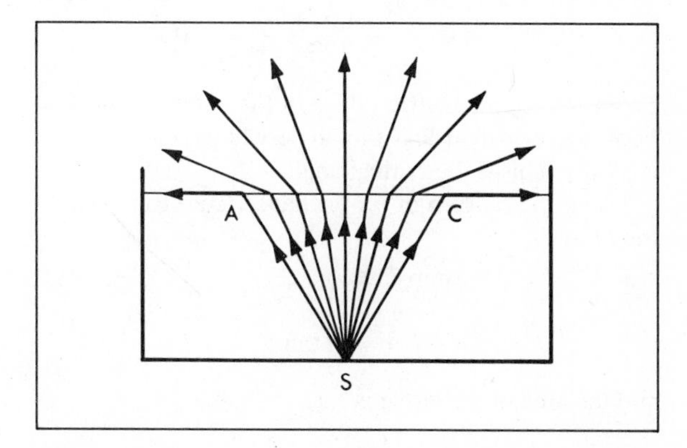

Figure 5

whose radius r is AP and whose area you wish to find (Fig. 6). To find r you need to know the tangent of the critical angle θ_c. You can find θ_c from Snell's law, $n_1 \sin \theta_1 = n_2 \sin \theta_2$, for the special case in which θ_1 is the critical angle and θ_2 is 90° (see Fig. 7).

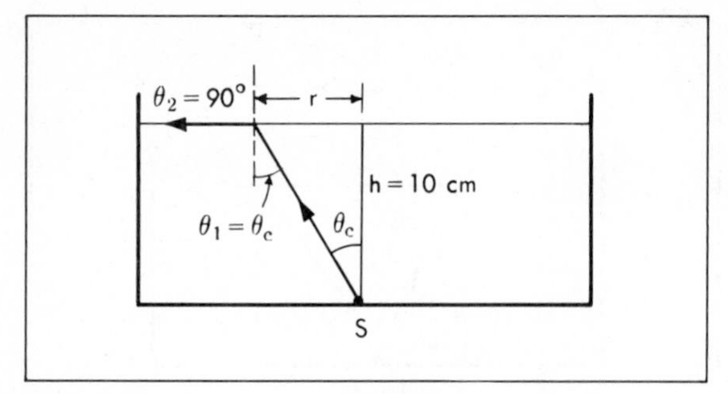

Figure 6

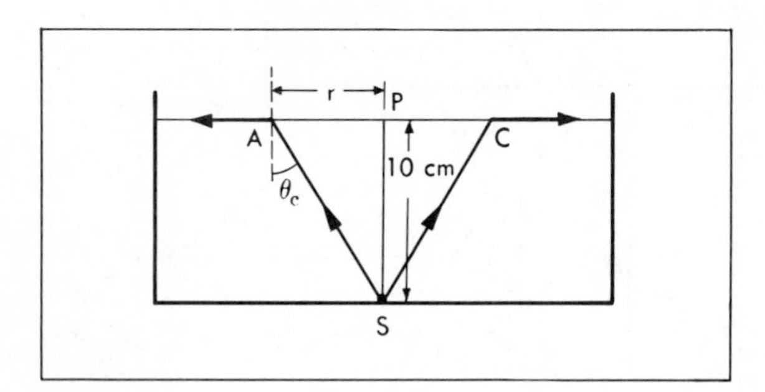

Figure 7

You know from the question that $n_1 = 1.63$, so

$$1.63 \sin \theta_c = 1.00 \sin 90°$$

$$\sin \theta_c = \frac{1.00 \times 1.00}{1.63} = 0.613.$$

From the trigonometric tables in the appendix of this book you will find that the angle whose sine is 0.613 is 37.8°. This is the critical angle.

Using the tangent of θ_c, we have (from Fig. 6) for the radius r

$$\tan \theta_c = \frac{r}{h}$$

$$r = h \tan \theta_c$$

and the area of the circle is

$$A = \pi r^2 = \pi(h \tan \theta_c)^2$$

$$= 3.14(10.0 \text{ cm} \times 0.776)^2 = 189 \text{ cm}^2.$$

Chapter 3, Problem 26

First search for applications among the situations you have so far studied: mirrors, both plane and curved; prisms, both rectangular and triangular; and lenses.

Here pencils of light are parallel in the original beam and remain so in each emergent beam. Curved mirrors and lenses seem to be ruled out. Since the beam is monochromatic, we are not concerned with dispersion and can retain prisms for consideration.

Let us examine case (d) to see how to go about solving this kind of problem. Could a pair of mirrors have changed the direction of the beam as shown? No; you can see that the pencils have not been interchanged right for left, as is the case in reflection from a plane surface (Fig. 8). What device can alter the

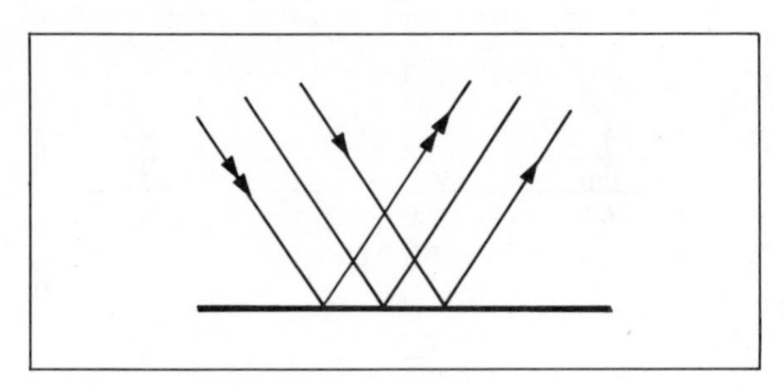

Figure 8

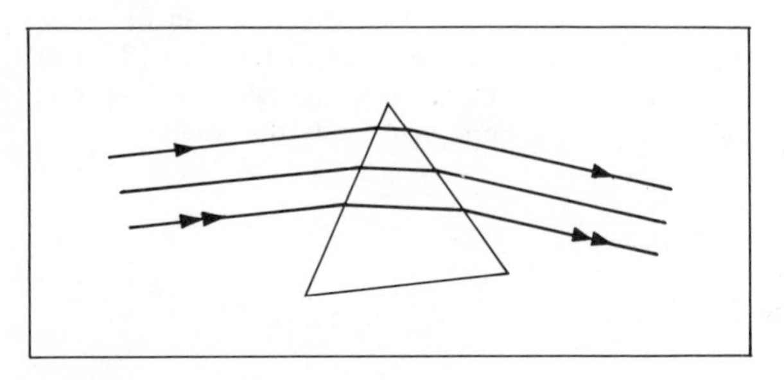

Figure 9

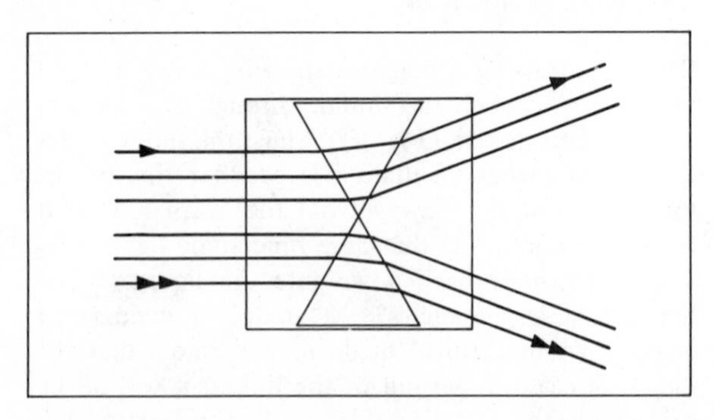

Figure 10

direction of a beam without interchanging the light pencils? What happens in a prism? You can see in Fig. 9 that it is possible to have a beam strike one surface of a prism so as to pass through it without interchanging the pencils of the beam.

Now you can visualize a possible model for this case: two prisms placed point to point (Fig. 10).

cording to the inverse-square law. This would mean that the particles gradually changed size as they moved away from a light source.

You can only conclude that the intensity of light sources, partial absorption, and the inverse-square law are better explained by changing numbers of particles than by a change in their size.

Chapter 4, Problem 3

If you want to test an assumption, think of what would have to be true according to the assumption, and then see if this is in agreement with your experience. In this case, can what you know about the behavior of light be explained by different-sized particles of light?

This would mean that two candles emit particles of a given size when they are separate from each other, but particles of another size when they are close together, since the intensity of light is greater when they are together. This does not seem reasonable.

The decrease in intensity caused by inserting a partially absorbing sheet of matter in a light beam would mean that the particles changed size as they went through the sheet. This too seems unreasonable.

You know that the intensity of light decreases ac-

Chapter 4, Problem 21

Even though you could measure the speed of light, there would be a limit to the accuracy with which you could measure it, as there is for any physical measurement. If a decrease in speed on reflection actually took place, the decrease might be so slight that you could not detect it. It would be necessary to amplify the effect before you could hope to measure it. Hence, your experimental procedure would have to use many reflections in succession. A variety of arrangements could be made. For example: you could reflect the light back and forth between two long, parallel mirrors, silvered on their facing surfaces so that the light would not have to pass through any glass (Fig. 11).

You would have to determine the length of the light path through this apparatus, then arrange to measure the speed of light over a path of equal length,

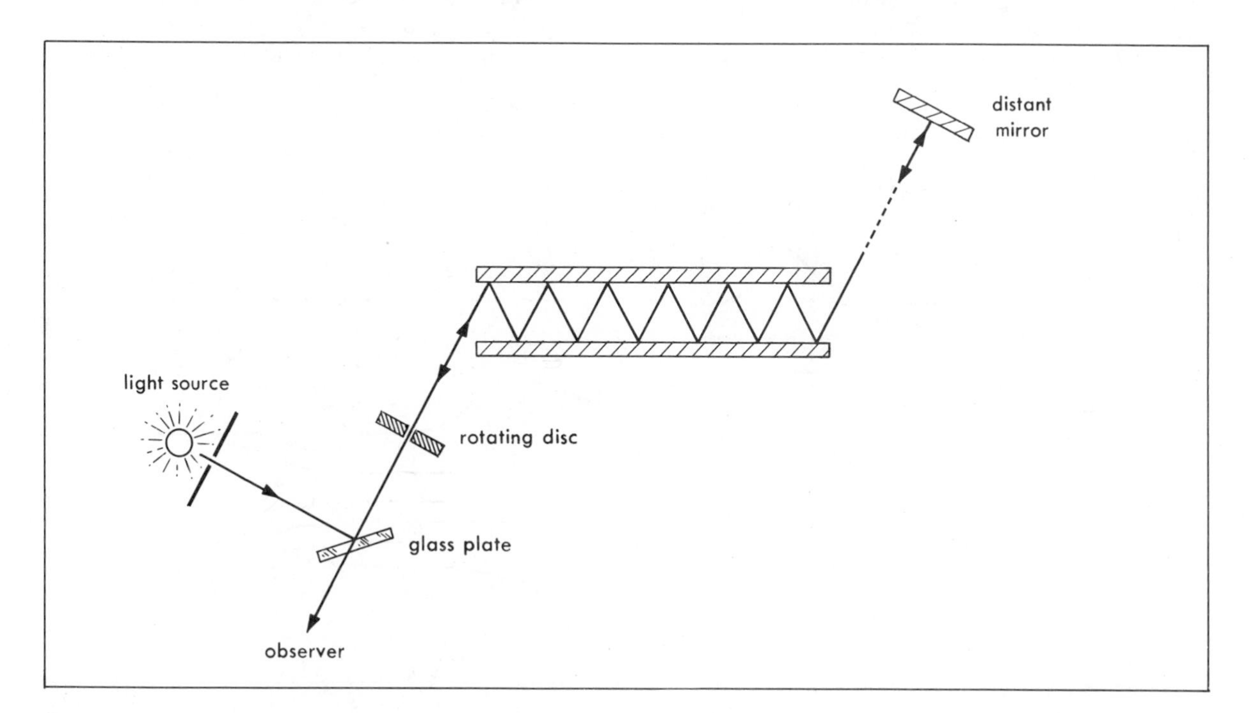

Figure 11

with only the reflection at the distant mirror taking place as shown in Fig. 11.

Could you be sure that the comparison of the two speeds you determined would answer the problem completely? Not necessarily! Nothing has been said about the mechanism by which light is reflected. You could as readily suppose that a slight time delay occurs during each reflection without a change in speed, rather than in a slight change in speed without a time delay. Could you modify the procedure so as to distinguish between these possibilities? Suppose you put the part of the path where the light undergoes many reflections between the source of light and the strobe instead of between the strobe and the distant mirror. How could you use the measurements made in this additional step to determine what happens to the light on reflection?

Chapter 5, Problem 19

In (b) you can see that the pulse in (a) has been partially reflected and partially transmitted. Therefore, there must be a junction in the region between the two pulses. How can you locate the junction exactly? The problem states that there are equal intervals of time between diagrams. The pulse, then, has to move the same distance in the time interval between diagrams (a) and (b) as it does between (b) and (c). Therefore, the distance l'' in Fig. 12 must equal $l' + 2\,l$ and the junction is located at J, a distance $l''/2 - l'2$ from the top of the pulse in (a).

The reflected pulse in (b) is right side up. What does this tell you about the rope to the right of the junction J? If you do not remember, refer to Figs. 5–10 through 5–13. The rope to the right of the junction must be less thick than the rope to the left. The relative thickness could also be determined by noting that the transmitted pulse has a larger speed than the original pulse.

In (c) you will notice that the pulse traveling to the right in (b) has been partially reflected and partially transmitted. Again, this implies a junction between the two pulses. Since there is no diagram (d) at a known later time interval, you cannot find the distance traveled by this reflected pulse in one time interval. Therefore, all you can say is that there is a junction somewhere in the region between P and P'.

In (c) the second reflected pulse is upside down; the second transmitted pulse is longer, and thus has a lower speed, than the transmitted pulse in (b). Therefore, the rope to the right of the junction must be thicker than the rope to the left.

Chapter 5, Problem 23

(a) Think of your experiment with waves on a coil spring. When were you able to neglect the decrease in size of the pulse? You could ignore this decrease

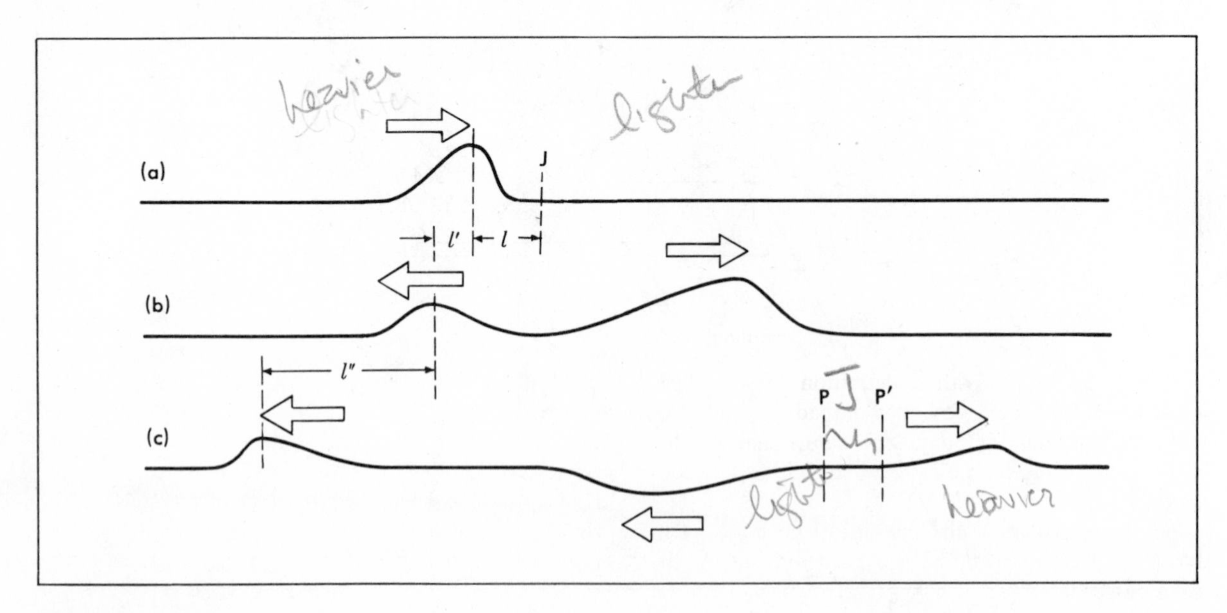

Figure 12

as long as you considered the motion of the pulse only for short periods of time, so that the decrease was hardly noticeable.

(b) The advantage in neglecting the decrease is that it is much easier to describe and predict the behavior of an "ideal" spring than that of a real spring.

If you want to think more about this problem, it would be beneficial to read Section 5–5 again.

Chapter 6, Problem 7

(a) Recall what you have observed thus far in the ripple tank. Within limits, a straight pulse remains straight and a circular pulse remains circular, as they travel. From this you can infer that the speed is everywhere the same, and a few measurements would verify that the speed is constant as long as the depth of the tank does not change.

In this question, when a pulse originating at point A is brought to focus at point B, you know that every part of the pulse reflected from the ellipse has taken the same time to reach B and hence has covered the same distance regardless of the point on the ellipse from which a segment of the pulse was reflected. In Fig. 13, the path APB must have the same length no matter what the position of P on the ellipse.

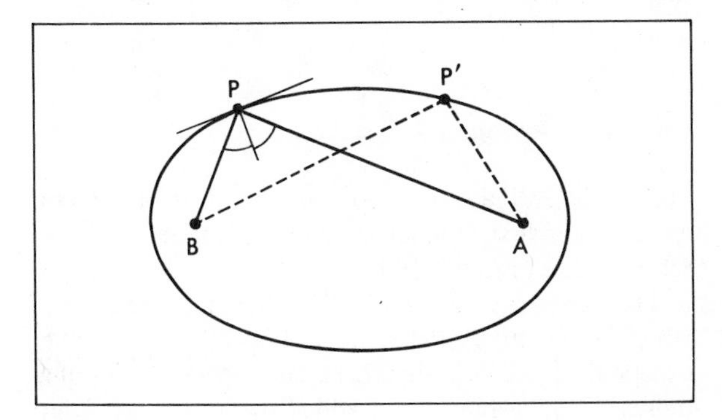

Figure 13

This is in accord with a definition of an ellipse from geometry which you may already know: "An ellipse is the locus of points the sum of whose distances from two fixed points (called the foci) is constant."

For every part of a circular pulse starting at A to arrive at B after reflection, the perpendicular to the tangent to the ellipse at any point P must bisect the angle APB. This is indeed a property of ellipses.

(b) You would expect, from symmetry, that a pulse originating at B would focus at A.

(c) The best way to answer this part of the question is to perform the experiment in the ripple tank. However, if you would like to make a prediction before doing the experiment, you can readily construct an ellipse on paper. You can do this with the aid of a pencil, two pins, and a loop of string. Set the two pins firmly any desired distance apart; pass the loop of string around the pins and the pencil, then move the pencil so as to keep the string taut (Fig. 14). Evidently the pins are the foci of the ellipse. Using the ellipse you have drawn, you can draw several paths from some point different from either focus. Draw the paths so that they obey the laws of reflection, and see if they converge at a single point after reflection.

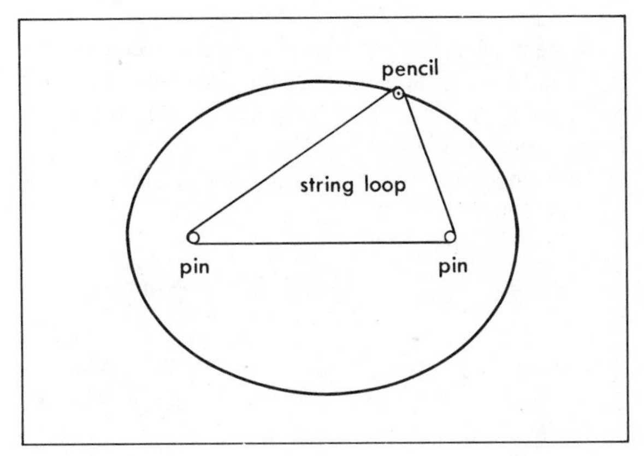

Figure 14

Chapter 6, Problem 24

(a) Begin with a sketch of the situation presented (Fig. 15). Since the speed is smaller in the shallow section, the wave front will bend so that the angle between the wave front and the line dividing the tank is less than 60°. Show the wave crests in each section.

You recall that the stroboscope "stops" waves in both sections of a ripple tank simultaneously, so the frequency is the same in both sections. Hence the time for a segment of the wave front in the deep section to travel the wavelength λ is the same as that for a segment of the wave front in the shallow section to travel the wavelength λ'.

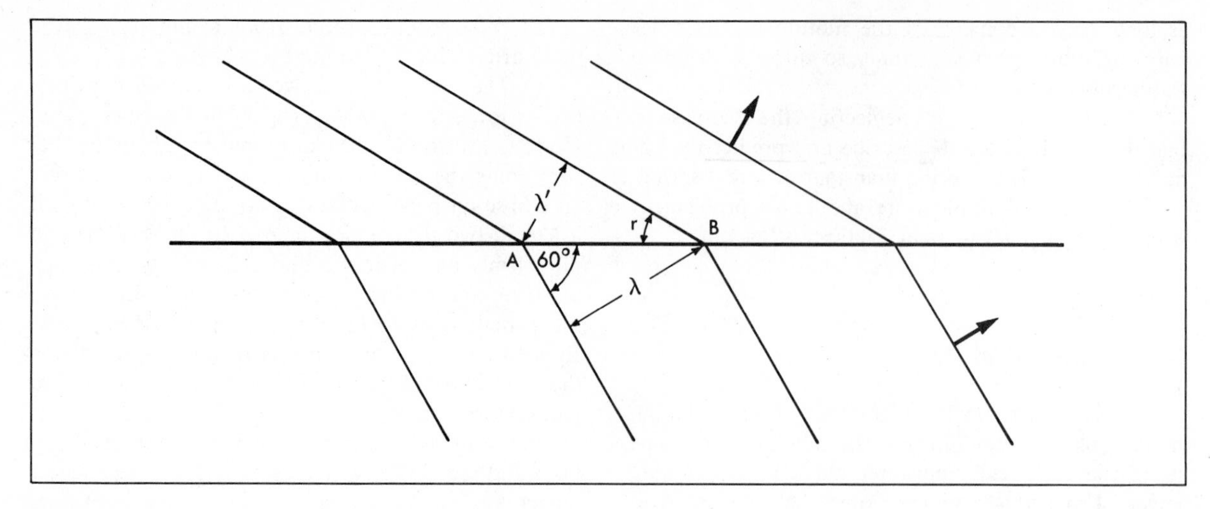

Figure 15

In particular, $\lambda = vt$, where v is the speed of the wave in the deep section and t the time for a segment to advance one wavelength, and $\lambda' = v't$, where v' is the speed of the wave in the shallow section and t is the same time. Now $\lambda' = AB \sin r$, and $\lambda = AB \sin 60°$; hence,

$$AB = \frac{vt}{\sin 60°} = \frac{v't}{\sin r} \text{ and } \sin r = \frac{v'}{v} \sin 60°.$$

Therefore, for case 1, $\sin r = 24/34 \times 0.866 = 0.612$, and $r = 38°$; and for case 2, $\sin r = 24/32 \times 0.866 = 0.649$, and $r = 41°$.

It would be tempting to thumb through the chapter, hunting for a formula that might work; however, this is a good opportunity to see if you can reconstruct some reasoning.

(b) This really is a question for discussion, rather than one that requires a decisive answer. Probably your work with the ripple tank suggests that it is easy to observe qualitatively that an increase in frequency is accompanied by a decrease in index of refraction: for when the frequency is high, you have trouble observing any refraction, whereas when it is low, refraction is easy to observe.

However, a quantitative determination of the small difference in speeds given in the problem would be difficult by any method. There are fewer measurements to take to determine the difference in the angle of refraction, so that if you can keep the rippler frequency constant under each of the two conditions and have the tank adjusted to cast sharp images, you probably will be able to measure the difference between 38° and 41°.

Otherwise, you might try to stop the waves with a stroboscope while measuring the distances across several wavelengths. You would have not only the same problem with steadiness of the rippler and sharpness of the images of the waves, but also the problem of keeping the stroboscope steady and aligned. It would require real skill with a hand stroboscope to make such refined measurements.

(c) With light, angles are easy to measure, whereas measurements of the speed of light are difficult.

Chapter 7, Problem 3

Since this problem asks you to extend an argument found in the text, it would be best to familiarize yourself with this argument first.

Then draw a diagram like that shown in Fig. 16(a) which is essentially the same as Fig. 7–3, in the text, except that two additional, equally spaced incoming pulses 4 and 5 have been added. At a time T later than Fig. 7–3, pulses 1 and 4 will each have moved a distance λ as shown in Fig. 16(b), and will have met at P_2, where they cancel each other. At a time $2T$ after Fig. 16(a), pulses 2 and 5 will meet and cancel at P_2 as shown in Fig. 16(c). Thus, equally spaced identical incident pulses will always superpose at P_2 with reflected pulses to give a node at this point.

(What other points will always be nodes?)

If you answer this question by using the ripple tank, you will need two separate wave generators. Actually doing the experiment will give you the best idea of how the pattern shifts.

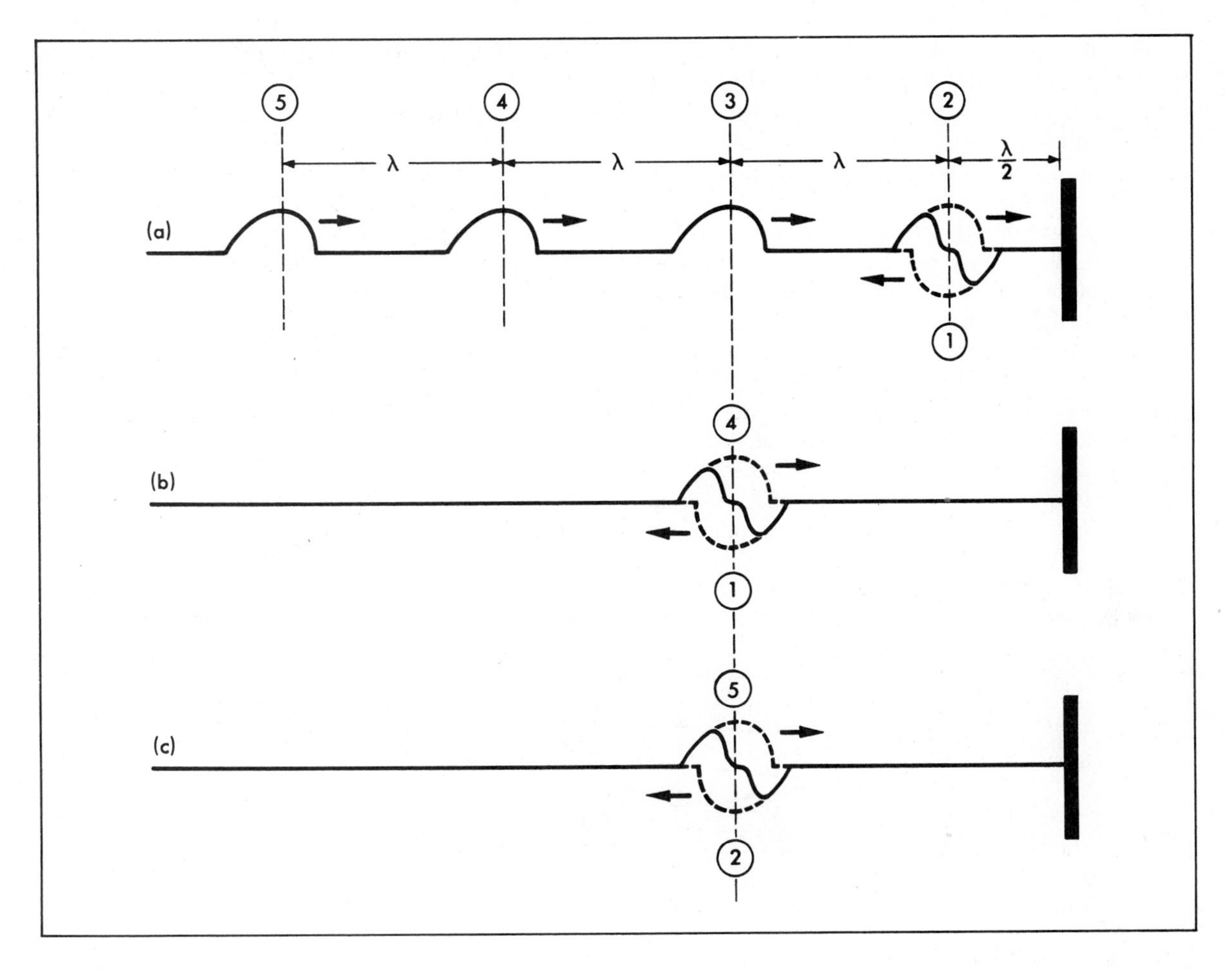

Figure 16

Chapter 7, Problem 24

Because the two sources are not operating at the same frequency, you must first realize that there is a continually changing phase delay which produces a changing interference pattern. You can see how the interference pattern changes either by doing an experiment in the ripple tank or by drawing sets of concentric circles to represent the wave crests from each source.

If you choose to make a drawing, you will first have to decide what scale to use. Since the problem only gives the frequencies of the sources, you are free to choose d and v. (Your choice of v will then determine the wavelengths.) The frequencies are 15 cycles per second and 16 cycles per second. (It will be easier then if your wave velocity is divisible by either 15 or 16.) For example, let $v = 15$ cm/sec and S_1 be 15 cycles per second and S_2 be 16 cycles per second; then $\lambda_1 = 1.0$ cm and $\lambda_2 = 0.94$ cm. Since the number of nodal lines is determined by d/λ, you want d to be large enough so that there will be several lines. In this case, you might let $d = 2.5$ cm.

Since the interference pattern is not constant, you will have to draw at least two different sets of crests from each source to see how the pattern changes with time. When you have done this, you will be able to see how the lines joining the double crests will sweep slowly across the tank toward the side of the lower-frequency source.

Chapter 8, Problem 6

If there is a dark bar near the surface of the plate, then both the direct light and the reflected light must travel very nearly parallel to the plate. Then the path of the reflected light will be very nearly the same as the light path from the source directly to the eye. In fact, if the source is far enough away, the difference in the two path lengths will be much less than the wavelength of the light. Since the light traveling the two paths comes from a common source, you would not expect a node. Hence, you conclude that there must be a phase delay of 1/2 due to the reflection to cause a dark bar close to the plane of the plate.

Why is this? Think back to what happened when you observed pulses traveling from one coiled spring to another of different size. Where the springs were joined, the pulses were partially reflected and partially transmitted. If the transmitted part of the pulse traveled slower than the incident and reflected part of the pulse, the reflected pulse was reflected upside down. This inversion of the reflected pulse corresponds to a phase delay of 1/2 in the case of periodic waves.

Since light waves travel slower in glass than in air, a phase delay of 1/2 is just what we would expect when light reflects from a plate of glass.

(b) The next dark line at P indicates an additional difference in phase of one period due to a difference in path length of λ from the sources. From Fig. 17,

$$\frac{x}{L} = \frac{\lambda}{d}$$

$$d = \frac{\lambda L}{x} = \frac{(5.4 \times 10^{-5})(60 \text{ cm})}{(0.09)} = 0.036 \text{ cm}$$

and the source is $d/2 = 0.018$ cm above the plane of the reflecting surface.

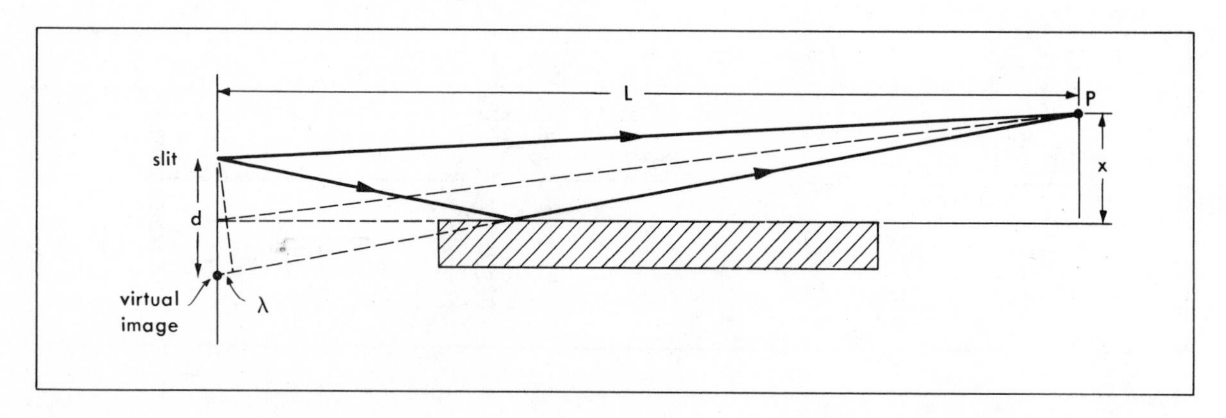

Figure 17

Chapter 9, Problem 16

(a) Since the velocity-time graphs for both cars are drawn on the same set of axes, you can see that the two cars will be moving with the same velocity where the two graphs intersect. The graph shows that this occurs at about 0.0065 hours.

(b) You have learned that the "area" under a velocity-time graph gives the displacement of an object, provided the units along the two axes are consistent. For example, if the vertical axis is marked off in units of meters/second, the horizontal axis should be marked off in units of seconds; then *meters/second* multiplied by *seconds* gives *meters*, a correct unit of displacement. You can apply this principle to Fig. 9–3. At the time in question, the velocity of each car is 40 mi/hr. The "area" under the graph of car B is a rectangle with base 0.0065 hours and altitude 40 mi/hr; hence the displacement of car B is 0.0065 hr $\times$ 40 mi/hr $=$ 0.26 mi. The area under the graph of car A is a triangle with base 0.0065 hr and altitude 40 mi/hr; hence the displacement of car A is $\frac{1}{2} \times$ 0.0065 hr $\times$ 40 mi/hr $=$ 0.13 mi. Hence, car B is 0.26 mi $-$ 0.13 mi $=$ 0.13 mi ahead.

(c) Again, you apply the same method as in part (b). The displacement of car B is represented by the rectangular area under its graph, whose base is 0.010 hr and whose altitude is 40 mi/hr. So the displacement of B is 0.010 hr $\times$ 40 mi/hr $=$ 0.40 mi. The displacement of car A is represented by the triangular area under its graph, whose base is 0.010 hr and whose altitude is now 60 mi/hr. So the displacement of A is ½ × 0.010 hr × 60 mi/hr = 0.30 mi. Hence car B is ahead by (0.40 − 0.30) mi = 0.10 mi.

(d) Obviously, car A will have caught up with car B when they are both at the same distance from the traffic light; so you can translate the question to read, "At what time will the areas under the graphs for A and B be equal?" There are a variety of ways of managing this. You know from part (c) that the time of this event will be greater than 0.010 hr. Perhaps the easiest way is to estimate the answer directly from the graph as shown in Fig. 18, where areas $a' = a$, $b' = b$, and $c' = c$.

Another way to solve this part of the problem is to first write down an equation for the displacements x'_A and x'_B of the cars. Taking the initial time t as 0.01 hr, we have for the cars moving at constant velocity after $t = 0.01$ hr

$$x'_A = x_A + v_A(t' - t)$$
$$x'_B = x_B + v_B(t' - t).$$

Since $x'_A = x'_B$ when car A catches up,

$$x_A + v_A(t' - t) = s_B + v_B(t' - t)$$
$$(v_A - v_B)(t' - t) = x_B - x_A$$
$$t' - t = \frac{x_B - x_A}{v_A - v_B}$$
$$t' = t + \frac{x_B - x_A}{v_A - v_B}.$$

From the graph, $v_A - v_B = 60 - 40 = 20$ mi/hr, and from part (c) $x_B - x_A = 0.13$ mi, so

$$t' = 0.01 \text{ hr} + \frac{0.10 \text{ mi}}{20 \text{ mi/hr}} = 0.015 \text{ hr}.$$

(e) It is most convenient to calculate the distance traveled by car B, as its speed is always constant. The total distance it travels from the traffic light is

$$x_B = vt$$
$$x_B = 40 \text{ mi/hr} \times 0.015 \text{ hr}$$
$$= 0.60 \text{ mi}.$$

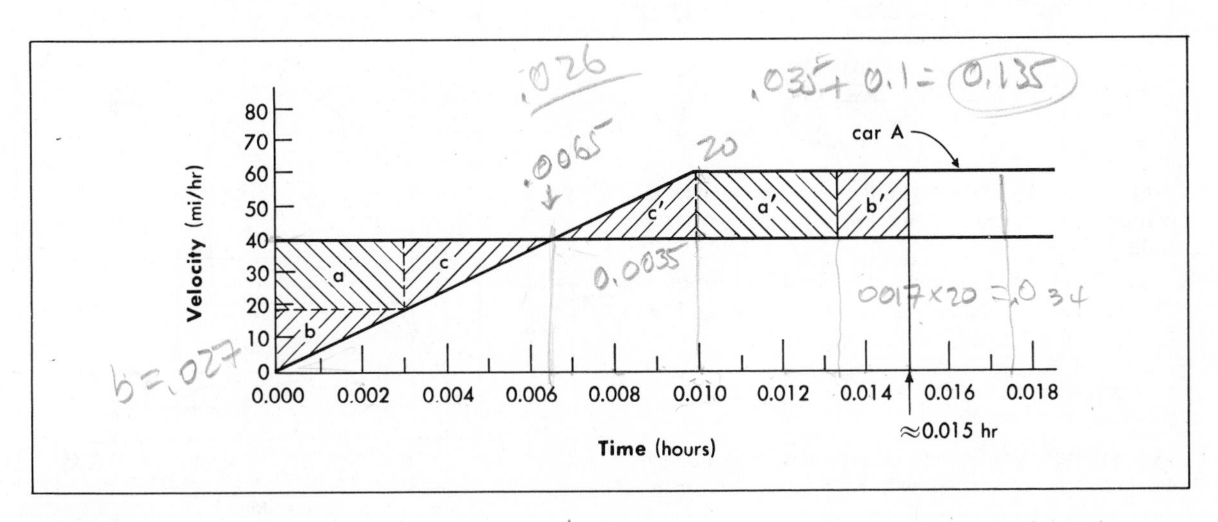

Figure 18

Chapter 9, Problem 28

(a) This problem involves uniform acceleration. Therefore, you may start from the general equation:

$$x' = x + v(t' - t) + \tfrac{1}{2}a(t' - t)^2.$$

For the express train, the final displacement $x' = 2$ miles; the initial displacement, $x = 0$; the initial velocity $v = 70$ miles per hour; and the initial time, $t = 0$. You are trying to find t', the time required for the train to stop. First you can simplify the equation by setting $x = 0$ at $t = 0$ and then eliminating the terms which equal zero:

$$x' = vt' + \tfrac{1}{2}at'^2. \tag{1}$$

Notice that in this equation you have two unknown quantities: t' and a. However, the acceleration may be expressed in terms of the initial velocity and t'

$$a = \frac{v' - v}{t' - t} = \frac{-v}{t'}. \qquad (2)$$

Substitute this value for the acceleration into equation (1):

$$x' = vt' + \tfrac{1}{2}\left(-\frac{v}{t'}\right)t'^2$$
$$= vt' - \tfrac{1}{2}vt'$$
$$= \tfrac{1}{2}vt'. \qquad (3)$$

Solve for the unknown, t'

$$t' = \frac{2x'}{v}$$

and substitute the values for x' and v'

$$t' = \frac{2(2 \text{ mi})}{70 \text{ mi/hr}} = 0.057 \text{ hr}.$$

(b) You can find the acceleration from equation (2).

$$a = -\frac{v}{t'} = \frac{-70 \text{ mi/hr}}{0.057 \text{ hr}}$$
$$a = -1.23 \times 10^3 \text{mi/hr}^2.$$

What is the significance of the negative sign? This confirms what you already knew, namely, that the train is slowing down.

(c) Equation (1) gives the displacement of the express train; substituting $v = 70$ mi/hr and $a = -1230$ mi/hr^2 gives

$$x'_e = 70 \text{ mi/hr} \times t' - \frac{1230 \text{ mi/hr}^2 \times t'^2}{2}.$$

You will not be interested in any time after 0.057 hr, when the express train will have stopped, or in a displacement greater than 2 miles, the stopping distance of the express. These values determine the limits of the axes on your graph.

Values for x'_e for the different values of t' can be calculated and are shown in the following table.

To find the displacement of the freight train as a function of the time, you note first that all displacements are measured from the express train's initial position. Since the freight is initially a mile down the track from this position, its initial displacement x is not zero. Therefore you use the equation for motion at constant velocity.

$$x'_f = x + vt'$$

t'	$70t'$	$-615t'^2$	x'_e
0	0	0	0
0.01	0.7	−0.06	0.64
0.02	1.4	−0.24	1.16
0.04	2.8	−0.96	1.84
0.0572	—*	—*	2.00

* *You can calculate these values, and if their sum is 2.00 miles you have a good check on your answers to (a) and (b).*

Since the graph of this function is a straight line, you need plot only two points, at $t' = 0$, $x'_f = x = 1$ mi. For $x'_f = 2$ mi, the limit of the displacement axis on the graph

$$t' = \frac{x'_f - x}{v} = \frac{2 \text{ mi} - 1 \text{ mi}}{40 \text{ mi/hr}} = 0.025 \text{ hr}.$$

The above two points and the points for the express are shown plotted on the same axes in Fig. 19.

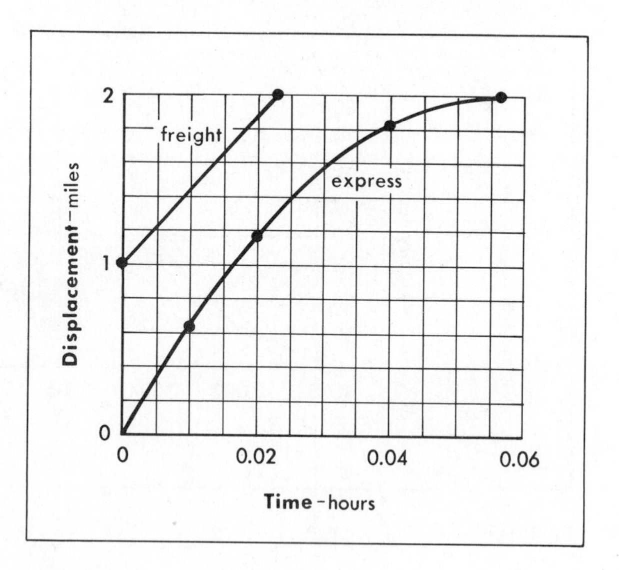

Figure 19

Since the two lines do not meet, the two trains can never have equal displacements at the same time and thus be at the same place at the same time. No collision takes place.

To solve this question algebraically, you will have to find if there is a time t when the trains are at the same position. In other words, is x'_f ever equal to x'_e?

Using equations $x'_f = 1 + 40t$ and $x'_e = 70t - \frac{1230}{2}t^2$ for x'_f and x'_e

$$1 + 40t = 70t - \frac{1230}{2}t^2.$$

To solve this quadratic equation, rearrange it so that you may use the formula: $x = \frac{-b \pm \sqrt{b^2 - 4ac}}{2a}$

$$\frac{1230}{2}t^2 - 30t + 1 = 0.$$

Since $\sqrt{b^2 - 4ac}$ is a negative quantity, there is no real solution for the equation. This tells you that the trains would not collide.

Chapter 10, Problem 2

Three quantities (a,b,c) may be arranged in six different ways: (a,b,c); (a,c,b); (b,a,c); (b,c,a); (c,a,b); and (c,b,a). To find the greatest sum of the three displacement vectors, you will have to try all six possible arrangements. In terms of the given vectors, the possible arrangements are:

(3mE, 2mN, 3mW);
(3mE, 3mW, 2mN);
(2mN, 3mE, 3mW);
(2mN, 3mW, 3mE);
(3mW, 3mE, 2mN); and
(3mW, 2mN, 3mE).

Adding the displacement vectors, head to tail, in the six different orders on graph paper you will always have, as Fig. 20 shows, the same net displacement—2 meters north, regardless of the order in which the vectors are added. In other words, these displacement vectors obey the commutative law for addition. (Can you show that all vector displacements, regardless of direction, obey the commutative law for addition?)

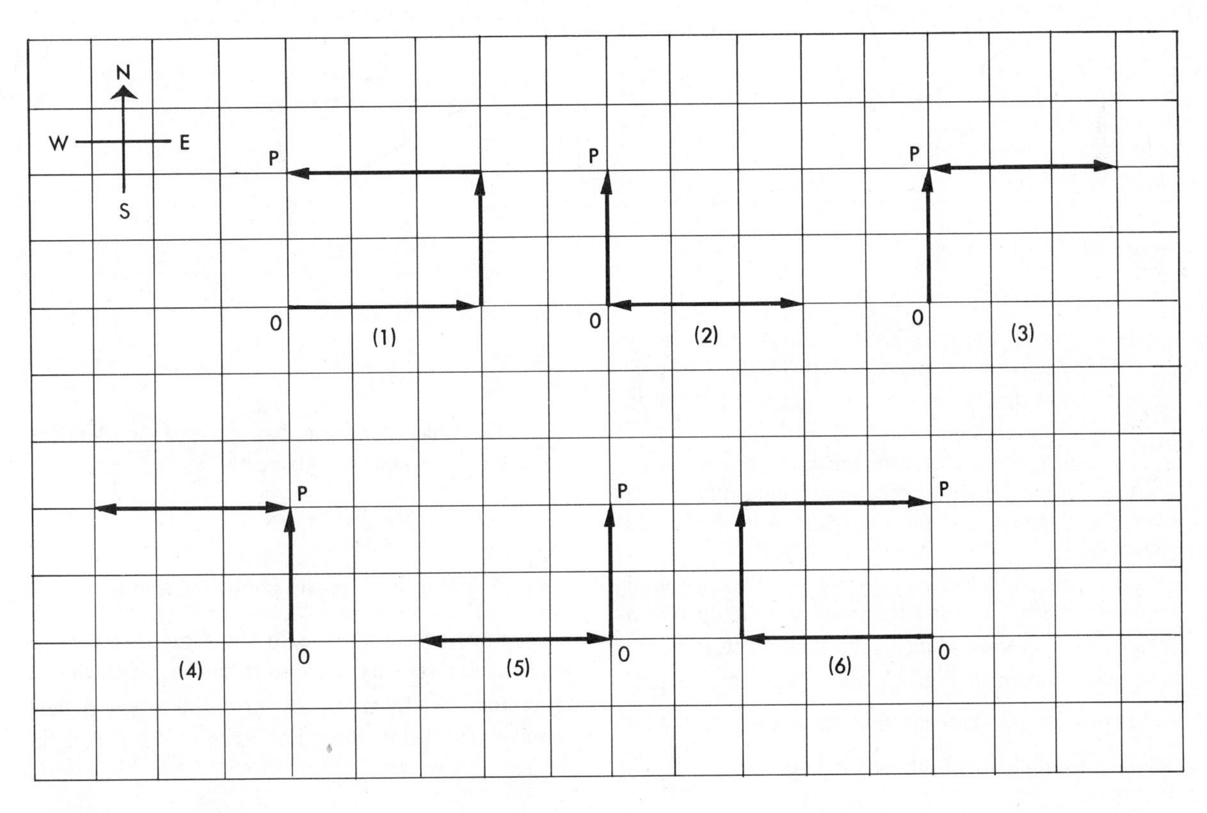

Figure 20

(a) Assume that the object is moving in a clockwise direction and that its initial velocity is $\vec{v}_1$. Then its velocity 3.0 sec. later is equal in magnitude, but different in direction, as shown in Fig. 21(a). To find the change in velocity $\Delta\vec{v} = \vec{v}_2 - \vec{v}_1$, you can redraw the two vectors with their tails at a common point, as shown in Fig. 21(b). You can make $\vec{v}_1$ and $\vec{v}_2$ each 10.0 cm long so that 1 cm on the drawing represents 0.20 cm/sec. Then the change $\Delta\vec{v}$ is just the vector drawn from the head of $\vec{v}_1$ to the head of $\vec{v}_2$.

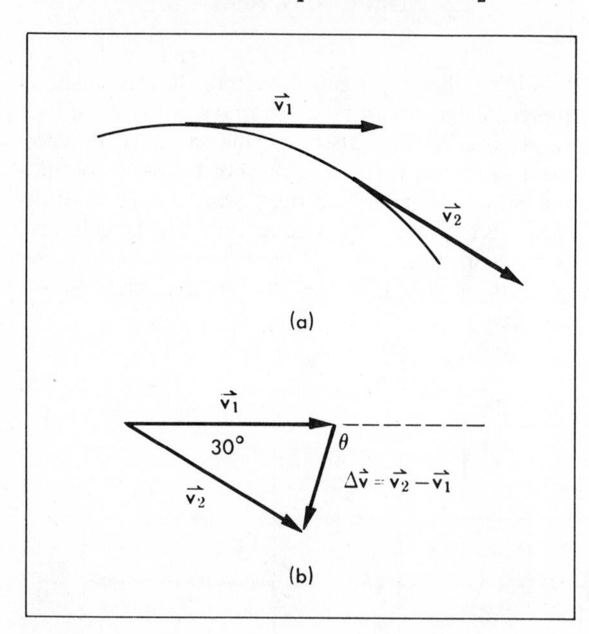

Figure 21

Measuring on the lines gives the magnitude of $\Delta\vec{v}$ as 1.05 m/sec. to the nearest 0.05 m/sec. A protractor shows that $\Delta\vec{v}$ makes an angle $\theta = 105°$ with $\vec{v}_1$.

The data given in the problem is expressed to an accuracy of only about 5 percent, so a carefully drawn scale drawing will give the change in velocity to this accuracy.

The change in velocity $(\vec{v}_2 - \vec{v}_1)$ can also be found by trigonometry. To use trigonometry, roughly sketch the triangle corresponding to the three vectors $\vec{v}_1$, $\vec{v}_2$ and $\Delta\vec{v}$ as shown in Fig. 22 with a dashed line perpendicular to Δv. You can find the angles and $\frac{\Delta v}{2}$ by recognizing that this is an isosceles triangle.

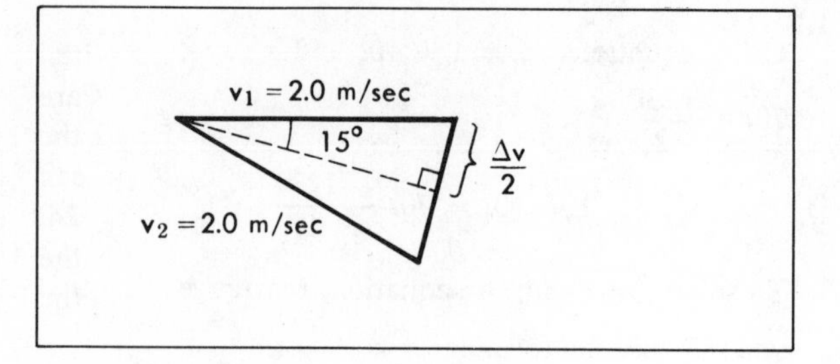

Figure 22

From the diagram, $\sin 15° = \dfrac{\frac{\Delta v}{2}}{v_1}$

$$\frac{\Delta v}{2} = v_1 \sin 15°$$

$$\Delta v = 2v_1 \sin 15° = 2 \times 2.0 \text{ m/sec} \times 0.259$$

$$\Delta v = 1.04 \text{ m/sec.}$$

This gives you the magnitude of the velocity change, but you also need its direction. This you can find from the isosceles triangle (see Fig. 23).

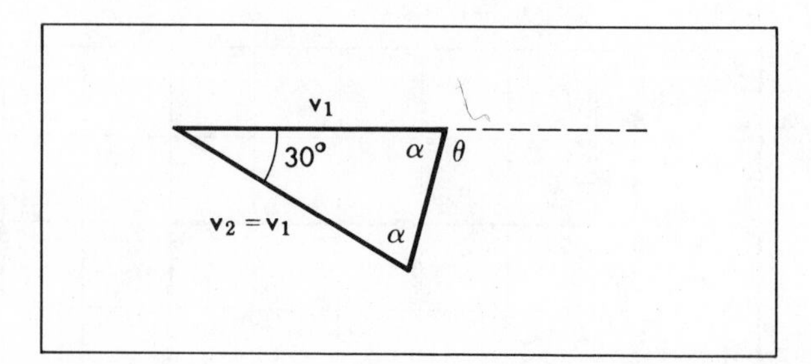

Figure 23

(b) The magnitude of the average acceleration during the 3.0 seconds is given by

$$\vec{a}_{av} = \frac{\Delta v}{\Delta t} = \frac{1.05 \text{ m/sec}}{3.0 \text{ sec}}$$

$$\vec{a}_{av} = 0.35 \text{ m/sec}^2, \text{ in the same direction as } \Delta\vec{v}.$$

The average acceleration is in the same direction as the change in velocity because to find $\vec{a}_{av}$, you divide the vector, $\Delta\vec{v}$, by a scalar, Δt. This gives a new vector in the same direction as $\Delta\vec{v}$. (See Section 10.4.)

Chapter 11, Problem 22

(a) You are given the displacement 9 m, the initial velocity v_o, and the final velocity $v_o/2$ and you are asked for the time. Since the force is constant, the motion is one of constant acceleration, and you can draw a velocity-time graph like the one in Fig. 24.

The displacement x of the object is equal to the shaded "area" in the figure. It is the area of the rectangle plus the area of the triangle or

$$x = \frac{v_o}{2}t + \frac{1}{2}\frac{v_o}{2}t$$

$$= \tfrac{3}{4}v_o t. \qquad (1)$$

The acceleration is the slope of the graph. It is

$$a = \frac{\frac{v_o}{2} - v_o}{t - 0}$$

$$= \frac{-v_o}{2t}. \qquad (2)$$

Combining equations (1) and (2) so as to eliminate v_o gives

$$t = \sqrt{\frac{-2x}{3a}}.$$

From Newton's law, $a = F/m$. Substituting this for a in the above equation gives

$$t = \sqrt{\frac{-2xm}{3F}}.$$

Substituting the given values of x, m, and F gives $t = 1$ sec (note that since the block is slowing down, F is negative.)

(b) You can now find the value of v_o from either equation (1) or equation (2). Choosing equation (2) gives

$$v_o = 2at = \frac{-2Ft}{m}$$

$$= \frac{-2 \times (-18) \times 1}{3}$$

$$= 12\ \text{m/sec}.$$

Both t and v_o can also be found using Newton's law and the two equations

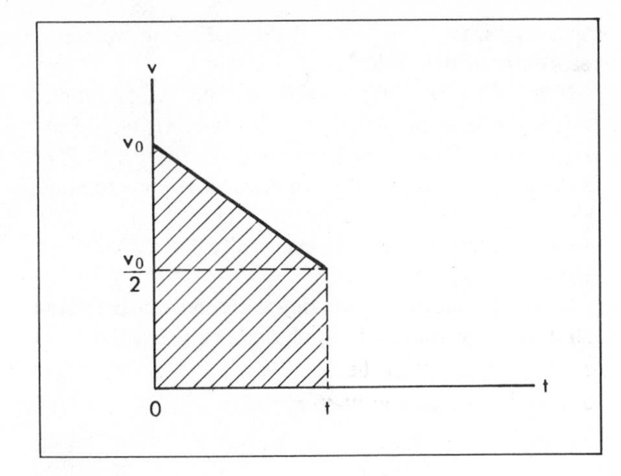

Figure 24

$$a = \frac{\Delta v}{\Delta t}$$

$$v'^2 - v^2 = 2a\,\Delta x.$$

If you first replace v' by $v_o/2$ and v by v_o, the two equations can then be combined to eliminate v_o and solved for t.

Chapter 11, Problem 23

In this problem neither the magnitude nor the direction of the applied force is specified; this gives you an infinite number of forces to choose from. Are there any forces you need not consider? If $\vec{F}$ has a component which is perpendicular to $\vec{v}_o$, then m_1 and m_2 can never have the same velocity. The velocity of m_2 would have a component perpendicular to $\vec{F}$, while the velocity of m_1 would be in the same direction as $\vec{F}$. Therefore, you need only consider forces which are in the same or opposite direction to $\vec{v}_o$, and you can drop the vector signs.

If $m_1 < m_2$, and the force F is in the same direction as v_o, m_1 will have a greater acceleration than m_2 and there will be a time when both masses have the same velocity—some velocity greater than v_o. If F is in the opposite direction to v_o, m_1 will move to the left, m_2 will slow down, stop, and then start moving to the left. Since m_1 has the greater acceleration, m_1 and m_2 will never have the same velocity.

If $m_1 = m_2$, F will produce the same acceleration for both masses. Because m_2 has an initial velocity, the two masses can never have the same velocity, regardless of the direction of F.

If $m_1 > m_2$, m_2 will have the greater acceleration. If F is in the same direction as v_o, then m_1 would always have a lower speed than m_2. However, if F is in the opposite direction from v_o, then m_2 has to slow down to stop before it begins moving to the left. This allows the two masses to eventually have the same velocity.

Since m_1 and m_2 are moving with constant acceleration, their velocities at a time t can be given by

$$v_1 = a_1 t = \frac{F}{m_1} t$$

$$v_2 = a_2 t + v_o = \frac{F}{m_2} t + v_o.$$

Is there a time t for which $v_1 = v_2$? Set $v_1 = v_2$ and solve for t:

$$\frac{F}{m_1} t = \frac{F}{m_2} t + v_o$$

$$t\left(\frac{F}{m_1} - \frac{F}{m_2}\right) = v_o$$

$$t = \frac{v_o}{F} \Big/ \left(\frac{1}{m_1} - \frac{1}{m_2}\right)$$

$$t = \frac{v_o}{F}\left(\frac{m_1 m_2}{m_2 - m_1}\right).$$

Now interpret this equation. If v_o and F are in the same direction $\left(\frac{v_o}{F}\right)$ is positive, and if $m_2 > m_1$, t will be positive. What does it mean if $m_2 < m_1$ and t is negative? This is a satisfactory solution mathematically, but in terms of the physical situation it means that m_1 and m_2 would have been moving at the same velocity sometime in the past.

If v_o and F are in opposite directions, then $\left(\frac{v_o}{F}\right)$ is negative and t can be positive only if $m_1 > m_2$.

If $m_1 = m_2$, there is no solution to the equation for t. These qualitative results support the discussion in the first part of the problem.

Chapter 12, Problem 10

Measuring displacements from the ground as the origin, you have for the displacements of the ball x'_b and the stone x_s from the ground:

$$x'_b = v_i t + \tfrac{1}{2} g t^2$$

$$x'_s = h + \tfrac{1}{2} g t^2.$$

When the ball and the stone meet, $x'_b = x'_s$. This gives

$$v_i t + \tfrac{1}{2} g t^2 = h + \tfrac{1}{2} g t^2$$

and

$$t = \frac{h}{v_i}.$$

Chapter 13, Problem 26

The question, "What are the accelerations of m_1 and m_2?" does not specify the frame of reference. Therefore, you can choose any frame of reference attached to m_1.

If you are not careful, you will apply Newton's law and the universal law of gravitation to calculate the acceleration and find

$$a_2 = \frac{F}{m_2} = \frac{G_1 m_1 m_2}{m_2 r^2} = \frac{G m_1}{r^2}.$$

Similarly, the acceleration of m_1 in a frame of reference attached to m_2 would be

$$a_1 = \frac{F}{m_1} = \frac{G m_1 m_2}{m_1 r^2} = \frac{G m_2}{r^2}$$

in the opposite direction.

This results in $a_1 \neq a_2$, contrary to your expectation. What you overlooked is that both m_1 and m_2 are at rest in accelerated frames and Newton's law does not hold in either of them. The accelerations a_1 and a_2 are those seen by an observer in a nonaccelerated frame of reference in which Newton's law holds. Suppose this observer is located somewhere between m_1 and m_2. He will see both masses accelerating in opposite directions toward him. The relative acceleration between them is, therefore, $a_1 + a_2$.

Chapter 14, Problem 20

When two masses interact only with each other, the total momentum is constant. Also, the velocity of the center of mass does not change. This can be a very useful way of looking at an interaction problem.

(a) The velocity of the center of mass $\vec{v}_c = 0.5$ m/sec does not change as a result of the collision.

(b) You can find the velocity of the other cart from the conservation of momentum. Thus:

Initial total momentum = Final total momentum

$$(m_1 + m_2)\vec{v}_c = m_1 \vec{v}_1{}' + m_2 \vec{v}_2{}'.$$

Since all velocities are along the same line,

$$(m_1 + m_2)v_c = m_1 v_1' + m_2 v_2'$$

and

$$v_2' = \frac{m_1 + m_2}{m_2} v_c - \frac{m_1}{m_2} v_1'$$

$$= 2 \times 0.5 \text{ m/sec} - 0.7 \text{ m/sec}$$

$$= 0.3 \text{ m/sec.}$$

Chapter 16, Problem 23

You have seen in Section 16.4 that changes in gravitational potential energy near the earth's surface are given by

$$\Delta U = mg\,\Delta r.$$

Now you are asked to show that this relationship can be derived from the more general expression for gravitational potential energy:

$$\Delta U = -GMm\left(\frac{1}{r'} - \frac{1}{r}\right) \qquad (1)$$

if $\frac{\Delta r}{r} \ll 1$. The problem you have to solve here is a purely mathematical one.

Since Δr does not appear in equation (1) you must do some algebraic manipulation to get it in a form that includes Δr. You can rewrite the terms inside the parentheses as follows:

$$\frac{1}{r'} - \frac{1}{r} = \frac{r - r'}{rr'}.$$

Substituting $r' = r + \Delta r$ gives

$$\frac{1}{r'} - \frac{1}{r} = \frac{-\Delta r}{r(r + \Delta r)} = \frac{-\Delta r}{r^2\left(1 + \frac{\Delta r}{r}\right)}.$$

If $\Delta r \ll r$, then $\frac{\Delta r}{r} \ll 1$. We can replace $1 + \frac{\Delta r}{r}$ by 1, giving

$$\frac{1}{r'} - \frac{1}{r} = \frac{-\Delta r}{r^2}.$$

Using this equality in equation (1) we find

$$\Delta U = \frac{GMm}{r^2}\Delta r.$$

The term $\frac{GMm}{r^2}$ is just the gravitational force mg on the object so

$$\Delta U = mg\,\Delta r$$

if $\Delta r \ll r$.

Chapter 17, Problem 2

The gas moves the piston, doing an amount of work $F\Delta x$. To show that $F\Delta x = P\Delta V$, first try to find an expression for F in terms of P.

By definition, the pressure $P = \frac{F}{A}$, the force per unit area. Then

$$F = PA$$

and the work done is

$$F\Delta x = PA\Delta x.$$

The volume change of the gas is equal to the surface area of the face of the piston times the distance the piston moves, or $\Delta V = A\Delta x$. Thus

$$PA\Delta x = P\Delta V$$

and you have shown that $F\Delta x = P\Delta V$.

Chapter 19, Problem 7

(a) From Coulomb's law, you know that the electric force between two charged spheres varies inversely with the square of the distance between the two charges. In this problem,

$$F_e = \frac{K}{x^2}. \qquad (1)$$

(b) Note that x and l are contained in one triangle and you are trying to find their relationship with F_e and mg contained in another triangle. Frequently, similar triangles may be used to find such a relationship: recall that the corresponding sides of similar triangles are proportional. If you can construct a triangle containing F_e and mg which is similar to the triangle containing x and l, the problem is solved.

There are three forces acting on A: $\vec{\mathbf{F}}_e$, $\vec{m\mathbf{g}}$, and the tension in the string, $\vec{\mathbf{F}}_t$. Since A is in equilibrium (it is not accelerating), the vector sum of these three forces must be zero:

$$\vec{\mathbf{F}}_e + \vec{m\mathbf{g}} + \vec{\mathbf{F}}_t = 0$$
$$\vec{\mathbf{F}}_e + \vec{m\mathbf{g}} = -\vec{\mathbf{F}}_t.$$

Because $\vec{\mathbf{F}}_t$ is a tension, it must act along the string. This means that $\vec{\mathbf{F}}_e + \vec{m\mathbf{g}}$ must also act along the direction of the string. You can make a sketch of these forces (Fig. 25) in an extension of the diagram given for the problem. The relative sizes of the forces are determined by choosing an arbitrary length for one of them.

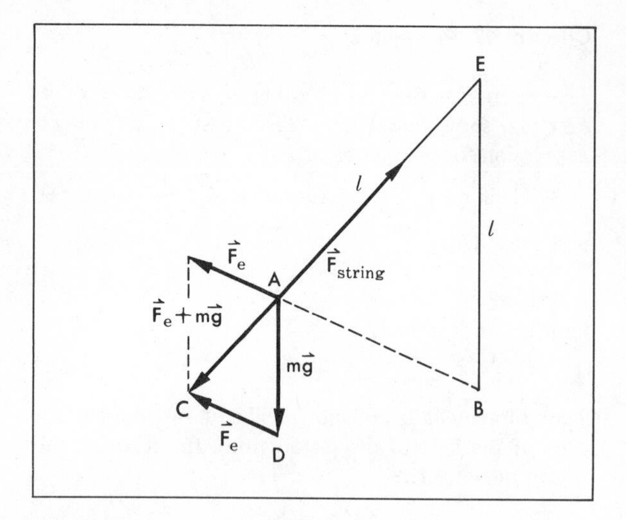

Figure 25

Now, to see if the two triangles ACD and ABE are similar. Since each side of the smaller triangle is parallel to a side in the larger triangle, the triangles are equiangular. Thus, they are similar and

$$\frac{F_e}{mg} = \frac{x}{l}. \quad (2)$$

(c) You will need to use equations (1) and (2) to answer this question. From equation (1), $F_e = \frac{K}{x^2}$, you see that if x is halved, F_e must be four times greater. Using this in equation (2) and calling the new weight $m'g$,

$$\frac{4F_e}{m'g} = \frac{x/2}{l}. \quad (3)$$

In order to compare equation (3) with equation (2) to find the new weight, divide both sides of equation (3) by 4.

$$\frac{F_e}{m'g} = \frac{x}{8l}.$$

Thus, $m'g = 8\ mg$; the weight must have increased by a factor of 8.

Chapter 21, Problem 6

The key to solving this problem is the recognition of the fact that when a current divides to follow two or more identical paths, it divides so that the current is the same through each path and the sum of these currents equals the current before it divides. It helps to sketch a labeled circuit diagram for each case. (See Fig. 26 which is the labeled circuit for part (c) of this problem.)

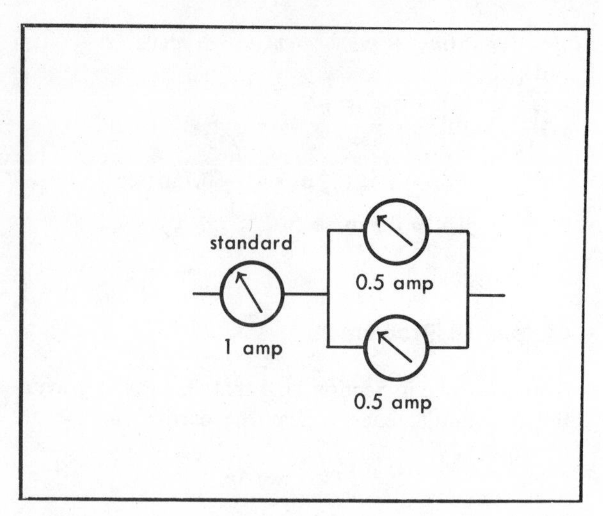

Figure 26

(a) You vary the voltage until each of the two meters in parallel stands at its calibration point, 1.0 ampere. Then the meter at the left will be carrying 2 amperes. (If you try this, it is important that the connecting wires be identical and that contacts be very clean.)

(b) To mark one scale at 3 amperes, add another identical meter in parallel and adjust the voltage until each of the three meters in parallel indicates 1 ampere.

(c) Figure 26 shows how to calibrate one meter at ½ ampere. To calibrate one meter at ⅓ ampere connect another calibrated meter in parallel with the other two.

Chapter 21, Problem 13

You know that as electrons flow through the coil of wire, they gain energy in an amount equal to the electrical work done on them. However, they almost immediately transfer this energy to the atoms of the metal (Sec. 21–3), increasing the internal energy of the metal and hence its temperature. The temperature of the wire being higher than the temperature of the surrounding water, the wire loses internal energy and the water gains it. This exchange rapidly reaches a steady state in which the temperature of the wire is constant and the energy transferred to the electrons by the source of EMF gives rise to an equivalent amount of energy transferred to the water.

(a) Consider what happens in one second (after a steady state has been reached). In one second, one gram of water passes through. The temperature of the gram of water is increased by 1°C, so the energy transferred to the water is 4.2 joules. The work done on the charge q is $qV = 4.2$ joules, so the charge q passing in one second must have been $q = 4.2$ joules/2.09 volts $= 2.0$ coulombs. A flow of 2.0 coulombs per second is a current of 2.0 amperes.

(b) To calibrate an ammeter we connect it in series as shown in the figure. The ammeter will carry the same current as the coil. By varying the applied EMF (not shown) which causes the current, we can vary the potential difference across the coil. For each V we measure with the voltmeter, we can calculate the current.

Chapter 21, Problem 25

(a) Let the current in the circuit be I, and the resistances of the elements be R_1 and R_2. You know from the conditions given that $I = V_1/R_1$, and also that $I = V_2/R_2$; or $V_1 = IR_1$, and $V_2 = IR_2$. In addition, $V = IR_e$, where R_e is the equivalent resistance of the combination. From the given conditions you also know that $V = V_1 + V_2$; so $IR_e = IR_1 + IR_2$. Hence $R_e = R_1 + R_2$.

(b) Similarly, it follows that $R_e = R_1 + R_2 + R_3$.

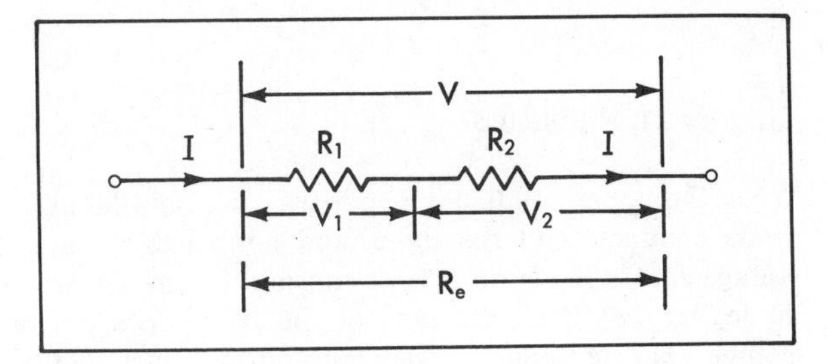

Figure 27

Chapter 21, Problem 26

(a) From conservation of charge, you know that the sum of the currents in the branches is equal to the total current; the potential difference is the same across both branches. Then for the elements, $I_1 = V/R_1$ and $I_2 = V/R_2$; for the combination $I = V/R_e$. Since $I = I_1 + I_2$,

$$V/R_e = V/R_1 + V/R_2$$

so

$$1/R_e = 1/R_1 + 1/R_2 \text{ and } R_e = R_1R_2/(R_1 + R_2).$$

(b) Similarly, $1/R_e = 1/R_1 + 1/R_2 + 1/R_3$
$= (R_2R_3 + R_1R_3 + R_1R_2)/R_1R_2R_3$

and

$$R_e = R_1R_2R_3/(R_2R_3 + R_1R_3 + R_1R_2).$$

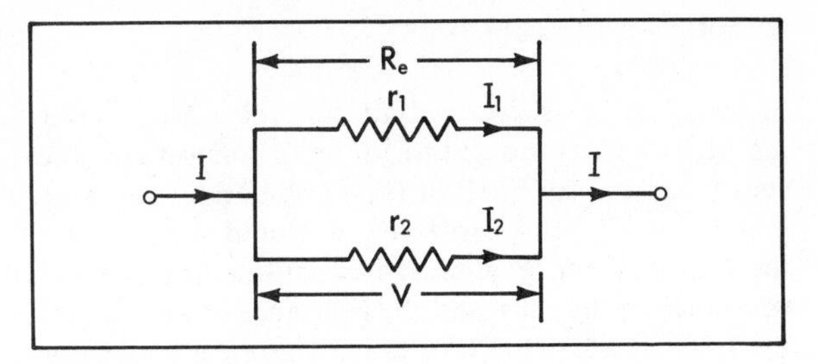

Figure 28

Chapter 21, Problem 27

In part (a) simply use the result of Problem 26(b) of this chapter; letting $r_1 = 20$ ohms, $r_2 = 40$ ohms, and $r_3 = 30$ ohms.

$$R = \frac{r_1r_2r_3}{r_2r_3 + r_1r_3 + r_1r_2} = \frac{20 \times 40 \times 30}{30 \times 40 + 20 \times 30 + 20 \times 40} = 9.2 \text{ ohms.}$$

The combination of resistors in part (b) is made up of several combinations of series and parallel connections. It is a more complex case than that of part (a). In this kind of resistive network, involving series strings and series-parallel strings (the three parallel strings on the right in part (b)), one starts "inside" and works "out." Begin with the series combination in the middle of the parallel set on the right, consisting of a 1.0 ohm and a 2.0 ohm resistor. These resistances simply add (from Problem 26(a)) and the resistance of this combination is 3.0 ohms. Then, if $r_1 = 2.0$ ohms, $r_2 = 3.0$ ohms, $r_3 = 3$ ohms, the combined resistance of the parallel group of three resistors is

$$\frac{2.0 \times 3.0 \times 3.0}{3.0 \times 3.0 + 2.0 \times 3.0 + 2.0 \times 3.0} = 0.86 \text{ ohm.}$$

The combined resistance of the parallel group of two resistors on the left is $\frac{2.0 \times 3.0}{2.0 + 3.0} = 1.20$ ohms. The two groups of parallel resistors are in series, so the total resistance of the whole combination $= 1.20 + 0.86 = 2.06$ ohms.

Chapter 21, Problem 29

Since the current I_r in the light bulb is sufficiently small compared to the current I_R, the current I_r in the lower section of R (below the sliding contact) will be nearly the same as the current I_R in the upper section of R. In this case, $I_R = V_i/R$ and $V_0 = I_Rr = (V_i/R)r = \frac{r}{R} V_i$.

Potentiometers are used extensively as voltage controls. Most of the controls on a television set are potentiometers.

Chapter 21, Problem 33

Since 12 volts is required to turn over the starter motor, your first thought might be to connect 8 of the cells in series; the EMF of the combination would be $8 \times 1.5 = 12$ volts. However, it would not operate the starter motor. You must also consider the internal resistance of the cells, and the resistance of the motor; the latter is $12/100 = 0.12$ ohm. The 8 cells in series have an internal resistance of $8 \times 0.1 = 0.8$ ohm; the total current would be only $12/(0.8 + 0.12) = 13$ amperes, nowhere near the 100 amperes required. Furthermore, the terminal voltage of the 8 cells delivering this current would be $12 - 13 \times 0.8 = 1.6$ volts, far less than the 12 volts required. Hence it seems evident that you must have more than 8 cells in series, so that the terminal voltage can be 12 volts while current is being delivered. A combination of resistors in parallel has a lower resistance than any of the individual resistors. This suggests that you should make up several sets of the same number of cells in series, then connect a number of these sets in parallel.

How should you decide on the proper arrangement? You might approach the problem by trial and error, a process likely to prove lengthy and tedious. It would be better to try to find a general solution first, as it so often is in solving problems. For this purpose imagine you have n_s cells in series in each set and that you have connected n_p such sets in parallel.

The terminal voltage V_t is the EMF minus the internal voltage drop in the battery:

$$V_t = \text{EMF} - Ir_i$$

where r_i is the effective internal resistance of the battery whose EMF is $1.5n_s$ where n_s is the number of cells in each series-connected string of cells. The total internal resistance of each series string is $0.1n_s$. Since all the series strings are connected in parallel, the total effective internal resistance of the battery is

$$r_i = \frac{0.1\, n_s}{n_p}$$

where n_p is the number of parallel-connected series strings. Thus

$$V_t = 1.5\, n_s - I \times \frac{0.1\, n_s}{n_p}.$$

Since $V_t = 12$ volts and $I = 100$ amp:

$$12 = 1.5\, n_s - \frac{10\, n_s}{n_p} \text{ and } n_p = \frac{10}{1.5 - 12/n_s}.$$

There are a number of solutions to this equation, having the conditions that n_s and n_p are integral numbers greater than 1. Table A shows all the possible integral numbers.

TABLE A

n_s	n_p
9	60
12	20
18	12
24	10
48	8

There are other solutions in which one of the numbers is a positive integral, the other number being close to an integral number. Can you find them?

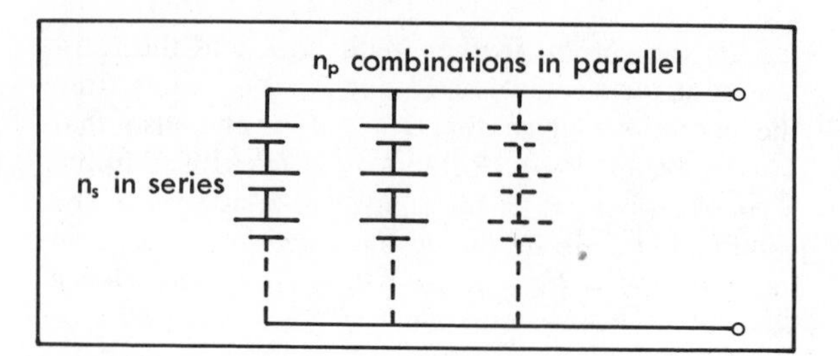

Figure 29

Chapter 21, Problem 37

When the upper conductor is positive, the right-hand diode conducts and the right-hand lamp lights; the voltage across the lamp is kept constant by the Zener diode, so that the brightness of the lamp does not change. The left-hand diode does not conduct and the left-hand lamp does not light. The situation is reversed when the upper conductor is negative.

Two ordinary diodes are required to determine which of the two lights will be lighted; and a Zener diode is required for each of the lamps, to maintain a constant voltage across it and hence to maintain a constant brightness.

Chapter 22, Problem 20

To answer this question, you must find out how the angle through which the beam of ions in a mass spectrograph is deflected depends on the mass of the ions in the beam. You can start with the equation $m = \frac{B^2 q}{2v} r^2$ derived in Section 22–7 in the text. Examining this equation, you see that if the magnetic field B, the charge q of the ions, and the accelerating potential V are kept fixed, the mass of the ions is proportional to the square of the radius.

$$m = kr^2$$

and

$$r = \frac{\sqrt{m}}{\sqrt{k}}. \qquad (1)$$

You should expect that the angle through which the ions are deflected depends on r the radius. To find the relationship between these two quantities, draw a diagram like that in Fig. 30 of the path of the ions between the pole pieces of the magnet.

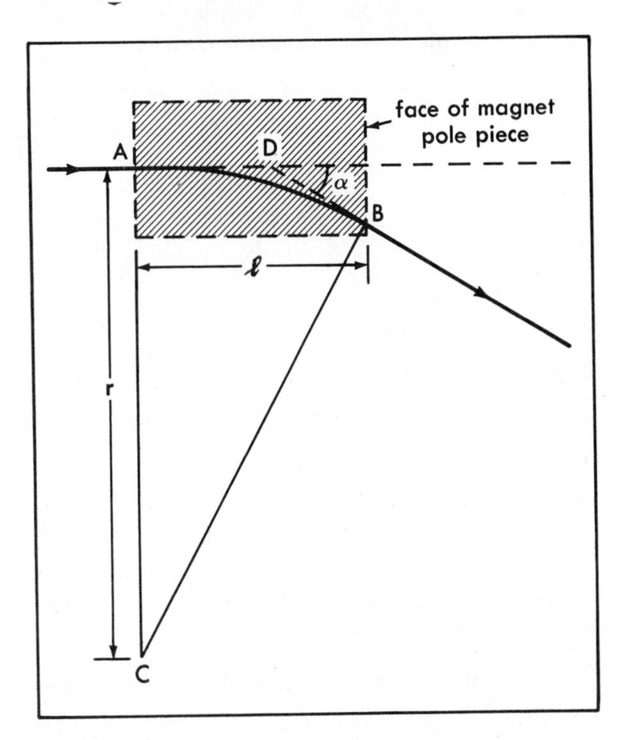

Figure 30

Since AC is perpendicular to AD and CB is perpendicular to DB, angle ACB equals angle α, the angle of deflection of the ions by the magnetic field.

In radians, angle ACB equals the ratio of the circular arc AB to the radius r. If α is small (as it usually is in a mass spectrograph) the arc AB is nearly equal to l, the length of the face of the pole piece. Therefore

$$\alpha = \measuredangle ACB = \frac{l}{r}$$

and

$$r = \frac{l}{\alpha}.$$

Substituting this value into equation (1) gives

$$\frac{l}{\alpha} = \frac{\sqrt{m}}{\sqrt{k}}$$

$$\alpha = \frac{l\sqrt{k}}{\sqrt{m}}.$$

The quantity l, of course, is constant for a given mass spectrograph.

If the mass spectrograph is to separate the isotopes of an element, the difference in the angle of deflection of ions of mass m amu and $m + 1$ amu must be measurable. This is because the smallest mass difference of two isotopes of an element is the mass of one neutron which is about 1 amu. You can call this difference $\Delta\alpha$ and write

$$\Delta\alpha = l\sqrt{k}\left(\frac{1}{\sqrt{m}} - \frac{1}{\sqrt{m+1}}\right).$$

As m increases the quantity $\left(\frac{1}{\sqrt{m}} - \frac{1}{\sqrt{m+1}}\right)$ decreases, resulting in smaller angular separation.

Chapter 23, Problem 21

The first thing to realize is that when the magnet falls through the loop there is a change in the magnetic flux through the loop. This change in flux will induce a current in the loop. The flux is zero when the magnet is far above the loop, rises to a maximum when the magnet is in the center of the loop, and then decreases to zero again when the magnet is far beyond the loop.

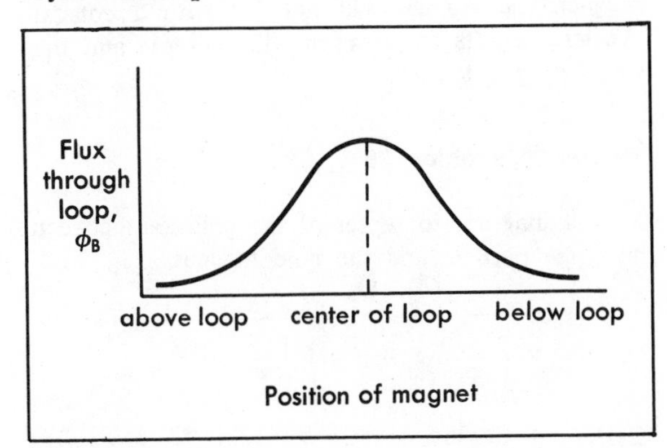

Figure 31

To get some idea of how the induced current depends on the position of the magnet, you can first make a sketch of the graph of the induced EMF resulting from the changing magnetic flux, assuming that the magnet falls at a nearly steady speed. (Fig. 31.)

Since $\mathcal{E} = \frac{\Delta\phi_B}{\Delta t}$, $\mathcal{E}$ is greatest where ϕ_B changes most rapidly and is zero where ϕ_B changes most slowly, you can use Fig. 31 to make a sketch of the graph of $\mathcal{E}$ as a function of position (Fig. 32). Do you understand why $\mathcal{E}$ changes direction? When the loop is above the magnet, $\Delta\phi_B$ is increasing, so $\mathcal{E} = \frac{\Delta\phi_B}{\Delta t}$ is positive. When the flux is decreasing $\Delta\phi_B$ is negative and E is negative.

The graph of the current will look like the graph for the EMF, since $I = \frac{E}{R}$ and R is constant.

To find the direction of the current in the loop at the different positions, you can use Lenz's law: the magnetic flux produced by the induced current must oppose the magnetic flux which is causing the current.

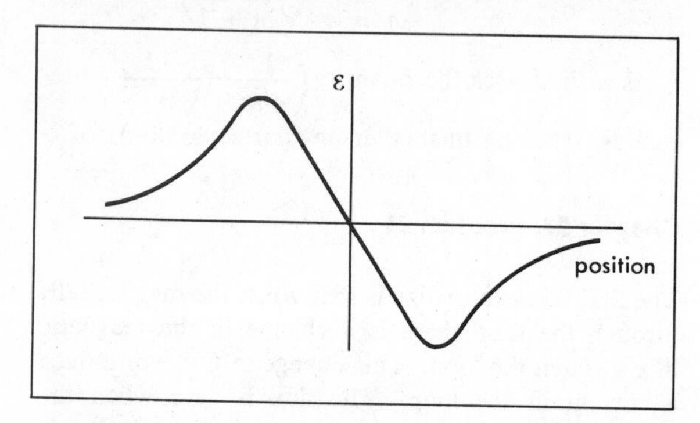

Figure 32

(b) Since the induced current produces a magnetic field which opposes the magnetic field of the falling magnet, the magnet will not fall with a constant acceleration. Its acceleration will decrease and then increase again to "g."

Chapter 24, Problem 28

Draw a diagram to represent the collision between the alpha particle and the gold nucleus (Fig. 33).

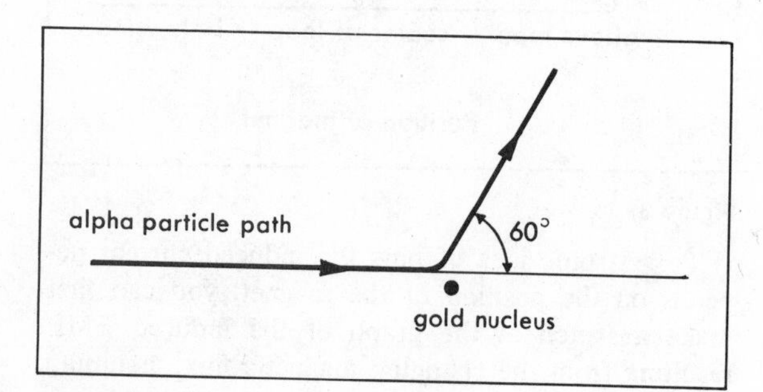

Figure 33

The purpose of this problem is to test the assumption that very little kinetic energy is transferred to the gold nucleus during the collision. You can find the actual amount of energy transferred by applying the law of conservation of momentum.

(a) Since momentum is conserved,

$$\vec{p}_{before} = \vec{p}_{after}$$

or

$$\vec{p}_\alpha = \vec{p}_\alpha{}' + \vec{p}_N{}'. \qquad (1)$$

From the assumption that the speed of the alpha particle does not change, it follows that $p_\alpha' = p_\alpha$. You know that the angle between $\vec{p}_\alpha$ and $\vec{p}_\alpha{}'$ is 60°. Therefore, you can draw a vector diagram (Fig. 34)

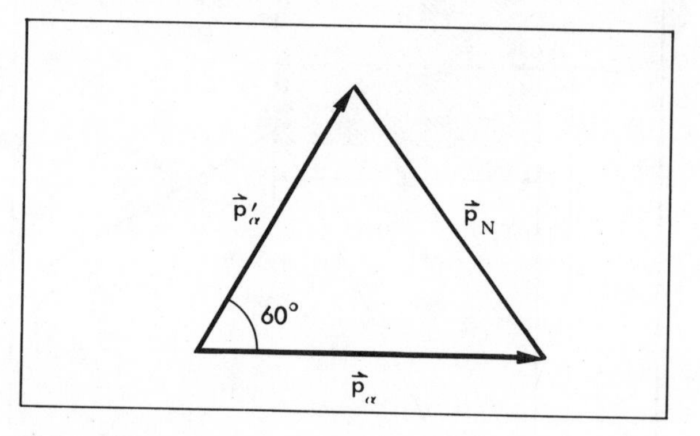

Figure 34

to represent equation (1). Since this is an isosceles triangle, with an angle of 60° at its apex, it is also equilateral. Hence:

$$p_N = p_\alpha = (6.62 \times 10^{-27}\text{ kg})(1.00 \times 10^7\text{ m/sec})$$
$$= 6.6 \times 10^{-20}\text{ kg-m/sec.}$$

(b) You know the momentum and mass of the nucleus, and, from these, you can find its velocity.

$$p_N = m_N v_N$$

$$v_N = \frac{p_N}{m_N} = \frac{6.6 \times 10^{-20}\text{ kg-m/sec}}{3.3 \times 10^{-25}\text{ kg}}$$
$$= 2.0 \times 10^5\text{ m/sec.}$$

(c) You can find the kinetic energy of the nucleus from its mass and speed.

$$E_{KN} = \tfrac{1}{2} m_N v_N{}^2$$

$$E_{KN} = \tfrac{1}{2}(3.3 \times 10^{-25}\text{ kg})(2.0 \times 10^5\text{ m/sec})^2$$

$$E_{KN} = 6.6 \times 10^{-15}\text{ joule.}$$

(d) The ratio of kinetic energy of the alpha particle to that of the nucleus is:

$$\frac{E_{KN}}{E_{K\alpha}} = \frac{\frac{1}{2}m_N v_N^2}{\frac{1}{2}m_\alpha v_\alpha^2} = \frac{6.6 \times 10^{-15}\text{ joule}}{(6.6 \times 10^{-27})(1.0 \times 10^7)^2}$$

$$= 0.02.$$

The original assumption was justified.

Chapter 25, Problem 15

(a) This question can be answered by using the equations given in the chapter for the energy and momentum of photons.

$$E = h\nu = \frac{hc}{\lambda}.$$

For an "average" photon of visible light you can choose $\lambda = 5000$ A. Then

$$E = (6.6 \times 10^{-34}\text{ joule-sec})(3.0 \times 10^8\text{ m/sec})/(5000\text{ A})(10^{-10}\text{ m/A})$$

$$= 4.0 \times 10^{-19}\text{ joule}$$

and

$$p = \frac{E}{c} = \frac{4.0 \times 10^{-19}\text{ joule}}{3.0 \times 10^8\text{ m/sec}}$$

$$= 1.3 \times 10^{-27}\text{ kg-m/sec.}$$

Or, you could also find the momentum from

$$p = \frac{h}{\lambda} = \frac{6.6 \times 10^{-34}\text{ joule-sec}}{(5000\text{ A})(10^{-10}\text{ m/A})}$$

$$= 1.3 \times 10^{-27}\text{ kg-m/sec.}$$

(b) There would be 1 watt = 1 joule/sec emitted in the visible region. Using the result from part (a) that one "average" photon of visible light has an energy of 4.0×10^{-19} joule, the number of photons per second is

$$\frac{1\text{ joule/sec}}{4.0 \times 10^{-19}\text{ joule/photon}} = 2.5 \times 10^{18}\text{ photons/sec.}$$

(c) The pressure on the black body is given by the change of momentum of the photons hitting per square meter per second (Section 17–2). At a distance of 2 meters this number of photons will be a fraction of $\frac{1}{4\pi 2^2} = \frac{1}{16\pi}$ of all the photons emitted per second by the light bulb. Since the photons are absorbed by the black body, the change in momentum equals the original momentum in magnitude. Using the results of parts (a) and (b), this yields a pressure

$$p = \frac{2.5 \times 10^{18} \times 1.3 \times 10^{-27}}{16\pi}$$

$$= 6.5 \times 10^{-11}\text{ newton/m}^2.$$

In Section 4–2, the pressure of light was first discussed. It might interest you to reread this now. You knew the pressure of light was small, but this calculation shows you how small. Compare the pressure of light from a 100-watt bulb to atmospheric pressure, which is about 10^5 newtons/m^2.

Appendix 3

Table of the Elements Based on the relative atomic mass of $^{12}C = 12$

ELEMENT	SYMBOL	ATOMIC NUMBER	ATOMIC MASS (in amu)
Actinium	Ac	89	(227.0278)
Aluminum	Al	13	26.9815
Americium	Am	95	(243.0614)
Antimony	Sb	51	121.75
Argon	Ar	18	39.948
Arsenic	As	33	74.9216
Astatine	At	85	(210)
Barium	Ba	56	137.34
Berkelium	Bk	97	(247.0702)
Beryllium	Be	4	9.0122
Bismuth	Bi	83	208.9806
Boron	B	5	10.81
Bromine	Br	35	79.904
Cadmium	Cd	48	112.40
Calcium	Ca	20	40.08
Californium	Cf	98	(251)
Carbon	C	6	12.011
Cerium	Ce	58	140.12
Cesium	Cs	55	132.9055
Chlorine	Cl	17	35.453
Chromium	Cr	24	51.996
Cobalt	Co	27	58.9332
Copper	Cu	29	63.546
Curium	Cm	96	(247)
Dysprosium	Dy	66	162.50
Einsteinium	Es	99	(254.0881)
Erbium	Er	68	167.26
Europium	Eu	63	151.96
Fermium	Fm	100	(253)
Fluorine	F	9	18.9984
Francium	Fr	87	(211.996)
Gadolinium	Gd	64	157.25
Gallium	Ga	31	69.72
Germanium	Ge	32	72.59
Gold	Au	79	196.9665
Hafnium	Hf	72	178.49
Helium	He	2	4.0026
Holmium	Ho	67	164.9303
Hydrogen	H	1	1.0080
Indium	In	49	114.82
Iodine	I	53	126.9045
Iridium	Ir	77	192.22
Iron	Fe	26	55.847
Krypton	Kr	36	83.80
Lanthanum	La	57	138.9055
Lawrencium	Lr	103	(257)
Lead	Pb	82	207.2
Lithium	Li	3	6.941
Lutetium	Lu	71	174.97
Magnesium	Mg	12	24.305
Manganese	Mn	25	54.9380
Mendelevium	Md	101	(256)
Mercury	Hg	80	200.59

ELEMENT	SYMBOL	ATOMIC NUMBER	ATOMIC MASS (in amu)
Molybdenum	Mo	42	95.94
Neodymium	Nd	60	144.24
Neon	Ne	10	20.179
Neptunium	Np	93	237.0482
Nickel	Ni	28	58.71
Niobium	Nb	41	92.9064
Nitrogen	N	7	14.0067
Nobelium	No	102	(255)
Osmium	Os	76	190.2
Oxygen	O	8	15.9994
Palladium	Pd	46	106.4
Phosphorus	P	15	30.9738
Platinum	Pt	78	195.09
Plutonium	Pu	94	(244)
Polonium	Po	84	(209)
Potassium	K	19	39.102
Praseodymium	Pr	59	140.0977
Promethium	Pm	61	(145)
Protactinium	Pa	91	(231.0359)
Radium	Ra	88	(226.0254)
Radon	Rn	86	(222.0175)
Rhenium	Re	75	186.2
Rhodium	Rh	45	102.9055
Rubidium	Rb	37	85.4678
Ruthenium	Ru	44	101.07
Samarium	Sm	62	150.4
Scandium	Sc	21	44.9559
Selenium	Se	34	78.96
Silicon	Si	14	28.086
Silver	Ag	47	107.868
Sodium	Na	11	22.9898
Strontium	Sr	38	87.62
Sulfur	S	16	32.06
Tantalum	Ta	73	180.9479
Technetium	Tc	43	98.9062
Tellurium	Te	52	127.60
Terbium	Tb	65	158.9254
Thallium	Tl	81	204.37
Thorium	Th	90	232.0381
Thulium	Tm	69	168.9342
Tin	Sn	50	118.69
Titanium	Ti	22	47.90
Wolfram	W	74	183.85
Uranium	U	92	238.029
Vanadium	V	23	50.9414
Xenon	Xe	54	131.30
Ytterbium	Yb	70	173.04
Yttrium	Y	39	88.9059
Zinc	Zn	30	65.37
Zirconium	Zr	40	91.22

Note: The figures in parentheses are the atomic masses of the most stable isotopes of radioactive elements. The other masses are the average for all the isotopes together, as they occur in their natural abundances.

Appendix 4

Table of Trigonometric Functions

sin (read down)

	.0	.1	.2	.3	.4	.5	.6	.7	.8	.9		
0°	.0000	.0017	.0035	.0052	.0070	.0087	.0105	.0122	.0140	.0157	.0175	**89°**
1°	.0175	.0192	.0209	.0227	.0244	.0262	.0279	.0297	.0314	.0332	.0349	**88°**
2°	.0349	.0366	.0384	.0401	.0419	.0436	.0454	.0471	.0488	.0506	.0523	**87°**
3°	.0523	.0541	.0558	.0576	.0593	.0610	.0628	.0645	.0663	.0680	.0698	**86°**
4°	.0698	.0715	.0732	.0750	.0767	.0785	.0802	.0819	.0837	.0854	.0872	**85°**
5°	.0872	.0889	.0906	.0924	.0941	.0958	.0976	.0993	.1011	.1028	.1045	**84°**
6°	.1045	.1063	.1080	.1097	.1115	.1132	.1149	.1167	.1184	.1201	.1219	**83°**
7°	.1219	.1236	.1253	.1271	.1288	.1305	.1323	.1340	.1357	.1374	.1392	**82°**
8°	.1392	.1409	.1426	.1444	.1461	.1478	.1495	.1513	.1530	.1547	.1564	**81°**
9°	.1564	.1582	.1599	.1616	.1633	.1650	.1668	.1685	.1702	.1719	.1736	**80°**
10°	.1736	.1754	.1771	.1788	.1805	.1822	.1840	.1857	.1874	.1891	.1908	**79°**
11°	.1908	.1925	.1942	.1959	.1977	.1994	.2011	.2028	.2045	.2062	.2079	**78°**
12°	.2079	.2096	.2113	.3130	.2147	.2164	.2181	.2198	.2115	.2233	.2250	**77°**
13°	.2250	.2267	.2284	.2300	.2317	.2334	.2351	.2368	.2385	.2402	.2419	**76°**
14°	.2419	.2436	.2453	.2470	.2487	.2504	.2521	.2538	.2554	.2571	.2588	**75°**
15°	.2588	.2605	.2622	.2639	.2656	.2672	.2689	.2706	.2723	.2740	.2756	**74°**
16°	.2756	.2773	.2790	.2807	.2823	.2840	.2857	.2874	.2890	.2907	.2924	**73°**
17°	.2924	.2940	.2957	.2974	.2990	.3007	.3024	.3040	.3057	.3074	.3090	**72°**
18°	.3090	.3107	.3123	.3140	.3156	.3173	.3190	.3206	.3223	.3239	.3256	**71°**
19°	.3256	.3272	.3289	.3305	.3322	.3338	.3355	.3371	.3387	.3404	.3420	**70°**
20°	.3420	.3437	.3453	.3469	.3486	.3502	.3518	.3535	.3551	.3567	.3584	**69°**
21°	.3584	.3600	.3616	.3633	.3649	.3665	.3681	.3697	.3714	.3730	.3746	**68°**
22°	.3746	.3762	.3778	.3795	.3811	.3827	.3843	.3859	.3875	.3891	.3907	**67°**
23°	.3907	.3923	.3939	.3955	.3971	.3987	.4003	.4019	.4035	.4051	.4067	**66°**
24°	.4067	.4083	.4099	.4115	.4131	.4147	.4163	.4179	.4195	.4210	.4226	**65°**
25°	.4226	.4242	.4258	.4274	.4289	.4305	.4321	.4337	.4352	.4368	.4384	**64°**
26°	.4384	.4399	.4415	.4431	.4446	.4462	.4478	.4493	.4509	.4524	.4540	**63°**
27°	.4540	.4555	.4571	.4586	.4602	.4617	.4633	.4648	.4664	.4679	.4695	**62°**
28°	.4695	.4710	.4726	.4741	.4756	.4772	.4787	.4802	.4818	.4833	.4848	**61°**
29°	.4848	.4863	.4879	.4894	.4909	.4924	.4939	.4955	.4970	.4985	.5000	**60°**
30°	.5000	.5015	.5030	.5045	.5060	.5075	.5090	.5105	.5120	.5135	.5150	**59°**
31°	.5150	.5165	.5180	.5195	.5210	.5225	.5240	.5255	.5270	.5284	.5299	**58°**
32°	.5299	.5314	.5329	.5344	.5358	.5373	.5388	.5402	.5417	.5432	.5446	**57°**
33°	.5446	.5461	.5476	.5490	.5505	.5519	.5534	.5548	.5563	.5577	.5592	**56°**
34°	.5592	.5606	.5621	.5635	.5650	.5664	.5678	.5693	.5707	.5721	.5736	**55°**
35°	.5736	.5750	.5764	.5779	.5793	.5807	.5821	.5835	.5850	.5864	.5878	**54°**
36°	.5878	.5892	.5906	.5920	.5934	.5948	.5962	.5976	.5990	.6004	.6018	**53°**
37°	.6018	.6032	.6046	.6060	.6074	.6088	.6101	.6115	.6129	.6143	.6157	**52°**
38°	.6157	.6170	.6184	.6198	.6211	.6225	.6239	.6252	.6266	.6280	.6293	**51°**
39°	.6293	.6307	.6320	.6334	.6347	.6361	.6374	.6388	.6401	.6414	.6428	**50°**
40°	.6428	.6441	.6455	.6468	.6481	.6494	.6508	.6521	.6534	.6547	.6561	**49°**
41°	.6561	.6574	.6587	.6600	.6613	.6626	.6639	.6652	.6665	.6678	.6691	**48°**
42°	.6691	.6704	.6717	.6730	.6743	.6756	.6769	.6782	.6794	.6807	.6820	**47°**
43°	.6820	.6833	.6845	.6858	.6871	.6884	.6896	.6909	.6921	.6934	.6947	**46°**
44°	.6947	.6959	.6972	.6984	.6997	.7009	.7022	.7034	.7046	.7059	.7071	**45°**
		.9	.8	.7	.6	.5	.4	.3	.2	.1	.0	

cos (read up)

Table of Trigonometric Functions

sin (read down)

	.0	.1	.2	.3	.4	.5	.6	.7	.8	.9		
45°	.7071	.7083	.7096	.7108	.7120	.7133	.7145	.7157	.7169	.7181	.7193	**44°**
46°	.7193	.7206	.7218	.7230	.7242	.7254	.7266	.7278	.7290	.7302	.7314	**43°**
47°	.7314	.7325	.7337	.7349	.7361	.7373	.7385	.7396	.7408	.7420	.7431	**42°**
48°	.7431	.7443	.7455	.7466	.7478	.7490	.7501	.7513	.7524	.7536	.7547	**41°**
49°	.7547	.7559	.7570	.7581	.7593	.7604	.7615	.7627	.7638	.7649	.7660	**40°**
50°	.7660	.7672	.7683	.7694	.7705	.7716	.7727	.7738	.7749	.7760	.7771	**39°**
51°	.7771	.7782	.7793	.7804	.7815	.7826	.7837	.7848	.7859	.7869	.7880	**38°**
52°	.7880	.7891	.7902	.7912	.7923	.7934	.7944	.7955	.7965	.7976	.7986	**37°**
53°	.7986	.7997	.8007	.8018	.8028	.8039	.8049	.8059	.8070	.8080	.8090	**36°**
54°	.8090	.8100	.8111	.8121	.8131	.8141	.8151	.8161	.8171	.8181	.8192	**35°**
55°	.8192	.8202	.8211	.8221	.8231	.8241	.8251	.8261	.8271	.8281	.8290	**34°**
56°	.8290	.8300	.8310	.8320	.8329	.8339	.8348	.8358	.8368	.8377	.8387	**33°**
57°	.8387	.8396	.8406	.8415	.8425	.8434	.8443	.8453	.8462	.8471	.8480	**32°**
58°	.8480	.8490	.8499	.8508	.8517	.8526	.8536	.8545	.8554	.8563	.8572	**31°**
59°	.8572	.8581	.8590	.8599	.8607	.8616	.8625	.8634	.8643	.8652	.8660	**30°**
60°	.8660	.8669	.8678	.8686	.8695	.8704	.8712	.8721	.8729	.8738	.8746	**29°**
61°	.8746	.8755	.8763	.8771	.8780	.8788	.8796	.8805	.8813	.8821	.8829	**28°**
62°	.8829	.8838	.8846	.8854	.8862	.8870	.8878	.8886	.8894	.8902	.8910	**27°**
63°	.8910	.8918	.8926	.8934	.8942	.8949	.8957	.8965	.8973	.8980	.8988	**26°**
64°	.8988	.8996	.9003	.9011	.9018	.9026	.9033	.9041	.9048	.9056	.9063	**25°**
65°	.9063	.9070	.9078	.9085	.9092	.9100	.9107	.9114	.9121	.9128	.9135	**24°**
66°	.9135	.9143	.9150	.9157	.9164	.9171	.9178	.9184	.9191	.9198	.9205	**23°**
67°	.9205	.9212	.9219	.9225	.9232	.9239	.9245	.9252	.9259	.9265	.9272	**22°**
68°	.9272	.9278	.9285	.9291	.9298	.9304	.9311	.9317	.9323	.9330	.9336	**21°**
69°	.9336	.9342	.9348	.9354	.9361	.9367	.9373	.9379	.9385	.9391	.9397	**20°**
70°	.9397	.9403	.9409	.9415	.9421	.9426	.9432	.9438	.9444	.9449	.9455	**19°**
71°	.9455	.9461	.9466	.9472	.9478	.9483	.9489	.9494	.9500	.9505	.9511	**18°**
72°	.9511	.9516	.9521	.9527	.9532	.9537	.9542	.9548	.9553	.9558	.9563	**17°**
73°	.9563	.9568	.9573	.9578	.9583	.9588	.9593	.9598	.9603	.9608	.9613	**16°**
74°	.9613	.9617	.9622	.9627	.9632	.9636	.9641	.9646	.9650	.9655	.9659	**15°**
75°	.9659	.9664	.9668	.9673	.9677	.9681	.9686	.9690	.9694	.9699	.9703	**14°**
76°	.9703	.9707	.9711	.9715	.9720	.9724	.9728	.9732	.9736	.9740	.9744	**13°**
77°	.9744	.9748	.9751	.9755	.9759	.9763	.9767	.9770	.9774	.9778	.9781	**12°**
78°	.9781	.9785	.9789	.9792	.9796	.9799	.9803	.9806	.9810	.9813	.9816	**11°**
79°	.9816	.9820	.9823	.9826	.9829	.9833	.9836	.9839	.9842	.9845	.9848	**10°**
80°	.9848	.9851	.9854	.9857	.9860	.9863	.9866	.9869	.9871	.9874	.9877	**9°**
81°	.9877	.9880	.9882	.9885	.9888	.9890	.9893	.9895	.9898	.9900	.9903	**8°**
82°	.9903	.9905	.9907	.9910	.9912	.9914	.9917	.9919	.9921	.9923	.9925	**7°**
83°	.9925	.9928	.9930	.9932	.9934	.9936	.9938	.9940	.9942	.9943	.9945	**6°**
84°	.9945	.9947	.9949	.9951	.9952	.9954	.9956	.9957	.9959	.9960	.9962	**5°**
85°	.9962	.9963	.9965	.9966	.9968	.9969	.9971	.9972	.9973	.9974	.9976	**4°**
86°	.9976	.9977	.9978	.9979	.9980	.9981	.9982	.9983	.9984	.9985	.9986	**3°**
87°	.9986	.9987	.9988	.9989	.9990	.9990	.9991	.9992	.9993	.9993	.9994	**2°**
88°	.9994	.9995	.9995	.9996	.9996	.9997	.9997	.9997	.9998	.9998	.9998	**1°**
89°	.9998	.9999	.9999	.9999	.9999	1.000	1.000	1.000	1.000	1.000	1.000	**0°**
		.9	.8	.7	.6	.5	.4	.3	.2	.1	.0	

cos (read up)

Table of Trigonometric Functions

tan (read down)

	.0	.1	.2	.3	.4	.5	.6	.7	.8	.9		
0°	.0000	.0017	.0035	.0052	.0070	.0087	.0105	.0122	.0140	.0157	.0175	**89°**
1°	.0175	.0192	.0209	.0227	.0244	.0262	.0279	.0297	.0314	.0332	.0349	**88°**
2°	.0349	.0367	.0384	.0402	.0419	.0437	.0454	.0472	.0489	.0507	.0524	**87°**
3°	.0524	.0542	.0559	.0577	.0594	.0612	.0629	.0647	.0664	.0682	.0699	**86°**
4°	.0699	.0717	.0734	.0752	.0769	.0787	.0805	.0822	.0840	.0857	.0875	**85°**
5°	.0875	.0892	.0910	.0928	.0945	.0963	.0981	.0998	.1016	.1033	.1051	**84°**
6°	.1051	.1069	.1086	.1104	.1122	.1139	.1157	.1175	.1192	.1210	.1228	**83°**
7°	.1228	.1246	.1263	.1281	.1299	.1317	.1334	.1352	.1370	.1388	.1405	**82°**
8°	.1405	.1423	.1441	.1459	.1477	.1495	.1512	.1530	.1548	.1566	.1584	**81°**
9°	.1584	.1602	.1620	.1638	.1655	.1673	.1691	.1709	.1727	.1745	.1763	**80°**
10°	.1763	.1781	.1799	.1817	.1835	.1853	.1871	.1890	.1908	.1926	.1944	**79°**
11°	.1944	.1962	.1980	.1998	.2016	.2035	.2053	.2071	.2089	.2107	.2126	**78°**
12°	.2126	.2144	.2162	.2180	.2199	.2217	.2235	.2254	.2272	.2290	.2309	**77°**
13°	.2309	.2327	.2345	.2364	.2382	.2401	.2419	.2438	.2456	.2475	.2493	**76°**
14°	.2493	.2512	.2530	.2549	.2568	.2586	.2605	.2623	.2642	.2661	.2679	**75°**
15°	.2679	.2698	.2717	.2736	.2754	.2773	.2792	.2811	.2830	.2849	.2867	**74°**
16°	.2867	.2886	.2905	.2924	.2943	.2962	.2981	.3000	.3019	.3038	.3057	**73°**
17°	.3057	.3076	.3096	.3115	.3134	.3153	.3172	.3119	.3211	.3230	.3249	**72°**
18°	.3249	.3269	.3288	.3307	.3327	.3346	.3365	.3385	.3404	.3424	.3443	**71°**
19°	.3443	.3463	.3482	.3502	.3522	.3541	.3561	.3581	.3600	.3620	.3640	**70°**
20°	.3640	.3659	.3679	.3699	.3719	.3739	.3759	.3779	.3799	.3819	.3839	**69°**
21°	.3839	.3859	.3879	.3899	.3919	.3939	.3959	.3979	.4000	.4020	.4040	**68°**
22°	.4040	.4061	.4081	.4101	.4122	.4142	.4163	.4183	.4204	.4224	.4245	**67°**
23°	.4245	.4265	.4286	.4307	.4327	.4348	.4369	.4390	.4411	.4431	.4452	**66°**
24°	.4452	.4473	.4494	.4515	.4536	.4557	.4578	.4599	.4621	.4642	.4663	**65°**
25°	.4663	.4684	.4706	.4727	.4748	.4770	.4791	.4813	.4834	.4856	.4877	**64°**
26°	.4877	.4899	.4921	.4942	.4964	.4986	.5008	.5029	.5051	.5073	.5095	**63°**
27°	.5095	.5117	.5139	.5161	.5184	.5206	.5228	.5250	.5272	.5295	.5317	**62°**
28°	.5317	.5340	.5362	.5384	.5407	.5430	.5452	.5475	.5498	.5520	.5543	**61°**
29°	.5543	.5566	.5589	.5612	.5635	.5658	.5681	.5704	.5727	.5750	.5774	**60°**
30°	.5774	.5797	.5820	.5844	.5867	.5890	.5914	.5938	.5961	.5985	.6009	**59°**
31°	.6009	.6032	.6056	.6080	.6104	.6128	.6152	.6176	.6200	.6224	.6249	**58°**
32°	.6249	.6273	.6297	.6322	.6346	.6371	.6395	.6420	.6445	.6469	.6494	**57°**
33°	.6494	.6519	.6544	.6569	.6594	.6619	.6644	.6669	.6694	.6720	.6745	**56°**
34°	.6745	.6771	.6796	.6822	.6847	.6873	.6899	.6924	.6950	.6976	.7002	**55°**
35°	.7002	.7028	.7054	.7080	.7107	.7133	.7159	.7186	.7212	.7239	.7265	**54°**
36°	.7265	.7292	.7319	.7346	.7373	.7400	.7427	.7454	.7481	.7508	.7536	**53°**
37°	.7536	.7563	.7590	.7618	.7646	.7673	.7701	.7729	.7757	.7785	.7813	**52°**
38°	.7813	.7841	.7869	.7898	.7926	.7954	.7983	.8012	.8040	.8069	.8098	**51°**
39°	.8098	.8127	.8156	.8185	.8214	.8243	.8273	.8302	.8332	.8361	.8391	**50°**
40°	.8391	.8421	.8451	.8481	.8511	.8541	.8571	.8601	.8632	.8662	.8693	**49°**
41°	.8693	.8724	.8754	.8785	.8816	.8847	.8878	.9810	.8941	.8972	.9004	**48°**
42°	.9004	.9036	.9067	.9099	.9131	.9163	.9195	.9228	.9260	.9293	.9325	**47°**
43°	.9325	.9358	.9391	.9424	.9457	.9490	.9523	.9556	.9590	.9623	.9657	**46°**
44°	.9657	.9691	.9725	.9759	.9793	.9827	.9861	.9896	.9930	.9965	1.000	**45°**
		.9	.8	.7	.6	.5	.4	.3	.2	.1	.0	

cot (read up)

Table of Trigonometric Functions

tan (read down)

	.0	.1	.2	.3	.4	.5	.6	.7	.8	.9		
45°	1.000	1.003	1.007	1.011	1.014	1.018	1.021	1.025	1.028	1.032	1.036	**44°**
46°	1.036	1.039	1.043	1.046	1.050	1.054	1.057	1.061	1.065	1.069	1.072	**43°**
47°	1.072	1.076	1.080	1.084	1.087	1.091	1.095	1.099	1.103	1.107	1.111	**42°**
48°	1.111	1.115	1.118	1.122	1.126	1.130	1.134	1.138	1.142	1.146	1.150	**41°**
49°	1.150	1.154	1.159	1.163	1.167	1.171	1.175	1.179	1.183	1.188	1.192	**40°**
50°	1.192	1.196	1.200	1.205	1.209	1.213	1.217	1.222	1.226	1.230	1.235	**39°**
51°	1.235	1.239	1.244	1.248	1.253	1.257	1.262	1.266	1.271	1.275	1.280	**38°**
52°	1.280	1.285	1.289	1.294	1.299	1.303	1.308	1.313	1.317	1.322	1.327	**37°**
53°	1.327	1.332	1.337	1.342	1.347	1.351	1.356	1.361	1.366	1.371	1.376	**36°**
54°	1.376	1.381	1.387	1.392	1.397	1.402	1.407	1.412	1.418	1.423	1.428	**35°**
55°	1.428	1.433	1.439	1.444	1.450	1.455	1.460	1.466	1.471	1.477	1.483	**34°**
56°	1.483	1.488	1.494	1.499	1.505	1.511	1.517	1.522	1.528	1.534	1.540	**33°**
57°	1.540	1.546	1.552	1.558	1.564	1.570	1.576	1.582	1.588	1.594	1.600	**32°**
58°	1.600	1.607	1.613	1.619	1.625	1.632	1.638	1.645	1.651	1.658	1.664	**31°**
59°	1.664	1.671	1.678	1.684	1.691	1.698	1.704	1.711	1.718	1.725	1.732	**30°**
60°	1.732	1.739	1.746	1.753	1.760	1.767	1.775	1.782	1.789	1.797	1.804	**29°**
61°	1.804	1.811	1.819	1.827	1.834	1.842	1.849	1.857	1.865	1.873	1.881	**28°**
62°	1.881	1.889	1.897	1.905	1.913	1.921	1.929	1.937	1.946	1.954	1.963	**27°**
63°	1.963	1.971	1.980	1.988	1.997	2.006	2.014	2.023	2.032	2.041	2.050	**26°**
64°	2.050	2.059	2.069	2.078	2.087	2.097	2.106	2.116	2.125	2.135	2.145	**25°**
65°	2.145	2.154	2.164	2.174	2.184	2.194	2.204	2.215	2.225	2.236	2.246	**24°**
66°	2.246	2.257	2.267	2.278	2.289	2.300	2.311	2.322	2.333	2.344	2.356	**23°**
67°	2.356	2.367	2.379	2.391	2.402	2.414	2.426	2.438	2.450	2.463	2.475	**22°**
68°	2.475	2.488	2.500	2.513	2.526	2.539	2.552	2.565	2.578	2.592	2.605	**21°**
69°	2.605	2.619	2.633	2.646	2.660	2.675	2.689	2.703	2.718	2.733	2.747	**20°**
70°	2.747	2.762	2.778	2.793	2.808	2.824	2.840	2.856	2.872	2.888	2.904	**19°**
71°	2.904	2.921	2.937	2.954	2.971	2.989	3.006	3.024	3.042	3.060	3.078	**18°**
72°	3.078	3.096	3.115	3.133	3.152	3.172	3.191	3.211	3.230	3.251	3.271	**17°**
73°	3.271	3.291	3.312	3.333	3.354	3.376	3.398	3.420	3.442	3.465	3.487	**16°**
74°	3.487	3.511	3.534	3.558	3.582	3.606	3.630	3.655	3.681	3.706	3.732	**15°**
75°	3.732	3.758	3.785	3.812	3.839	3.867	3.895	3.923	3.952	3.981	4.011	**14°**
76°	4.011	4.041	4.071	4.102	4.134	4.165	4.198	4.230	4.264	4.297	4.331	**13°**
77°	4.331	4.366	4.402	4.437	4.474	4.511	4.548	4.586	4.625	4.665	4.705	**12°**
78°	4.705	4.745	4.787	4.829	4.872	4.915	4.959	5.005	5.050	5.097	5.145	**11°**
79°	5.145	5.193	5.242	5.292	5.343	5.396	5.449	5.503	5.558	5.614	5.671	**10°**
80°	5.671	5.730	5.789	5.850	5.912	5.976	6.041	6.107	6.174	6.243	6.314	**9°**
81°	6.314	6.386	6.460	6.535	6.612	6.691	6.772	6.855	6.940	7.026	7.115	**8°**
82°	7.115	7.207	7.300	7.396	7.495	7.596	7.700	7.806	7.916	8.028	8.144	**7°**
83°	8.144	8.264	8.386	8.513	8.643	8.777	8.915	9.058	9.205	9.357	9.514	**6°**
84°	9.514	9.677	9.845	10.02	10.20	10.39	10.58	10.78	10.99	11.20	11.43	**5°**
85°	11.43	11.66	11.91	12.16	12.43	12.71	13.00	13.30	13.62	13.95	14.30	**4°**
86°	14.30	14.67	15.06	15.46	15.89	16.35	16.83	17.34	17.89	18.46	19.08	**3°**
87°	19.08	19.74	20.45	21.20	22.02	22.90	23.86	24.90	26.03	27.27	28.64	**2°**
88°	28.64	30.14	31.82	33.69	35.80	38.19	40.92	44.07	47.74	52.08	57.29	**1°**
89°	57.29	63.66	71.62	81.85	95.49	114.6	143.2	191.0	286.5	573.0	∞	**0°**
		.9	.8	.7	.6	.5	.4	.3	.2	.1	.0	

cot (read up)

Appendix 5

PHYSICAL CONSTANTS AND CONVERSION FACTORS

Physical Constants

Avogadro's number: $N_0 = 6.02 \times 10^{23}$

Speed of light: $c = 2.99793 \times 10^8$ m/sec

Gravitational constant: $G = 6.670 \times 10^{-11} \dfrac{\text{m}^3}{\text{kg-sec}^2}$

Boltzmann's constant: $k = 1.3805 \times 10^{-23} \dfrac{\text{joule}}{{}^\circ\text{K}}$

Constant in Coulomb's law: $k = 2.306 \times 10^{-28} \dfrac{\text{newton-m}^2}{(\text{elem. ch.})^2} = 8.988 \times 10^9 \dfrac{\text{newton-m}^2}{\text{coulomb}^2}$

Mass of electron: $m_e = 9.109 \times 10^{-31}$ kg

Mass of proton: $m_p = 1.672 \times 10^{-27}$ kg

Constant in Ampère's circuital law: $K = 2 \times 10^{-7} \dfrac{\text{newton}}{\text{amp}^2}$ (exact, by definition)

Planck's constant: $h = 6.626 \times 10^{-34}$ joule-sec $= 4.136 \times 10^{-15}$ ev-sec

Conversion Factors

1 atomic mass unit $= 1.66 \times 10^{-27}$ kg

1 electron volt $= 1.602 \times 10^{-19}$ joule

1 coulomb $= 6.242 \times 10^{18}$ elem. ch.

Answers to Short Problems

Marked with an asterisk (*)

CHAPTER 1

1. Firefly, flash bulb, and electric stove heating element.
2. Look at it in the dark.
13. Yes; if the screen is moved back far enough, there will be only a penumbra on it.
16. It would get narrower.

CHAPTER 2

1. Angle of incidence = ∡ 3.
   Angle of reflection = ∡ 2.
2. An infinite number of rays, which together form a hollow cone whose axis is perpendicular to the mirror surface.
3. One.
7. $f = 12$ mm.
8. The rays would be reflected back to the parabolic mirror, where they would then be reflected parallel to the axis.
9. Two rays are enough, because the others pass through almost the same spot.
11. No, none of the light from the image will pass into your eyes.
17. Yes; whenever $S_0 < f$.
18. No.

CHAPTER 3

1. Angle of incidence = ∡ 2.
   Angle of refraction = ∡ 4.
2. In the photo of Fig. 3–2.
4. On the left.
5. $r = \frac{3}{4} i$.
7. $n = \sqrt{3}/\sqrt{2} = 1.2$.
8. Yes, since the measured n is 1.3 to 1.4. The material is not necessarily water.
9. Diamond.
13.

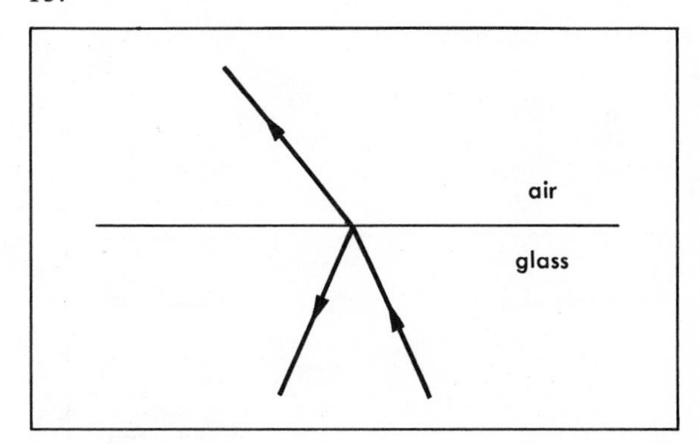

17. $n = \dfrac{\sin \theta_A}{\sin \theta_B} < 1$.
22. 63°.
30. 0.25 cm.

CHAPTER 4

1. Isolate each one in a dark room where there are no reflections, and see whether each gives the same reading on a light meter placed the same distance from it.
2. At $r/\sqrt{2} = 0.71\, r$.
6. 2 meters.
11. The smaller Ping-Pong ball is deflected by small bumps and holes that would not affect a basketball.
13. The speed on the lower level is the same for all angles of incidence.
14. The speed of light should be greater in a refractive material than in a vacuum.

18. The light is able to travel through the hole to the distant mirror and back through the hole before the disc has turned far enough to move the hole out of position.

26. The particle model predicts that the speed of light is greater in water than in air, while actual measurements show that it is less in water than in air.

CHAPTER 5

1. The waves all travel through some medium, but the medium itself does not travel with the wave.

3. At the point of maximum displacement.

4. It would still move up first.

7. No.

8. 7 cm.

12.

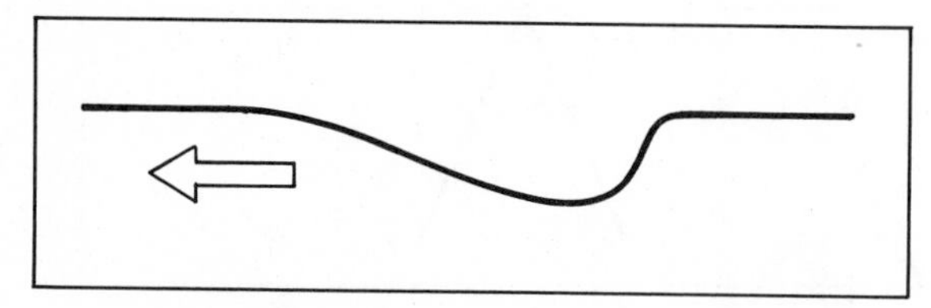

16. The incident pulse.

17. 3. (The speeds are proportional to the distances the pulses move between successive pictures.)

22. The pulse is smaller in the bottom frames than in the top frames.

CHAPTER 6

1. About 20 cm long.

2. $r' = 25°$.

3. (a) Incident pulse: (2); reflected pulse: (1). (b) $i = 70°$.

10. (a) 85 m. (b) It decreases.

11. (a) $T = 1/4$ sec. (b) $f = 4$ per sec.

12. Decrease the frequency.

17. Near the fifth crest from the bottom.

18. $n \approx 1.8$.

20. No, not unless one frequency was a whole-number multiple of the other.

CHAPTER 7

1.

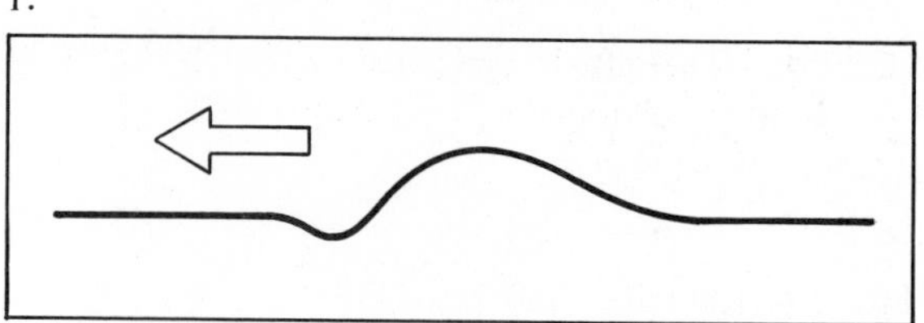

2. A distance λ from the wall.

4. "Double crest": A; "double trough": C; nodal point: B.

7. No; along this line are double crests separated by double troughs.

9. The second nodal line to the left of the central line.

10. On which nodal line the point lies—that is, the value of n.

13. Yes; all λ such that $\lambda > 2d$.

20. $p = 0.36$.

21. Yes. In both cases the two sources dip simultaneously.

CHAPTER 8

1. Try to find places where the boat does not bob up and down. Along the nodal lines the water does not move, while to either side of them the boat will bob up and down.

7. 0.9 meter.

9. $\Delta x = 0.68$ mm.

10. The nodes for red light are 1.4 times as far apart as those for blue light.

12. All the sources producing the central maximum are in phase. Of the sources producing the partial reinforcement in Fig. 18–10, two thirds cancel each other, and the remaining one third are not in phase but are separated by phase delays of less than 1/2.

18. The magnification is the same. Check the distance between the centers of the images.

22. It would increase.

23. $\lambda = 4200$ A.

24. No; the wavelength within each material depends on the index of refraction of the material.

25. The bars would be thinner and only half as far apart.

CHAPTER 9

1. (1) and (3); (4) and (6).
2. No. The graph indicates that between $t = 2$ and $t = 3$ hours, the car is in two different places at the same time.
8. $v_{ave} = 0.48$ m/sec.
11. (a) 120 miles. (b) 20 miles.
19. 6.7 miles per hour per second.
20. 0.45 m/sec^2.

CHAPTER 10

1. Length of displacement $2\sqrt{2} = 2.83$. Direction at an angle of 45° with the x axis.
8. −86.6 mi/hr north and −50 mi/hr east.
9. Zero.
12. Each velocity vector would have to be turned around to point in the opposite direction, but its tail would remain in the same place. Its magnitude would be unchanged.
13. If the motion is along a *curved* path, the *direction* of the velocity changes, requiring that there be an acceleration.
22. The flag will hang limply.
23. It appears not to move.

CHAPTER 11

1. Kinematics is the study of the motion of objects, without considering the cause of the motion; dynamics is the study of how forces cause motion.
2. No. Time was not involved in his reasoning.
4. The ball would orbit forever.
6. $v = 20$ cm/sec.
7. $v = 20.9$ cm/flash.
8. The force is constant.
12. Toward the right.
13. $a = 1.5$ m/sec^2.
16. (a) 3.0 (b) 3.0.
17. $a = 4$ m/sec^2.
18. $F = 2$ newtons.
19. $F = 15$ newtons.
25. Along the direction of the net force.
26. $a = 8$ m/sec^2 (to the right).
27. $\vec{\mathbf{F}} = 3.7$ newtons, to the right along the dotted line.
31. $\vec{\mathbf{F}} = 200$ newtons, in the direction opposite to that of your push.

CHAPTER 12

1. $m_g = 5.00$ kg.
2. (a) Weight on moon: 1.1×10^2 newtons. Weight on earth: 6.9×10^2 newtons. (b) Mass on earth: 70 kg. Mass on moon: 70 kg.
4. $g = 9.81$ newtons/kg.
7. 4.9 meters long.
13. The horizontal component of displacement of the ball is constant over equal time intervals; on the other hand, the vertical component of displacement is not constant.
16. $F = 24$ newtons.
17. The speed does not change, but the direction of motion changes in the direction of the force.
18. $\vec{\mathbf{v}} = 0.52$ m/sec, toward the left if the puck is moving clockwise, as in Fig. 12–11.
19. $\vec{\mathbf{a}} = 0.62$ m/sec^2, toward the center of the circle.
26. $\vec{\mathbf{F}} = -k\vec{\mathbf{x}}$, for some positive constant k.
27. At the points where the circle intersects the y axis.
28. $m = 0.2$ kg.
29. $l = 0.25$ meter.

CHAPTER 13

2. (a) 1.05. (b) About 6.
6. 4.75×10^{23} m^2.
8. (a) When it is closest to the sun. (b) When it is farthest from the sun.
9. 3.35×10^{18} m^3/sec^2.
10. $(8^3)^{1/2} = 23$ times longer.
13. 3.6×10^3 times greater.
14. (a) $K_e = 9.9 \times 10^{12}$ m^3/sec^2. (b) $m_e/m_s = 3.0 \times 10^{-6}$.
15. At a height equal to $(\sqrt{2} - 1) = 0.41$ times the earth's radius.

CHAPTER 14

1. 1.5 newton-sec.
2. 32.0 kg-m/sec = 32.0 newton-sec.
3. (a) Four: the throw, the bounce, the catch, and the continuous impulse due to gravity.

   (b) The bounce. Here the ball experienced about twice the change in velocity that it did during the throw or catch, and the gravitational impulse was the smallest of all.
7. $\vec{\mathbf{F}} = 500$ newtons, in the opposite direction to the momentum.
10. Zero. The net change in momentum is always in the direction of the total impulse.
11. Each velocity vector in (a) is multiplied by the mass of the body whose motion it represents, and a new scale for the units is chosen to represent these new vectors.
12. (Mass of faster)/(mass of slower) = 2/5.
19. The direction of motion of the center of mass after the collision is the same as it was before the collision: it is upward, at an angle of about a degree counterclockwise from the direction of motion of the ball entering from the bottom of the picture.
27. Most of the momentum is transferred to the earth (through the road) and a small amount to the air molecules surrounding the car.
29. No; the vector sum of the impulses due only to the forces of interaction cannot be zero, because the total impulse, which includes the momentum of the light, must equal zero.

CHAPTER 15

1. 200 joules.
2. The kinetic energy increases by a factor of four.
5. They will be equal, since they depend only on the force and the kinetic energy.
6. Yes; no energy is transferred when a force acts on a moving body at right angles to its motion.
12. They repel each other during the interaction.
14. When the force between the colliding bodies depends only on the separation of the bodies.
15. On the way in, the putty ball experiences a repulsive force at small separation, which slows it down; after the putty gets to the wall, the force suddenly becomes attractive, so it is not the same as on the way in. Hence we know the force does not depend on distance alone.
16. Probably. Since kinetic energy is conserved, we may assume that in this collision the interaction force is a function only of separation.
17. Mass of $m_1 <$ mass of m_2.

CHAPTER 16

1. In the middle picture, where the force is greatest.
2. The slope of the graph is k.
9. Each would move in a straight line in the direction of the force acting on it.
16. 3.4×10^3 joules.
25. An additional E_K.
26. $GMm/(10r_e)$.
27. They are equal.
32. $F = 0.20$ newton.

CHAPTER 17

1. P increases by a factor of 373/293 = 1.27.
3. The impulse given to the wall is $2mv$, by conservation of momentum.
10. Zero, if the container is not moving.
12. The total kinetic energy is proportional to PV/N.
13. E_K increases by a factor of 373/298 = 1.25.
15. (a) 2.07×10^{-23} joule.
    (b) No.
16. No.

CHAPTER 18

1. The sign is the same, but one cannot tell whether it is positive or negative.
2. Negative.
4. (a) Charge flows through the metal rod onto your hand and through your body, so no charge can accumulate on the rod. (b) Insulate it from everything except what you are charging it with.

CHAPTER 19

1. $F = 1.0 \times 10^{-5}$ newton.

9. 19–5(a) and 19–6(a)
19–5(b) and 19–6(b)
19–5(c) and 19–6(c)
19–5(d) has no analog in 19–6.

10. No. If they did, at the point where they cross, the *net* force on a movable charge would have to be in two directions at once, which is impossible.

15. The potential energy at r is $U = kq_1 q_2/r$. All this will be transferred to the moving charge as kinetic energy at infinity.

19. $F = 2.3 \times 10^{-16}$ newton.

23. $\vec{E} = 9 \times 10^{-18}$ newton/elementary charge, pointing toward the object.
$V = -5 \times 10^{-18}$ joule/elementary charge.

CHAPTER 20

1. In a good vacuum there are hardly any ions that can remove charge from the rod.

3. 14 ev.

4. $F = 4.2 \times 10^{-14}$ newton.

8. 45 ev, or 7.2×10^{-18} joule.

9. Between the curved plate and the first of the parallel plates, the electron coasts with a constant kinetic energy of 45 ev. Between the parallel plates the kinetic energy drops uniformly with distance from 45 ev to 15 ev as it strikes the final plate.

10. 2.9×10^{-17} joule.

11. $m = 2.7 \times 10^{-26}$ kg.

14. Positive.

15. 5.8×10^{-17}

22. 7.64×10^{-6} sec.

CHAPTER 21

2. $I = 5.00$ amp.

3. 0.67 ampere.

16. 1.7 joules/sec = 1.7 watts.

17. EMF = 1.9×10^{-18} joule/el. ch. = 12 volts.

19. $R = 4.4 \times 10^{4}$ ohm. (The potential difference across the resistor is (90 − 37) volts.)

22. $V = 4.9$ volts.

31. (a) 1.36 amperes.
(b) 1.36 volts.

38. (a) Increases.
(b) Decreases.

CHAPTER 22

1. The filings would try to stand up at right angles to the paper.

2. No; they are all closed loops, of various shapes.

3. Clockwise, in both cases.

4. The magnitude of the resultant field is $\sqrt{2}$ times the magnitude of the field from one.

5. The field is proportional to the magnitude of the current.

9. Parallel to the field, in either the same or opposite direction as the field.

10. Downward.

11. $F = 2.1$ newtons.

14. The force is directly proportional to the angle of rotation.

16. Switch the connections to the battery terminals, or reverse the direction of the magnetic field—but not both.

17. The radius increases as the square root of the accelerating potential.

18. The velocity of the charge, the sign and magnitude of the charge, and the component of magnetic field perpendicular to the velocity.

22. 360 electron volts = 5.76×10^{-17} joule.

28. $2\pi KI$.

29. Zero.

30. $B = 2\pi \times 10^{-3}$ newton/amp–m $= 6 \times 10^{-3}$ newton/amp–m.

CHAPTER 23

1. The positive charges will be at Q, the negative charges at P.

2. (a) The force on positive charges is directed outward from the page; the force on negative charges is directed into the page. (b) No; charges are not free to move in that direction.

4. There would be no current as long as the entire loop was in the region of uniform field.
5. (a) When the plane of the loop is parallel to the magnetic field. (b) When the plane of the loop is perpendicular to the magnetic field.
6. The flux through the loop will not change.
7. $\phi_B = 1 \times 10^{-2}$ newton–m/amp.
9. Yes. The force you must exert on the loop is proportional to the current, which would be reduced if the electrical resistance of the loop were increased.
10. At high speed, since the pulling force would have to be greater.
11. $\mathcal{E} = 1 \times 10^{-4}$ volt.
12. $\mathcal{E} = 0.5$ volt.
18. No; the magnetic force on the charges in the moving loop, which is proportional to the speed of the loop, would increase until it exactly canceled the applied force, and the loop would then move with constant velocity.
23. No; the lines are closed loops, since there are no free charges on which they can begin or end.
25. The rate at which charge is flowing onto either plate, which in turn determines the rate of change of electric flux through the region between the plates.
26. (a) Yes; with a constant current. (b) No; it is impossible to change the magnetic flux within a region at a constant rate indefinitely.
27. (a) Negative charges are on the top plate, positive charges on the bottom plate. (b) The electric field points upward.
29. (a) 3×10^8 m/sec. (b) 3×10^8 m/sec. (c) 3×10^8 m/sec.

CHAPTER 24

4. The large number of particles that come through the foil undeflected, compared with the small number that are deflected, tells us that the chance for one collision is very small; hence, the chance that one alpha particle will have two collisions is much, much smaller.
5. An atom must be mostly empty space, with something in it which can exert a strong force on an alpha particle.
6. It is the angle between the original and final directions of motion of a scattered particle.
10. We cannot measure b directly, because we do not know where each nucleus is, nor do we know the exact trajectories of the incident alpha particles.
11. N_θ is proportional to the square of the aiming error.
12. It is a Coulomb field; that is, it varies as $1/r^2$.
13. Through angles between 90° and 180°.
18. About 6 times as close.
19. $r_0 = 7.0 \times 10^{-15}$ meter.
20. The law of conservation of energy.

CHAPTER 25

1. The time at which a first jump occurred would always be later for weak light than for strong light.
4. (a) No. (b) Yes.
8. The point corresponding to a frequency of 7.4×10^{14} sec^{-1}.
9. 1.9ev.

CHAPTER 26

1. Light from an electron orbiting around a nucleus should give rise to a continuous spectrum in contradiction to the discrete spectral lines observed in practice.
2. 4.2 ev.
6. 6.67 ev, 4.86 ev, 1.81 ev.
7. 1.81 ev.
12. The atom absorbs 4.86 ev from the electron. It absorbs no energy from the photon. [See Fig. 26–3(b) and 26–6(c).]

CHAPTER 27

1. $\lambda = 0.73$ angstrom.
2. $2d \sin \theta = 2\lambda$.
12. No; since the only restriction is that $2d$ equal a whole number of wavelengths; d determines the lowest energy levels and also the energy values of the higher levels which can occur.
13. 10^{-17} joule.

Index

A

B

C

F

G

H

I

N

O

P

Q

R

S

T

U

V

W

X

Y

Z

4 5 6 7 8 9 0